Fungal Taxonomy, Phylogeny, and Ecology: A Themed Issue Dedicated to Academician Wen-Ying Zhuang

Fungal Taxonomy, Phylogeny, and Ecology: A Themed Issue Dedicated to Academician Wen-Ying Zhuang

Editors

Cheng Gao
Lei Cai

MDPI • Basel • Beijing • Wuhan • Barcelona • Belgrade • Manchester • Tokyo • Cluj • Tianjin

Editors
Cheng Gao
State Key Laboratory of Mycology
Institute of Microbiology
Chinese Academy of Sciences
Beijing
China

Lei Cai
State Key Laboratory of Mycology
Institute of Microbiology
Chinese Academy of Sciences
Beijing
China

Editorial Office
MDPI
St. Alban-Anlage 66
4052 Basel, Switzerland

This is a reprint of articles from the Special Issue published online in the open access journal *Journal of Fungi* (ISSN 2309-608X) (available at: www.mdpi.com/journal/jof/special_issues/Ecology_Fungi).

For citation purposes, cite each article independently as indicated on the article page online and as indicated below:

LastName, A.A.; LastName, B.B.; LastName, C.C. Article Title. *Journal Name* **Year**, *Volume Number*, Page Range.

ISBN 978-3-0365-6449-4 (Hbk)
ISBN 978-3-0365-6448-7 (PDF)

Contents

Journal of *Fungi*

MDPI

Editorial

Fungal Taxonomy, Phylogeny, and Ecology: A Themed Issue Dedicated to Academician Wen-Ying Zhuang

Cheng Gao * and Lei Cai *

State Key Laboratory of Mycology, Institute of Microbiology, Chinese Academy of Sciences, Beijing 100101, China
* Correspondence: gaoc@im.ac.cn (C.G.); cail@im.ac.cn (L.C.)

Citation: Gao, C.; Cai, L. Fungal Taxonomy, Phylogeny, and Ecology: A Themed Issue Dedicated to Academician Wen-Ying Zhuang. *J. Fungi* **2022**, *8*, 1294. https://doi.org/10.3390/jof8121294

Received: 9 December 2022
Accepted: 9 December 2022
Published: 11 December 2022

Publisher's Note: MDPI stays neutral with regard to jurisdictional claims in published maps and institutional affiliations.

We are honored and privileged to edit this Special Issue, "Fungal Taxonomy, Phylogeny, and Ecology: A Themed Issue Dedicated to Academician Wen-Ying Zhuang".

Professor Wen-Ying Zhuang is an outstanding mycologist in China and worldwide. Over the past 46 years, she has comprehensively investigated fungal biodiversity in forests, deserts, and plateaus habitats from 26 provinces and districts of China and accumulated numerous priceless resources for scientific research and economic utilization. She has discovered and described 1 new family, 13 new genera, and 360 new species of fungi, and resolved numerous taxonomic and nomenclature problems. Her collaboration with international colleagues contributed greatly to the selection of fungal DNA barcodes and the phylogenetic reconstruction of Leotiomycetes, Helotiales, and Hypocreales. She is the senior author of 280 articles, editor and co-editor of 18 monographs, independent editor of worldwide monographs of 3 important fungal genera, co-editor of Dictionary of the Fungi (v9), and co-Editor-in-Chief of Flora Sporophytae Sinicae. Prof. Wen-Ying Zhuang has been elected as a CAS academician, TWAS academician, IMA executive committee member, and MSA fellow. In honor of her outstanding contribution, one fungal genus *Wenyingia* and one bacterial genus *Wenyingzhuangia* have been named after her. More recently, her work on *Trichoderma* has renewed our understanding of its biodiversity and helped discover many potentially highly capable strains with significant enzyme profiles in degrading agricultural wastes.

In this Special Issue, we are pleased to publish a comprehensive assemblage of 23 papers covering fungal taxonomy, phylogeny, and ecology, in which 76 new taxa from a broad taxonomic group and different ecological habitats are reported.

Prof Wenying Zhuang and colleagues reported three new species of *Clonostachys* (Hypocreales, Ascomycota) from China, namely, *Clonostachys chongqingensis* sp. nov., *Clonostachys leptoderma* sp. nov., and *Clonostachys oligospora* sp. nov. [1]. Using contaminated substrates of edible fungi from North China, Cao et al. [2] detected 10 *Trichoderma* species, including 3 new species in Harzianum clade, *T. auriculariae* sp. nov., *T. miyunense* sp. nov., and *T. pholiotae* sp. nov. From the green mold diseased fruitbody of *Ganoderma sichuanense*, An et al. [3] described *Trichoderma ganodermatigerum* as a new species and reported *Trichoderma koningiopsis* as a new fungal pathogen on *Ganoderma sichuanense* fruitbodies. From 526 strains isolated from diseased watermelon in 8 growing provinces in China, Guo et al. [4] detected 12 known species of *Colletotrichum*, with *Colletotrichum kaifengense* and *Colletotrichum magnum* being the most aggressive to watermelon. Focusing on the large-spored *Alternaria* associated with Compositae plants in China, Zhao et al. [5] discovered five new species, namely, *Alternaria anhuiensis*, *A. coreopsidis*, *A. nanningensis*, *A. neimengguensis*, and *A. sulphureus*. From 353 *Calonectria* strains isolated from leaf blight pathogen of *Eucalyptus* plantations and adjacent plantings, as well as natural forests, Liu et al. [6] identified six known *Calonectria* taxa and one new species, *Calonectria minensis*. In a similar study, Zhang et al. [7] identified five new species, namely, *Calonectria cassiae*, *C. guangdongensis*, *C. melaleucae*, *C. shaoguanensis*, and *C. strelitziae*, based on isolates from leaf spots, stem blights, and root rots of species of *Arachis, Cassia, Callistemon, Eucalyptus, Heliconia, Melaleuca,* and *Strelitzia* plants in Guangdong province. Relying on newly collected *Otidea* specimens from northern China and herbarium specimens deposited in three important Chinese fungus herbaria (HMAS, HKAS, HMJAU), Xu et al. [8] recognized 16 species of *Otidea* in China, of which 7 new species were described, namely, *Otidea aspera, O. cupulata, O. filiformis, O. khakicolorata, O. parvula, O. plicara,* and *O. purpureobrunnea*. Based on a combination of morphology, phylogeny, and holomorphic states, Ou et al. [9] discovered four new species of *Orbilia*, i.e., *O. baisensis, O. hanzhongensis, O. nanningensis,* and *O. pinea*. In two other studies on plant-associated fungi from Hainan island of China, Liu et al. [10] described three new species of *Microdochium*, i.e., *M. hainanense, M. miscanthi,* and *M. sinense* from plant hosts *Miscanthus sinensis* and *Phragmites australis*; Wang et al. [11] described four new species of *Acrodictys*, namely, *A. bawanglingensis, A. diaoluoshanensis, A. ellisii,* and *A. pigmentosa,* from dead branches. Focusing on insect pathogenic fungi in Thailand, Thanakitpipattana et al. [12] discovered one new genus, *Neohyperdermium*, and five new species of *Ascopolyporus*, namely, *A. albus* sp. nov., *A. galloides, A. griseoperitheciatus, A. khaoyaiensis,* and *A. purpuratus*. Moreover, both macroscopic and phylogenetic evidence suggested that *Hyperdermium* is congeneric with *Ascopolyporus* [12]. In arid and semi-arid regions of Northwest China, Zhang et al. [13] discovered and described four new species of Verrucariaceae lichen, namely, *Clavascidium sinense, Placidium nigrum, Placidium nitidulum,* and *Placidium varium*. From the lichenized fungal genus *Astrothelium*, Jiang et al. [14] described five new species, namely, *A. jiangxiense, A. luminothallinum, A. pseudocrassum, A. subeustominspersum,* and *A. subrufescens*. These excellent works contributed significantly to recognizing fungal species diversity in Ascomycota.

Moving to Basidiomycota, from a residential area of Jiangxi Province in China, Zhang et al. [15] reported an unusual new bryophilous basidiolichen species in the genus *Omphalina*, namely, *O. licheniformis*, providing new insights and evidence for understanding the significance of lichenization during the evolution of Agaricomycetes. From the genus of edible mushrooms *Cantharellus*, Zhang et al. [16] discovered four new species, namely, *C. chrysanthus, C. convexus, C. neopersicinus,* and *C. sinocinnabarinus*. Demonstrated by both morphological and molecular analysis, Hu et al. [17] discovered eight new species of *Gymnopus* from Northeast China, namely, *G. changbaiensis, G. globulosus, G. longisterigmaticus, G. longus, G. macrosporus, G. striatus, G. tiliicola,* and *G. tomentosus*. Mao et al. [18] described two new species of *Mallocybe* and three new species of *Pseudosperma* from North China, namely, *M. depressa, P. gilvum, P. laricis, M. picea,* and *P. pseudoniveivelatum*. Song et al. [19] revealed the phylogenetic relationships in the genus *Phellodon* (Bankeraceae, Thelephorales) based on multi-locus sequences and described three new species, namely, *P. crassipileatus, P. griseofuscus,* and *P. perchocolatus*. Focusing on wood decomposers from a subtropical region of Yunnan Province in China, Luo and Zhao [20] reported three new species of *Trechispora*, namely *T. murina, T. odontioidea,* and *T. olivacea*. Kewessa et al. [21] recorded 64 wild edible fungal species belonging to 31 genera and 21 families from the plots established in the natural and plantation forests in Ethiopia, including ecologically and economically important fungal species such as *Agaricus campestroides, Tylopilus niger, Suillus luteus, Tricholoma portentosum,* and *Morchella americana*. The fungal community composition based on sporocarp observation was mainly correlated with the organic matter, available phosphorus, total nitrogen content of the soil, and daily minimum temperature [21].

Beyond the Dikarya and to the early fungal lineage *Conidiobolus* (Ancylistaceae; Entomophthorales; Zoopagomycota), Gryganskyi et al. [22] resolved the phylogeny, lifestyle, and evolution direction of parasitic in *Conidiobolus* group using molecular and genomic data. Gryganskyi et al. [22] found that parasitism evolved multiple times in the Conidiobolus group and suggested that the evolution of ballistic conidia preceded the evolution of the parasitic lifestyle. Focusing on cellular slime molds (Dictyostelids), Cavender et al. [23] discovered four new species, namely, *Cavenderia helicoidea, C. parvibrachiata, C. protumula,* and *C. ungulate*, from soil systems in northern Thailand.

Finally, we would like to thank all the contributors to this Special Issue and warm-heartedly wish Prof. Wen-Ying Zhuang all the best in the future.

Author Contributions: C.G. prepared a draft and L.C. reviewed and edited the definitive version. All authors have read and agreed to the published version of the manuscript.

Conflicts of Interest: The authors declare no conflict of interest.

References

1. Zeng, Z.-Q.; Zhuang, W.-Y. Three New Species of *Clonostachys* (Hypocreales, Ascomycota) from China. *J. Fungi* **2022**, *8*, 1027. [CrossRef] [PubMed]
2. Cao, Z.-J.; Qin, W.-T.; Zhao, J.; Liu, Y.; Wang, S.-X.; Zheng, S.-Y. Three New *Trichoderma* Species in Harzianum Clade Associated with the Contaminated Substrates of Edible Fungi. *J. Fungi* **2022**, *8*, 1154. [CrossRef] [PubMed]
3. An, X.-Y.; Cheng, G.-H.; Gao, H.-X.; Li, X.-F.; Yang, Y.; Li, D.; Li, Y. Phylogenetic Analysis of *Trichoderma* Species Associated with Green Mold Disease on Mushrooms and Two New Pathogens on *Ganoderma Sichuanense*. *J. Fungi* **2022**, *8*, 704. [CrossRef] [PubMed]
4. Guo, Z.; Luo, C.-X.; Wu, H.-J.; Peng, B.; Kang, B.-S.; Liu, L.-M.; Zhang, M.; Gu, Q.-S. *Colletotrichum* Species Associated with Anthracnose Disease of Watermelon (*Citrullus lanatus*) in China. *J. Fungi* **2022**, *8*, 790. [CrossRef]
5. Zhao, L.; Luo, H.; Cheng, H.; Gou, Y.-N.; Yu, Z.-H.; Deng, J.-X. New Species of Large-Spored *Alternaria* in Section Porri Associated with Compositae Plants in China. *J. Fungi* **2022**, *8*, 607. [CrossRef]
6. Liu, Q.; Wingfield, M.J.; Duong, T.A.; Wingfield, B.D.; Chen, S. Diversity and Distribution of *Calonectria* Species from Plantation and Forest Soils in Fujian Province, China. *J. Fungi* **2022**, *8*, 811. [CrossRef]
7. Zhang, Y.; Chen, C.; Chen, C.; Chen, J.; Xiang, M.; Wanasinghe, D.N.; Hsiang, T.; Hyde, K.D.; Manawasinghe, I.S. Identification and Characterization of *Calonectria* Species Associated with Plant Diseases in Southern China. *J. Fungi* **2022**, *8*, 719. [CrossRef]
8. Xu, Y.-Y.; Mao, N.; Yang, J.-J.; Fan, L. New Species and New Records of *Otidea* from China Based on Molecular and Morphological Data. *J. Fungi* **2022**, *8*, 272. [CrossRef]
9. Ou, X.-Y.; Shao, Y.-Y.; Zheng, H.-F.; Liu, B. Four New Species and New Records of *Orbilia* from China Based on Molecular and Morphological Data. *J. Fungi* **2022**, *8*, 1188. [CrossRef]
10. Liu, S.; Liu, X.; Zhang, Z.; Xia, J.; Zhang, X.; Meng, Z. Three New Species of *Microdochium* (Sordariomycetes, Amphisphaeriales) on Miscanthus sinensis and Phragmites australis from Hainan, China. *J. Fungi* **2022**, *8*, 577. [CrossRef]
11. Wang, S.; Liu, R.; Liu, S.; Zhang, Z.; Xia, J.; Li, D.; Liu, X.; Zhang, X. Morphological and Phylogenetic Analyses Reveal Four New Species of *Acrodictys* (Acrodictyaceae) in China. *J. Fungi* **2022**, *8*, 853. [CrossRef] [PubMed]
12. Thanakitpipattana, D.; Mongkolsamrit, S.; Khonsanit, A.; Himaman, W.; Luangsa-ard, J.J.; Pornputtapong, N. Is *Hyperdermium* Congeneric with *Ascopolyporus*? Phylogenetic Relationships of *Ascopolyporus* spp. (Cordycipitaceae, Hypocreales) and a New Genus *Neohyperdermium* on Scale Insects in Thailand. *J. Fungi* **2022**, *8*, 516. [CrossRef] [PubMed]
13. Zhang, T.; Zhang, X.; Yang, Q.; Wei, X. Hidden Species Diversity was Explored in Two Genera of Catapyrenioid Lichens (Verrucariaceae, Ascomycota) from the Deserts of China. *J. Fungi* **2022**, *8*, 729. [CrossRef] [PubMed]
14. Jiang, S.-H.; Zhang, C.; Xue, X.-D.; Aptroot, A.; Wei, J.-C.; Wei, X.-L. Morphological and Phylogenetic Characterizations Reveal Five New Species of *Astrothelium* (Trypetheliales, Ascomycota) from China. *J. Fungi* **2022**, *8*, 994. [CrossRef] [PubMed]
15. Zhang, T.; Zhu, X.; Vizzini, A.; Li, B.; Cao, Z.; Guo, W.; Qi, S.; Wei, X.; Zhao, R. New Insights into Lichenization in Agaricomycetes Based on an Unusual New Basidiolichen Species of *Omphalina* s. str. *J. Fungi* **2022**, *8*, 1033. [CrossRef]
16. Zhang, M.; Wang, C.-Q.; Gan, M.-S.; Li, Y.; Shao, S.-C.; Qin, W.-Q.; Deng, W.-Q.; Li, T.-H. Diversity of *Cantharellus* (Cantharellales, Basidiomycota) in China with Description of Some New Species and New Records. *J. Fungi* **2022**, *8*, 483. [CrossRef]
17. Hu, J.; Zhao, G.; Tuo, Y.; Rao, G.; Zhang, Z.; Qi, Z.; Yue, L.; Liu, Y.; Zhang, T.; Li, Y.; et al. Morphological and Molecular Evidence Reveal Eight New Species of *Gymnopus* from Northeast China. *J. Fungi* **2022**, *8*, 349. [CrossRef]
18. Mao, N.; Xu, Y.-Y.; Zhao, T.-Y.; Lv, J.-C.; Fan, L. New Species of *Mallocybe* and *Pseudosperma* from North China. *J. Fungi* **2022**, *8*, 256. [CrossRef]
19. Song, C.-G.; Chen, Y.-Y.; Liu, S.; Xu, T.-M.; He, X.-L.; Wang, D.; Cui, B.-K. A Phylogenetic and Taxonomic Study on *Phellodon* (Bankeraceae, Thelephorales) from China. *J. Fungi* **2022**, *8*, 429. [CrossRef]
20. Luo, K.; Zhao, C. A Molecular Systematics and Taxonomy Research on *Trechispora* (Hydnodontaceae, Trechisporales): Concentrating on Three New Trechispora Species from East Asia. *J. Fungi* **2022**, *8*, 1020. [CrossRef]
21. Kewessa, G.; Dejene, T.; Alem, D.; Tolera, M.; Martín-Pinto, P. Forest Type and Site Conditions Influence the Diversity and Biomass of Edible Macrofungal Species in Ethiopia. *J. Fungi* **2022**, *8*, 1023. [CrossRef] [PubMed]
22. Gryganskyi, A.P.; Nie, Y.; Hajek, A.E.; Hodge, K.T.; Liu, X.-Y.; Aadland, K.; Voigt, K.; Anishchenko, I.M.; Kutovenko, V.B.; Kava, L.; et al. The Early Terrestrial Fungal Lineage of *Conidiobolus*-Transition from Saprotroph to Parasitic Lifestyle. *J. Fungi* **2022**, *8*, 789. [CrossRef] [PubMed]
23. Cavender, J.C.; Vadell, E.M.; Perrigo, A.L.; Landolt, J.C.; Stephenson, S.L.; Liu, P. Four New Species of Dictyostelids from Soil Systems in Northern Thailand. *J. Fungi* **2022**, *8*, 593. [CrossRef] [PubMed]

Journal of *Fungi*

MDPI

Article

Three New Species of *Clonostachys* (Hypocreales, Ascomycota) from China

Zhao-Qing Zeng * and **Wen-Ying Zhuang ***

State Key Laboratory of Mycology, Institute of Microbiology, Chinese Academy of Sciences, Beijing 100101, China
* Correspondence: zengzq@im.ac.cn (Z.-Q.Z.); zhuangwy@im.ac.cn (W.-Y.Z.)

Abstract: Three new species of *Clonostachys* are introduced based on specimens collected from China. *Clonostachys chongqingensis* sp. nov. is distinguished by pale yellow to pale orange-yellow perithecia with a very low papilla, clavate to subcylindrical asci possessing ellipsoidal to elongate-ellipsoidal spinulose ascospores 13–16 × 4.5–5.5 μm; it has acremonium- to verticillium-like conidiophores and ellipsoidal to rod-shaped conidia. *Clonostachys leptoderma* sp. nov. has pinkish-white subglobose to globose perithecia on a well-developed stroma and with a thin perithecial wall, clavate to subcylindrical asci with ellipsoidal to elongate-ellipsoidal spinulose ascospores 7.5–11 × 2.5–3.5 μm; it produces verticillium-like conidiophores and ellipsoidal to subellipsoidal conidia. *Clonostachys oligospora* sp. nov. features solitary to gregarious perithecia with a papilla, clavate asci containing 6–8 smooth-walled ascospores 9–17 × 3–5.5 μm; it forms verticillium-like conidiophores and sparse, subfusiform conidia. The morphological characteristics and phylogenetic analyses of combined nuclear ribosomal DNA ITS1-5.8S-ITS2 and beta-tubulin sequences support their placement in *Clonostachys* and their classification as new to science. Distinctions between the novel taxa and their close relatives are compared herein.

Keywords: Bionectriaceae; morphology; sequence analyses; taxonomy

Citation: Zeng, Z.-Q.; Zhuang, W.-Y. Three New Species of *Clonostachys* (Hypocreales, Ascomycota) from China. *J. Fungi* **2022**, *8*, 1027. https://doi.org/10.3390/jof8101027

Academic Editor: Philippe Silar

Received: 24 August 2022
Accepted: 26 September 2022
Published: 28 September 2022

1. Introduction

Clonostachys Corda, typified by *C. araucaria* Corda, is characterized by solitary to gregarious, subglobose or globose to ovoid perithecia that are white, yellow, pale orange, tan, or brown; perithecial walls are KOH− and LA−; there are narrowly clavate to clavate asci containing eight ascospores; it produces penicillium-, verticillium-, gliocladium-, or acremonium-like conidiophores, cylindrical to narrowly flask-shaped phialides, and ellipsoidal to subfusiform conidia [1]. Members of the genus usually have a broad range of lifestyles and occur on the bark of recently dead trees, decaying leaves, and less frequently on other fungi, nematodes, and insects [1–3]. They are economically important in the fields of pharmaceutics and agriculture [4]. For instance, the secondary metabolites produced by *C. byssicola* Schroers exhibited antibacterial activities [5], and strains of *C. rosea* (Link) Schroers, Samuels, Seifert & W. Gams have been widely used as biocontrol agents [6].

About 100 names have been created under the genus *Clonostachys* (www.indexfungorum.org (accessed on 1 July 2022)), among which 65 species are commonly accepted [2,3,7–20]. Twenty-four species are known from China [2,7,21]. In this study, three additional taxa are introduced based on morphological characteristics and sequence analyses of combined nuclear ribosomal DNA ITS1-5.8S-ITS2 (ITS) and beta-tubulin (BenA) regions. Comparisons between these novel species and their close relatives are performed.

2. Materials and Methods

2.1. Sampling and Morphological Studies

Specimens were collected from Chongqing and Yunnan Province and were deposited in the Herbarium Mycologicum Academiae Sinicae (HMAS). Cultures were obtained by single ascospore isolation from fresh perithecium and are preserved in the State Key Laboratory of Mycology, Institute of Microbiology, Chinese Academy of Sciences. The methods of Hirooka et al. [22] were generally followed for morphological observations. Perithecial wall reactions were tested in 3% potassium hydroxide (KOH) and 100% lactic acid (LA). Longitudinal sections through the perithecia were made with a freezing microtome (YD-1508-III, Jinhua, China) at a thickness of 6–8 μm. Photographs were taken using a Canon G5 digital camera (Tokyo, Japan) connected to a Zeiss Axioskop 2 plus microscope (Göttingen, Germany). For colony characteristics and growth rates, strains were grown on potato dextrose agar (PDA) (200 g potato + 2% (*w*/*v*) dextrose + 2% (*w*/*v*) agar) and synthetic low-nutrient agar (SNA) [23] in 90 mm plastic Petri dishes at 25 °C for 2 weeks with alternating periods of light and darkness (12 h/12 h).

2.2. DNA Extraction, PCR Amplification, and Sequencing

The genomic DNA was extracted from fresh mycelium following the method of Wang and Zhuang [24]. Five primer pairs, namely ITS5/ITS4 [25], T1/T22 [26], acl1-230up/acl1-1220low [27], Crpb1a/rpb1c [28], and EF1-728F/EF2 [29,30], were used to amplify the sequences of ITS, BenA, ATP citrate lyase (ACL1), the largest subunit of RNA polymerase II (RPB1), and translation elongation factor 1-α (TEF1), respectively. PCR reactions were performed on an ABI 2720 Thermal Cycler (Applied Biosciences, Foster City, CA, USA) with a 25 μL reaction system consisting of 12.5 μL Taq MasterMix, 1 μL of each primer (10 μM), 1 μL template DNA, and 9.5 μL ddH_2O. DNA sequencing was carried out in both directions with the same primer pairs using an ABI 3730 XL DNA Sequencer (Applied Biosciences, Foster City, CA, USA). Newly achieved sequences and those retrieved from GenBank are listed in Table 1. *Fusarium acutatum* Nirenberg & O'Donnell and *Nectria cinnabarina* (Tode) Fr. were chosen as outgroup taxa.

Table 1. Sequences used in this study.

Species	Herbarium/Strain Numbers	GenBank Accession Numbers	
		ITS	BenA
C. agrawalii	CBS 53381	AF358241	AF358187
C. apocyni	CBS 13087	AF210688	AF358168
C. aranearum	QLS 0625	NR164542	KU212400
C. aureofulvella	CBS 19593	AF358226	AF358181
C. buxi	CBS 69693	KM231840	KM232111
C. byssicola	CBS 36478	MH861151	AF358153
C. candelabrum	CBS 50467/CML 2313	MH859044	KF871186
C. capitata	CBS 21893	AF358240	AF358188
C. chlorina	CBS 28790	NR137651	-
C. chloroleuca	CML 1941	KC806286	KF871172
C.chongqingensis	HMAS 290894	**OP205475** [a]	**OP205324**
C. coccicola	HD 2016	KU720552	KU720552
C. compactiuscula	CBS 72987	AF358242	AF358193
C. divergens	CBS 96773b	NR137532	AF358191
C. epichloe	CBS 101037	AF210675	AF358209
C. eriocamporesiana	MFLUCC 17-2620	NR168235	MN699965
C. eriocamporesii	MFLUCC 190486	NR168236	-
C. grammicospora	CBS 20993	NR137650	AF358206
C. grammicosporopsis	CBS 11587	AF210679	AF358204
C. impariphialis	HMAS 275560	KX096609	-
C. indica	RKV 2015	KT291441	KT291441
C. intermedia	CBS 50882	NR137652	AF358205
C. kowhai	CBS 46195	AF358250	AF358170
C. krabiensis	MFLUCC 160254	NR168189	-
C. leptoderma	HMAS 255834	**OP205474**	**OP205323**
C. levigata	CBS 94897	AF210680	AF358196
C. lucifer	CBS 100008	AF210683	AF358208
C. miodochialis	CBS 99769	NR137649	AF358210
C. oblongispora	CBS 100285	AF358248	AF358169
C.oligospora	HMAS 290895	**OP205473**	**OP205322**
C. phyllophila	CBS 92197	NR137531	-
C. pityrodes	CBS 102033	AF210672	AF358212
C. pseudochroleuca	CML 2513	KJ499909	KF871188
C. pseudostriata	CBS 12087	MH862056	AF358184
C. ralfsii	CBS 102845	AF358253	AF358219
C. rhizophaga	CBS 36177	AF358228	AF358158
C. rogersoniana	CBS 58289/CML 2557	AF210691	KX185047
C. rosea	CBS 71086	MH862010	AF358161
C. rossmaniae	CBS 21093	AF358227	AF358213
C. samuelsii	CBS 69997	AF358236	AF358190
C. saulensis	BRFM 2782	MK635054	-
C. sesquicillii	CBS 18088	AF210666	AF358214
C. setosa	CBS 83491	AF210670	AF358211
C. solani	CBS 101926	AF358230	AF358179
C. sporondochialis	CBS 101921	AF210685	AF358149
C. subquaternata	CBS 100003/CBS 10787	MT537603	AF358207
C. vesiculosa	HMAS 183151	NR119828	-
C. viticola	CAA 944	MK156282	MK156290
C. wenpingii	HMAS 172156	EF612465	HM054127
C. zelandiaenovae	CBS 23280	AF210684	AF358185
F. acutatum	CBS 40297	NR111142	KU603870
N. cinnabarina	CBS 18987	HM484699	HM484835

[a] Numbers in bold indicate the newly provided sequences.

2.3. Sequence Alignment and Phylogenetic Analyses

Sequences were assembled and aligned with BioEdit 7.0.5 [31] and converted to nexus files by ClustalX 1.8 [32]. To confirm the taxonomic positions of the new species, ITS and BenA sequences were combined and analyzed with Bayesian inference (BI), maximum likelihood

(ML), and maximum parsimony (MP) methods. A partition homogeneity test (PHT) was performed with 1000 replicates in PAUP*4.0b10 [33] to evaluate the statistical congruence between the two loci. The BI analysis was conducted by MrBayes 3.1.2 [34] using a Markov chain Monte Carlo (MCMC) algorithm. Nucleotide substitution models were determined by MrModeltest 2.3 [35]. Four Markov chains were run simultaneously for 1,000,000 generations with the trees sampled every 100 generations. A 50% majority rule consensus tree was computed after excluding the first 2500 trees as "burn-in". Bayesian inference posterior probability (BIPP) was determined from the remaining trees. The ML analysis was performed via IQ-Tree 1.6.12 [36] using the best model for each locus, as chosen by ModelFinder [37]. The MP analysis was performed with PAUP 4.0b10 [33] using heuristic searches with 1000 replicates of random addition of sequences and subsequent TBR (tree bisection and reconnection) branch swapping. The topological confidence of the resulting trees and statistical support of the branches were tested in maximum parsimony bootstrap proportion (MPBP) with 1000 replications and each with 10 replicates of the random addition of taxa. Trees were examined by TreeView 1.6.6 [38]. The maximum likelihood bootstrap (MLBP) values, MPBP values greater than 70%, and BIPP values greater than 90% were shown at the nodes.

3. Results

3.1. Phylogeny

The sequences of ITS and BenA from 50 representative species of *Clonostachys* were analyzed. The PHT ($p = 0.05$) indicated that the individual partitions were not highly incongruent [39]; thus, the two loci were combined for phylogenetic analyses. In the MP analysis, the datasets included 1159 nucleotide characters, of which 543 bp were constant, 154 were variable and parsimony-uninformative, and 462 were parsimony-informative. The MP analysis resulted in 123 most parsimonious trees (tree length = 2794, consistency index = 0.4098, homoplasy index = 0.5902, retention index = 0.4677, rescaled consistency index = 0.1917). One of the MP trees generated is shown in Figure 1. The topologies of the BI and ML trees were similar to that of the MP tree. The isolates 12581, 12672, and 11691 were grouped with the other *Clonostachys* taxa investigated (MPBP/MLBP/BIPP = 100%/100%/100%), which confirmed their taxonomic positions. The isolate 12581 was grouped with *C. agrawalii* (Kushwaha) Schroers and *C. capitata* Schroers, with low statistical support. The isolate 12672 clustered with *C. zelandiaenovae* Schroers (MPBP/MLBP/BIPP = 94%/96%/100%), and 11691 formed a separate lineage.

Figure 1. A maximum parsimony tree generated from analyses of combined ITS and BenA sequences of *Clonostachys* species. The MP analysis was performed using heuristic searches with 1000 replicates. MPBP (**left**) and MLBP (**middle**) values greater than 70% and BIPP (**right**) values greater than 90% were shown at the nodes. *Fusarium acutatum* and *Nectria cinnabarina* were chosen as outgroup taxa.

3.2. Taxonomy

Clonostachys chongqingensis Z.Q. Zeng and W.Y. Zhuang, sp. nov. (Figure 2).

Fungal Names: FN571276.

Etymology: The specific epithet refers to the type locality of the fungus.

Typification: China, Chongqing City, Jinfo Mountain, 29°2′50″ N 107°11′0″ E, on rotten bark of *Alnus* sp., 25 October 2020, Z.Q. Zeng, H.D. Zheng, X.C. Wang, C. Liu 12672 (holotype HMAS 290894).

DNA barcodes: ITS OP205475, BenA OP205324, ACL1 OP493559, TEF1 OP493562.

The mycelium was not visible on the natural substratum. Perithecia were superficial, solitary to gregarious, non-stromatic or with a basal stroma, subglobose to globose, with very low papilla and slightly roughened surface; they mostly did not collapse upon drying, and a few were slightly pinched at the apical portion, colored pale yellow to pale orange-yellow. There was no color change in 3% KOH or 100% LA, and the size was 304–353 × 294–392 μm. Perithecial walls were two-layered, 40–70 μm thick; the outer layer was of textura globulosa to textura angularis, 30–45 μm thick, with cells 5–15 × 3–12 μm and cell walls 0.8–1 μm thick. The inner layer was of textura prismatica, 10–25 μm thick, with cells 8–14 × 2.5–3.5 μm and cell walls 1–1.2 μm thick. Asci were clavate to subcylindrical, eight-spored, with a round and simple apex, and 60–85 × 6–13 μm. Ascospores were ellipsoidal to elongate-ellipsoidal, uniseptate, hyaline, spinulose, and uniseriate or irregular-biseriate, and 13–16 × 4.5–5.5 μm.

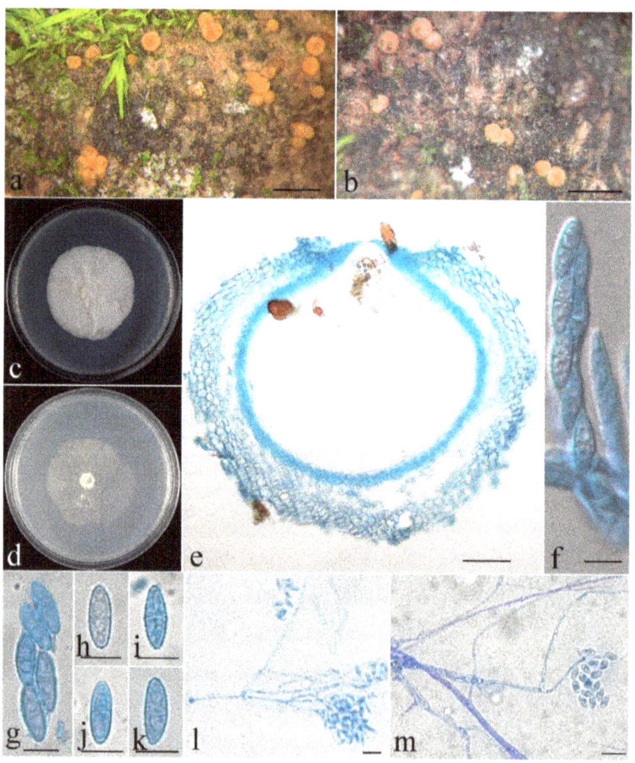

Figure 2. *Clonostachys chongqingensis* (holotype). (**a**,**b**) Ascomata on natural substratum; (**c**) colony after 2 weeks at 25 °C on PDA; (**d**) colony after 2 weeks at 25 °C on SNA; (**e**) median section through the perithecium; (**f**) asci with ascospores; (**g**–**k**) ascospore; (**l**,**m**) conidiophores, phialides, and conidia. Bars: (**a**,**b**) = 1 mm; (**e**) = 50 μm; (**f**–**m**) = 10 μm.

Colonies on PDA were 53 mm in diam. in average after 2 weeks at 25 °C; their surface was cottony, with a dense, whitish aerial mycelium. Colonies on SNA were 50 mm in diam. in average after 2 weeks at 25 °C; their surface was velvet, with a sparse, whitish aerial mycelium. Conidiophores were acremonium- to verticillium-like, arising from aerial hyphae and septate. Phialides were subulate tapered toward the apex, 15–74 μm long, 1.6–2.5 μm wide at the base, and 0.3–0.4 μm wide at the tip. Conidia were ellipsoidal to rod-shaped, unicellular, smooth-walled, hyaline, and 4–10 × 2.5–4 μm.

Notes: Morphologically, the species most resembles *C. sesquicillii* Schroers in having superficial, solitary to gregarious ascomata and clavate to subcylindrical asci with eight ellipsoidal, single-septate, spinulose ascospores [1]. However, the perithecia of the latter are often laterally or apically pinched when dry and have shorter asci (35–63 μm long) and smaller ascospores (8.2–14.4 × 2.2–4.4 μm) [1]. The two-locus phylogeny indicated the two fungi are remotely related (Figure 1).

Phylogenetically, *C. chongqingensis* is closely related to *C. zelandiaenovae* (Figure 1). The latter differs in its well-developed stroma, narrowly clavate asci with an apex ring, and wider ascospores (3.8–7.4 μm wide) [1].

Clonostachys leptoderma Z.Q. Zeng and W.Y. Zhuang, sp. nov. (Figure 3)

Fungal Names: FN571277.

Etymology: The specific epithet refers to the thin-walled perithecia.

Typification: China, Chongqing City, Jinyun Mountain, 29°49′46″ N 106°22′49″ E, on rotten bark, 23 October 2020, Z.Q. Zeng, H.D. Zheng, X.C. Wang, C. Liu 12581 (holotype HMAS 255834).

DNA barcodes: ITS OP205474, BenA OP205323, RPB1 OP493564, TEF1 OP493561.

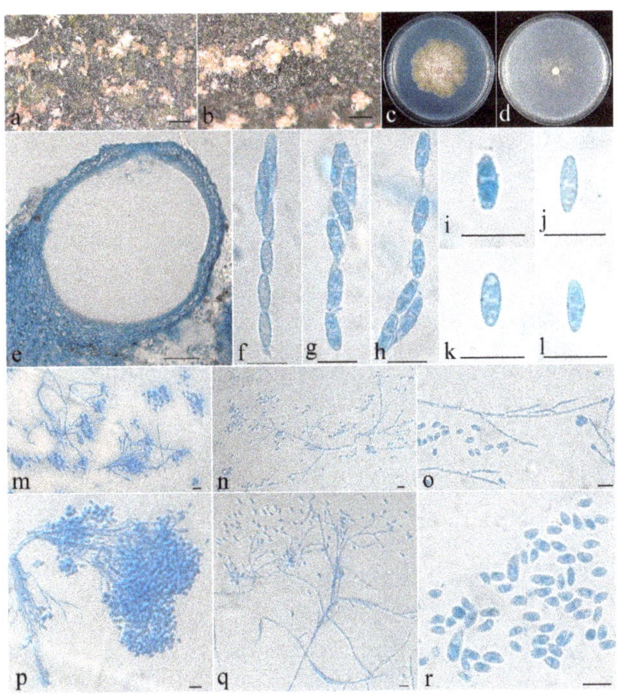

Figure 3. *Clonostachys leptoderma* (holotype). (**a**,**b**) Ascomata on natural substratum; (**c**) colony after 2 weeks at 25 °C on PDA; (**d**) colony after 2 weeks at 25 °C on SNA; (**e**) median section through the perithecium; (**f–h**) asci with ascospores; (**i–l**) ascospore; (**m–q**) conidiophores, phialides, and conidia; (**r**) conidia. Bars: (**a**,**b**) = 1 mm; (**e**) = 50 μm; (**f–r**) = 10 μm.

The mycelium was not visible on the natural substratum. Perithecia were superficial, solitary to gregarious, with a well-developed stroma, subglobose to globose, non-papillate, with surface slightly roughened, and did not collapse upon drying. They were pinkish-white, did not change color in 3% KOH or 100% LA, and were with a size of 216–284 × 206–265 μm. Perithecial walls were two-layered, 13–45 μm thick; the outer layer was of textura globulosa to textura angularis, 8–23 μm thick, with cells 5–10 × 4–8 μm and cell walls 1–1.2 μm thick; the inner layer was of textura prismatica, 5–22 μm thick, with cells 5–12 × 2–3 μm and cell walls 0.8–1 μm thick. Asci were clavate to subcylindrical, 6–8-spored, with a round and simple apex, and 53–63 × 4.8–7 μm. Ascospores were ellipsoidal to elongate-ellipsoidal, uniseptate, hyaline, spinulose, uniseriate or irregular-biseriate, and 7.5–11 × 2.5–3.5 μm.

Colonies on PDA was 31 mm in diam. in average after 2 weeks at 25 °C; their surface was cottony, with a dense, whitish aerial mycelium, and it produced a yellowish-brown pigment in medium. Colony on SNA were 18 mm in diam. in average after 2 weeks at 25 °C, with a sparse, whitish aerial mycelium. Conidiophores were verticillium-like, arising from aerial hyphae; they were septate, with dense phialides. Phialides were subulate, tapering toward the apex, 9–18 μm long, 1.5–2.5 μm wide at the base, and 0.2–0.3 μm wide at the tip. Conidia were subglobose, ellipsoidal to subellipsoidal, unicellular, smooth-walled, hyaline, and 2–7 × 2–5 μm.

Notes: Morphologically, the fungus is most similar to *C. epichloe* Schroers in having solitary to gregarious perithecia and ellipsoidal, bi-cellular, spinulose ascospores of a similar size [1]. Nevertheless, the latter differs in its smaller perithecia (140–240 × 140–200 μm) that is pinched when dry, its wider asci (5–10 μm wide) [1], and the presence of 36 bp and 132 bp divergences in the ITS and BenA regions. Obviously, they are not conspecific.

Phylogenetically, *C. leptoderma* is closely related to *C. capitata* and *C. agrawalii* (Figure 1). *Clonostachys capitata* can be differentiated by its thicker perithecial wall (45–60 μm thick), wider asci (7–12 μm wide), and larger ascospores (11.6–18.8 × 3.6–5.8 μm) [1]. *Clonostachys agrawalii*, which is known to have only an asexual stage, can be easily distinguished by its bi- to quarter-verticillate conidiophores and its cylindrical to flask-shaped, somewhat larger phialides (7–42 × 1.4–3.4 μm) [1].

Clonostachys oligospora Z.Q. Zeng and W.Y. Zhuang, sp. nov. (Figure 4)

Fungal Names: FN571278.

Etymology: The specific epithet refers to the very few conidia produced.

Typification: China, Yunnan Province, Chuxiong Prefecture, Zixi Mountain, Xianrengu, 25°54′0″ N 101°24′46″ E, on a rotten twig, 23 September 2017, Y. Zhang, H.D. Zheng, X.C. Wang, Y.B. Zhang 11691 (holotype HMAS 290895).

DNA barcodes: ITS OP205473, BenA OP205322, ACL1 OP493560, RPB1 OP493563.

The mycelium was not visible on the natural substratum. Perithecia were superficial, solitary to gregarious, either with a basal stroma or non-stromatic. They were subglobose to globose and papillate, with surface slightly warted; the warts were 6–25 μm high. They did not collapse upon drying, were colored pale yellow to light yellow, did not change color in 3% KOH or 100% LA, and were 225–274 × 225–265 μm. The perithecial walls were two-layered, 25–48 μm thick; the outer layer was of textura globulosa to textura angularis, 15–38 μm thick, with cells 8–18 × 9–15 μm and cell walls 0.5–0.8 μm thick; the inner layer was of textura prismatica, 10–15 μm thick, with cells 5–8 × 1.5–2.5 μm and cell walls 0.8–1 μm thick. Asci were clavate, 6–8-spored, with a rounded and simple apex, and 45–65 × 7.5–11 μm. Ascospores were ellipsoidal, uniseptate, hyaline, smooth-walled, uniseriate or irregular biseriate, and 9–17 × 3–5.5 μm.

Colonies on PDA were 50 mm in diam. in average after 2 weeks at 25 °C; their surface was cottony, with a dense, whitish aerial mycelium. Colonies on SNA were 25 mm in diam. in average after 2 weeks at 25 °C, with a sparse, whitish aerial mycelium. Conidiophores were verticillium-like, arising from aerial hyphae, and septate. Phialides were subulate, tapering toward the apex, 9–15 μm long; they were 1.5–2.5 μm wide at the base and

0.2–0.3 μm wide at the tip. Conidia were sparse, subfusiform, unicellular, smooth-walled, hyaline, and 5–13 × 1.8–2.2 μm.

Notes: Among the known species of *Clonostachys*, this fungus resembles *C. setosa* (Vittal) Schroers in terms of solitary to gregarious perithecia and ascospores that are ellipsoidal, bi-cellular, smooth-walled, and of a similar size [1]. However, the latter fungus is distinguished by asci with an apical ring, as well as by conidiophores that are penicillium-like and cylindrical conidia that are slightly larger (8.6–19.2 × 2–3.2 μm) [1]. In addition, there are 47 bp and 128 bp divergences in the ITS and BenA regions between HMAS 290895 and CBS 834.91. Both morphology and DNA sequence data support their distinction at the species level.

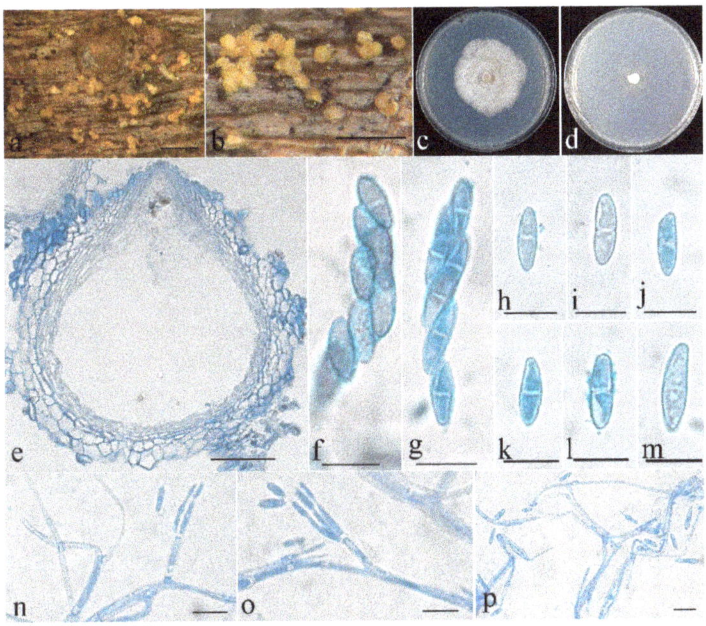

Figure 4. *Clonostachys oligospora* (holotype). (**a**,**b**) Ascomata on natural substratum; (**c**) colony after 2 weeks at 25 °C on PDA; (**d**) colony after 2 weeks at 25 °C on SNA; (**e**) median section through the perithecium; (**f**,**g**) asci with ascospores; (**h**–**m**) ascospore; (**n**–**p**) conidiophores, phialides, and conidia. Bars: (**a**,**b**) = 1 mm; (**e**) = 50 μm; (**f**–**p**) = 10 μm.

4. Discussion

Although *Clonostachys* was established in 1839, the name was not commonly used until Schroers' monographic treatment of the genus, from which 44 species were accepted [1]. The generic name *Bionectria* Speg. was introduced later [40], and the genus was reviewed by Rossman et al. [41]; in that work, the species included those previously placed in the *Nectria ochroleuca* group, the *N. ralfsii* group, and the *N. muscivora* group, as well as those having *Sesquicillium* W. Gams asexual stages. *Clonostachys* and *Bionectria* are of anamorph and teleomorph connections [1,41]. According to the current International Code of Nomenclature for algae, fungi, and plants [42], under the principle that one fungus requires one name, *Clonostachys* was recommended as the preferable name [43].

The previous phylogenetic overview of *Clonostachys* that was based on two-locus (ITS and BenA) sequence analyses showed that the genus is monophyletic [19,20]. Our analyses provided a similar tree topology, and species of the genus formed a well-supported clade (MPBP/MLBP/BIPP = 100%/100%/100%), including the three new taxa (Figure 1). *Clonostachys oligospora* is a well-separated lineage in between *C. indica* Prasher & R. Chauhan and *C. samuelsii* Schroers. *Clonostachys chongqingensis* clustered with *C. zelandiaenovae*, receiving relatively high statistical support (MPBP/MLBP/BIPP = 94%/96%/100%), and had moderate sequence divergences, i.e., 11/518 bp (2.1%) for ITS and 15/557 bp (2.7%) for BenA. *Clonostachys leptoderma* was grouped with *C. agrawalii* and *C. capitata*, which is poorly supported. Compared with the previously demonstrated phylogenies [19,20], minor changes were detected. For example, *C. pseudostriata* Schroers formerly constituted a separate lineage by itself [19,20]; whereas, with the joining of the new species, the fungus seemed to be closely related to *C. krabiensis* Tibpromma & K.D. Hyde and *C. viticola* C. Torcato & A. Alves, with low statistical support. Comparisons between each new species and closely related taxa are provided in Table 2. Along with the discovery of additional new species, the relationships among the species of the genus will become well-established.

Table 2. Morphological comparisons of the new species and their close relatives.

Species	Sexual Morph			Asexual Morph			Reference Source
	Perithecia	Asci	Ascospores	Conidiophores	Phialides	Conidia	
C. chongqingensis	Subglobose to globose, 304–353 × 294–392 μm.	Clavate to subcylindrical, 60–85 × 6–13 μm.	Ellipsoidal to elongate-ellipsoidal, spinulose, 13–16 × 4.5–5.5 μm.	Acremonium- to verticillium-like.	Subulate, 15–74 × 1.6–2.5 μm.	Ellipsoidal to rod-shaped, 4–10 × 2.5–4 μm.	This study
C. krabiensis	N.A.	N.A.	N.A.	Aggregated into sporodochia.	Subulate, 10–13 × 1.5–2.5 μm.	Cylindrical to oblong, 5–7 × 1–2 μm.	[15]
C. sesquicillii	Globose, 250–300 μm diam.	Narrowly clavate to cylindrical, 35–63 × 5–13 μm.	Ellipsoidal, warted, 8.2–14.4 × 2.2–4.4 μm.	Penicillate conidiophores bi- to quarter-verticillate, verticillate conidiophores sparsely formed.	Cylindrical to flask-shaped, 6.4–18.8 × 1.4–3.6 μm.	Ellipsoidal to oblong ellipsoidal, 4.2–9.6 × 1.6–3 μm.	[1]
C. viticola	N.A.	N.A.	N.A.	Primary conidiophores verticillate, secondary conidiophores bi- to ter-verticillate, 45.3–64.7 × 2.1–3.7 μm.	Cylindrical, 10.4–32.8 × 1.7–2.7 μm.	Ellipsoidal to oval, 4.5–6.7 × 2.4–3.4 μm.	[3]
C. zelandiaenovae	Subglobose to globose, 290–550 μm diam.	Narrowly clavate, 60–104 × 7–15.5 μm.	Ellipsoidal, spinulose, rarely smooth, 11.6–21.4 × 3.8–7.4 μm.	Primary conidiophores verticillate, secondary conidiophores penicillate, ter- to quinquies-verticillate.	Cylindrical to narrowly flask-shaped, 4.8–20.6 × 1.6–3.4 μm.	Distally broadly rounded, 4–13.2 × 2.4–4.2 μm.	[1]
C. leptoderma	Subglobose to globose, 216–284 × 206–265 μm.	Clavate to subcylindrical, 53–63 × 4.8–7 μm.	Ellipsoidal to elongate-ellipsoidal, spinulose, 7.5–11 × 2.5–3.5 μm.	Verticillium-like.	Subulate, 9–18 × 1.5–2.5 μm.	Subglobose, ellipsoidal to subellipsoidal, 2–7 × 2–5 μm.	This study
C. agrawalii	N.A.	N.A.	N.A.	Primary conidiophores irregularly branched to ter-verticillate, secondary conidiophores bi- to quarter-verticillate.	Cylindrical to flask-shaped, 7–42 × 1.4–3.4 μm.	Ends broadly rounded, 3.8–5.8 × 2.2–3 μm.	[1]
C. capitata	Subglobose to oval, around 300 μm diam.	Narrowly clavate, 50.5–89.5 × 7–12 μm.	Ellipsoidal to oblong-ellipsoidal, spinulose to warted, 11.6–18.8 × 3.6–5.8 μm.	Primary conidiophores verticillium-like, secondary conidiophores ter- to quinquies-verticillium-like.	Cylindrical to narrowly flask-shaped, 8.8–46.6 × 1.4–3.6 μm.	Ends broadly rounded, 4.6–12.4 × 2.2–4.2 μm.	[1]
C. epichloe	140–240 × 140–200 μm.	Clavate to narrowly clavate, 32–65 × 5–10 μm.	Ellipsoidal, smooth to spinulose, 7.2–13 × 2.4–4.4 μm.	Divergently branched to adpressed.	Cylindrical to narrowly flask-shaped, 7–29 × 2.2–3.2 μm.	Ellipsoidal to narrowly clavate, 4.8–9.6 × 1.6–3.6 μm.	[1]
C. oligospora	Subglobose to globose, 225–274 × 225–265 μm.	Clavate, 45–65 × 7.5–11 μm.	Ellipsoidal, smooth, 9–17 × 3–5.5 μm.	Verticillium-like.	Subulate, 9–15 × 1.5–2.5 μm.	Sparse, subfusiform, 5–13 × 1.8–2.2 μm.	This study
C. indica	N.A.	N.A.	N.A.	Primary conidiophores verticillium-like, secondary conidiophores bi- to quarter-verticillate.	Cylindrical, 10.5–35.9 × 1.9–3.9 μm.	Ovoid to subglobose, 3.9–7.4 × 2–3.7 μm.	[12]
C. samuelsii	Subglobose to oval, 250–350 μm diam.	Narrowly clavate, 43–71 × 5–11.5 μm.	Ellipsoidal to broadly ellipsoidal, warted, 7.8–15.4 × 2.8–5.6 μm.	Penicillate or irregularly penicillate, ter- to quarter-verticillate.	Phialides cylindrical to narrowly flask-shaped, 3.8–20.6 × 1.8–5 μm.	Conidia ellipsoidal, 4.4–11.6 × 2.2–3.8 μm.	[1]
C. setosa	Globose, around 200 μm diam.	Narrowly clavate, 45–53 × 6.5–9 μm.	Ellipsoidal, smooth to striate, 8.8–13 × 2.4–3.8 μm.	Penicillate, mono- to ter-verticillate.	Cylindrical, 5.4–13.4 × 2.4–4.2 μm.	Cylindrical, 8.6–19.2 × 2–3.2 μm.	[1]

More than 220 secondary metabolites have been reported from species of the genus. For example, *C. byssicola*, *C. candelabrum* (Bonord.) Schroers, *C. compactiuscula* (Sacc.) D. Hawksw. & W. Gams, *C. grammicospora* Schroers & Samuels, *C. pityrodes* Schroers, *C. rogersoniana* Schroers, and *C. rosea* were demonstrated to have the potential for biocontrol application [4,44–47]. Meanwhile, strains of *C. rosea* were occasionally reported as an opportunistic phytopathogen [48,49]. Therefore, studies on the biodiversity of *Clonostachys* are of theoretical and practical importance and need to be carried out continuously and extensively. China is diverse in climate, vegetation, and geographic structures and has rich niches for organisms [50,51]. Large-scale surveys in unexplored regions will significantly improve our knowledge of fungal species diversity.

5. Conclusions

The species diversity of the genus *Clonostachys* was investigated, and three new species were discovered. With the joining of the new species, the phylogenetic relationships among species of the genus are updated.

Author Contributions: Conceptualization, W.-Y.Z.; methodology, Z.-Q.Z.; software, Z.-Q.Z.; validation, Z.-Q.Z.; formal analysis, Z.-Q.Z.; investigation, Z.-Q.Z.; resources, W.-Y.Z. and Z.-Q.Z.; data curation, Z.-Q.Z.; writing—original draft preparation, Z.-Q.Z.; writing—review and editing, W.-Y.Z. and Z.-Q.Z.; visualization, Z.-Q.Z.; supervision, Z.-Q.Z. and W.-Y.Z.; project administration, W.-Y.Z.; funding acquisition, W.-Y.Z. and Z.-Q.Z. All authors have read and agreed to the published version of the manuscript.

Funding: This research was funded by the National Natural Science Foundation of China (31870012, 31750001), the National Project on Scientific Groundwork, Ministry of Science and Technology of China (2019FY100700), and the Frontier Key Program of the Chinese Academy of Sciences (QYZDY-SSWSMC029).

Institutional Review Board Statement: Not applicable.

Informed Consent Statement: Not applicable.

Data Availability Statement: The names of the new species were formally registered in the database Fungal Names (https://nmdc.cn/fungalnames (accessed on 11 August 2022)). Specimens were deposited in the Herbarium Mycologicum Academiae Sinicae (HMAS). The newly generated sequences were deposited in GenBank (https://www.ncbi.nlm.nih.gov/genbank (accessed on 22 September 2022)).

Acknowledgments: The authors would like to thank H.D. Zheng, X.C. Wang, Y.B. Zhang, Y. Zhang, and C. Liu for collecting specimens jointly for this study.

Conflicts of Interest: The authors declare no conflict of interest.

References

1. Schroers, H.J. A monograph of *Bionectria* (Ascomycota, Hypocreales, Bionectriaceae) and its *Clonostachys* anamorphs. *Stud. Mycol.* **2001**, *46*, 1–214.
2. Chen, W.H.; Han, Y.F.; Liang, J.D.; Zou, X.; Liang, Z.Q.; Jin, D.C. A new araneogenous fungus of the genus *Clonostachys*. *Mycosystema* **2016**, *35*, 1061–1069. (In Chinese)
3. Torcato, C.; Gonçalves, M.F.M.; Rodríguez-Gálvez, E.; Alves, A. *Clonostachys viticola* sp. nov., a novel species isolated from *Vitis vinifera*. *Int. J. Syst. Evol. Microbiol.* **2020**, *70*, 4321–4328. [CrossRef] [PubMed]
4. Han, P.P.; Zhang, X.P.; Xu, D.; Zhang, B.W.; Lai, D.W.; Zhou, L.G. Metabolites from *Clonostachys* fungi and their biological activities. *J. Fungi* **2020**, *6*, 229.
5. Zheng, C.J.; Kim, C.J.; Bae, K.S.; Kim, Y.H.; Kim, W.G. Bionectins A-C, epidithiodioxopiperazines with anti-MRSA activity, from *Bionectria byssicola* F120. *J. Nat. Prod.* **2006**, *69*, 1816–1819. [CrossRef] [PubMed]
6. Iqbal, M.; Dubey, M.; McEwan, K.; Menzel, U.; Franko, M.A.; Viketoft, M.; Jensen, D.F.; Karlsson, M. Evaluation of *Clonostachys rosea* for control of plant-parasitic *Nematodes* in soil and in roots of carrot and wheat. *Phytopathology* **2018**, *108*, 52–59. [CrossRef]
7. Zhuang, W.Y. (Ed.) *Flora Fungorum Sinicorum. Vol. Nectriaceae et Bionectriaceae*; Science Press: Beijing, China, 2013; p. 162. (In Chinese)
8. Rossman, A.Y. Lessons learned from moving to one scientific name for fungi. *IMA Fungus* **2014**, *5*, 81–89. [CrossRef]
9. Lombard, L.; Van der Merwe, N.A.; Groenewald, J.Z.; Crous, P.W. Generic concepts in Nectriaceae. *Stud. Mycol.* **2015**, *80*, 189–245. [CrossRef]
10. Dao, H.T.; Beattie, G.A.C.; Rossman, A.Y.; Burgess, L.W.; Holford, P. Four putative entomopathogenic fungi of armoured scale insects on *Citrus* in Australia. *Mycol. Prog.* **2016**, *15*, 47. [CrossRef]
11. Moreira, G.M.; Abreu, L.M.; Carvalho, V.G.; Schroers, H.J.; Pfenning, L.H. Multilocus phylogeny of *Clonostachys* subgenus *Bionectria* from Brazil and description of *Clonostachys chloroleuca* sp. nov. *Mycol. Prog.* **2016**, *15*, 1031–1039. [CrossRef]
12. Prasher, I.B.; Chauhan, R. *Clonostachys indicus* sp. nov. from India. *Kavaka* **2017**, *48*, 22–26.
13. Lechat, C.; Fournier, J. *Clonostachys spinulosispora* (Hypocreales, Bionectriaceae), a new species on palm from French Guiana. *Ascomycete.org* **2018**, *10*, 127–130.
14. Lechat, C.; Fournier, J. Two new species of *Clonostachys* (Bionectriaceae, Hypocreales) from Saül (French Guiana). *Ascomycete.org* **2020**, *12*, 61–66.

15. Tibpromma, S.; Hyde, K.D.; McKenzie, E.H.C.; Bhat, D.J.; Phillips, A.J.L.; Wanasinghe, D.N.; Samarakoon, M.C.; Jayawardena, R.S.; Dissanayake, A.J.; Tennakoon, D.S.; et al. Fungal diversity notes 840–928: Micro-fungi associated with Pandanaceae. *Fungal Divers.* **2018**, *93*, 1–160. [CrossRef]
16. Lechat, C.; Fournier, J.; Chaduli, D.; Lesage-Meessen, L.; Favel, A. *Clonostachys saulensis* (Bionectriaceae, Hypocreales), a new species from French Guiana. *Ascomycete.org* **2019**, *11*, 65–68.
17. Lechat, C.; Fournier, J.; Gasch, A. *Clonostachys moreaui* (Hypocreales, Bionectriaceae), a new species from the island of Madeira (Portugal). *Ascomycete.org* **2020**, *12*, 35–38.
18. Forin, N.; Vizzini, A.; Nigris, S.; Ercole, E.; Voyron, S.; Girlanda, M.; Baldan, B. Illuminating type collections of nectriaceous fungi in Saccardo's fungarium. *Persoonia* **2020**, *45*, 221–249. [CrossRef]
19. Hyde, K.D.; Dong, Y.; Phookamsak, R.; Jeewon, R.; Bhat, D.J.; Jones, E.B.G.; Liu, N.; Abeywickrama, P.D.; Mapook, A.; Wei, D.; et al. Fungal diversity notes 1151–1276: Taxonomic and phylogenetic contributions on genera and species of fungal taxa. *Fungal Divers.* **2020**, *100*, 5–277. [CrossRef]
20. Perera, R.H.; Hyde, K.D.; Maharachchikumbura, S.S.N.; Jones, E.B.G.; McKenzie, E.H.C.; Stadler, M.; Lee, H.B.; Samarakoon, M.C.; Ekanayaka, A.H.; Camporesi, E.; et al. Fungi on wild seeds and fruits. *Mycosphere* **2020**, *11*, 2108–2480. [CrossRef]
21. Zeng, Z.Q.; Zhuang, W.Y. Three new Chinese records of Hypocreales. *Mycosystema* **2017**, *36*, 654–662.
22. Hirooka, Y.; Kobayashi, T.; Ono, T.; Rossman, A.Y.; Chaverri, P. *Verrucostoma*, a new genus in the Bionectriaceae from the Bonin islands, Japan. *Mycologia* **2010**, *102*, 418–429. [CrossRef] [PubMed]
23. Nirenberg, H.I. Studies on the morphologic and biologic differentiation in *Fusarium* section *Liseola*. *Mitt. Biol. Bundesanst. Land-Forstw.* **1976**, *169*, 1–117.
24. Wang, L.; Zhuang, W.Y. Designing primer sets for amplification of partial calmodulin genes from penicillia. *Mycosystema* **2004**, *23*, 466–473.
25. White, T.J.; Bruns, T.; Lee, S.; Taylor, J. Amplification and direct sequencing of fungal ribosomal RNA genes for phylogenetics. In *PCR Protocols: A Guide to Methods and Applications*; Innis, M.A., Gelfland, D.H., Sninsky, J.J., White, T.J., Eds.; Academic Press: San Diego, CA, USA, 1990; pp. 315–322.
26. O'Donnell, K.; Cigelnik, E. Two divergent intragenomic rDNA ITS2 types within a monophyletic lineage of the fungus *Fusarium* are nonorthologous. *Mol. Phylogenet. Evol.* **1997**, *7*, 103–116. [CrossRef] [PubMed]
27. Gräfenhan, T.; Schroers, H.J.; Nirenberg, H.I.; Seifert, K.A. An overview of the taxonomy, phylogeny and typification of nectriaceous fungi in *Cosmospora*, *Acremonium*, *Fusarium*, *Stilbella* and *Volutella*. *Stud. Mycol.* **2011**, *68*, 79–113. [CrossRef] [PubMed]
28. Castlebury, L.A.; Rossman, A.Y.; Sung, G.H.; Hyten, A.S.; Spatafora, J.W. Multigene phylogeny reveals new lineage for *Stachybotrys chartarum*, the indoor air fungus. *Mycol. Res.* **2004**, *108*, 864–872. [CrossRef]
29. Carbone, I.; Kohn, L.M. A method for designing primer sets for speciation studies in filamentous ascomycetes. *Mycologia* **1999**, *91*, 553–556. [CrossRef]
30. O'Donnell, K.; Kistler, H.C.; Cigelnik, E.; Ploetz, R.C. Multiple evolutionary origins of the fungus causing Panama disease of banana: Concordant evidence from nuclear and mitochondrial gene genealogies. *Proc. Natl. Acad. Sci. USA* **1998**, *95*, 2044–2049. [CrossRef]
31. Hall, T.A. BioEdit: A user-friendly biological sequence alignment editor and analysis program for Windows 95/98/NT. *Nucleic Acids Symp. Ser.* **1999**, *41*, 95–98.
32. Thompson, J.D.; Gibson, T.J.; Plewniak, F.; Jeanmougin, F.; Higgin, D.G. The ClustalX windows interface: Flexible strategies for multiple sequences alignment aided by quality analysis tools. *Nucleic Acids Res.* **1997**, *25*, 4876–4883. [CrossRef]
33. Swofford, D.L. *PAUP 4.0b10: Phylogenetic Analysis Using Parsimony (* and Other Methods)*; Sinauer Associates: Sunderland, MA, USA, 2002.
34. Ronquist, F.; Huelsenbeck, J.P. MrBayes 3: Bayesian phylogenetic inference under mixed models. *Bioinformatics* **2003**, *19*, 1572–1574. [CrossRef] [PubMed]
35. Nylander, J.A.A. *MrModeltest v2.0. Program Distributed by the Author*; Evolutionary Biology Centre, Uppsala University: Uppsala, Sweden, 2004.
36. Nguyen, L.T.; Schmidt, H.A.; von Haeseler, A.; Minh, B.Q. IQ-TREE: A fast and effective stochastic algorithm for estimating maximum likelihood phylogenies. *Mol. Biol. Evol.* **2015**, *32*, 268–274. [CrossRef] [PubMed]
37. Chernomor, O.; von Haeseler, A.; Minh, B.Q. Terrace aware data structure for phylogenomic inference from supermatrices. *Mol. Syst. Biol.* **2016**, *65*, 997–1008. [CrossRef] [PubMed]
38. Page, R.D. TreeView: An application to display phylogenetic trees on personal computers. *Comp. Appl. Biosci.* **1996**, *12*, 357–358. [PubMed]
39. Cunningham, C.W. Can three incongruence tests predict when data should be combined? *Mol. Biol. Evol.* **1997**, *14*, 733–740. [CrossRef]
40. Spegazzini, C. Fungi costaricenses nonnulli. *Bol. Acad. Nac. Cienc. Córdoba* **1919**, *23*, 541–609.
41. Rossman, A.Y.; Samuels, G.J.; Rogerson, C.T.; Lowen, R. Genera of Bionectriaceae, Hypocreaceae and Nectriaceae (Hypocreales, Ascomycetes). *Stud. Mycol.* **1999**, *42*, 1–248.
42. Turland, N.J.; Wiersema, J.H.; Barrie, F.R.; Greuter, W.; Hawksworth, D.L.; Herendeen, P.S.; Knapp, S.; Kusber, W.H.; Li, D.Z.; Marhold, K.; et al. *International Code of Nomenclature for Algae, Fungi, and Plants (Shenzhen Code)*; Koeltz Scientific Books: Berlin, Germany, 2018; p. 254.

43. Rossman, A.Y.; Seifert, K.A.; Samuel, G.J.; Minnis, A.M.; Schroers, H.J.; Lombard, L.; Crous, P.W.; Põldma, K.; Cannon, P.F.; Summerbel, R.C.; et al. Genera in Bionectriaceae, Hypocreaceae, and Nectriaceae (Hypocreales) proposed for acceptance or rejection. *IMA Fungus* **2013**, *4*, 41–51. [CrossRef]
44. Zhu, S.; Ren, F.; Guo, Z.; Liu, J.; Liu, X.; Liu, G.; Che, Y. Rogersonins A and B, imidazolone *N*-oxide-incorporating indole alkaloids from a *ver*G disruption mutant of *Clonostachys rogersoniana*. *J. Nat. Prod.* **2019**, *82*, 462–468. [CrossRef]
45. Ouchi, T.; Watanabe, Y.; Nonaka, K.; Muramatsu, R.; Noguchi, C.; Tozawa, M.; Hokari, R.; Ishiyama, A.; Koike, R.; Matsui, H.; et al. Clonocoprogens A, B and C, new antimalarial coprogens from the Okinawan fungus *Clonostachys compactiuscula* FKR-0021. *J. Antibiot.* **2020**, *73*, 365–371. [CrossRef]
46. Sun, Z.B.; Li, S.D.; Ren, Q.; Xu, J.L.; Lu, X.; Sun, M.H. Biology and applications of *Clonostachys rosea*. *J. Appl. Microbiol.* **2020**, *129*, 486–495. [CrossRef] [PubMed]
47. Huang, T.; Xiang, L.G.; Zhuang, W.Y.; Zeng, Z.Q.; Yu, Z.H. Screening of antagonistic *Closnostachys* strains against *Phytophthora nicotianae*. *Microbiol. China* **2022**, *49*, 3860–3872.
48. Lee, S.A.; Kang, M.J.; Kim, T.D.; Park, E.J. First report of *Clonostachys rosea* causing root rot of *Gastrodia elata* in Korea. *Plant Dis.* **2020**, *104*, 3069. [CrossRef]
49. Haque, M.E.; Parvin, M.S. First report of *Clonostachys rosea* causing root rot of *Beta vulgaris* in North Dakota, USA. *New Dis. Rep.* **2020**, *42*, 21. [CrossRef]
50. Zeng, Z.Q.; Zhuang, W.Y. Current understanding of the genus *Hypomyces* (Hypocreales). *Mycosystema* **2015**, *34*, 809–816.
51. Yu, Y.N.; Mao, X.L. (Eds.) *100 Years of Mycology in China*; Science Press: Beijing, China, 2015; p. 763. (In Chinese)

Journal of
Fungi

MDPI

Article

Four New Species and New Records of Orbilia from China Based on Molecular and Morphological Data

Xiao-Yun Ou [1], Yuan-Yuan Shao [2], Hai-Fu Zheng [1,3] and Bin Liu [1,*]

1 Institute of Applied Microbiology, College of Agriculture, Guangxi University, Nanning 530005, China
2 Key Laboratory of Beibu Gulf Environment Change and Resources Utilization of Ministry of Education, Nanning Normal University, Nanning 530001, China
3 Guangxi Forest Inventory & Planning Institute, Nanning 530011, China
* Correspondence: liubin@gxu.edu.cn

Abstract: This study reports four new species and three new record species of Orbiliaceous fungi from China. *Orbilia baisensis*, *O. hanzhongensis*, *O. nanningensis* and *O. pinea* are described as new species and *O. crenatomarginata*, *O. vinosa* and *O. vitalbae* are described as new record species. All the studied species were identified by morphological characteristics and phylogenetic analysis of internal transcribed spacer (ITS) and large subunit (LSU) sequences. Four new species are described based on their sexual and asexual states, and their differences with the close relatives were compared and discussed.

Keywords: *Orbilia*; new species; morphology; phylogenetic; taxonomy

Citation: Ou, X.-Y.; Shao, Y.-Y.; Zheng, H.-F.; Liu, B. Four New Species and New Records of Orbilia from China Based on Molecular and Morphological Data. *J. Fungi* **2022**, *8*, 1188. https://doi.org/10.3390/jof8111188

Academic Editors: Cheng Gao and Lei Cai

Received: 31 August 2022
Accepted: 6 November 2022
Published: 11 November 2022

1. Introduction

The family *Orbiliaceae* is characterized by producing tiny, waxy, translucent, light-colored, sessile to sub-stipitate apothecia with small ascospores, which are asymmetrically globose to sub-fusoid [1]. Members of family *Orbiliaceae* are widely distributed in the environment and sporadically in arid habitats as saprophytic, parasitic or superficial on tree bark, deadwood, withered leaf and animals' excrement [2,3]. The most prominent feature of the family *Orbiliaceae* is the presence of a plasmatic spore body which is a strongly refractive vacuolar in the ascospore and is only visible in the living state [4]. The genus of *Orbilia* was established to accommodate *Orbilia leucostigma* [5], and the family *Orbiliaceae* was recognized by Nannfeldt [6] and assigned to the order Helotiales, which was revised to the class Leotiomycetes [4,7]. Attributed to the morphological features and molecular phylogenetic evidence, *Orbiliaceae* was transferred to the order *Orbiliales* and the class *Orbiliomycetes*, comprising two teleomorphic genera, *Orbilia* Fr. and *Hyalorbilia* Baral [8]. The third teleomorphic genus *Pseudorbilia* includes only one species carrying the characteristics intermediate between *Orbilia* and *Hyalorbilia* [9,10]. In the past decade, there is a continuous documentation of new species and taxonomic reforms, depicting the evolutionary changes in the diversity of the genus *Orbilia* [1,4,11–20]. Seven sexual-type genera and three asexual-type genera are accepted in the family *Orbiliaceae*; the teleomorph-typified include *Amphosoma*, *Bryorbilia*, *Liladisca*, *Lilapila*, *Pseudorbilia*, *Hyalorbilia* and *Orbilia*, while the anamorph-typified include *Lecophagus*, *Mycoceros*, *Retiarius*, and 415 species of *Orbilia* have been assigned among these genera [3].

The internal transcribed spacer (ITS) [21] and the large subunit gene of the rDNA (LSU) [22] region are extensively employed in phylogeny studies of fungi and these two markers have been proven to be effective to study the phylogenetic relationship in Orbiliaceous fungi. More recently, in order to overcome the ambiguities associated with *Orbilia leucostigma* and *Orbilia xanthostigma*, Baral revealed the high variation of ITS and LSU and presented distinct genotypes [23]. However, some studies have also made use of several other genes, which include the translation elongation factor 1-alpha (TEF) [24], beta-tubulin (TUB) [25,26], RNA polymerase second largest subunit (RPB2) [27,28] and chitin synthase 1 (CHS-1) [29], etc.

The concept of sexual and asexual states in Orbiliomycetes was first established by Brefeld [30] but it was well explained by Pfister [31]. The asexual states of *Orbiliaceae* include *Arthrobotrys, Dactylella, Dactylellina, Dicranidion, Drechslerella, Helicoon, Tridentaria, Trinacrium*, etc., while the *Arthrobotrys, Dactylellina* and *Drechslerella* belong to nematode-trapping fungi. Harkness established *Dicranidion* based on *Dicranidion fragile* Harkness [32], while the genus *Dicranidion* was placed into the section 2 of Hyphomycetes by Hughes [33]. The conidia of *Dicranidion* consist of two or three lobes, and in some species, conidia have multiple lobes, the lobes are equal or unequal, parallel or unparallel, septate or non-septate [34]. Brefeld firstly reported *Dicranidion* sp. that isolated from *Orbilia*, Berthet described the conidia of *Orbilia xanthostigma* with illustrations [35].

China is rich in endemic species resources and biological diversity owing to its varying environmental and geographic regions. During surveys of the orbiliaceous fungi from Guangxi Province of Southwestern China and Shaanxi province of Northwestern China, seven species of *Orbilia* were found and identified based on morphological evidence together with LSU and ITS sequence data. Among them, *Orbilia baisensis, Orbilia hanzhongensis, Orbilia nanningensis* and *Orbilia pinea* are described as new species, and *Orbilia crenatomarginata, Orbilia vinosa* and *Orbilia vitalbae* are described as new Chinese record.

2. Materials and Methods

2.1. Morphological Studies

Fresh specimens were collected from decayed and fallen tree branches and wood logs from Guangxi and Shaanxi provinces, China. In the description, the symbols were adopted as follows: * = living state, † = dead state. The specimens were dried and deposited in the GXU (Herbarium of Institute of Applied Microbiology, Guangxi University, China).

To obtain a pure culture, a fresh apothecium was fixed to the lid of a petri dish with the hymenia facing downward, allowing the ascospores to shoot on the surface of the water agar (18 g agar, 1 L distilled water). After germination, the ascospores deposits were transferred onto PDA (Potato Dextrose Agar) plates [36], MEA (Malt Extract Agar) plates, CMA (Corn Meal Agar) plates and LY (Lactose-Yeast Extract Agar) plates and incubated for 5–10 days at 25 °C. Cultures was deposited in the Institute of Applied Microbiology, Guangxi University, China. Observations and photographs were taken with a Nikon Eclipse 80i microscope (Nikon Corporation, Tokyo, Japan) equipped with Nikon Digital Sight DS-L1 microphotographic system. All the morphological measurements were recorded from 20 elements in water mounts employing Spot32 software v4.0.8 (Diagnostic Instruments, Sterling Heights, MI, USA).

2.2. DNA Extraction, PCR Amplification and Sequencing

Mycelia from the fresh cultures were inoculated in the potato dextrose broth (PDB) and were cultured under dark conditions in a thermostable shaker at 25 °C. After 2 weeks of shaking, mycelia were collected and washed with sterile distilled water and were used to extract DNA by CTAB method [37]. For those species without pure cultures, the sequencing DNA was directly amplified from the hydrated apothecia as described by Vitória et al. [38]. Briefly, the apothecia were placed in a PCR tube using a needle and stored at −80 °C for 12–24 h. One apothecium was transferred to a PCR tube containing 3 μL Cell Lysis Buffer, and vortexed for 2 min at maximum speed followed by incubation at 80 °C for 15 min. The samples were preserved at −20 °C for later use or directly used for PCR amplification.

Sequence data were generated from the internal transcriber spacer region of nuclear ribosomal DNA (ITS) and the large subunit of the rDNA genes (LSU) using primer pairs ITS1/ITS4 [21] and LROR/LR5 [22]. PCR amplification was performed in a reaction mixture of 50 μL containing 25 μL 2X Taq-Plus PCR Master Mix, 1 μL of each primer, 22 μL of doubly distilled H_2O, 1 μL of DNA template and the total. PCR reaction conditions were as follows: initial denaturation at 94 °C for 5 min; followed by 35 cycles of denaturation at 94 °C for 1 min, annealing at 56 °C for 1 min and extension at 72 °C for 1 min; and a final extension at 72 °C for 5 min. Amplified PCR products were separated on 1% agarose gel and examined under the UV light. PCR products were sequenced from the Beijing Genomics Institute (BGI).

2.3. Phylogenetic Analysis

Thirteen new sequences were generated in this study. To establish the preliminary identification of the studied species, the acquired sequences were first carefully examined for intactness and then blasted in the NCBI nucleotide sequence blast and were compared with the already published data. The obtained sequences of ITS and LSU were then used for phylogenetic studies. Related sequences of similar species were downloaded from NCBI GenBank and the sequences data sets were aligned using Clustal X 1.83 [39] and converted to FASTA files and constructed maximum likelihood tree by MEGA version 6.06 using the Kimura 2-parameter model [40]. The sequences were converted to NEXUS files by Phylosuite [41] and the partition homogeneity test was performed with 1000 replicates in PAUP*4 [42]. Nucleotide substitution models were selected by MrModeltest 2.31 [43]. The corresponding phylogenetic trees were constructed using the maximum likelihood and Bayesian inference analyses. Maximum likelihood analyses were performed with MEGA version 6.06, and Bayesian inference analyses were carried out using MrBayes v3.2.2 [44]. The tree was viewed in Fig Tree v1.4.4 [45]. The maximum likelihood bootstrap proportions (MLBP) were above 50% and Bayesian inference posterior probabilities (BIPP) greater than 0.95 at nodes. GenBank accession numbers are given in Table 1.

Table 1. GenBank accession numbers of taxa used in phylogenetic analyses.

Species	Strain Number	GenBank Accession Number	
		ITS	LSU
Dactylella clavata	YNUCC 5628	AY515568	AY261174
Dicranidion fissile	NBRC 31823	LC146730	LC146730
Hyalorbilia inflatula	H.B. 9080	KT222442	KT222442
Orbilia amarilla	TFC Mic. 23767	MH221071	MH221071
Orbilia baisensis *****	**DL17 (GXU2279)**	**OP225323**	**OP231636**
Orbilia baisensis *****	**BY44 (GXU2373)**	**OP585655**	**OP591332**
Orbilia cladodes	D.H.P. 90	U72592	U72592
Orbilia crenatomarginata ******	**(GXU2342)**	**OP225327**	**OP231640**
Orbilia crenatomarginata ******	**(GXU2343)**	**OP585656**	**OP591333**
Orbilia crenatomarginata ******	**(GXU2383)**	**OP585657**	**OP591334**
Orbilia crenatomarginata	H.B. 9452	KM248771	KM248771
Orbilia crenatomarginata	H.B. 9265	KM248772	KM248772
Orbilia eucalypti	G.M. 2015-10-02.1	MK473434	MK473434
Orbilia farnesianae	H.B. 8997h	KT222421	KT222421
Orbilia aff. *farnesianae*	B.L. 4090 (HMAS 139700)	DQ656643	DQ656688
Orbilia fissilis	CBS 117019	KT596781	KT596781
Orbilia gambelii	CBS 140815	KT215249	KT215249
Orbilia gambelii	G.M. 2018-09-12.12	OP237019	OP237019
Orbilia hanzhongensis *****	**BY35 (GXU2365)**	**OP225324**	**OP231637**
Orbilia leucostigma	H.B. 9958a	KY419187	KY419187
Orbilia nanningensis *****	**NN01 (GXU2466)**	**OP225326**	**OP231639**
Orbilia cf. *paracaudata*	H.B. 8685	KT222422	KT222422
Orbilia pilifera	G.F. 20110193	MK473413	MK473413
Orbilia pinea *****	**BY38 (GXU2368)**	**OP225325**	**OP231638**
Orbilia rectispora	M.E. 02-20-01	KT215289	KT215289
Orbilia renispora	GXU1487	MG742403	MG742404
Orbilia vinosa ******	**(GXU2394)**	**OP225328**	**OP231641**
Orbilia vinosa ******	**(GXU2397)**	**OP585658**	**OP591335**

Table 1. *Cont.*

Species	Strain Number	GenBank Accession Number	
		ITS	LSU
Orbilia vinosa **	**(GXU2421)**	**OP585659**	**OP591336**
Orbilia vinosa	G.M. 2014-02-14	KT380089	KT380089
Orbilia vinosa	CBS 116215	KT215266	KT215266
Orbilia vitalbae **	**(GXU2438)**	**OP225329**	**OP231642**
Orbilia vitalbae **	**(GXU2442)**	**OP585660**	**OP591337**
Orbilia vitalbae	H.B. 9905a	KT380075	KT380075
Orbilia xanthostigma	G.M. 2015-08-15-4b	KY419181	KY419181
Orbilia cf. *xanthostigma*	D.H.P. 120	U72593	U72593
Orbilia xinjiangensis	CBS 232.51	MH856835	MH856835
Orbilia xinjiangensis	H.B. 9646	KT222435	KT222435

Note: * new species, ** new Chinese record, specimen numbers are shown in parentheses, sequences newly generated in this study are in bold.

3. Results

3.1. Taxonomic Description

3.1.1. New Species

Orbilia baisensis X.Y. Ou and Bin Liu, sp. nov. (Figure 1).

MycoBank: MB 846093.

Etymology: from the geographical origin, Baise (Guangxi).

Holotype: CHINA, Guangxi province, Baise city, Dawangling drift scenic spot, from deadwood of *Castanea mollissima* on the ground, 11 July 2016, X.Y. Ou, GXU2279. Strain DL17 was isolated from GXU2279.

Sexual state: Apothecia superficial on the deadwood of *Castanea mollissima*, 0.2–1.4 mm in diameter, gregarious in groups or scattered, waxy, translucent, smooth, disc slight concave to flat, margin not protruding, sessile, yellow when fresh or rehydrated, turned yellow to orange when dry. Asci †12.2–39.3 × 2.3–4.6 µm, cylindric-clavate, 8-spored, pars sporifera †18.5–23.6 µm long, truncate to hemispherical at the apex, gradually narrowing towards to the base, flexuous stalk, forked to L-, T- or Y-shaped. Ascospores †2.9–4.6 × 1.6–2.1 µm, hyaline, non-septate, smooth, pronounced reniform, strongly curved ~90–150°, one end round to obtuse, the other end small with a short pointed base, pairs of ascospores arranged in the ascus; containing one refractive globose SB (spore body) at the end close to the wall in alive mature ascospores, 0.6–1.1 µm diameter. Paraphyses apically inflated to mostly capitate (-clavate), †21.9–29.0 × 0.8–1.9 µm, basally branched and expanding to 1.8–3.1 µm in diameter at the apex. Hymenium 70.5–192.3 µm thick; medullary excipulum 25.8–40.6 µm thick, of ± loose textura intricata with ± inflated cells, sharply delimited. Ectal excipulum composed of textura globulosa-angularis, thin-walled to slightly gelatinized, 38.8–146.6 µm thick, cells †3.2–7.9 × 2.0–5.5 µm diam., ovate to spherical.

Asexual state: *Dicranidion*-like.

Colonies beige-white on PDA, 34.8 mm in diameter at 25 °C after 10 d, strongly keratinized, aerial hyphae absent; beige-white on MEA, 24.8 mm diam., aerial hyphae sparse; white on CMA and LY, 11.8–12.0 mm diam., aerial mycelium absent; Hyphae hyaline, septate, branched, smooth and 1.4–3.2 µm wide. Conidiophores hyaline, erected or slightly bent, septate, unbranched, *5.7–33.1 µm long, *2.0–2.8 µm wide at the base and gradually tapering to *0.9–1.5 µm wide at the tip where bearing 1 apical spore. Conidia thallic, hyaline, Y-shaped, consisted of a spindle and two equal or unequal lobes; the spindle *6.0–11.5 µm long, 1–2 septate; the lobe *2.0–6.7 × 1.5–2.3 µm, 1 septate; in addition, columnar conidia *6.3–20.0 × 1.8–2.6 µm, 1–3 septate.

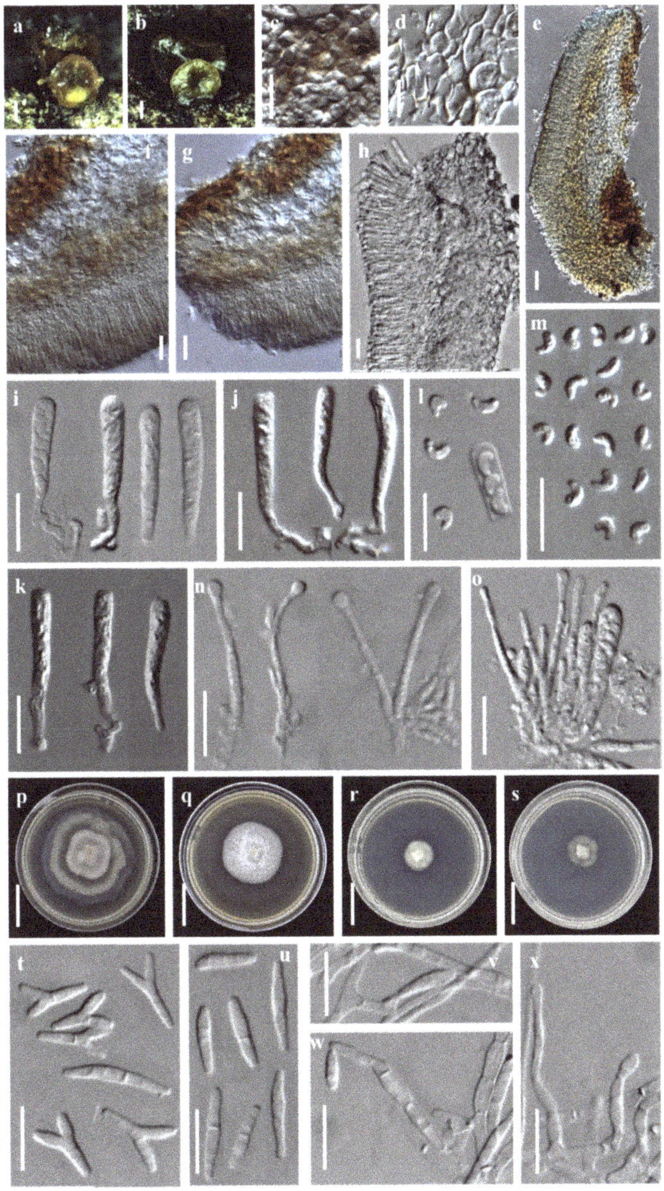

Figure 1. *Orbilia baisensis* and *dicranidion*-like asexual morph (strain DL17 was isolated from GXU2279). (**a**,**b**) apothecia; (**c**,**d**) basal excipular cells; (**e–h**) vertical section of apothecium; (**i–k**) ascus; (**l**,**m**) ascospores; (**n**) paraphyses; (**o**) asci and paraphyses; (**p–s**) colony after 10 d at 25 °C, (**p**) on PDA, (**q**) on MEA, (**r**) on CMA, (**s**) on LY; (**t**,**u**) conidia; (**v–x**) conidiophores with conidia. Scale bars: (**a**,**b**) = 0.5 mm; (**c**,**d**,**f**,**g**,**i–o**,**t–x**) = 10 μm; (**e**,**h**) = 20 μm; (**p–s**) = 10 mm.

Additional specimen examined: China, Shaanxi province, Hanzhong city, Foping County, Wangjiawan, from deadwood, 23 July 2017, X.Y. Ou, GXU2373.

Notes: *Orbilia baisensis* is clustered with *O. renispora* Y.Y. Shao, Quijada, Baral, Haelew. and Bin Liu, *O. leucostigma* (Fr.) Fr., *O. xanthostigma* (Fr.) Fr. and *O.* cf. *xanthostigma* (as

O. delicatula) by having reniform to nephroid or C-shaped ascospores and their asexual states are belonging to *Dicranidion*. However, *O. baisensis* features on 8-spored asci, flexuous base and forked to L-, T- or Y-shaped, pronounced reniform ascospores, round and larger at one end, the other end with a short pointed base, the middle section being the widest and capitate paraphyses. *Orbilia renispora* differs from *O. baisensis* by the smaller (3.0–3.6 × 1.5–1.8) and lower curl ascospores. *O. leucostigma* and *O. xanthostigma* differ from *O. baisensis* by the equant end of ascospores. In addition, molecular analyses reveal that *O. baisensis* shares less than 91.20% similarity with *O. renispora* in ITS sequence, and 89.91% similarity with *O. xanthostigma* and 89.53% similarity with *O. leucostigma*, respectively. Both the morphological and the molecular evidence support their separation at the species level.

Orbilia hanzhongensis X.Y. Ou & Bin Liu, sp. nov. (Figure 2)

MycoBank: MB 846094.

Etymology: according to the geographical origin, Hanzhong (Shaanxi).

Holotype: CHINA, Shaanxi province, Hanzhong city, Foping County, Wangjiawan, from fallen branch, 23 July 2017, X.Y. Ou, holotype GXU2365. Strain BY35 was isolated from GXU2365.

Sexual state: Apothecia rehydrated 0.1–1.5 mm diam., superficial on the deadwood, gregarious, waxy, medium translucent, smooth, disc round and slight to strongly concave, sometimes flat, margin not protruding, broadly sessile, pale to light yellow when fresh or rehydrated, yellow when dry. Asci †21.7–39.7 × 2.3–3.6 μm, cylindric-clavate, 8-spored, spores uniseriate, pars sporifera †20.5–30.7 μm long, truncate to hemispherical at the apex, the base with short to medium long, flexuous stalk, forked to L-, T-, H- or Y-shaped. Ascospores †2.5–4.4 × 1.7–2.2 μm, hyaline, non-septate, smooth, fusoid to ellipsoid, to ovoid, to lemon-shaped, one end subacute to acute, other end round or often acute, straight; SBs †0.2–0.4 μm diam., globose, at the end close to the wall in alive mature ascospores. Paraphyses straight to slightly capitate at the apex, †18.8–40.3 × 1.5–1.9 μm, the base branched and expanded to 2.0–3.0 μm in diameter at the apex. Hymenium 48.2–73.8 μm thick; medullary excipulum 16.6–22.2 μm thick, composed of medium dense textura intricata with inflated cells, sharply delimited. Ectal excipulum composed of textura globulosa-angularis from the base to the flanks, thin-walled, slightly gelatinized, 30.2–51.4 μm thick, cells †3.8–10.8 × 3.2–9.9 μm diam., ovate to spherical.

Asexual state: *Dactylella*-like.

Colonies beige-white on PDA, 50.5 mm diam. at 25 °C after 10 d, aerial hyphae dense; beige-white on MEA, 31.7 mm diam., aerial hyphae rarely sparse; white on CMA, 60 mm diam., aerial hyphae absent; white on LY, 22.7 mm diam., aerial mycelium absent. Hyphae hyaline, septate, branched, smooth, *1.0–3.5 μm wide. Conidiophores hyaline, erected, septate, unbranched, *10.6–72.8 μm long, *2.0–4.2 μm wide at the base and gradually tapering to *0.9–1.5 μm wide at the tip where bearing 1–3 apical spore. Conidia thallic, hyaline, unbranched, cylindric-ellipsoid (-clavate), obtuse at one end, truncate at the other end, straight when mature, slightly inflect when the immature, *10.1–20.3 × 2.6–4.2 μm, 1 septate.

Additional specimen examined: China, Shaanxi province, Hanzhong city, Foping County, Wangjiawan, from rotten branches, 23 July 2017, X.Y. Ou, GXU2379.

Notes: *Orbilia hanzhongensis* is characterized by ellipsoid to ovoid ascospores having one end subacute to acute, and mostly acute on the other end, globose SBs, capitate paraphyses. It is related to *O. rectispora* (Boud.) Baral and *O. xinjiangensis* (J. Chen, L.L. Xu, B. Liu and Xing Z. Liu) E. Weber, Baral and Helleman, but *O. rectispora* differs in having narrowly cylindrical to fusoid-clavate and larger ascospores (†5–9 × 0.9–1.2 μm), and *O. xinjiangensis* differs in distinctly larger spores (†7–9 × 1.8–2 μm) and larger conidia (*45–54 × 8–11 μm) with more septa [3]. Moreover, there is 51 bp (8.46%) divergence in the ITS region between *O. hanzhongensis* and *O. xinjiangensis*, and 90 bp (17.82%) divergence in the ITS region of *O. rectispora*. Both the morphology and DNA sequence data distinguish them as different species.

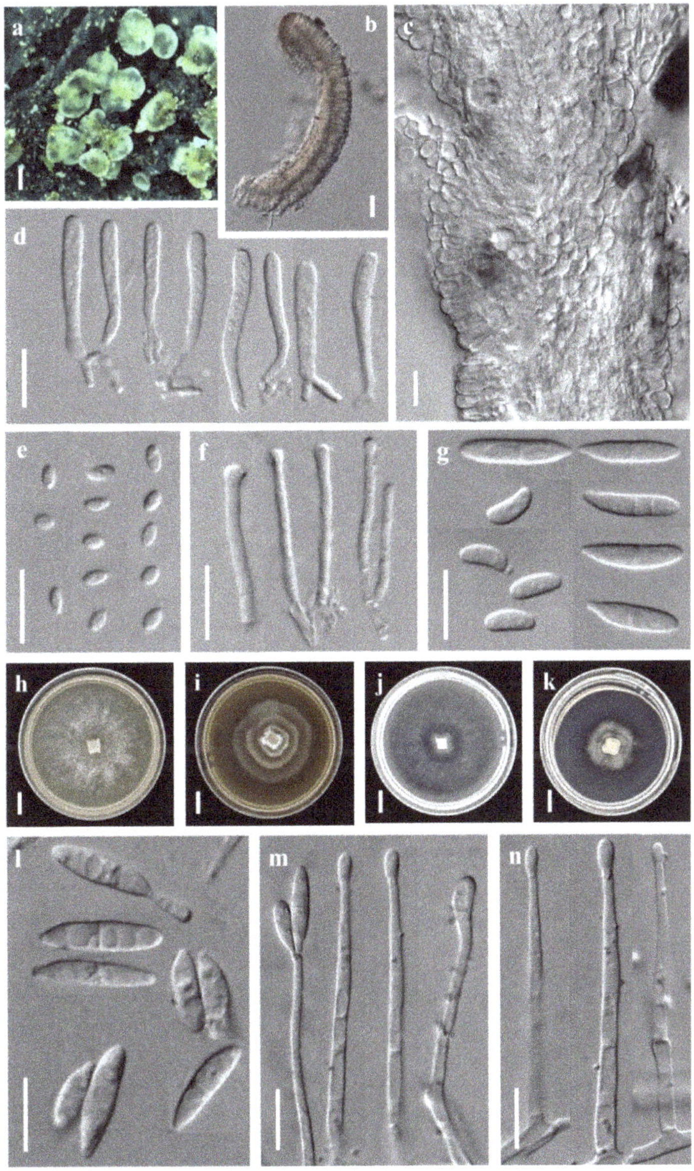

Figure 2. *Orbilia hanzhongensis* and *dactylella*-like asexual morph (strain BY35 isolated from GXU2365). (**a**) apothecia; (**b**) vertical section of apothecium; (**c**) basal excipular cells; (**d**) ascus; (**e**) ascospores; (**f**) paraphyses; (**h–k**) colony after 10 d at 25 °C, (**h**) on PDA, (**i**) on MEA, (**j**) on CMA, (**k**) on LY; (**g**,**l**) conidia; (**m–n**) conidiophores with conidia. Scale bars: (**a**) = 0.5 mm; (**b**) = 20 μm; (**c–g**,**l–n**) = 10 μm; (**h–k**) = 10 mm.

Orbilia nanningensis X.Y. Ou & Bin Liu, sp. nov. (Figure 3).

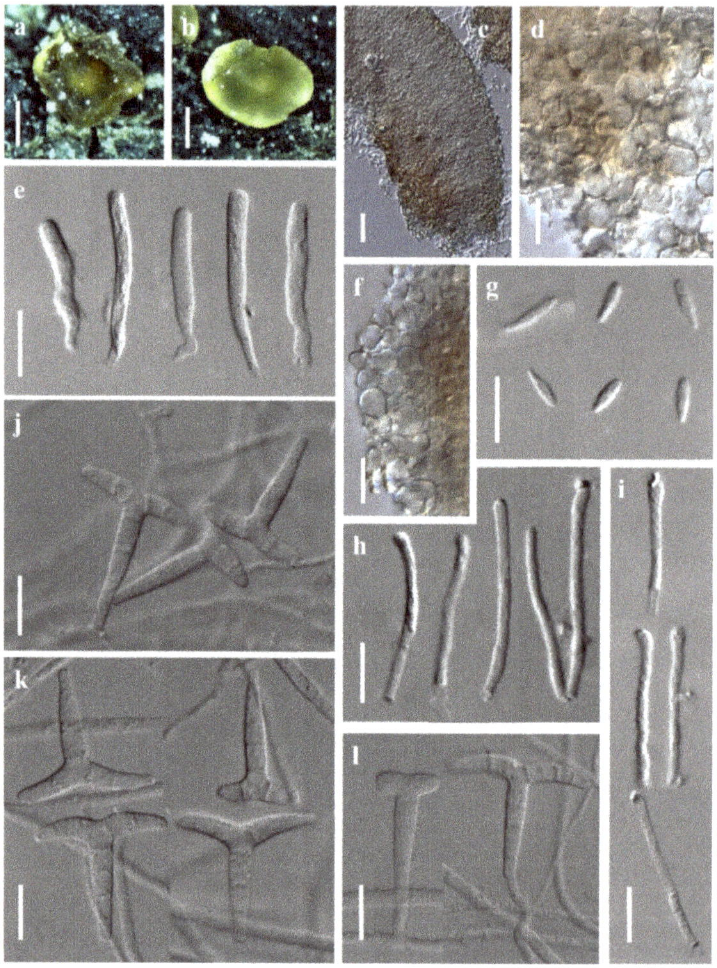

Figure 3. *Orbilia nanningensis* and *trinacrium*-like asexual morph (strain NN01 was isolated from GXU2466). (**a**,**b**) apothecia; (**c**) vertical section of apothecium; (**d**,**f**) basal excipular cells; (**e**) ascus; (**g**) ascospores; (**h**,**i**) paraphyses; (**j**,**k**) conidia; (**l**) conidiophores with conidia. Scale bars: (**a**,**b**) = 0.2 mm; (**c**) = 20 μm; (**d**–**l**) = 10 μm.

MycoBank: MB 846095.

Etymology: name after the geographical origin, Nanning (Guangxi).

Holotype: CHINA, Guangxi province, Nanning city, Xixiangtang District, Shibu Town, from deadwood on the ground, 1 January 2020, X.Y. Ou, holotype GXU2466. Strain NN01 was isolated from GXU2466.

Sexual state: Apothecia rehydrated 0.2–0.5 mm diam., scattered, round, light yellow-orange, translucent, sessile, superficial on dead branches on the ground, orange when fresh, dry deep yellow to orange, disc strongly concave, wet light yellow, disc flat, smooth, margin thin or thick. Asci 20.9–55.7 × 2.4–4.6 μm, pars sporifera 14.0–37.2 μm, cylindric to clavate, 8-spored, the apex truncate to hemispherical, thin-walled, gradually narrowing towards the base, flexuous stalk, unforked. Ascospores †6.2–7.5 × 1.6–2.1 μm, clavate to fusoid with a distinct short tapered, rarely ellipsoid, one end obtuse, tapered 1.0–2.5 µm long, straight or slightly curved. Paraphyses straight to slightly capitate at the apex, †17.8–37.5 × 1.2–2.0 µm, basally unbranched and expanding to 1.9–3.2 µm in diameter, exudate 0.8–1.3 µm thick, over paraphyses. Hymenium 101.8–151.2 µm thick; medullary excipulum 56.8–79.0 µm thick, always composed of dense textura intricata with many inflated cells, sharply delimited. Ectal excipulum 53.6–87.4 µm thick, of thin-walled, composed of oriented textura globulosa-angularis from the base to the flanks or margin, cells †5.7–12.1 × 3.7–8.3 µm diameter.

Asexual state: *Trinacrium*-like.

Colonies white on PDA. Mycelium *1.7–3.2 µm wide. Conidiophores unbranched, erected or slightly bent, septate, *7.1–15.9 µm long, the base *1.8–2.0 µm wide, the tip *1.0–1.4 µm wide where bearing 1 apical spore. Conidia thallic, T-shaped, consisting of one stipe and two arms, the two arms bent downwards, total size *20.7–32.7 × 14.9–30.8 µm, the stipe *16.9–28.7 × 2.8–4.1 µm, 3–5 septate, the arms *6.0–13.8 × 2.3–3.6 µm, 1–3 septate.

Additional specimens examined: China, Guangxi province, Nanning city, Lewan farm, from deadwood, 1 January 2020, X.Y. Ou, GXU2467.

Notes: *Orbilia nanningensis* is clustered with *O.* cf. *paracaudata* Baral and G. Marson, *O. farnesianae* Baral, *O. pilifera* Baral and R. Galán, *O.* aff. *farnesianae* and *O. amarilla* Quijada and Baral. Their ascospores were fusoid to clavate with a short tapered, and with similar T-shaped of conidia, but *O. nanningensis* differs from other related species by smooth margin of apothecia and straight to slightly capitate paraphyses at the apex. Among the known species of *Orbilia*, *O.* cf. *paracaudata* is the most closely related to *O. nanningensis* in the phylogenetic tree, there is only a distance of 3.80% in the ITS region between *O. nanningensis* and *O.* cf. *paracaudata*. However, *O. paracaudata* can be separated from *O. nanningensis* by its longer and narrower ascospores (5.8–8.5 × 1.6–1.8 µm) and a distinctly protruding apothecial margin.

Orbilia pinea X.Y. Ou & Bin Liu, sp. nov. (Figure 4).

MycoBank: MB 846096.

Etymology: named after the host from which it was collected, Pinus.

Holotype: CHINA, Shaanxi province, Hanzhong city, Foping County, Wangjiawan, from deadwood of pinus on the ground, 23 July 2017, X.Y. Ou, holotype GXU2368. Strain BY38 was isolated from GXU2368.

Sexual state: Apothecia rehydrated 1.0 mm diam., yellowish to orange, translucent, round, superficial and scattered, waxy, smooth, disc flat, margin thin and not protruding, sessile, dry orange or honey-yellow when fresh. Asci †26.5–41.0 × 2.6–4.2 µm, cylindric-clavate, 8-spored, spores uniseriate, ~3-seriate, ~4 lower spores inverted (sometimes mixed), pars sporifera †16.9–26.8 µm long, the apex strongly truncate or round to hemispherical, the base gradually narrowing with short to medium long and flexuous stalk, forked to L-, H- or Y-shaped. Ascospores †2.5–3.3 × 1.5–2.2 µm, hyaline, non-septate, smooth, pronounced reniform, strongly curved, ~48–158°, end round, rarely obtuse, middle largest; SBs globose, †0.4–0.6 µm diameter, usually close to one end in alive mature ascospores. Paraphyses apically inflated to capitate at the apex, sometimes uninflated or slightly inflated to sublageniform, †17.3–45.3 × 1.3–2.8 µm, branched at the base and expanded to 2.3–5.4 µm in diameter at the apex. Hymenium 77.6–139.0 µm thick; medullary excipulum 46.6–65.0 µm thick, subhyaline, composed of dense loose textura intricata, sharply delimited. Ectal excipulum 53.2–92.4 µm thick, hyaline, composed of thin-walled, textura globulosa-angularis from the base to the margin, cells †6.2–22.2 × 5.0–13.7 µm diameter.

Asexual state: *Dicranidion*-like.

Colonies beige-white on PDA, 20.0 mm diam. at 25 °C after 10 d, aerial hyphae absent; beige-white on MEA, 23.7 mm diam., aerial hyphae sparse; grow very slowly on CMA, only 15 mm diam. at 25 °C after 30 d, and could not grow on LY. Hyphae hyaline, septate, branched, smooth. Conidiophores hyaline, erected or slightly bent, septate, unbranched at the base, the tip where bearing 1 apical spore. Conidia thallic, hyaline, Y-shaped, consisted of a stipe and two equal or unequal arms; the stipe *5.9–10.2 × 2.2–2.7 µm, 1 septate; the arms *2.1–4.5 × 1.6–2.4 μm, 1 septate; in addition, columnar conidia *13.4–13.7 × 2.5–2.7 μm, 1–3 septate.

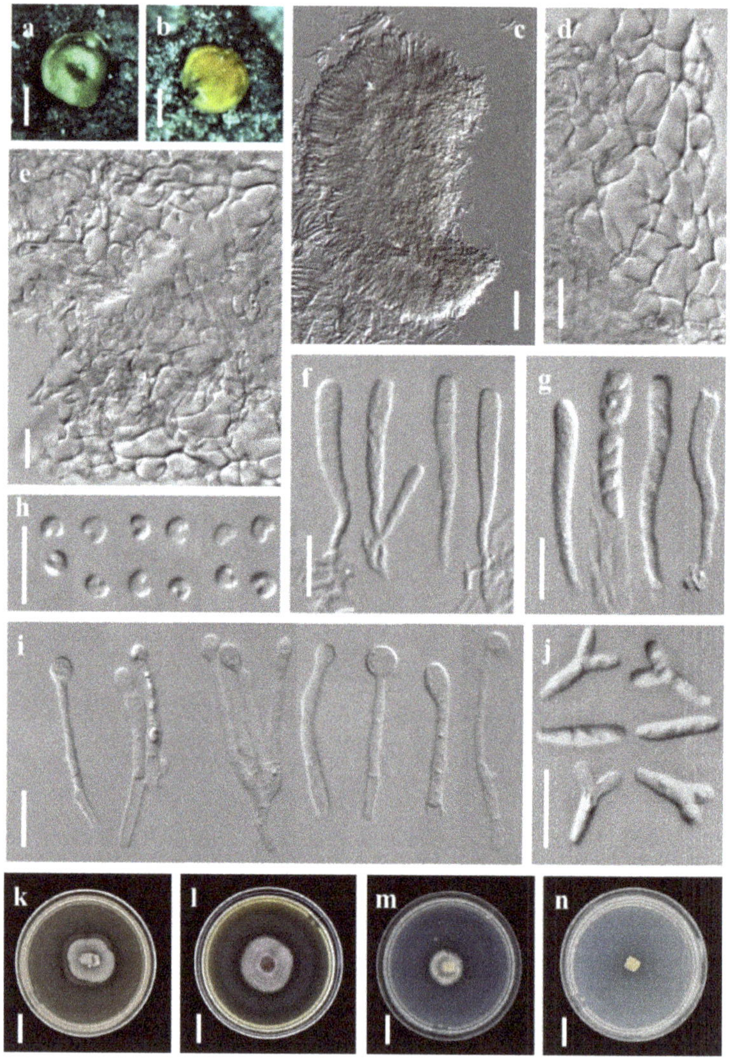

Figure 4. *Orbilia pinea* and *dicranidion*-like asexual morph (strain BY38 was isolated from GXU2368). (**a**,**b**) apothecia; (**c**) vertical section of apothecium; (**d**,**e**) basal excipular cells; (**f**,**g**) ascus; (**h**) ascospores; (**i**) paraphyses; (**j**) conidia; (**k**–**n**) colony at 25 °C, (**h**) on PDA after 10 d, (**i**) on MEA after 10 d, (**j**) on CMA after 30 d, (**k**) on LY after 30 d; Scale bars: (**a**,**b**) = 0.5 mm; (**c**–**j**) = 10 μm; (**k**–**n**) = 10 mm.

Notes: *Orbilia pinea* is most similar to *O. fissilis* (K. Ando and Tubaki) E. Weber and Baral, the most remarkable feature of *O. pinea* is pronounced reniform and strongly curved ascospores, *O. fissilis* differs in broadly ellipsoid to subglobose ascospores and frequently 4-armed of the *Dicranidion*-like conidia. There is only a distance of 2.36% in the ITS region between *O. pinea* and the type strain of *O. fissilis*. Obviously, they are not conspecific.

3.1.2. New Record Species

Orbilia crenatomarginata (Höhn.) Sacc. & Trotter, Syll. Fung. 22: 725 (1913) (Figure 5).

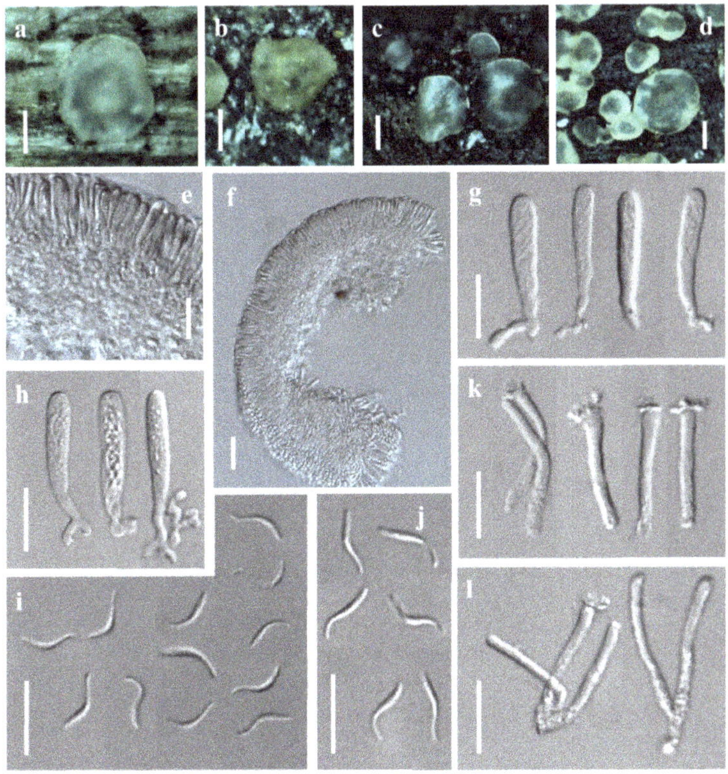

Figure 5. *Orbilia crenatomarginata* ((**a**,**e**,**f**,**g**,**j**,**k–l**) from GXU2342; (**b**,**c**) from GXU2343; (**d**,**h**,**i**) from GXU2383). (**a–d**) apothecia; (**e**,**f**) vertical section of apothecium; (**g**,**h**) ascus; (**i**,**j**) ascospores; (**k**,**l**) paraphyses. Scale bars: (**a**,**b**) = 0.2 mm; (**e**,**g–l**) = 10 μm; (**f**) = 20 μm.

Sexual state: Apothecia 0.1–0.5 mm in diameter, scattered on the surface of rotten wood, superficial, flat or slightly convex, smooth, sessile, margin protruding (or denticulate) and filamentous, with small and distinct triangular teeth, dry pale or light yellow to cream-carneous, rehydrated and fresh greyish to white. Asci †20.8–37.5 × 2.6–3.9 μm, clavate, pars sporifera †11.4–19.7 μm, 8-spored, spores strongly spirally and closely twine within asci, truncate to hemispherical at the apex, base gradually narrowed, flexuous stalk, forked to L-, or Y-shaped. Ascospores †8.8–10.2 × 0.9–1.0 μm wide, hyaline, non-septate, helicoid or S-shaped, sickle-shaped or falculate from profile, cylindrical at the one end, tapered at the other end, the four lower spores inversely oriented; spore bodies tear-shaped. Paraphyses †15.1–37.7 × 1.5–2.1 μm, cylindrical to claviform or slightly capitate, unbranched or occasionally branched at the base, slightly enlarged at the apex, 1.7–3.1 μm, covered with waxy exudates, 0.5–1.3 μm thick. Hymenium 41.9–67.9 μm thick, ectal excipulum composed of textura globulosa-angularis.

Specimens examined: China, Shaanxi province, Baoji city, Meixian County, Taibai mountain forest park, from branch of deciduous tree lying on the ground, 21 July 2017, X.Y. Ou, GXU2342. China, Shaanxi province, Baoji city, Meixian County, Taibai mountain forest park, from branch of deciduous tree lying on the ground, 21 July 2017, X.Y. Ou,

GXU2343. China, Shaanxi province, Ankang city, Ningshan County, Huoditang of Qinling, from branch of deciduous tree lying on the ground, 24 July 2017, X.Y. Ou, GXU2383.

Notes: *Orbilia crenatomarginata* features on strongly helicoid, worm or S-shaped ascospores, cylindrical but round gradually at the one end, strongly attenuated at the other end, cylindrical to claviform or slightly capitate paraphyses, apothecia margin denticulate with small and distinct triangular teeth. Our three specimens (GXU2342, GXU2343, GXU2383) corresponded to *O. crenatomarginata* H.B. 9452 and *O. crenatomarginata* H.B. 9265 (MLBP/BIPP = 100%/100%).

Orbilia vinosa (Alb. and Schwein.) P. Karst., Bidr. Känn. Finl. Nat. Folk 19: 101 (1871) (Figure 6).

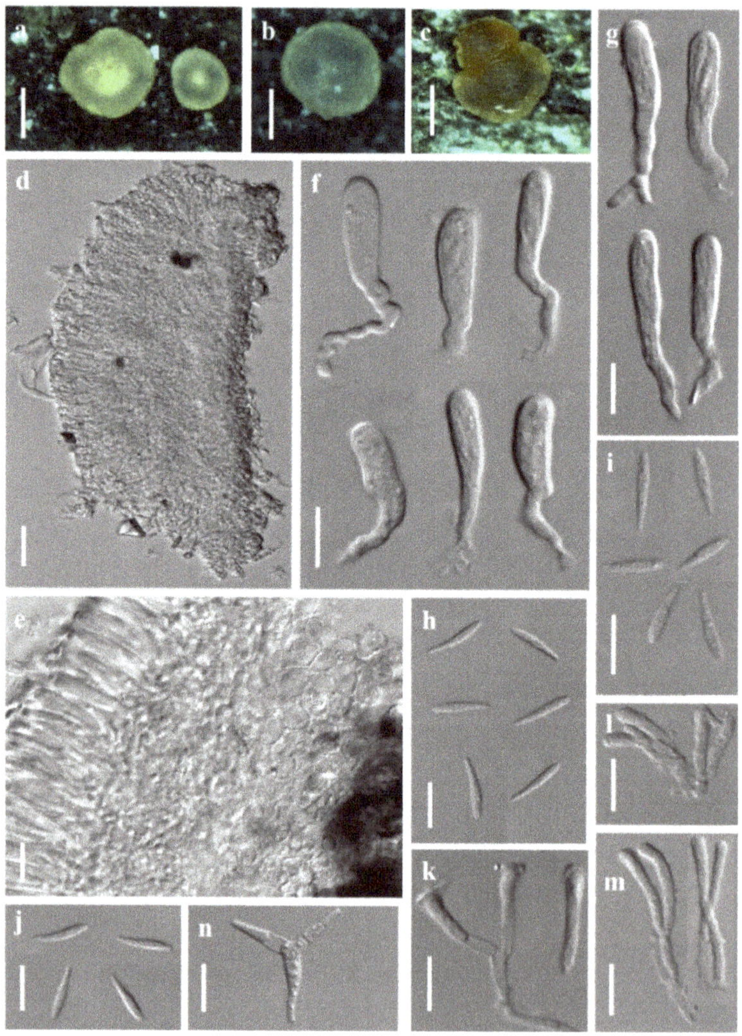

Figure 6. *Orbilia vinosa* ((**a**,**d**–**f**,**h**,**j**,**k**,**n**) from GXU2394; (**b**,**m**) from GXU2397; (**g**,**i**) from GXU2415; (**c**,**l**) from GXU2421). (**a**–**c**) apothecia; (**d**,**e**) vertical section of apothecium; (**f**,**g**) ascus; (**h**–**j**) ascospores; (**k**–**m**) paraphyses; (**n**) conidia (from apothecium). Scale bars: (**a**–**c**) = 0.2 mm; (**d**) = 20 μm; (**e**–**n**) = 10 μm.

Sexual state: Apothecia 0.1–0.4 mm in diam., scattered or gregarious on the surface of bark, disc flat to concave, sessile, waxy, translucent, round, fresh pale or light yellow to orange, sometimes cream-ochraceous or greyish, margin with crenulate, the back of disc with white glassy filament. Asci †16.2–53.4 × 5.5–6.0 μm, cylindric-clavate, pars sporifera †14.0–31.2 μm, 8-spores, spores seriate, lower spores inversely oriented, the apex hemispherical to truncate, the base gradually narrowing, flexuous stalk, forked to T-, L- or Y-shaped. Ascospores †7.3–14.1 × 1.1–2.3 μm, hyaline, non-septate, clavate, sometimes fusoid, one end obtuse or round, the other end slightly curved and smaller, strongly attenuated; spore bodies tear-shaped. Paraphyses †14.3–35.2 × 1.2–2.4 μm, cylindrical to slightly clavate-capitate, unbranched or occasionally branched at the base, slightly enlarged at the apex, terminal inflated, 1.9–3.3 μm in diameter. Hymenium 56.3–86.6 μm thick, ectal excipulum composed of textura globulosa-angularis, cell 3.5–10.2 × 2.4–8.0 μm and globose.

Specimens examined: China, Shaanxi province, Xi'an city, Cuihua Mountain, from rotten branch lying on the ground, 25 July 2017, X.Y. Ou, GXU2394. China, Shaanxi province, Xi'an city, Cuihua Mountain, from rotten wood lying on the ground, 25 July 2017, X.Y. Ou, GXU2397. China, Shaanxi province, Baoji city, Meixian County, Taibai mountain forest park, from branch of deciduous tree lying on the ground, 21 July 2017, X.Y. Ou and B. Liu, GXU2415. China, Shaanxi province, Hanzhong city, Foping County, Wangjiawan, from deadwood lying on the ground, 23 July 2017, X.Y. Ou and B. Liu, GXU2421.

Notes: *Orbilia vinosa* is characterized by clavate-fusoid ascospores, straight or slightly curved, one end obtuse, the other end strongly tapered. The gross morphology of our collections is similar to the original description, according to the detailed description and illustrations of the species provided by Baral et al. [3]. Sequence comparisons also revealed that the three specimens (GXU2394, GXU2397, GXU2421) corresponded to *O. vinosa* G.M. 2014-02-14 and *O. vinosa* CBS 116215 (MLBP/BIPP = 100%/100%).

Orbilia vitalbae Rehm, in Ade, Hedwigia 64: 315 (1923) (Figure 7).

Sexual state: Apothecia 0.1–0.4 mm in diameter, superficial on the rotten branch, scattered or gregarious, disc flat or slightly convex, round, translucent, sessile, pale to yellowish when fresh or rehydrated, dry deep cream to orange-yellow, margin slightly crenulate. Asci †20.0–51.0 × 3.1–5.2 μm, pars sporifera †19.1–26.2 μm, cylindric-clavate, 8-spored, the apex obtuse or strongly truncate, the base gradually thin, flexuous stalk, the lower part bifurcate to L- or Y-shaped. Ascospores †5.1–7.7 × 1.8–2.5 μm, fusoid to clavate, one end round to obtuse or subacute, the other end gradually attenuated, solely fastigiate arrangement in the ascus; SBs tear-shaped, in the end of ascospores. Paraphyses †19.5–29.8 × 1.2–2.5 μm, capitate, unbranched, enlarged to globose at the apex, 2.4–4.5 μm, a waxy exudate over terminal cell of paraphyses. Hymenium 66.6–102.1 μm thick, ectal excipulum composed of textura globulosa-angularis.

Specimens examined: China, Shaanxi province, Xi'an city, Cuihua Mountain, from rotten branch lying on the ground, 25 July 2017, X.Y. Ou and B. Liu, GXU2438. China, Shaanxi province, Xi'an city, Cuihua Mountain, from deadwood lying on the ground, 25 July 2017, X.Y. Ou and B. Liu, GXU2442.

Notes: *Orbilia vitalbae* featured on unipolar and straight, fusoid to clavate ascospores, round to obtuse at the one end and attenuated at the other end. In this study, our two specimens (GXU2438, GXU2442) corresponded to *O. vitalbae* H.B. 9905a (MLBP/BIPP = 99%/100%).

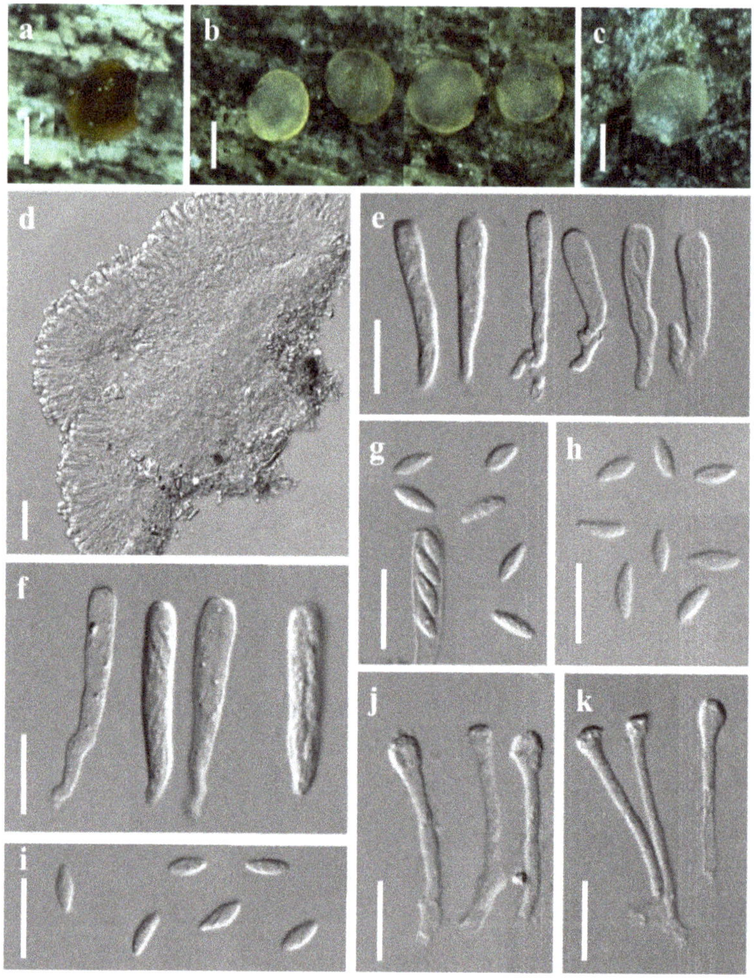

Figure 7. *Orbilia vitalbae* ((**a**,**b**,**d**,**e**,**h**,**j**) from GXU2438; (**c**,**f**,**g**,**i**,**k**) from GXU2442). (**a**–**c**) apothecia; (**d**) vertical section of apothecium; (**e**,**f**) ascus; (**g**–**i**) ascospores; (**j**,**k**) paraphyses. Scale bars: (**a**–**c**) = 0.2 mm; (**d**) = 20 μm; (**e**–**k**) = 10 μm.

3.2. Phylogenetic Analysis

The phylogenetic tree (Figure 8) was inferred from maximum likelihood analyses and Bayesian inference analyses with the combined ITS and LSU (528 bp from ITS and 561 bp from LSU) sequences. The analysis involved 38 nucleotide sequences that belonged to 25 species, 13 sequences were recognized in this study. The tree was composed of 37 strains as ingroup. *Hyalorbilia inflatula* (P. Karst.) Baral and G. Marson was used as the outgroup taxon. Maximum likelihood and Bayesian inference analyses generated semblable tree topologies. In the phylogenetic tree, five clades corresponding to sections of *Orbilia*, including *Arthrobotrys*, *Aurantiorubrae*, *Habrostictis*, *Hemiorbilia* and *Orbilia*, were revealed (Figure 8).

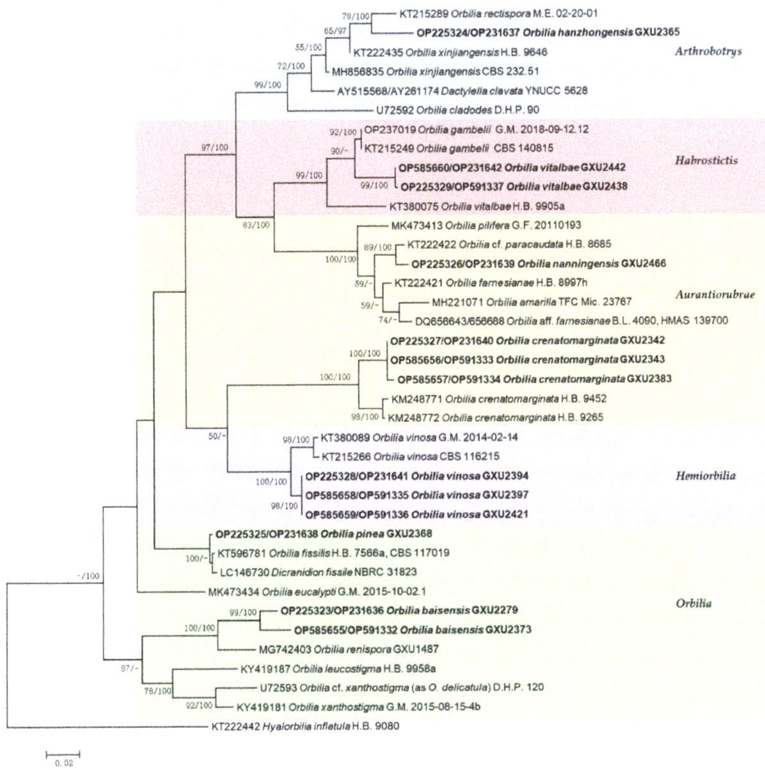

Figure 8. Phylogenetic tree generated from maximum likelihood analyses based on the combined ITS and LSU sequences expressing relationship of *Orbilia* species. Maximum likelihood bootstrap support ≥ 50% (left) and Bayesian posterior probability values ≥ 95% (right) are indicated at nodes (BIBP/MLBP). *Hyalorbilia inflatula* H.B. 9080 was used as outgroup. Bold names represent new species and new Chinese record.

In the phylogenetic tree inferred from combined sequences, our 13 samples were considered as four new species and three new record species in *Orbilia*. The new species *Orbilia baisensis* were located in a clade with high statistical support (MLBP/BIPP = 100%/100%) with *O. renispora*. The two specimens (GXU2279, GXU2373) formed a subclade and designated as *O. baisensis* (MLBP/BIPP = 99%/100%). *Orbilia hanzhongensis* and *O. rectispora* received medium statistical support (MLBP/BIPP = 78%/100%). *Orbilia nanningensis* was related to *O.* cf. *paracaudata*, *O. nanningensis* and *O.* cf. *paracaudata* clustered together in a high supported subclade (MLBP/BIPP = 89%/100%). *Orbilia pinea* and *O. fissilis*, *Dicranidion fissile* were located in a clade with high support (MLBP = 100%, BIPP < 95%). *O. pinea* clustered with *O. fissilis* in a high support (MLBP = 100%, BIPP < 95%).

New records species *Orbilia crenatomarginata* and *O. vinosa* clustered in a clade, which was divided into two strong supported monophyletic subclades. Our three specimens (GXU2342, GXU2343, GXU2383) formed a subclade corresponded to *O. crenatomarginata* (MLBP/BIPP = 100%/100%) and the three specimens (GXU2394, GXU2397, GXU2421) formed a subclade corresponding to *O. vinosa* (MLBP/BIPP = 100%/100%). *Orbilia vitalbae* and *O. gambelii* were located in a clade with high statistical support (MLBP/BIPP = 99%/100%). Our two specimens (GXU2438, GXU2442) formed a subclade corresponding to *O. vitalbae* (MLBP/BIPP = 99%/100%).

4. Discussion

The genus *Orbilia* is diversely and widely distributed in China. The morphological characteristic of the specific ascus and the polymorphic ascospores, especially the strongly refractive spore body, makes *Orbilia* distinctly unique to the other discomycetes. So far, only several species of *Orbilia* have been reported from Guangxi and Shaanxi province, China. In this study, sixteen specimens were collected from Guangxi and Shaanxi province, China. Seven species of *Orbilia* were identified based on morphological characters and phylogenetic analyses, containing four new species, viz. *Orbilia baisensis*, *O. hanzhongensis*, *O. nanningensis*, *O. pinea*, and three newly recorded species to China, viz. *O. crenatomarginata*, *O. vinosa* and *O. vitalbae*. Asexual states of the four new species are confirmed by obtaining pure cultures from the fresh apothecium, which connected to the anamorphic genus of *Dicranidion*, *Dactylella* and *Trinacrium*.

Orbilia baisensis is clustered with *O. renispora*, *O. leucostigma* and *O. xanthostigma* by having reniform to nephroid or C-shaped ascospores. *O. baisensis* features on 8-spored asci, flexuous base and forked to L-, T- or Y-shaped at the base, pronounced reniform ascospores, round and larger at the one end, small pointed base at the other end, the middle section being the widest and capitate paraphyses. *Orbilia renispora* differs from *O. baisensis* by the smaller and lower curl ascospores [18]. It can be confused with species of *O. xanthostigma-leucostigma* complex. However, *O. leucostigma* and *O. xanthostigma* differ from *O. baisensis* by the equant end of ascospores. Ascospores of *xanthostigma-leucostigma* complex are smaller, and with verrucose granule on the dorsal side [11,16]. The distinct warts on the dorsal side of ascospores were reported for the first time by Spooner [7], whereas Baral treated *O. delicatula* as the synonymy of *O.* cf. *xanthostigma* [3]. It was problematic to identify as *O. xanthostigma* and *O. leucostigma* only drawing on different color of apothecia by previous research, actually they contained different species, so they were arranged into *xanthostigma-leucostigma* complex. Baral revealed the high genovariation and represented multiple invariable genotypes in *Orbilia xanthostigma* and *Orbilia leucostigma* [23].

Orbilia hanzhongensis is characterized by fusoid to ellipsoid, to ovoid, to lemon-shaped ascospores with subacute to acute at the one end, round or often acute at the other end, globose SBs. It was related to *O. rectispora*, but differed in having ovoid-fusoid and smaller ascospores. Meanwhile, *O. hanzhongensis* deviated from *O. rectispora* [46] by a 9.47% distance in the ITS region. The sequences taken from the pure culture of *Orbilia nanningensis* comprised ITS and LSU regions, it was closed to *O.* cf. *paracaudata*, whereas *O. nanningensis* was deviated from *O.* cf. *paracaudata* [3] by a 5.8% distance in the ITS region and it had smooth margin of apothecia and capitate paraphyses. *Orbilia pinea* was related to *O. fissilis* and *D. fissile*, it differed from *O. fissilis* by smooth and pronounced reniform ascospores. *Orbilia crenatomarginata* was described and illustrated in detailed under the name of *Orbilia crystallina* [47]. The species is distinguished by white to cream apothecia and margin with the crystalline tooth, flexuous and forked to L-, or Y-shaped asci, helicoid or S-shaped to sickle-shaped ascospores.

Orbilia vinosa has been reported in Africa, America, Asia and Europe, growing on gymnosperms and angiosperms [16], but was first reported in China. *Peziza vinosa* is the primitive name of *O. vinosa* and described poorly [48], Spooner supplemented descriptions in detail and solved some problems with the type of *Peziza vinosa* [7]. *O. vinosa* clustered in a clade with *O. crenatomarginata*, but the former one differs by clavate ascospores.

Orbilia vitalbae can grow on rotten branches of various trees (Clematis et al.), decayed wood or herbaceous plants (Sideritis et al.) [49], and it is illustrated by asci †(27–)30–50(–54) × (3.5–)4–5.3(–5.5) μm and ascospores †5–8(–10) × 1.4–1.6 or 1.8–2.5(–2.7) μm [3]. In this study, the sizes of asci (†20.0–51.0 × 3.1–5.2 μm) and ascospores (†5.1–7.7 × 1.8–2.5 μm), the shape of ascospores, are well in agreement with the previous findings of Rehm.

Members of *Orbilia* are often found on dead twigs and branches hanging on trees, distributed in tropical, subtropical and temperate regions. There are 470 species currently known in the family *Orbiliaceae* [3], of which more than 100 species have been reported in China. Surveys of fungal resources in various regions with different climates and geographic structures will improve our understanding of the species diversity of orbiliaceous fungi in the country. It is necessary to investigate fungal resources in various regions in the future.

Author Contributions: B.L. supervised the project, conceived and designed the study, edited and revised the manuscript; X.-Y.O. collected samples, extracted DNA and PCR and wrote the manuscript; Y.-Y.S. conducted the experiments and phylogenetic analyses; H.-F.Z. conducted morphological observation. All authors have read and agreed to the published version of the manuscript.

Funding: This research was supported by the National Natural Science Foundation of China (No. 30860006) and Ministry of Science and Technology of China for Fundamental Research (No. 2013FY110400).

Institutional Review Board Statement: Not applicable.

Informed Consent Statement: Not applicable.

Data Availability Statement: The sequencing data were submitted to GenBank.

Acknowledgments: The authors are thankful to the project for Fundamental Research on Science and Technology that provided funding, and we thank Usman Rasheed for corrections of the language, and also thanks for the anonymous reviewers for their valuable comments and suggestions.

Conflicts of Interest: The authors declare no conflict of interest.

References

1. Ekanayaka, A.H.; Hyde, K.D.; Jones, E.B.G.; Zhao, Q. *Orbiliaceae* from Thailand. *Mycosphere* **2018**, *9*, 155–168. [CrossRef]
2. Marchal, E. *Champignons coprophiles* de Belgique. *Bull. Soc. R. Bot. Belg.* **1885**, *24*, 56–77.
3. Baral, H.O.; Weber, E.; Marson, G. *Monograph of Orbiliomycetes (Ascomycota) Based on Vital Taxonomy*; National Museum of Natural History: Luxembourg, 2020.
4. Baral, H.O.; Weber, E.; Gams, W.; Hagedorn, G.; Liu, B.; Liu, X.Z.; Marson, G.; Marvanová, L.; Stadler, M.; Weiβ, M. Generic names in the *Orbiliaceae* (Orbiliomycetes) and recommendations on which names should be protected or suppressed. *Mycol. Prog.* **2018**, *17*, 5–31. [CrossRef]
5. Fires, E.M. *Corpus florarum Provincialium Sueciae. I. Floram scanicum. XXIV*; Sebell & C: Upasla, Sweden, 1835; 394p.
6. Nannfeldt, J.Z. Studien über die Morphologie und Systematik der nich-lichenisierten inoperculaten Discomyceten. *Nova Acta Regiae Soc. Sci. Ups. Ser.* **1932**, *48*, 368.
7. Spooner, M.P. Helotiales of Australasia: *Geoglossaceae, Orbiliaceae, Sclerotiniaceae, Hyaloscyphaceae. Biblioth Mycol.* **1987**, *116*, 1–711.
8. Eriksson, O.E.; Baral, H.O.; Currah, R.S.; Hansen, K.; Kurtzman, C.P.; Læssøe, T.; Rambold, G. Notes on ascomycete systematics Nos 3580–3623. *Myconet* **2003**, *9*, 91–103.
9. Zhang, Y.; Yu, Z.F.; Baral, H.O.; Qiao, M.; Zhang, K.Q. *Pseudorbilia* gen. nov. (*Orbiliaceae*) from Yunnan, China. *Fungal Divers.* **2007**, *26*, 305–312.
10. Kumar, T.A.; Healy, R.; Spatafora, J.W.; Blackwell, M.; Mclanughlin, D.J. *Orbilia* ultrastructure, character evolution and phylogeny of *Pezizomycotina*. *Mycologia* **2012**, *104*, 462–476. [CrossRef]
11. Guo, J.W.; Li, S.F.; Yang, L.F.; Yang, J.; Ye, T.Z.; Yang, L. New records and new distribution of known species in the family *Orbiliaceae* from China. *Afr. J. Microbiol. Res.* **2014**, *8*, 3178–3190.
12. Qiao, M.; Li, J.Y.; Baral, H.O.; Zhang, Y.; Qiao, W.Y.; Su, H.Y.; Yu, Z.F. *Orbilia yuanensis* sp. nov. and its anamorph. *Mycol. Prog.* **2015**, *14*, 1. [CrossRef]
13. Quijada, L.; Baral, H.O.; Beltrán-Tejera, E. New species of *Orbilia* (Orbiliales) from arid ecosystems of the Canary Islands (Spain). *Nova Hedwig.* **2012**, *96*, 237–248. [CrossRef]
14. Quijada, L.; Baral, H.O.; Jaen-Molina, R.; Weiss, M.; Caujapé-Castelis, J.; Beltrán-Tejera, E. Phylogenetic and morphological circumscription of the *Orbilia aurantiorubra* group. *Phytotaxa* **2014**, *175*, 30. [CrossRef]
15. Quijada, L.; Baral, H.O.; Beltrán-Tejera, E. Diversity of *Hyalorbilia* (Orbiliales) in the Macaronesian Region. *Nova Hedwig.* **2015**, *100*, 1–14. [CrossRef]
16. Quijada, L.; Baral, H.O.; Beltrán-Tejera, E. A revision of the genus *Orbilia* in the Canary Islands. *Phytotaxa* **2016**, *284*, 231–262. [CrossRef]
17. Quijada, L.; Baral, H.O. *Orbilia beltraniae*, a new succulenticolous species from the Canary Islands. *MycoKeys* **2017**, *25*, 1–12. [CrossRef]
18. Shao, Y.Y.; Baral, H.O.; Ou, X.Y.; Wu, H.; Huang, F.C.; Zheng, H.F.; Liu, B. New species and records of orbiliaceous fungi from Georgia, USB. *Mycol. Prog.* **2018**, *17*, 1225–1235. [CrossRef]
19. Deng, C.; Yu, Z.F. The complete mitochondrial genomes of *Dactylellina leptospora* (Orbiliales, *Orbiliaceae*). *Mitochondrial DNA Part B Resour.* **2019**, *4*, 1615–1616. [CrossRef]

20. Wang, S.J.; Fang, M.L.; Xu, J.P.; Jiang, L.L.; Zhou, D.Y.; Zhang, K.Q.; Zhang, Y. Complete mitochondrial genome and phylogenetic analysis of *Orbilia dorsalia*, a species producing mature sexual structures on culture. *Mitochondrial DNA B* **2019**, *4*, 573–574. [CrossRef]
21. White, T.J.; Bruns, T.; Lee, S.; Taylor, J. *Amplification and Direct Sequencing of Fungal Ribosomal RNA Genes for Phylogenetics*; Academic Press: San Diego, CA, USA, 1990; pp. 315–322.
22. Vilgalys, R.; Hester, M. Rapid genetic identification and mapping of enzymatically amplified ribosomal DNA from several Cryptococcus species. *J. Bacteriol.* **1990**, *172*, 4239–4246. [CrossRef]
23. Baral, H.O.; Johnston, P.; Quijada, L.; Healy, R.; Pfister, D.H.; Lobuglio, K.F.; Rodriguez, V.; Weber, E. Cryptic speciation in *Orbilia xanthostigma* and *O. leucostigma* (Orbiliomycetes): An aggregate with worldwide distribution. *Mycol. Prog.* **2021**, *20*, 1503–1537. [CrossRef]
24. O'Donnell, K.; Kistler, H.C.; Cigelnik, E.; Ploetz, R.C. Multiple evolutionary origins of the fungus causing Panama disease of banana: Concordant evidence from nuclear and mitochondrial gene genealogies. *Proc. Natl. Acad. Sci. USA* **1998**, *95*, 2044–2049. [CrossRef] [PubMed]
25. O'Donnell, K.; Cigelnik, E. Two divergent intragenomic rDNA ITS2 types within a monophyletic lineage of the fungus *Fusarium* are nonorthologous. *Mol. Phylogenet. Evol.* **1997**, *7*, 103–116. [CrossRef] [PubMed]
26. Woudenberg, J.H.C.; Aveskamp, M.M.; De-Gruyter, J.; Spiers, A.G.; Crous, P.W. Multiple Didymella teleomorphs are linked to the *Phoma clematidina* morphotype. *Persoonia* **2009**, *22*, 56–62. [CrossRef]
27. Liu, Y.J.; Whelen, S.; Hall, B.D. Phylogenetic relationships among ascomycetes: Evidence from an RNA polymerase II subunit. *Mol. Biol. Evol.* **1999**, *16*, 1799–1808. [CrossRef] [PubMed]
28. Sung, G.H.; Sung, J.M.; Hywel-Jones, N.L.; Spatafora, J.W. A multigene phylogeny of *Clavicipitaceae* (Ascomycota, fungi): Identification of localized incongruence using a combinational bootstrap approach. *Mol. Phylogenet. Evol.* **2007**, *44*, 1204–1223. [CrossRef] [PubMed]
29. Carbone, I.; Kohn, L.M. A method for designing primer sets for speciation studies in filamentous ascomycetes. *Mycologia* **1999**, *91*, 553–556. [CrossRef]
30. Brefeld, O. AmericaUntersuchungen aus dem Gesammtgebiete der Mykologie, Ascomyceten II. *Mycology* **1891**, *7*, 50–51.
31. Pfister, D.H. Castor, Pollux, and life histories of fungi. *Micologia* **1997**, *89*, 1–23. [CrossRef]
32. Harkness, H.W. Fungi of the Pacific Coast. *Bull. Calif. Acad. Sci.* **1885**, *1*, 159–176.
33. Hughes, S.J. Conidiophores, conidia, and classification. *Can. J. Bot.* **1953**, *31*, 577–659. [CrossRef]
34. Peek, C.A.; Solheim, W.G. The hyphomycetous genera of H. W. Harkness and the ascomycetous genus *Cleistosama* Harkn. *Mycologia* **1958**, *50*, 844–861. [CrossRef]
35. Berthet, P. Forrnes conidiennes de divers Discomycetes. *Bull. Soc. Mycol. Fr.* **1964**, *80*, 125–149.
36. Liu, B.; Liu, X.Z.; Zhuang, W.Y. *Orbilia querci* sp. nov. and its knob-forming nematophagous anamorph. *FEMS Microbiol. Lett.* **2005**, *245*, 99–105. [CrossRef] [PubMed]
37. Doyle, J.J. A rapid DNA isolation procedure for small quantities of fresh leaf tissue. *Phytochem. Bull.* **1987**, *19*, 11–15.
38. Vitória, N.S.; Bezerra, J.L.; Gramacho, K.P. A simplified DNA Extraction Method for PCR Analysis of *Camarotella* spp. *Braz. Arch. Biol. Technol.* **2010**, *55*, 249–252. [CrossRef]
39. Thompson, J.D.; Gibson, T.J.; Plewniak, F.; Jeanmougin, F.; Higgins, D.G. The Clustal_X windows interface: Flexible strategies for multiple sequence alignment aided by quality analysis tools. *Nucleic Acids Res.* **1997**, *25*, 4876–4882. [CrossRef]
40. Tamura, K.; Stecher, G.; Peterson, D.; Filipski, A.; Kumar, S. MEGA6: Molecular Evolutionary Genetics Analysis version 6.0. *Mol. Biol. Evol.* **2013**, *30*, 2725–2729. [CrossRef]
41. Zhang, D.; Gao, F.L.; Jakovlic, I.; Zhou, H.; Zhang, J.; Li, W.X.; Wang, G.T. Phylosuite: An integrated and scalable desktop platform for streamlined molecular sequence data management and evolutionary phylogenetics studies. *Mol. Ecol. Resour.* **2020**, *20*, 348–355. [CrossRef]
42. Swofford, D.L. *PAUP 4.0b10: Phylogenetic Analysis Using Parsimony (* and Other Methods)*; Sinauer Associates: Sunderland, MA, USA, 2002.
43. Nylander, J.A.A. *MrModeltest v2. Program Distributed by the Author*; Evolutionary Biology Centre, Uppsala University: Uppsala, Sweden, 2004.
44. Ronquist, F.; Huelsenbeck, J.P. MrBayes 3: Bayesian phylogenetic inference under mixed models. *Bioinformatics* **2003**, *19*, 1572–1574. [CrossRef]
45. Rambaut, A. FigTree v1.4.3 2016. Available online: http://tree.bio.ed.ac.uk/software/figtree/ (accessed on 1 August 2022).
46. Liu, B.; Liu, X.Z.; Zhuang, W.Y.; Baral, H.O. Orbiliaceous fungi from Tibet, China. *Fungal Divers.* **2006**, *22*, 107–120.
47. Priou, J.P.; Poncelet, A. *Orbilia crystallina* et Tapesina griseovitellina, deux rare ascomycètes nouveaux pour la Bretagne. *Cah. Mycol. Nantais* **2006**, *18*, 3–8.
48. Albertini, I.B.; Schweinitz, L.D. *Conspectus Fungorum in Lusatiae Superioris Agro Niskiensi Crescentium e Methodo Persooniana*; Sumtibus Kummerianis: Lipsiae, Germany, 1805; 376p.
49. Ade, A. Mykologische Beiträge. *Hedwigia* **1923**, *64*, 286–320.

Journal of *Fungi*

MDPI

Article

Three New *Trichoderma* Species in Harzianum Clade Associated with the Contaminated Substrates of Edible Fungi

Zi-Jian Cao [1,2], Wen-Tao Qin [2,*], Juan Zhao [2], Yu Liu [2], Shou-Xian Wang [2] and Su-Yue Zheng [1,*]

1 School of Landscape and Ecological Engineering, Hebei University of Engineering, Handan 056038, China
2 Institute of Plant Protection, Beijing Academy of Agriculture and Forestry Sciences, Beijing 100097, China
* Correspondence: qinwentao@ipepbaafs.cn (W.-T.Q.); zhengsuyue@sina.com (S.-Y.Z.)

Abstract: *Trichoderma* is known worldwide as biocontrol agents of plant diseases, producers of enzymes and antibiotics, and competitive contaminants of edible fungi. In this investigation of contaminated substrates of edible fungi from North China, 39 strains belonging to 10 *Trichoderma* species isolated from four kinds of edible fungi were obtained, and three novel species belonging to the Harzianum clade were isolated from the contaminated substrates of *Auricularia heimuer* and *Pholiota adipose*. They were recognized based on integrated studies of phenotypic features, culture characteristics, and molecular analyses of RNA polymerase II subunit B and translation elongation factor 1-α genes. *Trichoderma auriculariae* was strongly supported as a separate lineage and differed from *T. vermifimicola* due to its larger conidia. *Trichoderma miyunense* was closely related to *T. ganodermatigerum* but differed due to its smaller conidia and higher optimum mycelial growth temperature. As a separate lineage, *T. pholiotae* was distinct from *T. guizhouense* and *T. pseudoasiaticum* due to its higher optimum mycelial growth temperature and larger conidia. This study extends the understanding of *Trichoderma* spp. contaminating substrates of edible fungi and updates knowledge of species diversity in the group.

Keywords: Hypocreaceae; *Trichoderma*; phylogeny; morphology; taxonomy

Citation: Cao, Z.-J.; Qin, W.-T.; Zhao, J.; Liu, Y.; Wang, S.-X.; Zheng, S.-Y. Three New *Trichoderma* Species in Harzianum Clade Associated with the Contaminated Substrates of Edible Fungi. *J. Fungi* **2022**, *8*, 1154. https://doi.org/10.3390/jof8111154

Academic Editors: Lei Cai and Cheng Gao

Received: 10 September 2022
Accepted: 27 October 2022
Published: 31 October 2022

1. Introduction

Trichoderma Pers. is ubiquitous in various niches and around the world. The genus contains at least eight infrageneric clades, of which the Harzianum clade is one of the largest [1]. According to our investigated statistics, the Harzianum clade consists of more than 95 accepted species, which are morphologically heterogeneous and phylogenetically complicated. They play important roles in agriculture, industry, and other fields and are employed as biocides or biofertilizers for plant growth [2–4], act as producers of enzymes and antibiotics, and are endophytic in plants that can resist both physiological stress and pathogen invasion [5,6].

Green mold contamination caused by *Trichoderma* spp. in the cultivation and various growth stages of edible fungi has been one of the biggest biological constraints in the industry since the 1980s [7], with the economic losses accounting for 10–20% of total production [8]. At present, green mold is one of the most devastating diseases in nearly all production areas of cultivated edible fungi due to its high disease incidence and serious economic loss [9,10]. Mycelia of *Trichoderma* spp. show stronger competitiveness than those of edible fungi, and thus they can inhibit mycelial growth or decrease the fruiting rate of edible fungi. Lots of green conidia of *Trichoderma* will gradually cover the contaminated substrates or fruiting bodies, and the contaminated fruiting bodies will eventually shrivel and rot.

In order to better understand the *Trichoderma* species contaminating substrates of edible fungi and preserve biological control resources, substrates of edible fungi contaminated by green mold in North China were investigated, and three undescribed species belonging to the Harzianum clade were found on contaminated substrates of *Auricularia heimuer* and *Pholiota adipose*. Their phylogenetic positions were determined based on sequence analyses of the combined translation elongation factor 1-alpha (*tef1-α*) and the second largest nuclear RNA polymerase subunit (*rpb2*) genes. Similarities and differences in morphological characteristics between the new species and their closely related species were investigated and compared in detail.

2. Materials and Methods

2.1. Isolates and Specimens

Specimens were separately collected from contaminated substrate of edible fungi in North China from 2020 to 2022 (Table S1), and strains were isolated following the method of a previous study [11]. The ex-type strains were deposited in the culture collection of Institute of Plant Protection, Beijing Academy of Agriculture and Forestry Sciences (JZB culture collection).

2.2. Morphology and Growth Characterization

For morphological studies, growth rates were determined on three different media: potato dextrose agar (PDA; 200 g potato, 18 g dextrose, 18 g agar, and 1 L distilled water), cornmeal dextrose agar (CMD; 40 g cornmeal, 20 g glucose, 18 g agar, and 1 L distilled water), and synthetic low nutrient agar (SNA; 1 g KH_2PO_4, 1 g KNO_3, 0.5 g $MgSO_4 \cdot 7H_2O$, 0.5 g KCl, 0.2 g glucose, 0.2 g sucrose, 18 g agar, and 1 L distilled water) at 25, 30, and 35 °C in darkness. Mycelial discs (5 mm diameter) were incubated in Petri dishes (90 mm diameter) with three replicates for each isolate. Colony diameters were measured after 3 days. The time when mycelia entirely covered the surface of the plate and the morphological characteristics of colonies, such as colony appearance, color, and spore production, were recorded [12]. For microscopic morphology, photographs were taken with an Axio Imager Z2 microscope (Carl Zeiss, Jena, Germany). Microscopic characteristics and micromorphological data were examined on the cultures grown on SNA and PDA for 7–9 days at 25 °C.

2.3. DNA Extraction, PCR Amplification, and Sequencing

Genomic DNA was extracted from the cultures grown on PDA for 7 days using a plant genomic DNA Kit (DP305, TIANGEN Biotech, Beijing, China). Fragments of *tef1-α* and *rpb2* were amplified with the primer pairs EF1-728F [13] and TEF1LLErev [14] and fRPB2-5f/7cr [15], respectively. Each PCR reaction consisted of 12.5 μL Premix Taq™ (TaKaRa, Dalian, China), 1.0 μL of forward primer (10 μM), 1.0 μL of reverse primer (10 μM), 1.5 μL of DNA, and 9 μL of double-sterilized water. Polymerase chain reaction (PCR) conditions followed Zhu and Zhuang [16]. The products were purified and subjected to sequencing on an ABI 3730 DNA sequencer (Applied Biosystems, Bedford, MA, USA) at SinoGenoMax company. Sequences generated from this study and those retrieved from GenBank are listed in Table 1.

Table 1. Materials including strain numbers and GenBank accessions of sequences used for phylogenetic analyses.

Species	Voucher	GenBank Accession Number	
		rpb2	*tef1-α*
T. achlamydosporum	YMF 1.6177	MT052180	MT070156
T. afarasin	CBS 130755	–	AF348093
T. afarasin	DIS 314F	FJ442778	FJ463400
T. afroharzianum	CBS 124620 ET	FJ442691	FJ463301
T. afroharzianum	GJS 04-193	FJ442709	FJ463298
T. aggregatum	HMAS 248863	KY688001	KY688062
T. aggregatum	HMAS 248864	KY688002	KY688063
T. aggressivum	CBS 100525	AF545541	AF348095

Table 1. *Cont.*

Species	Voucher	GenBank Accession Number	
		rpb2	*tef1-α*
T. aggressivum	DAOM 222156 [ET]	FJ442752	AF348098
T. alni	CBS 120633 [ET]	EU498349	EU498312
T. alpinum	HMAS 248821 [T]	KY687958	KY688012
T. amazonicum	IB95	HM142368	HM142377
T. amazonicum	CBS126898 [ET]	HM142367	HM142376
T. anaharzianum	YMF 1.00241	MH262577	MH236493
T. anaharzianum	YMF 1.00383 [T]	MH158995	MH183182
T. asiaticum	YMF 1.00168	MH262575	MH236492
T. asiaticum	YMF 1.00352 [T]	MH158994	MH183183
T. atrobrunneum	GJS90-254	FJ442735	AF443943
T. atrobrunneum	GJS 05-101	FJ442745	FJ463392
T. atrogelatinosum	CBS 237.63 [ET]	KJ842201	–
T. atroviride	CBS 119499	FJ860518	FJ860611
T. auriculariae	JZBQT1Z7 [T]	**ON649949**	**ON649896**
T. auriculariae	JZBQT1Z8	**ON649950**	**ON649897**
T. auriculariae	JZBQT1Z9	**ON649951**	**ON649898**
T. austroindianum	BAFC 3583	–	MH352421
T. azevedoi	CEN1422	MK696821	MK696660
T. bannaense	HMAS 248840 [T]	KY687979	KY688037
T. bannaense	HMAS 248865	KY688003	KY688038
T. botryosum	COAD 2422	MK044212	MK044119
T. botryosum	COAD 2401	MK044181	MK044088
T. breve	HMAS 248844 [T]	KY687983	KY688045
T. breve	HMAS 248845	KY687984	KY688046
T. brunneoviride	CBS 121130	EU498357	EU498316
T. brunneoviride	CBS 120928	EU498358	EU498318
T. caeruloviride	COAD 2416	MK044201	MK044108
T. caeruloviride	COAD 2415	MK044202	MK044109
T. camerunense	GJS 99-230	–	AF348107
T. catoptron	GJS 02-76 [T]	AY391900	AY737726
T. ceraceum	GJS 95-159	AF545508	AY937437
T. ceratophylletum	YMF 1.04621 [T]	MK327580	MK327579
T. cerinum	DAOM 230012	KJ842184	KJ871242
T. christiani	CBS 132572 [ET]	KJ665244	KJ665439
T. cinnamomeum	GJS 96-128	AY391916	AY391977
T. cinnamomeum	GJS 97-233	AY391919	AY391978
T. compactum	CBS 121218	KF134789	KF134798
T. concentricum	HMAS 248833	KY687971	KY688027
T. confertum	HMAS 248893	MF371203	MF371218
T. confertum	HMAS 248896	MF371205	MF371220
T. corneum	GJS 97-82 [ET]	KJ665252	KJ665455
T. dacrymycellum	WU29044	FJ860533	FJ860633
T. endophyticum	CBS 130753	FJ442722	FJ463326
T. endophyticum	CBS 130733	FJ442690	FJ463330
T. epimyces	CBS120534 [ET]	EU498360	EU498320
T. ganodermatigerum	CCMJ5245 [T]	ON567189	ON567195
T. ganodermatigerum	CCMJ5246	ON567190	ON567196
T. ganodermatigerum	CCMJ5247	ON567191	ON567197
T. globoides	HMAS 248747	KX026963	KX026955
T. guizhouense	HGUP0038 [T]	JQ901400	JN215484
T. guizhouense	S278	KF134791	KF134799
T. guizhouense	DAOM 231435	–	EF191321
T. harzianum	CBS 226-95	AF545549	AF348101
T. harzianum	GJS 05 107	FJ442708	FJ463329
T. hausknechtii	CBS 133493	KJ665276	KJ665515
T. helicolixii	CBS 133499	KJ665278	KJ665517
T. hengshanicum	HMAS 248852 [T]	KY687991	KY688054

Table 1. *Cont.*

Species	Voucher	GenBank Accession Number	
		rpb2	*tef1-α*
T. hirsutum	HMAS 248834 [T]	KY687972	KY688029
T. hortense	BMCC LU994	–	KJ871185
T. ingratum	HMAS 248822	KY687973	KY688018
T. inhamatum	CBS 273-78	FJ442725	AF348099
T. italicum	CBS 132567	KJ665282	KJ665525
T. koreanum	SFC20131005-S066	MH025988	MH025979
T. lentiforme	DIS 253B	FJ442756	FJ851875
T. lentiforme	DIS 94D	FJ442749	FJ463379
T. lentinulae	HMAS 248256	MN605867	MN605878
T. lentinulae	CGMCC 3.19848	MN605868	MN605879
T. liberatum	HMAS 248831 [T]	KY687969	KY688025
T. linzhiense	HMAS 248846 [T]	KY687985	KY688047
T. lixii	CBS 110080	KJ665290	FJ716622
T. longifialidicum	LESF 552	KT278955	KT279020
T. miyunense	JZBQF5	**ON649968**	**ON649915**
T. miyunense	JZBQF7 [T]	**ON649969**	**ON649916**
T. miyunense	JZBQF9	**ON649970**	**ON649917**
T. neotropicale	LA11 [ET]	–	HQ022771
T. paratroviride	S385	KJ665321	KJ665627
T. parepimyces	CBS 122769 [ET]	FJ860562	FJ860664
T. peberdyi	CEN1426	MK696825	MK696664
T. peruvianum	CP15-2	MW480153	MW480145
T. peruvianum	CP15-9	MW480154	MW480146
T. perviride	HMAS 273786	KX026962	KX026954
T. phayaoense	SDBR-CMU349	MW002074	MW002073
T. pholiotae	JZBQH11	**ON649971**	**ON649918**
T. pholiotae	JZBQH12 [T]	**ON649972**	**ON649919**
T. pholiotae	JZBQH13	**ON649973**	**ON649920**
T. pinicola	KACC 48486 [ET]	MH025993	MH025981
T. pinicola	SFC20130926-S014	MH025991	MH025978
T. pleuroti	CBS 124387 [ET]	HM142372	HM142382
T. pleuroticola	CBS 124383 [ET]	HM142371	HM142381
T. pleuroticola	TRS70 [ET]	KP009172	KP008951
T. pollinicola	LC11682 = LF1542 [ET]	MF939604	MF939619
T. pollinicola	LC11686 = LF2050	MF939605	MF939620
T. polypori	HMAS 248855 [T]	KY687994	KY688058
T. priscilae	CBS 131487 [ET]	KJ665333	KJ665691
T. propepolypori	YMF 1.06224 [T]	MT052181	MT070158
T. propepolypori	YMF 1.06199	MT052182	MT070157
T. pseudoasiaticum	YMF 1.06200 [T]	MT052183	MT070155
T. pseudodensum	HMAS 248828 [T]	KY687967	KY688023
T. pseudogelatinosum	CNUN309 [ET]	HM920173	HM920202
T. pseudopyramidale	COAD 2419	MK044206	MK044113
T. pseudopyramidale	COAD 2506	MK044207	MK044114
T. purpureum	HMAS 273787 [T]	KX026961	KX026953
T. pyramidale	CBS 135574 [ET]	KJ665334	KJ665699
T. rifaii	CBS 130746	–	FJ463324
T. rifaii	DIS 337F	FJ442720	FJ463321
T. rufobrunneum	HMAS 266614 [T]	KF730010	KF729989
T. rugulosum	SFC20180301-001 [T]	MH025986	MH025984
T. rugulosum	SFC20180301-002	MH025987	MH025985
T. simile	YMF 1.06201 [T]	MT052184	MT070154
T. simile	YMF 1.06202	MT052185	MT070153
T. simmonsii	CBS 130431	FJ442757	AF443935
T. simmonsii	S7	KJ665337	KJ665719
T. simplex	HMAS 248842 [T]	KY687981	KY688041
T. solum	HMAS 248848 [T]	KY687987	KY688050

Table 1. *Cont.*

Species	Voucher	GenBank Accession Number	
		rpb2	*tef1-α*
T. stramineum	GJS 02-84	AY391945	AY391999
T. subalni	HMAS 275683	MH612371	MH612377
T. subalni	HMAS 275684	MH612370	MH612376
T. syagri	BAFC 4357	–	MG822711
T. tawa	CBS 114233 ET	AY391956	FJ463313
T. tawa	DAOM 232841	KJ842187	EU279972
T. tenue	HMAS 273785 ET	KX026960	KX026952
T. tomentosum	DAOM 178713a	AF545557	AY750882
T. velutinum	CPK 298	KF134794	KJ665769
T. velutinum	DAOM 230013 ET	JN133569	AY937415
T. vermifimicola	CGMCC 3.19850	MN605870	MN605881
T. vermifimicola	HMAS 248255	MN605871	MN605882
T. xixiacum	HMAS 248253 T	MN605874	MN605885
T. xixiacum	CGMCC 3.19698	MN605875	MN605886
T. zayuense	HMAS 248835 T	KY687974	KY688031
T. zelobreve	HMAS 248254 T	MN605872	MN605883
T. zelobreve	CGMCC 3.19696	MN605873	MN605884
T. zeloharzianum	YMF 1.00268	MH158996	MH183181

Numbers in bold indicate newly submitted sequences in this study. T: type strains. ET: ex-type strains.

2.4. Phylogenetic Analyses

Sequences for all isolates generated in this study were blasted against the NCBIs GenBank nucleotide datasets (https://blast.ncbi.nlm.nih.gov/Blast.cgi) and MIST (http://mmit.china-cctc.org/index.php) [17] to obtain an initial identification. To identify the phylogenetic positions of *Trichoderma* species isolated from contaminated substrates of edible fungi, *rpb2* and *tef1-α* sequences of all *Trichoderma* species belonging to the Harzianum clade were combined for the analyses, with *T. atroviride* and *T. paratroviride* selected as outgroup taxa. Their sequences of type or ex-type strains based on previous publications were downloaded from NCBI database and assembled using BioEdit 7.0.5.3 [18]. Alignment was generated and converted to nexus files with Clustal X 1.83 [19].

Maximum parsimony (MP) analysis was performed with PAUP 4.0b10. Starting trees were obtained via random stepwise addition with 1000 replicates and subsequent branch-swapping algorithm using tree bisection–reconnection (TBR) [20]. Analyses were performed with all characters treated as unordered and unweighted, and gaps treated as missing data. MaxTrees was set to 1000, and branches collapsed when maximum branch length was zero. Maximum parsimony bootstrap proportion (MPBP) was calculated to test topological confidence of the resulting trees.

Bayesian inference (BI) trees were calculated using MrBayes v. 3.1.2 [21]. The best-fit nucleotide substitution model GTR+I+G was selected using MrModeltest 2.3 [22]. Four chains were run from random trees for 6,000,000 generations and sampled every 100 generations. The first 25% of trees were discarded as the burn-in phase of the analyses, and Bayesian inference posterior probability (BIPP) was determined from the remaining trees. Trees were visualized in FigTree v1.4.3 [23].

3. Results

3.1. Phylogenetic Analyses

The partition homogeneity test of *rpb2* and *tef1-α* sequences indicated that the individual partitions were generally congruent ($p = 0.01$). The combined *rpb2* and *tef1-α* dataset was subsequently used for phylogenetic analysis to determine the positions of the new species. In MP analysis, the dataset contained 140 taxa and 2307 characters, of which 1468 characters were constant, 150 variable characters were parsimony uninformative, and 689 were parsimony informative. Five most parsimonious trees with the same topology were generated, and one of them is shown in Figure 1 (tree length = 3091, CI = 0.3999, HI = 0.6001, RC = 0.3039, and RI = 0.7600). The BI tree topology was generally the same as that of the MP tree.

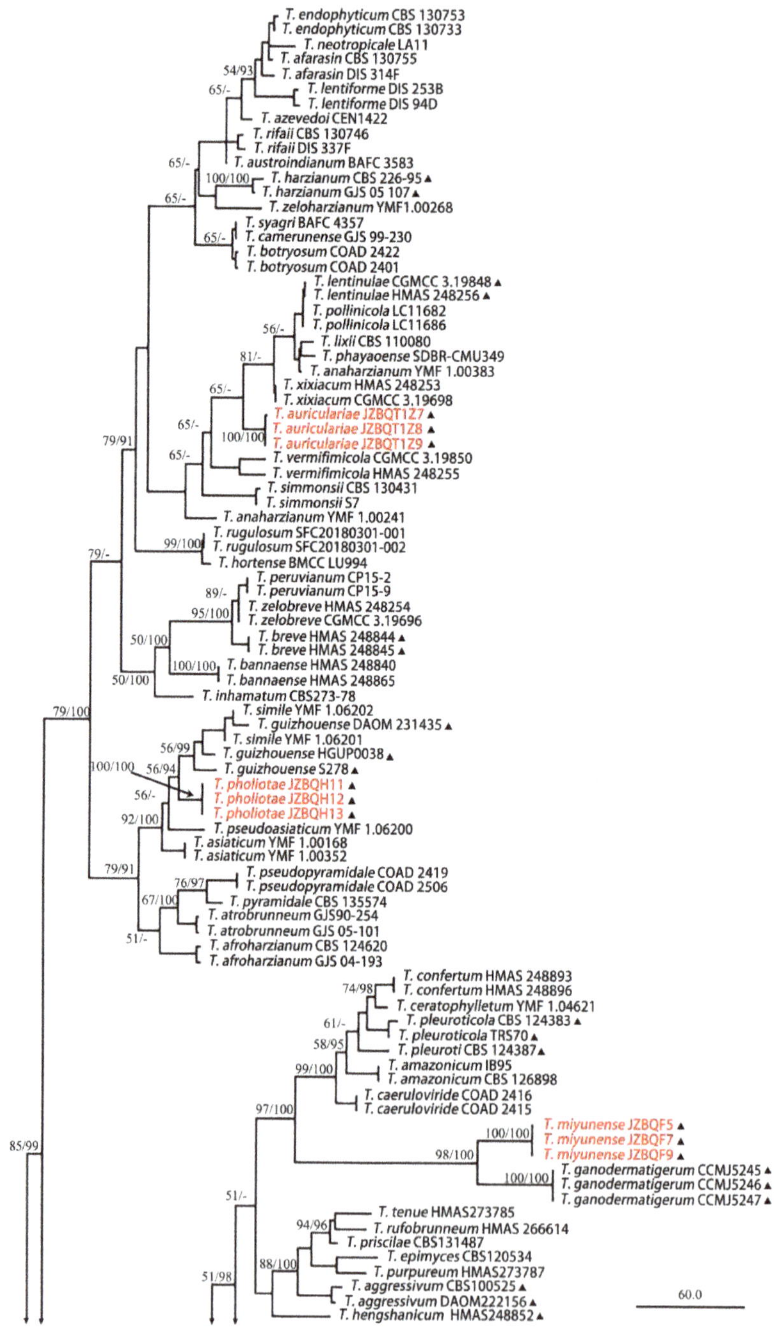

Figure 1. *Cont.*

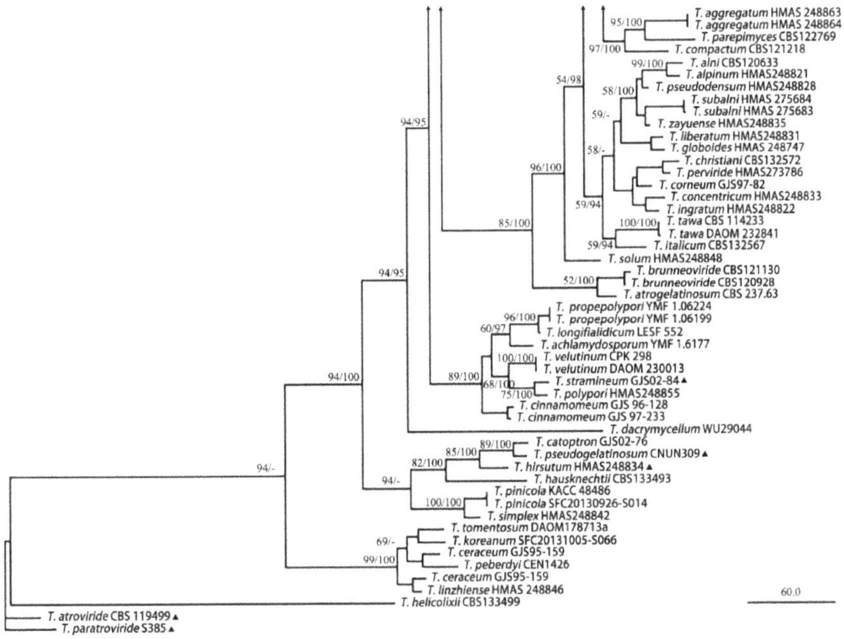

Figure 1. Maximum parsimony phylogram of the investigated *Trichoderma* species inferred from the combined sequences of *rpb2* and *tef1-α*. MPBP above 50% (left) and BIPP above 90% (right) are indicated at the nodes. New species proposed are indicated in red font. *Trichoderma* species isolated from substrate or fruiting bodies of edible fungi are marked with ▲.

A total of 140 sequences representing 95 *Trichoderma* species, including our three new species, were used for constructing the phylogenetic tree, and *T. atroviride* and *T. paratroviride* were used as outgroups. Results showed that all the investigated *Trichoderma* species formed a strongly supported group (MPBP/BIPP = 100%/100%), which was generally congruent with the previous studies [24].

In the phylogenetic tree (Figure 1), *T. auriculariae*, *T. miyunense*, and *T. pholiotae* were newly added to the *T. harzianum* clade. *Trichoderma auriculariae* was distributed as a separate terminal branch (MPBP/BIPP = 100%/100%) among *T. vermifimicola* and *T. xixiacum*. *Trichoderma miyunense* was a sister of *T. ganodermatigerum* (MPBP/BIPP = 98%/100%). *Trichoderma pholiotae* formed a linage with *T. asiaticum*, *T. guizhouense*, *T. pseudoasiaticum*, and *T. simile* with high support value (MPBP/BIPP = 92%/100%), and our three strains of *T. pholiotae* were distributed as a highly supported separate terminal branch (MPBP/BIPP = 100%/100%) among *T. pseudoasiaticum* and *T. guizhouense*.

3.2. Taxonomy

Trichoderma auriculariae Z. J. Cao and W.T. Qin, sp. nov.

MycoBank MB845141 (Figure 2).

Etymology: The specific epithet refers to the host from which the fungus was isolated.

Typification: China, Beijing, Tongzhou, from the contaminated substrates of *Auricularia heimuer*, 26 August 2021, W.T. Qin, Z.J. Cao, L. Gao, J. Li (ex-type strain JZBQT1Z7).

DNA barcodes: ITS = ON653396, *rpb2* = ON649949, *tef1-α* = ON649896.

On CMD after 72 h, colony radius 65–66 mm at 25 °C, 69–70 mm at 30 °C, and 8–10 mm at 35 °C. Colony hyaline and radial, not zonate. Aerial hyphae rare in colony center. A large number of white pustules formed after 2 days. Conidiation formed on aerial hyphae and in pustules, abundant, spreaded throughout the colony, then gradually turned green. No diffusing pigment noted.

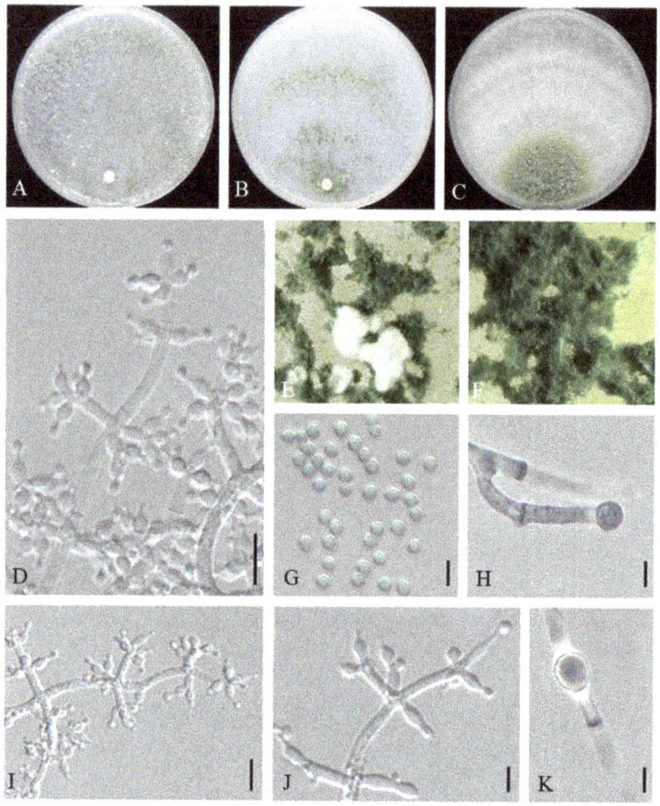

Figure 2. *Trichoderma auriculariae* (JZBQT1Z7). Cultures at 25 °C after 7 days on (**A**) CMD, (**B**) SNA, and (**C**) PDA; (**D**,**I**,**J**) conidiophores and phialides; (**E**,**F**) conidiation pustules on CMD after 7 days; (**G**) conidia; (**H**,**K**) chlamydospores. Scale bars: (**D**,**I**) = 10 μm, (**G**,**H**,**J**,**K**) = 5 μm.

On PDA after 72 h, colony radius 47–49 mm at 25 °C, 66–68 mm 30 °C, and 5–7 mm at 35 °C. Colony regularly circular, distinctly zonate. Aerial mycelium dense and radial, forming a dense, zonate, floccose mat. Conidial production noted after 2 days, starting around the original inoculum, effuse in aerial hyphae, more abundant along the original inoculum. No diffusing pigment noted, odor fruity.

On SNA after 72 h, colony radius 47–49 mm at 25 °C, 51–55 mm at 30 °C, and 5–7 mm at 35 °C. Colony hyaline, mycelium loose. Conidial production noted after 2 days, starting around the inoculum, effuse in the aerial hyphae, forming a few inconspicuous rings. Small pustules formed around the inoculum, first white, turning green after 3 days, with hairs protruding beyond the surface. No diffusing pigment.

Conidiophores pyramidal, with opposing branches borne on a conspicuously broad spindle, less solitary. The main axis and branches terminating in 3–5 cruciate to nearly verticillate disposed phialides. Hyphal septa clearly visible. Phialides ampulliform, sometime lageniform, 4.6–9.9 × (2.2–) 2.7–3.8 μm, l/w 1.4–3.5 (–4.4), 1.4–2.7 μm wide at the base (n = 50). Conidia green, globose or subglobose, sometimes ellipsoidal, smooth, 2.7–3.8 × 2.3–3.1 μm, l/w 1.0–1.3 (n = 50). Chlamydospores common, intercalary or terminal, variable in shape, ellipsoid, globose or oblong, 4.6–7.5 × 3.8–6.3 μm (n = 20).

Additional strains examined: China, Beijing, Tongzhou, from the contaminated substrates of *A. heimuer*, 26 August 2021, W.T. Qin, Y. Liu, S.X. Wang, JZBQT1Z8; *ibid.*, JZBQT1Z9.

Notes: Phylogenetically, *T. auriculariae* formed a separate group (MPBP/BIPP = 100%/100%) in the Harzianum clade among *T. vermifimicola* and *T. xixiacum*. The *tef1-α* sequences between *T. auriculariae* and *T. vermifimicola* were very similar, but they shared 28 bp divergent among 1117 bp for *rpb2* sequences (97.49%). Phylogenetically, *T. auriculariae* shared a common ancestor with *T. xixiacum*, *T. vermifimicola*, and *T. simmonsi*. *Trichoderma auriculariae* shared typical characteristics of the Harzianum clade in pyramidal conidiophores comprising a long main axis, and 3–5 phialides in whorls arose at the tips of the branches. However *T. auriculariae* had longer phialides and grew much slower at 35 °C on PDA than *T. simmonsi* [5.2–6.5 mm, 25–55 mm] [25] and had larger conidia than that of *T. vermifimicola* [2.3–2.6 × 2.0–2.4 μm] and *T. xixiacum* [2.3–2.7 × 2.0–2.6] [24]. Meanwhile, chlamydospores were common in *T. auriculariae* (Table S1).

Trichoderma miyunense Z. J. Cao and W.T. Qin, sp. nov.

MycoBank MB845142 (Figure 3).

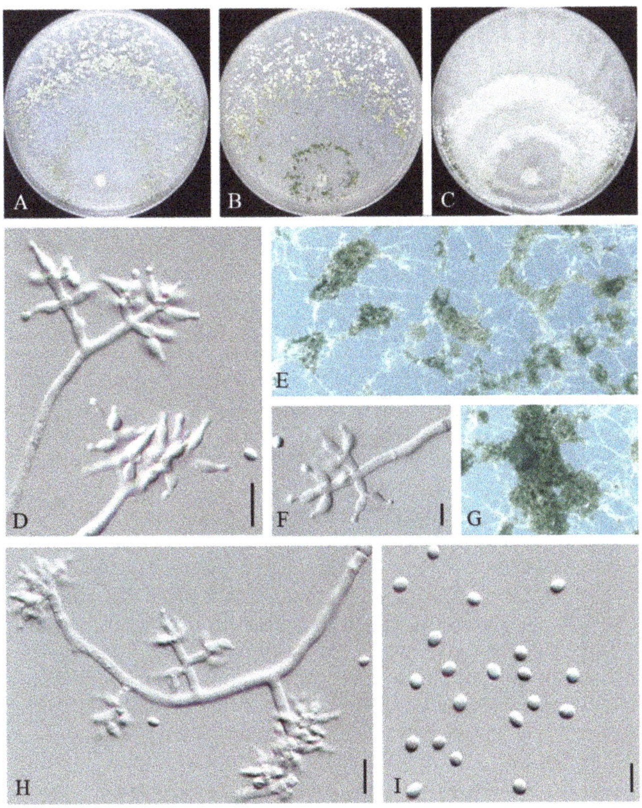

Figure 3. *Trichoderma miyunense* (JZBQF9). Cultures at 25 °C after 7 days on (**A**) CMD, (**B**) SNA, and (**C**) PDA; (**D**,**F**,**H**) conidiophores and phialides; (**E**,**G**) conidiation pustules on SNA after 7 days; (**I**) chlamydospores. Scale bars: (**D**,**H**) = 10 μm, (**F**,**I**) = 5 μm.

Etymology: The specific epithet refers to the type locality.

Typification: China, Beijing, Miyun, from the contaminated substrates of *Auricularia heimuer*, 9 September 2020, Y. Liu, W.T. Qin, S. Song (ex-type strain JZBQF9).

DNA barcodes: ITS = ON653404, *rpb2* = ON649970, *tef1-α* = ON649917.

On CMD after 72 h, colony radius 51–52 mm at 25 °C and 65–66 mm at 30 °C. No growth at 35 °C. Colony hyaline, weak, regularly circular, distinctly zonate. Conidiation first formed in white pustules on aerial hyphae, turned green after a few days. No diffusing pigment noted, odor slightly fruity.

On PDA after 72 h, colony radius 42–43 mm at 25 °C and 51–54 mm at 30 °C. No growth at 35 °C. Mycelium white, aerial along the edge, irregularly circular, less with sporulation. No diffusing pigment noted, odor slightly fruity.

On SNA after 72 h, colony radius 30–33 mm at 25 °C and 25–29 mm at 30 °C. No growth at 35 °C. Mycelium hyaline and smooth, dark green to light green pustules, irregular in shape, relatively abundant in the zonation regions, with the formation of 2–3 concentric rings. Aerial hyphae short and inconspicuous. No diffusing pigment, no distinct odor.

Conidiophores pyramidal, with a relatively obvious main axis, multiple branches unpaired, with the longest branches near the base of the main axis. Branches perpendicular to the main axis or at acute angles with the main axis, with septa conspicuous and producing barrel-shaped or cylindrical metulae. Phialides densely disposed at the terminal of branches, often formed in whorls of 2–4, variable in shape and size, ampulliform to lageniform, (5.2–) 5.6–9.7 (–10.3) × 1.9–3.2 (–3.7) μm, l/w 1.9–4.4, 1.0–2.1 (–2.6) wide at the base (n = 80). Conidia green, smooth, ellipsoid, sometimes globose to subglobose, 2.2–3.4 × (1.8–) 2–2.9 μm, l/w 1–1.3 (–1.4) (n = 80). Chlamydospores unobserved.

Additional strains examined: China, Beijing, Miyun, from the contaminated substrates of *Auricularia heimuer*, 9 September 2020, W.T. Qin, Y. Liu, S. Song, JZBQF5; *ibid.*, JZBQF7.

Notes: Phylogenetically, *T. miyunense* formed a sister group with *T. ganodermatigerum* (Figure 1). They shared 36 bp divergent among 1132 bp for *rpb2* sequences (96.82%) and 35 bp divergent among 1102 bp for *tef1-α* sequences (96.82%). Morphologically, compared to *T. miyunense*, *T. ceratophylletum* possessed shorter phialides (4.1–8.4 μm) and lesser l/w of phialides [(1.0–) 1.2–2.8 (–3.2) μm] [26], while *T. ganodermatigerum* had larger conidia [(3.4–) 3.6–4.8 (–5.3) × (2.9–) 3.2–4.3 (–4.6)], and the optimum temperature was 25 °C [27]. *T. miyunense* was distinctly different from *T. caeruloviride*, which possessed abundant chlamydospores on CMD after 4 days with no concentric rings present [28]. In contrast, *T. confertum* had slightly larger phialides [8.3–12.5 × 2.5–4.2 μm] [29], *T. amazonicum* had distinctly wider phialides [3.3–3.5 μm] and chlamydospore-like structures in the clusters, and *T. pleuroticola* featured diffuse brown pigment and yellow crystals on PDA [30] (Table S2).

Trichoderma pholiotae Z.J. Cao & W.T. Qin, sp. nov.

MycoBank MB845143 (Figure 4).

Etymology: The specific epithet refers to the host from which the fungus was isolated.

Typification: China, Beijing, Haidian, from the contaminated substrates of *Pholiota adipose*, 25 September 2020, W.T. Qin, Z.J. Cao, L. Gao, J. Li (ex-type strain JZBQH12).

DNA barcodes: ITS = ON653405, *rpb2* = ON649972, *tef1-α* = ON649919.

On CMD after 72 h, colony radius 71–72 mm at 25 °C, 73–74 mm at 30 °C, and 13–18 mm at 35 °C. Colonies hyaline, fan-shaped, tending to aggregate toward the distal parts of the colony. Aerial hyphae loose, sparse, radial. Conidiation effuse in aerial hyphae or in loosely disposed pustules. Pustules minute, irregular in shape, relatively abundant in the zonation regions, formed concentric rings around the outer ring, white at first, then gradually green. No diffusing pigment noted, odor slightly fruity.

On PDA after 72 h, colony radius 67–68 mm at 25 °C, 70–72 mm at 30 °C, and 8–10 mm at 35 °C. Colonies white in the center, with the zone around the central part of the colony forming a distinct circular and green part. Aerial hyphae distinctly radial, abundant, dense, floccose to cottony. Light diffusing yellow pigment, odor slightly fruity.

On SNA after 72 h, colony radius 49–50 mm at 25 °C, 54–55 mm at 30 °C, and 8–10 mm at 35 °C. Colonies translucent and round-like. Aerial hyphae short, radial distribution. Pustules abundant, irregular in shape, from white to green, with the formation of concentric rings. No diffusing pigment noted.

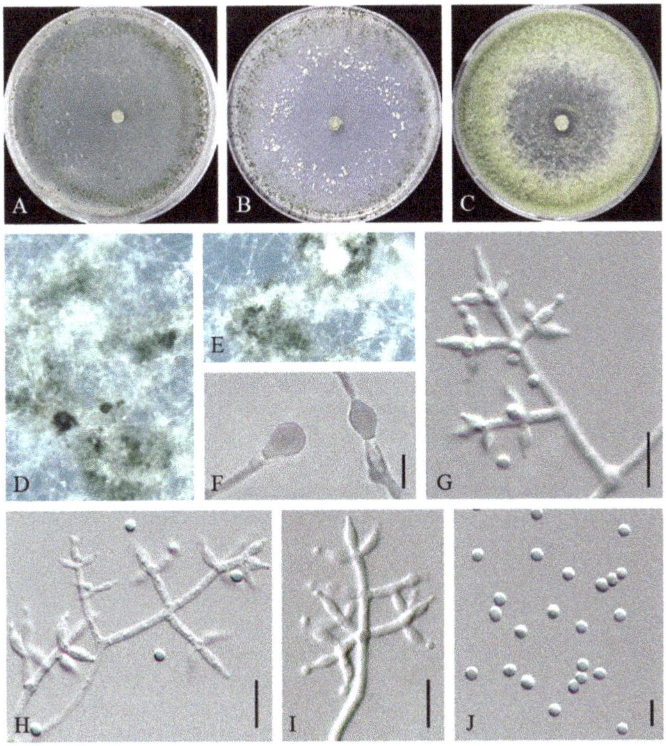

Figure 4. *Trichoderma pholiotae* (JZBQH12). Cultures at 25 °C after 7 days on (**A**) CMD, (**B**) SNA, and (**C**) PDA; (**D**,**E**) conidiation pustules on SNA after 7 days; (**F**) chlamydospores; (**G**–**I**) conidiophores and phialides; (**J**) conidia. Scale bars: (**F**,**J**) = 5 μm, (**G**–**I**) = 10 μm.

Conidiophores typically pyramidal with opposing branches, formed densely intricate reticulum, with one terminal whorl of generally 3–4 phialides and mostly paired side branches, less frequently solitary. Branches mostly perpendicular to the main axis with septa conspicuous. Phialides varied, borne in regular levels around the axis, some regular ampulliform or lageniform and others apex and inequilateral to curved, (4.1–) 4.9–10.9 (–11.6) × 2.4–4.2 (–5.0) μm, l/w 1.4–3.4 (–3.9), (1.3–) 1.4–3.1 (–3.4) μm wide at the base (n = 100). Conidia elliptic to subspheroidal, less globose, green, smooth, 2.6–3.8 (–4.2) × 2.4–3.3 (–3.5) μm, l/w 1–1.3 (n = 80). Chlamydospores common, intercalary or terminal, ellipsoid, globose, 5.0–7.4 (8.3) × (3.9–) 4.9–7.0 μm (n = 25).

Additional strains examined: China, Beijing, Haidian, from the contaminated substrates of *Pholiota adipose*, 25 September 2020, W.T. Qin, Z.J. Cao, L. Gao, J. Li, JZBQH11; *ibid.*, JZBQH13.

Notes: Phylogenetically, *T. pholiotae* formed a linage with *T. asiaticum*, *T. guizhouense*, *T. pseudoasiaticum*, and *T. simile* with high support value (MPBP/BIPP = 92%/100%), and our three strains of *T. pholiotae* were distributed as a highly supported separate terminal branch (MPBP/BIPP = 100%/100%) among *T. pseudoasiaticum* and *T. guizhouense* in the Harzianum clade. However, compared to *T. pholiotae*, *T. guizhouense* possessed thinner phialides [2.0–3.0 μm] and globose conidia [31]. *T. simile* had distinct lower optimum growth temperature (25 °C) in the three media, and *T. asiaticum* had shorter phialides [(3.0–) 4.0–6.0 (–7.0) μm] [12]. In addition, *T. pholiotae* and *T. pseudoasiaticum* could be distinguished by the branching pattern, with *T. pholiotae* being pyramidal and *T. pseudoasiaticum* being verticillium-like (Table S3).

4. Discussion

During exploration of contaminated substrates of edible fungi in North China, 39 strains representing 10 *Trichoderma* species were isolated from four kinds of edible fungi and examined, and three new species were recognized based on integrated studies of phenotypic and molecular data (Table S1). To explore their taxonomic positions, a phylogenetic tree containing all species of the Harzianum clade was constructed based on analyses of the combined sequences of *rpb2* and *tef1-α*. The three new species were well located in the Harzianum clade with separate terminal branches and were clearly distinguishable from any of the existing species. The results of this study have a number of practical implications to identify and diagnose *Trichoderma* species contaminating edible fungi. This work provides useful information on the epidemiological and geographical distribution of *Trichoderma*, which will help in the development of targeted interventions aimed at comprehensive management and control of green mold contamination of edible fungi.

With further study of *Trichoderma* classification, researchers have reached a consensus that accurate identification of *Trichoderma* species cannot depend only on the morphological identification as sometimes there is high ambiguity in the morphological features of *Trichoderma* spp. [32,33]. *Trichoderma* spp. isolated from the fruiting bodies or substrates of edible fungi is usually anamorph with high morphological similarity with many species, which is not conducive to identification. With DNA-based techniques gradually perfected and widely used, the integrative (polyphasic) taxonomy approach for species delimitation is recommended, including the combination of genealogy and multiparametric phenotypes [34,35], especially for examining the presence of species complexes and cryptic species [31]. Therefore, we hypothesized that *T. harzianum*, which was originally identified by ITS sequence and morphology in previous studies, probably belonged to the *T. harzianum* complex. However, the present study showed that the complex still contained many taxa, indicating that the previous identification was not accurate. Furthermore, it is also difficult to identify species of the Harzianum clade according to exclusive *tef1-α* or *rpb2* sequence data [24,25]. Therefore, the combination of *tef1-α* and *rpb2* sequences for phylogenetic analysis is highly recommended to identify species in the Harzianum clade.

Taxonomy of *Trichoderma* dates back to the late 18th century [36], and some of them cause economic losses in commercial mushroom farms [37]. Over more than a century, successive findings have brought the number of known species of the genus to over 441 [1,23,38]. *Trichoderma* species are located throughout the world, and more than 30 of them are mushroom inhabiting (Figure 1, Table 2). They are isolated from the substrate or fruiting bodies of *Agaricus bisporus*, *Lentinula edodes*, *Pleurotus ostreatus*, *Ganoderma lingzhi*, etc. and are mainly located in the Harzianum, Longibrachiatum, and Viride clades [39]. There may still be many unknown *Trichoderma* species associated with the growth of edible fungi and their related living environment. The phylogenetical difference between *Trichoderma* spp. on edible fungi substrates and from other sources deserves further analysis.

Analysis of the biological characteristics of *Trichoderma* species from contaminated substrates showed that the optimum growth temperature of many *Trichoderma* species was generally around 30 °C, which was consistent with the phenomenon that contamination of *Trichoderma* on edible fungi is more likely to occur at high temperatures. Therefore, reasonable control the growth environment temperature of edible fungi may be a reasonable approach to prevent or delay the outbreak of *Trichoderma* contamination during production. More broadly, research is also needed to analyze the mechanism of occurrence of *Trichoderma* spp. contamination, such as the correlation between contamination occurrence and the growth environment of edible fungi.

With the increased number of species joining the Harzianum clade, understanding of *Trichoderma* spp. will become more sophisticated and intelligible, and reasonable species concepts will be firmly established. Accumulated knowledge of *Trichoderma*, especially the Harzianum clade, will provide useful information for sufficient utilization of resources and for the prevention of contamination of edible fungi.

Table 2. *Trichoderma* spp. associated with the contaminated substrates of edible fungi.

Species	Cultivated Mushroom	Reference
T. aggressivum	*Agaricus bisporus*	[40,41]
T. asperellum	*A. bisporus*	[9,42]
T. atroviride	*L. edodes, Pleurotus ostreatus, A. bisporus, Ganoderma lingzhi*	[8,9,43]
T. aureoviride	*Auricularia heimuer, Flammulina filiformis, L. edodes*	[44]
T. breve	*L. edodes*	[45]
T. capillare	*Agaricus* sp.	[46]
T. citrinviride	*L. edodes, P. ostreatus*	[43,47]
T. deliquescens	*L. edodes*	[11]
T. ganodermatigerum	*G. sichuanense*	[27]
T. ghanense	*A. bisporus*	[9]
T. guizhouense	*P. ostreatus*	[48]
T. hamatum	*A. bisporus*	[49]
T. harzianum	*L. edodes, A. bisporus, P. ostreatus, Agrocybe aegerita*	[43,50]
T. hengshanicum	*G. lingzhi*	[51]
T. hirsutum	*L. edodes*	[45]
T. koningii	*P. ostreatus, A. bisporus*	[37,40]
T. koningiopsis	*Dictyophora rubrovolvata, P. eryngii*	[52,53]
T. lentinulae	*L. edodes*	[24]
T. longibrachiatum	*L. edodes, P. ostreatus, A. aegerita*	[9,43,50]
T. oblongisporum	*L. edodes*	[54]
T. patella	*P. ostreatus*	[55]
T. pleuroti	*P. ostreatus*	[56]
T. pleuroticola	*P. ostreatus, L. edodes, G. lingzhi*	[50,54,56]
T. polysporum	*L. edodes*	[57]
T. pseudogelatinosum	*L. edodes*	[58]
T. pseudokoningii	*P. ostreatus*	[37]
T. pseudolacteum	*L. edodes*	[59]
T. pseudostramineum	*L. edodes*	[58]
T. reesei	*P. ostreatus*	[60]
T. stramineum	*L. edodes*	[57]
T. stromaticum	*A. bisporus*	[49]
T. virens	*P. ostreatus, A. bisporus*	[37,40]
T. viride	*L. edodes*	[54]

5. Conclusions

In this study, 39 strains belonging to 10 *Trichoderma* species isolated from four kinds of edible fungi in North China were obtained, and three novel species belonging to the Harzianum clade were isolated from the contaminated substrates of *Auricularia heimuer* and *Pholiota adipose*. More than 30 mushroom-inhabiting *Trichoderma* species throughout the world mainly located in the Harzianum, Longibrachiatum, and Viride clades were indicated. This study enrich the biodiversity of *Trichoderma* and provide important support for systematic development of the Harzianum clade.

Supplementary Materials: The following supporting information can be downloaded at: https://www.mdpi.com/article/10.3390/jof8111154/s1. Table S1: Strain information and their accession numbers. Table S2: Comparison of the morphological characteristics of *Trichoderma auriculariae* and its relatives. Table S3: Comparison of the morphological characteristics of *Trichoderma miyunense* and its relatives. Table S4: Comparison of the morphological characteristics of *Trichoderma pholiotae* and its relatives. Table S5: The growth rate of three new species in this study incubated at different temperatures and media.

Author Contributions: Conceptualization, W.-T.Q. and S.-Y.Z.; methodology, W.-T.Q.; software, Z.-J.C. and J.Z.; validation, W.-T.Q. and Y.L.; formal analysis, Z.-J.C.; investigation, Z.-J.C. and S.-X.W.; data curation, Z.-J.C. and J.Z.; writing—original draft preparation, Z.-J.C.; writing—review and editing, W.-T.Q. and S.-Y.Z.; visualization, Z.-J.C. and J.Z.; supervision, W.-T.Q. and S.-X.W.; project administration, W.-T.Q. and Y.L.; funding acquisition, W.-T.Q. All authors have read and agreed to the published version of the manuscript.

Funding: This study was funded by the Beijing Academy of Agriculture and Forestry Sciences, China (KJCX20220415), the National Natural Science Foundation of China (32002106), and the Rural Revitalization Project of Beijing Municipal Bureau of Agriculture (BJXCZX20221229).

Institutional Review Board Statement: Not applicable.

Informed Consent Statement: Not applicable.

Data Availability Statement: Not applicable.

Acknowledgments: The authors are thankful to Xing-Hong Li and Wei Zhang for technical assistance and thankful to all the sample collectors in this study.

Conflicts of Interest: The authors declare no conflict of interest.

References

1. Cai, F.; Druzhinina, I.S. In honor of John Bissett: Authoritative guidelines on molecular identification of *Trichoderma*. *Fungal Divers.* **2021**, *107*, 1–69. [CrossRef]
2. Druzhinina, I.S.; Seidl-Seiboth, V.; Herrera-Estrella, A.; Horwitz, B.A.; Kenerley, C.M.; Monte, E.; Mukherjee, P.K.; Zeilinger, S.; Grigoriev, I.V.; Kubicek, C.P. *Trichoderma*: The genomics of opportunistic success. *Nat. Rev. Microbiol.* **2011**, *9*, 749–759. [CrossRef] [PubMed]
3. Nuangmek, W.; Aiduang, W.; Kumla, J.; Lumyong, S.; Suwannarach, N. Evaluation of a newly identified endophytic fungus, *Trichoderma phayaoense* for plant growth promotion and biological control of gummy stem blight and wilt of muskmelon. *Front. Microbiol.* **2021**, *12*, 634772. [CrossRef]
4. Carillo, P.; Woo, S.L.; Comite, E.; El-Nakhel, C.; Vinale, F. Application of *Trichoderma harzianum*, 6-pentyl-α-pyrone and plant biopolymer formulations modulate plant metabolism and fruit quality of plum tomatoes. *Plants* **2020**, *9*, 771. [CrossRef]
5. Wei, H.; Wu, M.; Fan, A.; Su, H. Recombinant protein production in the filamentous fungus *Trichoderma*. *Chin. J. Chem. Eng.* **2021**, *30*, 74–81. [CrossRef]
6. Gupta, S.; Smith, P.; Boughton, B.; Twt, R.; Sha, N.; Roessner, U. Inoculation of barley with *Trichoderma harzianum* T-22 modifies lipids and metabolites to improve salt tolerance. *J. Exp. Bot.* **2021**, *72*, 7229–7246. [CrossRef]
7. Samuel, G.J.; Dodd, S.L.; Gams, W.; Castlebury, L.A.; Petrini, O. *Trichoderma* species associated with the green mold epidemic of commercially grown *Agaricus bisporus*. *Mycologia* **2002**, *94*, 146–170. [CrossRef]
8. Yan, Y.; Zhang, C.; Moodley, O.; Zhang, L.; Xu, J. Green mold caused by *Trichoderma atroviride* on the lingzhi medicinal mushroom, *Ganoderma lingzhi* (Agaricomycetes). *Int. J. Med. Mushrooms* **2019**, *21*, 515–521. [CrossRef]
9. Hatvani, L.; Antal, Z.; Manczinger, L.; Szekeres, A.; Druzhinina, I.S.; Kubicek, C.P.; Nagy, A.; Nagy, E.; Vagvolgyi, C.; Kredics, L. Green mold diseases of *Agaricus* and *Pleurotus* spp. are caused by related but phylogenetically different *Trichoderma* species. *Phytopathology* **2007**, *97*, 532–537. [CrossRef]
10. Komon-Zelazowska, M.; Bissett, J.; Zafari, D.; Hatvani, L.; Manczinger, L.; Woo, S.; Lorito, M.; Kredics, L.; Kubicek, C.P.; Druzhinina, I.S. Genetically closely related but phenotypically divergent *Trichoderma* species cause green mold disease in oyster mushroom farms worldwide. *Appl. Environ. Microb.* **2007**, *73*, 7415–7426. [CrossRef]
11. Kim, J.Y.; Yun, Y.H.; Hyun, M.W.; Kim, M.H.; Kim, S.H. Identification and characterization of *Gliocladium viride* isolated from mushroom fly infested oak log beds used for shiitake cultivation. *Mycobiology* **2010**, *38*, 7–12. [CrossRef] [PubMed]
12. Zheng, H.; Qiao, M.; Lv, Y.F.; Du, X.; Zhang, K.Q.; Yu, Z.F. New species of *Trichoderma* isolated as endophytes and saprobes from southwest China. *J. Fungi* **2021**, *7*, 467. [CrossRef] [PubMed]
13. Ignazio, C.; Linda, M.K. A method for designing primer sets for speciation studies in filamentous ascomycetes. *Mycologia* **1999**, *91*, 553–556.
14. Jaklitsch, W.M.; Komon, M.; Kubicek, C.P.; Druzhinina, I.S. *Hypocrea voglmayrii* sp. nov. from the Austrian Alps represents a new phylogenetic clade in *Hypocrea/Trichoderma*. *Mycologia* **2005**, *97*, 1365–1378. [CrossRef] [PubMed]
15. Liu, Y.J.; Whelen, S.; Hall, B.D. Phylogenetic relationships among ascomycetes: Evidence from an RNA polymerse II subunit. *Mol. Biol. Evol.* **1999**, *16*, 1799–1808. [CrossRef]
16. Zhu, Z.X.; Zhuang, W.Y. *Trichoderma* (*Hypocrea*) species with green ascospores from China. *Persoonia* **2015**, *34*, 113–129. [CrossRef]
17. Dou, K.; Lu, Z.; Wu, Q.; Ni, M.; Yu, C.; Wang, M.; Li, Y.; Wang, X.; Xie, H.; Chen, J.; et al. MIST: A multilocus identification system for *Trichoderma*. *Appl. Environ. Microb.* **2020**, *86*, e01532-20. [CrossRef]

18. Hall, T.A. BioEdit: A user-friendly biological sequence alignment editor and analysis program for windows 95/98/NT. *Nucleic Acids Symp.* **1999**, *41*, 95–98.
19. Higgins, D.G.; Jeanmougin, F.; Gibson, T.J.; Plewniak, F.; Thompson, J.D. The Clustal X windows interface: Flexible strategies for multiple sequence alignment aided by quality analysis tools. *Nucleic Acids Symp.* **1997**, *25*, 4876–4882.
20. Swofford, D.L. *PAUP*. Phylogenetic Analysis Using Parsimony (*and Other Methods)*; Version 4.0; Sinauer Associates: Sunderland, MA, USA, 2002.
21. Ronquist, F.; Huelsenbeck, J.P. MrBayes 3: Bayesian phylogenetic inference under mixed models. *Bioinformatics* **2003**, *19*, 1572–1574. [CrossRef]
22. Nylander, J.A.A. MrModeltest v2. Program Distributed by the Author. 2004. Available online: http://paup.csit.fsu.edu (accessed on 1 January 2018).
23. Rambaut, A. FigTree. Tree Figure Drawing Tool, v. 1.4.3. 2016. Available online: http://tree.bio.ed.ac.uk/ (accessed on 1 January 2018).
24. Gu, X.; Wang, R.; Sun, Q.; Wu, B.; Sun, J.-Z. Four new species of *Trichoderma* in the Harzianum clade from northern China. *Mycokeys* **2020**, *73*, 109–132. [CrossRef] [PubMed]
25. Chaverri, P.; Branco-Rocha, F.; Jaklitsch, W.; Gazis, R.; Degenkolb, T.; Samuels, G.J. Systematics of the *Trichoderma harzianum* species complex and the re-identification of commercial biocontrol strains. *Mycologia* **2015**, *107*, 558–590. [CrossRef] [PubMed]
26. Yuan, H.S.; Lu, X.; Dai, Y.C.; Hyde, K.V.D.; Kan, Y.H.; Kusan, I.; He, S.H.; Liu, N.G.; Sarma, V.V.; Zhao, C.L.; et al. Fungal diversity notes 1277–1386: Taxonomic and phylogenetic contributions to fungal taxa. *Fungal Divers.* **2020**, *104*, 1–266. [CrossRef]
27. An, X.Y.; Cheng, G.H.; Gao, H.X.; Li, D.; Li, Y. Phylogenetic analysis of *trichoderma* species associated with green mold disease on mushrooms and two new pathogens on *Ganoderma sichuanense*. *J. Fungi* **2022**, *8*, 704. [CrossRef] [PubMed]
28. Rodriguez, M.D.H.; Evans, H.C.; De Abreu, L.M.; De Macedo, D.M.; Ndacnou, M.K.; Bekele, K.B.; Barreto, R.W. New species and records of *Trichoderma* isolated as mycoparasites and endophytes from cultivated and wild coffee in Africa. *Sci. Rep.* **2021**, *11*, 5671. [CrossRef]
29. Chen, K.; Zhuang, W.Y. Seven new species of *Trichoderma* from soil in China. *Mycosystema* **2017**, *36*, 1441–1462.
30. Chaverri, P.; Gazis, R.O.; Samuels, G.J. *Trichoderma amazonicum*, a new endophytic species on *Hevea brasiliensis* and *H. guianensis* from the Amazon basin. *Mycologia* **2011**, *103*, 139–151. [CrossRef]
31. Li, Q.R.; Tan, P.; Jiang, Y.L.; Hyde, K.D.; Mckenzie, E.H.C.; Bahkali, A.H.; Kang, J.C.; Wang, Y. A novel *Trichoderma* species isolated from soil in Guizhou, *T. guizhouense*. *Mycol. Prog.* **2013**, *12*, 167–172. [CrossRef]
32. Chaverri, P.; Samuels, G.J. *Hypocrea/Trichoderma* (Ascomycota, Hypocreales, Hypocreaceae): Species with green ascospores. *Stud. Mycol.* **2003**, *48*, 1–116.
33. Jaklitsch, W.M.; Samuels, G.J.; Dodd, S.L.; Lu, B.S.; Druzhinina, I.S. *Hypocrea rufa/Trichoderma viride*: A reassessment, and description of five closely related species with and without warted conidia. *Stud. Mycol.* **2006**, *56*, 135–177. [CrossRef]
34. Luecking, R.; Aime, M.C.; Robbertse, B.; Miller, A.N.; Ariyawansa, H.A.; Aoki, T.; Cardinali, G.; Crous, P.W.; Druzhinina, I.S.; Geiser, D.M.; et al. Unambiguous identification of fungi: Where do we stand and how accurate and precise is fungal DNA barcoding? *IMA Fungus* **2020**, *11*, 14. [CrossRef] [PubMed]
35. Druzhinina, I.; Kubicek, C.P. Species concepts and biodiversity in *Trichoderma* and *Hypocrea*: From aggregate species to species clusters? *J. Zhejiang Univ. Sci. B* **2005**, *6*, 100–112. [CrossRef] [PubMed]
36. Persoon, C.H. *Neurospora* genetic nomenclature. *Romers Neues Mag. Bot.* **1794**, *1*, 81–128.
37. Jhune, C.S.; Leem, H.T.; Park, H.S.; Lee, C.J.; Weon, H.Y.; Seok, S.J.; Yoo, K.H.; Sung, G.H. Identification of oyster mushroom green mold pathogen that causes and pathological characteristics. *J. Mushrooms* **2014**, *12*, 132–137. [CrossRef]
38. Barrera, V.A.; Iannone, L.; Romero, A.I.; Chaverri, P. Expanding the *Trichoderma harzianum* species complex: Three new species from Argentine natural and cultivated ecosystems. *Mycologia* **2021**, *113*, 1136–1155. [CrossRef]
39. Allaga, H.; Zhumakayev, A.; Buchner, R.; Kocsube, S.; Szucs, A.; Vagvolgyi, C.; Kredics, L.; Hatvani, L. Members of the *Trichoderma harzianum* species complex with mushroom pathogenic potential. *Agronomy* **2021**, *11*, 2434. [CrossRef]
40. Kosanovic, D.; Potocnik, I.; Duduk, B.; Vukojevic, J.; Stajic, M.; Rekanovic, E.; Milijasevic-Marcic, S. *Trichoderma* species on *Agaricus bisporus* farms in Serbia and their biocontrol. *Ann. Appl. Biol.* **2013**, *163*, 218–230. [CrossRef]
41. Marik, T.; Urban, P.; Tyagi, C.; Szekeres, A.; Leitgeb, B.; Vagvolgyi, M.; Manczinger, L.; Druzhinina, I.S.; Vagvolgyi, C.; Kredics, L. Diversity profile and dynamics of peptaibols produced by green mould *Trichoderma* species in interactions with their hosts *Agaricus bisporus* and *Pleurotus ostreatus*. *Chem. Biodivers.* **2017**, *14*, e1700033. [CrossRef]
42. Wu, X.J.; Hu, F.P.; He, H.Z.; Xie, B.G. Identification of *Trichoderma* species associated with cultivated edible fungi. *J. Agric. Biotechnol.* **2008**, *16*, 1048–1055.
43. Kim, C.S.; Park, M.S.; Kim, S.C.; Maekawa, N.; Yu, S.H. Identification of *Trichoderma*, a competitor of shiitake mushroom (*Lentinula edodes*), and competition between *Lentinula edodes* and *Trichoderma* species in Korea. *Plant Pathol. J.* **2012**, *28*, 137–148. [CrossRef]
44. Cui, L.H. Isolation, Identification and Diversity Analysis of the Contaminating Fungi from the Edible Mushroom-Growing Synthetic Wood Logs. Master's Thesis, Liaoning Normal University, Dalian, China, 2017.
45. Wang, Y. Identification of the Pathogen of *Lentinus edodes* Sticks Rot and Preliminary Study on Its Occurrence. Master's Thesis, Guizhou University, Guiyang, China, 2021.

46. Samuels, G.J.; Ismaiel, A.; Mulaw, T.B.; Szakacs, G.; Druzhinina, I.S.; Kubicek, C.P.; Jaklitsch, W.M. The Longibrachiatum clade of *Trichoderma*: A revision with new species. *Fungal Divers.* **2012**, *55*, 77–108. [CrossRef] [PubMed]
47. Park, M.S.; Seo, G.S.; Bae, K.S.; Yu, S.H. Characterization of *Trichoderma* spp. associated with green mold of oyster mushroom by PCR-RFLP and sequence analysis of ITS regions of rDNA. *Plant Pathol. J.* **2005**, *21*, 229–236. [CrossRef]
48. Innocenti, G.; Montanari, M.; Righini, H.; Roberti, R. *Trichoderma* species associated with green mould disease of *Pleurotus ostreatus* and their sensitivity to prochloraz. *Plant Pathol.* **2019**, *68*, 392–398. [CrossRef]
49. Song, X.X.; Wang, Q.; Jun, J.X.; Zhang, J.J.; Chen, H.; Chen, M.J.; Huang, J.C.; Xie, B.Q. Study on accurate identification of four *Trichoderma* diseases in *Agaricus bisporus* in factory cultivation. *Edible Fungi* **2019**, *41*, 67–72.
50. Choi, I.Y.; Choi, J.N.; Hyu, L.W.; Sharma, P.K. Isolation and identification of mushroom pathogens from *Agrocybe aegerita*. *Mycobiology* **2010**, *38*, 310–315. [CrossRef]
51. Cai, M.; Idrees, M.; Zhou, Y.; Zhang, C.; Xu, J. First report of green mold disease caused by *Trichoderma hengshanicum* on *Ganoderma lingzhi*. *Mycobiology* **2020**, *48*, 427–430. [CrossRef]
52. Chen, X.; Zhou, X.; Zhao, J.; Tang, X.; Pasquali, M.; Migheli, Q.; Berg, G.; Cernava, T. Occurrence of green mold disease on *Dictyophora rubrovolvata* caused by *Trichoderma koningiopsis*. *J. Plant Pathol.* **2021**, *103*, 981–984. [CrossRef]
53. Kim, S.W.; Kim, S.; Lee, H.J.; Park, J.W.; Ro, H.S. Isolation of fungal pathogens to an edible mushroom, *Pleurotus eryngii*, and development of specific ITS primers. *Mycobiology* **2013**, *41*, 252–255. [CrossRef]
54. Wang, G.; Cao, X.; Ma, X.; Guo, M.; Liu, C.; Yan, L.; Bian, Y. Diversity and effect of *Trichoderma* spp. associated with green mold disease on *Lentinula edodes* in China. *Microbiologyopen* **2016**, *5*, 709–718. [CrossRef]
55. Hua, R.; Li, J.Y.; Liu, S.X.; Luo, X.K.; Wang, X.Y.; Liu, C.L.; Zhang, L.Y.; Sun, D.F. Investigation on common dieases of *Pleurotus ostreatus* and identification of pathogens. *Edible Fungi China* **2021**, *40*, 80–86.
56. Park, M.S.; Bae, K.S.; Yu, S.H. Two new species of *Trichoderma* associated with green mold of oyster mushroom cultivation in Korea. *Mycobiology* **2006**, *34*, 111–113. [CrossRef] [PubMed]
57. Miyazaki, K.; Tsuchiya, Y.; Okuda, T. Specific PCR assays for the detection of *Trichoderma harzianum* causing green mold disease during mushroom cultivation. *Mycoscience* **2009**, *50*, 94–99. [CrossRef]
58. Kim, C.S.; Yu, S.H.; Nakagiri, A.; Shirouzu, T.; Sotome, K.; Kim, S.C.; Maekawa, N. Re-evaluation of *Hypocrea pseudogelatinosa* and *H. pseudostraminea* isolated from shiitake mushroom (*Lentinula edodes*) cultivation in Korea and Japan. *Plant Pathol. J.* **2012**, *28*, 341–356. [CrossRef]
59. Kim, C.S.; Shirouzu, T.; Nakagiri, A.; Sotome, K.; Maekawa, N. *Trichoderma eijii* and *T. pseudolacteum*, two new species from Japan. *Mycol. Prog.* **2013**, *12*, 739–753. [CrossRef]
60. Lan, B.M. Isolation and identification of *Trichoderma* on oyster mushroom in Quanzhou. *Chin. J. Trop. Agric.* **2022**, *42*, 81–85.

Journal of *Fungi*

MDPI

Article

Phylogenetic Analysis of *Trichoderma* Species Associated with Green Mold Disease on Mushrooms and Two New Pathogens on *Ganoderma sichuanense*

Xiao-Ya An [1,2], Guo-Hui Cheng [1,2], Han-Xing Gao [2], Xue-Fei Li [2], Yang Yang [3], Dan Li [2,*] and Yu Li [1,2,*]

[1] Department of Plant Protection, Shenyang Agricultural University, Shenyang 110866, China; anxiaoya2016@163.com (X.-Y.A.); chengguohui2016@163.com (G.-H.C.)
[2] Engineering Research Center of Chinese Ministry of Education for Edible and Medicinal Fungi, Jilin Agricultural University, Changchun 130118, China; gaohanxing168@163.com (H.-X.G.); lixuefei2020@163.com (X.-F.L.)
[3] Environment and Plant Protection Institute, Chinese Academy of Tropical Agricultural Sciences, Haikou 571101, China; yyjob1992@163.com
* Correspondence: lidan@jlau.edu.cn (D.L.); liyu@jlau.edu.cn (Y.L.)

Abstract: Edible and medicinal mushrooms are extensively cultivated and commercially consumed around the world. However, green mold disease (causal agent, *Trichoderma* spp.) has resulted in severe crop losses on mushroom farms worldwide in recent years and has become an obstacle to the development of the *Ganoderma* industry in China. In this study, a new species and a new fungal pathogen on *Ganoderma sichuanense* fruitbodies were identified based on the morphological characteristics and phylogenetic analysis of two genes, the translation elongation factor 1-α (TEF1) and the second-largest subunit of RNA polymerase II (RPB2) genes. The new species, *Trichoderma ganodermatigerum* sp. nov., belongs to the Harzianum clade, and the new fungal pathogen was identified as *Trichoderma koningiopsis*. Furthermore, in order to better understand the interaction between *Trichoderma* and mushrooms, as well as the potential biocontrol value of pathogenic *Trichoderma*, we summarized the *Trichoderma* species and their mushroom hosts as best as possible, and the phylogenetic relationships within mushroom pathogenic *Trichoderma* species were discussed.

Keywords: taxonomy; green mold disease; one new taxon; mycoparasites; biological agents

Citation: An, X.-Y.; Cheng, G.-H.; Gao, H.-X.; Li, X.-F.; Yang, Y.; Li, D.; Li, Y. Phylogenetic Analysis of *Trichoderma* Species Associated with Green Mold Disease on Mushrooms and Two New Pathogens on *Ganoderma sichuanense*. *J. Fungi* **2022**, *8*, 704. https://doi.org/10.3390/jof8070704

Academic Editors: Cheng Gao and Lei Cai

Received: 8 June 2022
Accepted: 1 July 2022
Published: 3 July 2022

1. Introduction

Mushrooms have been used by humans for millennia and are consumed for their nutritive and medicinal values [1,2]. Most of them are appreciated as delicacies and are extensively cultivated and commercially consumed in many countries. Some mushrooms also have high pharmacological activities, especially *Ganoderma* spp. [3,4]. *Ganoderma sichuanense*, described from China and previously confused with *G. lucidum*, an oriental fungus, has a long history in China, Japan, and other Asian countries for promoting health and longevity [5,6]. The mushroom is famous for its pharmacological effects [7,8] and is widely cultivated in northeastern China. However, *Trichoderma* green mold diseases have increased and pose a serious threat to its production [9–11].

Trichoderma has been studied for more than 200 years since it was established by Persoon in 1794 [12], while sharp development occurred in the past few decades, when a large number of taxonomic articles were published [13–26]. At present, similar to *Fusarium*, *Aspergillus*, or *Penicillium*, *Trichoderma* is a species-rich genus [15] and has been segregated into many groups or clades based on the phylogenetic relationships within the genus [27–29]. Moreover, the rapid development of *Trichoderma* is inseparable from its various uses. For example, it can not only be used as a highly efficient producer of plant biomass-degrading enzymes for biofuel and other industries, but also as a very effective biological agent for plant disease management [30–33]. Furthermore, *Trichoderma* has also been an initially produce white and dense mycelia highly similar to mushroom mycelia, which makes it difficult to distinguish them, causing the best period of control to be missed. Thus, it is particularly important to explore the specificity of *Trichoderma* species and the interaction between *Trichoderma* and its host for disease control.

Between 2020 and 2021, during fieldwork at mushroom cultivation bases, we found that green mold disease occurred continuously in *G. sichuanense* production areas in the following provinces of China: Heilongjiang, Jilin, and Shandong, leading to a significant negative effect on the development of fruitbodies. We collected diseased specimens and isolated the pathogens from several bases and identified them based on molecular and morphological characteristics. A new *Trichoderma* species and a new host record were confirmed. In addition, we summarized the *Trichoderma* species reported on mushrooms as best as possible and provided their recorded hosts. The relationships among these species were also discussed by constructing a phylogeny tree with multi-locus data, which is expected to help us know more about the relationships between *Trichoderma* species and their hosts, and to help search for *Trichoderma* species with potential biocontrol value.

2. Materials and Methods

2.1. Fungal Isolation

Diseased samples of *G. sichuanense* were collected from Jilin, Heilongjiang, and Shandong Provinces, China, and deposited in the Herbarium of Mycology, Jilin Agricultural University (HMJAU). Diseased tissues were cut into small pieces (5 mm × 5 mm × 5 mm) using a sterilized scalpel, immersed in 75 percent alcohol for 45 s before being rinsed three times with sterilized water, and placed onto Potato Dextrose Agar (PDA, BD, USA) plates containing 100 mg/L of streptomycin sulfate (Solarbio, Bejing, China), and then incubated at room temperature. Pure cultures were obtained using single-spore isolates following the method described by Chomnuti et al. [34]. Germinated spores were transferred to fresh PDA plates and incubated at 25 °C for one or two weeks. Living cultures were deposited in the Engineering Research Center of Edible and Medicinal Fungi, Ministry of Education, Jilin Agricultural University (Changchun, Jilin, China).

2.2. Growth Characterization

Colony characteristics, growth rates, and optimum temperatures for growth were determined according to the methods of Jaklitsch [18,19] by using agar media cornmeal dextrose agar (CMD, 40 g cornmeal + 2% (*w*/*v*) dextrose (Genview, Beijing, China) + 2% (*w*/*v*) agar (Genview, Beijing, China)), PDA, and synthetic low nutrient agar (SNA, pH adjusted to 5.5) [35]. Colonies were incubated in 9 cm diameter Petri dishes at 25 °C with alternating 12 h/12 h fluorescent light/darkness and measured daily until the dishes were covered with mycelia. The influence of temperature on growth was studied by growing isolates on PDA, SNA, and CMD at 15 °C, 20 °C, 25 °C, 30 °C, and 35 °C under dark conditions. Each temperature was repeated for five plates, and the experiment was repeated three times.

2.3. Morphological Study

The characteristics of asexual states were described following the methods of Jaklitsch [36] and Rifai [37]. Microscopic observations were conducted using a Zeiss Axio Lab A1 light microscope (Göttingen, Germany) (objectives 10, 20, 40, and 100 oil immersion). All measurements and photographs were performed using a Zeiss Imager A2 microscope with an Axiocam 506 color camera and integrated software. Microscopically, the characteristics of 50 conidia and conidiophores from the isolates were observed. The effects of *Trichoderma* on *Ganoderma* morphology were studied using a Hitachi, model SU8010, Field Emission Scanning Electron Microscope (FESEM) at Jilin Agricultural University.

2.4. DNA Extraction, PCR, and Sequencing

Mycelia were harvested from three-day-old cultures on PDA for DNA extraction according to the manufacturer's instructions (NuClean Plant Gen DNA Kit, CWBIO, Taizhou,

China). Sequences of ITS, TEF1, and RPB2 genes were amplified by polymerase chain reaction (PCR) with the pairs of primers ITS4 (5′-TCCTCCGCTTATTGATATGC-3′) and ITS5 (5′-GGAAGTAAAAGTCGTAACAAGG-3′) [38], primers EF1-728F (5′-CATCGAGAAGT TCGAGAAGG-3′) [39] and TEF1-LLErev (5′-GCCATCCTTGGAGATACCAGC-3′) [40], and primers RPB2-5F (5′-GAYGAYMGWGATCAYTTYGG-3′) and RPB2-7CR (5′-CCCATRGCTT GYTTRCCCA-3′) [41], respectively.

PCR was carried out in a 25 μL reaction mixture containing 1 μL of DNA sample, 12.5 μL 2 × SanTaq PCR Mix (Sangon Biotech, Shanghai, China), 1 μL of each primer (10 μM), and 9.5 μL nuclease-free water. The PCR conditions were as follows: initial denaturation at 94 °C for 3 min, then denaturation at 94 °C for 30 s, annealing for 45 s with the corresponding temperatures (56 °C for TEF1, and 55 °C for RPB2), extension at 72 °C for 1 min, followed by 35 cycles, then a final extension at 72 °C for 10 min, using an Applied Biosystems S1000 TM Thermal Cycler machine. PCR products were sent to the Changchun Branch of Sangon Biotech Co., Ltd. (Changchun, China) for paired-end sequencing, and the results were first assembled using BioEdit [42] and then confirmed by BLAST on NCBI (https://blast.ncbi.nlm.nih.gov/Blast.cgi, accessed on 21 June 2021).

2.5. Phylogenetic Analyses

BLASTn searches with the sequences were performed against NCBI to detect the most closely related species (http://www.blast.ncbi.nlm.nih.gov/, accessed on 22 December 2021). Phylogenetic trees were constructed using TEF1 and RPB2 sequences, and phylogenetic analyses were performed with the Maximum Likelihood (ML), Maximum Parsimony (MP), and Bayesian Inference (BI) methods. New sequences were generated from the new species in this study, along with reference sequences retrieved from GenBank (Table 1). The *Trichoderma* sequences associated with mushroom green mold are listed in Table 2. Multiple alignments of all common sequences and reference sequences were automatically generated using MAFFT V.7.471 [43], with manual improvements made using BioEdit when necessary [42], and converted to nexus and NEX format through the software Aliview [44]. In the analysis, ambiguous areas were excluded and gaps were regarded as missing data.

Table 1. Strain information and GenBank accession numbers of sequences used for phylogenetic analyses for new species.

Species	Strains	GenBank Accession Number		References
		TEF1	RPB2	
T. afarasin	GJS 99-227	AF348093	—	[45]
T. afroharzianum	LESF229	KT279013	KT278945	[46]
T. afroharzianum	GJS04-186 (T)	FJ463301	FJ442691	In GenBank
T. aggregatum	HMAS248864	KY688063	KY688002	[47]
T. aggressivum	CBS100525	AF534614	AF545541	[48]
T. aggressivum	DAOM222156	AF348098	FJ442752	[45]
T. alni	CPK2494	EU498313	EU498350	[49]
T. alni	CBS120633 = CPK1982 (T)	EU498312	EU498349	[49]
T. alpinum	HMAS248870	KY688017	KY687963	[47]
T. alpinum	HMAS248821 (T)	KY688012	KY687958	[47]
T. amazonicum	IB95	HM142377	HM142368	[50]
T. asperellum	CBS433.97 = TR3 (T)	AF456907	EU248617	[51]
T. atrobrunneum	S3	KJ665376	KJ665241	[20]
T. atrobrunneum	GJS92-110 (T)	AF443942	—	[16]
T. atrogelatinosum	CBS237.63 (T)	—	KJ842201	In GenBank
T. azevedoi	CEN1403	MK696638	MK696800	[52]
T. azevedoi	CEN1422	MK696660	MK696821	[52]
T. bannaense	HMAS248865	KY688038	KY688003	[47]
T. bannaense	HMAS248840 (T)	KY688037	KY687979	[47]
T. breve	HMAS248845	KY688046	KY687984	[47]
T. breve	HMAS248844 (T)	KY688045	KY687983	[47]

Table 1. *Cont.*

Species	Strains	GenBank Accession Number		References
		TEF1	RPB2	
T. brunneoviride	CBS121130 = CPK2014	EU498316	EU498357	[49]
T. camerunense	GJS99-231	AF348108	—	[45]
T. camerunense	GJS99-230 (T)	AF348107	—	[45]
T. catoptron	GJS02-76 = CBS114232 (T)	AY391963	AY391900	[53]
T. christiani	CBS132572 = S442 (T)	KJ665439	KJ665244	[20]
T. cinnamomeum	GJS97-237 (T)	AY391979	AY391920	[53]
T. compactum	CBS121218	KF134798	KF134789	[54]
T. concentricum	HMAS248858	KY688028	KY687997	[47]
T. concentricum	HMAS248833 (T)	KY688027	KY687971	[47]
T. endophyticum	DIS220J	FJ463330	FJ442690	[55]
T. endophyticum	DIS221E	FJ463316	FJ442775	In GenBank
T. epimyces	CPK1980	EU498319	EU498359	[49]
T. epimyces	CBS120534 = CPK1981 (T)	EU498320	EU498360	[49]
T. ganodermatigerum	**CCMJ5245 (T)**	**ON567195**	**ON567189**	**This study**
T. ganodermatigerum	**CCMJ5246**	**ON567196**	**ON567190**	**This study**
T. ganodermatigerum	**CCMJ5247**	**ON567197**	**ON567191**	**This study**
T. ganodermatigerum	**CCMJ5248**	**ON567198**	**ON567192**	**This study**
T. ganodermatigerum	**CCMJ5249**	**ON567199**	**ON567193**	**This study**
T. ganodermatigerum	**CCMJ5250**	**ON567200**	**ON567194**	**This study**
T. guizhouense	S278	KF134799	KF134791	[54]
T. guizhouense	S628	KJ665511	KJ665273	[20]
T. harzianum	GJS05-107	FJ463329	FJ442708	In GenBank
T. harzianum	GJS04-71	FJ463396	FJ442779	In GenBank
T. harzianum	Thaum12	MT081433	MT118248	In GenBank
T. harzianum	CBS226.95 (T)	AF534621	AF545549	[48]
T. hausknechtii	Hypo649 = CBS133493 (T)	KJ665515	KJ665276	[20]
T. helicolixii	S640 = CBS133499 (T)	KJ665517	KJ665278	[20]
T. hengshanicum	HMAS248853	KY688055	KY687992	[47]
T. hengshanicum	HMAS248852 (T)	KY688054	KY687991	[47]
T. hirsutum	HMAS248859	KY688030	KY687998	[47]
T. hirsutum	HMAS248834 (T)	KY688029	KY687972	[47]
T. ingratum	HMAS248824	KY688019	KY687964	[47]
T. ingratum	HMAS248873	KY688022	KY688010	[47]
T. ingratum	HMAS248822 (T)	KY688018	KY687973	[47]
T. inhamatum	CBS273.78 (T)	AF348099	FJ442725	[45]
T. italicum	S131 = CBS132567 (T)	KJ665525	KJ665282	[20]
T. lentiforme	DIS167C	FJ463309	FJ442689	In GenBank
T. lentiforme	GJS98-6 (T)	AF469195	—	[16]
T. liberatum	HMAS248832	KY688026	KY687970	[47]
T. liberatum	HMAS248831 (T)	KY688025	KY687969	[47]
T. linzhiense	HMAS248874	KY688048	KY688011	[47]
T. linzhiense	HMAS248846 (T)	KY688047	KY687985	[47]
T. lixii	CBS110080 = GJS97-96	FJ716622	KJ665290	[20]
T. neocrassum	DAOM164916 = CBS336.93 (T)	AF534615	AF545542	[48]
T. neotropicale	LA11	HQ022771	—	[56]
T. peberdyi	CEN1387	MK696619	MK696781	[52]
T. peberdyi	CEN1388	MK696620	MK696782	[52]
T. pleuroticola	T1295	EU279973	—	[57]
T. pleuroticola	CBS124383 (T)	HM142381	HM142371	[50]
T. pleuroti	CBS124387 (T)	HM142382	HM142372	[50]
T. polypori	HMAS248855	KY688058	KY687994	[47]
T. polypori	HMAS248861	KY688059	KY688000	[47]
T. priscilae	S129	KJ665689	KJ665332	[20]
T. pseudodensum	HMAS248829	KY688024	KY687968	[47]
T. pseudodensum	HMAS248828 (T)	KY688023	KY687967	[47]

Table 1. *Cont.*

Species	Strains	GenBank Accession Number		References
		TEF1	RPB2	
T. pseudogelatinosum	TUFC60186 (T)	JQ797397	JQ797405	[58]
T. pyramidale	S573	KJ665698	—	[20]
T. pyramidale	S73 = CBS135574 (T)	KJ665699	KJ665334	[20]
T. rifaii	DIS337F	FJ463321	FJ442720	In GenBank
T. rifaii	DIS355B (T)	FJ463324	—	In GenBank
T. simmonsii	GJS90-22	AY391984	AY391925	[53]
T. simmonsii	GJS92-100	AF443937	FJ442710	[16]
T. simmonsii	GJS91-138	AF443935	FJ442757	[16]
T. simplex	HMAS248860	KY688042	KY687999	[47]
T. simplex	HMAS248842 (T)	KY688041	KY687981	[47]
T. solum	HMAS248848	KY688050	KY687987	[47]
T. solum	HMAS248847 (T)	KY688049	KY687986	[47]
T. spirale	DAOM183974	EU280049	—	[57]
T. spirale	LESF107	KT279022	KT278956	[46]
T. stramineum	GJS02-84 = CBS114248 (T)	AY391999	AY391945	[53]
T. tawa	GJS97-174 = CBS114233 (T)	AY392004	AY391956	[53]
T. tomentosum	S33	KF134801	KF134793	[54]
T. tomentosum	DAOM178713A (T)	AF534630	AF545557	[48]
T. velutinum	DAOM230013 = CPK298	AY937415	KF134794	[59]
T. virens	DIS162	FJ463367	FJ442696	In GenBank
T. zayuense	HMAS248836	KY688032	KY687975	[47]
T. zayuense	HMAS248835 (T)	KY688031	KY687974	[47]

New sequences are shown in bold. The type sequences are marked with (T).

Table 2. Isolates and GenBank accession numbers of *Trichoderma* species associated with green mold on mushrooms.

Species	Host Range	Isolates	GenBank Accession Number		References
			TEF1	RPB2	
T. aggressivum	*Agaricus bisporus*	CBS100525	AF534614	AF545541	[48]
T. aggressivum f. aggressivum	*Agaricus bisporus*	GJS99-30	AF348109	—	[60]
		DAOM222156	AF348098	FJ442752	[45]
T. aggressivum f. europaeum	*Agaricus bisporus*	CBS100526 (T)	KP008993	KP009166	[45]
	—	TRS27	KP008994	KP009163	In GenBank
	—	CBS435.95	KP008998	KP009169	In GenBank
T. alni	*Macrotyphula* cf. *contorta*	CBS120633	EU498312	EU498349	[49]
		CPK2494	EU498313	EU498350	
T. asperellum	*Pleurotus ostreatus*	T11 (ACCC32725)	MF049065	—	[61]
	Pleurotus eryngii	—	—	—	[62]
	—	CGMCC6422	KF425756	KF425755	[63]
	—	CBS433.97 = TR3 (T)	AF456907	EU248617	In GenBank
T. atrobrunneum	*Ganoderma sichuanense*	CGMCC3.19070	MH464779	—	[64]
	—	T17-27	MW232537	MW232508	[65]
T. atroviride	*Pleurotus ostreatus*	CPK3277	EU918154	—	[66]
	Ganoderma sichuanense	2015005	—	—	[10]
	Agaricus bisporus	T33	—	—	[67]
	Lentinula edodes	T25	—	—	[68]
	Pleurotus eryngii	—	—	—	[69]
	—	PARC1011	MT454114	MT454130	[70]
	—	PARC1018	MT454121	MT454137	

Table 2. *Cont.*

Species	Host Range	Isolates	GenBank Accession Number		References
			TEF1	RPB2	
	—	DAOM222144	AF456889	FJ442754	[71]
	—	Th002	AB558906	AB558915	[72]
T. aureoviride	*Pleurotus ostreatus*	HMAS266607	KF923280	KF923306	[73]
T. austriacum	*Peziza* sp.	CBS122494 (T)	FJ860619	FJ860525	[19]
T. capillare	*Agaricus bisporus*	CPK2883	JN182283	JN182312	[74]
		GJS99-3	JN175584	JN175529	
T. catoptron	*Aphyllophorales* s. l.	GJS02-76 (T)	AY391963	AY391900	[53]
T. cerinum	*Lentinula edodes*	S357	KF134797	KF134788	[75]
T. chromospermum	black mycelium and black pyrenomycete	GJS95-196	AY391975	AY391914	[53]
		GJS98-73	AY391976	AY391915	
		GJS94-68 = CBS114577	—	AY391913	
	—	HMAS252537	KF729986	KF730004	[25]
	—	HMAS252539	KF923287	KF923314	
	—	HMAS252535	KF923292	KF923315	
T. citrinoviride	*Lentinula edodes*	TAMA0154	AB807641	AB807653	[76]
	Pleurotus ostreatus	GJS92-8	JN175595	JN175544	[77]
	Pleurotus eryngii	GJS01-364	AY225860	AF545565	[69]
	Polypore mushroom	TAMA0188	AB807644	AB807656	[76]
	—	HZA9	MK850831	MK962804	[78]
T. epimyces	*Polyporus umbellatus*	CPK1980	EU498319	EU498359	[49]
		CBS120534 (T)	EU498320	EU498360	
T. erinaceum	—	DIS7	DQ109547	EU248604	[79]
T. fasciculatum	*Hypocrea* ascospores	CBS118.72	—	—	[80]
	—	DAOM172827	AF534628	AF545555	[48]
T. fomiticola	*Fomes fomentarius*	CBS121136	FJ860639	FJ860538	[18]
		CPK3137	FJ860640	FJ860539	
T. ghanense	*Agaricus bisporus*	NBRC30902	AB807638	AB807650	[76]
T. ganodermatis	*Ganoderma sichuanense*	HMAS248856	KY688060	KY687995	[47]
		HMAS248869	KY688061	KY688007	[47]
T. ganodermatigerum	***Ganoderma sichuanense***	**CCMJ5245(T)**	**ON567195**	**ON567189**	**This study**
		CCMJ5246	**ON567196**	**ON567190**	
		CCMJ5247	**ON567197**	**ON567191**	
		CCMJ5248	**ON567198**	**ON567192**	
		CCMJ5249	**ON567199**	**ON567193**	
		CCMJ5250	**ON567200**	**ON567194**	
T. ghanense	*Agaricus bisporus*	NBRC30902	AB807638	—	[76]
T. hamatum	*Agaricus bisporus*	Tham20-3	—	—	[81]
	Lentinula edodes	—	—	—	[82]
	—	DAOM167057 (T)	EU279965	AF545548	[57]
	—	Hypo647 = WU31629	KJ665513	KJ665274	[20]
	—	Hypo648 = CBS132565	KJ665514	KJ665275	[20]
T. harzianum	*Pleurotus ostreatus*	KACC40558	—	—	[66]
	Cyclocybe aegerita	JB1	—	—	[73]
	Lentinula edodes	T50	—	—	[83]
	Pleurotus eryngii *Pleurotus ostreatus*	KACC40784	—	—	[69]
	Agaricus bisporus	—	—	—	[45]
	Pleurotus ostreatus	—	—	—	[84]
	Polypores/Corticiaceous	—	—	—	[18]

Table 2. *Cont.*

Species	Host Range	Isolates	GenBank Accession Number		References
			TEF1	RPB2	
	Pleurotus tuoliensis *Tremella fuciformis* *Flammulina filiformis*	—	—	—	[85]
	—	CBS226.95	AF348101	AF545549	[48]
	—	Thaum12	MT081433	MT118248	[86]
	—	CBS227.95	AF348100	—	[45]
	—	GJS05-107	FJ463329	FJ442708	In GenBank
	—	GJS04-71	FJ463396	FJ442779	In GenBank
T. hengshanicum	*Ganoderma sichuanense*	1009	—	—	[87]
	—	HMAS248852 (T)	KY688054	KY687991	[47]
	—	HMAS248853	KY688055	KY687992	
T. inhamatum	*Agaricus bisporus*	CBS273.78 (T)	AF348099	FJ442725	[81]
	Pleurotus tuoliensis	—	—	—	[85]
T. koningii	*Pleurotus eryngii*	—	—	—	[69]
	Agaricus bisporus	—	—	—	[88]
	Lentinula edodes *Pleurotus ostreatus* *Pleurotus tuoliensis* *Flammulina filiformis* *Volvariella volvacea* *Hypsizygus marmoreus*	—	—	—	[85]
	Ganoderma sichuanense	TFl040917	—	—	[75]
	Tremella fuciformis	TGy040604	—	—	
	—	7723	KJ634753	KJ634720	[89]
	—	GJS90-18	DQ289007	EU248600	[23]
	—	CBS979.70	AY665703	EU248601	In GenBank
	—	S22	KC285595	KC285749	[90]
T. koningiopsis	*Phaiius rubrovolvata*	CXYL	MN135988	MT038997	[91]
	Ganoderma sichuanense	**CCMJ5253**	**ON567187**	**ON567201**	**This study**
		CCMJ5254	**ON567188**	**ON567202**	
T. kunigamense	*Lentinula edodes*	TAMA193	AB807645	AB807657	[76]
T. leguminosarum	dark corticiaceous fungus	S391	KJ665548	KJ665287	[20]
		CBS130014	KJ665551	KJ665288	
		S503	KJ665552	KJ665289	
T. lieckfeldtiae	*Moniliophthora roreri*	GJS00-14 = CBS123049 (T)	EU856326	EU883562	[92]
T. longibrachiatum	*Pleurotus ostreatus*	TUFC61535 = CBS816.68(T)	EU401591	DQ087242	[40]
	Agrocybe aegerita	JB4	—	—	[73]
	Lentinula edodes	T57	—	—	[83]
	Ganoderma sichuanense	TFl040921	—	—	[75]
	Pleurotus eryngii	—	—	—	[93]
	Agaricus bisporus	—	—	—	[81]
	Pleurotus tuoliensis *Hypsizygus marmoreus* *Volvariella volvacea*	—	—	—	[85]

Table 2. *Cont.*

Species	Host Range	Isolates	GenBank Accession Number		References
			TEF1	RPB2	
T. mienum	*Lentinula edodes*	TUFC61517	JQ621975	JQ621965	[94]
T. orientale	*Ganoderma applanatum*	LESF516	KT279041	KT278976	[46]
	Ganoderma applanatum	LESF540	KT279042	KT278977	
	Ganoderma applanatum	LESF544	KT279043	KT278978	
	Ganoderma applanatum	TRS707	KP008888	KP009202	
T. oblongisporum	*Lentinula edodes*	T37	—	—	[83]
	—	DAOM167085	AF534623	AF545551	[48]
T. parareesei	*Pleurotus eryngii*	TAMA0153	AB807640	AB807652	[76]
T. parestonica	*Hymenochaete tabacina*	CBS120636 (T)	FJ860667	FJ860565	[18]
T. pleuroticola	*Pleurotus ostreatus*	CBS124383 (T)	HM142381	HM142371	[66]
		CPK2885	EU918161	EU918141	
	Pleurotus eryngii	CAF-TP3	—	—	[69]
	Lentinula edodes	T22	—	—	[83]
	Cyclocybe aegerita	JB7	—	—	[73]
	—	T1295	EU279973	—	[57]
T. pleuroti	*Pleurotus ostreatus*	KACC44537	—	—	[69]
	Pleurotus eryngii var. *ferulae*	—	—	—	[95]
	—	CBS124387 (T)	HM142382	HM142372	[50]
T. polypori	*Lentinula edodes*	HMAS248861	KY688059	KY688000	[47]
	Polyporus sp.	HMAS248855 (T)	KY688058	KY687994	
T. polysporum	*Lentinula edodes*	—	—	—	[96]
	—	8232	KJ634779	KJ634746	[89]
	—	8147	KJ634771	KJ634738	
T. priscilae	*Crepidotus* sp.	S168 = CBS131487 (T)	KJ665691	KJ665333	[20]
	Stereum sp.	S129	KJ665689	KJ665332	
	—	HMAS245002	KT343760	KT343764	In GenBank
T. protopulvinatum	*Fomitopsis pinicola*	CPK2434	FJ860677	FJ860574	[18]
T. pulvinatum	*Fomitopsis pinicola*	CBS121279	FJ860683	FJ860577	[18]
T. pseudokoningii	*Lentinula edodes*	DUCC4021	KX431217	—	[77]
	Cyclocybe aegerita	TGc050619	—	—	[75]
	Ganoderma sichuanense	TFl040926	—	—	
	Pleurotus eryngii	—	—	—	[97]
	Flammulina filiformis	—	—	—	[98]
	Pleurotus tuoliensis *Volvariella volvacea* *Hypsizygus marmoreus*	—	—	—	[85]
	—	DAOM167678	AY865641	KJ842214	[99]
	—	GJS99-149	JN175589	JN175536	[17]
	—	GJSNS19	JN175588	JN175535	
T. pseudolacteum	*Lentinula edodes*	TUFC61496	JX238494	JX238479	[100]
		TUFC61502	JX238480	JX238471	
T. samuelsii	*Hymenochaete* sp.	S5 = CBS130537	JN715651	JN715599	[101]
		S42	JN715652	JN715598	

Table 2. *Cont.*

Species	Host Range	Isolates	GenBank Accession Number		References
			TEF1	RPB2	
T. songyi	*Tricholoma matsutake*	TC556 TC480	KX266244 KX266243	KX266250 KX266249	[102]
T. stilbohypoxyli	*Stilbohypoxylon moelleri*	Hypo256 = CPK1977	FJ860702	FJ860592	[23]
T. stromaticum	*Agaricus bisporus*	GJS97-181	AY937447	HQ342227	[59]
	—	GJS07-88	HQ342195	HQ342258	[103]
	—	GJS03-47	HQ342201	HQ342264	
	—	GJS00-107	HQ342202	HQ342265	
T. sulphureum	*Laetiporus sulphureus*	CBS119929 CPK1593	FJ860710 FJ860709	FJ179620 FJ860599	[18]
	Thelephora sp.	GJS95-135 = CBS114237	AY392006	AY391958	[53]
T. tsugarense	*Lentinula edodes*	TAMA203 (T)	AB807647	AB807659	[76]
T. viride	*Lentinula edodes*	T13	—	—	[83]
	Pleurotus ostreatus	—	—	—	[82]
	Tremella fuciformis	TGc040905	—	—	[75]
	Ganoderma sichuanense	TFl080706	—	—	[75]
	Flammulina filiformis	TFj10010	—	—	[75]
	Cyclocybe aegerita	TGc040905	—	—	[75]
	Phallus indusiatus	TFl080706	—	—	[75]
	Tremella fuciformis	TGc040905	—	—	[75]
	Agaricus bisporus	—	—	—	[88]
	Pleurotus eryngii	—	—		[69]
	—	TRS575	KP008931	KP009081	In GenBank
	—	LESF115	KT278989	KT278921	[46]
T. virens	*Agaricus bisporus* *Pleurotus eryngii*	— —	— —	— —	[88]
	—	DIS162	FJ463367	FJ442696	In GenBank
	—	DIS328A	FJ463363	FJ442738	In GenBank
T. cf. *virens*	*Pleurotus eryngii*	KACC40783	—	—	[69]
	Pleurotus ostreatus	TUCIM2558	KX655776	—	[104]
T. viridarium	*Steccherinum ochraceum*	GJS89-142	AY376049	EU241495	[51]
	Nemania sp.	GJS98-182	DQ307511	EU252011	[23]
Protocrea farinosa *Protocrea pallida*	— —	CBS121551 CBS121552	EU703889 EU703897	EU703935 EU703944	[105]

The type sequences are marked with (T), the new sequences are shown in bold.

An MP phylogram was constructed with PAUP 4.0b10 [106] from the combined sequences of TEF1 and RPB2, using 1000 replicates of a heuristic search with random addition of sequences and subsequent tree bisection and reconnection (tbr) branch swapping. Analyses were performed with all characters treated as unordered and unweighted, and gaps treated as missing data. The topological confidence of the resulting trees was tested by maximum parsimony bootstrap proportion (MPBP) with 1000 replications, each with 10 replicates of random addition of taxa. An ML phylogram was constructed with Raxml-gui 2.0 [107] with the sequence after alignment. The ML + Rapid bootstrap program and 1000 repeats of the GTRGAMMAI model were used to evaluate the bootstrap proportion (BP) of each branch for constructing the phylogenetic tree. The BI analysis was conducted using MrBayes 3.2.7 [108] using a Markov Chain Monte Carlo (MCMC) algorithm. Nucleotide substitution models were determined using MrModeltest 2.3 [109]. The best model for combined sequences was HKY + I + G.

3. Results

3.1. Molecular Phylogeny

Species recognition: The dataset for the new species phylogenetic analyses included sequences from 100 taxa (Table 1). Multi-locus data were concatenated, which comprised 2321 characters, with TEF1 1293 characters and RPB2 1028 characters. Estimated base frequencies were as follows: A = 0.231650, C = 0.281772, G = 0.234671, and T = 0.251907; substitution rates were as follows: AC = 1.069464, AG = 4.197119, AT = 0.935747, CG = 0.993621, CT = 4.979475, and GT = 1.000000. The MP and ML trees showed similar topologies with high statistical support values. The MP tree was selected as the representative phylogeny. In Bayesian analysis, the average standard deviation of split frequencies at the end of the total MCMC generations was calculated as 0.008946, which is less than 0.01. Most of the tree topologies resulting from three analyses were nearly the same. In the resulting tree (Figure 1), the combined phylogenetic analyses using TEF1-α and RPB2 showed that the six strains of *T. ganodermatigerum* represent phylogenetically distinct species with high statistical supports (MPBP/MLBP/BIBP = 100%/100%/1.0), and clustered together with the species in the Harzianum clade [16]. The new species is most related to the clade that contains *T. amazonicum*, *T. pleuroticola*, *T. hengshanicum*, and *T. pleuroti*. Two collections of CCMJ5253 and CCMJ5254 clustered with *T. koningiopsis* with high support (MPBP/MLBP = 100/100) (Figure 2).

Phylogenetic structure: Some sections could be found among the *Trichoderma* strains associated with mushrooms and are mainly concentrated in the Harzianum clade (Figure 2). *Trichoderma longibrachiatum*, *T. citrinoviride*, *T. pseudokoningii*, and *T. ghanense* are from section *Longibrachiatum*, whose members are best known as producers of cellulose-hydrolyzing enzymes [74,110,111]. *Trichoderma atroviride*, *T. viride*, *T. koningii*, *T. hamatum*, *T. minutisporum*, *T. polysporum*, *T. viride*, and *T. asperellum* are from section *Trichoderma* or the Viride clade [36,111].

The phylogenetic structure according to ecology: Species in the Harzianum clade are commonly fungicolous, living in different types of habitats [112,113]. They are most commonly isolated from soil or found on decomposing plant material where they occur cryptically or parasitize other fungi [18,53,114], and those species are possibly the most common endophytic "species" in wild trees [115,116]. There is usually no apparent host specialization [117]. However, some exceptions to this trend exist. Clade I in the Harzianum clade of the tree is a collection of species with relatively narrow host ranges, or in other words, a strong host preference. *Trichoderma atrobrunneum* was found in soil or on decaying wood, clearly or cryptically parasitizing other fungi. *Trichoderma pleuroti*, just like *T. aggressivum*, has thus far never been isolated from areas outside of mushroom farms [118]. Furthermore, *T. epimyces* has only been reported on *Polyporus umbellatus* [49], and *T. priscilae* has been reported from basidiomes of *Crepidotus* and *Stereum* [20].

Some other species such as *T. atroviride*, *T. asperellum*, *T. harzianum*, and *T. longibrachiatum* were also found in significant proportions in *Agaricus* compost [119]. *Trichoderma stromaticum* and its *Hypocrea* teleomorph are only known from cocoa and are often associated with tissue infected with the basidiomycetous pathogen *Crinipellis perniciosa* [55].

Although some of these pathogenic *Trichoderma* species (e.g., species gathered in or near Clade II) have been explored as biocontrol agents for plant diseases, *T. atroviride*, *T. viride*, *T. koningii*, *T. koningiopsis*, and *T. asperellum* serve as pathogens with broad host ranges on mushrooms. *Trichoderma sulphureum*, *T. protopulvinatum*, *T. pulvinatum*, and *T. austriacum* coalesce into a subclade (Clade III), and each of these species has been reported on a particular fungus [18,19].

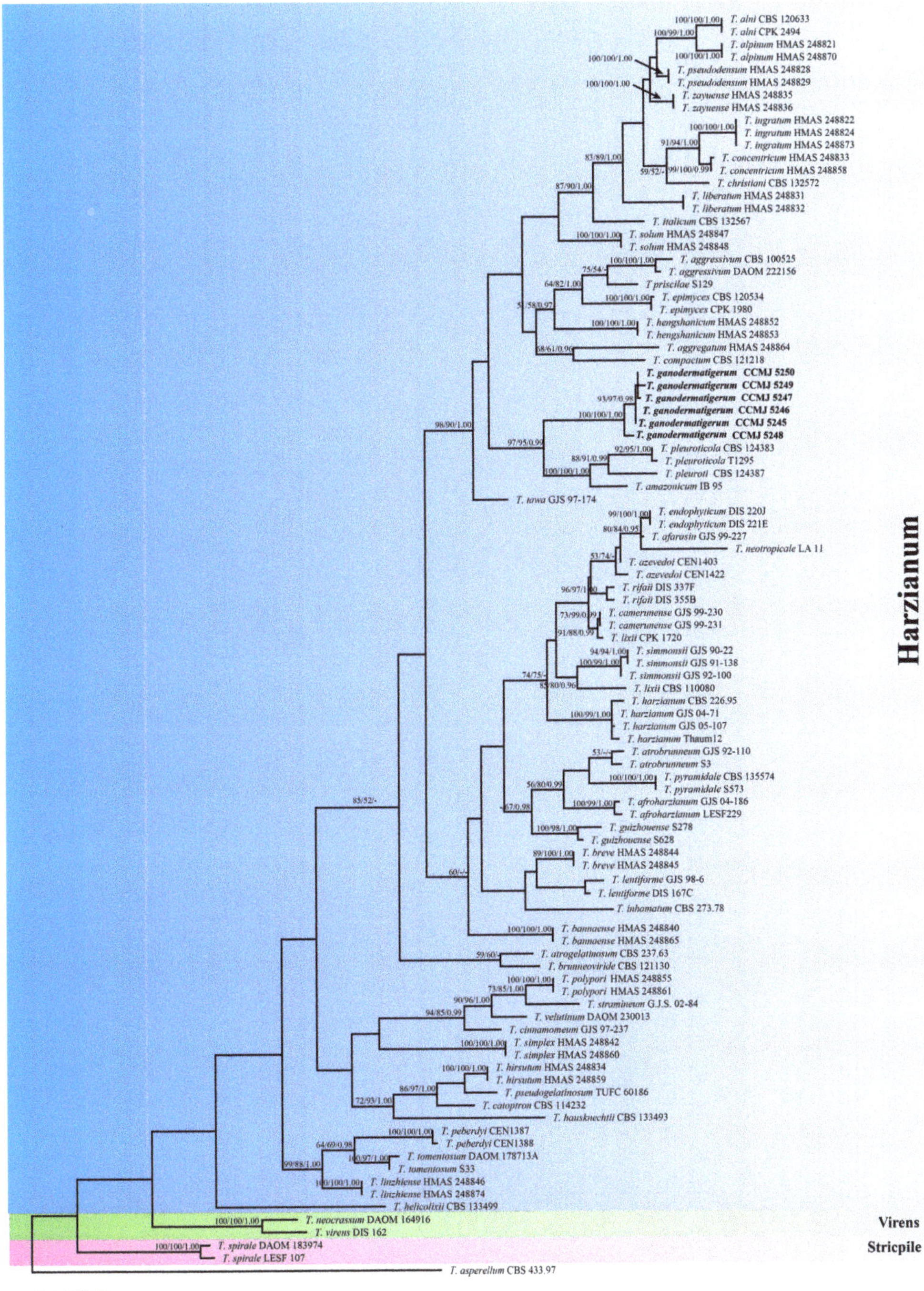

Figure 1. Phylogeny of *Trichoderma* using MP analysis based on combined TEF1 and RPB2 sequences. MPBP $\geq$ 50%, MLBP $\geq$ 50%, and BIPP $\geq$ 0.9 are shown on the branches (MPBP/MLBP/BIPP). The sequences in bold are the new species.

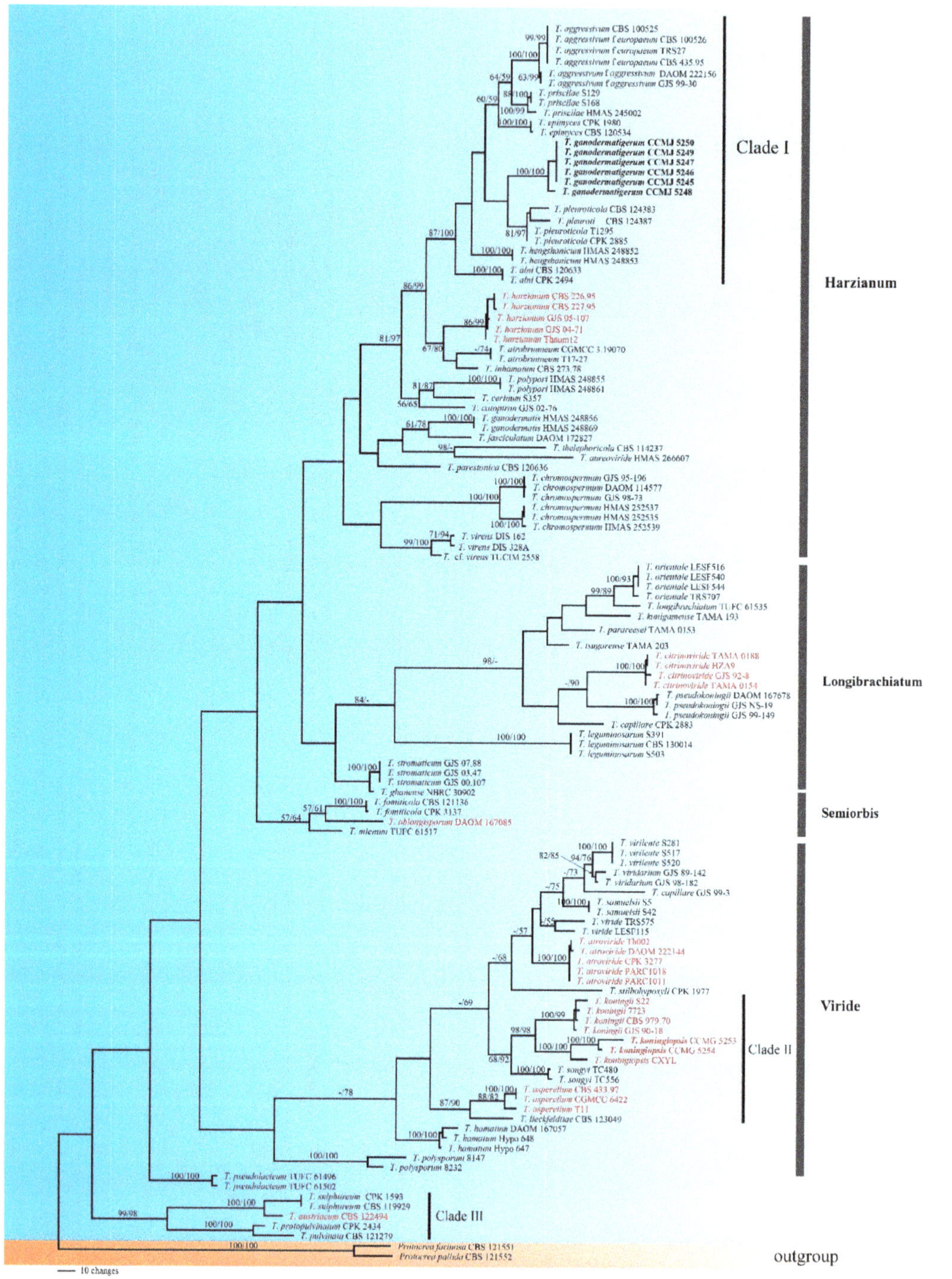

Figure 2. Phylogeny of *Trichoderma* associated with mushrooms using MP analysis based on concatenated TEF1 and RPB2 sequences. Branches are labeled with MPBP ≥ 50% and MLBP ≥ 50%. The biological agents are marked in red, and the new sequences in this study are in bold.

3.2. *Taxonomy*

Trichoderma ganodermatigerum X.Y. An & Y. Li, sp. nov. Figure 3A–L.

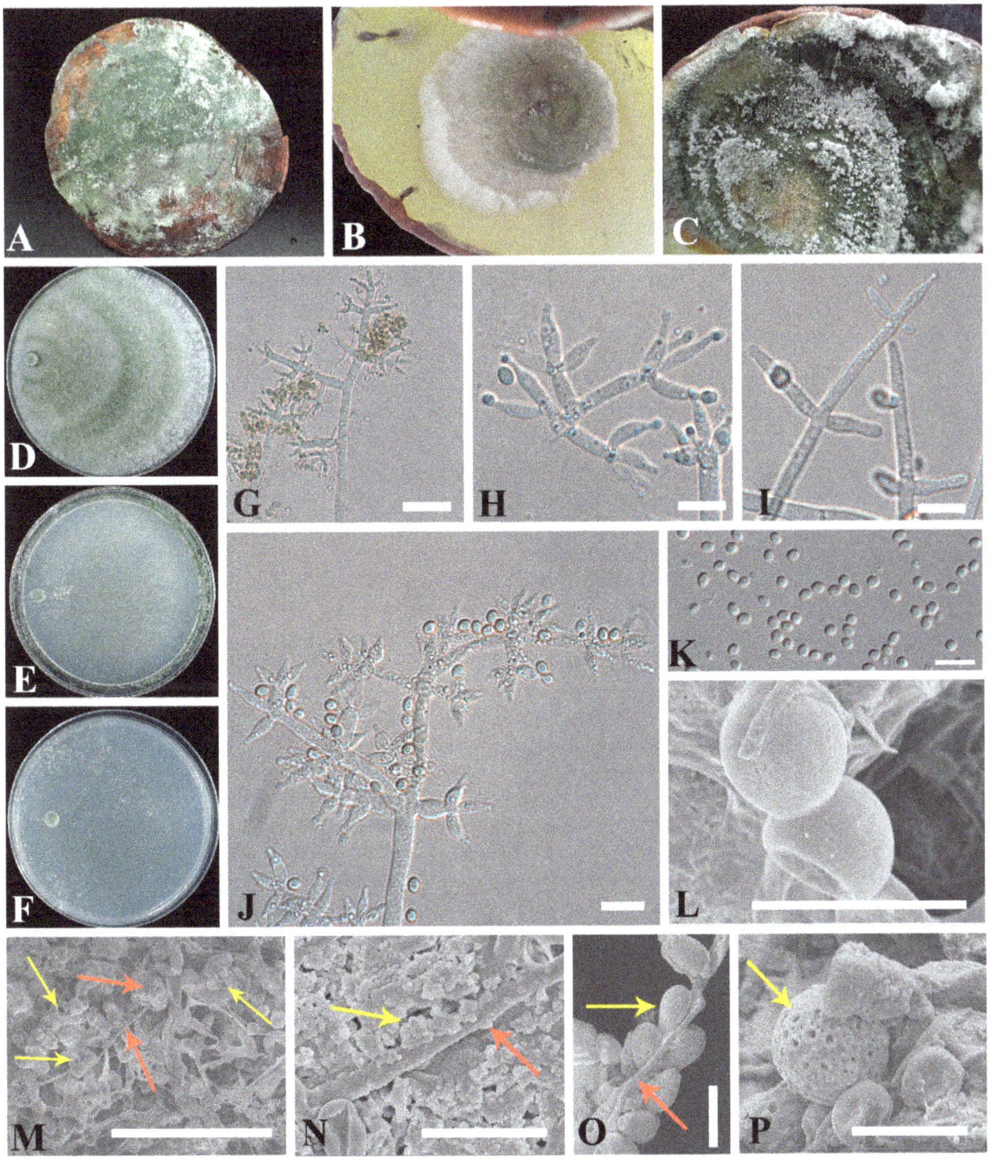

Figure 3. Morphological characteristics of *T. ganodermatigerum*. (**A–C**) diseased fruitbody; (**D–F**) colony on PDA, CMD, and SNA; (**G–J**) conidiophores and phialides; (**K**,**L**) conidia; (**M–P**) interactions of *G. sichuanense* and *T. ganodermatigerum*; (**M**) *Trichoderma* hyphae and conidia are filled in the *Ganoderma* tissue, causing the tissue to become rough or even depressed; (**N**) *Trichoderma* hyphae covered with *Ganoderma* tissue; (**O**) clinged *Trichoderma* hyphae and healthy *Ganoderma* spores; (**P**) abnormal *Ganoderma* spores in diseased tissue. Bars: G, Q = 20 μm; H–J, M–P = 10 μm; K = 50 μm; L = 5 μm. The yellow arrows indicate the tissues and spores of *G. sichuanense*, and the red arrows indicate the hyphae and spores of *T. ganodermatigerum*.

MycoBank: MB 843898.

Diagnosis: Phylogenetically, *T. ganodermatigerum* formed a distinct clade and is related to *T. amazonicum* (Figure 1). Both *T. amazonicum* and *T. ganodermatigerum* form dense concentric rings, pyramidal branching patterns, and branches toward the tip; mycelium grows slowly or does not grow at 35 °C; conidia globose, smooth, and green. As for *T. amazonicum*, there is no diffusing pigmentation on CMD media and a slightly fruity odor; a brown diffusing pigmentation of the agar is formed in some strains on PDA media [50]. Phylogenetic analysis of TEF1 and RPB2 gene sequences also revealed that *T. ganodermatigerum* was phylogenetically distinct not only from *T. amazonicum* but also from other previously reported *Trichoderma* species.

Etymology: The name refers to the host genus "*Ganoderma*" from which it was isolated.

Typification: CHINA. Jilin Province, Panshi City, Songshan County, from *Ganoderma sichuanense*, alt. 310 m, 126°56′ E, 42°77′ N, 18 August 2021, *Xiaoya An*, HMJAU59014, preserved in Engineering Research Center of Chinese Ministry of Education for Edible and Medicinal Fungi of Jilin Agricultural University. Ex-type culture CCMJ5245. Sexual morph: Undetermined. (ITS: ON399102, TEF1: ON567195, and RPB2: ON567189).

Teleomorph: Unknown.

Description: The optimum temperature was 25 °C, and the colony radius on CMD was 7–9 mm at 15 °C, 19–23 mm at 20 °C, 43–52 mm 25 °C, and 32–36 mm at 30 °C, with no growth at 35 °C, and mycelium covering the plate after ten days at 25 °C (Figure 3E). Colony hyaline, thin, and radiating, white in the initial stage, and gradually turned to light green with slight zonate. Mycelia were sparse and delicate, hard to be observed, and aerial hyphae were inconspicuous. Conidiation starting after six days, formed in pustules. Pustules were spreading near the original inoculum or at the edge of the colony, distributed loosely in the plate, white in the initial stage and then turned green. No chlamydospores were observed. No distinct odor and no diffusing pigment were observed.

Colony radius on SNA after 72 h 5–8 mm at 15 °C,13–15 mm at 20 °C, 42–43 mm at 25 °C, and 25–28 mm at 30 °C, and can hardly see the growth at 35 °C. Mycelium covering the plate after six days at 25 °C (Figure 3F). Colony hyaline, thin, irregular, surface mycelium scant. Aerial hyphae are inconspicuous and short. Conidiation starting after three days, formed in loose pustules. Pustules initially white, loose distribution, later turn aggregated and green. No chlamydospores were observed. No distinct odor and no diffusing pigment were observed.

On PDA, the colony radius was 9–12 mm at 15 °C, 22–28 mm at 20 °C, 38–44 mm at 25 °C, and 30–40 mm at 30 °C, with no growth at 35 °C after 72 h, and mycelium covering the plate after 5–6 days at 25 °C (Figure 3D). The colony was circular, spreading in several concentric rings; aerial hyphae were common, dense, and green; the margin was relatively loose and whitish under the alternative light situations. However, mycelia were aerated and white, and only green appeared near the inoculation site under the condition of total darkness. Conidiation starting after 3–4 days, formed on aerial hyphae, spreading in a circle around the original inoculum. Conidiophores are typically tree-like, straight, or slightly curved, comprising a distinct main axis with side branches paired or unilateral and often terminating in whorls of 3–4 divergent phialides, rarely with a terminal solitary phialide (Figure 3G–J), branches densely disposed, arising at mostly vertical angles upwards, rebranching 1–3 times; the distance between two neighboring branches is (6.6–) 10.0–30.0 (–35.6) μm. Phialides formed paired or in whorls of 3–5, lageniform, spindly, usually arising at an acute angle to the axis, rarely solitary (Figure 3F), (1.1–) 2.8–12.3 (–16) μm× (0.2–) 1.9–3.4 (–3.6) μm, *l/w* ratio (1.6–) 1.7–5.9 (–7.0), (0.2–) 1.4–2.6 (–2.8) μm wide at the base. Conidia one-celled, green, smooth-walled, globose to subglobose, sometimes ellipsoid, (3.4–) 3.6–4.8 (–5.3) μm× (2.9–) 3.2–4.3 (–4.6) μm, *l/w* ratio 1.1–1.5. No chlamydospores were observed. No distinct odor and no diffusing pigment were observed.

Distribution: Jilin, Shandong, and Heilongjiang Provinces, China.

Additional specimen examined: China, Jilin Province, Panshi city, Songshan County, from *Ganoderma sichuanense*, alt. 310 m, 126°56′ E, 42°77′ N, 11 Oct. 2021, *Xiaoya An*, HMJAU59013.

Notes: Fungicolous on the fruiting body of *G. sichuanense* in terrestrial habitats. It produces extremely tree-like main axes and branches and green, globose conidia (Figure 3N). The results of the phylogenetic tree strongly support its status as a new taxon (Figure 1), indicating its affinity to the Harzianum clade [16]. The species was related to *T. amazonicum* and *T. pleuroticola*. Regarding *T. amazonicum*, it is a host-specific endophyte and might have potential for biocontrol of *Hevea* diseases [50]. Phylogenetically, *T. ganodermatigerum* is related to *T. pleuroticola* in the mycoparasite group. Morphologically, both species grow rapidly and form broad concentric rings on PDA. Conidiation formed small pustules, and the green spores cause the colony to change from light to dark green [120]. The difference is that the new species starts with white, aerial mycelia and spores are more spherical or nearly spherical, with obvious green color, while the spores of *T. pleuroticola* are light green, subglobose to broadly ellipsoidal conidia, slightly smaller than *T. ganodermatigerum*, and reported more on *Pleurotus ostreatus*, *Pleurotus eryngii* var. *ferulae*, *Lentinula edodes*, and *Cyclocybe aegerita* [69,73,83,120].

Trichoderma koningiopsis Samuels, Carm. Suárez & H.C. Evans 2006.

Description: Fungicolous, colonized the fruiting body of *G. sichuanense*, causing green mold disease and occurring mostly from June to September. It is very difficult to distinguish the mycelium in the early stage, and only scattered spots present under the cap. Then, white mycelium appeared, with radiating growth. The edge of the colony is often accompanied by a yellow or brown line. A large number of green spores were produced in the late stage. Young basidiomes were inoculated with *T. koningiopsis*, which reproduced the original signs; the same pathogen was isolated again from the diseased fruitbody.

On PDA, the colony was radial, first whitish, became dark green with fluffy hyphae after ten days. Aerial hyphae were common and dense, but no concentric rings were observed. Mycelia often appear white in complete darkness, and light stimulates spore production, resulting in a green colony. Conidia formed in pustules, spreading near the original inoculum, white, turning green later. On CMD, mycelium covering the plate after ten days at 25 °C, loose and slim, aerial hyphae were absent. Conidia were formed in pustules, which were only produced at the edge of a colony. On SNA media, concentric rings of light yellow or green appeared, and spores were produced in four days. Conidiophore branches arose at right angles, and primary branches arose singly or in pairs. Conidia were ellipsoidal to oblong-shaped, green, 2.8–7.3 × 2.5–7.0 μm. No chlamydospores, no distinct odor, and no diffusing pigment were observed.

Material examined: CHINA, Jilin Province, on a fruiting body of *Ganoderma*, 4 August 2020; *Xiaoya An*, HMJAU59012, living culture CCMJ5253, CCMJ5254 (ITS: ON385996, ON385947; TEF1: ON567187, ON567188, and RPB2: ON567201, ON567202, respectively).

Notes: *Trichoderma koningiopsis* is found throughout tropical America, as well as East Africa, Europe, Canada, and eastern North America [23]. This species is mainly found in soil, twigs, and decayed leaves, and the sexual type is mostly found in wood. At present, *T. koningiopsis* has been reported to cause green mold of *Phaiius rubrovolvata* [91], and to our knowledge, this is the first time that it has caused green mold on *G. sichuanense*. Our sequences had high similarity to the *T. koningiopsis* sequence after BLAST, and the results of the phylogenetic tree also confirmed the correctness of the classification (Figure 2).

4. Discussion

Edible and medicinal mushrooms have become a very important crop and are grown commercially in many countries [1,121], but the production, including the yield and quantity, is challenged by fungal diseases [2,24]. *Trichoderma ganodermatigerum* is a new species of *Trichoderma*. The results from the phylogenetic analyses separate the new species from other closely related and morphologically similar species. The sequences indicate it belongs to the Harzianum clade. To date, more than forty *Trichoderma* species have been reported to be associated with mushroom green mold disease. *Trichoderma atroviride*, *T. harzianum*, *T. koningii*, *T. longibrachiatum*, *T. pseudokoningii*, and *T. viride* are the six most commonly cited species causing disease on edible mushrooms (Table 2), all of which could infect six to eleven species of cultivated mushrooms [61,64,68,73,83,91,119,122,123]. Before this study, there were seven known species that could cause *G. sichuanense* diseases, namely, *T. koningii*, *T. longibrachiatum*, *T. pseudokoningii*, *T. viride*, *T. atrobrunneum*, *T. ganodermatis* [47], and *T. hengshanicum* [87], while *T. orientale* can cause disease on *G. applanatum* [124].

Trichoderma green mold infection in edible basidiomycetes has a long history [125]. There are many types of interactions between mushrooms and *Trichoderma* [126–129]. Similar to *T. aggressivum*, the causal agent of *Agaricus* green mold disease [130], no obvious biting phenomenon was observed between pathogen and mushroom in this study. Through SEM observation, in the interaction zone between *G. sichuanense* and *T. ganodermatigerum*, the tissue surface of *Ganoderma* became uneven with irregular holes (Figure 3K), the pores on the *Ganoderma* spores became larger, and the double-layer structure was damaged, resulting in spore invagination (Figure 3L), which was similar to the interaction between *Trichoderma* and shiitake [83]. We can at least suspect that the cell-wall-degrading enzymes play an important role in the process according to the symptoms of soft tissue with holes or even oozing liquid of *Ganoderma*. In addition, *T. songyi* could have great biological potential because it is closely related to the biological agents (Figure 2, Clade II).

The application of the *Trichoderma* species as biocontrol agents began in 1934 when Weindling first discovered that *Trichoderma* could be parasitic on the hyphae of *Rhizoctonia solani*, and since then, an increasing amount of research has focused on this field [131]. Because many *Trichoderma* species are symbiotic and fungal parasitoids, they need to produce degradation enzymes or secondary metabolites to obtain nutrients from the host, so they have been developed as biocontrol agents for plant diseases [50,55,112,132,133]. Among the species associated with mushrooms, nine species are used as biological agents already. *Trichoderma koningiopsis*, the new pathogen for *G. sichuanense* in this study, has been a biocontrol agent for a long time [134]. Since *T. ganodermatigerum* can infect cultivated *Ganoderma*, leading to growth stagnation or the cessation of sporulation of *Ganoderma*, it could be a potential biocontrol agent for plant disease. Therefore, the parasitic characteristics and compounds should be further studied.

Author Contributions: X.-Y.A., D.L. and Y.L. conceived and designed the study. X.-Y.A., G.-H.C. and X.-F.L. collected specimens from China. X.-Y.A., G.-H.C. and H.-X.G. generated the DNA sequence data, checked the specimens, and analyzed the data. X.-Y.A., Y.Y., D.L. and Y.L. checked issues related to nomenclatural articles. X.-Y.A. wrote the manuscript draft. X.-Y.A., G.-H.C., H.-X.G., D.L. and Y.L. revised the draft, and all authors approved the final manuscript. All authors have read and agreed to the published version of the manuscript.

Funding: This research was funded by the National Natural Science Foundation of China (No. U20A2046), China Agriculture Research System (No. CARS-20), Central Public-interest Scientific Institution Basal Research Fund (No.1630042022003), the Creation of Ganoderma Germplasm resources and breeding and development of new varieties under the grant (No. GF20190034), Central Public-interest Scientific Institution Basal Research Fund (No. 1630042022020), and Overseas Expertise Introduction Project for Discipline Innovation (111 Center) (No. D17014).

Institutional Review Board Statement: Not applicable.

Informed Consent Statement: Not applicable.

Acknowledgments: We would like to express our gratitude to the staff of the Engineering Research Center of Edible and Medicinal Fungi, Ministry of Education, Jilin Agricultural University, including Lan Yao, Yu-Kun Ma, and Ye-Tong Li for their help during molecular experiments, Meng-Le Xie for his help during the phylogenetic analyses and taxonomy process, and Chang-Tian Li and Yong-Ping Fu (Plant Protection College of Jilin Agricultural University) for the sample collection in Jilin and Heilongjiang. We also thank Zhuang Li (Plant Protection College of Shandong Agricultural University, China) for his kind help during the sample collection in Shandong.

Conflicts of Interest: The authors declare no conflict of interest.

References

1. Dong, C.H.; Guo, S.P.; Wang, W.F.; Liu, X.Z. *Cordyceps* industry in China. *Mycology* **2015**, *6*, 121–129. [CrossRef] [PubMed]
2. Zhang, J.X.; Chen, Q.; Huang, C.Y.; Gao, W.; Qu, J.B. History, current situation and trend of edible mushroom industry development. *Mycosystema* **2015**, *34*, 524–540. [CrossRef]
3. Andri Wihastuti, T.; Sargowo, D.; Heriansyah, T.; Eka Aziza, Y.; Puspitarini, D.; Nur Iwana, A.; Astrida Evitasari, L. The reduction of aorta histopathological images through inhibition of reactive oxygen species formation in hypercholesterolemia rattus norvegicus treated with polysaccharide peptide of *Ganoderma lucidum*. *Iran. J. Basic Med. Sci.* **2015**, *18*, 514–519. [PubMed]
4. Nguyen, V.T.; Tung, N.T.; Cuong, T.D.; Hung, T.M.; Kim, J.A.; Woo, M.H.; Choi, J.S.; Lee, J.H.; Min, B.S. Cytotoxic and anti-angiogenic effects of lanostane triterpenoids from *Ganoderma lucidum*. *Phytochem. Lett.* **2015**, *12*, 69–74. [CrossRef]
5. Cao, Y.; Wu, S.; Dai, Y. Species clarification of the prize medicinal *Ganoderma* mushroom "Lingzhi". *Fungal Divers.* **2012**, *56*, 49–62. [CrossRef]
6. Kim, S.; Song, J.; Choi, H.T. Genetic transformation and mutant isolation in *Ganoderma lucidum* by restriction enzyme-mediated integration. *FEMS Microbiol. Lett.* **2004**, *233*, 201–204. [CrossRef]
7. Vitak, T.Y.; Wasser, S.P.; Nevo, E.; Sybirna, N.O. The effect of the medicinal mushrooms *Agaricus brasiliensis* and *Ganoderma lucidum* (higher Basidiomycetes) on the erythron system in normal and streptozotocin-induced diabetic rats. *Int. J. Med. Mushrooms* **2015**, *17*, 277–286. [CrossRef]
8. Zhao, L.Y.; Dong, Y.H.; Chen, G.T.; Hu, Q.H. Extraction, purification, characterization and antitumor activity of polysaccharides from *Ganoderma lucidum*. *Carbohydr. Polym.* **2010**, *80*, 783–789. [CrossRef]
9. Lu, B.; Zuo, B.; Liu, X.; Feng, J.; Wang, Z.; Gao, J. *Trichoderma harzianum* causing green mold disease on cultivated *Ganoderma lucidum* in Jilin province, China. *Plant Dis.* **2016**, *100*, 2524. [CrossRef]
10. Yan, Y.H.; Zhang, C.L.; Moodley, O.; Zhang, L.; Xu, J.Z. Green mold caused by *Trichoderma atroviride* on the Lingzhi medicinal mushroom, *Ganoderma lingzhi* (Agaricomycetes). *Int. J. Med. Mushrooms* **2019**, *21*, 515–521. [CrossRef]
11. Zuo, B.; Lu, B.; Liu, X.; Wang, Y.; Ma, G.; Gao, J. First report of *Cladobotryum mycophilum* causing cobweb on *Ganoderma lucidum* cultivated in Jilin province, China. *Plant Dis.* **2016**, *100*, 1239. [CrossRef]
12. Persoon, C.H. Neuer versuch einer systematischen eintheilung der schwämme. *Römer's Neues Mag. Bot.* **1794**, *1*, 63–128. [CrossRef]
13. Barrera, V.A.; Iannone, L.; Romero, A.I.; Chaverri, P. Expanding the *Trichoderma harzianum* species complex: Three new species from Argentine natural and cultivated ecosystems. *Mycologia* **2021**, *113*, 1136–1155. [CrossRef] [PubMed]
14. Bissett, J.; Gams, W.; Jaklitsch, W.; Samuels, G.J. Accepted *Trichoderma* names in the year 2015. *IMA Fungus* **2015**, *6*, 263–295. [CrossRef] [PubMed]
15. Cai, F.; Druzhinina, I.S. In honor of John Bissett: Authoritative guidelines on molecular identification of *Trichoderma*. *Fungal Divers.* **2021**, *107*, 1–69. [CrossRef]
16. Chaverri, P.; Castlebury, L.A.; Samuels, G.J.; Geiser, D.M. Multilocus phylogenetic structure within the *Trichoderma harzianum/Hypocrea lixii* complex. *Mol. Phylogenet. Evol.* **2003**, *27*, 302–313. [CrossRef]
17. Druzhinina, I.S.; Komoń-Zelazowska, M.; Ismaiel, A.; Jaklitsch, W.; Mullaw, T.; Samuels, G.J.; Kubicek, C.P. Molecular phylogeny and species delimitation in the section *Longibrachiatum* of *Trichoderma*. *Fungal Genet. Biol.* **2012**, *49*, 358–368. [CrossRef]
18. Jaklitsch, W.M. European species of *Hypocrea* Part I. The green-spored species. *Stud. Mycol.* **2009**, *63*, 1–91. [CrossRef]
19. Jaklitsch, W.M. European species of *Hypocrea* part II: Species with hyaline ascospores. *Fungal Divers.* **2011**, *48*, 1–250. [CrossRef]
20. Jaklitsch, W.M.; Voglmayr, H. Biodiversity of *Trichoderma* (Hypocreaceae) in Southern Europe and Macaronesia. *Stud. Mycol.* **2015**, *80*, 1–87. [CrossRef]
21. Kullnig-Gradinger, C.M.; Szakacs, G.; Kubicek, C.P. Phylogeny and evolution of the genus *Trichoderma*: A multigene approach. *Mycol. Res.* **2002**, *106*, 757–767. [CrossRef]
22. Pani, S.; Kumar, A.; Sharma, A. *Trichoderma harzianum*: An overview. *Bull. Environ. Pharmacol. Life Sci.* **2021**, *10*, 32–39.
23. Samuels, G.J.; Dodd, S.L.; Lu, B.S.; Petrini, O.; Schroers, H.J.; Druzhinina, I.S. The *Trichoderma koningii* aggregate species. *Stud. Mycol.* **2006**, *56*, 67–133. [CrossRef] [PubMed]
24. Sun, J.Z.; Liu, X.Z.; McKenzie, E.H.C.; Jeewon, R.; Liu, J.K.; Zhang, X.L.; Zhao, Q.; Hyde, K.D. Fungicolous fungi: Terminology, diversity, distribution, evolution, and species checklist. *Fungal Divers.* **2019**, *95*, 337–430. [CrossRef]
25. Zhu, Z.X.; Zhuang, W.Y. *Trichoderma* (*Hypocrea*) species with green ascospores from China. *Persoonia* **2015**, *34*, 113–129. [CrossRef]
26. Zhu, Z.X.; Zhuang, W.Y.; Li, Y. A new species of the Longibrachiatum Clade of *Trichoderma* (Hypocreaceae) from Northeast China. *Nova Hedwigia* **2018**, *106*, 441–453. [CrossRef]
27. Atanasova, L.; Druzhinina, I.S.; Jaklitsch, W.M. Two hundred *Trichoderma* species recognized on the basis of molecular phylogeny. In *Trichoderma: Biology and Applications*; Mukherjee, P.K., Horwitz, B.A., Singh, U.S., Mukherjee, M., Schmoll, M., Eds.; CABI Publishing: Croydon, UK, 2013; pp. 10–42.
28. Bissett, J. A revision of the genus *Trichoderma*. II. Infrageneric classification. *Can. J. Bot.* **1991**, *69*, 2357–2372. [CrossRef]
29. Bissett, J. A revision of the genus *Trichoderma*. IV. Additional notes on section *Longibrachiatum*. *Can. J. Bot.* **1991**, *69*, 2418–2420. [CrossRef]

30. Aamir, M.; Kashyap, S.P.; Zehra, A.; Dubey, M.K.; Singh, V.K.; Ansari, W.A.; Upadhyay, R.S.; Singh, S. *Trichoderma erinaceum* bio-priming modulates the WRKYs defense programming in tomato against the *Fusarium oxysporum* f. sp. *lycopersici (Fol)* challenged condition. *Front. Plant Sci.* **2019**, *10*, 911. [CrossRef]
31. Erazo, J.G.; Palacios, S.A.; Pastor, N.; Giordano, F.D.; Rovera, M.; Reynoso, M.M.; Venisse, J.S.; Torres, A.M. Biocontrol mechanisms of *Trichoderma harzianum* ITEM 3636 against peanut brown root rot caused by *Fusarium solani* RC 386. *Biol. Control* **2021**, *164*, 104774. [CrossRef]
32. John, R.P.; Tyagi, R.D.; Prévost, D.; Brar, S.K.; Pouleur, S.; Surampalli, R.Y. Mycoparasitic *Trichoderma viride* as a biocontrol agent against *Fusarium oxysporum* f. sp. *adzuki* and *Pythium arrhenomanes* and as a growth promoter of soybean. *Crop Prot.* **2010**, *29*, 1452–1459. [CrossRef]
33. Wonglom, P.; Ito, S.; Sunpapao, A. Volatile organic compounds emitted from endophytic fungus *Trichoderma asperellum* T1 mediate antifungal activity, defense response and promote plant growth in lettuce (*Lactuca sativa*). *Fungal Ecol.* **2020**, *43*, 100867. [CrossRef]
34. Chomnunti, P.; Hongsanan, S.; Aguirre-Hudson, B.; Tian, Q.; Peršoh, D.; Dhami, M.K.; Alias, A.S.; Xu, J.C.; Liu, X.Z.; Stadler, M.; et al. The sooty moulds. *Fungal Divers.* **2014**, *66*, 1–36. [CrossRef]
35. Nirenberg, H.I. Neuerscheinung. Untersuchungen über die morphologische und biologische Differenzierung in der *Fusarium*-Sektion Liseola, von Dr. Helgard Nirenberg (Inst. f. Mykologie). *Z. Pflanzenernahr. Bodenkd.* **1977**, *140*, 243. [CrossRef]
36. Jaklitsch, W.M.; Samuels, G.J.; Dodd, S.L.; Lu, B.S.; Druzhinina, I.S. *Hypocrea rufa/Trichoderma viride*: A reassessment, and description of five closely related species with and without warted conidia. *Stud. Mycol.* **2006**, *56*, 135–177. [CrossRef]
37. Rifai, M.A. A revision of the genus *Trichoderma*. *Mycol. Pap.* **1969**, *116*, 1–56.
38. White, T.J.; Bruns, T.; Lee, S.; Taylor, J. Amplification and direct sequencing of fungal ribosomal RNA genes for phylogenetics. In *PCR Protocols: A Guide to Methods And Applications*; Innis, M.A., Gelfand, D.H., Sninsky, J.J., White, T.J., Eds.; Academic Press: San Diego, CA, USA, 1990; pp. 315–322.
39. Carbone, I.; Kohn, L.M. A method for designing primer sets for speciation studies in filamentous ascomycetes. *Mycologia* **1999**, *91*, 553–556. [CrossRef]
40. Jaklitsch, W.M.; Komon, M.; Kubicek, C.P.; Druzhinina, I.S. *Hypocrea voglmayrii* sp. nov. from the Austrian Alps represents a new phylogenetic clade in *Hypocrea/Trichoderma*. *Mycologia* **2005**, *97*, 1365–1378. [CrossRef]
41. Liu, Y.J.; Whelen, S.; Hall, B.D. Phylogenetic relationships among ascomycetes: Evidence from an RNA polymerse II subunit. *Mol. Biol. Evol.* **1999**, *16*, 1799–1808. [CrossRef]
42. Hall, T.A. BioEdit: A user-friendly biological sequence alignment editor and analysis program for Windows 95/98/NT. *Nucleic Acids Symp. Ser.* **1999**, *41*, 95–98. [CrossRef]
43. Katoh, K.; Standley, D.M. MAFFT multiple sequence alignment software version 7: Improvements in performance and usability. *Mol. Biol. Evol.* **2013**, *30*, 772–780. [CrossRef] [PubMed]
44. Larsson, A. AliView: A fast and lightweight alignment viewer and editor for large datasets. *Bioinformatics* **2014**, *30*, 3276–3278. [CrossRef] [PubMed]
45. Samuels, G.J.; Dodd, S.L.; Gams, W.; Castlebury, L.A.; Petrini, O. *Trichoderma* species associated with the green mold epidemic of commercially grown *Agaricus bisporus*. *Mycologia* **2002**, *94*, 146–170. [CrossRef] [PubMed]
46. Montoya, Q.V.; Meirelles, L.A.; Chaverri, P.; Rodrigues, A. Unraveling *Trichoderma* species in the attine ant environment: Description of three new taxa. *Antonie Van Leeuwenhoek* **2016**, *109*, 633–651. [CrossRef]
47. Chen, K.; Zhuang, W.Y. Discovery from a large-scaled survey of *Trichoderma* in soil of China. *Sci. Rep.* **2017**, *7*, 9090. [CrossRef]
48. Chaverri, P.; Castlebury, L.A.; Overton, B.E.; Samuels, G.J. *Hypocrea/Trichoderma*: Species with conidiophore elongations and green conidia. *Mycologia* **2003**, *95*, 1100–1140. [CrossRef]
49. Jaklitsch, W.M.; Kubicek, C.P.; Druzhinina, I.S. Three European species of *Hypocrea* with reddish brown stromata and green ascospores. *Mycologia* **2008**, *100*, 796–815. [CrossRef]
50. Chaverri, P.; Gazis, R.O.; Samuels, G.J. *Trichoderma amazonicum*, a new endophytic species on *Hevea brasiliensis* and *H. guianensis* from the Amazon basin. *Mycologia* **2011**, *103*, 139–151. [CrossRef]
51. Hanada, R.E.; de Jorge Souza, T.; Pomella, A.W.V.; Hebbar, K.P.; Pereira, J.O.; Ismaiel, A.; Samuels, G.J. *Trichoderma martiale* sp. nov., a new endophyte from sapwood of *Theobroma cacao* with a potential for biological control. *Mycol. Res.* **2008**, *112*, 1335–1343. [CrossRef]
52. Inglis, P.W.; Mello, S.C.M.; Martins, I.; Silva, J.B.T.; Macêdo, K.; Sifuentes, D.N.; Valadares-Inglis, M.C. *Trichoderma* from Brazilian garlic and onion crop soils and description of two new species: *Trichoderma azevedoi* and *Trichoderma peberdyi*. *PLoS ONE* **2020**, *15*, e0228485. [CrossRef]
53. Chaverri, P.; Samuels, G.J. *Hypocrea/Trichoderma* (Ascomycota, Hypocreales, Hypocreaceae): Species with green ascospores. *Stud. Mycol.* **2003**, *48*, 1–116.
54. Jaklitsch, W.M.; Lechat, C.; Voglmayr, H. The rise and fall of *Sarawakus* (Hypocreaceae, Ascomycota). *Mycologia* **2014**, *106*, 133–144. [CrossRef] [PubMed]
55. Chaverri, P.; Branco-Rocha, F.; Jaklitsch, W.; Gazis, R.; Degenkolb, T.; Samuels, G.J. Systematics of the *Trichoderma harzianum* species complex and the re-identification of commercial biocontrol strains. *Mycologia* **2015**, *107*, 558–590. [CrossRef] [PubMed]
56. Gazis, R.; Rehner, S.; Chaverri, P. Species delimitation in fungal endophyte diversity studies and its implications in ecological and biogeographic inferences. *Mol. Ecol.* **2011**, *20*, 3001–3013. [CrossRef] [PubMed]

57. Hoyos-Carvajal, L.; Orduz, S.; Bissett, J. Genetic and metabolic biodiversity of *Trichoderma* from Colombia and adjacent neotropic regions. *Fungal Genet. Biol.* **2009**, *46*, 615–631. [CrossRef]
58. Kim, C.S.; Yu, S.H.; Nakagiri, A.; Shirouzu, T.; Sotome, K.; Kim, S.C.; Maekawa, N. Re-evaluation of *Hypocrea pseudogelatinosa* and *H. pseudostraminea* isolated from shiitake mushroom (*Lentinula edodes*) cultivation in Korea and Japan. *Plant Pathol. J.* **2012**, *28*, 341–356. [CrossRef]
59. Samuels, G. *Trichoderma*: Systematics, the sexual state, and ecology. *Phytopathology* **2006**, *96*, 195–206. [CrossRef]
60. Ospina-Giraldo, M.D.; Royse, D.J.; Chen, X.; Romaine, C.P. Molecular phylogenetic analyses of biological control strains of *Trichoderma harzianum* and other biotypes of *Trichoderma* spp. associated with mushroom green mold. *Phytopathology* **1999**, *89*, 308–313. [CrossRef]
61. Qiu, Z.H.; Wu, X.L.; Zhang, J.X.; Huang, C.Y. High temperature enhances the ability of *Trichoderma asperellum* to infect *Pleurotus ostreatus* mycelia. *PLoS ONE* **2017**, *12*, e0187055. [CrossRef]
62. Liu, X.M.; Wu, X.L.; Chen, Q.; Qiu, Z.H.; Zhang, J.X.; Huang, C.Y. Effects of heat stress on *Pleurotus eryngii* mycelial growth and its resistance to *Trichoderma asperellum*. *Mycosystema* **2017**, *36*, 1566–1574. [CrossRef]
63. Jiang, H.; Zhang, L.; Zhang, J.Z.; Ojaghian, M.R.; Hyde, K.D. Antagonistic interaction between *Trichoderma asperellum* and *Phytophthora capsici in vitro*. *J. Zhejiang Univ. Sci. B* **2016**, *17*, 271–281. [CrossRef]
64. Sun, J.Z.; Liu, X.Z.; Jeewon, R.; Li, Y.L.; Lin, C.G.; Tian, Q.; Zhao, Q.; Xiao, X.P.; Hyde, K.D.; Nilthong, S. Fifteen fungicolous Ascomycetes on edible and medicinal mushrooms in China and Thailand. *Asian J. Mycol.* **2019**, *2*, 129–169. [CrossRef]
65. Rees, H.J.; Bashir, N.; Drakulic, J.; Cromey, M.G.; Bailey, A.M.; Foster, G.D. Identification of native endophytic *Trichoderma* spp. for investigation of *in vitro* antagonism towards *Armillaria mellea* using synthetic- and plant-based substrates. *J. Appl. Microbiol.* **2021**, *131*, 392–403. [CrossRef] [PubMed]
66. Kredics, L.; Kocsubé, S.; Nagy, L.; Komoń-Zelazowska, M.; Manczinger, L.; Sajben, E.; Nagy, A.; Vágvölgyi, C.; Kubicek, C.P.; Druzhinina, I.S.; et al. Molecular identification of *Trichoderma* species associated with *Pleurotus ostreatus* and natural substrates of the oyster mushroom. *FEMS Microbiol. Lett.* **2009**, *300*, 58–67. [CrossRef] [PubMed]
67. Kosanović, D.; Potocnik, I.; Duduk, B.; Vukojevic, J.; Stajic, M.; Rekanovic, E.; Milijašević-Marčić, S. *Trichoderma* species on *Agaricus bisporus* farms in Serbia and their biocontrol. *Ann. Appl. Biol.* **2013**, *163*, 218–230. [CrossRef]
68. Ma, X.L.; Fan, X.L.; Wang, G.Z.; Xu, R.P.; Yan, L.L.; Zhou, Y.; Gong, Y.H.; Xiao, Y.; Bian, Y.B. Enhanced expression of thaumatin-like protein gene (*LeTLP1*) endows resistance to *Trichoderma atroviride* in *Lentinula edodes*. *Life* **2021**, *11*, 863. [CrossRef]
69. Lee, S.H.; Jung, H.J.; Hong, S.B.; Choi, J.I.; Ryu, J.S. Molecular markers for detecting a wide range of *Trichoderma* spp. that might potentially cause green mold in *Pleurotus eryngii*. *Mycobiology* **2020**, *48*, 313–320. [CrossRef]
70. Úrbez-Torres, J.R.; Tomaselli, E.; Pollaed-Flamand, J.; Boule, J.; Gerin, D.; Pollastro, S. Characterization of *Trichoderma* isolates from southern Italy, and their potential biocontrol activity against grapevine trunk disease fungi. *Phytopathol. Mediterr.* **2020**, *59*, 425–439. [CrossRef]
71. Dodd, S.L.; Lieckfeldt, E.; Samuels, G.J. *Hypocrea atroviridis* sp. nov., the teleomorph of *Trichoderma atroviride*. *Mycologia* **2003**, *95*, 27–40. [CrossRef]
72. Smith, A.; Beltrán, C.A.; Kusunoki, M.; Cotes, A.M.; Motohashi, K.; Kondo, T.; Deguchi, M. Diversity of soil-dwelling *Trichoderma* in Colombia and their potential as biocontrol agents against the phytopathogenic fungus *Sclerotinia sclerotiorum* (Lib.) de Bary. *J. Gen. Plant Pathol.* **2013**, *79*, 74–85. [CrossRef]
73. Choi, I.Y.; Choi, J.N.; Sharma, P.K.; Lee, W.H. Isolation and identification of mushroom pathogens from *Agrocybe aegerita*. *Mycobiology* **2010**, *38*, 310–315. [CrossRef] [PubMed]
74. Samuels, G.J.; Ismaiel, A.; Mulaw, T.B.; Szakacs, G.; Druzhinina, I.S.; Kubicek, C.P.; Jaklitsch, W.M. The Longibrachiatum Clade of *Trichoderma*: A revision with new species. *Fungal Divers.* **2012**, *55*, 77–108. [CrossRef] [PubMed]
75. Yan, Y.H. Research on Identification of *Trichoderma* of Mushrooms and Control of *Trichoderma*, *Mycogone cervine*. Master Thesis, Fujian Agriculture and Forestry University, Fuzhou, China, 2011.
76. Yabuki, T.; Miyazaki, K.; Okuda, T. Japanese species of the Longibrachiatum Clade of *Trichoderma*. *Mycoscience* **2014**, *55*, 196–212. [CrossRef]
77. Kim, J.Y.; Kwon, H.W.; Lee, D.H.; Ko, H.K.; Kim, S.H. Isolation and characterization of airborne mushroom damaging *Trichoderma* spp. from indoor air of cultivation houses used for oak wood mushroom production using sawdust media. *Plant Pathol. J.* **2019**, *35*, 674–683. [CrossRef] [PubMed]
78. Tomah, A.A.; Abd Alamer, I.S.; Li, B.; Zhang, J.Z. A new species of *Trichoderma* and gliotoxin role: A new observation in enhancing biocontrol potential of *T. virens* against *Phytophthora capsici* on chili pepper. *Biol. Control* **2020**, *145*, 104261. [CrossRef]
79. Samuels, G.J.; Suarez, C.; Solis, K.; Holmes, K.A.; Thomas, S.E.; Ismaiel, A.; Evans, H.C. *Trichoderma theobromicola* and *T. paucisporum*: Two new species isolated from cacao in South America. *Mycol. Res.* **2006**, *110*, 381–392. [CrossRef]
80. Bissett, J. A revision of the genus *Trichoderma*. III. Section *Pachybasium*. *Can. J. Bot.* **1991**, *69*, 2373–2417. [CrossRef]
81. Górski, R.; Sobieralski, K.; Siwulski, M.; Frąszczak, B.; Sas-Golak, I. The effect of *Trichoderma* isolates, from family mushroom growing farms, on the yield of four *Agaricus bisporus* (Lange) Imbach strains. *J. Plant Prot. Res.* **2014**, *54*, 102–105. [CrossRef]
82. Xu, Y.B.; Wen, C.J. Studies of contaminative fungi in the substituted cultivation of *Pleurotus ostreatus* and *Lentinus edodes*. *Edible Fungi China* **2004**, *23*, 45–47.
83. Wang, G.Z.; Cao, X.T.; Ma, X.L.; Guo, M.P.; Liu, C.H.; Yan, L.L.; Bian, Y.B. Diversity and effect of *Trichoderma* spp. associated with green mold disease on *Lentinula edodes* in China. *MicrobiologyOpen* **2016**, *5*, 709–718. [CrossRef]

84. Reper, C.; Penninckx, M.J. Inhibition of Pleurotus ostreatus growth and fructification by a diffusible toxin from *Trichoderma hamatum*. *J. Food Sci. Technol.* **1987**, *20*, 291–292.
85. He, Z.D.; Sun, H.Q.; Gao, Y.F. Identification of *Trichoderma* species on mushrooms. *J. Hebei Norm. Univ. Sci. Technol.* **2008**, *22*, 41–45.
86. Zhou, Y.; Wang, J.; Laying, Y.; Guo, L.J.; He, S.T.; Zhou, W.; Huang, J.S. Development of species/genus specific primers for identification of three *Trichoderma* species and for detection of *Trichoderma* genus. *Research Square* 2021, *preprint*. [CrossRef]
87. Cai, M.Z.; Idrees, M.; Zhou, Y.; Zhang, C.L.; Xu, J.Z. First report of green mold disease caused by *Trichoderma hengshanicum* on *Ganoderma lingzhi*. *Mycobiology* **2020**, *48*, 427–430. [CrossRef] [PubMed]
88. Song, X.X.; Wang, Q.; Jun, J.X.; Zhang, J.J.; Chen, H.; Chen, M.J.; Huang, J.C.; Xie, B.Q. Study on accurate identification of four *Trichoderma* diseases in *Agaricus bisporus* in factory cultivation. *Edible Fungi* **2019**, *41*, 67–72.
89. Zhu, Z.X.; Zhuang, W.Y. Three new species of *Trichoderma* with hyaline ascospores from China. *Mycologia* **2015**, *107*, 328–345. [CrossRef]
90. Jaklitsch, W.M.; Samuels, G.J.; Ismaiel, A.; Voglmayr, H. Disentangling the *Trichoderma viridescens* complex. *Persoonia* **2013**, *31*, 112–146. [CrossRef]
91. Chen, X.Y.L.; Zhou, X.H.; Zhao, J.; Tang, X.L.; Pasquali, M.; Migheli, Q.; Berg, G.; Cernava, T. Occurrence of green mold disease on *Dictyophora rubrovolvata* caused by *Trichoderma koningiopsis*. *J. Plant Pathol.* **2021**, *103*, 981–984. [CrossRef]
92. Samuels, G.J.; Ismaiel, A. *Trichoderma evansii* and *T. lieckfeldtiae*: Two new *T. hamatum*-like species. *Mycologia* **2009**, *101*, 142–156. [CrossRef]
93. Choi, I.Y.; Joung, G.T.; Ryu, J.; Choi, J.S.; Choi, Y.G. Physiological characteristics of green mold (*Trichoderma* spp.) isolated from oyster mushroom (*Pleurotus* spp.). *Mycobiology* **2003**, *31*, 139–144. [CrossRef]
94. Kim, C.S.; Shirouzu, T.; Nakagiri, A.; Sotome, K.; Nagasawa, E.; Maekawa, N. *Trichoderma mienum* sp. nov., isolated from mushroom farms in Japan. *Antonie Van Leeuwenhoek* **2012**, *102*, 629–641. [CrossRef]
95. Błaszczyk, L.; Siwulski, M.; Sobieralski, K.; Fružyńska-Jóźwiak, D. Diversity of *Trichoderma* spp. causing *Pleurotus* green mould diseases in Central Europe. *Folia Microbiol.* **2013**, *58*, 325–333. [CrossRef] [PubMed]
96. Lee, H.M.; Bak, W.C.; Lee, B.H.; Park, H.; Ka, K.H. Breeding and screening of *Lentinula edodes* strains resistant to *Trichoderma* spp. *Mycobiology* **2008**, *36*, 270–273. [CrossRef] [PubMed]
97. Al-Rubaiey, W.; Al-Juboory, H.H. Molecular identification of *Trichoderma longibrachiatum* causing green mold in *Pleurotus eryngii* culture media. *Plant Archives* **2020**, *20*, 181–184. [CrossRef]
98. Choi, I.Y.; Lee, W.; Choi, J.S. Forest green mold disease caused by *Trichoderma pseudokoningii* in winter mushroom, *Flammulina velutipes*. *Korean J. Mycol.* **1998**, *26*, 531–537.
99. Druzhinina, I.S.; Kopchinskiy, A.G.; Komoń, M.; Bissett, J.; Szakacs, G.; Kubicek, C.P. An oligonucleotide barcode for species identification in *Trichoderma* and *Hypocrea*. *Fungal Genet. Biol.* **2005**, *42*, 813–828. [CrossRef] [PubMed]
100. Kim, C.S.; Shirouzu, T.; Nakagiri, A.; Sotome, K.; Maekawa, N. *Trichoderma eijii* and *T. pseudolacteum*, two new species from Japan. *Mycol. Prog.* **2012**, *12*, 739–753. [CrossRef]
101. Jaklitsch, W.M.; Stadler, M.; Voglmayr, H. Blue pigment in *Hypocrea caerulescens* sp. nov. and two additional new species in sect. *Trichoderma*. *Mycologia* **2012**, *104*, 925–941. [CrossRef]
102. Park, M.S.; Oh, S.Y.; Cho, H.J.; Fong, J.J.; Cheon, W.J.; Lim, Y.W. *Trichoderma songyi* sp. nov., a new species associated with the pine mushroom (*Tricholoma matsutake*). *Antonie Van Leeuwenhoek* **2014**, *106*, 593–603. [CrossRef]
103. Samuels, G.J.; Ismaiel, A.; de Souza, J.; Chaverri, P. *Trichoderma stromaticum* and its overseas relatives. *Mycol. Prog.* **2012**, *11*, 215–254. [CrossRef]
104. Choi, I.Y.; Hong, S.B.; Yadav, M.C. Molecular and morphological characterization of green mold, *Trichoderma* spp. isolated from oyster mushrooms. *Mycobiology* **2003**, *31*, 74–80. [CrossRef]
105. Jaklitsch, W.M.; Põldmaa, K.; Samuels, G.J. Reconsideration of *Protocrea* (Hypocreales, Hypocreaceae). *Mycologia* **2008**, *100*, 962–984. [CrossRef] [PubMed]
106. Swofford, D.L. *PAUP*: Phylogenetic Analysis Using Parsimony (*and other methods)*; Version 4.0b10; Sinauer Associates: Sunderland, UK, 2002.
107. Edler, D.; Klein, J.; Antonelli, A.; Silvestro, D. raxmlGUI 2.0: A graphical interface and toolkit for phylogenetic analyses using RAxML. *Methods Ecol. Evol.* **2021**, *12*, 373–377. [CrossRef]
108. Ronquist, F.; Teslenko, M.; Van Der Mark, P.; Ayres, D.L.; Darling, A.; Höhna, S.; Larget, B.; Liu, L.; Suchard, M.A.; Huelsenbeck, J.P. MrBayes 3.2: Efficient Bayesian phylogenetic inference and model choice across a large model space. *Syst. Biol.* **2012**, *61*, 539–542. [CrossRef]
109. Nylander, J.A.A. MrModeltest v2. Program Distributed by the Author. Ph.D. Thesis, Uppsala University, Uppsala, Sweden, 2004.
110. Harman, G.E.; Kubicek, C.P. *Trichoderma and Gliocladium, Volume 2: Enzymes, Biological Control and Commercial Applications*, 1st ed.; CRC Press: London, UK, 1998.
111. Kubicek, C.P.; Mikus, M.; Schuster, A.; Schmoll, M.; Seiboth, B. Metabolic engineering strategies for the improvement of cellulase production by *Hypocrea jecorina*. *Biotechnol. Biofuels* **2009**, *2*, 19. [CrossRef] [PubMed]
112. Chaverri, P.; Samuels, G.J. Evolution of habitat preference and nutrition mode in a cosmopolitan fungal genus with evidence of interkingdom host jumps and major shifts in ecology. *Evolution* **2013**, *67*, 2823–2837. [CrossRef] [PubMed]
113. Samuels, G.J. *Trichoderma*: A review of biology and systematics of the genus. *Mycol. Res.* **1996**, *100*, 923–935. [CrossRef]

114. Li, Q.R.; Tan, P.; Jiang, Y.L.; Hyde, K.D.; McKenzie, E.; Bahkali, A.; Kang, J.C.; Wang, Y. A novel *Trichoderma* species isolated from soil in Guizhou, *T. guizhouense*. *Mycol. Prog.* **2012**, *12*, 167–172. [CrossRef]
115. Evans, H.C.; Holmes, K.A.; Thomas, S.E. Endophytes and mycoparasites associated with an indigenous forest tree, *Theobroma gileri*, in Ecuador and a preliminary assessment of their potential as biocontrol agents of cocoa diseases. *Mycol. Prog.* **2003**, *2*, 149–160. [CrossRef]
116. Gazis, R.; Chaverri, P. Diversity of fungal endophytes in leaves and stems of wild rubber trees (*Hevea brasiliensis*) in Peru. *Fungal Ecol.* **2010**, *3*, 240–254. [CrossRef]
117. Rossman, A.Y.; Samuels, G.J.; Rogerson, C.T.; Lowen, R. Genera of bionectriaceae, Hypocreaceae and Nectriaceae (Hypocreales, Ascomycetes). *Stud. Mycol.* **1999**, *42*, 1–248.
118. Hatvani, L. Mushroom Pathogenic *Trichoderma* Species, Occurrence, Biodiversity, Diagnosis and Extracellular Enzyme Production. Ph.D. Thesis, University of Szeged, Szeged, Hungary, 2008.
119. Castle, A.; Speranzini, D.; Rghei, N.; Alm, G.; Rinker, D.; Bissett, J. Morphological and molecular identification of *Trichoderma* isolates on North American mushroom farms. *Appl. Environ. Microbiol.* **1998**, *64*, 133–137. [CrossRef] [PubMed]
120. Park, M.S.; Bae, K.S.; Yu, S.H. Two new species of *Trichoderma* associated with green mold of oyster mushroom cultivation in Korea. *Mycobiology* **2006**, *34*, 111–113. [CrossRef] [PubMed]
121. Fletcher, J.T.; Gaze, R.H. *Mushroom Pest and Disease Control: A Color Handbook*, 1st ed.; CRC Press: London, UK, 2007.
122. Mamoun, M.L.; Lapicco, R.; Savoie, J.M.; Olivier, J.M. Green mould disease in France: *Trichoderma harzianum* Th2 and other species causing damages on mushroom farms. In Proceedings of the 15th International Congress on the Science and Cultivation of Edible Fungi, Maastricht, The Netherlands, 15–19 May 2000.
123. Kim, C.S.; Park, M.S.; Kim, S.C.; Maekawa, N.; Yu, S.H. Identification of *Trichoderma*, a competitor of shiitake mushroom (*Lentinula edodes*), and competition between *Lentinula edodes* and *Trichoderma* species in Korea. *Plant Pathol. J.* **2012**, *28*, 137–148. [CrossRef]
124. Jaklitsch, W.M.; Voglmayr, H. New combinations in *Trichoderma* (Hypocreaceae, Hypocreales). *Mycotaxon* **2013**, *126*, 143–156. [CrossRef]
125. Sinden, J.W.; Hauser, E. Nature and control of three mildew diseases of mushrooms in America. *Mushroom Sei.* **1953**, *2*, 177–180.
126. Elad, Y. Biocontrol of foliar pathogens: Mechanisms and application. *Commun Agric. Appl. Biol. Sci.* **2003**, *68*, 17–24.
127. Mumpuni, A.; Sharma, H.S.S.; Brown, A.E. Effect of metabolites produced by *Trichoderma harzianum* biotypes and *Agaricus bisporus* on their respective growth radii in culture. *Appl. Environ. Microbiol.* **1998**, *64*, 5053–5056. [CrossRef]
128. Neethling, D.; Nevalainen, H. Mycoparasitic species of *Trichoderma* produce lectins. *Can. J. Microbiol.* **1996**, *42*, 141–146. [CrossRef]
129. Williams, J.; Clarkson, J.M.; Mills, P.R.; Cooper, R.M. Saprotrophic and mycoparasitic components of aggressiveness of *Trichoderma harzianum* groups toward the commercial mushroom *Agaricus bisporus*. *Appl. Environ. Microbiol.* **2003**, *69*, 4192–4199. [CrossRef]
130. Abubaker, K.S.; Sjaarda, C.; Castle, A.J. Regulation of three genes encoding cell-wall-degrading enzymes of *Trichoderma aggressivum* during interaction with *Agaricus bisporus*. *Can. J. Microbiol.* **2013**, *59*, 417–424. [CrossRef]
131. Weindling, R. Studies on a lethal principle effective in the parasitic action of *Trichoderma lignorum* on *Rhizoctonia solani* and other soil fungi. *Phytopathology* **1934**, *24*, 1153–1179.
132. Castrillo, M.L.; Bich, G.A.; Zapata, P.D.; Villalba, L.L. Biocontrol of *Leucoagaricus gongylophorus* of leaf-cutting ants with the mycoparasitic agent *Trichoderma koningiopsis*. *Mycosphere* **2016**, *7*, 810–819. [CrossRef]
133. Kubicek, C.P.; Herrera-Estrella, A.; Seidl-Seiboth, V.; Martínez, D.; Druzhinina, I.S.; Thon, M.R.; Zeilinger, S.; Casas-Flores, S.; Horwitz, B.A.; Mukherjee, P.K.; et al. Comparative genome sequence analysis underscores mycoparasitism as the ancestral life style of *Trichoderma*. *Genome Biol.* **2011**, *12*, R40. [CrossRef] [PubMed]
134. Yu, C.; Luo, X. *Trichoderma koningiopsis* controls *Fusarium oxysporum* causing damping-off in *Pinus massoniana* seedlings by regulating active oxygen metabolism, osmotic potential, and the rhizosphere microbiome. *Biol. Control* **2020**, *150*, 104352. [CrossRef]

Journal of *Fungi*

Article

New Insights into Lichenization in Agaricomycetes Based on an Unusual New Basidiolichen Species of *Omphalina* s. str.

Tingting Zhang [1,2,†], Xinyu Zhu [1,3,†], Alfredo Vizzini [4], Biting Li [1,5], Zhenghua Cao [6], Wenqing Guo [6], Sha Qi [1], Xinli Wei [1,2,*] and Ruilin Zhao [1,2,*]

1 State Key Laboratory of Mycology, Institute of Microbiology, Chinese Academy of Sciences, Beijing 100101, China
2 College of Life Sciences, University of Chinese Academy of Sciences, Beijing 100049, China
3 Center of Excellence in Fungal Research, Mae Fah Luang University, Chiang Rai 57100, Thailand
4 Dipartimento di Scienze Della Vita e Biologia Dei Sistemi, Università di Torino, Viale P.A. Mattioli 25, 10125 Torino, Italy
5 College of Life Science and Technology, Xinjiang University, Urumqi 830000, China
6 The Wanan County Second Middle School, Wanan 343800, China
* Correspondence: weixl@im.ac.cn (X.W.); zhaorl@im.ac.cn (R.Z.)
† These authors contributed equally to this work.

Abstract: The genus *Omphalina* is an ideal genus for studying the evolutionary mechanism of lichenization. Based on molecular phylogeny using ITS and nuLSU sequences by means of Bayesian and maximum likelihood analyses and morphological examination, combining the existence of green algae in basidiomata stipe and a *Botrydina*-type vegetative thallus, we described a bryophilous new basidiolichen species, *Omphalina licheniformis*, from a residential area of Jiangxi Province, China. This finding of unusual new basidiolichen species updated our understanding of the delimitation of *Omphalina*, indicating that both non-lichen-forming and lichen-forming fungal species are included simultaneously. The presence of algal cells in the basidiomata should receive more attention, as this would be helpful to distinguish more potential basidiolichens and explore the cryptic species diversity. This work provides new insights and evidence for understanding the significance of lichenization during the evolution of Agaricomycetes.

Keywords: agaricales; basidiolichen; basidiomycota; fruiting body; green algae; phenotype; systematics; new taxon

Citation: Zhang, T.; Zhu, X.; Vizzini, A.; Li, B.; Cao, Z.; Guo, W.; Qi, S.; Wei, X.; Zhao, R. New Insights into Lichenization in Agaricomycetes Based on an Unusual New Basidiolichen Species of *Omphalina* s. str.. *J. Fungi* **2022**, *8*, 1033. https://doi.org/10.3390/jof8101033

Academic Editor: Philippe Silar

Received: 23 August 2022
Accepted: 28 September 2022
Published: 29 September 2022

1. Introduction

Lichens are symbionts of fungi (mycobionts) and algae and/or cyanobacteria (photobionts), among which only 0.9% species belong to the Basidiomycota [1]. *Omphalina* Quél. is an undisputedly important genus when talking about basidiolichen species, because it included both non-lichenized and lichenized species originally, and was subsequently separated into non-lichenized *Omphalina* s. str. and lichenized genera such as *Lichenomphalia* Redhead, Lutzoni, Moncalvo & Vilgalys [2] and *Agonimia* Zahlbr. (syn. *Marchandiomphalina* Diederich, Manfr. Binder & Lawrey [3]). In addition, *Omphalina* also includes saprophytic, parasitic, and bryophilous species [4]. Therefore, *Omphalina* has been regarded as an ideal genus for studying the evolutionary mechanisms associated with lichenization [4,5].

Recently, molecular phylogenetic analyses pointed out that the classical concept of *Omphalina*, mainly based on morphological features [6–9], includes several omphalinoid genera nested inside the order Agaricales [2,10–14], as well as Hymenochaetales Oberw. [15–18]. Inside the Agaricales there are omphalinoid taxa in the suborders Hygrophorineae (family Hygrophoraceae, subfamily Lichenomphaloideae [13]), Marasmiineae (family Porotheleaceae [19,20]) and Tricholomatineae (family Omphalinaceae [21]). Rickenellaceae is the family encompassing omphalinoid taxa in the Hymenochaetales [17,18,22]. *Omphalina* was restricted to the species phylogenetically related to *O. pyxidata* (the conserved lectotype of *Omphalina* [2,10,23,24]), which typically show reddish brown, rusty, or orange-brown tinges on the pileus and stipe, and a non-concolorous hymenophore [2]. Micromorphological features, such as non-amyloid spores, sub-regular to irregular hymenophoral trama and pileipellis with encrusting pigment, are shared by all members of the genus *Omphalina* s. stricto [2,9,25–29]. This genus is sister to *Infundibulicybe* [14,21,30–32], and together they form the family Omphalinaceae [33] in the suborder Tricholomatineae.

Basidiolichens are mainly distributed in five orders of Agaricomycetes, viz. Agaricales, Atheliales, Lepidostromatales, Cantharellales and Corticiales, among which Agaricales and the family Hygrophoraceae within this order accommodate most of the basidiolichen species [1]. In China, thirteen basidiolichen species, including five *Dictyonema* spp. (Hygrophoraceae, Agaricales), four *Lichenomphalia* spp. (Hygrophoraceae, Agaricales), and four *Sulzbacheromyces* spp. (Lepidostromataceae, Lepidostromatales), have been discovered [34–36].

A cluster of bryophilous *Omphalina* basidiomata was found in a residential area of Jiangxi Province, China. The phylogenetic analyses of nrDNA ITS and nuLSU sequences also confirmed it to be an unknown *Omphalina* species. However, interestingly, it was found in the stipe existence of green algae; moreover, far fewer vegetative thalli consisting of green tiny globules (*Botrydina*-type) were also seen near to the hairs at the base of stipe. Therefore, an unusual new basidiolichen species of *Omphalina* is described and reported here. This finding indicates that *Omphalina* s. str. still consists of both non-lichenized and lichenized species with very a close phylogenetic relationship, and further provides new insights into the lichenization in Agaricomycetes.

2. Materials and Methods

2.1. Taxon Sampling and Morphological Examination

Five basidiomata specimens were collected from a residential area of Wan'an County, Jiangxi Province of China (Figure 1), and are preserved in the Herbarium Mycologicum Academiae Sinicae, Beijing, China (HMAS). Morphology and anatomy were examined using a MOTIC SMZ-168 stereomicroscope and a LEICA M125 dissecting microscope equipped with a Leica DFC450 camera.

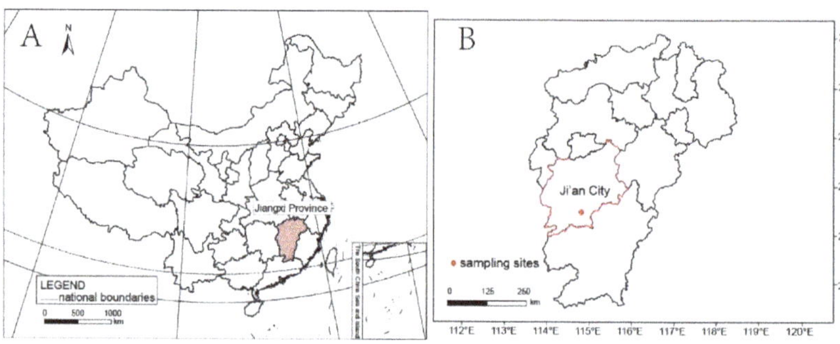

Figure 1. Collection site. (**A**). Jiangxi Province is colored with pink. (**B**). Ji'an City is circled with a red line and the collection site is marked with a red solid circle.

Photographs of fresh specimens were taken immediately in the residential area, and the basidiomes were gathered. The morphological characteristics, including cap, stipe, pileus, and odor, were recorded. Specimens were dried in an electrical food drier at 55 °C to ensure that no moisture was left, and then were sealed in plastic bags.

To observe anatomic characteristics, parts of dried specimens were cut and mounted in 5% KOH and stained with 1% Congo Red. Anatomic characteristics including basidiospores, basidia, cystidia and pileipellis were observed under an Olympus CX31RTSF microscope (Made in Philippines, Tokyo, Japan) with at least 20 records. Data were analyzed and recorded as X = the mean of length by width ± SD, Q = the quotient of basidiospore length to width, and Qm = the mean of Q values ± SD. All of the protocols of the morphological study followed Largent's methodology [37].

2.2. DNA Extraction, PCR, and Sequencing

DNA was extracted from two fresh basidiomata (Table S1) by means of the modified CTAB method [38]. PCR was performed to amplify two gene loci: nuclear ribosomal DNA internal transcribed spacer (ITS) and large subunit (nuLSU), using primers ITS4 and ITS5 [39], and LR0R and LR5 [40], respectively. The PCR procedure followed Yang et al. [41].

2.3. Sequence Alignment and Phylogenetic Analysis

A total of 136 DNA sequences including four new sequences were used in this study (Table S1). Representative species of the lichenized genera of Hygrophoraceae (Agaricales), non-lichenized genera of Agaricales, and other related orders in the Agaricomycetes were chosen in the phylogenetic analyses. *Multiclavula* spp. (Clavulinaceae, Cantharellales) were taken as the outgroup.

Raw sequences were firstly assembled and edited with SeqMan [42], and then aligned using MAFFT v.7 [43]. We used the program Gblocks v.0.19b [44,45] to remove ambiguously aligned sites. The congruence of the two loci (ITS and nuLSU) was tested as described previously [46,47]. All maximum likelihood (ML) and Bayesian inference (BI) analyses were performed using the GTR + I + G model selected by jModelTest 2 [48]. The ML analysis involved 1000 pseudoreplicates with RAxML v.8.2.6 [49]. The BI analysis was performed using MrBayes v. 3.2.7 [50,51] with two parallel Markov chain Monte Carlo (MCMC), each using 5 million generations and sampling every 1000th generation. We used TRACER v.1.7.2 [52] to examine the standard deviation of split frequencies less than 0.01, reflecting the fact that the two trees differed very little, and the parameters converged. The 50% majority rule consensus tree was generated after discarding the first 25% as burn-in.

Phylogenetic analysis was run on the Cipres Science Gateway (http://www.phylo.org, (accessed on 18 July 2011)) and visualized using FigTree v.1.4.3 (http://tree.bio.ed.ac.uk/software/figtree, (accessed on 28 August 2014)). The clades with bootstrap (BP) values above 75% or posterior probability (PP) values above 0.95 were considered highly supportive.

3. Results

3.1. Phylogenetic Analysis

The aligned matrix contained 2011 unambiguous nucleotide (1129 ITS and 882 nuLSU) position characteristics for the full dataset of 93 members. BI and ML phylogenetic trees were constructed, and they had similar topological structures. The RAxML tree is shown in Figure 2 with both bootstrap support (BS) and posterior probability (PP) values of BI analysis. In the tree, all of the *Omphalina* species clustered into a well-supported monophyletic clade (BS 100/PP 1.00), obviously separated from other groups, in which the two samples (Coll. Nos. JX001 and ZRL20220005) are included and co-formed into a separate branch (BS 100/PP 0.99), indicating that this is a new species also supported by the morphological characteristics (see below). The BI phylogenetic tree and two single-gene-locus RAxML trees are shown in Figures S1–S3.

3.2. Taxonomy

The genus *Omphalina* is known to be a non-lichenized basidiolichen genus, because the original lichenized species contained in this genus have been separated out and formed totally different other genera such as *Lichenomphalia* [2]. However, an *Omphalina* new species was found to be lichenized in this study, and is described below. Therefore, the definition of the genus *Omphalina* also needs to be redefined as the genus with lichenized species in some cases.

Omphalina Quél., Enchir. fung. (Paris): 42 (1886)

Type species: *O. pyxidata* (Bull.) Que'l., *incertae sedis*, Agaricales, Agaricomycetes, worldwide. This genus is characterized by small basidiomes, pileus convex to umbilicate, reddish brown tinged, smooth, without scales; lamellae decurrent, paler and well-developed; stipe central, reddish brown tinged; hymenial and pileal cystidia absent or sparse, and presence of clamp-connections [2,53,54]. Sometimes lichenized.

Omphalina licheniformis X.L. Wei, Z.H. Cao & R.L. Zhao, **sp. nov.** (Figures 3 and 4)

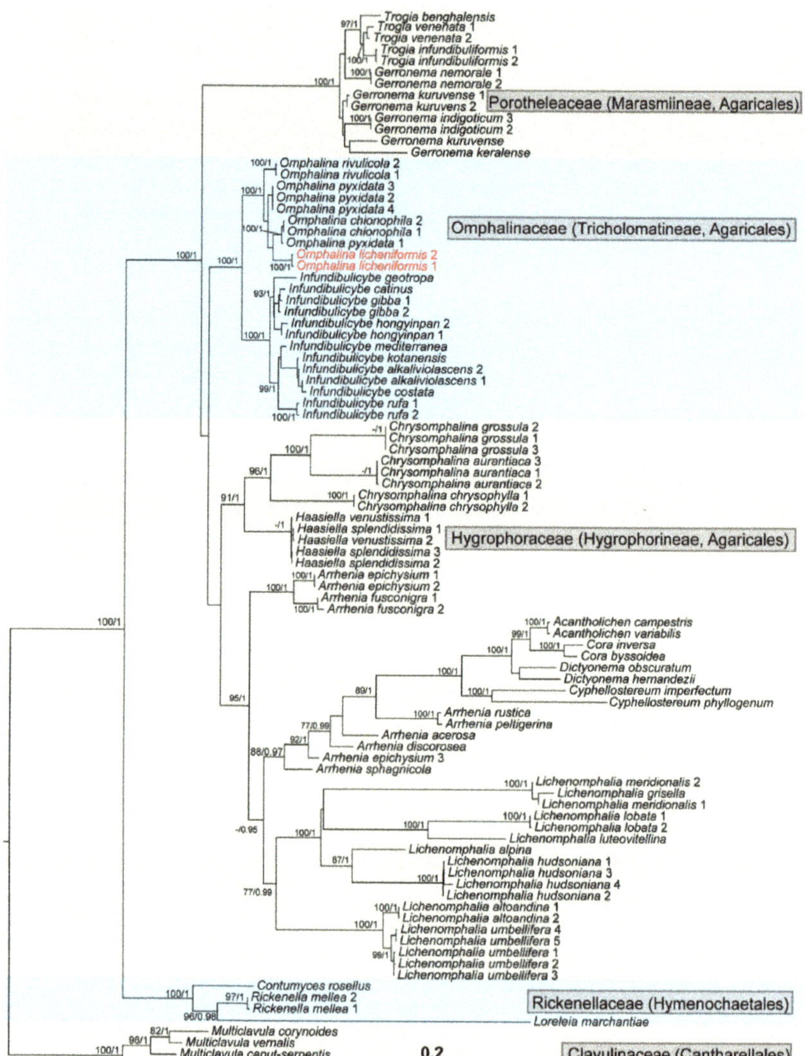

Figure 2. The RAxML tree of omphalinoid species based on the concatenated ITS + nuLSU dataset. The numbers in each node represent bootstrap support (BS) and posterior probability (PP) values. BS values ≥ 75% and PP values ≥ 0.95 were plotted on the branches of the tree. The samples corresponding to the new species are in red. Scale in 0.2 substitution per site.

Figure 3. The habitat and habits of *Omphalina licheniformis* sp. nov (holotype HMAS–L 154705). (**A**) Community balcony in a residential area—the habitat of basidiomata is marked by a red circle. (**B**,**C**) The basidiomata in situ. (**D**) Lamellae. (**E**) Hair. (**F**) Young basidiomata. (**G**) Microscopic observation of a young basidiomata with green algae cells in the stipe marked by the black box. (**H**) Zoom in on the black box. (**I**) Tiny and very few *Botrydina*-type vegetative thalli. (**J**) Green algae cells and hyphae in the thallus. (**K**) Leaf cells and chloroplasts inside of the moss *Hyophila involute*. Bars: D-F = 1 mm, G = 200 µm, H-I = 100 µm, J = 20 µm, K = 10 µm.

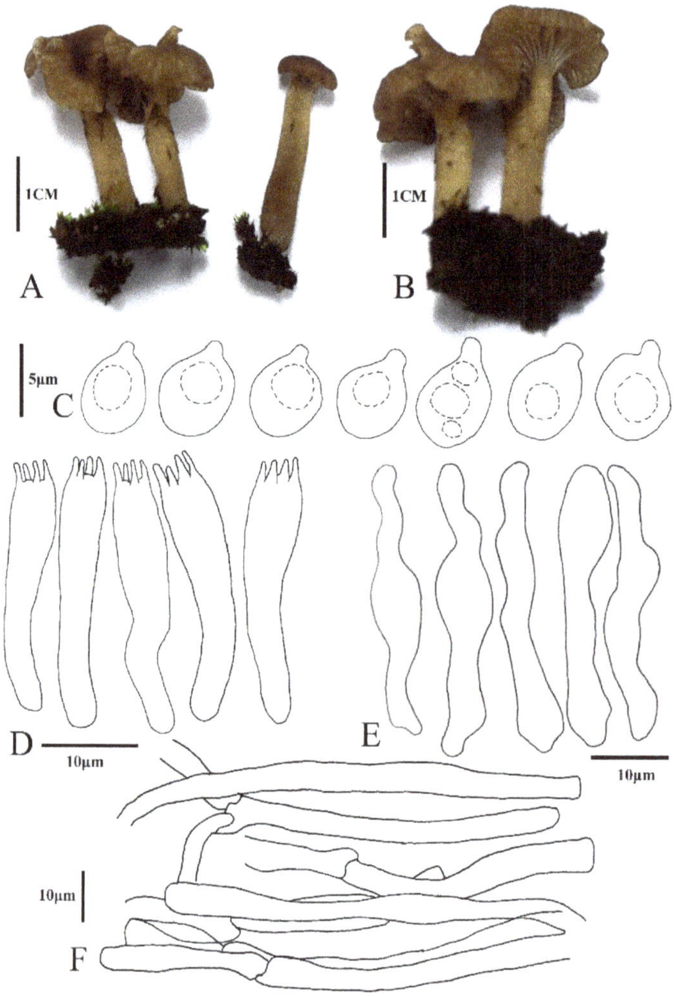

Figure 4. The basidiomes' habits and the anatomic structure of *Omphalina licheniformis* sp. nov (ZRL 20220005, HMAS 281952, isotype). (**A**,**B**) Basidiomes. (**C**) Basidiospores. (**D**) Basidia. (**E**) Cheilocystidia. (**F**) Pileipellis hyphae.

Fungal Names No.: FN571066

Etymology: The epithet '*licheniformis*' indicates that this species is a lichen-forming fungus.

Typus: China. Jiangxi Province, Ji'an City, Wan'an County, 26.465265° N 114.798753° E, 65 m alt., on moss *Hyophila involuta*, 10 March 2022, Z.H. Cao & W.Q. Guo JX001 (HMAS–L 154705, holotype), ZRL20220005 (HMAS 281952, isotype).

Diagnosis: *Omphalina licheniformis* is distinguished from other species of this genus by having smaller basidiospores (4.9–5.5 × 3.8–4.6 μm) and presenting distinctive cheilocystidia. It is also characterized by the presence of far fewer vegetative thalli consisting of green tiny globules (*Botrydina*-type) near to the hairs at the base of the stipe, and green algae in the stipe.

Description: Thalli not obvious, *Botrydina*-type, globular, with globules clustered, very few and tiny, plump when wet, yellow-green to green, c. 80 μm in diam.; globules consist of unicellular green alga enveloped by the hyaline hyphae, hyphae 2.5–5 μm wide, algal cells 6–10 μm in diam.

Basidiomes small, omphalinoid. *Pileus* 7–15 mm in diam., subhemispherical when young and subfunnel when mature, distinctly depressed, hygrophanous, red brown to yellowish brown, edge paler, becoming alutaceous when drying, margin involute when young, wavy and striate when mature. *Context* thin, up to 1 mm, concolorous with pileus surface. *Lamellae* decurrent, distant, thick, forked and anastomosing, concolorous with pileus, but paler. *Stipe* 15–20 × 3–5 mm, cylindrical, hollow, smooth to white fibrillose, concolorous with the pileus, darker at the middle and lower parts with white mycelium-like cilia. *Smell* and *taste* indistinct.

Basidiospores 4.9–5.5 × 3.8–4.6 μm, [x = 5.2 ± 0.2 × 4.1 ± 0.2, Q = 1.2–1.3, Q_m = 1.3, n = 20], broadly ellipsoid to ellipsoid, rough with granular contents, smooth, thick-walled, hyaline. *Basidia* 25.0–27.4 × 3.3–4.7 μm, clavate, hyaline, (2-)4-spored, smooth. *Cheilocystidia* 27.1–35.7 × 3.7–7.0 μm, cylindrical and flexuose, narrowly clavate or lageniform, thin-walled, hyaline. *Hymenophoral trama* irregular, consisting of 3.0–7.6 μm-wide hyphae. Pileipellis a cutis composed of hyphae of 3.3–5.3 μm in diam., smooth, cylindrical, hyaline; pigment epiparietal, minutely to strongly encrusting.

Habitat and distribution: This species is bryophilous, growing on the moss *Hyophila involuta* (Hook.) A. Jaeger mixing with soil located on a community balcony in a residential area of Wan'an County, Jiangxi Province of China, which is characterized by a subtropical monsoon climate, and is the only known distribution up to now.

Additional specimens examined: CHINA. Jiangxi Province, Ji'an City, Wan'an County, 26.465265° N 114.798753° E, 65 m alt., on moss, 30 March 2022, Z.H. Cao & W.Q. Guo JX002 (HMAS–L 154704), ZRL20220006 (HMAS 281953); 7 April 2022, Z.H. Cao & W.Q. Guo JX003 (HMAS–L 154706).

Notes: The vegetative thalli of this species are so tiny and few in number that they can very easily be ignored. The coexistence of algal cells in the base of the stipe near to the hairs is a very new finding, because previously, algal cells were only reported in the vegetative thallus of basidiolichens and known as green algae *Coccomyxa* [2,34,35]. The algal cells found in the new species are also unicellular and green, and unfortunately, this algal species has not been identified. However, the possibility that they are moss chloroplasts can be excluded, although moss chloroplasts are also unicellular and green, because chloroplasts are organelles within the cells of moss and need to live in the cytoplasm [55], and so it seems unlikely that the moss chloroplasts would escape from the moss cells and exist separately, trapped in the fungal hyphae. Furthermore, the moss chloroplasts are oval and 2.5–3 × 5–6 μm in size (Figure 3K), different from the algal cells (6–10 μm in diam., Figure 3H). Inconspicuous or absent thalli are common in the basidiolichens of Agaricales, for example in the bryophilous or phycophilous basidiolichen genus *Lichenomphalia*, and the thalli are not obvious in some species such as *L. umbellifera* (L.) Redhead, Lutzoni, Moncalvo & Vilgalys and *L. velutina* (Quél.) Redhead, Lutzoni, Moncalvo & Vilgalys [2,34]. The *Botrydina*-type globular thallus of the new species (Figure 3I) is also similar to *L. meridionalis* (Contu & La Rocca) P.A. Moreau & Courtec. [56], and the algal cells and hyphae are also observed (Figure 3J). However, the new species is distant from *Lichenomphalia* spp., but clusters within the genus *Omphalina*, close to *O. pyxidata* (Bull.) Quél. and *O. chionophila* Lamoure in phylogeny (Figure 2), which also have brown caps, small agarics, and decurrent lamellae [57], but are not known to have a lichenized form.

4. Discussion

Phylogenetic and morphological analyses support this new species as a member of *Omphalina* s. stricto. In the phylogenetic tree (Figure 2), *O. chionophile* is sister to *O. licheniformis*. However, in morphology, they can be distinguished by the size of the basidospores—that of *O. chionophile* is 8–10 × 5–6 μm [25,54,58], and that of *O. licheniformis* is 4.9–5.5 × 3.8–4.6 μm. Furthermore *O. chionophile* lacks cheilocystidia. The type species *O. pyxidata* is also phylogenetically close to *O. licheniformis*, sharing similar basidiome features which remain difficult to distinguish in situ. However, they can be distinctly separated under a microscope according to the basidiospore and cheilocystidia, as the basidiospores of *O. pyxidata* (7–8 × 5–6 μm) are larger than those of *O. licheniformis* (4.9–5.5 × 3.8–4.6 μm), and the cheilocystidia of *O. pyxidata* are often branched, while the cheilocystidia of *O. licheniformis* are cylindrical and flexuose, narrowly clavate or lageniform, and not branched [53,54].

The following species differ significantly from *O. licheniformis* (4.9–5.5 × 3.8–4.6 μm) in spore size: *O. rivulicola* (8–10.5 × 5.5–7 μm), *O. mutila* (6.5–10 × 4–6 μm), and *O. subhepatica* (6–8 × 4–5 μm) [54,59]. In terms of the presence or absence of cheilocystidia, *O. licheniformis* possesses cheilocystidia, but none of the following species have cheilocystidia: *O. mutila*, *O. demissa*, *O. subhepatica*, *O. chionophile*, *O. galericolor*, *O. kuehneri*, *O.arctia*, *O. rivulicola* [54,60]. In terms of habitat, *O. licheniformis* is found on mosses, but the following species are found in other habitats: *O. mutila* grows on humid soil in heathland and marshes with *Calluna*, *Erica* and *Mokunia*; *O. chionophile* grows on naked solifluction soil; and *O. pyxidata* and *O. rivulicola* grow on dry sandy soil [54]. In terms of the color of basidiomes, *O. licheniformis* is reddish brown to yellowish brown, but *O. mutila* forms white basidiomes [54,61]. Therefore, *O. licheniformis* is a distinguished species in both molecular composition and morphology.

Lichen-forming fungi are an important component of the kingdom Fungi, making up nearly 20% of the known fungal species, among which over 99% of species belong to Ascomycota [1]. Previous studies showed that more losses than gains of lichenization have occurred during the evolution of Ascomycota, resulting in lichen-forming fungi becoming the ancestors of major lineages of non-lichen-forming fungi in Ascomycota [62]. Compared with ascolichens, basidiolichens are much rarer, and comprise less than 1% of species of the known lichen-forming fungi [1]; however, lichenization in the evolution of Basidiomycota is also very important in related research, and is even treated as one of necessary models to study the evolution of lichens [63].

As early as 1995, Gargas et al. found three independent origins of lichenization in Basidiomycota, i.e., coral *Multiclavula* (Coker) R.H. Petersen as the basal origin, gilled mushroom *Lichenomphalia umbellifera* (syn. *Omphalina umbellifera*), and cyanolichen *Dictyonema pavonium* (Weber & D. Mohr) Parmasto. Basidiolichens are known to consist of far more than three genera nowadays, as mentioned above, including 5 orders, 5 families, 15 genera and 172 species, and six to seven independent lichenization events happened in the Basidiomycota, among which over 85% (147 species) belong to eight genera of Hygrophoraceae Lotsy in Agaricales, viz. *Acantholichen* P.M.Jørg., *Arrhenia* Fr., Cora Fr., *Corella* Vain., *Cyphellostereum* D.A.Reid, *Dictyonema* C.Agardh ex Kunth, *Lichenomphalia*, and *Semiomphalina* Redhead [1]. Except for *Corella* and *Semiomphalina*, all of the other six genera of Hygrophoraceae were included in our phylogenetic analyses (Figure 2), among which the genus *Lichenomphalia* was included in *Omphalina* [2].

The genus *Omphalina* has been taken as a model system to study lichenization since the late 1990s due to its variable nutritional modes [4]; however, after a series of species such as *Omphalina umbellifera*, etc., were transferred to other genera, no lichen-forming species were reported in *Omphalina* s.str., until the finding of *Omphalina licheniformis* in this study. Our study indicates that within the Agaricales, the lichenization process also occurred in the Omphalinaceae of the suborder Tricholomatineae. In the previous reports on basidiolichens, algal cells have never been found in the fruiting body structures [2,34,35,63], but indeed existed in the stipe of *Omphalina licheniformis* (Figure 3). This finding of unusual new species updated our understanding of the delimitation of *Omphalina*, indicating that both non-lichen-forming and lichen-forming fungal species are included simultaneously. Moreover, these results provide new insights and evidence for understanding the significance of lichenization during the evolution of Basidiomycota. Through this study, it should be noted that we need to pay more attention to the Basidiomycota fungi, especially to whether the algal cells are present in the fruiting bodies, which would be very helpful to distinguish more potential basidiolichens and explore the cryptic species diversity through these algal cell examinations.

The presence of algal cells as lichen photobionts is well-known to provide a carbon source for the mycobiont [64], which is relatively easy to understand in ascolichens, because fruiting bodies such as apothecia and pycnidia are closely connected parts of the lichen thallus, and the photobiont can be found both in the thallus and fruiting bodies, except the lecideine-type apothecia and pycnidia without hymenial algae. The algal cells in basidiolichens are assumed to be similar in function [5,65], but there is still an absence of strong evidence, especially due to the fruiting bodies of basidiolichens, which often look separable from the thallus in most cases, and algal cells have only been reported in the thallus previously [2,34,35,63]; moreover, sometimes the thallus is not obvious [2,34].

Supplementary Materials: The following supporting information can be downloaded at: https://www.mdpi.com/article/10.3390/jof8101033/s1, Table S1: Specimens used for DNA extraction and GenBank accession number of all samples used in this study, Figure S1: The Bayesian tree based on the concatenated ITS + nuLSU (two genes) dataset, Figure S2: The maximum likelihood tree based on the ITS dataset, Figure S3: The maximum likelihood tree based on the nuLSU dataset.

Author Contributions: X.W. and R.Z. conceived and designed the study. Z.C. and W.G. discovered and collected specimens. T.Z., X.Z. and B.L. generated the DNA sequence data, X.Z., S.Q., X.W. and R.Z. performed the phenotypic analysis, T.Z., X.Z., A.V., X.W. and R.Z. analyzed the DNA data and checked issues related to nomenclatural articles. X.W. wrote the manuscript draft. A.V. and R.Z. All authors have read and agreed to the published version of the manuscript.

Funding: This research was funded by the National Natural Science Foundation of China (32070096, 31961143010, 31970010), National Science and Technology Fundamental Resources Investigation Program of China (Grant No. 2019FY101808), Biological Resources Programme, Chinese Academy of Sciences (KFJ-BRP-009), Beijing Innovative Consortium of Agriculture Research System (Project ID: BAIC05-2022).

Data Availability Statement: Publicly available datasets were analyzed in this study. All newly generated sequences were deposited in GenBank (accessed on 9 June 2022, https://www.ncbi.nlm.nih.gov/genbank/; Table S1). All new taxa were deposited in Fungal names (accessed on 14 July 2022, https://www.fungalinfo.im.ac.cn).

Acknowledgments: The authors sincerely thank L. Zhang at the Fairy Lake Botanical Garden (Shenzhen), Chinese Academy of Sciences, for his great help in the identification of the moss in this study. Special thanks go to Q. X. Yang for her assistance during our studies in HMAS–L.

Conflicts of Interest: The authors declare no conflict of interest.

References

1. Lücking, R.; Hodkinson, B.P.; Leavitt, S.D. The 2016 classification of lichenized fungi in the Ascomycota and Basidiomycota—Approaching one thousand genera. *Bryologist* **2017**, *119*, 361–416. [CrossRef]
2. Redhead, S.; Lutzoni, F.; Moncalvo, J.M.; Vilgalys, R. Phylogeny of agarics: Partial systematics solutions for core Omphalinoid genera in the Agaricales (Euagarics). *Mycotaxon* **2002**, *83*, 19–57.
3. Lücking, R.; Moncada, B. Dismantling Marchandiomphalina into *Agonimia* (Verrucariaceae) and *Lawreymyces* gen. nov. (Corticiaceae): Setting a precedent to the formal recognition of thousands of voucherless fungi based on type sequences. *Fungal Divers.* **2017**, *84*, 119–138. [CrossRef]
4. Lutzoni, F.; Vilgalys, R. *Omphalina* (Basidiomycota, Agaricales) as a model system for the study of coevolution in lichens. *Cryptogam. Bot.* **1995**, *5*, 71–81.
5. Lutzoni, F.M. Phylogeny of lichen- and non-lichen-forming omphalinoid mushrooms and the utility of testing for combinability among multiple data sets. *Syst. Biol.* **1997**, *46*, 373–406. [CrossRef] [PubMed]
6. Bigelow, H.E. *Omphalina* in North America. *Mycologia* **1970**, *62*, 1–32. [CrossRef]
7. Clémençon, H. Kompendium der Blätterpilze Europäische omphalinoide Tricholomataceae. *Zeitschr. F. Mykol.* **1982**, *48*, 195–237.
8. Norvell, L.L.; Redhead, S.A.; Ammirati, J.F. *Omphalina* sensu lato in North America 1-2. 1: *Omphalina wynniae* and the genus *Chrysomphalina*. 2: *Omphalina* sensu Bigelow. *Mycotaxon* **1994**, *50*, 379–407.
9. Bon, M. Flore mycologique d'Europe 4. Les Clitocybes, Omphales et ressemblants. *Doc. Mycologia. Mém. Hors Sér* **1997**, *4*, 1–181.
10. Moncalvo, J.-M.; Vilgalys, R.; Redhead, S.; Johnson, J.; James, T.; Aime, M.; Valerie, H.; Verduin, S.; Larsson, E.; Baroni, T.; et al. One Hundred and Seventeen Clades of Eu-agarics. *Mol. Phylogenet. Evol.* **2002**, *23*, 357–400. [CrossRef]
11. Matheny, P.; Curtis, J.; Valerie, H.; Aime, M.; Moncalvo, J.-M.; Ge, Z.-W.; Slot, J.; Ammirati, J.; Baroni, T.; Bougher, N.; et al. Major clades of Agaricales: A multilocus phylogenetic overview. *Mycologia* **2006**, *98*, 982–995. [CrossRef] [PubMed]

12. Lawrey, J.; Lücking, R.; Sipman, H.; Chaves, J.; Redhead, S.; Bungartz, F.; Sikaroodi, M.; Gillevet, P. High concentration of basidiolichens in a single family of agaricoid mushrooms (Basidiomycota: Agaricales: Hygrophoraceae). *Mycol. Res.* **2009**, *113*, 1154–1171. [CrossRef] [PubMed]
13. Lodge, D.; Padamsee, M.; Matheny, P.; Aime, M.; Cantrell, S.; Boertmann, D.; Kovalenko, A.; Vizzini, A.; Dentinger, B.; Kirk, P.; et al. Molecular phylogeny, morphology, pigment chemistry and ecology in Hygrophoraceae (Agaricales). *Fungal Divers.* **2013**, *64*, 1–99. [CrossRef]
14. Varga, T.; Krizsan, K.; Földi, C.; Dima, B.; Sanchez-Garcia, M.; Sanchez-Ramirez, S.; Szollosi, G.; Szarkándi, J.; Papp, V.; Albert, L.; et al. Megaphylogeny resolves global patterns of mushroom evolution. *Nat. Ecol. Evol.* **2019**, *3*, 668–678. [CrossRef]
15. Redhead, S.; Moncalvo, J.-M.; Vilgalys, R.; Lutzoni, F. Phylogeny of agarics: Partial systematics solutions for bryophilous omphalinoid agarics outside of the Agaricales (Euagarics). *Mycotaxon* **2002**, *82*, 151–168.
16. Larsson, K.-H.; Parmasto, E.; Fischer, M.; Langer, E.; Nakasone, K.; Redhead, S. Hymenochaetales: A molecular phylogeny for the hymenochaetoid clade. *Mycologia* **2006**, *98*, 926–936. [CrossRef]
17. Korotkin, H.; Swenie, R.; Miettinen, O.; Budke, J.; Chen, K.-H.; Lutzoni, F.; Smith, M.; Matheny, P. Stable isotope analyses reveal previously unknown trophic mode diversity in the Hymenochaetales. *Am. J. Bot.* **2018**, *105*, 1869–1887. [CrossRef]
18. Olariaga, I.; Huhtinen, S.; Læssøe, T.; Petersen, J.; Hansen, K. Phylogenetic origins and family classification of typhuloid fungi, with emphasis on *Ceratellopsis*, *Macrotyphula* and *Typhula* (*Basidiomycota*). *Stud. Mycol.* **2020**, *96*, 155–184. [CrossRef]
19. Vizzini, A.; Consiglio, G.; Marchetti, M.; Borovička, J.; Campo, E.; Cooper, J.; Lebeuf, R.; Ševčíková, H. New data in Porotheleaceae and Cyphellaceae: Epitypification of *Prunulus scabripes* Murrill, the status of *Mycopan* Redhead, Moncalvo & Vilgalys and a new combination in *Pleurella* Horak emend. *Mycol. Prog.* **2022**, *21*, 44. [CrossRef]
20. Vizzini, A.; Picillo, B.; Luigi, P.; Dovana, F. *Chrysomycena perplexa* gen. et sp. nov. (Agaricales, Porotheleaceae), a new entity from the Lazio region (Italy). *Riv. Micol. Romana* **2019**, *107*, 96–107.
21. Sanchez-Garcia, M.; Matheny, P. Is the switch to an ectomycorrhizal state an evolutionary key innovation in mushroom-forming fungi? A case study in the Tricholomatineae (Agaricales). *Evolution* **2016**, *71*, 51–65. [CrossRef] [PubMed]
22. Vizzini, A. Segnalazioni di muscinputa laevis (Basidiomycota, Agaricomycetes) per il Nord Italia. *Micol. Veg. Mediterr.* **2010**, *25*, 141–148.
23. Redhead, S.; Norvell, L.L. Notes on *Bondarzewia*, *Heterobasidion* and *Pleurogala*, new genus. *Mycotaxon* **1993**, *48*, 371–380.
24. Moncalvo, J.-M.; Lutzoni, F.; Rehner, S.; Johnson, J.; Vilgalys, R. Phylogenetic Relationships of Agaric Fungi Based on Nuclear Large Subunit Ribosomal DNA Sequences. *Syst. Biol.* **2000**, *49*, 278–305. [CrossRef]
25. Lamoure, D. Agaricales de la zone alpine. Genre *Omphalina* (1ère partie). *Trav. Sci. Parc. Natl. Vanoise* **1974**, *5*, 149–164.
26. Lamoure, D. Alpine and circumpolar *Omphalina* species. In *Arctic and Alpine Mycology 1*; Laursen, G.A., Ammirati, J.F., Eds.; University of Washington Press: Seattle, WA, USA, 1982; pp. 201–215.
27. Bigelow, H.E. The *Clitocybe pyxidata* group. In *Travaux Mycologiques Dédiés à R. Kühner, Numéro spécial du Bulletin de la Société Linnéenne de Lyon*; Société Linnéenne de Lyon: Lyon, France, 1974; pp. 39–46.
28. Bigelow, H.E. North American species of *Clitocybe*. Part I. *Beih. Nova Hedwigia* **1982**, *72*, 1–280.
29. Bigelow, H.E. North American species of *Clitocybe*. Part II. *Beih. Nova Hedwigia* **1985**, *81*, 281–471.
30. Sanchez-Garcia, M.; Ryberg, M.; Khan, F.; Varga, T.G.; Nagy, L.; Hibbett, D. Fruiting body form, not nutritional mode, is the major driver of diversification in mushroom-forming fungi. *Proc. Natl. Acad. Sci. USA* **2020**, *117*, 32528–32534. [CrossRef]
31. Vizzini, A.; Para, R.; Fontenla, R.; Ghignone, S.; Ercole, E. A preliminary ITS phylogeny of *Melanoleuca* (Agaricales), with special reference to European taxa. *Mycotaxon* **2012**, *118*, 361–381. [CrossRef]
32. Vizzini, A.; Marco, C.; Musumeci, E.; Ercole, E. A new taxon in the Infundibulicybe gibba complex (Basidiomycota, Agaricales, Tricholomataceae) from Sardinia (Italy). *Mycologia* **2010**, *103*, 203–208. [CrossRef]
33. Vizzini, A.; Consiglio, G.; Marchetti, M.; Alvarado, P. Insights into the Tricholomatineae (Agaricales, Agaricomycetes): A new arrangement of Biannulariaceae and Callistosporium, Callistosporiaceae fam. nov., *Xerophorus* stat. nov., and *Pleurocollybia* incorporated into *Callistosporium*. *Fungal Divers.* **2020**, *101*, 211–259. [CrossRef]
34. Liu, D.; Goffinet, B.; Wang, X.Y.; Hur, J.S.; Shi, H.X.; Zhang, Y.Y.; Yang, M.X.; LI, l.J.; Yin, A.C.; Wang, L.S. Another lineage of basidiolichen in China, the genera *Dictyonema* and *Lichenomphalia* (Agaricales: Hygrophoraceae). *Mycosystema* **2018**, *37*, 849–864. [CrossRef]
35. Liu, D.; Goffinet, B.; Ertz, D.; De Kesel, A.; Wang, X.; Hur, J.S.; Shi, H.; Zhang, Y.; Yang, M.; Wang, L. Circumscription and phylogeny of the Lepidostromatales (lichenized Basidiomycota) following discovery of new species from China and Africa. *Mycologia* **2017**, *109*, 730–748. [CrossRef] [PubMed]
36. Wei, J.C. *The Enumeration of Lichenized Fungi in China*; China Forestry Publishing House: Beijing, China, 2020; pp. 1–606.
37. Largent, D.L. *How to Identify Mushrooms to Genus*; Mad River Press: Eureka, CA, USA, 1986; Volume 1–5.
38. Rogers, S.O.; Bendich, A.J. Extraction of DNA from plant tissues. In *Plant Molecular Biology Manual A6*; Gelvin, S.B., Schilperoort, R.A., Eds.; Kluwer Academic Publishers: Boston, MA, USA, 1988; pp. 1–10.
39. White, T.; Bruns, T.; Lee, S.; Taylor, J.; Innis, M.; Gelfand, D.; Sninsky, J. Amplification and Direct Sequencing of Fungal Ribosomal RNA Genes for Phylogenetics. *PCR Protoc. Guide Methods Appl.* **1990**, *31*, 315–322.
40. Vilgalys, R.; Hester, M. Rapid genetic identification and mapping of enzymatically amplified ribosomal DNA from several *Cryptococcus* species. *J. Bacteriol.* **1990**, *172*, 4238–4246. [CrossRef]

41. Yang, Q.; Cheng, X.; Zhang, T.; Liu, X.; Wei, X. Five new species of the lichen-forming fungal genus *Peltula* from China. *J. Fungi* **2022**, *8*, 134. [CrossRef]
42. Swindell, S.R.; Plasterer, T.N. SEQMAN. In *Sequence Data Analysis Guidebook*; Swindell, S.R., Ed.; Humana Press: Totowa, NJ, USA, 1997; pp. 75–89.
43. Katoh, K.; Asimenos, G.; Toh, H. Multiple alignment of DNA sequences with MAFFT. *Methods Mol. Biol.* **2009**, *537*, 39–64. [CrossRef] [PubMed]
44. Castresana, J. Selection of conserved blocks from multiple alignments for their use in phylogenetic analysis. *Mol. Biol. Evol.* **2000**, *17*, 540–552. [CrossRef]
45. Talavera, G.; Castresana, J. Improvement of phylogenies after removing divergent and ambiguously aligned blocks from protein sequence alignments. *Syst. Biol.* **2007**, *56*, 564–577. [CrossRef]
46. Swofford, D. *PAUP*. Phylogenetic Analysis Using Parsimony (*and Other Methods)*; Version 4.0b10; Sinauer Associates, Inc.: Sunderland, UK.
47. Zhang, T.; Zhang, X.; Yang, Q.; Wei, X. Hidden species diversity was explored in two genera of catapyrenioid lichens (Verrucariaceae, Ascomycota) from the deserts of China. *J. Fungi* **2022**, *8*, 729. [CrossRef]
48. Darriba, D.; Taboada, G.L.; Doallo, R.; Posada, D. jModelTest 2: More models, new heuristics and parallel computing. *Nat. Methods* **2012**, *9*, 772. [CrossRef] [PubMed]
49. Stamatakis, A. RAxML Version 8: A Tool for Phylogenetic Analysis and Post-Analysis of Large Phylogenies. *Bioinformatics* **2014**, *30*, 1312–1313. [CrossRef]
50. Ronquist, F.; Teslenko, M.; van der Mark, P.; Ayres, D.L.; Darling, A.; Höhna, S.; Larget, B.; Liu, L.; Suchard, M.A.; Huelsenbeck, J.P. MrBayes 3.2: Efficient Bayesian phylogenetic inference and model choice across a large model space. *Syst. Biol.* **2012**, *61*, 539–542. [CrossRef]
51. Huelsenbeck, J.P.; Ronquist, F. MRBAYES: Bayesian inference of phylogenetic trees. *Bioinformatics* **2001**, *17*, 754–755. [CrossRef] [PubMed]
52. Rambaut, A.; Drummond, A.J.; Xie, D.; Baele, G.; Suchard, M.A. Posterior Summarization in Bayesian Phylogenetics Using Tracer 1.7. *Syst. Biol.* **2018**, *67*, 901–904. [CrossRef] [PubMed]
53. Vizzini, A.; Curti, M.; Contu, M.; Ercole, E. A new cystidiate variety of *Omphalina pyxidata* (Basidiomycota, tricholomatoid clade) from Italy. *Mycotaxon* **2012**, *120*, 361–371. [CrossRef]
54. Vesterholt, J. Funga Nordica. Agaricoid, boletoid, cyphelloid and gasteroid genera. In *Funga Nordica*; Knudsen, H., Vesterholt, J., Eds.; Nordsvamp: Copenhagen, Denmark, 2012; pp. 485–487.
55. Kutschera, U.; Niklas, K.J. Endosymbiosis, cell evolution, and speciation. *Theory Biosci.* **2005**, *124*, 1–24. [CrossRef]
56. Masumoto, H.; Ohmura, Y.; Degawa, Y. *Lichenomphalia meridionalis* (Hygrophoraceae, lichenized Basidiomycota) new to Asia. *Opusc. Philolichenum* **2019**, *18*, 379–389.
57. Voitk, A.; Saar, I.; Moncada, B.; Lickey, E. Circumscription and typification of sphagnicolous omphalinoid species of *Arrhenia* (Hygrophoraceae) in Newfoundland and Labrador: Three obligate and one facultative species. *Mycol. Prog.* **2022**, *21*, 57. [CrossRef]
58. Oberwinkler, F. Basidiolichens. In *Fungal Associations*; Hock, B., Ed.; Springer: Berlin/Heidelberg, Germany, 2001; pp. 211–225.
59. Gulden, G.; Senn-Irlet, B.; Jenssen, K.M.; Stordal, J. *Arctic and Alpine Fungi Volume 3: Agarics of the Swiss Alps*; Soppkonsulenten: Drammen, Norway, 1990; pp. 1–58.
60. Courtecuisse, R.; Duhem, B. *Guide des Champignons de France et d'Europe*; Delachaux et Niestle: Paris, France, 1994; pp. 1–544.
61. Courtecuisse, R.; Duhem, B. *Les Champignons de France*; Eclectis, La Maison d'édition de la CAMIF: Paris, France, 1994.
62. Lutzoni, F.; Pagel, M.; Reeb, V. Major fungal lineages are derived from lichen symbiotic ancestors. *Nature* **2001**, *411*, 937–940. [CrossRef]
63. Gargas, A.; DePriest, P.; Grube, M.; Tehler, A. Multiple origins of lichen symbioses in fungi suggested by SSU rDNA phylogeny. *Science* **1995**, *268*, 1492–1495. [CrossRef] [PubMed]
64. Honegger, R. Functional aspects of the lichen symbiosis. *Annu. Rev. Plant Physiol. Plant Mol. Biol.* **1991**, *42*, 553–578. [CrossRef]
65. Tschermak-Woess, E. The algal partner. In *Handbook of Lichenology*; Galum, M., Ed.; CRC Press: Boca Raton, FL, USA, 1988; pp. 39–92.

Journal of *Fungi*

Article

Forest Type and Site Conditions Influence the Diversity and Biomass of Edible Macrofungal Species in Ethiopia

Gonfa Kewessa [1,2,3], Tatek Dejene [1,4], Demelash Alem [5], Motuma Tolera [3,4] and Pablo Martín-Pinto [1,*]

1 Sustainable Forest Management Research Institute, University of Valladolid, Avda. Madrid 44, 34071 Palencia, Spain
2 Department of Forestry, Ambo University, Ambo P.O. Box 19, Ethiopia
3 Wondo Genet College of Forestry and Natural Resources, Hawassa University, Shashemene P.O. Box 128, Ethiopia
4 Ethiopian Forestry Development, Addis Ababa P.O. Box 3243, Ethiopia
5 Ethiopian Forest Development Bahir Dar Center, Bahir Dar P.O. Box 12627, Ethiopia
* Correspondence: pmpinto@pvs.uva.es; Tel.: +34-979-108-340

Abstract: Ethiopian forests are rich in valuable types of non-wood forest products, including mushrooms. However, despite their nutritional, economic, and ecological importance, wild edible mushrooms have been given little attention and are rarely documented in Ethiopia. In this study, we assessed mushroom production levels in natural and plantation forests and the influence of climate and environmental variables on mushroom production. Sporocarps were sampled weekly from July to August 2019 at a set of permanent plots (100 m^2) in both forest systems. We analyzed 63 plots to quantify sporocarp species' richness and fresh weight as well as to elucidate the degree of influence of forest types and site conditions, including soil and climate. Morphological analyses were used to identify fungi. In total, we recorded 64 wild edible fungal species belonging to 31 genera and 21 families from the plots established in the natural and plantation forests. A significantly greater total number of edible fungi were collected from natural forests (n = 40 species) than from plantations. Saprotrophs (92.19%) were the dominant guild whereas ectomycorrhizal fungi represented only 6.25% of species. Ecologically and economically important fungal species such as *Agaricus campestroides*, *Tylopilus niger*, *Suillus luteus*, *Tricholoma portentosum*, and *Morchella americana* were collected. The sporocarp yield obtained from plantation forests (2097.57 kg ha^{-1} yr^{-1}) was significantly greater than that obtained from natural forests (731.18 kg ha^{-1} yr^{-1}). The fungal community composition based on sporocarp production was mainly correlated with the organic matter, available phosphorus, and total nitrogen content of the soil, and with the daily minimum temperature during collection. Accordingly, improving edible species' richness and sporocarp production by maintaining ecosystem integrity represents a way of adding economic value to forests and maintaining biological diversity, while providing wood and non-wood forest products; we propose that this approach is imperative for managing Ethiopian forests.

Keywords: edaphic variables; edible mushrooms; natural forests; plantation forests; sporocarp yield

Citation: Kewessa, G.; Dejene, T.; Alem, D.; Tolera, M.; Martín-Pinto, P. Forest Type and Site Conditions Influence the Diversity and Biomass of Edible Macrofungal Species in Ethiopia. *J. Fungi* **2022**, *8*, 1023. https://doi.org/10.3390/jof8101023

Academic Editors: Lei Cai and Cheng Gao

Received: 26 August 2022
Accepted: 24 September 2022
Published: 28 September 2022

1. Introduction

The Ethiopian highlands were once covered with dense natural high forests [1,2]. However, anthropogenic influences associated with the expansion of farming and human habitation have brought about deforestation and forest degradation [2], resulting in dramatic changes in the abundance and extent of native forest types. What remains is highly fragmented and frequently modified by non-native fast-growing trees such as *Eucalyptus* and *Pinus* species [3]. However, an important feature of these natural and plantation forests is their richness in valuable types of non-timber forest products (NTFPs) [4–6]. Various studies have revealed long lists of NWFPs, indicating their importance, contribution, and use by rural communities in Ethiopia [7,8]. If managed and conserved properly, NTFPs could potentially support the livelihoods of rural people by providing a source of income as well as food and medicine [7]. The most important NTFPs include honey, bees-wax, bamboo, coffee, spices, edible plant products such as fruits, seeds, and fodder, medicinal plants, various extractives and flavorings [7,9], and wild edible mushrooms [10,11].

More than 2000 fungi are known to produce edible sporocarps [12]. Edible wild mushrooms are not only a source of nutrition but also can be an important income generator [12,13], helping rural people to reduce their vulnerability to poverty and strengthen their livelihoods by providing a reliable source of revenue in many regions of the world [14–16]. In addition, edible mushrooms also form the basis of many manufactured products, including medicine, and are the focus of a new wave of tourism resulting from recreational programs linked to nature. Furthermore, wild mushrooms in general are important components of biodiversity in forest systems [17], playing an essential ecological role in forest communities. In particular, mycorrhizal associations have significant effects on nutrient and water uptake, growth, and plant survival [18,19], improving soil aeration and porosity [20] and resistance to pathogens [21], and also providing food for many organisms [22]. Saprotrophic fungi are essential for the decomposition of dead matter and, therefore, for nutrient cycling in forest ecosystems [23]. They are also important in the cycling of various elements, such as carbon, nitrogen, and oxygen [17].

There has been a tradition of ethnomycological usage among communities that dwell in Ethiopian forests [10,11,24]. However, despite all the benefits, wild mushrooms are the most neglected of Ethiopia's NTFP resources. To date, there have been few studies of wild mushrooms and they are rarely documented [25,26], which may reflect that forest resource management has been primarily based on the production of wood products [8,9,27]. Wild mushroom formation is influenced by various factors, such as the forest type, host plant, nutritional status of the mycelium, and environmental factors [28–30]. Climatic variables also influence mushroom production because the development of fungal fruit bodies is dependent on soil temperature and the availability of surface water [31,32]. In addition, sporocarp composition is strongly determined by soil nutrients and chemical properties [33,34]. This is particularly the case for saprotrophic fungi, which are more dependent on their respective substrates than mycorrhizal fungi [35]. Forest management can, therefore, play a crucial part in shaping macrofungal communities because it can modify vegetation parameters such as tree density, canopy cover, understory plant communities, and soil conditions [30,33].

The negative impact of plantation forests is a common narrative in Ethiopia. Among these criticisms, the most cited is the lack of plant diversity in plantation forests [36,37]. However, our knowledge and understanding of fungal communities in plantation forests are limited compared to our knowledge of those in natural forests. Such information is essential to encourage forest management practices that include the conservation and production of valuable mushrooms through the adoption of a mycosilvicultural approach [38,39] for plantation and natural forest systems in Ethiopia. However, few studies have assessed the different environmental factors required for the production of wild edible mushrooms in these forest systems. To date, most have been performed at local scales and investigated a limited set of environmental variables. To gain a deeper understanding of the different environmental factors driving fungal communities as well as host plants, we should consider a wide range of variables from broad geographic areas. This should enable the development of strategies to manage different forests in different landscapes, to promote sporocarps production in the country. Thus, the main goal of this study was to assess the diversity of valuable macrofungal species in different areas of the country and compare the edible macrofungal species' richness, sporocarp production, and community assemblages in native mixed forests and non-native fast-growing plantation forests. We hypothesized (1) that forest type is important for structuring macrofungal communities and the production of their sporocarps at small spatial scales in Ethiopian forest ecosystems. We expected that the macrofungal community would differ between the two forest types, resulting in an overall higher diversity value for study sites in natural forests because fungal diversity would be driven mainly by vegetation type, substrate availability, and other environmental variables [40–42]. However, the shorter rotation period of trees in plantation forests would also result in nutrient stress in these forests [43], which would favor only certain macrofungal species, such as ectomycorrhizal species [44]. Thus, we hypothesized that the overall sporocarp biomass produced in plantation forests would be higher than that in natural mixed forests, particularly those of potentially marketable species such as *Suillus luteus*. Differences in environmental variables such as edaphic and climate variables are the main driving forces for edible mushroom communities and, hence, would be important for the maintenance of their local production and diversity [45,46]. However, in the dry Afromontane region, water is scarce for a long period each year, implying that sporocarp development by some species would be limited. Thus, we also expected (2) that climate variables would govern fungal communities more than edaphic variables in our study areas [47]. When we set out, our specific objectives were: (i) to analyze the richness and production of sporocarps according to forest types, and (ii) to assess the influence of edaphoclimatic variables on taxa composition. The information generated from this study should highlight the economic value of edible mushrooms as NTFPs and encourage forest managers to manage these forests sustainably to promote sporocarp development and maintain the biological diversity of Ethiopian forests.

2. Materials and Methods

2.1. Study Sites

The study was conducted in five dry Afromontane areas in Ethiopia: the Taragedam, Alemsaga, Banja, Wondo Genet, and Menagesha Suba forests (Figure 1). Comprehensive descriptions of the forests are provided in Table 1. Some pictures of the forest types where the sporocarps collection was carried out in the study areas are provided in Figures 2 and 3.

Table 1. Comprehensive descriptions of the forests and the study sites.

Forest	Geographical Coordinates	Altitude (m asl)	MAP (mm)	MAT (°C)	Vegetation Types	Reference
Taragedam	12°06′59″–12°07′25″ N and 37°46′14″–37°47′02″ E	2062–2457	1300	20.4	Natural forests	[48]
Alemsaga	11°54′30″–11°56′00″ N and 37°55′00″–37°57′00″ E	2100–2470	1484	16.4	Natural forests	[49]
Banja	10°57′17″–11°03′05″ N and 36°39′09″–36°48′25″ E	1870–2570	1215.3	17.7	Natural forests	[50]
Wondo Genet	7°06′–7°07′ N and 38°37′–38°42′ E	1600–2580	1210	20	Natural forests and *Pinus patula* and *Eucalyptus grandis* plantation forests	[51]
Menagesha Suba	8°56′–9°03′ N and 38°28′–38°36′ E	2200–3385	1100	16	*Pinus radiata* plantation forests	[52]

Note: MAP, mean annual precipitation; MAT, mean annual temperature.

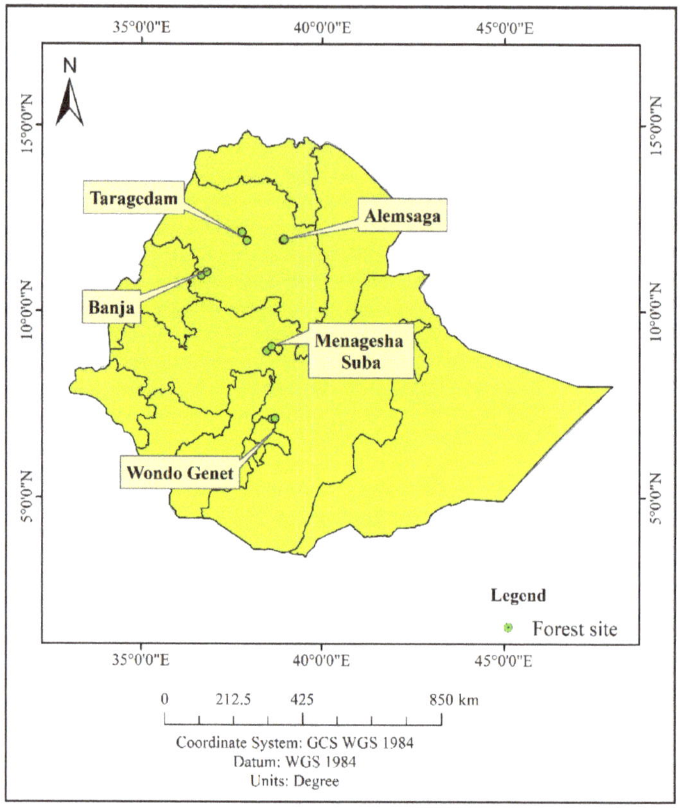

Figure 1. Map showing the locations of forests in which the study plots were located.

Figure 2. Plantation of *Pinus patula* where sporocarps collection was carried out in Wondo Genet (photo credit: Tatek Dejene).

Figure 3. Dry Afromontane forests in (**A**) Wondo Genet, (**B**) Taragedam, (**C**) Alemsaga, and (**D**) Banja where sporocarps collections were conducted (photo credits: 3A, Tatek Dejene; 3B–3D, Demelash Alem).

2.2. Experimental Design and Sporocarp Sampling

In total, 63 sample plots were established at seven sites, nine plots in each of the natural and plantation forests described in [53–55]. The plantation forests are located in the Wondo Genet and Menagesha Suba areas, while the natural forests are found in the Wondo Genet, Taragedam, Banja, and Alemsaga areas. The plantations mainly constitute the *Eucalyptus grandis*, *Pinus patula*, and *Pinus radiata* tree species (Table 1). The main tree species in the dry Afromontane natural forests include *Juniperus procera*, *Podocarpus falcatus*, *Hagenia abyssinica*, and *Olea africana* in the Wondo Genet natural forest area [27]. The dominant tree species in the natural forests of Taragedam, Banja, and Alemsaga, meanwhile, include *Maytenus obscura*, *Carissa edulis*, *Olea* sp., *Acacia abyssinica*, *Buddleja polystachya*, *Acacia nilotica*, *Albizia gummifera*, *Prunus africana*, and *Brucea antidysenterica*. These trees serve as the main sources of timber for the country [27] and thus indicate a need for the sustainable management of these forests. Each plot in each forest type was rectangular in shape (2 m × 50 m) and covered an area of 100 m^2. Within each of the selected forest stands, we studied three different blocks with three plots per block. The plots were established at least 500 m apart and were laid out randomly in the forests to avoid confounding spatial effects inherent to such a plot-based design [56,57] and to reduce environmental heterogeneity [58]. The plots were analyzed as independent samples as suggested by Ruiz-Almenara [59]. All fungal fruit bodies found were harvested weekly during the major rainy season through July and August 2019. Fresh weight measurements were taken in situ using a digital sensitive balance (SF-400) to determine the fruit body production in kilograms per hectare per year. The number of sporocarps of each species in each plot was also recorded. Specimens were photographed in the field, and their morphological features and ecological characteristics were noted to facilitate taxonomic identification in the laboratory [60]. Specimens of each macrofungal species were taken to the laboratory and dried to preserve as herbaria specimens, then used for species identification purposes.

2.3. Species Identification and Characterization

In this study, morphological analyses were used for taxa identification. In the laboratory, fruit body tissues and spores were examined using an Optika B-350PL microscope (Optika, Pontenarica, Italy). Monographs were also used for taxa identification [61–69]. Up-to-date fungal taxa names and authors' names were obtained from the Mycobank database (http://mycobank.org (accessed on 1 November 2020). Trophic levels were assigned to species using the recent classification compiled by Põlme et al. [70]. Fungal species' edibility classification was accomplished by assessing the commercial importance of the collected species [71,72].

2.4. Environmental Data Collection and Analysis

To relate the composition of edible fungal taxa to edaphic variables, soil samples were collected from each of the sample plots established in each forest. After clearing and removing plant matter and debris from the soil surface, five soil cores were extracted from the center and the four corners of each plot using an auger (2 cm radius, 20 cm depth, and 250 cm^3). The soil cores collected from each plot were pooled, and a composite, relatively homogeneous subsample of approximately 500 g from each plot was placed in a plastic bag for analysis. The soil pH was determined by analyzing a soil:water (1:2.5) suspension with the aid of a pH meter [73]. The organic carbon (C) content of the soil was determined using wet digestion [74]. The Kjeldahl procedure was used to determine the total nitrogen (N) content of the soil samples [75]. Sodium bicarbonate (0.5 M NaHCO3) was used as an extraction solution to determine the available phosphorus (P) [76]. The soil analysis was conducted by Water Works Design and Supervision Enterprises, a laboratory service sub-process, the soil fertility section at Addis Ababa, and the Amhara Water Works Design and Supervision Works Enterprise at Bahir Dar, Ethiopia.

In addition to the soil samples, the following climate variables were obtained for each forest from nearby meteorological stations: daily, mean, minimum, and maximum temperatures (°C), total annual precipitation (mm), and the average temperature (°C) and precipitation (mm) values for July and August of 2019 (i.e., the sporocarp collection season).

2.5. Data Analysis

Data were transformed when necessary to achieve the parametric criteria of normality and homoscedasticity. Macrofungal data were normalized by rarefying the abundance data to the smallest number of macrofungi per plot. In addition, data from environmental variables were scaled using base R and used for subsequent statistical analyses. The sporocarp biomass (kg ha^{-1} yr^{-1}) was estimated for each forest. Differences in sporocarp production levels across forests were assessed using linear mixed-effects models (LME) [77], where a block (a set of plots at the same site in each forest) was defined as random and the forest was defined as a fixed factor. LME analyses were used to prevent false positive associations due to the relatedness structure in the sampling. Tukey's test was later used to check for significant differences ($p \leq 0.05$) between forests when needed.

Relationships between sporocarp composition and environmental parameters were visualized using non-metric multidimensional scaling (NMDS) based on an absence and presence species data matrix and environmental scaled data. A permutation-based non-parametric MANOVA (PerMANOVA) [78] using the Euclidean distance was performed to analyze differences in sporocarp communities between forest types and across the five sites. Isolines were also plotted on the NMDS ordinations for rainfall using the ordisurf function. Correlations between NMDS axes scores with explanatory variables were assessed using the envfit function in R. To assess the influence of edaphic, climate, and location variables on the fungal community, we performed a Mantel test (Bray–Curtis distance) on the total species matrix and scaled environmental parameters.

3. Results

3.1. Edible Fungal Richness

In total, 64 edible fungal species belonging to 31 genera and 21 families were identified from the plots established in the natural and plantation forests. All these species belonged to the Basidiomycota, except for *Morchella* spp., which belonged to the Ascomycota. There was a significant difference in the total number of edible species found in each of the forest types ($F = 7.23$, $p = 0.002$). Forty edible species belonging to 22 genera and 15 families were found in the natural forests, while 16 species belonging to 16 genera and 12 families were found in plantation forests (Table 2). Eight edible species were common to both forest types.

Saprotrophs were the dominant guild (92.19%; n = 59 species) followed by ectomycorrhizal fungi (6.25%; n = 4 species), with other guilds comprising 1.56% of the species (Table 2). Ecologically and economically important edible fungal species such as *Agaricus campestroides*, *Agaricus subedulis*, *Tylopilus niger*, *Suillus luteus*, *Tricholoma portentosum*, *Tricholoma saponaceum*, *Morchella americana*, and *Morchella anatolica* were collected (Table 2).

Table 2. List of wild edible fungal species collected from study sites in natural and plantation forests, Ethiopia.

Species	Family	T	N	P
Agaricus augustus Fr.	Agaricaceae	S	x	
Agaricus bitorquis (Quél.) Sacc.	Agaricaceae	S	x	
Agaricus campestris L.	Agaricaceae	S	x	
Agaricus campestroides Heinem. and Gooss. -Font.	Agaricaceae	S	x	x
Agaricus moelleri Wasser	Agaricaceae	S	x	
Agaricus murinaceus Bull.	Agaricaceae	S	x	
Agaricus subedulis Heinem.	Agaricaceae	S	x	x
Ampulloclitocybe clavipes (Pers.) Redhead, Lutzoni, Moncalvo and Vilgalys	Hygrophoraceae	S	x	
Armillaria heimii Pegler	Physalacriaceae	P	x	
Auricularia auricula-judae (Bull.) Quél.	Auriculariaceae	S	x	

Table 2. *Cont.*

Species	Family	T	N	P
Calvatia cyathiformis (Bosc) Morgan	Lycoperdaceae	S	x	
Calvatia gigantea (Batsch) Lloyd	Lycoperdaceae	S	x	
Calvatia subtomentosa Dissing and M. Lange	Lycoperdaceae	S		x
Clitocybe carolinensis H.E. Bigelow and Hesler	Tricholomataceae	S	x	
Clitocybe cistophila Bon and Contu	Tricholomataceae	S	x	
Clitocybe foetens Melot	Tricholomataceae	S	x	
Clitocybe fragrans (With.) P. Kumm.	Tricholomataceae	S	x	
Clitocybe geotropa (Bull. ex DC.) Quél.	Tricholomataceae	S	x	
Coprinellus domesticus (Bolton) Vilgalys, Hopple and Jacq. Johnson	Psathyrellaceae	S	x	x
Coprinopsis nivea (Pers.) Redhead, Vilgalys and Moncalvo	Psathyrellaceae	S	x	
Coprinus comatus (O.F. Müll.) Pers.	Agaricaceae	S	x	
Coprinus lagopus (Fr.) Fr.	Psathyrellaceae	S	x	
Coprinus micaceus (Bull.) Fr.	Psathyrellaceae	S	x	
Coprinus niveus (Pers.) Fr.	Psathyrellaceae	S	x	
Craterellus ignicolor (R.H. Petersen) Dahlman, Danell and Spatafora	Hydnaceae	S	x	
Crepidotus applanatus (Pers.) P. Kumm.	Crepidotaceae	S	x	
Crepidotus mollis (Schaeff.) Staude	Crepidotaceae	S	x	
Gymnopilus pampeanus (Speg.) Singer	Strophariaceae	S		x
Hygrocybe chlorophana (Fr.) Wünsche	Hygrophoraceae	S	x	
Hygrophoropsis aurantiaca (Wulfen) Maire	Hygrophoropsidaceae	S	x	x
Laetiporus sulphureus (Bull.) Murrill	Laetiporaceae	S	x	
Lentinellus cochleatus (Pers.) P. Karst.	Auriscalpiaceae	S	x	
Lepista sordida (Schumach.) Singer	Tricholomataceae	S		x
Lepista sordida var. lilacea (Quél.) Bon	Tricholomataceae	S		x
Leucoagaricus americanus (Peck) Vellinga	Agaricaceae	S	x	
Leucoagaricus holosericeus (J.J. Planer) M.M. Moser	Agaricaceae	S	x	x
Leucoagaricus leucothites (Vittad.) Wasser	Agaricaceae	S	x	x
Leucoagaricus purpureolilacinus Huijsman	Agaricaceae	S	x	
Leucoagaricus rubrotinctus (Peck) Singer	Agaricaceae	S	x	x
Leucocoprinus birnbaumii (Corda) Singer	Agaricaceae	S	x	
Leucocoprinus cepistipes (Sowerby) Pat.	Agaricaceae	S	x	x
Lycoperdon perlatum Pers.	Lycoperdaceae	S		x
Lycoperdon umbrinum Pers.	Lycoperdaceae	S		x
Macrolepiota africana (R. Heim) Heinem.	Agaricaceae	S		x
Macrolepiota procera (Scop.) Singer	Agaricaceae	S	x	
Morchella cf americana Clowez and C. Matherly	Morchellaceae	S		x
Morchella anatolica Isiloglu, Spooner, Alli and Solak	Morchellaceae	S		x
Omphalotus illudens (Schwein.) Bresinsky and Besl	Omphalotaceae	S		x
Pholiota aurivella (Batsch) P. Kumm.	Strophariaceae	S	x	
Pleurotus luteoalbus Beeli	Pleurotaceae	S	x	
Pleurotus populinus O. Hilber and O.K. Mill.	Pleurotaceae	S	x	
Pleurotus pulmonarius (Fr.) Quél.	Pleurotaceae	S	x	
Polyporus badius (Pers.) Schwein.	Polyporaceae	S		x
Polyporus tenuiculus (P. Beauv.) Fr.	Polyporaceae	S		x
Polyporus tuberaster (Jacq. ex Pers.) Fr.	Polyporaceae	S		x
Schizophyllum commune Fr.	Schizophyllaceae	S		x
Suillus luteus (L.) Roussel	Suillaceae	EM		x
Termitomyces clypeatus R. Heim	Lyophyllaceae	S	x	
Termitomyces microcarpus (Berk. and Broome) R. Heim	Lyophyllaceae	S	x	
Termitomyces robustus (Beeli) R. Heim	Lyophyllaceae	S	x	
Termitomyces schimperi (Pat.) R. Heim	Lyophyllaceae	S	x	
Tricholoma portentosum (Fr.) Quél.	Tricholomataceae	EM	x	
Tricholoma saponaceum (Fr.) P. Kumm.	Tricholomataceae	EM	x	
Tylopilus niger (Heinem. and Gooss. -Font.) Wolfe	Boletaceae	EM		x

Note: T, trophic groups; S, saprotrophic; P, parasitic; EM, ectomycorrhizal; N, natural forest; P, plantation forest.

3.2. Sporocarp Production

We found significant differences in the total edible sporocarp production between the two forest types (F = 4.293 $p = 0.04$; Figure 4A), with significantly greater mean sporocarp production levels in plantation forests (2097.57 kg ha^{-1} yr^{-1}) than in natural forests (731.18 kg ha^{-1} yr^{-1}).

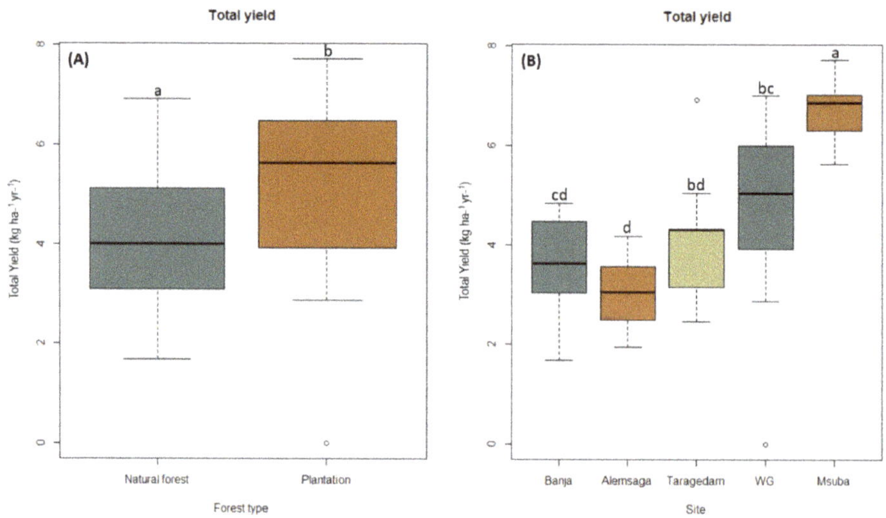

Figure 4. Production of sporocarps at five study sites in Ethiopia according to the forest type (**A**) and site (**B**). Data are presented as boxplots showing the maximum and minimum values. The bar in the box is the standard deviation of the mean. Values with the same letter are not significantly different. Msuba, Menagesha Suba; WG, Wondo Genet.

The average sporocarp production levels of edible species differed significantly among the five studied sites (F = 9.24; $p = 0.0001$), with the highest mean production levels recorded in the Menagesha Suba forest (7.49 kg ha^{-1}). This value was significantly higher than that of the Wondo Genet ($p_{\text{-MS}} - p_{\text{-WG}} = 0.004$), Banja ($p_{\text{-MS}} - p_{\text{-Ba}} = 0.0001$), Alemsaga ($p_{\text{-MS}} - p_{\text{-Al}} = 0.000$), and Taragedam forests ($p_{\text{-MS}} - p_{\text{-TG}} = 0.001$). Mean sporocarp production in Wondo Genet forests (3.29 kg ha^{-1}) was also significantly higher than that in Alemsaga forests ($p_{\text{-WG}} - p_{\text{-Al}} = 0.02$) but was not significantly different from that recorded for the Taragedam ($p_{\text{-WG}} - p_{\text{-TA}} = 0.775$) and Banja forests ($p_{\text{-WG}} - p_{\text{-TG}} = 0.2626$). Mean sporocarp production levels in Alemsaga, Banja, and Taragedam forests did not differ significantly ($p > 0.05$; Figure 3B).

3.3. Sporocarp Composition and Environmental Variables

The perMANOVA analyses indicated that the two forest types differed significantly in their sporocarp composition (F = 5.343, R2 = 0.14, $p = 0.001$; Figure 5A). Explanatory variables categorized as edaphic, climate, and location parameters were correlated with fungal community composition ($p < 0.05$; Table 3). Of these, a Mantel test confirmed that location variables had a significantly stronger aggregate effect on the sporocarp composition of both forest types ($p = 0.000$) than climate ($p = 0.0001$) or edaphic variables ($p = 0.0001$). The significance of each explanatory variable and their aggregated contribution to differences in sporocarp composition are shown in Table 3.

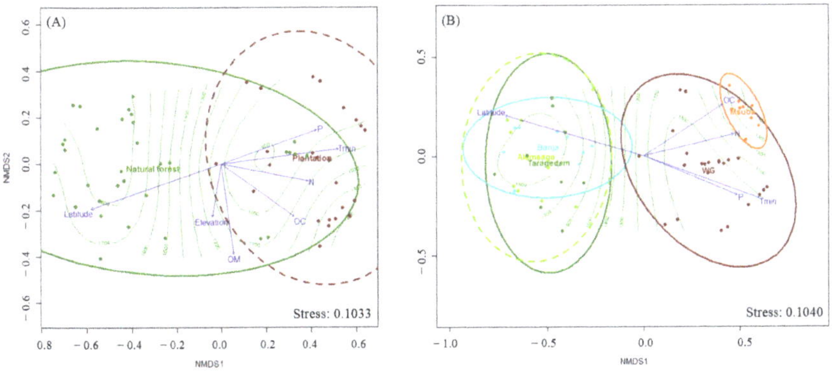

Figure 5. Non − metric multidimensional scaling (NMDS) ordination graph with fitted explanatory variables based on dissimilarities calculated using the Euclidean distance of sporocarp composition from plots in the two forest types (**A**) and by study sites (**B**), with rainfall displayed as isolines (fitted LOESS model R2 = 0.86). Arrows represent the environmental variables that were most significantly ($p < 0.005$) related to ordination. The five ellipses indicate the plots in the five study areas (Msuba, Menagesha Suba; WG, Wondo Genet). Explanatory variables are shown in blue: OM, organic matter; Tmin, minimum daily temperature; N, total nitrogen; P, available phosphorus; OC, organic carbon.

Table 3. Significance of explanatory variables for sporocarp composition in the two forest types. The numbers in bold indicate a highly significant effect of environmental variables ($p < 0.001$).

Sources	Contribution	Variables	Pseudo-F	*p*-Values
Edaphic variables	32.33%	OM	0.2824	0.001
		N	0.3057	0.001
		P	0.4184	0.001
		OC	0.3122	0.001
Climate	60.76%	Elevation	0.0762	0.088
		Tmin	0.5217	0.001
Spatial factors	64.19%	Latitude	0.7777	0.001

Note: OM, organic matter; N, total nitrogen; P, available phosphorus; OC, organic carbon; Tmin, minimum daily temperature.

When sites were analyzed separately, environmental variables such as N, P, OC, Tmin, and latitude were significantly correlated ($p < 0.05$) with the composition of edible sporocarps in the study areas. Of these variables, the Mantel test confirmed that location variables had a significantly stronger aggregate effect on sporocarp composition (r = 0.6419; $p = 0.0001$) than climate (r = 0.4378; $p = 0.0001$) and edaphic variables (r = 0.3166; $p = 0.0001$) when analyzed by site (Figure 4B).

4. Discussion

In the most rural parts of Ethiopia, the local people are dependent on forest resources, either in the form of subsistence or as a cash income derived from NTFPs [7]. The collection of wild edible mushrooms by local people is a common practice, particularly in the southwestern parts of the country [11]. However, wild mushrooms are not considered important sources of food and medicine by rural communities in the northern part of the country [79]. This might be due to the continuing exodus of people from the countryside, which has meant that local communities are gradually losing their traditional knowledge, particularly about wild mushroom species. Furthermore, although a limited number of studies have reported the availability of wild mushrooms in Ethiopia and their importance as sources of food, medicine, and to some extent, income for local communities [10,11], information about the type of wild edible fungal species that are available, their potential production, and their status in different forest systems is scant [80]. This study is the first systematic survey focused on wild edible mushrooms, which was carried out in forests located in central and northern Ethiopia, where remnants of natural forests and plantations of exotic trees exist [81]. We collected a total of 64 edible fungal species from the study sites. The majority of the edible species (n = 40) were collected only from study sites in natural forests.

We collected wild edible species that have both economic and ecological significance belonging to the genera *Morchella*, *Suillus*, and *Tylopilus* in plantation forests and *Tricholoma* in natural forests [82]. In addition, we found some *Agaricus* species and *Termitomyces*, which are known to be used by rural people in the southwest part of the country [26,83], and a *Schizophyllum* species, which is eaten by local people in southern Ethiopia [10,24]. Of these species, *Suillus luteus* is consumed by local people and is also sold in markets at a good commercial price in different developing countries, along with other NTFPs [44]. Furthermore, these kinds of mushrooms could help to sustain communities during periods of food scarcity, serving as an important source of nutrients for local people [82]. The use of wild mushrooms by rural people as a food source during lean periods has been documented in Ethiopian ethnomycological literature. In most cases, these species are collected for subsistence use [84]. However, in some places, mushrooms can provide households with additional income when sold in the markets. For example, in local markets in the southern and southwestern parts of the country, *Agaricus* sp. and *Termitomyces* sp. are available occasionally in association with other vegetal products, which the local people sell to earn some extra money to supplement their household income [83]. Therefore, the conservation and development of these kinds of valuable species deserve special attention given their possible role in increasing food security and income generation to subsidize rural household economies. In addition, as mushroom collection from wild habitats is seasonal, maintaining some of these edible species by means of local small-scale cultivation practices or in private forest areas would be very remunerative. Therefore, a strategy is needed for the adoption and cultivation of important species from the wild, which will not only increase their utilization but also create new sources of income for rural people and contribute to food security.

Although we collected more edible species from plots in natural forests than in plantation forests, sporocarp production levels were higher in plots in plantation forests than in natural forests. The greater number of species but lower sporocarp production levels in natural forests is unsurprising given that almost all the species found were saprotrophic and they were mainly composed of singleton taxa, with a small number of frequent species, which is in agreement with the findings of previous studies [85]. The saprophytic fungal species collected were also characterized by low levels of biomass production, in accordance with Gassibe et al. and Mediavilla et al. [54,86]. Nonetheless, they are relevant for decomposition processes and ecosystem functioning [87], particularly in tropical forest systems such as the natural forests in this study, where decomposition is rapid [88]. This may reflect the accumulation of favorable substrates, which is likely to enhance the richness [89] of these systems. The conspicuous sporocarps produced by these saprophytic fungi may also have favored the collection of this particular fungal class, although basidiomycete mycelia are reported to be everywhere in forests [90]. Overall, the sporocarp yield obtained from plantation forests (2097.57 kg $ha^{-1}yr^{-1}$) was significantly higher than that from natural forests (731.18 kg $ha^{-1}yr^{-1}$). Interestingly, almost 25% of the species collected from plantation forests were marketable species and have economic significance, including *Morchella* sp., *Suillus* sp., and *Tylopilus* sp. [12]. These species were characterized by their high levels of biomass production. For example, in Peru and Mexico, *Suillus* and *Morchella* species are commercial NTFPs produced in plantation forests. They guarantee the economic performance of those forests [44,91] and the livelihoods of local communities [92], thus providing incentives for farmers to plant and manage more plantations in their surroundings. Furthermore, in Mexico, *Morchella* species are also exported to generate income [44]. Although the overall biomass produced by sites in this study was low, the most productive site (Menagesha Suba) had mean production levels per stand of 7.49 kg ha^{-1}, which suggests the potential production levels of edible sporocarp species in forests with similar conditions in this area. This also provides a starting point in terms of broadening the management of forests for the production of NTFPs such as edible mushrooms in Ethiopia, depending on the location and type of forest.

Although natural forests produced lower sporocarp yields than plantation forests and had fewer marketable species, the overall yields and species were still valuably enhanced by plantations of exotic conifer species. A recent study [85] in the northern part of Ethiopia indicated that the overall land connectivity of natural forests with that of plantations provided important ectomycorrhizal species such as *Tricholoma* and *Suillus* in natural forest systems, indicating that such forest management activities could create important microniches with suitable resources and abiotic conditions to support more valuable mushroom species [93]. Thus, enrichment of natural forest systems through planting diverse tree species could potentially offer suitable habitats to enhance the richness and productivity of valuable edible species in natural forests in the study areas.

As we hypothesized, distinct fungal communities were observed in the two forest types. Fungal communities in natural forests were characterized by a large number of species, which may have been due to the greater spatial heterogeneity of the soil in natural forests compared with plantation forests. A heterogeneous soil environment and high rainfall levels create microhabitats in which saprotrophic species should be able to find the resources they require to survive in natural forests. The vegetation composition also impacts the composition of edible fungal species via the quantity and quality of the organic inputs, which mainly affect the saprotrophic community structure [94]. Most of the species found in the natural forests were associated with litter decomposition, which is typically characteristic of tropical forests [42]. However, some specific species such as *Termitomyces* and *Tricholoma* species were exclusively found in natural forests. The genus *Termitomyces* comprises a group of gilled mushrooms that have formed a termitophilic association with a particular family of termites, the Macrotermitinae (Isoptera), which are commonly found in Africa in places with a dry and humid climate [95]. This might be because our sampling sites in the natural forest areas were generally classified as dry Afromontane forest areas, characterized by high humidity and prolonged dry seasons [1]. These conditions might favor the occurrence and formation of distinct fungal community compositions in natural forest systems. Distinct fungal community compositions were also observed for plantation forests. Ectomycorrhizal fungi characterized the fungal composition of plantation forests along with some saprotrophic species. *Tylopilus* and *Suillus* species were site-exclusive species that were significantly more abundant in plantation forests comprised mainly *Pinus* than in other plantation forests. Some mushroom species, such as *Agaricus campestroides*, *Coprinellus domesticus*, *Leucoagaricus holosericeus*, *Hygrophoropsis aurantiaca*, and *Leucocoprinus cepistipes*, were common in both forest systems, indicating that these genera might be characterized as generalists.

Studies have shown that different environmental variables govern the composition of fungal species and that different fungal taxa are likely to respond to edaphic variables in different ways, depending on their characteristics [96,97] and, in turn, the composition of fungal communities is directly correlated with soil parameters [98]. In this study, organic matter, P, and N were significantly correlated with the whole edible fungal species community dataset. This is likely to be because organic matter influences the fungal community through its impact on the water-holding capacity of the soil and on nutrient availability [99]. Thus, organic matter may favor more fungal assembly in an area, particularly saprotrophic fungi. Furthermore, the finding that N was an important factor correlated with fungal taxa compositions is in accordance with previous studies [28,35,100] that noted the influence of N on fungal distribution patterns. These studies reported that fungi showed community specialization toward more N-rich soil sites. This might be because nitrogen can influence the formation of extraradical mycelium in the soil and play a vital role in sporocarp formation [101]. Other studies have also noted that fungal communities adapt to more nitrogen-rich sites [100,102]. Furthermore, the microclimate directly influences ecological processes and reflects subtle changes in ecosystem functioning, particularly in forests where the majority of the identified edible species are saprotrophic and depend on a suitable microclimate for their growth and production. In that context, a mosaic forest management scheme is needed that considers both timber production and edible wild mushroom production. Such as scheme must uphold the environment variables needed to create suitable habitats, with variable microclimates to promote diverse sporocarps. This could increase the value of Ethiopia's remnant natural forests and provide incentives for forest owners to sustainably manage and conserve the forests' resources in different forms.

5. Conclusions

Although our data were based on only one year of sampling during the rainy season in July and August 2019, we conclude that the richness of edible mushrooms is relatively higher in natural forests. The majority of these species are saprophytic and characterized by the production of small sporocarps. Although only a small number of edible species were recorded in plantation forests, sporocarp production levels in these forests were high. We also observed that some species specifically characterized each forest, and there was also a noticeable presence of valuable species in both forest types that could be potentially marketed in rural areas, providing forest managers and local people with supplementary incomes. Species such as *Termitomyces* and *Tricholoma* were exclusively found in natural forests, while *Suillus luteus* was found in plantation forests. However, environmental variables such as climate, spatial parameters, and edaphic parameters were found to affect the composition and, thus, sporocarp production in the study areas. In this regard, we found that the relative contributions of the spatial and climatic parameters were greater than those of the edaphic-specific variables, which significantly affected sporocarp production and species composition. However, a mosaic landscape, mixing natural and plantation forests, could provide timber and high levels of edible mushroom production. Accordingly, this could contribute to the conservation of remnant natural forests, which have high biodiversity values. Thus, a management scheme is needed that combines timber and mushroom production, which could provide economic and ecological benefits, especially in natural forest systems, which are characterized by excessive deforestation and forest degradation. Furthermore, the scheme should consider the influence of different environmental conditions, to create habitats that offer a suitable environment with variable microclimates for the promotion of diverse edible fungi and higher levels of sporocarp production, particularly in natural forests where the majority of the identified edible species were characterized by low levels of sporocarp production.

Author Contributions: Conceptualization, P.M.-P.; methodology, T.D., G.K. and P.M.-P.; formal analysis, P.M.-P., T.D. and G.K.; investigation, G.K., D.A. and T.D.; data curation, G.K. and D.A.; writing—original draft preparation, G.K.; writing—review and editing, P.M.-P., T.D. and M.T.; supervision, P.M.-P., T.D. and M.T.; project administration, P.M.-P.; funding acquisition, P.M.-P. All authors have read and agreed to the published version of the manuscript.

Funding: This work was supported by the projects SUSTIFUNGI_ET (Sustfungi_Eth: 2017/ACDE/002094) and MYCOPROED_ET (Mycoproed_Eth: 2019/ACDE/000921) and SUSTFUNGI_ET_III (Sustfungi_Eth_III: 2022/ACDE/000201), which were funded by the Spanish Agency for International Development and Cooperation. Gonfa Kewessa also received funds through the UVa 2021 call for predoctoral contracts, co-financed by Banco Santander.

Institutional Review Board Statement: Not applicable.

Informed Consent Statement: Not applicable.

Data Availability Statement: Not applicable.

Acknowledgments: We would like to express our gratitude to the people involved in the fieldwork.

Conflicts of Interest: The authors declare no conflict of interest.

References

1. Friis, B.; Demissew, S.; Breugel, P. *Atlas of the Potential Vegetation of Ethiopia*; The Royal Danish Academy of Sciences and Letters: Copenhagen, Denmark, 2010.
2. Badege, B. Deforestation and Land Degradation in the Ethiopian Highlands: A Strategy for Physical Recovery. *Northeast Afr. Stud.* **2001**, *8*, 7–26.
3. Bekele, M. Forest Plantations and Woodlots in Ethiopia. *Afr. For. forum* **2011**, *1*, 52.
4. Adilo, M. The Contribution of Non-Timber Forest Products to Rural Livelihood in Southwest Ethiopia. Master's Thesis, Wageningen University, Wageningen, The Netherlands, 2007.
5. Seyoum, A. Economic value of Afromontane Forests of Sheka: Forest of Sheka. In *Forest of Sheka*; Fetene, M., Ed.; MEHBER: Addis Ababa, Ethiopia, 2007; pp. 183–218.
6. Sultan, M. The Role of Non-Timber Forest Products to Rural Livelihoods and Forest Conservation: A Case Study at Harana Bulluk District Oromia National Regional State, Ethiopia. Master's Thesis, Wondo Genet College of Forestry and Natural Resource, Hawassa University, Wondo Genet, Ethiopia, 2009.
7. Lulekal, E.; Asfaw, Z.; Kelbessa, E.; Van Damme, P. Wild edible plants in Ethiopia: A review on their potential to combat food insecurity. *Afrika Focus* **2011**, *24*, 71–122. [CrossRef]
8. Melaku, E.; Ewnetu, Z.; Teketay, D. Non-timber forest products and household incomes in Bonga forest area, southwestern Ethiopia. *J. For. Res.* **2014**, *25*, 215–223. [CrossRef]
9. Asfaw, Z.; Tadesse, M. Prospects for sustainable use and development of wild food plants in Ethiopia. *Econ. Bot.* **2001**, *55*, 47–62. [CrossRef]
10. Sitotaw, R.; Lulekal, E.; Abate, D. Ethnomycological study of edible and medicinal mushrooms in Menge District, Asossa Zone, Benshangul Gumuz Region, Ethiopia. *J. Ethnobiol. Ethnomed.* **2020**, *16*, 11. [CrossRef]
11. Dejene, T.; Oria-de-Rueda, J.A.; Martín-Pinto, P. Edible wild mushrooms of Ethiopia: Neglected non-timber forest products. *Rev. Fitotec. Mex.* **2017**, *40*, 391–397. [CrossRef]
12. Boa, E. *Wild Edible Fungi: A Global Overview of Their Use and Importance to People*; FAO: Rome, Italy, 2004; ISBN 9251051577.
13. Barroetaveña, C.; La Manna, L.; Alonso, M.V. Variables affecting Suillus luteus fructification in ponderosa pine plantations of Patagonia (Argentina). *For. Ecol. Manag.* **2008**, *256*, 1868–1874. [CrossRef]
14. De Román, M.; Boa, E. International Seminar on Ethnomycology Collection, Marketing and Cultivation of Edible Fungi in Spain. *Micol. Apl. Int.* **2004**, *16*, 25–33.
15. Chang, Y.S.; Lee, S.S. Utilisation of macrofungi species in Malaysia. *Fungal Divers* **2004**, *15*, 15–22.
16. Mau, J.; Chang, C.; Huang, S.; Chen, C. Antioxidant properties of methanolic extracts from Grifola frondosa, Morchella esculenta and Termitomyces albuminosus mycelia. *Food Chem.* **2004**, *87*, 111–118. [CrossRef]
17. Chen, Y.; Yuan, Z.; Bi, S.; Wang, X.; Ye, Y.; Svenning, J.-C. Macrofungal species distributions depend on habitat partitioning of topography, light, and vegetation in a temperate mountain forest. *Sci. Rep.* **2018**, *8*, 13589. [CrossRef]
18. Brundrett, M.C. Mycorrhizal associations and other means of nutrition of vascular plants: Understanding the global diversity of host plants by resolving conflicting information and developing reliable means of diagnosis. *Plant Soil* **2009**, *320*, 37–77. [CrossRef]
19. Courty, P.-E.; Buée, M.; Diedhiou, A.G.; Frey-Klett, P.; Le Tacon, F.; Rineau, F.; Turpault, M.-P.; Uroz, S.; Garbaye, J. The role of ectomycorrhizal communities in forest ecosystem processes: New perspectives and emerging concepts. *Soil Biol. Biochem.* **2010**, *42*, 679–698. [CrossRef]
20. Fernández-Toirán, L.; Agreda, T.; Olano, J. Stand age and sampling year effect on the fungal fruit body community in Pinus pinaster forests in central Spain. *Can. J. Bot.* **2006**, *84*, 1249–1258. [CrossRef]
21. Martín-Pinto, P.; Vaquerizo, H.; Peñalver, F.; Olaizola, J.; Oria-De-Rueda, J.A. Early effects of a wildfire on the diversity and production of fungal communities in Mediterranean vegetation types dominated by *Cistus ladanifer* and Pinus pinaster in Spain. *For. Ecol. Manag.* **2006**, *225*, 296–305. [CrossRef]
22. Yanagihara, H.; Kamo, K.; Imori, S.; Satoh, K. Bias-corrected AIC for selecting variables in multinomial logistic regression models. *Linear Algebra Appl.* **2012**, *436*, 4329–4341. [CrossRef]
23. Ferris, R.; Peace, A.J.; Newton, A.C. Macrofungal communities of lowland Scots pine (*Pinus sylvestris* L.) and Norway spruce (*Picea abies* (L.) Karsten.) plantations in England: Relationships with site factors and stand structure. *For. Ecol. Manag.* **2000**, *131*, 255–267. [CrossRef]
24. Tuno, N. Mushroom utilization by the Majangir, an Ethiopian tribe. *Mycologist* **2001**, *15*, 78–79. [CrossRef]
25. Alemu, F. Assessment of wild mushrooms and wood decaying fungi in Dilla University, main campus, Ethiopia. *Int. J. Adv. Res.* **2013**, *1*, 458–467.
26. Muleta, D.; Woyessa, D.; Teferi, Y. Mushroom consumption habits of Wacha Kebele residents, southwestern Ethiopia. *Glob. Res. J. Agric. Biol. Sci.* **2013**, *4*, 6–16.
27. Kassa, H.; Campbell, B.; Sandewall, M.; Kebede, M.; Tesfaye, Y.; Dessie, G.; Seifu, A.; Tadesse, M.; Garedew, E.; Sandewall, K. Building future scenarios and uncovering persisting challenges of participatory forest management in Chilimo Forest, Central Ethiopia. *J. Environ. Manag.* **2009**, *90*, 1004–1013. [CrossRef] [PubMed]
28. Gassibe, P.V.; Oria-de-Rueda, J.A.; Martín-Pinto, P.P. pinaster under extreme ecological conditions provides high fungal production and diversity. *For. Ecol. Manag.* **2015**, *337*, 161–173. [CrossRef]

29. Bonet, J.A.; Fischer, C.R.; Colinas, C. The relationship between forest age and aspect on the production of sporocarps of ectomycorrhizal fungi in *Pinus sylvestris* forests of the central Pyrenees. *For. Ecol. Manag.* **2004**, *203*, 157–175. [CrossRef]
30. Smith, J.E.; Molina, R.; Huso, M.M.; Luoma, D.L.; McKay, D.; Castellano, M.A.; Lebel, T.; Valachovic, Y. Species richness, abundance, and composition of hypogeous and epigeous ectomycorrhizal fungal sporocarps in young, rotation-age, and old-growth stands of Douglas-fir (*Pseudotsuga menziesii*) in the Cascade Range of Oregon, U.S.A. *Can. J. Bot.* **2002**, *80*, 186–204. [CrossRef]
31. Bonet, J.A.; Palahí, M.; Colinas, C.; Pukkala, T.; Fischer, C.R.; Miina, J.; Martínez de Aragón, J. Modelling the production and species richness of wild mushrooms in pine forests of the Central Pyrenees in northeastern Spain. *Can. J. For. Res.* **2010**, *40*, 347–356. [CrossRef]
32. Pinna, S.; Gévry, M.F.; Côté, M.; Sirois, L. Factors influencing fructification phenology of edible mushrooms in a boreal mixed forest of Eastern Canada. *For. Ecol. Manag.* **2010**, *260*, 294–301. [CrossRef]
33. Santos-Silva, C.; Gonçalves, A.; Louro, R. Canopy cover influence on macrofungal richness and sporocarp production in montado ecosystems. *Agrofor. Syst.* **2011**, *82*, 149–159. [CrossRef]
34. Straatsma, G.; Ayer, F.; Egli, S. Species richness, abundance, and phenology of fungal fruit bodies over 21 years in a Swiss forest plot. *Mycol. Res.* **2001**, *105*, 515–523. [CrossRef]
35. Reverchon, F.; Del Ortega-Larrocea, P.M.; Pérez-Moreno, J. Saprophytic fungal communities change in diversity and species composition across a volcanic soil chronosequence at Sierra del Chichinautzin, Mexico. *Ann. Microbiol.* **2010**, *60*, 217–226. [CrossRef]
36. Jaleta, D.; Mbilinyi, B.; Mahoo, H.; Lemenih, M. Eucalyptus expansion as relieving and provocative tree in Ethiopia. *J. Agric. Ecol. Res. Int.* **2016**, *6*, 1–12. [CrossRef]
37. Teketay, D. The ecological effects of Eucalyptus: Ground for making wise and informed decision. In Proceedings of the Eucalyptus Dilemma, Addis Ababa, Ethiopia, 15 November 2000.
38. Castellano, M.; Molina, R. Mycorrhizae. In *The Biological Component: Nursery Pests and Mycorrhizal*; Landis, T.D., Tinus, R.W., Mc Donald, S.E., Barnett, J.P., Eds.; The Container Tree Nursery Manual. Agriculture Handbook 674, Forest Service; USDA: Washington, DC, USA, 1989; Volume 5, 171p.
39. Trappe, J. Selection of fungi for ectomycorrhizal inoculation in nurseries. *Annu. Rev. Phytopatho.* **1977**, *15*, 203–222. [CrossRef]
40. Glassman, S.I.; Wang, I.J.; Bruns, T.D. Environmental filtering by pH and soil nutrients drives community assembly in fungi at fine spatial scales. *Mol. Ecol.* **2017**, *26*, 6960–6973. [CrossRef] [PubMed]
41. Li, P.; Li, W.; Dumbrell, A.J.; Liu, M.; Li, G.; Wu, M.; Jiang, C.; Li, Z. Spatial Variation in Soil Fungal Communities across Paddy Fields in Subtropical China. *mSystems* **2020**, *5*, e00704-19. [CrossRef] [PubMed]
42. Tedersoo, L.; Bahram, M.; Polme, S.; Koljalg, U.; Yorou, N.S.; Wijesundera, R.; Ruiz, L.V.; Vasco-Palacios, A.M.; Thu, P.Q.; Suija, A.; et al. Global diversity and geography of soil fungi. *Science* **2014**, *346*, 1256688. [CrossRef]
43. Kranabetter, J.M.; Friesen, J.; Gamiet, S.; Kroeger, P. Ectomycorrhizal mushroom distribution by stand age in western hemlock—Lodgepole pine forests of northwestern British Columbia. *Can. J. For. Res.* **2005**, *35*, 1527–1539. [CrossRef]
44. Pérez-Moreno, J.; Martínez-Reyes, M.; Yescas-Pérez, A.; Delgado-Alvarado, A.; Xoconostle-Cázares, B. Wild mushroom markets in central Mexico and a case study at Ozumba. *Econ. Bot.* **2008**, *62*, 425–436. [CrossRef]
45. Collado, E.; Bonet, J.A.; Camarero, J.J.; Egli, S.; Peter, M.; Salo, K.; Martínez-Peña, F.; Ohenoja, E.; Martín-Pinto, P.; Primicia, I.; et al. Mushroom productivity trends in relation to tree growth and climate across different European forest biomes. *Sci. Total Environ.* **2019**, *689*, 602–615. [CrossRef] [PubMed]
46. Martínez de Aragón, J.; Bonet, J.A.; Fischer, C.R.; Colinas, C. Productivity of ectomycorrhizal and selected edible saprotrophic fungi in pine forests of the pre-Pyrenees mountains, Spain: Predictive equations for forest management of mycological resources. *For. Ecol. Manag.* **2007**, *252*, 239–256. [CrossRef]
47. Newsham, K.K.; Hopkins, D.W.; Carvalhais, L.C.; Fretwell, P.T.; Rushton, S.P.; O'Donnell, A.G.; Dennis, P.G. Relationship between soil fungal diversity and temperature in the maritime Antarctic. *Nat. Clim. Chang.* **2016**, *6*, 182–186. [CrossRef]
48. Worku, M. The Role of Forest Biodiversity Conservation Practices for Tourism Development in a Case of Tara Gedam Monastery, South Gonder Zone, Ethiopia. *J. Ecosyst. Ecography* **2017**, *7*, 2–7. [CrossRef]
49. Esubalew, E.; Giday, K.; Hishe, H.; Goshu, G. Carbon stock of woody species along Altitude gradient in Alemsaga Forest, South Gondar, North Western Ethiopia. *Agric. J. IJOEAR* **2019**, *5*, 13–21.
50. Abere, F.; Belete, Y.; Kefalew, A.; Soromessa, T. Carbon stock of Banja forest in Banja district, Amhara region, Ethiopia: An implication for climate change mitigation. *J. Sustain. For.* **2017**, *36*, 604–622. [CrossRef]
51. Alem, D.; Dejene, T.; Oria-de-Rueda, J.A.; Geml, J.; Castaño, C.; Smith, J.E.; Martín-Pinto, P. Soil fungal communities and succession following wildfire in Ethiopian dry Afromontane forests, a highly diverse underexplored ecosystem. *For. Ecol. Manag.* **2020**, *474*, 118328. [CrossRef]
52. Jemal, A.; Getu, E. Diversity of butterfly communities at different altitudes of Menagesha-suba state forest, Ethiopia. *J. Entomol. Zool. Stud.* **2018**, *6*, 2197–2202.
53. Dejene, T.; Oria-de-Rueda, J.A.; Martín-Pinto, P. Fungal diversity and succession under Eucalyptus grandis plantations in Ethiopia. *For. Ecol. Manag.* **2017**, *405*, 179–187. [CrossRef]

54. Gassibe, P.V.; Fabero, R.F.; Hernández-Rodríguez, M.; Oria-de-Rueda, J.A.; Martín-Pinto, P. Fungal community succession following wildfire in a Mediterranean vegetation type dominated by Pinus pinaster in Northwest Spain. *For. Ecol. Manag.* **2011**, *262*, 655–662. [CrossRef]
55. Hernández-Rodríguez, M.; Oria-de-Rueda, J.A.; Martín-Pinto, P. Post-fire fungal succession in a Mediterranean ecosystem dominated by *Cistus ladanifer* L. *For. Ecol. Manag.* **2013**, *289*, 48–57. [CrossRef]
56. Hiiesalu, I.; Bahram, M.; Tedersoo, L. Plant species richness and productivity determine the diversity of soil fungal guilds in temperate coniferous forest and bog habitats. *Mol. Ecol.* **2017**, *26*, 4846–4858. [CrossRef]
57. Rudolph, S.; Maciá-Vicente, J.G.; Lotz-Winter, H.; Schleuning, M.; Piepenbring, M. Temporal variation of fungal diversity in a mosaic landscape in Germany. *Stud. Mycol.* **2018**, *89*, 95–104. [CrossRef]
58. O'Hanlon, R.; Harrington, T.J. Macrofungal diversity and ecology in four Irish forest types. *Fungal Ecol.* **2012**, *5*, 499–508. [CrossRef]
59. Ruiz-Almenara, C.; Gándara, E.; Gómez-Hernández, M. Comparison of diversity and composition of macrofungal species between intensive mushroom harvesting and non-harvesting areas in Oaxaca, Mexico. *PeerJ* **2019**, *7*, e8325. [CrossRef] [PubMed]
60. Adeniyi, M.; Adeyemi, Y.; Odeyemi, Y.; Odeyemi, O. Ecology, diversity and seasonal distribution of wild mushrooms in a Nigerian tropical forest reserve. *Biodiversitas J. Biol. Divers* **2018**, *19*, 285–295. [CrossRef]
61. Antonin, V. Monograph of Marasmius, Gloiocephala, Palaeocephala and Setulipes in Tropical Africa. *Fungus Fl Trop Afr.* **2007**, *1*, 177.
62. Hama, O.; Maes, E.; Guissou, M.; Ibrahim, D.; Barrage, M.; Parra, L.; Raspe, O.; De Kesel, A. Agaricus subsaharianus, une nouvelle espèce comestible et consomméeau Niger, au Burkina Faso et en Tanzanie. *Crypto. Mycol.* **2010**, *31*, 221–234.
63. Pegler, D.N. Studies on African Agaricales: II *. *Kew Bull.* **1969**, *23*, 219–249. [CrossRef]
64. Morris, B. An annotated check-list of the macrofungi of Malawi. *Kirkia* **1990**, *13*, 323–364.
65. Pegler, D.N. Studies on African Agaricales: 1. *Kew Bull.* **1968**, *21*, 499–533. [CrossRef]
66. Pegler, D.; Rayner, R. A contribution to the Agaric flora of Kenya. *Kew Bull.* **1969**, *23*, 347–412. [CrossRef]
67. Rammeloo, J.; Walleyn, R. The edible fungi of Africa South of the Sahara: A literature survey. *Scr. Bot. Belg.* **1993**, *5*, 62.
68. Ryvarden, L.; Piearce, G.D.; Masuka, A.J. *An Introduction to the Larger Fungi of South Central Africa*; Baobab Books: Harare, Zimbabwe, 1994.
69. Singer, R. Marasmius. *Fl. Icon. Champ. Congo* **1965**, *14*, 253–278.
70. Põlme, S.; Abarenkov, K.; Nilsson, R.H.; Lindahl, B.D.; Clemmensen, K.E.; Kauserud, H.; Nguyen, N.; Kjøller, R.; Bates, S.T.; Baldrian, P.; et al. FungalTraits: A user-friendly traits database of fungi and fungus-like stramenopiles. *Fungal Divers* **2020**, *105*, 1–16. [CrossRef]
71. Gerhardt, E.; Vila, J.; Llimona, X. *Mushroom from Spain and Europe: Manual for Identification*; Omega SA Editions: Barcelona, Spain, 2000; ISBN 84-282-1120-5.
72. Moreno, G.; Manjón, J.L. *Guide to Fungi of the Iberian Peninsula*; Omega SA Editions: Barcelona, Spain, 2010.
73. Reeuwijk, L. *Procedures for Soil Analysis*, 6th ed.; International Soil Reference and Information Centre: Wageningen, The Netherlands, 2002.
74. Walkley, A.; Black, I.A. An examination of the digestion method for determining soil organic matter and a proposed modification of the chromic acid titration method. *Soil Sci.* **1934**, *34*, 29–38. [CrossRef]
75. Kim, J.; Kreller, C.R.; Greenberg, M.M. Preparation and Analysis of Oligonucleotides Containing the C4′-Oxidized Abasic Site and Related Mechanistic Probes. *J. Org. Chem.* **2005**, *70*, 8122–8129. [CrossRef]
76. Kim, H.T. *Soil Sampling, Preparation and Analysis*; CRC Press: Boca Raton, FL, USA, 1996; pp. 139–145.
77. Pinheiro, J.; Bates, D.; DebRoy, S.; Sarkar, D.; Team, R.C. Nlme: Linear and Nonlinear Mixed Effects Models. *R Packag. Version 3.1–126*. 2016. Available online: http://CRAN\T1\textgreater{}R-project.org/package=nlme (accessed on 1 August 2022).
78. Anderson, M.J. A new method for non parametric multivariate analysis of variance. *Austral Ecol.* **2001**, *26*, 32–46. [CrossRef]
79. Zeleke, G.; Dejene, T.; Tadesse, W.; Agúndez, D.; Martín-Pinto, P. Ethnomycological Knowledge of Three Ethnic Groups in Ethiopia. *Forests* **2020**, *11*, 875. [CrossRef]
80. Sitotaw, R.; Mulat, A.; Abate, D. Morphological and molecular studies on Termitomyces species of Menge District, Asossa Zone, Northwest Ethiopia. *Sci. Technol. Arts Res. J.* **2015**, *4*, 49–57. [CrossRef]
81. Faye, M.D.; Weber, J.C.; Abasse, T.A.; Boureima, M.; Larwanou, M.; Bationo, A.B.; Diallo, B.O.; Sigué, H.; Dakouo, J.M.; Samaké, O.; et al. Farmers' preferences for tree functions and species in the west African sahel. *For. Trees Livelihoods* **2011**, *20*, 113–136. [CrossRef]
82. Sundriyal, M.; Sundriyal, R. Wild edible plants of the Sikkim Himalaya: Nutritive values of selected species. *Econ. Bot.* **2001**, *55*, 377–390. [CrossRef]
83. Abate, D. Wild mushrooms and mushroom cultivation efforts in Ethiopia. *WSMBMP Bull.* **2014**, *11*, 1–3.
84. Yehuala, K. Contribution of Small Scale Mushroom Production for Food Security in the Amhara. In Proceedings of the Experience Sharing and Revitalization Workshop of Amhara Region Food Security Network, Bahir Dar, Ethiopia, 26–27 August 2010.
85. Alem, D.; Dejene, T.; Oria-de-Rueda, J.A.; Martín-Pinto, P. Survey of macrofungal diversity and analysis of edaphic factors influencing the fungal community of church forests in Dry Afromontane areas of Northern Ethiopia. *For. Ecol. Manag.* **2021**, *496*, 119391. [CrossRef]

86. Mediavilla, O.; Oria-de-Rueda, J.A.; Martín-Pinto, P. Changes in sporocarp production and vegetation following wildfire in a Mediterranean Forest Ecosystem dominated by Pinus nigra in Northern Spain. *For. Ecol. Manag.* **2014**, *331*, 85–92. [CrossRef]
87. Deacon, J. *Fungal Biology*, 4th ed.; Blackwell Publishing Oxford Press: Malden, MA, USA, 2006.
88. Powers, J.S.; Montgomery, R.A.; Adair, E.C.; Brearley, F.Q.; DeWalt, S.J.; Castanho, C.T.; Chave, J.; Deinert, E.; Ganzhorn, J.U.; Gilbert, M.E.; et al. Decomposition in tropical forests: A pan-tropical study of the effects of litter type, litter placement and mesofaunal exclusion across a precipitation gradient. *J. Ecol.* **2009**, *97*, 801–811. [CrossRef]
89. Wardle, D.A.; Bardgett, R.; Klironomos, J.; Setala, H.; van der Putten, W.; Wall, D. Ecological Linkages Between Aboveground and Belowground Biota. *Science* **2004**, *304*, 1629–1633. [CrossRef]
90. Cairney, J.W.G.; Bastias, B.A. Influences of fire on forest soil fungal communitiesThis article is one of a selection of papers published in the Special Forum on Towards Sustainable Forestry—The Living Soil: Soil Biodiversity and Ecosystem Function. *Can. J. For. Res.* **2007**, *37*, 207–215. [CrossRef]
91. Toivanen, T.; Markkanen, A.; Kotiaho, J.S.; Halme, P. The effect of forest fuel harvesting on the fungal diversity of clear-cuts. *Biomass Bioenergy* **2012**, *39*, 84–93. [CrossRef]
92. Elizabeth, M.-E.; Felipe, R.-S.; Maribel, I.-M. Conocimiento Popular Acerca De La *K'Allampa De Pino* (*Suillus luteus* (L) Roussel) En La Localidad De Alalay, Mizque (Cochabamba, Bolivia): Un Ejemplode Dilogo De Saberes. *Revista Etnobiologia* **2018**, *16*, 76–86.
93. Gómez-Hernández, M.; Williams-Linera, G. Diversity of macromycetes determined by tree species, vegetation structure, and microenvironment in tropical cloud forests in Veracruz, Mexico. *Botany* **2011**, *89*, 203–216. [CrossRef]
94. Read, D.; Perez-Moreno, J. Mycorrhizas and nutrient cycling in ecosystems—A journey towards relevance? *New Phytol.* **2003**, *157*, 475–492. [CrossRef]
95. Kirk, P.; Cannon, P.; Minter, D.; Stalpers, J. *Dictionary of the Fungi*, 10th ed.; The Centre for Agriculture and Bioscience International (CABI): Wallingford, UK, 2008.
96. Crowther, T.W.; Stanton, D.W.G.; Thomas, S.M.; A'Bear, A.D.; Hiscox, J.; Jones, T.H.; Voříšková, J.; Baldrian, P.; Boddy, L. Top-down control of soil fungal community composition by a globally distributed keystone consumer. *Ecology* **2013**, *94*, 2518–2528. [CrossRef]
97. Koide, R.T.; Fernandez, C.; Malcolm, G. Determining place and process: Functional traits of ectomycorrhizal fungi that affect both community structure and ecosystem function. *New Phytol.* **2014**, *201*, 433–439. [CrossRef] [PubMed]
98. Cozzolino, V.; Di Meo, V.; Monda, H.; Spaccini, R.; Piccolo, A. The molecular characteristics of compost affect plant growth, arbuscular mycorrhizal fungi, and soil microbial community composition. *Biol. Fertil. Soils* **2016**, *52*, 15–29. [CrossRef]
99. Harrington, T.J. Relationships between macrofungi and vegetation in the burren. *Biol. Environ.* **2003**, *103*, 147–159. [CrossRef]
100. Kranabetter, J.M.; Durall, D.M.; MacKenzie, W.H. Diversity and species distribution of ectomycorrhizal fungi along productivity gradients of a southern boreal forest. *Mycorrhiza* **2009**, *19*, 99–111. [CrossRef] [PubMed]
101. Trudell, S.A.; Edmonds, R.L. Macrofungus communities correlate with moisture and nitrogen abundance in two old-growth conifer forests, Olympic National Park, Washington, USA. *Can. J. Bot.* **2004**, *82*, 781–800. [CrossRef]
102. Toljander, J.F.; Eberhardt, U.; Toljander, Y.K.; Paul, L.R.; Taylor, A.F.S. Species composition of an ectomycorrhizal fungal community along a local nutrient gradient in a boreal forest. *New Phytol.* **2006**, *170*, 873–884. [CrossRef] [PubMed]

Journal of *Fungi*

MDPI

Article

A Molecular Systematics and Taxonomy Research on *Trechispora* (Hydnodontaceae, Trechisporales): Concentrating on Three New *Trechispora* Species from East Asia

Kaiyue Luo [1,2] and Changlin Zhao [1,2,3,4,*]

1 Yunnan Key Laboratory of Plateau Wetland Conservation, Restoration and Ecological Services, Southwest Forestry University, Kunming 650224, China
2 College of Biodiversity Conservation, Southwest Forestry University, Kunming 650224, China
3 Key Laboratory for Forest Resources Conservation and Utilization in the Southwest Mountains of China, Ministry of Education, Southwest Forestry University, Kunming 650224, China
4 Yunnan Key Laboratory for Fungal Diversity and Green Development, Kunming Institute of Botany, Chinese Academy of Sciences, Kunming 650201, China
* Correspondence: fungi@swfu.edu.cn

Abstract: *Trechispora* are an important genus of wood-inhabiting fungi that have the ability to decompose rotten wood in the forest ecosystem. In this study, we reported three new species of *Trechispora*: *T. murina*, *T. odontioidea*, *T. olivacea* from a subtropical region of Yunnan Province, China. Species descriptions were based on a combination of morphological features and phylogenetic analyses of the ITS and LSU region of nuclear ribosomal DNA. *Trechispora murina* is characterized by the resupinate basidiomata, grandinioid hymenial surface with a greyish tint, monomitic hyphal system and ellipsoid, thick-walled, ornamented basidiospores; *T. odontioidea* has an odontioid hymenial surface with cylindrical to conical, blunt aculei and subglobose to globose, colorless, slightly thick-walled, ornamented basidiospores; *T. olivacea* has a farinaceous hymenial surface with olivaceous tint, basidia clavate and thick-walled, ornamented, broadly ellipsoid to globose basidiospores. Sequences of the ITS and nLSU rDNA markers of the studied samples were generated, and phylogenetic analyses were performed with maximum likelihood, maximum parsimony, and Bayesian inference methods. After a series of phylogenetic analyses, the 5.8S+nLSU dataset was constructed to test the phylogenetic relationship of *Trechispora* with other genera of Hydnodontaceae. The ITS dataset was used to evaluate the phylogenetic relationship of the three new species with other species of *Trechispora*. Using ITS phylogeny, the new species *T. murina* was retrieved as a sister to *T. bambusicola* with moderate supports; *T. odontioidea* formed a single lineage and then grouped with *T. fimbriata* and *T. nivea*; while *T. olivacea* formed a monophyletic lineage with *T. farinacea*, *T. hondurensis*, and *T. mollis*.

Keywords: fungal diversity; morphology; southwest China; subtropical region; wood-inhabiting fungi

Citation: Luo, K.; Zhao, C. A Molecular Systematics and Taxonomy Research on *Trechispora* (Hydnodontaceae, Trechisporales): Concentrating on Three New *Trechispora* Species from East Asia. *J. Fungi* **2022**, *8*, 1020. https://doi.org/10.3390/jof8101020

Academic Editors: Cheng Gao and Lei Cai

Received: 20 August 2022
Accepted: 24 September 2022
Published: 27 September 2022

1. Introduction

Fungi form an essential branch of the tree of life, inferred from the important relationship with animals and plants [1], and it drives the carbon cycling in forest soils, mediate mineral nutrition of plants, and alleviates carbon limitations of other soil organisms as the decomposers and mutualists of plants and animals being the fundamental ecological roles [2]. Inferred from growing on a variety of the boreal, temperate, subtropical, and tropical divers vegetations, wood-inhabiting fungi have a rich diversity [3–13]. *Trechispora* P. Karst. belongs to Trechisporales, a small but strongly supported order in Agaricomycotina [14,15]. *Trechispora* (Hydnodontaceae Jülich) typified by *T. onusta* P. Karst., which is characterized by resupinate to effused basidiomata; a smooth to hydnoid to poroid hymenophore; ampullaceous septa; short cylindric basidia; and smooth to verrucose or aculeate basidiospores [5,16]. Currently, MycoBank and Index Fungorum have registered 121 specific and intraspecific names in

Trechispora. About 60 species are currently accepted in *Trechispora* worldwide [5,17–27], of which 18 species of the genus have been found in China [28–34].

The high phylogenetic diversity on the corticioid Agaricomycetes based on two genes, 5.8S and 28S in which nine taxa of *Trechispora* nested into trechisporoid clade [35]. The molecular systematics suggested that *Trechispora* belonged to Hydnodontaceae and was related to genera *Brevicellicium* K.H. Larss. & Hjortstam, *Porpomyces* Jülich, *Sistotremastrum* J. Erikss., and *Subulicystidium* Parmasto [36], the similar morphological characters of *Trechispora* to these genera are basidiomata resupinate, hyphal system monomitic, cystidia absent [5,37]. The phylogeny of Trechisporales was inferred from a combined ITS-nLSU sequences, which revealed that several related genera *Porpomyces*, *Scytinopogon* Singer, and *Trechispora* grouped closely together and nested within Hydnodontaceae [38].

Based on the ITS and nLSU datasets, the phylogenetic study of *Trechispora* reports two new *Trechispora* species: *T. cyatheae* Ordynets, Langer & K.H. Larss. and *T. echinocristallina* Ordynets, Langer & K.H. Larss., which were found in La Réunion Island [24]. Recently, a new species of *Trechispora* has been reported from North America and China [26,33,34].

During the investigations of the corticioid fungi, Yunnan Province, China, we collected three fungal taxa, which could not be assigned to any described species within Hydnodontaceae. We present morphological and molecular phylogenetic evidence that support them as the three new species in *Trechispora*.

2. Materials and Methods

2.1. Sample Collection and Herbarium Specimen Preparation

Fresh fruiting bodies of the fungi growing on fallen angiosperm branches were collected in 2019 from the Honghe and Wenshan of Yunnan Province, China. The samples were photographed in situ and macroscopic details were recorded. Field photographs were taken by a Jianeng 80D camera (Tokyo, Japan). All photographs were focus-stacked and merged using Helicon Focus Pro 7.7.5 software. Once the macroscopic details were recorded, the specimens were transported to a field station where the specimens were dried on an electronic food dryer at 45 °C. Once dried, the specimens were labeled and sealed in envelopes and plastic bags. The dried specimens were deposited in the herbarium of the Southwest Forestry University (SWFC), Kunming, Yunnan Province, China.

2.2. Morphology

The macromorphological descriptions were based on field notes and photos captured in the field and laboratory. Color, texture, taste and odor of fruit bodies were mostly based on the authors' field trip investigations. Color terminology follows Kornerup and Wanscher [39]. All materials were examined under a Nikon 80i microscope (Nikon Corporation, Tokyo, Japan). Drawings were made with the aid of a drawing tube. The measurements and drawings were made from slide preparations stained with cotton blue (0.1 mg aniline blue dissolved in 60 g pure lactic acid), Melzer's reagent (1.5 g potassium iodide, 0.5 g crystalline iodine, 22 g chloral hydrate, aq. dest. 20 mL), and 5% potassium hydroxide. Spores were measured from the sections of the basidiomata and when presenting spore size data, 5% of the measurements excluded from each end of the range are shown in parentheses [40]. The following abbreviations were used: KOH = 5% potassium hydroxide water solution, CB = cotton clue, CB– = acyanophilous, IKI = Melzer's reagent, IKI– = both inamyloid and indextrinoid, L = means spore length (arithmetic average for all spores), W = means spore width (arithmetic average for all spores), Q = variation in the L/W ratios between the specimens studied, and n = a/b ((a) number of spores were measured in total, coming from (b) number of specimen).

2.3. Molecular Phylogeny

The CTAB rapid plant genome extraction kit-DN14 (Aidlab Biotechnologies Co., Ltd., Beijing, China) was used to obtain genomic DNA from the dried specimens following the manufacturer's instructions [41]. The nuclear ribosomal ITS region was amplified with the primers ITS5 and ITS4 [42]. The nuclear nLSU region was amplified with the primer pairs LR0R and LR7 (http://lutzonilab.org/nuclear-ribosomal-dna/, accessed on 7 June 2019). The PCR procedure used for ITS was as follows: initial denaturation at 95 °C for 3 min, followed by 35 cycles at 94 °C for 40 s, 58 °C for 45 s, and 72 °C for 1 min, and a final extension of 72 °C for 10 min. The PCR procedure used for nLSU was as follows: initial denaturation at 94 °C for 1 min, followed by 35 cycles at 94 °C for 30 s, 48 °C for 1 min, and 72 °C for 1.5 min, and a final extension of 72 °C for 10 min. The PCR products were purified and sequenced at Kunming Tsingke Biological Technology Limited Company (Kunming, Yunnan Province, China). All the newly generated sequences were deposited in NCBI GenBank (https://www.ncbi.nlm.nih.gov/genbank/, accessed on 28 November 2021) (Table 1).

Table 1. List of species, specimens, and GenBank accession numbers of sequences used in this study, the newly generated sequences are in bold fonts.

Species Name	Specimen No.	GenBank Accession No.		References
		ITS	nLSU	
Brevicellicium exile	H (Spirin 8370)	MT002322	MT002338	[43]
B. olivascens	KHL 8571	HE963792	HE963793	[36]
Dextrinocystis calamicola	He 5700	MK204534	MK204547	[38]
Fibrodontia alba	TNMF 24944	NR153983	NG060401	[24]
F. brevidens	Wu 9807-16	KC928276	KC928277	[44]
Litschauerella gladiola	He 3171	MK204555	MK204556	[38]
Luellia cystidiata	JHP 09455	MW371211		Unpublished
Porpomyces mucidus	Dai 12692	KT157833	KT157838	[45]
P. submucidus	Cui 5183	KT152143	KT152145	[45]
Scytinopogon pallescens	He 5192		MK204553	[38]
S. havencampii	DED 8300	KT253946	KT253947	[46]
Sistotremastrum guttuliferum	He 3338	MK204540	MK204552	[38]
S. niveocremeum	CBS 42854	MH857381	MH868921	[47]
S. suecicum	H (Miettinen14550)	MT075860	MT002336	[43]
Sphaerobasidium minutum	KHL 11714	DQ873652	DQ873653	[48]
Subulicystidium brachysporum	KASL 1584b	MH041544	MH041610	[49]
S. cochleum	KHL 11200	MN207036	MN207024	[50]
S. longisporum	Ordynets 00146	MN207039	MN207032	[50]
S. meridense	Hjm 16400	MH041538	MH041604	[49]
Trechispora amianthina	CBS 202.54	MH857292		[47]
T. araneosa	KHL 8570	AF347084		[35]
T. bambusicola	CLZhao 3305	MW544022	MW520172	[33]
T. bispora	CBS 142.63	MH858241		[47]
T. byssinella	UC 2023068	KP814481		Unpublished
T. clancularis	FRDBI 4426619	MW487976		Unpublished
T. cohaerens	HHB 19445	MW740327		Unpublished
T. copiosa	AMO 453	MN701018		[27]
T. confinis	KHL 11197	AY463473	AY586719	[35]
T. daweishanensis	CLZhao 18255	MW302338		[34]
T. echinospora	MA Fungi 82486	JX392853		[36]
T. farinacea	MA Fungi 79474	JX392855	JX392856	[36]
T. fimbriata	CLZhao 9006	MW544025	MW520175	[33]
T. fissurata	CLZhao 4571	MW544027		[33]
T. gelatinosa	AMO 1139	MN701021		[27]
T. havencampii	DED 8300	NR154418		[46]
T. hondurensis	HONDURAS 19-F016	MT571523	MT636540	Unpublished
T. hymenocystis	KHL 8795	AF347090		[35]
T. incisa	GB 0090648	KU747095		Unpublished
T. invisitata	UC 2023088	KP814425		Unpublished
T. kavinioides	KGN 981002	AF347086		[35]
T. mollis	URM 85884	MK514945		[26]
T. mollusca	CBS 43948	MH856428		[47]
T. murina	**CLZhao 11736**	**OL615003**		**Present study**
T. murina	**CLZhao 11752**	**OL615004**	**OL615009**	**Present study**
T. nivea	MA Fungi 74044	JX392832		[36]
T.odontioidea	**CLZhao 17890**	**ON417458**		**Present study**
T.olivacea	**CLZhao 17826**	**ON417457**		**Present study**

Table 1. *Cont.*

Species Name	Specimen No.	GenBank Accession No.		References
		ITS	nLSU	
T. pallescens	FLOR 56186	MK458766		Unpublished
T. papillosa	AMO 713	MN701022		[27]
T. regularis	KHL 10881	AF347087		[35]
T. rigida	URM 85754	MT406381		[26]
T. stellulata	UC 2023096	KP814450		Unpublished
T. stevensonii	MA Fungi 70669	JX392841		[36]
T. subsphaerospora	KHL 8511	AF347080		[35]
T. termitophila	AMO 1169	MN701028		[27]
T. thelephora	1820 AMV	KF937369		[51]
T. torrendii	URM 85886	MK515148		[26]
T. xantha	CLZhao 17781	MW302340		[34]
T. yunnanensis	CLZhao 215	MN654923		[31]
Tubulicium raphidisporum	He 3191	MK204537	MK204545	[38]
T. vermiferum	KHL 8714	—	AY463477	[35]

The sequences and alignment were adjusted manually using AliView version 1.27 [52]. The datasets were aligned with Mesquite version 3.51. The 5.8S+nLSU sequences dataset was used to know the phylogenetic relationship of the three new species in *Trechispora* and related genera, and the ITS dataset was used to evaluate the phylogenetic relationships of the new species with known species of the genus. Sequences of *Porpomyces mucidus* (Pers.) that Jülich and *P. submucidus* F. Wu & C.L. Zhao retrieved from GenBank were used as the outgroup for the 5.8S+nLSU analysis (Figure 1) [34], and sequences of *Fibrodontia alba* that Yurchenko & Sheng H. Wu and *F. brevidens* (Pat.) Hjortstam & Ryvarden retrieved from GenBank were used as the outgroup for the ITS analysis (Figure 2) [24,34].

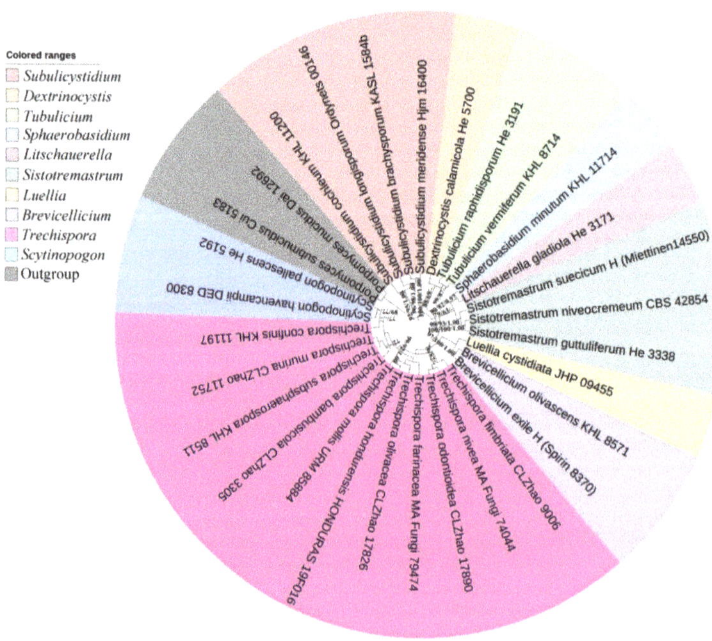

Figure 1. Maximum parsimony strict consensus tree illustrating the phylogeny of *Trechispora* and related genera in Trechisporales based on 5.8S+nLSU sequences. The genera represented by each color are indicated in the upper left of the phylogenetic tree. Branches are labelled with maximum likelihood bootstrap value ≧ 70%, parsimony bootstrap value ≧ 50%, and Bayesian posterior probabilities ≧ 0.95, respectively.

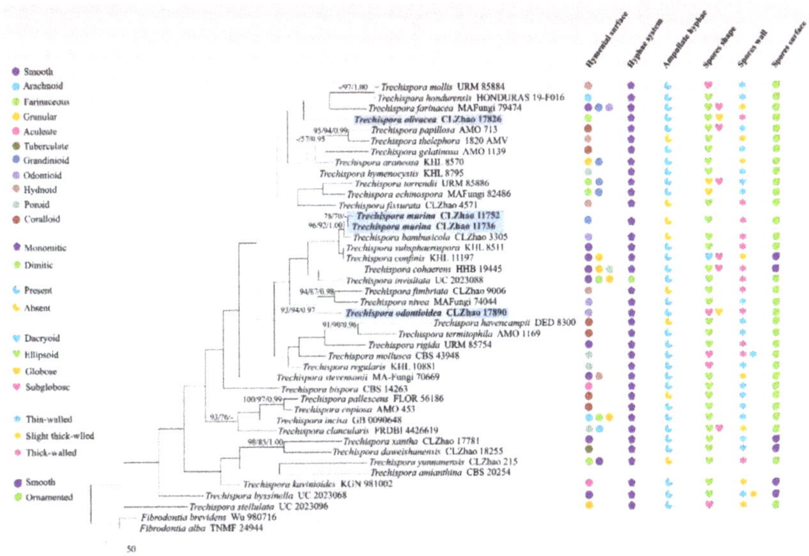

Figure 2. Maximum parsimony strict consensus tree illustrating the phylogeny of three new species and related species in *Trechispora* based on ITS sequences. Branches are labelled with maximum likelihood bootstrap value ≧ 70%, parsimony bootstrap value ≧ 50%, and Bayesian posterior probabilities ≧ 0.95, respectively. The new species are in bold.

The three combined datasets were analyzed using maximum parsimony (MP), maximum likelihood (ML), and Bayesian inference (BI), according to Zhao and Wu [41]. Maximum parsimony analyses were constructed using PAUP* version 4.0b10 [53]. All characters were equally weighted and gaps were treated as missing data. Trees were inferred using the heuristic search option with TBR branch swapping and 1000 random sequence additions. Max trees were set to 5000, branches of zero length were collapsed, and all parsimonious trees were saved. Clade robustness was assessed using bootstrap (BT) analysis with 1000 replicates [54]. Descriptive tree statistics—tree length (TL), consistency index (CI), retention index (RI), rescaled consistency index (RC), and homoplasy index (HI)—were calculated for each maximum parsimonious tree generated. Multiple sequence alignment was also analyzed using ML in RAxML-HPC2 through the Cipres Science Gateway [55]. Branch support (BS) for ML analysis was determined by 1000 bootstrap replicates.

MrModeltest 2.3 [56] was used to determine the best-fit evolution model for each dataset for Bayesian inference (BI), which was performed using MrBayes 3.2.7a with a GTR+I+G model of DNA substitution and a gamma distribution rate variation across sites [57]. A total of 4 Markov chains were run, each consisting of 1.6 million generations, with random starting trees for 5.8S+nLSU (Figure 1) and 1.2 million generations for ITS (Figure 2) with trees and parameters sampled every 1000 generations. The first one-fourth of all generations were discarded as burn-in. The majority rule consensus tree of all remaining trees was calculated. Branches were considered as significantly supported if they received a maximum likelihood bootstrap value (BS) ≥ 70%, maximum parsimony bootstrap value (BT) ≥ 70%, or Bayesian posterior probabilities (BPP) ≥ 0.95.

3. Results

3.1. Molecular Phylogeny

The 5.8S+nLSU dataset (Figure 1) included sequences from 30 fungal samples representing 30 species. The dataset had an aligned length of 1508 characters, of which 1141 characters are constant, 104 are variable and parsimony uninformative, and 263 are parsimony informative. Maximum parsimony analysis yielded 54 equally parsimonious trees (TL = 986, CI = 0.5172, HI = 0.4828, RI = 0.5211, and RC = 0.2695). The best model was GTR+I+G (lset nst = 6, rates = invgamma; prset statefreqpr = dirichlet (1,1,1,1)). Bayesian and ML analyses showed a topology similar to that of MP analysis with split frequencies equal to 0.022581 (BI), and the effective sample size (ESS) across the two runs is double of the average ESS (avg ESS) = 869.5.

The 5.8S+nLSU rDNA gene regions (Figure 1) include ten genera within Trechisporales, *Brevicellicium*, *Dextrinocystis* Gilb. & M. Blackw., *Litschauerella* Oberw., *Luellia* K.H. Larss. & Hjortstam, *Scytinopogon*, *Sistotremastrum* J. Erikss., *Sphaerobasidium* Oberw., *Subulicystidium* Parmasto, *Tubulicium* Oberw., and *Trechispora*, shows that all related genera cluster into Trechisporales and the three new species grouped into *Trechispora*.

The ITS-alone dataset (Figure 2) included sequences from 42 fungal specimens representing 41 species. The dataset had an aligned length of 580 characters, of which 178 characters are constant, 61 are variable and parsimony-uninformative, and 341 are parsimony-informative. Maximum parsimony analysis yielded 584 equally parsimonious trees (TL = 2802, CI = 0.3123, HI = 0.6877, RI = 0.2519, and RC = 0.0787). Best model for the ITS dataset estimated and applied in the Bayesian analysis was GTR+I+G (lset nst = 6, rates = invgamma; prset statefreqpr = dirichlet (1,1,1,1). Bayesian and ML analyses resulted in a topology similar to that of MP analysis with split frequencies equal to 0.025000 (BI), and the effective sample size (ESS) across the two runs is double of the average ESS (avg ESS) = 621.5.

The phylogram inferred from the ITS dataset (Figure 2) indicated that the three new species grouped into *Trechispora*, in which the new species *T. murina* was sister to *T. bambusicola* with higher supports (96% BS, 92% BP, and 1.00 BPP); *T. odontioidea* formed a unique position within the clade of *T. fimbriata* C.L. Zhao and *T. nivea* (Pers.) K.H. Larss; while *T. olivacea* shared a clade formed by the members of *T. farinacea* (Pers.) Liberta, *T. hondurensis* Schoutteten & Haelew., and *T. mollis*.

3.2. Taxonomy

Trechispora murina K.Y. Luo & C.L. Zhao, sp. nov. Figures 3 and 4.

Figure 3. Basidiomata of *Trechispora murina* (holotype CLZhao 11752): the front of the basidiomata (**A**), characteristic hymenophore (**B**). Bars: (**A**) = 5 mm and (**B**) = 1 mm.

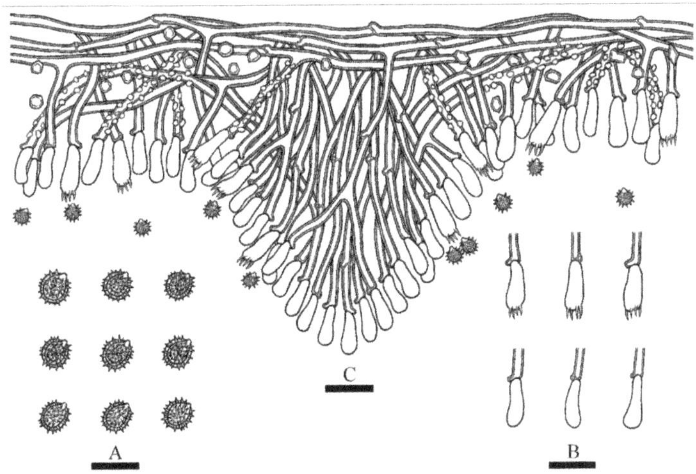

Figure 4. Microscopic structures of *Trechispora murina* (holotype CLZhao 11752): basidiospores (**A**), a cross-section of basidiomata (**B**), basidia and basidioles (**C**). Bars: (**A**) = 5 µm, (**B**,**C**) = 10 µm.

MycoBank no.: 842491.

Holotype—China, Yunnan Province, Wenshan, Funing County, Guying Village, GPS coordinates 23°44′ N, 105°56′ E, altitude 750 m asl., on a fallen angiosperm branch, leg. C.L. Zhao, 20 January 2019, CLZhao 11752 (SWFC).

Etymology—***murina*** (Lat.): Referring to the furry mouse-like hymenial surface.

Basidiomata—Annual, resupinate, thin, growing adnate but easily separable, up to 15 cm long, 3 cm wide, 100–500 µm thick. Hymenial surface grandinioid, pale greyish to grey when fresh, turn to greyish upon drying. Sterile margin concolorous with a hymenial surface, up to 2 mm wide.

Hyphal system—Monomitic; generative hyphae with clamp connections; colorless; thick-walled with a wide to lumen; richly branched; interwoven; encrusted; 2–3.5 µm in diameter; IKI–, CB–; tissues unchanged in KOH.

Hymenium—Cystidia and cystidioles absent; basidia more or less clavate, with four sterigmata and a basal clamp connection, 10.0–14.0 × 3.5–4.5 µm; basidioles dominant; basidioles in shape similar to basidia, but slightly smaller.

Spores—Basidiospores ellipsoid, colorless, thick-walled, ornamented, IKI–, CB–, (2.5–) 3–4 × (2–) 2.5–3 µm, L = 3.42 µm, W = 2.87 µm, Q = 1.17–1.20 (n = 60/2).

Additional specimen examined (paratype)—China, Yunnan Province, Wenshan, Funing County, Guying Village, GPS coordinates 23°39′ N, 105°59′ E, altitude 1400 m asl., on a fallen angiosperm branch, leg. C.L. Zhao, 20 January 2019, CLZhao 11736 (SWFC).

Trechispora odontioidea K.Y. Luo & C.L. Zhao, sp. nov. Figures 5 and 6.

MycoBank no.: 844493.

Holotype—China, Yunnan Province, Honghe, Pingbian County, Daweishan National Nature Reserve. GPS coordinates: 23°420′ N, 103°300′ E; altitude: 1500 m asl., on fallen angiosperm branches, leg. C.L. Zhao, 1 August 2019, CLZhao 17890 (SWFC).

Etymology—*odontioidea* (Lat.): Referring to the odontioid hymenophore.

Basidiomata—Annual, adnate, thin, up to 11 cm long, 2.5 cm wide, 50–200 µm thick. Hymenial surface odontioid, aculei cylindrical to conical, blunt, 0.3–0.6 mm long, pale buff when fresh, turn to buff upon drying. Sterile margin indistinct, cream to buff, 0.5–1 mm wide.

Hyphal system—Monomitic; generative hyphae with clamp connections; colorless, thin- to thick-walled; frequently branched; interwoven; 2–3.5 µm in diameter; ampullate hyphae frequently present; IKI−, CB−; tissues unchanged in KOH.

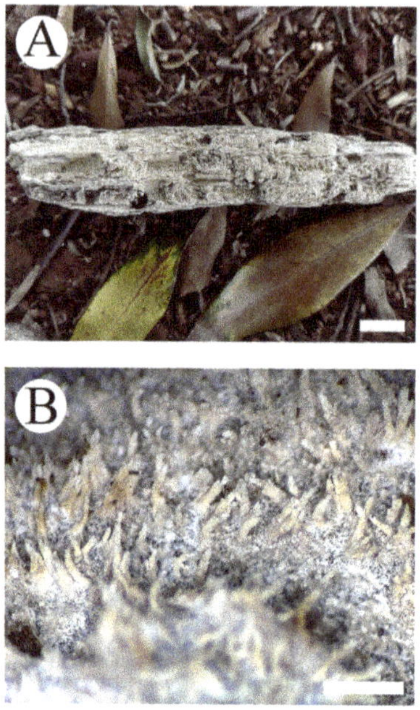

Figure 5. Basidiomata of *Trechispora odontioidea* (holotype CLZhao 17890): the front of the basidiomata (**A**), characteristic hymenophore (**B**). Bars: (**A**) = 1 cm and (**B**) = 1 mm.

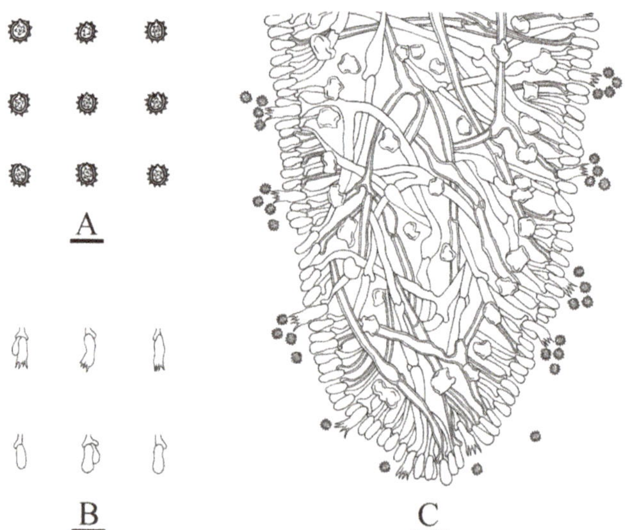

Figure 6. Microscopic structures of *Trechispora odontioidea* (holotype CLZhao 17890): basidiospores (**A**), basidia and basidioles (**B**), a cross section of basidiomata (**C**). Bars: (**A**) = 5 μm, (**B**,**C**) = 10 μm.

Hymenium—Cystidia and cystidioles absent.; basidia clavate, with four sterigmata and a basal clamp connection, 8.0–12.0 × 2.5–4 μm; basidioles dominant, in shape similar to basidia, but smaller.

Spores—Basidiospores subglobose to globose, colorless, slightly thick-walled, ornamented, IKI–, CB–, 2–3 × 1.5–2.5 μm, L = 2.53 μm, W = 2.00 μm, Q = 1.27 (n = 30/1).

Trechispora olivacea K.Y. Luo & C.L. Zhao, sp. nov. Figures 7 and 8.

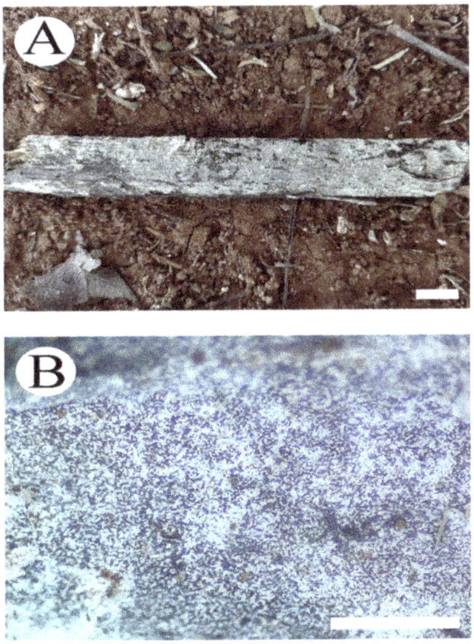

Figure 7. Basidiomata of *Trechispora olivacea* (holotype CLZhao 17826): the front of the basidiomata (**A**), characteristic hymenophore (**B**). Bars: (**A**) = 1 cm and (**B**) = 1 mm.

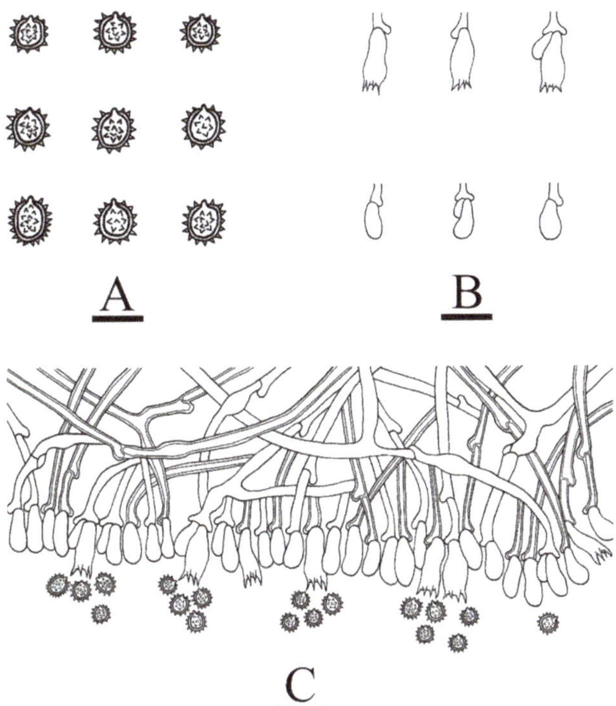

Figure 8. Microscopic structures of *Trechispora olivacea* (holotype CLZhao 17826): basidiospores (**A**), basidia and basidioles (**B**), a cross-section of basidiomata (**C**). Bars: (**A**) = 5 μm, (**B**,**C**) = 10 μm.

MycoBank no.: 844494.

Holotype—China, Yunnan Province, Honghe, Pingbian County, Daweishan National Nature Reserve. GPS coordinates: 23°420′ N, 103°300′ E; altitude: 1500 m asl., on fallen angiosperm branches, leg. C.L. Zhao, 1 August 2019, CLZhao 17826 (SWFC).

Etymology—*olivacea* (Lat.): Referring to the olivaceous hymenial surface.

Basidiomata—Annual, resupinate, thin, very hard to separate from substrate, up to 11 cm long, 2.5 cm wide, 30–80 µm thick. Hymenial surface farinaceous, pale white to slightly olivaceous when fresh, turn to olivaceous upon drying. Sterile margin indistinct, slightly olivaceous, 0.2–0.5 mm wide.

Hyphal system—Monomitic; generative hyphae with clamp connections; colorless; thin- to thick-walled; occasionally branched; interwoven; 1.5–3.0 µm in diameter; ampullate hyphae present; IKI–, CB–; tissues unchanged in KOH.

Hymenium—Cystidia and cystidioles absent; basidia clavate, with four sterigmata and a basal clamp connection, 10.0–12.0 × 4.5–5 µm; basidioles dominant, with the shape similar to basidia, but smaller.

Spores—Basidiospores broadly ellipsoid to globose, colorless, thick-walled, ornamented, IKI–, CB–, 2.5–4 × 1.5–2.5 µm, L = 3.30 µm, W = 2.65 µm, Q = 1.25 (n = 30/1).

4. Discussion

The classification of corticioid fungi revealed that two taxa of *Trechispora farinacea* and *T. hymenocystis* nested into *Trechispora* located in Hydnodontaceae (Trechisporales) [15]. In the present study (Figure 2), *Trechispora murina*, *T. odontioidea*, and *T. olivacea* are nested into *Trechispora*, in which *T. murina* was sister to *T. bambusicola*; *T. odontioidea* formed a monophyletic lineage and then grouped with *T. fimbriata* and *T. nivea*; while *T. olivacea* formed a monophyletic lineage and then grouped with *T. farinacea*, *T. hondurensis*, and *T. mollis*. However, *T. bambusicola* is morphologically distinguishable from *T. murina* by having the odontioid hymenophore with cream to buff the hymenial surface [33]. *Trechispora fimbriata* is distinguishable from *T. odontioidea* by having the hydnoid hymenial surface and longer basidiospores (3–3.6 × 2.4–3.2 µm) [33]; *T. nivea* differs from *T. odontioidea* by its thin-walled, larger basidiospores (3.5–4 × 2.5–3 µm) [5]. *Trechispora farinacea* is distinguishable from *T. olivacea* by its smooth to grandinioid or odontioid hymenophore with whitish to ochraceous hymenial surface and larger basidiospores (4–5 × 3.5–4 µm) [5]; *T. hondurensis* is separated from *T. olivacea* by having a hydnoid to partly irpicoid hymenial surface and thin-walled, wider basidiospores (3.6–3.8 × 2.7–2.9 µm) [58]; *T. mollis* is distinguishable from *T. olivacea* because it has white-yellow to pale yellow hydnoid hymenial surface, and wider ampullate septa at generative hyphae (reaching 8 µm in width) [26].

Morphologically, *Trechispora murina* is similar to *T. farinacea*, *T. rigida*, *T. subsphaerospora* (Litsch.) Liberta, and *T. torrendii* Chikowski & K.H. Larss. Based on the character of the grandinioid hymenial surface. However, *Trechispora farinacea* is separated from *T. murina* by having a whitish to ochraceous hymenial surface and larger, subglobose to broadly elliposid basidiospores (4–5 × 3.5–4 µm) [5]. *Trechispora rigida* differs from *T. murina* due to the presence of its dirty white to buff hymenophore [59] and having larger basidiospores (4.5–5.5 × 4 µm) [27]. *Trechispora subsphaerospora* differs from *T. murina* by having smooth basidiospores [34]. *Trechispora torrendii* differs in its pale yellow to yellow hymenophore [26] and has globose to subglobose basidiospores (2.8–3.5 × 3–3.5 µm) [27].

Trechispora murina is similar to *T. canariensis* Ryvarden & Liberta, *T. fastidiosa* (Pers.) Liberta, *T. praefocata* (Bourdot & Galzin) Liberta, *T. stevensonii* (Berk. & Broome) K.H. Larss., and *T. yunnanensis* C.L. Zhao due to the presence of the ellipsoid, ornamented basidiospores. However, *Trechispora canariensis* differs from *T. murina* because it has arachonoid to pelliculose hymenial surface and thin-walled, larger basidiospores (5–7 × 3–3.5 µm) [5]. *Trechispora fastidiosa* is separated from *T. murina* by having a membranaceous, whitish to cream hymenial surface and larger basidiospores (6–7 × 4.5–5.5 µm) [5]. *Trechispora praefocata* differs by having the farinaceous to arachnoid hymenial surface and larger basidiospores (5–6.5 × 4–5.5 µm) [5]. *Trechispora stevensonii* differs from *T. murina* by its hydnoid hymenophore and larger basidiospores (4–4.5 × 3–3.5 μm) [5]. *Trechispora yunnanensis* is separated from *T. murina* by having the farinaceous hymenial surface and larger basidiospores (7–8.5 × 5–5.5 μm) [31].

Trechispora odontioidea is similar to *T. bambusicola* C.L. Zhao and *T. nivea* in having an odontioid hymenial surface. However, *Trechispora bambusicola* differs from *T. odontioidea* because it has granulose basidiomata, and the absence of the ampullaceous septa [33]. *Trechispora nivea* differs from *T. odontioidea* due to the presence of white to ochraceous basidiomata and broadly ellipsoid to subglobose, thin-walled, larger basidiospores (3.5–4 × 2.5–3 μm) [5].

Trechispora odontioidea resembles *T. clancularis* (Park.-Rhodes) K.H. Larss., *T. cyatheae* Ordynets, Langer & K.H. Larss., *T. hymenocystis* (Berk. & Broome) K.H. Larsson, *T. invisitata* (H.S. Jacks.) Liberta, and *T. torrendii* Chikowski & K.H. Larss. due to the presence of ornamented or aculeate basidiospores. However, *Trechispora clancularis* is distinguishable from *T. odontioidea* due to the presence of its poroid to irpicoid hymenial surface and subglobose to ovoid, larger basidiospores (6–6.5 × 5–6 μm) [5]. *Trechispora cyatheae* differs from *T. odontioidea* in having a farinaceous to grandinioid hymenial surface, and broadly elliptical to slightly lacrymiform and adaxial side convex or straight, longer basidiospores (3–3.5 × 2–3 μm) [24]. *Trechispora hymenocystis* is distinguishable from *T. odontioidea* by its poroid basidiomata and broadly ellipsoidal to ellipsoidal, larger basidiospores (4.5–5.5 × 3.5–4.5 μm) [19]. *Trechispora invisitata* differs from *T. odontioidea* because it has a smooth to porulose, farinaceous to granulose hymenial surface and ellipsoid to ovate, larger basidiospores (4.5–5.5 × 3–4 μm) [5]. *Trechispora torrendii* differs from *T. odontioidea* because it has a farinose to grandinioid hymenial surface and larger basidiospores (3.2–3.5 × 2.8–3.2 μm) [26].

Trechispora olivacea is similar to *T. caucasica* (Parmasto) Liberta, *T. dimitica* Hallenb., *T. gelatinosa* Meiras-Ottoni & Gibertoni, *T. verruculosa* (G. Cunn.) K.H. Larss., and *T. yunnanensis* C.L. Zhao due to the presence of a farinaceous hymenial surface. However, *Trechispora caucasica* differs from *T. olivacea* by having a white to greyish hymenial surface and narrowly ellipsoid to reniform with a median constriction, larger basidiospores (5–5.5 × 4–4.5 μm) [5]. *Trechispora dimitica* differs from *T. olivacea* in its white to pale greyish hymenial surface, dimitic hyphal system, and shorter basidia (7–10 × 4.5–5.5 μm) [5]. *Trechispora gelatinosa* is distinguishable from *T. olivacea* by its coralloid basidiomata and wider basidiospores (3.2–4.5 × 2.5–3.5 μm) [27]. *Trechispora verruculosa* differs from *T. olivacea* because it has granulose to hydnoid with small cylindrical aculei, white to yellowish to ochraceous hymenial surface and larger basidiospores (4.5–5.5 × 3.5–4.5 μm) [5]. *Trechispora yunnanensis* can be delimited from *T. olivacea* by its larger basidiospores (7–8.5 × 5–5.5 μm) [31].

Trechispora olivacea resembles *T. hypogeton* (Maas Geest.) Hjortstam & K.H. Larss., *T. nivea*, *T. rigida*, and *T. thelephora* (Lév.) Ryvarden in having broadly ellipsoid to globose, ornamented basidiospores. However, *Trechispora hypogeton* is distinguishable from *T. olivacea* by its stipitate basidiomata and wider basidiospores (3.8–4.3 × 2.7–3.1 μm) [26]. *Trechispora nivea* differs from *T. olivacea* by the presence of a odontioid hymenial surface with white to pale ochraceous and wider basidiospores (3.5–4 × 2.5–3 μm) [5]. *Trechispora rigida* differs from *T. olivacea* due to the presence of a colliculose hymenial surface and larger basidiospores (4.5–5.5 × 4 μm) [26]. *Trechispora thelephora* differs from *T. olivacea* because it has a stipitate basidiomata and larger basidiospores (4.0–5.0 × 3.4–4.5 μm) [26].

Wood-rotting fungi are an extensively studied group of Basidiomycota [12,13,60–66] and the three taxa of *Trechispora* are a typical example group of wood-rotting fungi [15,33–35,67]. Based on our present morphology and phylogeny focusing on *Trechispora*, all taxa in this genus can be separated from the three new species.

Key to 21 accepted species of *Trechispora* in China

1. Basidiospores smooth ——— 2
1′ Basidiospores aculeate, verrucose or ornamented ——— 5
2. Ampullate hyphae > 5 μm in width, basidiospores angular ——— *T. subsphaerospora*
2′ Ampullate hyphae < 5 μm in width, basidiospores ellipsoid ——— 3
3. Basidiospores thick-walled ——— *T. cohaerens*
3′ Basidiospores thin-walled ——— 4

4. Hymenial surface tuberculate ———— *T. daweishanensis*
4′ Hymenial surface smooth ———— *T. xantha*
5. Hyphal system dimitic ———— *T. dimitica*
5′ Hyphal system monomitic ———— 6
6. Hyphae without ampullate septa ———— 7
6′ Hyphae with ampullate septa ———— 12
7. Basidiospores thin-walled, ovoid to subglobose ———— *T. suberosa*
7′ Basidiospores thick-walled, ellipsoid ———— 8
8. Basidiospores > 7 μm in length ———— *T. yunnanensis*
8′ Basidiospores < 7 μm in length ———— 9
9. Basidiomata margin greyish ———— ***T. murina***
9′ Basidiomata margin white to cream ———— 10
10. Hymenial surface odontioid ———— *T. bambusicola*
10′ Hymenial surface hydnoid ———— 11
11. Hymenophore with blunt aculei ———— *T. fimbriata*
11′ Hymenophore with sharp aculei ———— *T. fissurata*
12. Sphaerocysts present, hyphae inflated ———— *T. hymenocystis*
12′ Sphaerocysts absent, hyphae uninflated ———— 13
13. Ampullate septa > 6 μm in width ———— 14
13′ Ampullate septa < 6 μm in width ———— 15
14. Basidiospores sparsely verrucose ———— *T. polygonospora*
14′ Basidiospores densely aculeate ———— *T. mollusca*
15. Subhymenium with short-celled hyphae ———— 16
15′ Subhymenium with long-celled hyphae ———— 17
16. Basidiome thin, ochraceous ———— *T. farinacea*
16′ Basidiome thick, dirty white to buff ———— *T. rigida*
17. Basidiospores thin-walled ———— 18
17′ Basidiospores thick-walled ———— 19
18. Hymenophore with hydnoid ———— *T. nivea*
18′ Hymenophore without hydnoid ———— *T. microspora*
19. Basidiospores > 5 μm in length ———— *T. praefocata*
19′ Basidiospores < 5 μm in length ———— 20
20. Hymenial surface farinaceous with olivaceous ———— ***T. olivacea***
20′ Hymenial surface odontioid with buff ———— ***T. odontioidea***

Author Contributions: Conceptualization, C.Z.; methodology, C.Z. and K.L.; software, C.Z. and K.L.; validation, C.Z. and K.L.; formal analysis, C.Z. and K.L.; investigation, C.Z. and K.L.; resources, C.Z.; writing—original draft preparation, C.Z. and K.L.; writing—review and editing, C.Z. and K.L.; visualization, C.Z. and K.L.; supervision, C.Z.; project administration, C.Z.; funding acquisition, C.Z. All authors have read and agreed to the published version of the manuscript.

Funding: The research was supported by the National Natural Science Foundation of China (Project No. 32170004, U2102220), Yunnan Fundamental Research Project (Grant No. 202001AS070043), the High-level Talents Program of Yunnan Province (YNQR-QNRC-2018-111), and the Opening Foundation of Yunnan Key Laboratory of Plateau Wetland Conservation, Restoration and Ecological Services (202105AG070002).

Institutional Review Board Statement: Not applicable.

Informed Consent Statement: Not applicable.

Data Availability Statement: Publicly available datasets were analyzed in this study. This data can be found here: [https://www.ncbi.nlm.nih.gov/, accessed on 28 November 2021; https://www.mycobank.org/page/Simple%20names%20search].

Conflicts of Interest: The authors declare no conflict of interest.

References

1. James, T.Y.; Stajich, J.E.; Hittinger, C.T.; Rokas, A. Toward a fully resolved fungal tree of life. *Annu. Rev. Microbiol.* **2020**, *74*, 291–313. [CrossRef] [PubMed]
2. Tedersoo, L.; Bahram, M.; Põlme, S.; Koljalg, U.; Yorou, N.S.; Wijesundera, R.; Ruiz, L.V.; Vasco-Palacios, A.M.; Thu, P.Q.; Suija, A.; et al. Global diversity and geography of soil fungi. *Science* **2014**, *346*, 1256688. [CrossRef] [PubMed]
3. Gilbertson, R.L.; Ryvarden, L. *North American Polypores 1–2*; Fungiflora: Oslo, Norway, 1987; pp. 1–433.
4. Núñez, M.; Ryvarden, L. East Asian polypores 2. *Synop. Fungorum* **2001**, *14*, 165–522.
5. Bernicchia, A.; Gorjón, S.P. *Fungi Europaei 12: Corticiaceae s.l.*; Edizioni Candusso: Alassio, Italy, 2010.
6. Dai, Y.C. Polypore diversity in China with an annotated checklist of Chinese polypores. *Mycoscience* **2012**, *53*, 49–80. [CrossRef]
7. Ryvarden, L.; Melo, I. Poroid fungi of Europe. *Synop. Fungorum* **2014**, *31*, 1–455.
8. Dai, Y.C.; Cui, B.K.; Si, J.; He, S.H.; Hyde, K.D.; Yuan, H.S.; Liu, X.Y.; Zhou, L.W. Dynamics of the worldwide number of fungi with emphasis on fungal diversity in China. *Mycol. Prog.* **2015**, *14*, 62. [CrossRef]
9. Dai, Y.C.; Yang, Z.L.; Cui, B.K.; Wu, G.; Yuan, H.S.; Zhou, L.W.; He, S.H.; Ge, Z.W.; Wu, F.; Wei, Y.L.; et al. Diversity and systematics of the important macrofungi in Chinese forests. *Mycosystema* **2021**, *40*, 770–805.
10. Cui, B.K.; Yuan, H.S.; Zhou, L.W.; He, S.H.; Wei, Y.L. Diversity of wood-decaying fungi in conifer trees of the Greater and Lesser Khinggan Mountains. *Biodivers. Sci.* **2019**, *27*, 887–895. [CrossRef]
11. Wu, F.; Yuan, H.S.; Zhou, L.W.; Yuan, Y.; Cui, B.K.; Dai, Y.C. Polypore diversity in South China. *Mycosystema* **2020**, *39*, 653–682.
12. Luo, K.Y.; Chen, Z.Y.; Zhao, C.L. Phylogenetic and taxonomic analyses of three new wood-inhabiting fungi of *Xylodon* (Basidiomycota) in a forest ecological system. *J. Fungi* **2022**, *8*, 405. [CrossRef]
13. Qu, M.H.; Wang, D.Q.; Zhao, C.L. A phylogenetic and taxonomic study on *Xylodon* (Hymenochaetales): Focusing on three new *Xylodon* species from southern China. *J. Fungi* **2022**, *8*, 35. [CrossRef]
14. Hibbett, D.S.; Binder, M.; Bischoff, J.F.; Blackwell, M.; Cannon, P.F.; Eriksson, O.E.; Huhndorf, S.; James, T. A higher-level phylogenetic classification of the Fungi. *Mycol. Res.* **2007**, *111*, 509–547. [CrossRef]
15. Larsson, K.H. Re-thinking the classification of corticioid fungi. *Mycol. Res.* **2007**, *111*, 1040–1063. [CrossRef]
16. Karsten, P.A. Fragmenta mycologica XXIX. *Nova Hedwig.* **1890**, *29*, 147–149.
17. Liberta, A.E. On *Trechispora*. *Taxon* **1966**, *15*, 317–319. [CrossRef]
18. Liberta, A.E. The genus *Trechispora* (Basidiomycetes, Corticiaceae). *Can. J. Bot.* **1973**, *51*, 1871–1892. [CrossRef]
19. Larsson, K.H. Poroid species in *Trechispora* and the use of calcium oxalate crystals for species identification. *Mycol. Res.* **1994**, *98*, 1153–1172. [CrossRef]
20. Ryvarden, L. A note on the genus *Hydnodon* Banker. *Synop. Fungorum* **2002**, *15*, 31–33.
21. Trichiès, G.; Schultheis, B. *Trechispora antipus* sp. nov., une seconde espèce bisporique du genre *Trechispora* (Basidiomycota, Stereales). *Mycotaxon* **2002**, *82*, 453–458.
22. Miettinen, O.; Larsson, K.H. *Trechispora elongata* species nova from North Europe. *Mycotaxon* **2006**, *96*, 193–198.
23. Kirk, P.M.; Cannon, P.F.; David, J.C.; Minter, D.W.; Stalpers, J.A. *Ainsworth and Bisby's Dictionary of the Fungi*, 10th ed.; CAB International Press: Wallingford, UK, 2008; p. 783. [CrossRef]
24. Ordynets, A.; Larsson, K.H.; Langer, E. Two new *Trechispora* species from La Réunion Island. *Mycol. Prog.* **2015**, *14*, 113. [CrossRef]
25. Phookamsak, R.; Hyde, K.D.; Jeewon, R.; Bhat, D.J.; Jones, E.B.G.; Maharachchikumbura, S.S.N.; Raspe´, O.; Karunarathna, S.C.; Wanasinghe, D.N.; Hongsanan, S.; et al. Fungal diversity notes 929–1035: Taxonomic and phylogenetic contributions on genera and species of fungi. *Fungal Divers.* **2019**, *95*, 1–273. [CrossRef]
26. Chikowski, R.S.; Larsson, K.H.; Gibertoni, T.B. Taxonomic novelties in *Trechispora* from Brazil. *Mycol. Prog.* **2020**, *19*, 1403–1414. [CrossRef]
27. Meiras-Ottoni, A.; Larsson, K.H.; Gibertoni, T.B. Additions to *Trechispora* and the status of *Scytinopogon* (Trechisporales, Basidiomycota). *Mycol. Prog.* **2021**, *20*, 203–222. [CrossRef]
28. Dai, Y.C. A checklist of polypores in China. *Mycosystema* **2009**, *28*, 315–327.
29. Dai, Y.C. A revised checklist of corticioid and hydnoid fungi in China for 2010. *Mycoscience* **2011**, *52*, 69–79. [CrossRef]
30. Yuan, H.S.; Dai, Y.C. Wood-inhabiting fungi in southern China. 6. Polypores from Guangxi Autonomous region. *Ann. Bot. Fenn.* **2012**, *49*, 341–351. [CrossRef]
31. Xu, T.M.; Chen, Y.H.; Zhao, C.L. *Trechispora yunnanensis* sp. nov. from China. *Phytotaxa* **2019**, *424*, 253–261. [CrossRef]
32. Wang, K.; Zhao, M.J.; Su, J.H.; Yang, L.; Deng, H.; Wang, Y.H.; Wu, H.J.; Li, Y.; Wu, H.M.; Wei, X.D.; et al. The use of checklist of fungi in China database in the red list assessment of macrofungi in China. *Biodivers. Sci.* **2020**, *28*, 74–98. [CrossRef]
33. Zhao, W.; Zhao, C.L. The phylogenetic relationship revealed three new wood-inhabiting fungal species from genus *Trechispora*. *Front. Microbiol.* **2021**, *12*, 650195. [CrossRef] [PubMed]
34. Zong, T.K.; Liu, C.M.; Wu, J.R.; Zhao, C.L. *Trechispora daweishanensis* and *T. xantha* spp. nov. (Hydnodontaceae, Trechisporales) found in Yunnan Province of China. *Phytotaxa* **2021**, *479*, 147–159. [CrossRef]
35. Larsson, K.H.; Larsson, E.; Kõljalg, U. High phylogenetic diversity among corticioid homobasidiomycetes. *Mycol. Res.* **2004**, *108*, 983–1002. [CrossRef]
36. Telleria, M.T.; Melo, I.; Dueñas, M.; Larsson, K.H.; Paz, M.M.P. Molecular analyses confifirm *Brevicellicium* in Trechisporales. *IMA Fungus* **2013**, *4*, 21–28. [CrossRef]
37. Jülich, W. Notes on some Basidiomycetes (Aphyllophorales and Heterobasidiomycetes). *Persoonia* **1982**, *11*, 421–428.

38. Liu, S.L.; Ma, H.X.; He, S.H.; Dai, Y.C. Four new corticioid species in Trechisporales (Basidiomycota) from East Asia and notes on phylogeny of the order. *MycoKeys* **2019**, *48*, 97–113. [CrossRef]
39. Kornerup, A.; Wanscher, J. *Metheun Hand Book of Colour*, 3rd ed.; Metheun London Ltd.: London, UK, 1978; pp. 144–148.
40. Wu, F.; Zhou, L.W.; Vlasák, J.; Dai, Y.C. Global diversity and systematics of Hymenochaetaceae with poroid hymenophore. *Fungal Divers.* **2022**, *113*, 1–192. [CrossRef]
41. Zhao, C.L.; Wu, Z.Q. *Ceriporiopsis kunmingensis* sp. nov. (Polyporales, Basidiomycota) evidenced by morphological characters and phylogenetic analysis. *Mycol. Prog.* **2017**, *16*, 93–100. [CrossRef]
42. White, T.J.; Bruns, T.; Lee, S.; Taylor, J. Amplification and direct sequencing of fungal ribosomal RNA genes for phylogenetics. *PCR Protoc. A Guide Methods Appl.* **1990**, *18*, 315–322. [CrossRef]
43. Spirin, V.; Volobuev, S.; Viner, I.; Miettinen, O.; Vlasak, J.; Schoutteten, N.; Motato-Vásquez, V.; Kotiranta, H.; Hernawati; Larsson, K.H. On *Sistotremastrum* and similar-looking taxa (Trechisporales, Basidiomycota). *Mycol. Prog.* **2021**, *20*, 453–476. [CrossRef]
44. Yurchenko, E.; Wu, S.H. *Hyphoderma formosanum* sp. nov. (Meruliaceae, Basidiomycota) from Taiwan. *Sydowia* **2014**, *66*, 19–23.
45. Wu, F.; Yuan, Y.; Zhao, C.L. *Porpomyces submucidus* (Hydnodontaceae, Basidiomycota) from tropical China. *Phytotaxa* **2015**, *230*, 61–68. [CrossRef]
46. Desjardin, D.E.; Perry, B.A. A new species of *Scytinopogon* from the island of Príncipe, Republic of São Tomé and Príncipe, West Africa. *Mycosphere* **2015**, *6*, 434–441. [CrossRef]
47. Vu, D.; Groenewald, M.; de Vries, M.; Gehrmann, T.; Stielow, B.; Eberhardt, U.; Al-Hatmi, A.; Groenewald, J.Z.; Cardinali, G.; Houbraken, J.; et al. Large-scale generation and analysis of fifilamentous fungal DNA barcodes boosts coverage for kingdom fungi and reveals thresholds for fungal species and higher taxon delimitation. *Stud. Mycol.* **2019**, *92*, 135–154. [CrossRef]
48. Larsson, K.H.; Parmasto, E.; Fischer, M.; Langer, E.; Nakasone, K.K.; Redhead, S.A. Hymenochaetales: A molecular phylogeny for the hymenochaetoid clade. *Mycologia* **2006**, *98*, 926–936. [CrossRef] [PubMed]
49. Ordynets, A.; Scherf, D.; Pansegrau, F.; Denecke, J.; Langer, E. Short-spored *Subulicystidium* (Trechisporales, Basidiomycota): High morphological diversity and only partly clear species boundaries. *MycoKeys* **2018**, *35*, 41–99. [CrossRef]
50. Ordynets, A.; Liebisch, R.; Lysenko, L.; Scherf, D.; Volobuev, S.; Saitta, A.; Larsson, K.H.; Yurchenko, E.; Buyck, B.; Bolshakov, S. Morphologically similar but not closely related: The long-spored species of *Subulicystidium* (Trechisporales, Basidiomycota). *Mycol. Prog.* **2020**, *19*, 691–703. [CrossRef]
51. Vasco-Palacios, A.M.; Lopez-Quintero, C.; Franco-Molano, A.E.; Boekhout, T. *Austroboletus amazonicus* sp. nov. and *Fistulinella campinaranae* var. scrobiculata, two commonly occurring boletes from a forest dominated by *Pseudomonotes tropenbosii* (Dipterocarpaceae) in Colombian Amazonia. *Mycologia* **2014**, *106*, 1004–1014. [CrossRef] [PubMed]
52. Larsson, A. AliView: A fast and lightweight alignment viewer and editor for large data sets. *Bioinformatics* **2014**, *30*, 3276–3278. [CrossRef] [PubMed]
53. Swofford, D.L. *PAUP *: Phylogenetic Analysis Using Parsimony (* and Other Methods)*; Version 4.0b10; Sinauer Associates: Sunderland, MA, USA, 2002.
54. Felsenstein, J. Confidence intervals on phylogenetics: An approach using bootstrap. *Evolution* **1985**, *39*, 783–791. [CrossRef] [PubMed]
55. Miller, M.A.; Pfeiffer, W.; Schwartz, T. The CIPRES science gateway: Enabling high-impact science for phylogenetics researchers with limited resources. *Assoc. Comput. Mach.* **2012**, *39*, 1–8. [CrossRef]
56. Nylander, J.A.A. *MrModeltest V2. Program Distributed by the Author*; Evolutionary Biology Centre, Uppsala University: Uppsala, Sweden, 2004.
57. Ronquist, F.; Teslenko, M.; van der Mark, P.; Ayres, D.L.; Darling, A.; Hohna, S.; Larget, B.; Liu, L.; Suchard, M.A.; Huelsenbeck, J.P. Mrbayes 3.2: Efficient bayesian phylogenetic inference and model choice across a large model space. *Syst. Biol.* **2012**, *61*, 539–542. [CrossRef]
58. Haelewaters, D.; Dima, B.; Abdel-Hafiz, A.I.; Abdel-Wahab, M.A.; Abul-Ezz, S.R.; Acar, I.; Aguirre-Acosta, E.; Aime, M.C.; Aldemir, S.; Ali, M.; et al. Fungal systematics and evolution: Fuse 6. *Sydowia* **2020**, *72*, 231–356. [CrossRef]
59. Larsson, K.H. New species and combinations in *Trechispora* (Corticiaceae, Basidiomycotina). *Nord. J. Bot.* **1996**, *16*, 83–98. [CrossRef]
60. Zhang, M.; Wang, C.Q.; Gan, M.S.; Li, Y.; Shao, S.C.; Qin, W.Q.; Deng, W.Q.; Li, T.H. Diversity of *Cantharellus* (Cantharellales, Basidiomycota) in China with Description of Some New Species and New Records. *J. Fungi* **2022**, *8*, 483. [CrossRef]
61. Song, C.G.; Chen, Y.Y.; Liu, S.; Xu, T.M.; He, X.L.; Wang, D.; Cui, B.K. A Phylogenetic and Taxonomic Study on *Phellodon* (Bankeraceae, Thelephorales) from China. *J. Fungi* **2022**, *8*, 429. [CrossRef]
62. Hu, J.; Zhao, G.; Tuo, Y.; Rao, G.; Zhang, Z.; Qi, Z.; Yue, L.; Liu, Y.; Zhang, T.; Li, Y.; et al. Morphological and Molecular Evidence Reveal Eight New Species of *Gymnopus* from Northeast China. *J. Fungi* **2022**, *8*, 349. [CrossRef]
63. Mao, N.; Xu, Y.Y.; Zhao, T.Y.; Lv, J.C.; Fan, L. New Species of *Mallocybe* and *Pseudosperma* from North China. *J. Fungi* **2022**, *8*, 256. [CrossRef]
64. Wu, F.; Tohtirjap, A.; Fan, L.F.; Zhou, L.W.; Alvarenga, R.L.M.; Gibertoni, T.B.; Dai, Y.C. Global diversity and updated phylogeny of *Auricularia* (Auriculariales, Basidiomycota). *J. Fungi* **2021**, *7*, 933. [CrossRef]
65. Guan, Q.X.; Zhao, C.L. Taxonomy and phylogeny of the wood-inhabiting fungal genus *Hyphoderma* with descriptions of three new species from East Asia. *J. Fungi* **2021**, *7*, 308. [CrossRef]

66. Wang, D.Q.; Zhao, C.L. Morphological and Phylogenetic Evidence for Recognition of Two New Species of *Phanerochaete* from East Asia. *J. Fungi* **2021**, *7*, 1063. [CrossRef]
67. He, M.Q.; Zhao, R.L.; Hyde, K.D.; Begerow, D.; Kemler, M.; Yurkov, A.; McKenzie, E.H.C.; Raspe, O.; Kakishima, M.; Sanchez-Ramırez, S.; et al. Notes, outline and divergence times of Basidiomycota. *Fungal Divers.* **2019**, *99*, 105–367. [CrossRef]

Journal of *Fungi*

MDPI

Article

Morphological and Phylogenetic Characterizations Reveal Five New Species of *Astrothelium* (*Trypetheliales, Ascomycota*) from China

Shu-Hua Jiang [1], Chao Zhang [1,2], Xian-Dong Xue [1,3], André Aptroot [4], Jiang-Chun Wei [1,5] and Xin-Li Wei [1,5,*]

[1] State Key Laboratory of Mycology, Institute of Microbiology, Chinese Academy of Sciences, Beijing 100101, China
[2] Institute of Life Science and Green Development, College of Life Sciences, Hebei University, Baoding 071002, China
[3] College of Life Sciences, Shandong Normal University, Jinan 250014, China
[4] Laboratório de Botânica/Liquenologia, Instituto de Biociências, Universidade Federal de Mato Grosso do Sul, Avenida Costa e Silva s/n, Bairro Universitário, Campo Grande CEP 79070-900, Mato Grosso do Sul, Brazil
[5] University of Chinese Academy of Sciences, Beijing 100049, China
* Correspondence: weixl@im.ac.cn

Abstract: The lichenized fungal genus *Astrothelium* is an important element of crustose lichen communities in tropical to subtropical forests. Morphological and molecular phylogenetic approaches to investigate species diversity of *Astrothelium* (*Trypetheliaceae*) from Southern China were carried out in this study. Bayesian and maximum-likelihood (ML) analyses were generated based on the combined data set of internal transcribed spacer (ITS), partial regions of the nuclear ribosomal large subunit (LSU), and the largest subunit of RNA polymerase II gene sequences (RPB1). The morphological comparison with the known *Astrothelium* taxa and molecular phylogeny support five new species: *Astrothelium jiangxiense* sp. nov., *A. luminothallinum* sp. nov., *A. pseudocrassum* sp. nov., *A. subeustominspersum* sp. nov., and *A. subrufescens* sp. nov. All these species are described and illustrated in detail.

Keywords: diversity; morphology; new taxa; *Trypetheliaceae*; phylogeny

Citation: Jiang, S.-H.; Zhang, C.; Xue, X.-D.; Aptroot, A.; Wei, J.-C.; Wei, X.-L. Morphological and Phylogenetic Characterizations Reveal Five New Species of *Astrothelium* (*Trypetheliales, Ascomycota*) from China. *J. Fungi* **2022**, *8*, 994. https://doi.org/10.3390/jof8100994

Academic Editor: Philippe Silar

Received: 25 August 2022
Accepted: 19 September 2022
Published: 22 September 2022

1. Introduction

Astrothelium Eschw. is a genus of lichenized fungi, with the type species *Astrothelium conicum* Eschw., belonging to the family *Trypetheliaceae* in the order *Trypetheliales* in the class *Dothideomycetes* of the phylum *Ascomycota* [1–3]. Traditionally, *Astrothelium* included the species with fused lateral ostioles and transversely septate ascospores. Harris (1989, 1995) anticipated that the classification scheme employed for *Trypetheliaceae* was artificial and would result in the polyphyly of some genera [4,5]. Aptroot et al. (2008) subsequently echoed his contentions, further emphasizing the need for a revision of generic concepts within *Trypetheliaceae* [6]. Utilizing molecular data, Del Prado et al. (2006) and Nelsen et al. (2009) began assessing Harris's assertions and demonstrated the non-monophyly of *Trypethelium* and *Astrothelium* [3,7]. Further, ontogenetic studies of muriform-spored taxa revealed these spores initially form transverse septa with diamond-shaped lumina and subsequently develop muriform septation, thus suggesting a close evolutionary connection between species producing these different ascospore types [8]. Nelsen et al. (2014) explicitly studied relationships within *Trypetheliaceae* based on molecular analysis and showed that species from a number of genera together form a strongly supported group, referred to as the "Astrothelium" clade [9]. This result was supported by Lücking et al. [10,11] and Aptroot and Lücking [12]. With its modern circumscription [12], the genus, with a total of over 250 taxa [1,13–17], is the largest in the family and exhibits much variation in perithecial arrangement and ascospore septation, including species essentially referred to as the previously separate genera *Astrothelium*, *Bathelium*, *Campylothelium*, *Cryptothelium*, *Laurera*, and *Trypethelium* [1,2,9,10,12–15]. In both its original and revised delimitation, the genus has a pantropical distribution [1,2,15,18–21].

Astrothelium is distinguished by the corticate, usually olive-green thallus, the simple to aggregated or pseudostromatic ascomata with apical to lateral, separate or fused ostioles; the ascomata or pseudostromata can be immersed to prominent, often different from the thallus in structure and color. The ascospores in *Astrothelium* are distoseptate, with diamond-shaped lumina, which are best seen in the species with transversely septate ascospores, but ascospores with muriform septation are also common [1,12]. Thus far, most known species in the genus produce hyaline ascospores, except *Astrothelium fuscosporum* Soto-Medina, Aptroot and Lücking, producing very characteristic, multiseptate, and brown ascospores [12,22].

Astrothelium species with astrothelioid ascospores in China are poorly known; up to now, only four species have been recorded: *A. cinnamomeum* (Eschw.) Müll. Arg. [23], *A. sinense* S.H. Jiang and C. Zhang [17], *A. speciosum* Zahlbr. [24], and *A. variolosum* (Ach.) Müll. Arg. [24]. The aim of this study is to investigate the diversity of *Astrothelium* in Southern China, combining morphological, chemical, and molecular data.

2. Materials and Methods

2.1. Material Examined

All materials were collected from Guangdong, Guangxi, Guizhou, and Jiangxi Provinces, and are preserved in the Herbarium Mycologicum Academiae Sinicae-Lichenes, Beijing, China (HMAS–L) and Fungarium of the College of Life Sciences, Liaocheng University, Liaocheng, China (LCUF).

2.2. Phenotypic Analysis

Morphological characters were examined and photographed under a Motic SMZ-168 series stereo microscope and LEICA M125 and DFC450 dissecting microscopes. The anatomical characters were explored with a razor blade under an A2 dissecting stereoscope with an Axio Imager.

2.3. Chemical Analysis

The lichen substances were detected using thin layer chromatography (TLC) [25,26]. The particularly stable and reliable solvent C (toluene/acetic acid 170:30) was used in this study as it often provides the best discrimination of lichen substances [25]. Relative Rf values were determined on the plate by a control mixture: atranorin and norstictic acid, in *Lethariella cladonioides* (Nyl.) Krog. The controls were assigned invariant Rf values, and other spots were measured relative to them.

We utilized the normal procedure—to soak the lichen fragments firstly in c. 1 mL of acetone for 10 min in a small test tube. Then, this solution was used for spotting on the TLC plate. After that, the plate was preequilibrated with glacial acetic acid vapor and subsequently proceeded with elution in solvent C. The plate was dried and then examined under short wavelength (254 nm) ultraviolet light for pigments. Further, it was sprayed with 10% sulfuric acid and heated at 110° in an oven for 10 min to develop the spots. The Rf values and color of each lichen substance were recorded and immediately examined under long wavelength (365 nm) ultraviolet light. The Rf values, as well as the fluorescent properties, were compared and analyzed to confirm the identity of the substance [25,26].

2.4. Phylogenetic Analyses

2.4.1. DNA Extraction and PCR Amplification

Sixteen fresh specimens were chosen for DNA extraction (Supplementary file S1) using the modified CTAB method [27]. The partial region of the internal transcribed spacer (ITS) was amplified using the ITS4 and ITS5 primers [28]. The fungal nuclear ribosomal large subunit (LSU) was amplified using combinations of the primers: LROR-ACCCGCTGAACTTAAGC [29], LR3-GGTCCGTGTTTCAAGAC [29], 1F-CAGTCTGAGTGAATTGCTAA (in this study), and 1R-TTTCTTGACATTGGCATTTG (in this study). The largest subunit of RNA polymerase II gene sequence RPB1 was amplified using the primers RPB1-Af and RPB1-Cr [29].

PCR reactions were carried out in 25 μL containing 1 μL each primer solution (10 μM), 2 μL genomic DNA, 8 μL ddH_2O, and 13 μL 2 × Taq PCR MasterMix® (Cw-bio Inc., Jiangsu, China). Thermocycling of ITS conditions comprised initial denaturation at 95 °C for 5 min; followed by 31 cycles of denaturation at 94 °C for 30 s, annealing at 52 °C for 30 s, elongation at 72 °C for 50 s, and a final extension at 72 °C for 10 min. PCR amplification of LSU and RPB1 included: a 1 min initial denaturation at 94 °C, 38 cycles of 1 min denaturation at 94 °C, 45 s (for LSU) or 90 s (for RPB1) annealing step at 52 °C, 1 min extension at 72 °C, a final extension at 72 °C for 10 min. The target product of PCR was checked by 0.8% agarose electrophoresis gels and sequenced by Majorbio Sanger Inc. (Beijing, China). The new sequences derived in this study were deposited in GenBank (https://www.ncbi.nlm.nih.gov/, accessed on 25 August 2022; Supplementary file S1).

2.4.2. Phylogenetic Analyses

Sequences for each marker in this study were combined with those obtained from GenBank by Basic Local Alignment Search Tool (BLAST) (Supplementary file S1), generating a separate ITS and a concatenated three-locus (ITS, LSU, RPB1) dataset (Supplementary files S2 and S3). The top hits obtained after running a BLAST search of ITS were also included (Supplementary file S4), which can help us judge the novelty of species more easily based on quantitative measurement. Compared to ITS, LSU is relatively conservative, and difficult to distinguish species when they are closely related. However, when the two independent branches in the LSU tree were estimated for large evolutionary divergence, it also indicates that the two species are distinct. Therefore, considering that most of the known sequences of *Astrothelium* are about LSU, we generated a separate LSU dataset for analysis (Supplementary file S5). Estimates of evolutionary divergence between LSU sequences were conducted in MEGA v.7 (Kumar et al., Philadelphia, PA, USA) [30].

For constructing the phylogenetic tree, the genus *Bathelium*, belonging to the same family *Trypetheliaceae* within *Astrothelium*, was chosen as outgroup [10]. Sequences for each marker were firstly aligned independently with MAFFT v.7 (Katoh and Standley, Osaka, Japan) [31], and the combinability was tested as described previously [32]. Only when no significant conflict was detected, would the three markers ITS, LSU, and RPB1 be combined. Bayesian analyses were performed by using MrBayes v.3.2.7 (Ronquist et al., Stockholm, Sweden), as detailed in Jiang et al. [32]. Every 100th generation was sampled as a tree with 5,000,000 generations running. Maximum likelihood (ML) analyses involving 1000 pseudoreplicates were performed using IQ-TREE v2.0.6 (Minh et al., Canberra, Australia) [33]. The best-fit substitution model was selected using ModelFinder [34]. In the ML analyses of ITS sequences, the TIM2 + F + G4 model was selected as the best model according to BIC. TN + F + I + G4 and TIM2 + F + I + G4 were selected as the best models for LSU and the three-locus dataset, respectively. The phylogenetic tree was drawn by FigTree v.1.4.3 (http://tree.bio.ed.ac.uk/software/figtree/, accessed on 25 August 2022).

3. Results

3.1. Phylogenetic Analyses

The dataset included 16 ITS sequences, 10 LSU sequences, and 7 RPB1 sequences newly generated in this study. The ITS (Supplementary file S6) and LSU sequences were analyzed separately (Figure 1) and subsequently compared with the three-marker tree based on a concatenated alignment with 1880 bp (ITS: 455 bp; LSU: 587 bp; RPB1: 838 bp; Figure 2). No different relationships were revealed by the separate analyses for the ITS, LSU, and RPB1 datasets, all with reciprocal posterior probabilities (PP) of 0.99; therefore, these three markers can be combined.

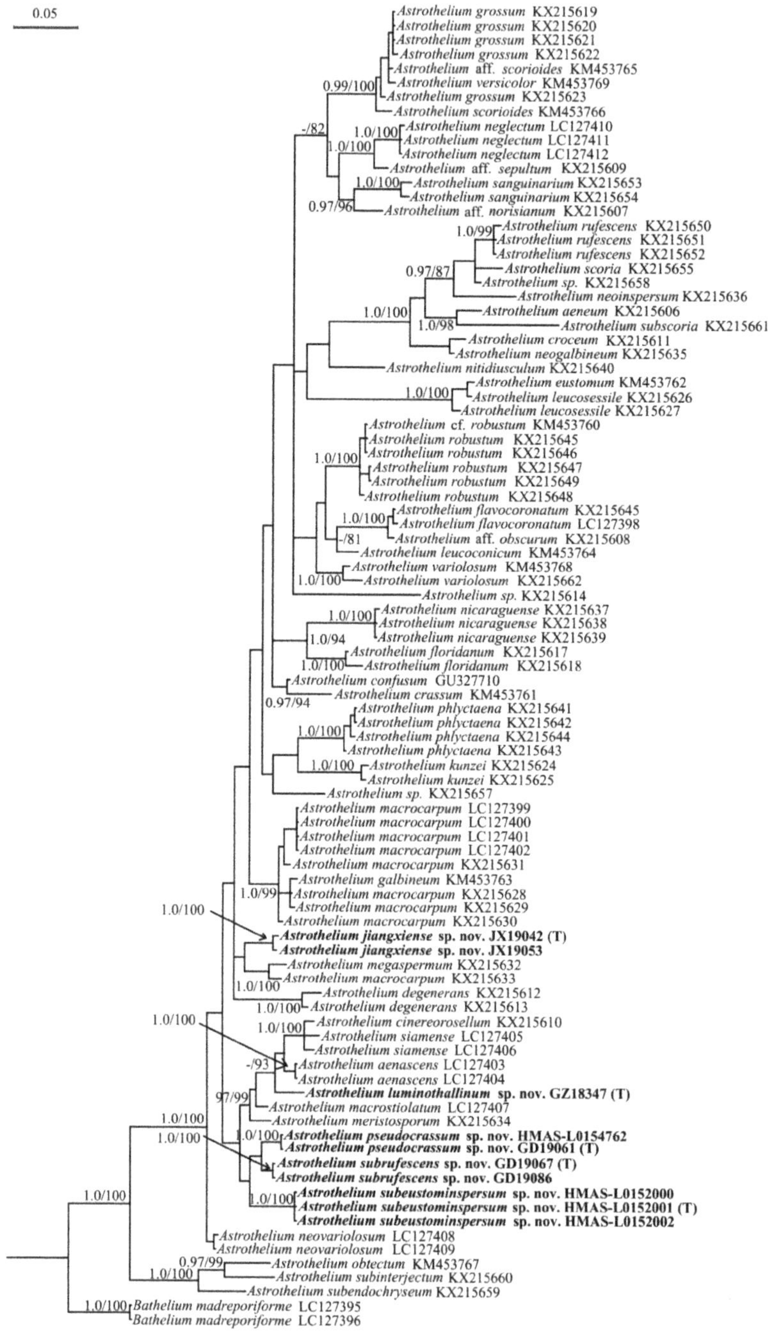

Figure 1. Bayesian tree showing the internal phylogeny of *Astrothelium* based on LSU (587 bp). Bayesian inference posterior probabilities above 0.95 (left) and maximum likelihood bootstrap support above 70% (right) are shown at nodes (B–PP/ML–BP). Terminals in boldface indicate newly generated sequences for this study. The type specimens in the phylogenetic tree were labeled by letter T.

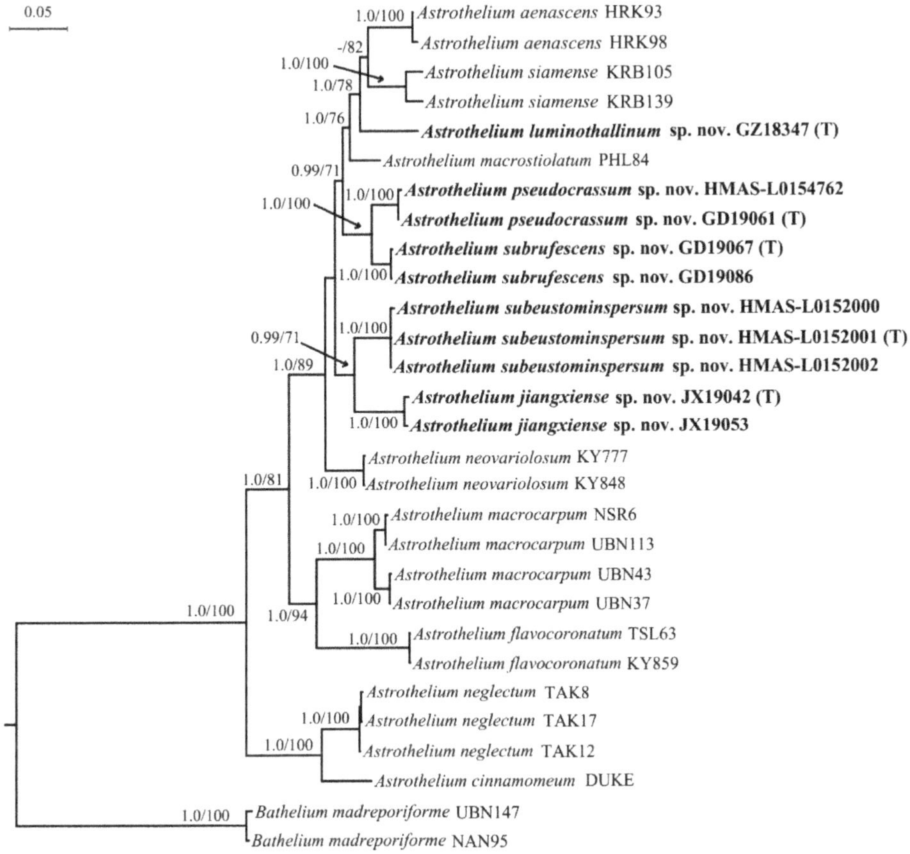

Figure 2. Bayesian tree showing the internal phylogeny of *Astrothelium* based on three markers (ITS, LSU, and RPB1) with an alignment length of 1880 bp. Bayesian inference posterior probabilities above 0.95 (left) and maximum likelihood bootstrap support above 70% (right) are shown at nodes (B–PP/ML–BP). Terminals in boldface indicate newly generated sequences for this study. The type specimens in the phylogenetic tree were labeled by letter T.

As we know, ITS was often used as barcoding to distinguish lichen species due to its high variability [35]. Compared to ITS, LSU is relatively conservative, and its sensitivity is lower at the species level. Thus, LSU is difficult to distinguish species with when they are closely related and similar, but when the two independent lineages in the LSU tree were estimated of large evolutionary divergence (Supplementary file S7), it can help us delimit them into different species.

The single ITS and LSU phylogeny and the combined sequence matrices revealed five new monophyletic lineages corresponding to five new species here: *Astrothelium jiangxiense* S.H. Jiang and C. Zhang sp. nov., *A. luminothallinum* S.H. Jiang and C. Zhang sp. nov., *A. pseudocrassum* S.H. Jiang and C. Zhang sp. nov., *A. subeustominspersum* S.H. Jiang and C. Zhang sp. nov., and *A. subrufescens* S.H. Jiang and C. Zhang sp. nov. Most clades in all trees had strong support through Bayesian and ML analyses. *Astrothelium luminothallinum* is such a unique species that no close relatives were found in all trees (Figures 1 and 2, and Supplementary file S6); they all supported it as a distinct clade. In addition, *A. pseudocrassum* clustered with *A. subrufescens* (Figures 1 and 2, and Supplementary file 6), but they can be distinguished easily in morphology. *A. jiangxiense* and *A. subeustominspersum* were grouped into one clade according to the three-gene combined tree (Figure 2). However, the single ITS and LSU phylogenetic trees did not support this relationship, showing that each species was highly supported and obviously separated from the others (Figure 1 and Supplementary file S6). This minor conflict makes it challenging to evaluate the exact relationships between different clades in *Astrothelium*; however, it does not affect their interpretation as distinct phylogenetic entities.

3.2. Taxonomy

Astrothelium jiangxiense S.H. Jiang & C. Zhang, sp. nov. (Figure 3)

Fungal Names FN 571292.

Etymology: The epithet refers to the discovery of this new species in Jiangxi province, China.

Typus: CHINA, Jiangxi, Shangrao City, Yushan County, Sanqing Mountain, alt. 1295 m, on bark, 17 March 2019, M. Li JX19042 (LCUF, holotype).

Diagnosis: The new species differs from the similar species *Astrothelium scoria* (Fée) Aptroot and Lücking and *A. subscoria* Flakus and Aptroot by the bullate thallus and usually single but not white-covered pseudostromata.

Description: Thallus crustose, corticolous, olive-green to bright green, verrucose to bullate, continuous, without prothallus, 40–90 µm thick, covering areas up to 5 cm diam. Algae trentepohlioid. Ascomata perithecia, conical to pyriform, black, 0.6–1 mm diam., erumpent to prominent, usually single, but sometimes 2–4 aggregated in pseudostromata. Pseudostromata blackish brown. Ostiole apical, black, except for a pale brown rim around 50–100 µm diam. Ascomata wall carbonized, black, 40–260 µm thick. Hamathecium hyaline, interspersed with oil droplets. Asci cylindrical to clavate, 120–140 × 12–15 µm. Ascospores eight per ascus, biseriate, hyaline, fusiform, transversely three-septate, 16–30 × 6–8 µm, lumina diamond-shaped, surrounded by a smooth gelatinous sheath, 2–8 µm wide. Pycnidia not seen.

Chemistry: Thallus UV-, pseudostromata UV-. TLC showed an unidentified substance at Rf five of solvent C, with UV+ red reaction (Supplementary file S8).

Habitat and distribution: The new species grows on the bark of tropical and subtropical regions and is currently only found in China.

Additional specimens examined: CHINA. Jiangxi, Shangrao City, Yushan County, Sanqing Mountain, alt. 1317 m, on bark, 17 March 2019, Z.T. Yao JX19053 (LCUF).

Notes: This taxon can be recognized by its olive-green to bright green bullate thallus (UV-), pale yellow to brown, somewhat conical pseudostromata, usually single ascomata, and apical ostiole. In TLC examination, we found an unknown substance at Rf five of solvent C, UV+ red (Supplementary file S8). It would key out in the recent world key [36] in key I at couplet 5 and 6 (Supplementary file S9) and differs by the bullate thallus and usually single but not white-covered pseudostromata. *Astrothelium scoria* resembles the new species in ascospore characters; however, its thallus is often yellowish or brownish and smooth; besides, ascomata are irregularly grouped to pseudostromata, and no substances can be detected by TLC [12]. Further, the new species and *A. scoria* form independent branches in the phylogenetic tree constructed by LSU (Figure 1) and show large evolutionary divergence (15.4%, Supplementary file S7), which supports them as different species. *A. subscoria* is another similar species but can be distinguished by ascomata with white cover [37]. In addition, the LSU analysis indicates their large genetic distance (16.7%; Figure 1, Supplementary file S7). It is also somewhat similar to *A. aenascens* Aptroot, but the latter has parietin detected by TLC, and ascospores are not surrounded by a gelatinous layer [38]. Furthermore, the molecular evidence support that they are two distinct species (Figures 1 and 2, Supplementary file S6). *A. diaphanocorticatum* Aptroot and Sipmanis is similar to the new species in the bullate thallus and three-septate ascospores, but its asci are wider (120–150 × 16–20 µm) and ascospores are not surrounded by a gelatinous layer [16].

Astrothelium luminothallinum S.H. Jiang and C. Zhang, sp. nov. (Figure 4)

Fungal Names FN 571293.

Etymology: The epithet refers to the thallus containing lichexanthone, contrasting with the pseudostromata, which are absent of lichexanthone.

Typus: CHINA, Guizhou, Libo County, Limingguanxiang, alt. 750 m, on bark, 23 October 2018, Z.F. Jia GZ18347 (LCUF, holotype).

Diagnosis: The new species was similar to *Astrothelium phlyctaena* (Fée) Aptroot and Lücking but can be distinguished by interspersed hamathecium, and UV-negative pseudostromata.

Description: Thallus crustose, corticolous, continuous, grey-green or light green, 2–4 cm diam., 300–600 µm thick, smooth to uneven. Algae trentepohlioid. Ascomata perithecia, conical to pyriform, elongated, black, 0.15–0.40 mm diam., immersed to erumpent, single or aggregated in pseudostromata. Pseudostromata surfaces grey-white or grey-yellow, not concolorous to the thallus, raised, top flattened. Ostiole apical, appearing as blackish dots, black, 0.1–0.3 mm diam. Ascomata wall carbonized, black, 32–72 µm thick. Hamathecium hyaline, interspersed with oil droplets. Asci cylindrical to clavate, colorless, 75–120 × 12–18 µm. Ascospores eight per ascus, biseriate to irregular, hyaline, fusiform, ends rounded, transversely three-septate, 18–29 × 6–10 µm, lumina diamond-shaped, surrounded by a smooth gelatinous sheath, 4–18 µm wide. Pycnidia not seen.

Chemistry: Thallus UV+ yellow, pseudostromata UV-. TLC: lichexanthone.

Habitat and distribution: The new species grows on the bark of tropical and subtropical regions and is currently only found in China.

Additional specimens examined: CHINA, Guizhou, Libo County, Limingguanxiang, alt. 750 m, on bark, 23 October 2018, Z.F. Jia 20191092 (HMAS-L 0154761).

Notes: This new species would key out in the recent world key [36] in key H at couplet 16 and 17 (Supplementary file S9) and is similar to *A. phlyctaena*, but the latter often has clear hamathecium and pseudostromata UV+ yellow [12]. In addition, they formed independent branches in the LSU phylogenetic tree, with large evolutionary divergence (10.2%; Figure 1, Supplementary file S7). Another related species is *A. leucosessile* Lücking, M.P. Nelsen and Aptroot, but it differs by having whitish and sessile pseudostromata [11]. Further, the LSU analysis reveals its large genetic distance from *A. luminothallinum* (14.3%, Supplementary file S7, Figure 1).

Astrothelium pseudocrassum S.H. Jiang and C. Zhang, sp. nov. (Figure 5)

Fungal Names FN 571294.

Etymology: The epithet conveys the potential confusion with *Astrothelium crassum*.

Typus: CHINA, Guangdong, Shenzhen City, Dapeng Peninsula National Geopark, alt. 167 m, on bark, 17 January 2019, F.Y. Liu GD19061 (LCUF, holotype).

Diagnosis: The new species differs from the similar species *Astrothelium crassum* (Fée) Aptroot and Lücking by the thin thallus and usually white-covered pseudostromata.

Description: Thallus crustose, corticolous, pale green, smooth to uneven, continuous, not surrounded by prothallus, 40–100 µm thick, covering areas up to 4 cm diam., not inducing gall formation of the host bark. Algae trentepohlioid. Ascomata perithecia, conical to pyriform, black, 0.3–0.6 mm diam., immersed to erumpent, sometimes single, but often 4–25 aggregated in pseudostromata. Pseudostromata surface white to grey, not concolorous to the thallus, raised, irregular to often linear in outline. Ostiole eccentric, fused, black, 0.2–0.5 mm diam., covered by a whitish rim. Ascomata wall carbonized, black, 50–160 µm thick. Hamathecium hyaline, clear, not interspersed. Asci cylindrical to clavate, 85–120 × 13–16 µm. Ascospores eight per ascus, biseriate to irregular, hyaline, fusiform, transversely three-septate, 20–27 × 6–9 µm, lumina diamond-shaped, surrounded by a smooth gelatinous sheath, 2–11 µm wide. Pycnidia not seen.

Figure 3. *Astrothelium jiangxiense* sp. nov. (**a**,**b**) Thallus with ascomata (holotype, LCUF JX19042); (**c**) ascus, with hamathecium interspersed with oil droplets (LCUF JX19053); (**d**,**e**) ascospores (holotype, LCUF JX19042). Scale bars: (**a**,**b**) = 0.1 mm; (**c**) = 20 μm; (**d**,**e**) = 10 μm.

Chemistry: Thallus UV-, pseudostromata UV-. TLC: no substances detected.

Habitat and distribution: The new species grows on the bark of tropical and subtropical regions and is currently only found in China.

Additional specimens examined: CHINA. Guangdong, Shenzhen City, Dapeng Peninsula National Geopark, alt. 167 m, on bark, 17 January 2019, F.Y. Liu 20191093 (HMAS-L 0154762).

Notes: This species would key out in the recent world key [36] in key K at couplet 16 (Supplementary file S9) and differs from *A. crassum* by the thin thallus and usually white-covered pseudostromata. The latter was often characterized by ascomata with a whitish rim bordering a wide, dark ostiolar area and thickened thallus [12]. Further, they form independent branches in the phylogenetic tree constructed by LSU (Figure 1) and show large evolutionary divergence (8.2%; Supplementary file S7), which supports them as different species. Another related species *A. subdissocians* (Nyl. ex Vain.) Aptroot and Lücking can be distinguished by ascomata with complete whitish cover, including the ostiolar area [12]. *A. nitidiusculum* (Nyl.) Aptroot and Lücking is another similar species in having whitish ascomata with dark ostiolar area, but its pseudostromata are not distinct, and the ostioles are apical [12]. In addition, they form independent branches in the phylogenetic tree constructed by LSU (Figure 1) and show large evolutionary divergence (11.4%; Supplementary file S7), which supports them as two different species.

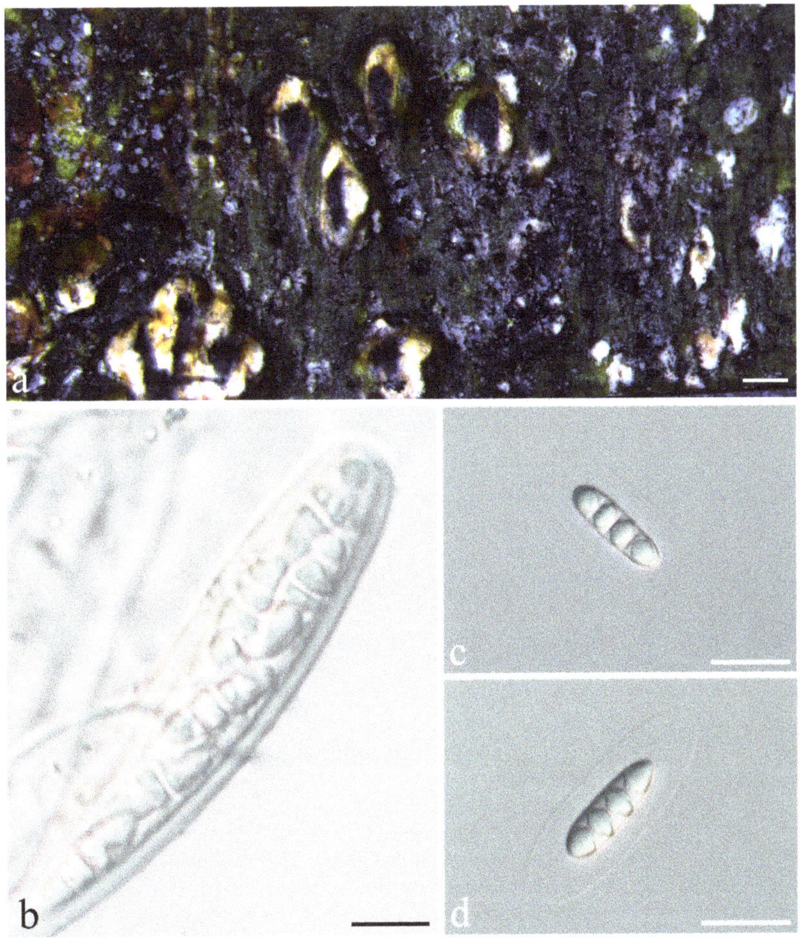

Figure 4. *Astrothelium luminothallinum* sp. nov. (holotype, LCUF GZ18347). (**a**) Thallus and pseudostromata; (**b**) ascus; (**c**,**d**) ascospores. Scale bars: (**a**) = 0.25 mm; (**b**) = 20 μm; (**c**,**d**) = 20 μm.

Astrothelium subeustominspersum S.H. Jiang and C. Zhang, sp. nov. (Figure 6).

Fungal Names FN 571295.

Etymology: The epithet refers to the potential confusion with *Astrothelium eustominspersum*.

Typus: CHINA. Guangxi, Shangsi County, Shiwandashan National Nature Reserve, alt. 290 m, on bark, 21 March 2019, S.H. Jiang and C. Zhang 20190856 (HMAS–L 0152001, holotype).

Diagnosis: The new species can be distinguished from the most related species *Astrothelium eustominspersum* Aptroot and Oliveira-Junior by the lichexanthone only contained in the thallus.

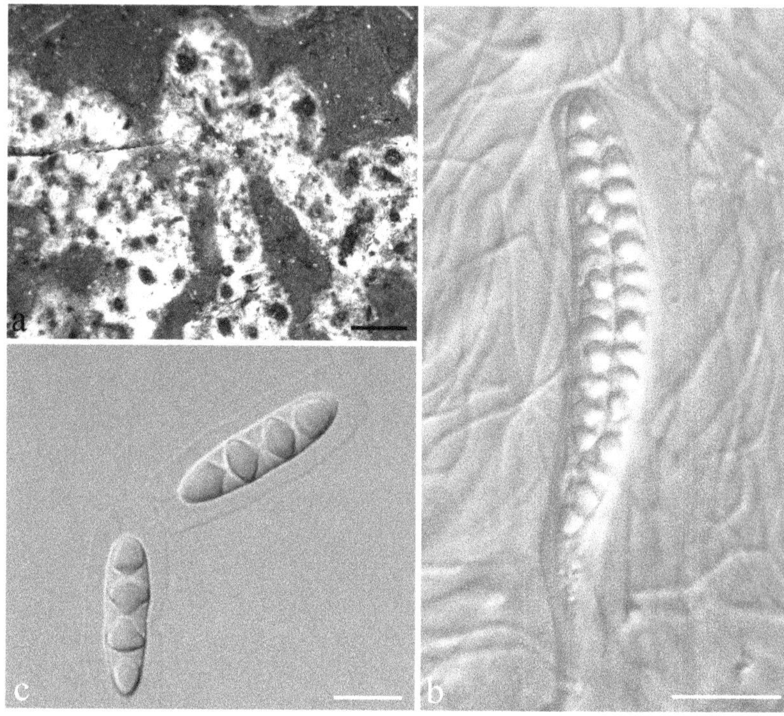

Figure 5. *Astrothelium pseudocrassum* sp. nov. (holotype, LCUF GD19061). (**a**) Thallus with ascomata; (**b**) ascus; (**c**) ascospores. Scale bars: (**a**) = 1 mm; (**b**) = 20 μm; (**c**) = 10 μm.

Description: Thallus crustose, corticolous, greenish-grey to olive-green, smooth to uneven, somewhat shiny, continuous, covering areas up to 6 cm diam., 0.2–0.4 mm thick, not surrounded by a prothallus. Algae trentepohlioid. Ascomata perithecia, globose to pyriform, black, 0.15–0.4 mm diam., single to 2–15 aggregated groups immersed in pseudostromata. Pseudostromata surfaces white-grey to brownish, not concolorous to the thallus, raised, rounded to irregular. Ostiole eccentric, fused, black, 0.2–0.6 mm diam. Ascomata wall carbonized, 30–100 μm thick. Hamathecium hyaline, interspersed with oil droplets. Asci cylindrical to clavate, 88–170 × 13–17 μm. Ascospores eight per ascus, biseriate to irregular, hyaline, fusiform, transversely three-septate, 17–28 × 6–8 μm, lumina diamond-shaped, surrounded by a smooth gelatinous sheath, 1–6 μm wide. Pycnidia not seen.

Chemistry: Thallus UV+, pseudostromata UV-. TLC: lichexanthone.

Habitat and distribution: The new species grows on the bark of tropical and subtropical regions and is currently only found in China.

Additional specimens examined: CHINA. Guangxi, Shangsi County, Shiwandashan National Nature Reserve, alt. 290 m, on bark, 21 March 2019, S.H. Jiang and C. Zhang 20190855 (HMAS–L 0152000), 20190873 (HMAS–L 0152002).

Notes: This species would key out in the recent world key [36] in key J at couplet 17 (Supplementary file S9) and differs from the remaining species keying out there in that it has lichexanthone only in the thallus and that it has an interspersed hamathecium. The most similar species is *Astrothelium eustominspersum*, but the lichexanthone is only in the ostiole [39]. *A. studerae* Aptroot and M. Cáceres resembles this new species, but it can be distinguished by apical ostiole and lichexanthone only in ascomata [16]. *A. neovariolosum* Luangsuph., Aptroot and Sangvichien is also similar to this new species in pseudostromata and ascospores, but it differs in black prothallus, apical ostiole, and smaller perithecia and asci [29]. In addition, they formed two distinct clades in both ITS and ITS-LSU-RPB1 phylogenetic trees (Figure 2 and Supplementary file S6).

Astrothelium subrufescens S.H. Jiang and C. Zhang, sp. nov. (Figure 7).

Fungal Names FN 571296.

Etymology: The epithet refers to the potential confusion with *Astrothelium rufescens*.

Typus: CHINA, Guangdong, Shenzhen City, Dapeng Peninsula National Geopark, alt. 167 m, on bark, 17 January 2019, F.Y. Liu GD19067 (LCUF, holotype).

Diagnosis: The new species can be distinguished from the somewhat similar species *Astrothelium rufescens* (Müll. Arg.) Aptroot and Lücking by the pseudostromata that contain few ascomata.

Description: Thallus crustose, corticolous, continuous, 1–9 cm diam., 30–90 μm thick, smooth to uneven, or bullate, olive-green. Algae trentepohlioid. Ascomata perithecia, black, conical to pyriform, 0.6–1.5 mm diam., immersed to erumpent in pseudostromata. Pseudostromata surfaces grey-white or grey-yellow, not concolorous to the thallus. Ostiole apical, black, with whitish surrounding area, 0.1–0.5 mm diam. Ascomata wall carbonized, black, 55–160 μm thick. Hamathecium is clear, not interspersed with oil droplets. Asci cylindrical to clavate, colorless, 85–117 × 13–16 μm. Ascospores eight per ascus, biseriate to irregular, hyaline, fusiform, ends rounded, transversely three-septate, 19–27 × 6–9 μm, lumina diamond-shaped, surrounded by a smooth gelatinous sheath, 1.5–11 μm wide. Pycnidia not seen.

Chemistry: Thallus UV-, pseudostromata UV-. TLC: no substance detected.

Habitat and distribution: The new species grows on the bark of tropical and subtropical regions and is currently only found in China.

Additional specimens examined: CHINA, Guangdong, Shenzhen City, Dapeng Peninsula National Geopark, alt. 167 m, on bark, 17 January 2019, F.Y. Liu GD19065 (LCUF), GD19086 (LCUF), GD19094 (LCUF). Hainan, Changjiang County, Bawangling National Nature Reserve, alt. 762 m, on bark, 8 December 2019, Z. Chao and Z.T. Yao HN20192176 (HMAS–L 0152007); Wuzhishan City, Wuzhishan National Nature Reserve, alt. 784 m, on bark, 12 December 2019, Z. Chao and Z.T. Yao HN20192312 (HMAS–L 0152041), HN20192317 (HMAS–L 0152042).

Notes: This species would key out in the recent world key [36] in key I at couplet 11–13 (Supplementary file S9). It is closely related to *Astrothelium rufescens*, but the latter species has pseudostromata containing more ascomata [12]. Further, from the LSU phylogenetic tree, they formed into two distinct clades (Figure 1), and their evolutionary divergence is far (13.6%; Supplementary file S7), which also supports them as different species. *A. crassum* also resembles this new species in whitish pseudostromata, but it differs in the ostiole lateral [12]. Furthermore, they formed independent lineages in the LSU tree (Figure 1), with a large genetic distance (7.3%; Supplementary file S7), indicating that they are two different species.

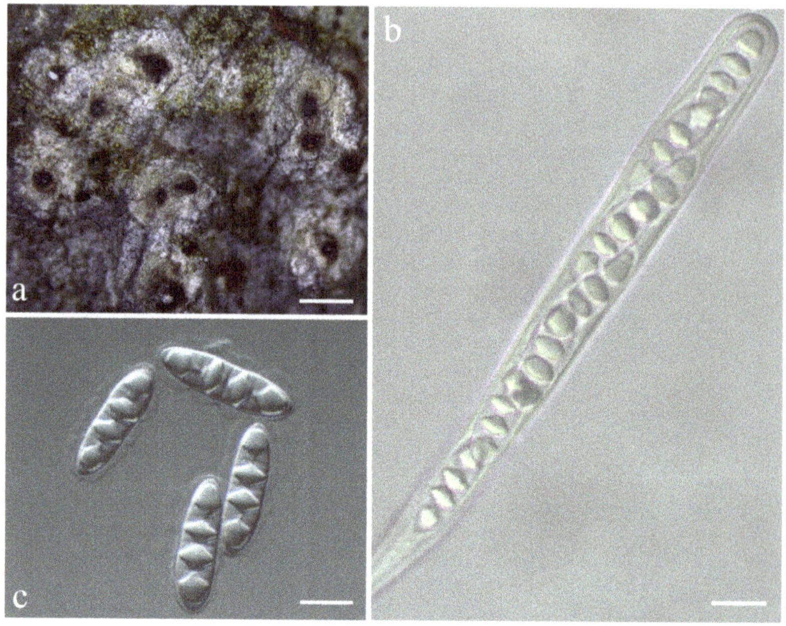

Figure 6. *Astrothelium subeustominspersum* sp. nov. (holotype, HMAS–L 0152001). (**a**) Thallus with ascomata; (**b**) ascus; (**c**) ascospores. Scale bars: (**a**) = 1 mm; (**b**,**c**) = 10 μm.

Figure 7. *Astrothelium subrufescens* sp. nov. (holotype, LCUF GD19067). (**a**) Thallus and pseudostromata; (**b**,**c**) ascus; (**d**–**f**) ascospores. Scale bars: (**a**) = 0.1 mm; (**b**–**f**) = 20 μm.

4. Discussion

The *Trypetheliaceae* is one of the oldest described families in the lichenized *Ascomycota* [2]. Based on molecular data, the core *Trypetheliaceae* were reorganized, with most species now in a single genus, *Astrothelium*, and additional lineages allocated in the other seventeen genera [1]. Species in *Astrothelium* often have a pantropical distribution [1,2,6,18,19]. In Southern China, there is abundant subtropical to tropical evergreen resources [40]. This habitat is favorable for the pyrenocarpous lichens, including *Astrothelium*. However, the genus has not been sufficiently studied; therefore, so far, four species have been recorded from China [17,23,24]. *A. speciosum* and *A. variolosum* were reported by Zahlbruckner (1933) for the first time from China [24]. After that, Aptroot and his colleagues contributed to the research on pyrenocarpous species and other microlichens in tropical China, especially in Hong Kong, Taiwan, and Xishuangbanna regions in Yunnan [23,41–44]. The vast majority of the species they encountered were new records for China. Furthermore, among those surveys of lichens in tropical China, *A. cinnamomeum* was reported in their annotated checklist of the lichens of Hong Kong [23]. During the investigation of lichenized fungi from (sub-)tropical China in our study, some samples were not able to be categorized as any previously described *Astrothelium* species. Based on morphological characteristics and phylogenetic analysis of the combined ITS, LSU, and RPB1 sequence datasets, there is sufficient evidence to verify five species new to science: *Astrothelium jiangxiense* sp. nov., *A. luminothallinum* sp. nov., *A. pseudocrassum* sp. nov., *A. subeustominspersum* sp. nov., and *A. subrufescens* sp. nov., even though the specimens were limited in this study.

In the modern circumscription [12], *Astrothelium* exhibits much variation in perithecial arrangement and ascospore septation, including septate and muriform ascospores. Five new species described here are all characterized by three-septate ascospores and surrounded by a smooth gelatinous sheath, indicating the diversity of *Astrothelium* is higher than previously understood. To better recognize the species relationship of *Astrothelium*, more taxonomic studies should be continuously carried out in the near future. More material will doubtlessly be found if the pyrenocarpous lichen flora in the area is investigated in more detail. Experience shows that with deeper studies, more rare species will be discovered, and there might be quite a number of yet undiscovered taxa. In addition, with further investigations, we expect to discover additional species within muriform ascospores in this genus from China.

Supplementary Materials: The following supporting information can be downloaded at: https://www.mdpi.com/article/10.3390/jof8100994/s1. Supplementary file S1: Specimens used in the phylogenetic analysis; Supplementary file S2: ITS_ alignment dataset; Supplementary file S3: Concatenated three-locus (ITS, LSU, RPB1) dataset; Supplementary file S4: The top hits obtained after running a BLAST search of ITS; Supplementary file S5: LSU_ alignment dataset; Supplementary file S6: The phylogenetic tree of ITS; Supplementary file S7: Estimates of evolutionary divergence between LSU sequences; Supplementary file S8: TLC test of the new speices *Astrothelium jiangxiense* sp. nov.; Supplementary file S9: World key to the species of *Pyrenulaceae* and *Trypetheliaceae*.

Author Contributions: S.-H.J., J.-C.W. and X.-L.W. conceived and designed the study. S.-H.J. and C.Z. generated the DNA sequence data. S.-H.J., C.Z., X.-D.X. and A.A. performed the phenotypic assessment of the material. S.-H.J., C.Z. and A.A. analyzed the data. S.-H.J. and A.A. checked issues related to nomenclatural articles. S.-H.J. wrote the manuscript draft. S.-H.J., A.A. and X.-L.W. revised the manuscript. All authors have read and agreed to the published version of the manuscript.

Funding: This research was supported by the National Natural Science Foundation of China (No. 31750001, 31800010).

Institutional Review Board Statement: Not applicable.

Informed Consent Statement: Not applicable.

Data Availability Statement: The newly generated sequences can be found in GenBank (https://www.ncbi.nlm.nih.gov/, accessed on 25 August 2022; Supplementary file S1). All new taxa were deposited in Fungal Names (https://nmdc.cn/fungalnames, accessed on 25 August 2022).

Acknowledgments: The authors are indebted to Z.F. Jia at the Liaocheng University for the loan of specimens.

Conflicts of Interest: The authors declare no conflict of interest.

References

1. Hongsanan, S.; Hyde, K.D.; Phookamsak, R.; Wanasinghe, D.N.; McKenzie, E.H.C.; Sarma, V.V.; Lücking, R.; Boonmee, S.; Bhat, J.D.; Liu, N.G.; et al. Refined families of *Dothideomycetes*: Orders and families *incertae sedis* in *Dothideomycetes*. *Fungal Divers.* **2020**, *105*, 17–318. [CrossRef]
2. Hyde, K.D.; Jones, E.B.G.; Liu, J.K.; Ariyawansa, H.; Boehm, E.; Boonmee, S.; Braun, U.; Chomnunti, P.; Crous, P.W.; Dai, D.Q.; et al. Families of *Dothideomycetes*. *Fungal Divers.* **2013**, *63*, 1–313. [CrossRef]
3. Nelsen, M.P.; Lücking, R.; Grube, M.; Mbatchou, J.S.; Muggia, L.; Plata, E.R.; Lumbsch, H.T. Unravelling the phylogenetic relationships of lichenised fungi in Dothideomyceta. *Stud. Mycol.* **2009**, *64*, 135–144. [CrossRef] [PubMed]
4. Harris, R.C. Working Keys to the Lichen-Forming Fungi of Puerto Rico. In *Tropical Lichen Workshop*; Catholic University of Puerto Rico: Ponce, Puerto Rico, 1989.
5. Harris, R.C. More Florida Lichens. In *Including the 10¢ Tour of the Pyrenolichens*; The New York Botanical Garden: New York, NY, USA, 1995; pp. 1–182.
6. Aptroot, A.; Lücking, R.; Sipman, H.J.M.; Umaña, L.; Chaves, J.L. Pyrenocarpous lichens with bitunicate asci. A first assessment of the lichen biodiversity inventory in Costa Rica. *Bibl. Lichenol.* **2008**, *97*, 1–162.
7. Del Prado, R.; Schmitt, I.; Kautz, S.; Palice, Z.; Lücking, R.; Lumbsch, H.T. Molecular data place *Trypetheliaceae* in *Dothideomycetes*. *Mycol. Res.* **2006**, *110*, 511–520. [CrossRef] [PubMed]
8. Sweetwood, G.; Lücking, R.; Nelsen, M.P.; Aptroot, A. Ascospore ontogeny and discharge in megalosporous *Trypetheliaceae* and *Graphidaceae* (*Ascomycota*: *Dothideomycetes* and *Lecanoromycetes*) suggest phylogenetic relationships and ecological constraints. *Lichenologist* **2012**, *44*, 277–296. [CrossRef]
9. Nelsen, M.P.; Lücking, R.; Aptroot, A.; Andrew, C.J.; Cáceres, M.; Plata, E.R.; Gueidan, C.; Silva Canêz, L.; Knight, A.; Ludwig, L.R.; et al. Elucidating phylogenetic relationships and genus-level classification within the fungal family *Trypetheliaceae (Ascomycota: Dothideomycetes)*. *Taxon* **2014**, *63*, 974–992. [CrossRef]
10. Lücking, R.; Nelsen, M.P.; Aptroot, A.; De Klee, R.B.; Bawingan, P.A.; Benatti, M.N.; Binh, N.Q.; Bungartz, F.; Cáceres, M.E.S.; Silva Canêz, L.; et al. A phylogenetic framework for reassessing generic concepts and species delimitation in the lichenized family *Trypetheliaceae* (*Ascomycota*: *Dothideomycetes*). *Lichenologist* **2016**, *48*, 739–762. [CrossRef]
11. Lücking, R.; Nelsen, M.P.; Aptroot, A.; Benatti, M.N.; Binh, N.Q.; Gueidan, C.; Guitiérrez, M.C.; Jungbluth, P.; Lumbsch, H.T.; Marcelli, M.P.; et al. A pot-pourri of new species of *Trypetheliaceae* resulting from molecular phylogenetic studies. *Lichenologist* **2016**, *48*, 639–660. [CrossRef]
12. Aptroot, A.; Lücking, R. A revisionary synopsis of the *Trypetheliaceae* (*Ascomycota*: *Trypetheliales*). *Lichenologist* **2016**, *48*, 763–982. [CrossRef]
13. Lücking, R.; Hodkinson, B.P.; Leavitt, S.D. The 2016 classification of lichenized fungi in the *Ascomycota* & *Basidiomycota*—Approaching one thousand genera. *Bryologist* **2017**, *119*, 361–416.
14. Cáceres, M.E.S.; Aptroot, A. Lichens from the Brazilian Amazon, with special reference to the genus *Astrothelium*. *Bryologist* **2017**, *120*, 165–181. [CrossRef]
15. Aptroot, A.; Weerakoon, G. Three new species and ten new records of *Trypetheliaceae* (*Ascomycota*) from Sri Lanka. *Cryptogam. Mycol.* **2018**, *39*, 373–377. [CrossRef]
16. Aptroot, A.; Sipman, H.J.M.; Barreto, F.M.O.; Nunes, A.D.; Cáceres, M.E.S. Ten new species and 34 new country records of *Trypetheliaceae*. *Lichenologist* **2019**, *51*, 27–43. [CrossRef]
17. Jiang, S.H.; Zhang, C.; Yao, Z.T.; Liu, H.J. A species new to science and a new Chinese record of *Astrothelium* (*Typetheliaceae*, *Ascomycota*). *Mycosystema* **2021**, *40*, 31–39.
18. Harris, R.C. The family *Trypetheliaceae* (*Loculoascomycetes*: Lichenized *Melanommatales*) in Amazonian Brazil. *Suppl. Acta Amaz.* **1984**, *14*, 55–80. [CrossRef]
19. Awasthi, D.D. A key to the microlichens of India, Nepal and Sri Lanka. *Bibl. Lichenol.* **1991**, *40*, 1–340.
20. Kirk, P.M.; Cannon, P.F.; Minter, D.W.; Stalpers, J.A. *Dictionary of the Fungi*, 10th ed.; Wallingford, CAB International: Wallingford, UK, 2008.
21. Aptroot, A. Trypetheliaceae. *Flora Aust.* **2009**, *57*, 535–552.
22. Medina, E.S.; Aptroot, A.; Lücking, R. *Aspidothelium silverstonei* and *Astrothelium fuscosporum*, two new corticolous lichen species from Colombia. *Cryptogam. Mycol.* **2017**, *38*, 253–258. [CrossRef]
23. Aptroot, A.; Seaward, M.R.D. Annotated checklist of Hongkong Lichens. *Trop. Bryol.* **1999**, *17*, 57–101.
24. Zahlbruckner, A. Flechten der Insel Formosa in Feddes, Lichenes in Feddes, *Repertorium* sp. *Feddes Repert.* **1933**, *31*, 194–224. [CrossRef]

25. Elix, J.A. *A Catalogue of Standardized Thin-Layer Chromatographic Data and Biosynthetic Relationships for Lichen Substances*, 3rd ed.; Australian National University: Canberra, Australia, 2014.
26. Orange, A.; James, P.W.; White, F.J. *Microchemical Methods for the Identification of Lichens*; British Lichen Society: London, UK, 2010.
27. Rogers, S.O.; Bendich, A.J. Extraction of DNA from Plant Tissues. In *Plant Molecular Biology Manual*; Gelvin, S.B., Schilperoort, R.A., Verma, D.P.S., Eds.; Kluwer Academic Publishers: Dordrecht, The Netherlands, 1988; pp. 1–10.
28. White, T.J.; Bruns, T.; Lee, S.J.; Taylor, J.L. Amplification and direct sequencing of fungal ribosomal RNA genes for phylogenetics. *PCR Protoc. A Guide Methods Appl.* **1990**, *18*, 315–322.
29. Luangsuphabool, T.; Lumbsch, H.T.; Aptroot, A.; Piapukiew, J.; Sangvichien, E. Five new species and one new record of *Astrothelium* (*Trypetheliaceae, Ascomycota*) from Thailand. *Lichenologist* **2016**, *48*, 727–737. [CrossRef]
30. Kumar, S.; Stecher, G.; Tamura, K. MEGA7: Molecular evolutionary genetics analysis version 7.0 for bigger datasets. *Mol. Biol. Evol.* **2016**, *33*, 1870–1874. [CrossRef] [PubMed]
31. Katoh, K.; Standley, D.M. MAFFT multiple sequence alignment software version 7: Improvements in performance and usability. *Mol. Biol. Evol.* **2013**, *30*, 772–780. [CrossRef] [PubMed]
32. Jiang, S.H.; Lücking, R.; Liu, H.J.; Wei, X.L.; Xavier-Leite, A.B.; Portilla, C.V.; Ren, Q.; Wei, J.C. Twelve new species reveal cryptic diversification in foliicolous lichens of *Strigula* s.lat. (*Strigulales, Ascomycota*). *J. Fungi* **2022**, *8*, 2. [CrossRef] [PubMed]
33. Minh, B.Q.; Schmidt, H.A.; Chernomor, O.; Schrempf, D.; Woodhams, M.D.; Haeseler, A.; Lanfear, R. IQ-TREE 2: New models and efficient methods for phylogenetic inference in the genomic era. *Mol. Biol. Evol.* **2020**, *37*, 1530–1534. [CrossRef]
34. Kalyaanamoorthy, S.; Minh, B.Q.; Wong, T.K.F.; von Haeseler, A.; Jermiin, L.S. ModelFinder: Fast model selection for accurate phylogenetic estimates. *Nat. Methods* **2017**, *14*, 587–589. [CrossRef]
35. Kelly, L.J.; Hollingsworth, P.M.; Coppins, B.J.; Ellis, C.J.; Harrold, P.; Tosh, J.; Yahr, R. DNA barcoding of lichenized fungi demonstrates high identification success in a floristic context. *New Phytol.* **2011**, *191*, 288–300. [CrossRef]
36. Aptroot, A. World key to the species of *Pyrenulaceae* and *Trypetheliaceae*. *Arch. Lichenol.* **2021**, *29*, 1–91.
37. Flakus, A.; Kukwa, M.; Aptroot, A. *Trypetheliaceae* of Bolivia: An updated checklist with descriptions of twenty-four new species. *Lichenologist* **2016**, *48*, 661–692. [CrossRef]
38. Aptroot, A.; Ertz, D.; Etayo, J.A.S.; Gueidan, C.; Mercado-Díaz, J.A.; Schumm, F.; Weerakoon, G. Forty-six new species of *Trypetheliaceae* from the tropics. *Lichenologist* **2016**, *48*, 609–638. [CrossRef]
39. Junior, I.O.; Aptroot, A.; Marcela, E.S.C. Lichens from Monte Pascoal, Bahia, Brazil, with some new pyrenocarpous species and a key to the *Pyrenula* species from Brazil. *Bryologist* **2021**, *124*, 552–568. [CrossRef]
40. Zhu, H. The tropical forests of Southern China and conservation of biodiversity. *Bot. Rev.* **2017**, *83*, 87–105. [CrossRef]
41. Aptroot, A.; Sparrius, L.B. New microlichens from Taiwan. *Fungal Divers.* **2003**, *14*, 1–50.
42. Aptroot, A. Pyrenocarpous lichens and related nonlichenized ascomycetes from Taiwan. *J. Hattort. Bot. Lab.* **2003**, *93*, 155–173.
43. Aptroot, A.; Ferraro, L.I.; Lai, M.J.; Sipman, H.J.M.; Sparrius, L.B. Foliicolous lichens and their lichenicolous ascomycetes from Yunnan and Taiwan. *Mycotaxon* **2003**, *88*, 41–47.
44. Aptroot, A.; Sipman, H.J.M. New Hong Kong lichens, ascomycetes and lichenicolous fungi. *J. Hattort. Bot. Lab.* **2001**, *91*, 317–343.

Journal of *Fungi*

MDPI

Article

Morphological and Phylogenetic Analyses Reveal Four New Species of *Acrodictys* (*Acrodictyaceae*) in China

Shi Wang [1,2], Rongyu Liu [2], Shubin Liu [2], Zhaoxue Zhang [2], Jiwen Xia [2], Duhua Li [2], Xiaoyong Liu [1,*] and Xiuguo Zhang [1,2,*]

1 College of Life Sciences, Shandong Normal University, Jinan 250358, China
2 Shandong Provincial Key Laboratory for Biology of Vegetable Diseases and Insect Pests, College of Plant Protection, Shandong Agricultural University, Taian 271018, China
* Correspondence: 622001@sdnu.edu.cn (X.L.); zhxg@sdau.edu.cn (X.Z.)

Abstract: During our ongoing survey of dematiaceous hyphomycetes associated with dead branches in tropical forests, eight *Acrodictys* isolates were collected from Hainan, China. Morphology from the cultures and phylogeny based on partial small subunit (SSU), entire internal transcribed spacer regions with intervening 5.8S (ITS), partial large subunit (LSU) of rRNA gene, partial beta-tubulin (tub2), and partial RNA polymerase II second largest subunit (rpb2) genes were employed to identify these isolates. As a result, four new species, namely *Acrodictys bawanglingensis* sp. nov., *A. diaoluoshanensis* sp. nov., *A. ellisii* sp. nov., and *A. pigmentosa* sp. nov., are introduced. Illustrations and descriptions of the four taxa are provided, along with comparisons with closely related taxa in the genus. For facilitating relative studies, an updated key to all accepted species of this genus is also compiled.

Keywords: morphology; new taxa; phylogeny; *Sordariomycetes*; taxonomy

Citation: Wang, S.; Liu, R.; Liu, S.; Zhang, Z.; Xia, J.; Li, D.; Liu, X.; Zhang, X. Morphological and Phylogenetic Analyses Reveal Four New Species of *Acrodictys* (*Acrodictyaceae*) in China. *J. Fungi* **2022**, *8*, 853. https://doi.org/10.3390/jof8080853

Academic Editors: Cheng Gao and Lei Cai

Received: 8 July 2022
Accepted: 5 August 2022
Published: 15 August 2022

1. Introduction

Acrodictys M.B. Ellis was erected by Ellis and typified by *A. bambusicola* M.B. Ellis [1]. In the protologue, this genus was characterized by globose to subglobose, uniformly pigmented, angularly, or obliquely septate conidia. This broad, generic circumscription of *Acrodictys* has been followed for nearly 40 years until recent segregations starting at the beginning of this century [2–4].

Acrodictyella W.A. Baker and Partr., typified by *A. obovata* W.A. Baker and Partr., was considered the first genus to accommodate those species similar to *Acrodictys* but characterized by producing hyaline, muriform conidia, which secede well before maturation and become pigmented sometime after their release [2]. Based on conidial morphology, conidiogenesis, and conidial secession, Baker et al. [3,5] and Baker and Morgan-Jones [4] divided the genus *Acrodictys* sensu lato into four genera, viz. *Acrodictys* sensu stricto, *Junewangia* W.A. Baker and Morgan-Jones, *Pseudoacrodictys* W.A. Baker and Morgan-Jones and *Rhexoacrodictys* W.A. Baker and Morgan-Jones. The *Acrodictys* sensu stricto is mainly characterized by macronematous, mononematous, cylindrical, unbranched or infrequently branched, percurrently proliferating conidiophores, and muriform conidia [6,7]. Based on morphological and phylogenetic analyses and typified by the genus *Acrodictys* sensu stricto, *Acrodictyaceae* J.W. Xia and X.G. Zhang was introduced [7].

Species in *Acrodictys* sensu stricto are saprobic on dead branches. They are thought to be wood-decaying fungi, promoting the carbon cycle in the ecosystem by converting cellulose, hemicellulose, and lignin into inorganic substances [7]. Although having a worldwide distribution, they are mainly recorded from tropical areas, such as Brazil, Thailand, and Mexico (https://www.gbif.org, accessed on 25 June 2022). Hainan Province (18°10′–20°10′ N,108°37′–111°05′ E) is an island in southern China. Its annual mean temperature is 22–27 °C, and its annual precipitation is 1000–2600 mm. It is a typical tropical rainforest climate, suitable for the growth and reproduction of wood-decaying fungi.

In China, 12 isolates representing 9 species (*A. bambusicola*, *A. elaeidicola*, *A. fluminicola*, *A. globulosa*, *A. hainanensis*, *A. liputii*, *A. malabarica*, *A. peruamazonensis*, and *A. porosiseptata*) have been documented from tropical areas in Hainan and Yunnan Provinces [6,7]. The samples in this study were collected from the Bawangling National Nature Reserve and Diaoluoshan National Nature Reserve in Hainan Province. Dead branches of unidentified trees were collected in tropical rain forests dominated by *Lauraceae* and *Fagaceae* trees, and the fungi in pure culture were isolated from these dead branches. Fungi from these axenic cultures were identified based on classic morphological and modern molecular approaches. As a result, four new species are described and illustrated herein in the genus *Acrodistys* sensu stricto.

2. Materials and Methods

2.1. Isolation and Morphological Observation

Samples were collected from the Bawangling National Nature Reserve and Diaoluoshan National Nature Reserve, Hainan Province, China (108°37′–117°50′ E, 3°58′–20°20′ N), and taken to the laboratory in plastic bags. The samples were placed in Petri dishes with moist filter paper and cultured in an incubator at 25 °C for 1–2 weeks. The samples were examined using a stereomicroscope (SZX10 model, Olympus, Tokyo, Japan, https://www.olympus-lifescience.com.cn/zh/microscopes/, accessed on 25 June 2022) to locate sporulating structures. Isolations were made with single spore methods according to the following steps: Spores were suspended with sterilized water and spread onto PDA (PDA: 200 g potato, 20 g dextrose, 20 g agar, 1000 mL distilled water, pH 7.0) plate and incubated for one day under a biochemical incubator. After germination, colonies were transferred to a new PDA plate to obtain a pure culture. After 30 days of incubation, morphological characters were recorded. Photographs of the colonies were taken on the 30th day using a digital camera (G7X model, Canon, Tokyo, Japan, https://www.canon.com.cn/, accessed on 25 June 2022). The micromorphological characteristics of colonies were observed using two stereomicroscopes (SZX10 model and BX53 model, Olympus, Tokyo, Japan, https://www.olympus-lifescience.com.cn/zh/microscopes/, accessed on 25 June 2022), both fitted with a digital camera (DP80 model, Olympus, Tokyo, Japan, https://www.olympus-lifescience.com.cn/zh/microscopes/, accessed on 25 June 2022). Sporulating structures were mounted in water for microscopic observation and photomicrography. At least 50 mature conidia and conidiophores were measured using the cellSens software (Olympus, Tokyo, Japan), and images were used for further processing with Photoshop ver. CS5 (Adobe Systems, San Jose, CA, USA, https://www.adobe.com/cn/, accessed on 25 June 2022). The cultures were stored in 10% sterilized glycerin and sterile water at 4 °C for further studies. Specimens were deposited in the Herbarium Mycologicum Academiae Sinicae, Institute of Microbiology, Chinese Academy of Sciences, Beijing, China (HMAS), and the Herbarium of the Department of Plant Pathology, Shandong Agricultural University (HSAUP). Living cultures were deposited in the Shandong Agricultural University Culture Collection (SAUCC). New taxa were submitted to the taxonomic database MycoBank (http://www.mycobank.org, accessed on 25 June 2022) and obtained relative deposition numbers.

2.2. DNA Extraction and Amplification

Genomic DNA was extracted from fungal mycelia grown on PDA, using a modified cetyltrimethylammonium bromide (CTAB) protocol as described in Guo et al. [8]. Five loci, namely the internal transcribed spacer regions with intervening 5.8S rRNA gene (ITS), partial large subunit (LSU) rRNA gene, the partial small subunit (SSU) rRNA gene, the partial RNA polymerase II second largest subunit (rpb2), and part of the beta-tubulin gene region (tub2) were amplified with the primer pairs and polymerase chain reaction (PCR) processes listed in Table 1.

PCR was performed using a thermocycler (Mastercycler X50, Eppendorf, Hamburg, Germany, https://www.eppendorf.com/CN-zh/, accessed on 25 June 2022). Amplifications were performed in a 25 μL reaction volume, which contained 12.5 μL 2 × Taq Plus Master Mix II (Vazyme, Nanjing, China), 1 μL of each forward and reverse primer (10 μM) (Tsingke Company Limited, Qingdao, China), 1 μL template genomic DNA (approximately 1 ng/μL), and 10 μL distilled deionized water. PCR products were visualized through 1% agarose gel electrophoresis. Sequencing was bidirectionally conducted by a company (Tsingke Company Limited, Qingdao, China), and a consensus was obtained using MEGA 7.0 [9]. All sequences generated in this study were deposited in GenBank under the accession numbers listed in Table 2.

2.3. Phylogenetic Analyses

Novel sequences were generated from the eight strains in this study, and all available reference sequences of *Acrodictys* species were downloaded from GenBank [7,10]. No rpb2 sequences are available in GenBank, and only two sequences are available for tub2. Multiple sequences for ITS, LSU, and SSU were aligned and concatenated using an online program (MAFFT v.7.11, http://mafft.cbrc.jp/alignment/server/, accessed on 25 June 2022) [11], with default settings and manually corrected where necessary (Supplementary File S1). To establish the identity of the isolates at a species level, phylogenetic analyses were conducted for each locus and for a combination of all three loci (ITS, LSU, and SSU), based on maximum likelihood (ML) and Bayesian inference (BI). The best evolutionary model for each partition was determined using MrModeltest v. 2.3 [12] and then incorporated into the BI analyses. ML and BI were run through the CIPRES Science Gateway portal (https://www.phylo.org, accessed on 25 June 2022) using RaxML-HPC2 on XSEDE 8.2.12 (Heidelberg, Germany) [13,14] and MrBayes on XSEDE 3.2.7a (Stockholm, Sweden), respectively [15–17]. ML was performed using the default parameters, and BI was carried out using the rapid bootstrapping algorithm with the automatic halt option. Bayesian analyses included four parallel runs of 5,000,000 generations and a sampling frequency of 500 generations. The burn-in fraction was set to 0.25, and posterior probabilities (PP) were determined from the remaining trees. The resulting trees were plotted using FigTree v. 1.4.4 (http://tree.bio.ed.ac.uk/software/figtree, accessed on 25 June 2022) and edited with Adobe Illustrator CS6.0 (https://www.adobe.com/cn/, accessed on 25 June 2022).

Table 1. Gene regions, respective primer pairs, and PCR programs used in the study.

Locus	Primer	Sequence (5′–3′)	PCR Program	References
ITS	ITS5 ITS4	GGA AGT AAA AGT CGT AAC AAG G TCC TCC GCT TAT TGA TAT GC	(94 °C for 30 s, 55 °C for 30 s, 72 °C for 45 s) × 29 cycles	[18]
LSU	LR0R LR5	GTA CCC GCT GAA CTT AAG C TCC TGA GGG AAA CTT CG	(94 °C for 30 s, 48 °C for 50 s, 72 °C for 90 s) × 35 cycles	[19,20]
SSU	NS1 NS4	GTA GTC ATA TGC TTG TCT C CTT CCG TCA ATT CCT TTA AG	(94 °C for 30 s, 50 °C for 45 s, 72 °C for 90 s) × 35 cycles	[18]
tub2	Bt-2a Bt-2b	GGT AAC CAA ATC GGT GCT GCT TTC ACC CTC AGT GTA GTG ACC CTT GGC	(95 °C for 30 s, 56 °C for 30 s, 72 °C for 60 s) × 35 cycles	[21]
rpb2	fRPB2-5F fRPB2-7R	GGG GWG AYC AGA AGA AGG C CCC ATR GCT TGY TTR CCC AT	(95 °C for 50 s, 56 °C for 50 s, 72 °C for 120 s) × 37 cycles	[22]

Table 2. Strains and GenBank accession numbers used in this study.

Species	Strain	GenBank Accession Number				
		ITS	LSU	SSU	rpb2	tub2
Acrodictys aquatica	MFLUCC 18-0356 *	MG835711	MG835712	–	–	–
A. bambusicola	CGMCC 3.18641	KU999973	KX033564	KX033535	–	KX036219

Table 2. *Cont.*

Species	Strain	GenBank Accession Number				
		ITS	LSU	SSU	rpb2	tub2
A. bawanglingensis	**SAUCC 1342 ***	**ON606324**	**ON614219**	**ON620164**	**ON859853**	**ON859845**
	SAUCC 1343	**ON606325**	**ON614220**	**ON620165**	**ON859852**	**ON859844**
A. diaoluoshanensis	**SAUCC 1601 ***	**ON645265**	**ON644407**	**ON645269**	**ON859847**	**ON859839**
	SAUCC 1602	**ON645266**	**ON644408**	**ON645270**	**ON859846**	**ON859838**
A. ellisii	**SAUCC 1471 ***	**ON645254**	**ON644405**	**ON645267**	**ON859851**	**ON859843**
	SAUCC 1472	**ON645255**	**ON644406**	**ON645268**	**ON859850**	**ON859842**
A. elaeidicola	CGMCC 3.18642	KU999977	KX033568	KX033539	–	–
	CGMCC 3.18643	KU999978	KX033569	KX033540	–	–
A. fluminicola	KUMCC 15-0240 *	MK828642	MK849786	–	–	–
A. globulosa	CGMCC 3.18644	KU999970	KX033562	KX033532	–	–
A. hainanensis	CGMCC 3.18645 *	KU999974	KX033565	KX033536	–	–
A. liputii	CGMCC 3.18646	KU999966	KX033558	KX033528	–	–
	CGMCC 3.18647	KU999979	KX033570	KX033541	–	–
	MFLUCC 18-0323	MZ412512	MZ412524	MZ413269	–	–
A. malabarica	CGMCC 3.18648	KU999968	KX033560	KX033530	–	–
A. peruamazonensis	CGMCC 3.18649	KU999969	KX033561	KX033531	–	–
A. pigmentosum	**SAUCC 1591 ***	**ON606326**	**ON614221**	**ON620166**	**ON859849**	**ON859841**
	SAUCC 1592	**ON606327**	**ON614222**	**ON620167**	**ON859850**	**ON859842**
A. porosiseptata	CGMCC 3.18650	KU999967	KX033559	KX033529	–	KX036220
Fluminicola saprophytica	MFLUCC 15-0976 *	MF374358	MF374367	MF374375	MF370954	–

Notes: New species established in this study are in bold. Ex-type strains are marked with "*".

3. Results

3.1. Phylogenetic Analyses

Eight strains of *Acrodictys* were isolated from the dead branches of unidentified trees in Hainan Province, China. The alignment of ITS, LSU, and SSU sequences was composed of 21 strains (including the 8 new strains) of *Acrodictys* and *Fluminicola saprophytica* (MFLUCC 15-0976) as the outgroup taxon, and 2419 characters, viz. 1–592 (ITS), 593–1419 (LSU), and 1420–2419 (SSU). Of these characters, 2089 were constant, 94 were variable and parsimony uninformative, and 236 were parsimony informative.

The Bayesian analysis lasted for 65,000 generations, resulting in 1301 total trees, of which 976 trees were used to calculate the posterior probabilities. The ML tree topology is consistent with that from the BI analyses, and therefore, the ML tree is presented with BI posterior probabilities being plotted together (Figure 1).

The 22 strains are assigned to 15 species clades based on the three-locus phylogeny (Figure 1). The involved eight strains represent four new species, namely *Acrodictys bawanglingensis* (Figure 2), *A. diaoluoshanensis* (Figure 3), *A. ellisii* (Figure 4), and *A. pigmentosa* (Figure 5), forming a clade together with *A. bambusicola* (MLBS = 96% and BYPP = 1.0).

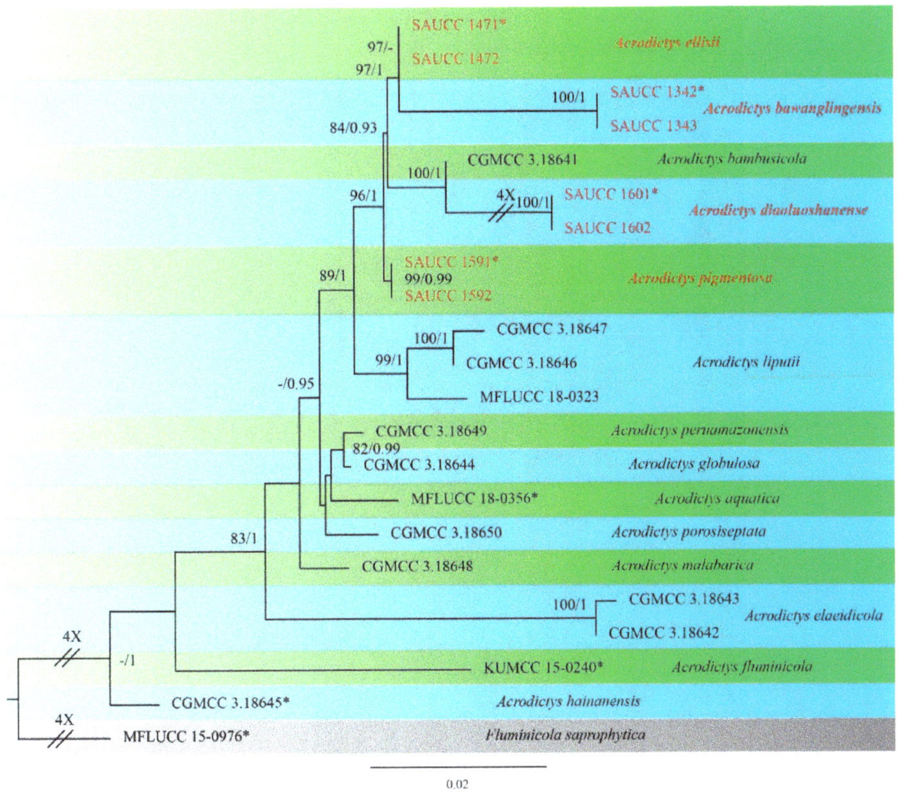

Figure 1. Phylogram generated from RAxML analysis based on combined ITS, LSU, and SSU sequence data, demonstrating phylogenetic relationships within *Acrodictys* with the *Fluminicola saprophytica* MFLUCC 15-0976 as outgroup. The maximum likelihood bootstrap value (MLBV ≥ 70%) and Bayesian inference posterior probability (BIPP ≥ 0.90) are shown at relative nodes at the first and second positions, respectively. Strains marked with "*" are ex-types. Strains from this study are shown in red. Three branches were shortened to fit the page size and are indicated by a double slash (//) with a fold number showing how many times they are shortened.

3.2. Taxonomy

3.2.1. *Acrodictys bawanglingensis* S. Wang, J.W. Xia, X.Y. Liu, and X.G. Zhang, sp. nov.

MycoBank No. 844582

Etymology—The epithet *bawanglingensis* is named after the Bawangling National Nature Reserve where the holotype was collected.

Type—China, Hainan Province: the Bawangling National Nature Reserve (109°03′–109°17′ E, 18°57′–19°11′ N), on the dead branches of unidentified trees collected in tropical rain forests dominated by *Lauraceae* and *Fagaceae* trees, 19 May 2021, R.Y. Liu, holotype HMAS 352233, ex-holotype living culture SAUCC 1342.

Description—Asexual morph on PDA: Mycelia are white to pale brown, floccose cottony, reverse black. Conidiophores are macronematous, mononematous, erect, unbranched, straight or flexuous, thick-walled, smooth, dark brown at the base, paler toward the apex, septate, 60.0–120.0 × 4.5–6.5 μm. Conidiogenous cells are monoblastic, integrated, terminal, determinate, cylindrical, pale brown to brown, and smooth. Conidia are solitary, muriform, obovoid to obpyriform, pale brown to brown, 18.0–26.0 × 10.0–16.0 μm, usually with 3 transverse septa and 1–3 longitudinal septa, slightly constricted at the septa, with conspicuous pores in the septa, and truncate at the base. Chlamydospores were not observed. Sexual morphs are unknown.

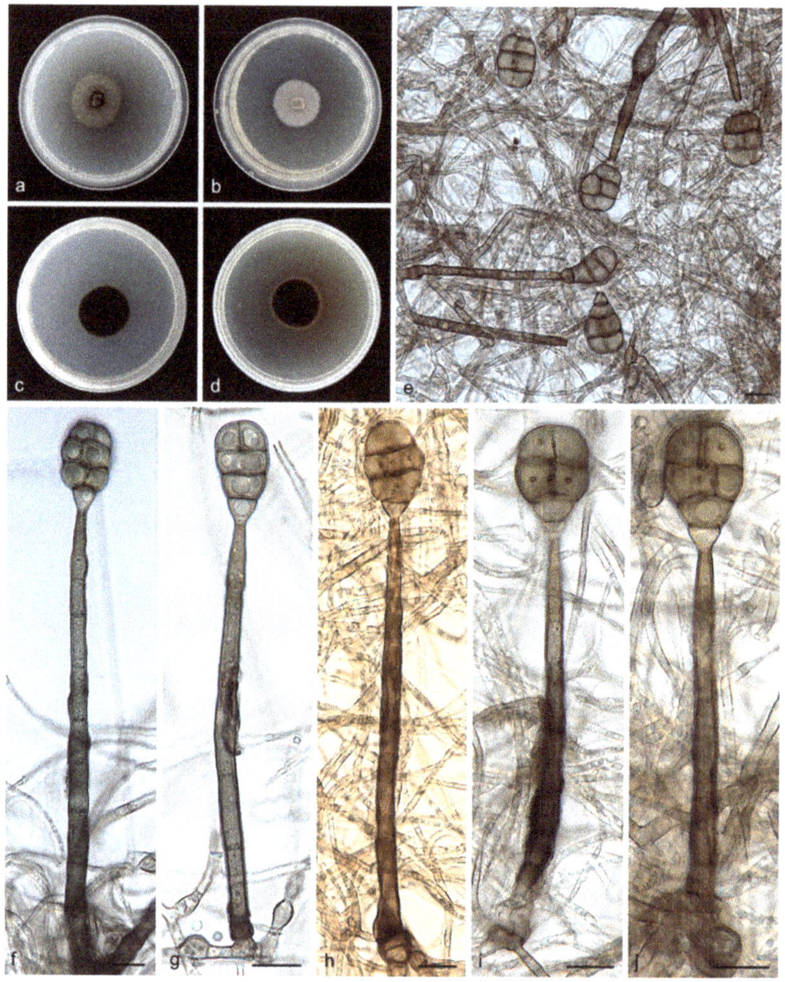

Figure 2. *Acrodictys bawanglingensis* (holotype HMAS 352233, ex-holotype SAUCC 1342): (**a**,**b**) surface sides of colony after incubation for 7 days on PDA (**a**) and MEA (**b**);(**c**,**d**) reverse sides of colony after incubation for 7 days on PDA (**c**) and MEA (**d**); (**e**–**j**) conidiophores, conidiogenous cells, and conidia. Petri dish diameter: 90 mm (**a**–**d**). Scale bars: 10 μm (**e**–**j**).

Culture characteristics—Colonies on PDA are flat with an entire margin, attaining 25.0–30.0 mm in diameter after 14 days at 25 °C, with a growth rate of 1.5–2.5 mm/day, a greenish-brown color, and with a layer of white aerial hyphae on the surface. Colonies on MEA are flat with an entire margin, have a generally mouse-gray color, and are covered with a layer of white-to-gray, dense, aerial mycelia that are floccose cottony; the reverse is black with a pale brown margin.

Additional specimen examined—China, Hainan Province: the Bawangling National Nature Reserve (109°03′–109°17′ E, 18°57′–19°11′ N), on the dead branches of unidentified trees collected in tropical rain forests dominated by *Lauraceae* and *Fagaceae* trees, 19 May 2021, R.Y. Liu, paratype HMAS 352234, ex-paratype living culture SAUCC 1343.

Notes—Strains SAUCC 1342 and SAUCC 1343 have similar morphological features and identical DNA sequences and form a monophyletic group with a long branch and robust support values (MLBV = 100% and BIPP = 1.00, Figure 1). Therefore, SAUCC 1342 and SAUCC 1343 are identified as the same new species: *Acrodictys bawanglingensis*.

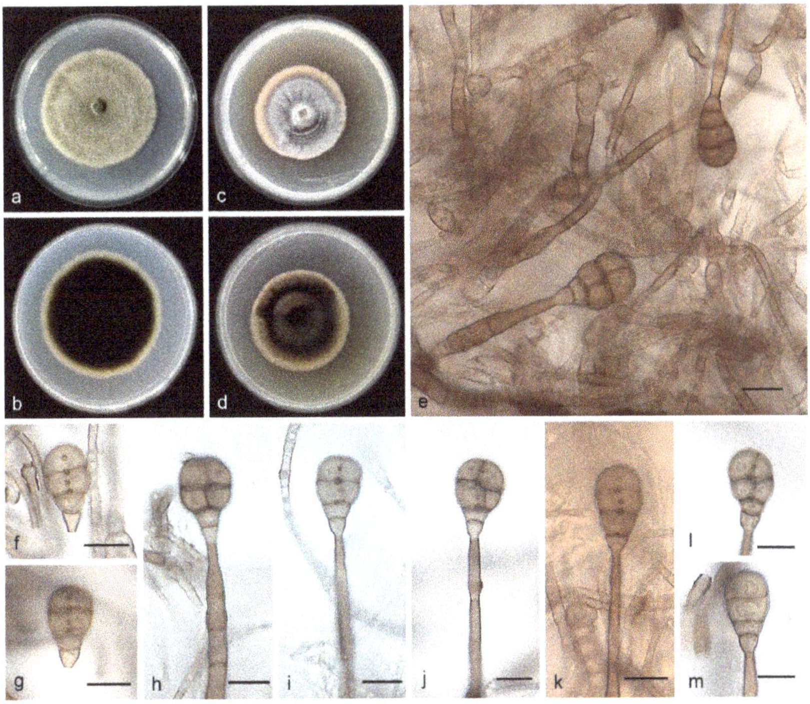

Figure 3. *Acrodictys diaoluoshanensis* (holotype HMAS 352235, ex-holotype SAUCC 1601): (**a**,**b**) surface sides of colony after incubation for 30 days on PDA (**a**) and MEA (**b**); (**c**,**d**) reverse sides of colony after incubation for 30 days on PDA (**c**) and MEA (**d**); (**e–m**) conidiophores, conidiogenous cells, and conidia. Petri dish diameter: 90 mm (**a–d**). Scale bars: 10 μm (**e–m**).

3.2.2. *Acrodictys diaoluoshanensis* S. Wang, J.W. Xia, X.Y. Liu, and X.G. Zhang, sp. nov.

MycoBank No. 844583

Etymology—The epithet *diaoluoshanensis* pertains to the Diaoluoshan National Nature Reserve, where the holotype was collected.

Type—China, Hainan Province: the Diaoluoshan National Nature Reserve (109°41′–110°4′ E, 18°38′–18°50′ N), on the dead branches of unidentified trees collected in tropical rain forests dominated by *Lauraceae* and *Fagaceae* trees, 21 May 2021, R.Y. Liu, holotype HMAS 352235, ex-holotype living culture SAUCC 1601.

Description—Asexual morph on PDA: Conidiophores are macronematous, mononematous, erect, unbranched, straight or flexuous, thick-walled, smooth, pale brown, septate, and 34.0–65.0 × 1.8–5.6 μm. Conidiogenous cells are monoblastic, integrated, terminal, determinate, cylindrical, pale brown, and smooth. Conidia are solitary, muriform, obovoid to obpyriform, pale brown to brown, and 18.0–22.0 × 10.0–13.0 μm, usually with 3 transverse septa and 1–2 longitudinal septa; they are slightly constricted at the septa, with conspicuous pores in the septa, and truncate at the base.

Culture characteristics—Colonies on PDA are flat with an entire margin, attaining 26.0–30.0 mm in diameter after 14 days at 25 °C, with a growth rate of 1.8–2.2 mm/day; they have aerial mycelia that are white to pale brown and floccose cottony; the reverse is black. Colonies on MEA are flat with an entire margin, with white-to-gray aerial mycelia that are floccose cottony; the reverse is black with a pale brown margin.

Additional specimen examined—China, Hainan Province: the Diaoluoshan National Nature Reserve (109°41′–110°4′ E, 18°38′–18°50′ N), on the dead branches of unidentified trees collected in tropical rain forests dominated by *Lauraceae* and *Fagaceae* trees, 21 May 2021, R.Y. Liu, paratype HMAS 352236, ex-paratype living culture SAUCC 1602.

Notes—Strains SAUCC 1601 and SAUCC 1602 have similar morphological features and identical DNA sequences and gather together with robust support values (MLBV = 100% and BIPP = 1.00, Figure 1). These two strains are, therefore, identified as the same new species: *Acrodictys diaoluoshanensis*. In the phylogenetical tree, based on a combined dataset of three genetic markers, *A. diaoluoshanensis* is closely related to *A. bambusicola* (MLBV = 100%, BIPP = 1.00), but they are different in conidia (*A. diaoluoshanensis* 18.0–22.0 × 10–13 μm vs. *A. bambusicola* 17.0–36.0 × 12.0–18.0 μm).

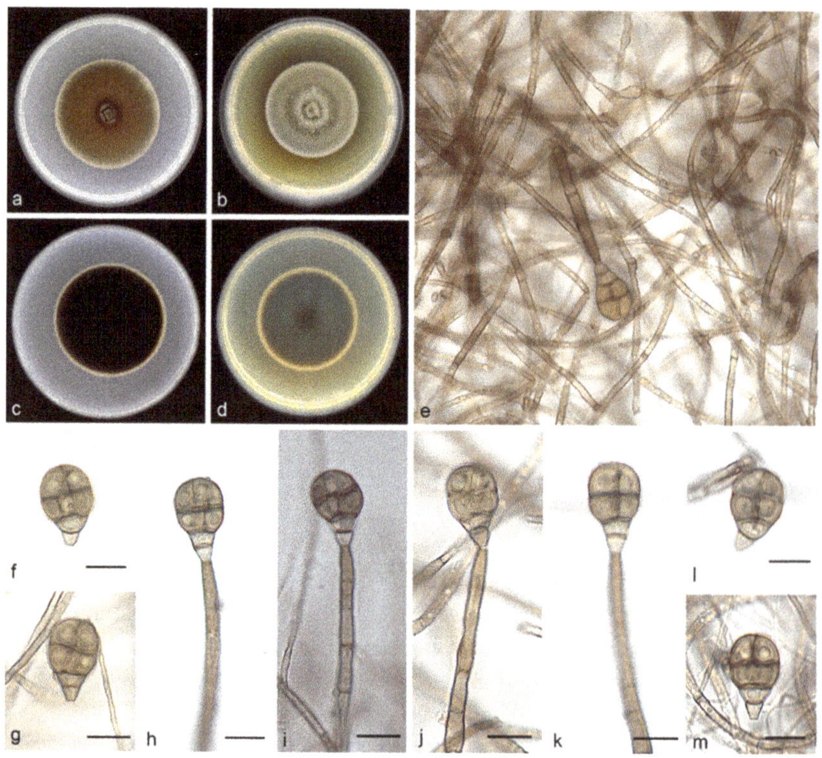

Figure 4. *Acrodictys ellisii* (holotype HMAS 352237, ex-holotype SAUCC 1471): (**a**,**b**) surface sides of colony after incubation for 30 days on PDA (**a**) and MEA (**b**); (**c**,**d**) reverse sides of colony after incubation for 30 days on PDA (**c**) and MEA (**d**); (**e**–**m**) conidiophores, conidiogenous cells. and conidia. Petri dish diameter: 90 mm (**a**–**d**). Scale bars: 10 μm (**e**–**m**).

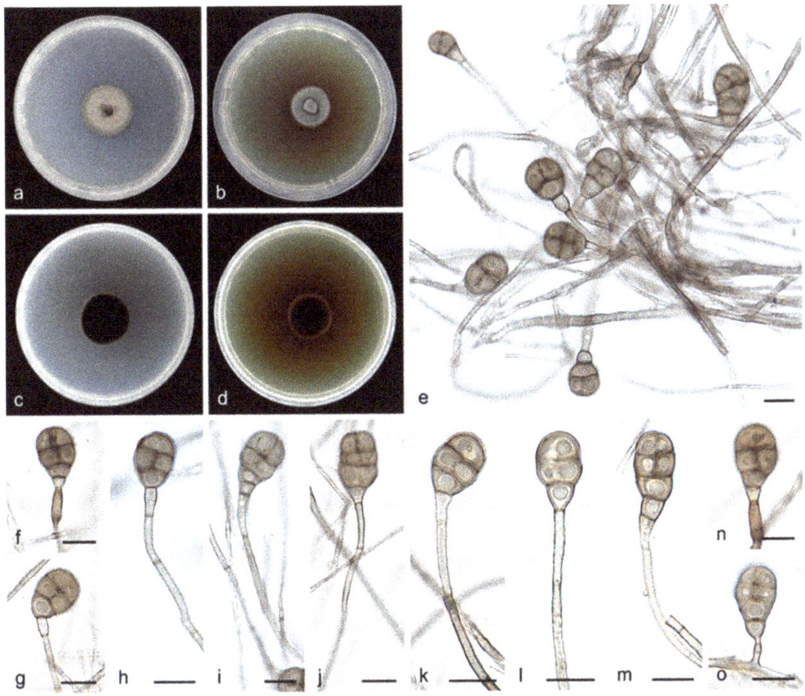

Figure 5. *Acrodictys pigmentosa* (holotype HMAS 352239, ex-holotype SAUCC 1591): (**a**,**b**) surface sides of colony after incubation for 7 days on PDA (**a**) and MEA (**b**); (**c**,**d**) reverse sides of colony after incubation for 7 days on PDA (**c**) and MEA (**d**); (**e**–**o**) conidiophores, conidiogenous cells, and conidia. Petri dish diameter: 90 mm (**a**–**d**). Scale bars: 10 µm (**e**–**o**).

3.2.3. *Acrodictys ellisii* S. Wang, J.W. Xia, X.Y. Liu, and X.G. Zhang, sp. nov.

MycoBank No. 844584

Etymology—The epithet *ellisii* is named in honor of the great English mycologist, M. B. Ellis.

Type—China, Hainan Province: the Bawangling National Nature Reserve (109°03′–109°17′ E, 18°57′–19°11′ N), on the dead branches of unidentified trees collected in tropical rain forests dominated by *Lauraceae* and *Fagaceae* trees, 19 May 2021, R.Y. Liu, holotype HMAS 352237, ex-holotype living culture SAUCC 1471.

Description—Asexual morph on PDA: Conidiophores are macronematous, mononematous, erect, unbranched, straight or flexuous, thick-walled, smooth, pale brown, septate, and 47.0–82.0 × 2.1–5.2 µm. Conidiogenous cells are monoblastic, integrated, terminal, determinate, cylindrical, pale brown, and smooth. Conidia are solitary, muriform, obovoid to obpyriform, pale brown to brown, and 17.0–22.0 × 11.0–14.0 µm, usually with 3 transverse septa and 1–3 longitudinal septa; they are slightly constricted at the septa and truncate at the base.

Culture characteristics—Colonies on PDA are flat with an entire margin, attaining 25.0–30.0 mm in diameter after 14 days at 25 °C, with a growth rate of 1.8–2.1 mm/day; they have aerial mycelia that are white to pale brown and floccose cottony; the reverse is black. Colonies on MEA are flat with an entire margin, with white-to-gray aerial mycelia that are floccose cottony; the reverse is black with a pale brown margin.

Additional specimen examined—China, Hainan Province: the Bawangling National Nature Reserve (109°03′–109°17′ E, 18°57′–19°11′ N), on the dead branches of unidentified trees collected in tropical rain forests dominated by *Lauraceae* and *Fagaceae* trees, 19 May 2021, R.Y. Liu, paratype HMAS 352238, ex-paratype living culture SAUCC 1472.

Notes—Strains SAUCC 1471 and SAUCC 1472 are similar in morphological features and identical DNA sequences and form a clade (MLBV = 97%, Figure 1). These two strains are, therefore, identified as the same new species: *Acrodictys ellisii*. Phylogenetically, *A. ellisii* is closely related to *A. bawanglingensis* (MLBV = 97%, BIPP = 1.00), having only 4 and 18 bp of dissimilarity in LSU and SSU, respectively. Morphologically, *Acrodictys bawanglingensis* also differs from *A. ellisii* in conidial size (60.0–120.0 × 4.5–6.5 μm vs. 47.0–82.0 × 2.1–5.2 μm).

3.2.4. *Acrodictys pigmentosa* S. Wang, J.W. Xia, X.Y. Liu, and X.G. Zhang, sp. nov.

MycoBank No. 844585

Etymology—The epithet *pigmentosa* originates from its pigmented colony on MEA.

Type—China, Hainan Province: the Bawangling National Nature Reserve (109°03′–109°17′ E, 18°57′–19°11′ N), on the dead branches of unidentified trees collected in tropical rain forests dominated by *Lauraceae* and *Fagaceae* trees, 19 May 2021, R.Y. Liu, holotype HMAS 352239, ex-holotype living culture SAUCC 1591.

Description—Asexual morph on PDA: Conidiophores are macronematous, mononematous, erect, unbranched, straight or flexuous, thick-walled, smooth, pale brown, septate, and 4.5–75.0 × 1.5–3.0 μm. Conidiogenous cells are monoblastic, integrated, terminal, determinate, cylindrical, pale brown, and smooth. Conidia are solitary, muriform, obovoid to obpyriform, pale brown to brown, 12.0–24.0 × 7.0–12.0 μm, usually with 1–4 transverse septa and 1–3 longitudinal septa, slightly constricted at the septa, and truncate at the base.

Culture characteristics—Colonies on PDA are flat with an entire margin, attaining 27.0–32.0 mm in diameter after 14 days at 25 °C, with a growth rate of 1.5–2.5 mm/day; they have white-to-pale-brown aerial mycelia that are floccose cottony; the reverse is black. Colonies on MEA are flat with an entire margin, with white-to-gray aerial mycelia that are floccose cottony; the reverse is black with a pale brown margin.

Additional specimen examined—China, Hainan Province: the Bawangling National Nature Reserve (109°03′–109°17′ E, 18°57′–19°11′ N), on the dead branches of unidentified trees collected in tropical rain forests dominated by *Lauraceae* and *Fagaceae* trees, 19 May 2021, R.Y. Liu, paratype HMAS 352240, ex-paratype living culture SAUCC 1592.

Notes—Strains SAUCC 1591 and SAUCC 1592 are similar in morphological features and identical DNA sequences and form a monophyletic group with robust support values (MLBV = 100%, BIPP = 1.00, Figure 1). These two strains are, therefore, identified as the same new species: *Acrodictys pigmentosa*. Phylogenetic analyses on a combined dataset of three genetic markers showed that *A. pigmentosa* is basal to the clade of *A. ellisii*, *A. bawanglingensis*, *A. bambusicola*, and *A. diaoluoshanensis* (MLBV = 96%, BIPP = 1.00), but they are different in conidia (12.0–24.0 × 7.0–12.0 μm vs. 17.0–22.0 × 11.0–14.0 μm vs. 18.0–26.0 × 10.0–16.0 μm vs. 17.0–36.0 × 12.0–18.0μm vs. 18.0–22.0 × 10.0–13.0 μm).

3.3. Key to the Species of *Acrodictys*

Together with the four new species proposed in this study, we currently accepted a worldwide total of 27 species in the genus *Acrodictys*. In order to facilitate identification in the future, a key to the species of *Acrodictys* is provided herein, updating the key compiled 11 years ago [6]. Since then, as many as eight new species have been added to this genus. The characteristics adopted in the updated key include perithecia, septa, asci, ascospores, conidiogenous cells, conidia, and chlamydospores.

1. Sexual morph known————————————————————————————2
1′ Sexual morph unknown———————————————————————————6
2. Maximum number of septa of ascospores > 3——————————————————3
2′ Maximum number of septa of ascospores ≤ 3——————————————————4
3. Conidia obovate——————————————————————————*A. satwalekeri*
3′ Conidia pyriform——————————————————————————*A. elaeidicola*
4. Conidia ellipsoid——————————————————————————*A. nigra*
4′ Conidia muriform——————————————————————————5
5. Conidia size 23.0–34.0 × 18.0–22.0 μm———————————————*A. peruamazonensis*
5′ Conidia size 15.0–22.0 × 7.0–13.0 μm————————————————*A. hainanensis*
6. Conidia with transverse septa only———————————————————7
6′ Conidia with transverse and longitudinal septa————————————————14
7. Conidia clavate——————————————————————————8
7′ Conidia rounded——————————————————————————10
8. Conidiogenous cells in groups————————————————————*A. similis*
8′ Conidiogenous cells singly——————————————————————9
9. Conidiogenous cells size 70.0–100.0 × 4.0–6.0 μm————————————*A. aquatica*
9′ Conidiogenous cells size 98.0–142.0 × 4.0–6.0 μm———————————*A. fluminicola*
10. Conidiogenous cells branched—————————————————————*A. caribensis*
10′ Conidiogenous cells unbranched————————————————————11
11. Conidiogenous cells septate———————————————————————12
11′ Conidiogenous cells aseptate——————————————————————13
12. Conidia spheroid—————————————————————————*A. brooksiae*
12′ Conidia ellipsoidal————————————————————————*A. sacchari*
13. Conidia 2–3 transverse septa——————————————————————*A. elliptica*
13′ Conidia 4–9 transverse septa——————————————————————*A. liputii*
14. Conidiophores in groups——————————————————————*A. furcata*
14′ Conidiophores singly———————————————————————————15
15. Conidia maximum length > 100 μm——————————————————*A. septosporioides*
15′ Conidia maximum length < 100 μm———————————————————16
16. Conidiogenous cells lageniform————————————————————17
16′ Conidiogenous cells cylindrical————————————————————22
17. Conidiogenous cells determinate proliferations————————————*A. balladynae*
17′ Conidiogenous cells percurrent proliferations———————————————18
18. Conidia subglobose or ellipsoidal——————————————————————19
18′ Conidia clavate or pyriform——————————————————————————20
19. Conidia size 12.0–22.0 × 8.0–16.0 μm————————————————*A. irregularis*
19′ Conidia size 27.0–32.0 × 12.0–16.0 μm———————————————*A. oblonga*
20. Conidiogenous cells maximum length > 200 μm ———————————*A. porosiseptata*
20′ Conidiogenous cells maximum length < 200 μm ——————————————21
21. Conidia size 20.0–30.0 × 13.0–16.0 μm———————————————*A. bambusicola*
21′ Conidia size 17.0–27.0 × 11.0–15.0 μm———————————————*A. atroapicula*
22. Conidiogenous cells with percurrent proliferations———————————————23
22′ Conidiogenous cells with determinate proliferations——————————————25
23. Conidiogenous cells maximum length > 60 μm———————————————*A. micheliae*
23′ Conidiogenous cells maximum length < 60 μm————————————————24
24. Conidia size 28.0–32.0 × 8.0–12.0 μm————————————————*A. lignicola*
24′ Conidia size 16.0–20.0 × 12.0–15.0 μm———————————————*A. papillata*
25. Conidiogenous cells maximum length > 80 μm—————————————————26
25′ Conidiogenous cells maximum length < 80 μm—————————————————27
26. Conidia size 18.0–26.0 × 10.0–16.0 μm, exceeding 23 and 13 um in length and width, respectively——————————————————————***A. bawanglingensis***
26′ Conidia size 18.0–22.0 × 10.0–13.0 μm————————————————***A. ellisii***
27. Conidia size 18.0–22.0 × 10.0–13.0 μm———————————————***A. diaoluoshanensis***
27′ Conidia size 12.0–24.0 × 7.0–12.0 μm————————————————***A. pigmentosa***

4. Discussion

Traditionally, *Acrodictys* sensu lato species have been characterized and identified based on conidial schizolytic/rhexolytic secession, conidiophores, conidiogenous cells, and conidia [2–5]. *Acrodictys* as a single genus in *Acrodictyaceae* was introduced by Xia et al. [7] based on ITS, LSU, SSU, and tub2 sequence data. In previous studies, *Acrodictys* species have been characterized and identified based on dictyoseptate pigmented conidia seceding schizolytically from monoblastic integrated terminal determinate or lageniform to doliiform percurrently extending conidiogenous cells [1,6]. Xia et al. [7] described eight *Acrodictys* species, including a new species based on morphology and ITS, LSU, SSU, and tub2 sequence data. This makes it possible for us to study *Acrodictys* species through molecular systematics. Subsequently, Luo et al. [10] introduced a new species and a new combination for this genus. To date, 10 *Acrodictys* species have molecular data, viz. *A. aquatica*, *A. bambusicola*, *A. elaeidicola*, *A. fluminicola*, *A. globulosa*, *A. hainanensis*, *A. liputii*, *A. malabarica*, *A. peruamazonensis*, and *A. porosiseptata* [7,10].

Tropical forest ecosystems offer suitable habitats for microfungi, among which the anamorphic species are the most abundant and diverse. Many anamorphic species have been recorded in rainforests, forest parks, and national nature reserves of Hainan Province, China [23–26]. As decomposers in the ecosystem, saprobic hyphomycetes can decompose a large amount of litter in tropical rain forests and convert them into inorganic substances, returning them to the soil. During our studies of saprobic hyphomycetes from the national nature reserves of Hainan Province, particularly the regions of the Bawangling National Nature Reserve and Diaoluoshan National Nature Reserve, several collections were made on dead branches. Studies of their morphological characteristics and phylogenetic data revealed four new species. These four species share a feature in conidiogenous cells, namely determinate proliferations. The conidial characteristics provide useful information for species delimitation (Table 3). In this study, we also uploaded rpb2 and tub2 sequence data (Table 2), although few were deposited in GenBank by other experts. All the four new species reported herein are different from one another in these two loci.

Table 3. Micromorphological comparison of phylogenetically related *Acrodictys* species.

Species	Shape	Conidial Size (μm)	Septa
Acrodictys aquatica	Clavate	20.0–27.0 × 10.0–17.0	3–4 transverse septa
A. bambusicola	Broadly clavate or pyriform	17.0–36.0 × 12.0–18.0	2–5 transverse and 1 or more longitudinal septa
A. bawanglingensis	**Obovoid to obpyriform**	**18.0–26.0 × 10.0–16.0**	**3 transverse and 1–3 longitudinal septa**
A. diaoluoshanensis	**Obovoid to obpyriform**	**18.0–22.0 × 10.0–13.0**	**3 transverse and 1–2 longitudinal septa**
A. ellisii	**Obovoid to obpyriform**	**17.0–22.0 × 11.0–14.0**	**3 transverse and 1–3 longitudinal septa**
A. elaeidicola	Turbinate, pyriform or clavate	17.0–26.0 × 11.0–19.0	3 transverse and 1–3 longitudinal septa
A. fluminicola	Broadly clavate, obovoid to pyriform	24.0–30.0 × 13.0–17.0	2–3 transverse and a few longitudinal septa
A. globulosa	Subglobose	22.0–27.0 × 17.0–23.0	2 transverse and several longitudinal and oblique septa
A. hainanensis	Oblong to obovoid	15.0–22.0 × 7.0–13.0	3–5 transverse and several longitudinal or oblique septa
A. liputii	Subglobose	18.5–22.5 × 13.5–17.5	2–3 parallel transverse and 2 perpendicular longitudinal septa
A. malabarica	Gangliar, somewhat top-shaped	16.0–21.0 × 14.0–17.0	3 transverse and 2–4 longitudinal septa

Table 3. *Cont.*

Species	Shape	Conidial Size (μm)	Septa
A. peruamazonensis	Ellipsoidal	28.0–36.0 × 17.0–21.0	3–4 transverse and some longitudinal or oblique septa
A. pigmentosa	**Obovoid to obpyriform**	**12.0–24.0 × 7.0–12.0**	**1–4 transverse septa and 1–3 longitudinal septa**
A. porosiseptata	Broadly clavate to pyriform	25.0–30.0 × 13.5–16.5	4–5 transverse and 3 perpendicular longitudinal septa

Notes: New species established in this study are in bold.

5. Conclusions

Hainan Province has a typical tropical rainforest climate, which is very suitable for the growth and reproduction of saprotrophic fungi. In this study, we chose the Bawangling National Nature Reserve and Diaoluoshan National Nature Reserve as representative sites for sample collection. Through morphological observations and molecular date analyses, we identified four new *Acrodictys* species, namely *A. bawanglingensis*, *A. diaoluoshanensis*, *A. ellisii*, and *A. pigmentosa*. The morphological descriptions and molecular data of *Acrodictys* in this study not only enrich the world's fungal resources and diversity but also contribute materials for studies of the effects of saprobic hyphomycetes on carbon cycling in ecosystems.

Supplementary Materials: The following supporting information can be downloaded at: https://www.mdpi.com/xxx/s1, Supplementary File S1: The combined ITS, LSU, and SSU multiple sequence alignment.

Author Contributions: Conceptualization, S.W. and R.L.; methodology, X.L.; software, Z.Z.; validation, J.X.; formal analysis, R.L.; investigation, S.W.; resources, S.L.; data curation, S.W.; writing—original draft preparation, S.W.; writing—review and editing, R.L.; visualization, S.L.; supervision, D.L.; project administration, X.L.; funding acquisition, X.Z. All authors have read and agreed to the published version of the manuscript.

Funding: This research was funded by the National Natural Science Foundation of China (nos. U2002203, 31750001, 31900014).

Institutional Review Board Statement: Not applicable for studies involving humans or animals.

Informed Consent Statement: Not applicable.

Data Availability Statement: The sequences from the present study were submitted to the NCBI database (https://www.ncbi.nlm.nih.gov/, accessed on 10 June 2022), and the accession numbers are listed in Table 2.

Conflicts of Interest: The authors declare no conflict of interest.

References

1. Ellis, M.B. Dematiaceous Hyphomycetes II. *Mycol. Pap.* **1961**, *79*, 1–23.
2. Baker, W.A.; Partridge, E.C.; Morgan-Jones, G. Notes on hyphomycetes LXXXI. *Acrodictyella obovata*, a new lignicolous, dematiaceous genus and species collected in Alabama. *Mycotaxon* **2001**, *78*, 29–35.
3. Baker, W.A.; Partridge, E.C.; Morgan-Jones, G. Notes on hyphomycetes LXXXV. *Junewangia*, a genus in which to classify four *Acrodictys* species and a new taxon. *Mycotaxon* **2002**, *81*, 293–319.
4. Baker, W.A.; Morgan-Jones, G. Notes on hyphomycetes. XCI. *Pseudoacrodictys*, a novel genus for seven taxa formerly placed in *Acrodictys*. *Mycotaxon* **2003**, *85*, 371–391.
5. Baker, W.A.; Partridge, E.C.; Morgan-Jones, G. Notes on hyphomycetes LXXXVII. *Rhexoacrodictys*, a new segregate genus to accommodate four species previously classified in *Acrodictys*. *Mycotaxon* **2002**, *82*, 95–113.
6. Zhao, G.Z.; Cao, A.X.; Zhang, T.Y.; Liu, X.Z. *Acrodictys* (Hyphomycetes) and related genera from China. *Mycol. Prog.* **2011**, *10*, 67–83. [CrossRef]
7. Xia, J.W.; Ma, Y.R.; Li, Z.; Zhang, X.G. Acrodictys-like wood decay fungi from southern china, with two new families *Acrodictyaceae* and *Junewangiaceae*. *Sci. Rep.* **2017**, *7*, 7888. [CrossRef]

8. Guo, L.D.; Hyde, K.D.; Liew, E.C.Y. Identification of endophytic fungi from *Livistona chinensis* based on morphology and rDNA sequences. *New Phytol.* **2000**, *147*, 617–630. [CrossRef]
9. Kumar, S.; Stecher, G.; Tamura, K. MEGA7: Molecular Evolutionary Genetics Analysis Version 7.0 for Bigger Datasets. *Mol. Biol. Evol.* **2016**, *33*, 1870–1874. [CrossRef]
10. Luo, Z.L.; Hyde, K.D.; Liu, J.K.; Maharachchikumbura, S.S.N.; Jeewon, R.; Bao, D.F.; Bhat, D.J.; Lin, C.G.; Li, W.L.; Yang, J.; et al. Freshwater Sordariomycetes. *Fungal Divers.* **2019**, *99*, 451–660. [CrossRef]
11. Katoh, K.; Rozewicki, J.; Yamada, K.D. MAFFT online service: Multiple sequence alignment, interactive sequence choice and visualization. *Brief. Bioinform.* **2017**, *20*, 1160–1166. [CrossRef] [PubMed]
12. Nylander, J.A.A. *MrModelTest v. 2. Program Distributed by the Author*; Evolutionary Biology Centre, Uppsala University: Uppsala, Sweden, 2004.
13. Miller, M.A.; Pfeiffer, W.; Schwartz, T. The CIPRES science gateway: Enabling high-impact science for phylogenetics researchers with limited resources. In Proceedings of the 1st Conference of the Extreme Science and Engineering Discovery Environment. Bridging from the Extreme to the Campus and Beyond, Chicago, IL, USA, 16 July 2012; Association for Computing Machinery: San Diego, CA, USA, 2012; p. 8. [CrossRef]
14. Stamatakis, A. RAxML Version 8: A tool for phylogenetic analysis and post-analysis of large phylogenies. *Bioinformatics* **2014**, *30*, 1312–1313. [CrossRef] [PubMed]
15. Huelsenbeck, J.P.; Ronquist, F. MRBAYES: Bayesian inference of phylogeny. *Bioinformatics* **2001**, *17*, 754–755. [CrossRef] [PubMed]
16. Ronquist, F.; Huelsenbeck, J.P. MrBayes 3: Bayesian phylogenetic inference under mixed models. *Bioinformatics* **2003**, *19*, 1572–1574. [CrossRef] [PubMed]
17. Ronquist, F.; Teslenko, M.; Van, D.M.P.; Ayres, D.L.; Darling, A.; Höhna, S.; Larget, B.; Liu, L.; Suchard, M.A.; Huelsenbeck, J.P. MrBayes 3.2: Efficient Bayesian phylogenetic inference and model choice across a large model space. *Syst. Biol.* **2012**, *61*, 539–542. [CrossRef] [PubMed]
18. White, T.J.; Bruns, T.; Lee, S.; Taylor, F.J.R.M.; Lee, S.H.; Taylor, L.; Shawe-Taylor, J. Amplification and direct sequencing of fungal ribosomal rna genes for phylogenetics. In *PCR Protocols: A Guide to Methods and Applications*; Innis, M.A., Gelfand, D.H., Sninsky, J.J., Eds.; Academic Press Inc.: New York, NY, USA, 1990; pp. 315–322. [CrossRef]
19. Rehner, S.A.; Samuels, G.J. Taxonomy and phylogeny of *Gliocladium* analysed from nuclear large subunit ribosomal DNA sequences. *Mycol. Res.* **1994**, *98*, 625–634. [CrossRef]
20. Vilgalys, R.; Hester, M. Rapid genetic identification and mapping of enzymatically amplified ribosomal DNA from several *Cryptococcus* species. *J. Bacteriol.* **1990**, *172*, 4238–4246. [CrossRef]
21. Glass, N.L.; Donaldson, G.C. Development of primer sets designed for use with the PCR to amplify conserved genes from filamentous ascomycetes. *Appl. Environ. Microbiol.* **1995**, *61*, 1323–1330. [CrossRef]
22. Liu, Y.J.; Whelen, S.; Hall, B.D. Phylogenetic Relationships among Ascomycetes: Evidence from an RNA polymerse II subunit. *Mol. Biol. Evol.* **1999**, *16*, 1799–1808. [CrossRef]
23. Wu, W.P.; Zhuang, W. *Sporidesmium, Endophragmiella and Related Genera from China*; Fungal Diversity Research Series; Fungal Diversity Press: Chiang Rai, Thailand, 2005; Volume 15, p. 351.
24. Jiang, N.; Voglmayr, H.; Bian, D.R.; Piao, C.G.; Wang, S.K.; Li, Y. Morphology and Phylogeny of *Gnomoniopsis* (Gnomoniaceae, Diaporthales) from Fagaceae Leaves in China. *J. Fungi.* **2021**, *7*, 792. [CrossRef]
25. Xia, J.W.; Mu, T.C.; Zhang, Z.X.; Li, Z.; Zhang, X.G. *Neoacrodictys elegans* gen. & sp. nov. from Hainan Province, China. *Mycotaxon* **2022**, *137*, 63–71. [CrossRef]
26. Zhang, Z.X.; Liu, R.Y.; Liu, S.B.; Mu, T.C.; Zhang, X.G.; Xia, J.W. Morphological and phylogenetic analyses reveal two new species of Sporocadaceae from Hainan, China. *MycoKeys* **2022**, *88*, 171–192. [CrossRef] [PubMed]

Journal of *Fungi*

MDPI

Article

Diversity and Distribution of *Calonectria* Species from Plantation and Forest Soils in Fujian Province, China

Qianli Liu [1,2], Michael J. Wingfield [1], Tuan A. Duong [1], Brenda D. Wingfield [1] and Shuaifei Chen [1,2,*]

1 Department of Biochemistry, Genetics and Microbiology, Forestry and Agricultural Biotechnology Institute (FABI), University of Pretoria, Pretoria 0028, South Africa; qianli.liu@fabi.up.ac.za (Q.L.); mike.wingfield@fabi.up.ac.za (M.J.W.); tuan.duong@fabi.up.ac.za (T.A.D.); brenda.wingfield@fabi.up.ac.za (B.D.W.)

2 Research Institute of Fast-Growing Trees (RIFT), Chinese Academy of Forestry (CAF), Zhanjiang 524022, China

* Correspondence: shuaifei.chen@gmail.com

Abstract: To meet the growing demand for wood and pulp products, *Eucalyptus* plantations have expanded rapidly during the past two decades, becoming an integral part of the southern China landscape. Leaf blight caused by various *Calonectria* spp., is a serious threat to these plantations. In order to explore the diversity and distribution of *Calonectria* spp. in Fujian Province soils, samples were collected in *Eucalyptus* plantations and adjacent plantings of *Cunninghamia lanceolata*, *Phyllostachys heterocycle* and *Pinus massoniana* as well as in natural forests. Three hundred and fifty-three *Calonectria* isolates were recovered from soil samples and they were identified based on a comparison of multilocus DNA sequence data for the *act* (actin), *cmdA* (calmodulin), *his3* (histone H3), *rpb2* (the second largest subunit of RNA polymerase), *tef1* (translation elongation factor 1-alpha) and *tub2* (β-tubulin) gene regions, as well as morphological characteristics. Six known taxa including *Calonectria aconidialis*, *Ca. hongkongensis*, *Ca. ilicicola*, *Ca. kyotensis*, *Ca. pacifica*, *Ca. pseudoreteaudii* and one novel species described here as *Ca. minensis* sp. nov. were identified. Of these, *Ca. aconidialis* and *Ca. kyotensis* were the most prevalent species, and found in eight and seven sites, and four and five forest types, respectively. *Calonectria* spp. were most abundant in soils from *Eucalyptus* stands, followed by *P. heterocycle* and natural forests. Relatively few species were found in the soils associated with *Cunninghamia lanceolata* and *Pinus massoniana*. The abundance of known *Calonectria* spp. suggests that these fungi have been relatively well sampled in Fujian. The results are also consistent with the fact that most Calonectria diseases are found on Angiosperm as opposed to Gymnosperm plants.

Keywords: Calonectria leaf blight; forest pathogens; fungal diversity; phylogeny; taxonomy

Citation: Liu, Q.; Wingfield, M.J.; Duong, T.A.; Wingfield, B.D.; Chen, S. Diversity and Distribution of *Calonectria* Species from Plantation and Forest Soils in Fujian Province, China. *J. Fungi* **2022**, *8*, 811. https://doi.org/10.3390/jof8080811

Academic Editors: Lei Cai and Cheng Gao

Received: 2 July 2022
Accepted: 25 July 2022
Published: 31 July 2022

Publisher's Note: MDPI stays neutral with regard to jurisdictional claims in published maps and institutional affiliations.

1. Introduction

Species of *Eucalyptus* are the most important trees used to establish plantations in the tropics and Southern Hemisphere, where they provide substantial resources for the global fibre market [1]. These trees were first introduced into China as ornamentals in 1890 and plantations of *Eucalyptus* spp. had reached 5.46 million hm^2 by 2018 [1]. Plantations of these trees are mainly distributed in 11 provinces of China, and over 75% can be found in the Guangxi, Guangdong, Yunnan and Fujian Provinces of southern China [1]. The *Eucalyptus* plantations in China have been established with a relatively narrow genetic base and consequently many disease problems, caused by a variety of pathogens, have emerged as threats to their sustainability [2–6].

Among the diseases threatening *Eucalyptus* plantations, leaf blight caused by species of *Calonectria* De Not. has become a major constraint in southern China [4,7–10]. Symptoms of infection are characterised by water-soaked spots on leaves in the lower and middle parts of the tree crowns. These coalesce and gradually develop into extended necrotic areas, which result in blight and often severe defoliation [9]. In China, Calonectria Leaf Blight (CLB) has been observed in *Eucalyptus* plantations in Fujian, Guangdong, Guangxi, Hainan and Yunnan Provinces [4,7,9–11]. This is similar to the situation in Australia, Brazil, Indonesia, Thailand and Vietnam where *Eucalyptus* plantations have also suffered significant damage due to CLB [12–16].

The genus *Calonectria* includes many aggressive plant pathogens. These species are extensively distributed particularly in sub-tropical and tropical regions of the world, and they have a wide host range including more than 335 plant species [17]. *Calonectria* species are generally considered as soil-borne fungi and they can survive in the soil for extended periods due to their thick-walled microsclerotia [17].

A recent taxonomic revision of *Calonectria* by Liu and co-authors [18] accepted 120 species. Of these, 65 have been reported from soils samples; the remaining species are known from infections on plant tissues [10,18–22]. To date, 27 species of *Calonectria* have been recorded in China, 18 of which have been isolated from soil samples [4,7,10,11,18,21,23–26].

Plantations of *Eucalyptus* spp. are commonly established alongside those of *Cunninghamia lanceolata*, *Phyllostachys heterocycle* and *Pinus massoniana* and can also be in mixed plantings in the Fujian Province (Figure 1). In recent years, leaf blight has become a serious threat to *Eucalyptus* plantations in this province [7,8]. *Calonectria* species including *Ca. crousiana*, *Ca. eucalypti*, *Ca. fujianensis*, *Ca. pauciromosa* and *Ca. pseudoreteaudii* [7,8,18] have been isolated from diseased *Eucalyptus* tissues and are regarded as the important causal agents of CLB in Fujian. *Calonectria* infections initially arise from inoculum in the soil but very little is known regarding the species diversity and distribution of these fungi in Fujian soils. The aim of this study was thus to determine the identity and distribution of *Calonectria* spp. from a wide variety of soils in Fujian, with a particular focus on *Eucalyptus* spp. but also including other trees that are found in the area.

Figure 1. Different forest plantations and natural forests in southern China. (**a**). mixed species plantations in Zhangzhou Region, Fujian Province, 1: *Eucalyptus* sp., 2: *Pinus massoniana*, 3: *Cunninghamia lanceolata*; (**b**). mixed species plantations in Jiangxi Province, 1: *Eucalyptus* sp., 3: *Cunninghamia lanceolata*; 4: *Phyllostachys heterocycle*; (**c**). *Eucalyptus* sp. in Yongan Region, Fujian Province; (**d**). *Cunninghamia lanceolata* in JiangXi Province; (**e**). *Phyllostachys heterocycle* in Nanping Region, Fujian Province; (**f**). natural forests in Nanping Region, Fujian Province. Soil samples in this study were collected from Fujian Province.

2. Materials and Methods

2.1. Sample Collection and Fungal Isolation

Soil samples were collected from *Eucalyptus* plantations and adjacent plantings, including those of *Cunninghamia lanceolata*, *Phyllostachys heterocycle* and *Pinus massoniana* as well as in natural forests (Figure 1). These plantations and forests were distributed in nine counties or districts in five regions of Fujian Province (one site in Nanping Region, two sites in Fuzhou Region, two sites in Sanming Region, three sites in Longyan Region, one site in Zhangzhou Region) of southern China (Figure 2). These forests typically have thick layers of leaf litter, which was removed before collecting soil samples from the upper 0–20 cm of the humid soil profile. Between three and 37 soil samples (Table 1) were collected randomly at each site. The soil samples were placed in re-sealable plastic bags to maintain moisture and transported to the laboratory for further study.

Soil samples were placed in plastic cups and moistened using distilled water. *Medicago sativa* (alfalfa) seeds were surface-disinfested in 75% ethanol for 30 s and scattered onto the surface of the moistened soil to bait for *Calonectria* spp. as described by Crous [17]. After eight to ten days at 25 °C, conidiophores typical of *Calonectria* spp. were observed with a Zeiss Stemi 2000C dissection microscope on the germinating alfalfa plants. Conidial masses were transferred to 2% MEA (Malt Extract Agar) using a sterile needle. After 12 h of incubation at 25 °C, single hyphal tips were transferred to fresh MEA plates using a sterile needle and these cultures were incubated at 25 °C for seven days. Cultures were sorted based on their morphological characteristics and one to five isolates were retained for each of the soil samples.

Cultures were deposited in the Culture Collection (CSF) at the Research Institute of Fast-growing Trees (RIFT) (previous institution: China Eucalypt Research Centre, CERC), Chinese Academy of Forestry (CAF), ZhanJiang, Guangdong Province, China. Representative isolates have also been maintained in the China General Microbiological Culture Collection Centre (CGMCC), Beijing, China. Dried specimens were deposited in the Mycological Fungarium of the Institute of Microbiology, Chinese Academy of Sciences (HMAS), Beijing, China.

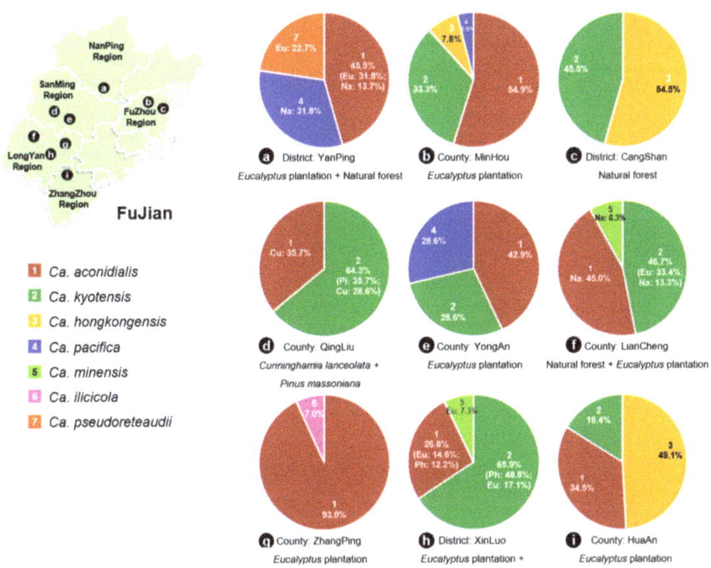

Figure 2. *Calonectria* species collected from nine counties (districts) in Fujian Province. (**a**–**i**). the percentage of each species in nine different counties (districts). Different species are indicated by numbers with different colours.

Table 1. Details of soils sampled, associated forest types and *Calonectria* spp. isolated.

Code	Sampling Site	Substrate	Number of Samples	Number of Samples Obtained *Calonectria*	Number of *alonectria* Isolates Obtained	*Calonectria* spp. (Number of Isolates)
a	Yanping District	*Eucalyptus* plantation	5	3	12	*Ca. aconidialis* (7); *Ca. pseudoreteaudii* (5)
		natural forest	13	3	10	*Ca. aconidialis* (3); *Ca. pacifica* (7)
		Cunninghamia lanceolata	10	0	0	N/A [a]
b	Minhou County	*Eucalyptus* plantation	15	12	51	*Ca. aconidialis* (28); *Ca. kyotensis* (17); *Ca. hongkongensis* (4); *Ca. pacifica* (2)
c	Cangshan District	natural forest	3	3	11	*Ca. kyotensis* (5); *Ca. hongkongensis* (6)
d	Qingliu County	natural forest	10	0	0	N/A
		Cunninghamia lanceolata	11	2	9	*Ca. aconidialis* (5); *Ca. kyotensis* (4)
		Pinus massoniana	10	1	5	*Ca. kyotensis* (5)
e	Yongan County	*Eucalyptus* plantation	27	7	28	*Ca. aconidialis* (12); *Ca. kyotensis* (8); *Ca. pacifica* (8)
f	Liancheng County	*Eucalyptus* plantation	20	4	20	*Ca. kyotensis* (20)
		natural forest	17	8	40	*Ca. aconidialis* (27); *Ca. kyotensis* (8); *Ca. minensis* (5)
g	Zhangping County	*Eucalyptus* plantation	20	15	71	*Ca. aconidialis* (66); *Ca. ilicicola* (5)
h	Xinluo District	*Eucalyptus* plantation	19	4	16	*Ca. aconidialis* (6); *Ca. kyotensis* (7); *Ca. minensis* (3)
		Phyllostachys heterocycle	14	5	25	*Ca. aconidialis* (5); *Ca. kyotensis* (20)
i	Hua'an County	*Eucalyptus* plantation	15	12	55	*Ca. aconidialis* (19); *Ca. kyotensis* (9); *Ca. hongkongensis* (27)
		In total	209	79	353	

[a] N/A refers to samples that did not yield *Calonectria* isolates.

2.2. DNA Extraction, PCR Amplifications and Sequencing

Mycelium was collected from axenic cultures grown on MEA for 5–7 days using a sterilised scalpel. Genomic DNA was extracted from the cultures using the CTAB method "5" described by Van Burik et al. [27]. Partial gene sequences were determined for the actin (*act*), calmodulin (*cmdA*), histone H3 (*his3*), the second largest subunit of RNA polymerase (*rpb2*), translation elongation factor 1-alpha (*tef1*) and β-tubulin (*tub2*) regions. Primer pairs ACT-512F/ACT-783R, CAL-228F/CAL-2Rd, CYLH3F/CYLH3R, fRpb2-5F/fRpb2-7cR, EF1-728F/EF2 and T1/CYLTUB1R [18] were used to amplify the six gene regions, respectively.

The PCR reaction mixtures contained 17.5 μL TopTaq^{TM} Master Mix, 1 μL of each primer (10 mM), 2 μL DNA sample and RNase-Free H_2O to a final volume of 35 μL. The amplifications were conducted under conditions described by Liu and co-authors [18]. All PCR products were sequenced in both directions using the same primers used for amplification. Raw sequences were inspected and manually corrected in Geneious v. 9.1.4 (Biomatters, Auckland, New Zealand) [28]. All sequences generated in this study were submitted to GenBank (http://www.ncbi.nlm.nih.gov; accessed on 24 July 2022) (Table 2, Appendix A Table A1).

2.3. Phylogenetic Analyses

To obtain the preliminary identification of the isolates, a standard nucleotide BLAST search was conducted using sequences of the six (*act*, *cmdA*, *his3*, *rpb2*, *tef1* and *tub2*) gene regions. Furthermore, sequences obtained in this study (Table 2) and sequences of other phylogenetically closely related *Calonectria* species downloaded from NCBI (http://www.ncbi.nlm.nih.gov; accessed on 24 July 2022) (Table 3) were used in the analyses. Sequence alignments were conducted online with MAFFT v. 7 (Suita, Janpan) [29] and were manually adjusted in MEGA v. 6.0.5 software (Auckland, New Zealand) [30] when necessary. The final alignments used in phylogenetic analyses were submitted to TreeBASE (http://treebase.org; accessed on 3 October 2021).

Genotypes of all the isolates were determined based on the sequences for the six gene regions. Representative isolates for all the genotypes were selected for the phylogenetic analyses. All the isolates of the novel species were used in the analyses. Maximum Parsimony (MP) and Maximum Likelihood (ML) approaches were used for phylogenetic analyses. The sequence datasets for the six individual gene regions and a concatenated dataset for those regions were used to determine the phylogenetic relatedness of all the isolates. PAUP v. 4.0 b10 [31] was used to perform the MP analyses, and PhyML v. 3.0 [32] was applied to conduct the ML analyses. A partition homogeneity test (PHT) [33] was performed to assess whether the datasets for the six gene regions could be combined.

For MP analyses, all characters were unordered and equally weighted. Gaps were regarded as fifth character and phylogenetic trees were obtained using a heuristic tree search criterion including 1000 random stepwise additions and tree-bisection-reconstruction (TBR) branch swapping. Branches of zero-length were collapsed. Supports for tree-branching points were determined using bootstrap analyses with 1000 replicates [34]. Tree length (TL), retention index (RI), consistency index (CI), rescaled consistency indexes (RC) and homoplasy index (HI) (Table 4) were calculated for parsimony trees. For ML analyses, the best substitution model for each dataset was determined using JModeltest 2.1.7 [35]. Sequence data for two isolates of *Curvicladiella cignea* (CBS 109167 and CBS 109168) were used as outgroup taxa (Table 3).

Table 2. Isolates sequenced in this study and used for phylogenetic analyses and morphological studies.

Species [a]	Isolate No. [b,c]	Genotype [d]	Substrate	Sampling Site	GPS Coordinate	Collector	GenBank Accession No. [e]					
							act	*cmdA*	*his3*	*rpb2*	*tef1*	*tub2*
Calonectria aconidialis	CSF9779	AAA-AA	Soil (*Eucalyptus* plantation)	Hua'an, Zhangzhou, Fujian, China	24°53′49.369″ N, 117°32′45.070″ E	S.F. Chen, Q.L. Liu and F.F. Liu	OK253064	OK253135	OK253279	N/A [f]	OK253491	OK253844
	CSF9857	AAAAAA	Soil (*Eucalyptus* plantation)	Zhangping, Longyan, Fujian, China	25°17′10.882″ N, 117°27′33.635″ E	S.F. Chen, Q.L. Liu and F.F. Liu	OK253065	OK253136	OK253280	OK253423	OK253492	OK253845
	CSF9937	AAAABA	Soil (*Eucalyptus* plantation)	Xinluo, Longyan, Fujian, China	25°07′08.597″ N, 116°44′42.257″ E	S.F. Chen, Q.L. Liu and F.F. Liu	OK253066	OK253137	OK253281	OK253424	OK253493	OK253846
	CSF9938	AAAABA	Soil (*Eucalyptus* plantation)	Xinluo, Longyan, Fujian, China	25°07′08.597″ N, 116°44′42.257″ E	S.F. Chen, Q.L. Liu and F.F. Liu	OK253067	OK253138	OK253282	OK253425	OK253494	OK253847
	CSF9939	AAAABA	Soil (*Eucalyptus* plantation)	Xinluo, Longyan, Fujian, China	25°07′08.597″ N, 116°44′42.257″ E	S.F. Chen, Q.L. Liu and F.F. Liu	OK253068	OK253139	OK253283	OK253426	OK253495	OK253848
	CSF9809	ABAAAA	Soil (*Eucalyptus* plantation)	Hua'an, Zhangzhou, Fujian, China	24°53′49.369″ N, 117°32′45.070″ E	S.F. Chen, Q.L. Liu and F.F. Liu	OK253069	OK253140	OK253284	OK253427	OK253496	OK253849
	CSF10105	ABAAAA	Soil (*Eucalyptus* plantation)	Minhou, Fuzhou, Fujian, China	26°15′04.285″ N, 119°02′38.917″ E	S.F. Chen, Q.L. Liu and F.F. Liu	OK253070	OK253141	OK253285	OK253428	OK253497	OK253850
	CSF9789	ABAAAB	Soil (*Eucalyptus* plantation)	Hua'an, Zhangzhou, Fujian, China	24°53′49.369″ N, 117°32′45.070″ E	S.F. Chen, Q.L. Liu and F.F. Liu	OK253071	OK253142	OK253286	OK253429	OK253498	OK253851
	CSF9839	ABAAAC	Soil (*Eucalyptus* plantation)	Zhangping, Longyan, Fujian, China	25°17′10.882″ N, 117°27′33.635″ E	S.F. Chen, Q.L. Liu and F.F. Liu	OK253072	OK253143	OK253287	OK253430	OK253499	OK253852
	CSF9844	ABAAAC	Soil (*Eucalyptus* plantation)	Zhangping, Longyan, Fujian, China	25°17′10.882″ N, 117°27′33.635″ E	S.F. Chen, Q.L. Liu and F.F. Liu	OK253073	OK253144	OK253288	OK253431	OK253500	OK253853
	CSF9882	ABAAAD	Soil (*Eucalyptus* plantation)	Zhangping, Longyan, Fujian, China	25°17′10.882″ N, 117°27′33.635″ E	S.F. Chen, Q.L. Liu and F.F. Liu	OK253074	OK253145	OK253289	OK253432	OK253501	OK253854

Table 2. *Cont.*

Species [a]	Isolate No. [b,c]	Genotype [d]	Substrate	Sampling Site	GPS Coordinate	Collector	GenBank Accession No. [e]					
							act	*cmdA*	*his3*	*rpb2*	*tef1*	*tub2*
	CSF9987	ABAAAD	Soil (natural forest area)	Liancheng, Longyan, Fujian, China	25°26′14.348″ N, 116°38′42.400″ E	S.F. Chen, Q.L. Liu and F.F. Liu	OK253075	OK253146	OK253290	OK253433	OK253502	OK253855
	CSF9813	ABAACA	Soil (*Eucalyptus* plantation)	Hua'an, Zhangzhou, Fujian, China	24°53′49.369″ N, 117°32′45.070″ E	S.F. Chen, Q.L. Liu and F.F. Liu	OK253076	OK253147	OK253291	OK253434	OK253503	OK253856
	CSF9841	ABAACA	Soil (*Eucalyptus* plantation)	Zhangping, Longyan, Fujian, China	25°17′10.882″ N, 117°27′33.635″ E	S.F. Chen, Q.L. Liu and F.F. Liu	OK253077	OK253148	OK253292	OK253435	OK253504	OK253857
	CSF9870	ABBAAA	Soil (*Eucalyptus* plantation)	Zhangping, Longyan, Fujian, China	25°17′10.882″ N, 117°27′33.635″ E	S.F. Chen, Q.L. Liu and F.F. Liu	OK253078	OK253149	OK253293	OK253436	OK253505	OK253858
	CSF9875	ABB-AA	Soil (*Eucalyptus* plantation)	Zhangping, Longyan, Fujian, China	25°17′10.882″ N, 117°27′33.635″ E	S.F. Chen, Q.L. Liu and F.F. Liu	OK253079	OK253150	OK253294	N/A	OK253506	OK253859
	CSF9957	ACBAAA	Soil (natural forest area)	Liancheng, Longyan, Fujian, China	25°26′14.348″ N, 116°38′42.400″ E	S.F. Chen, Q.L. Liu and F.F. Liu	OK253080	OK253151	OK253295	OK253437	OK253507	OK253860
Ca. hongkongensis	CSF7124	AAAAAA	Soil (natural forest area)	Cangshan, Fuzhou, Fujian, China	26°5′16.2″ N, 119°14′19.8″ E	S.F. Chen, Q.L. Liu and F.F. Liu	OK253081	OK253192	OK253336	OK253438	OK253669	OK253900
	CSF9784	AAAAAA	Soil (*Eucalyptus* plantation)	Hua'an, Zhangzhou, Fujian, China	24°53′49.369″ N, 117°32′45.070″ E	S.F. Chen, Q.L. Liu and F.F. Liu	OK253082	OK253193	OK253337	OK253439	OK253670	OK253901
	CSF9794	ABAAAA	Soil (*Eucalyptus* plantation)	Hua'an, Zhangzhou, Fujian, China	24°53′49.369″ N, 117°32′45.070″ E	S.F. Chen, Q.L. Liu and F.F. Liu	OK253083	OK253194	OK253338	OK253440	OK253671	OK253902
	CSF9799	ABAAAA	Soil (*Eucalyptus* plantation)	Hua'an, Zhangzhou, Fujian, China	24°53′49.369″ N, 117°32′45.070″ E	S.F. Chen, Q.L. Liu and F.F. Liu	OK253084	OK253195	OK253339	OK253441	OK253672	OK253903

Table 2. *Cont.*

Species [a]	Isolate No. [b,c]	Genotype [d]	Substrate	Sampling Site	GPS Coordinate	Collector	GenBank Accession No. [e]					
							act	*cmdA*	*his3*	*rpb2*	*tef1*	*tub2*
Ca. ilicicola	CSF9862	AAAAAA	Soil (*Eucalyptus* plantation)	Zhangping, Longyan, Fujian, China	25°17′10.882″ N, 117°27′33.635″ E	S.F. Chen, Q.L. Liu and F.F. Liu	OK253085	OK253202	OK253346	OK253442	OK253706	OK253910
	CSF9863	AAAAAA	Soil (*Eucalyptus* plantation)	Zhangping, Longyan, Fujian, China	25°17′10.882″ N, 117°27′33.635″ E	S.F. Chen, Q.L. Liu and F.F. Liu	OK253086	OK253203	OK253347	OK253443	OK253707	OK253911
Ca. kyotensis	CSF7130	AAAAAA	Soil (natural forest area)	Cangshan, Fuzhou, Fujian, China	26°5′16.2″ N, 119°14′19.8″ E	S.F. Chen, Q.L. Liu and F.F. Liu	OK253087	OK253207	OK253351	OK253444	OK253711	OK253915
	CSF10088	AAAAAA	Soil (*Eucalyptus* plantation)	Minhou, Fuzhou, Fujian, China	26°15′04.285″ N, 119°02′38.917″ E	S.F. Chen, Q.L. Liu and F.F. Liu	OK253088	OK253208	OK253352	OK253445	OK253712	OK253916
	CSF9834	AAA-AB	Soil (*Eucalyptus* plantation)	Hua'an, Zhangzhou, Fujian, China	24°58′22.263″ N, 117°31′09.708″ E	S.F. Chen, Q.L. Liu and F.F. Liu	OK253089	OK253209	OK253353	N/A	OK253713	OK253917
	CSF9910	AAAAAB	Soil (*Phyllostachys heterocycla*)	Xinluo, Longyan, Fujian, China	25°07′31.133″ N, 116°51′37.485″ E	S.F. Chen, Q.L. Liu and F.F. Liu	OK253090	OK253210	OK253354	OK253446	OK253714	OK253918
	CSF10014	AAAAAC	Soil (*Eucalyptus* plantation)	Yongan, Sanming, Fujian, China	25°55′10.860″ N, 117°16′39.591″ E	S.F. Chen, Q.L. Liu and F.F. Liu	OK253091	OK253211	OK253355	OK253447	OK253715	OK253919
	CSF10080	AAAAAD	Soil (*Eucalyptus* plantation)	Minhou, Fuzhou, Fujian, China	26°15′04.285″ N, 119°02′38.917″ E	S.F. Chen, Q.L. Liu and F.F. Liu	OK253092	OK253212	OK253356	OK253448	OK253716	OK253920
	CSF10086	AAAAAE	Soil (*Eucalyptus* plantation)	Minhou, Fuzhou, Fujian, China	26°15′04.285″ N, 119°02′38.917″ E	S.F. Chen, Q.L. Liu and F.F. Liu	OK253093	OK253213	OK253357	OK253449	OK253717	OK253921
	CSF10053	AAAABB	Soil (*Pinus massoniana*)	Qingliu, Sanming, Fujian, China	26°10′54.311″ N, 116°52′50.901″ E	S.F. Chen, Q.L. Liu and F.F. Liu	OK253094	OK253214	OK253358	OK253450	OK253718	OK253922
	CSF10054	AAAABB	Soil (*Pinus massoniana*)	Qingliu, Sanming, Fujian, China	26°10′54.311″ N, 116°52′50.901″ E	S.F. Chen, Q.L. Liu and F.F. Liu	OK253095	OK253215	OK253359	OK253451	OK253719	OK253923

Table 2. *Cont.*

Species [a]	Isolate No. [b,c]	Genotype [d]	Substrate	Sampling Site	GPS Coordinate	Collector	GenBank Accession No. [e]					
							act	*cmdA*	*his3*	*rpb2*	*tef1*	*tub2*
	CSF9922	AAAABF	Soil (*Phyllostachys heterocycla*)	Xinluo, Longyan, Fujian, China	25°07′31.133″ N, 116°51′37.485″ E	S.F. Chen, Q.L. Liu and F.F. Liu	OK253096	OK253216	OK253360	OK253452	OK253720	OK253924
	CSF9923	AAAABF	Soil (*Phyllostachys heterocycla*)	Xinluo, Longyan, Fujian, China	25°07′31.133″ N, 116°51′37.485″ E	S.F. Chen, Q.L. Liu and F.F. Liu	OK253097	OK253217	OK253361	OK253453	OK253721	OK253925
	CSF9949	AAAADB	Soil (*Eucalyptus* plantation)	Xinluo, Longyan, Fujian, China	25°07′08.597″ N, 116°44′42.257″ E	S.F. Chen, Q.L. Liu and F.F. Liu	OK253098	OK253218	OK253362	OK253454	OK253722	OK253926
	CSF9951	AAAADB	Soil (*Eucalyptus* plantation)	Xinluo, Longyan, Fujian, China	25°07′08.597″ N, 116°44′42.257″ E	S.F. Chen, Q.L. Liu and F.F. Liu	OK253099	OK253219	OK253363	OK253455	OK253723	OK253927
	CSF9932	AAAADG	Soil (*Eucalyptus* plantation)	Xinluo, Longyan, Fujian, China	25°07′08.597″ N, 116°44′42.257″ E	S.F. Chen, Q.L. Liu and F.F. Liu	OK253100	OK253220	OK253364	OK253456	OK253724	OK253928
	CSF9935	AAAADG	Soil (*Eucalyptus* plantation)	Xinluo, Longyan, Fujian, China	25°07′08.597″ N, 116°44′42.257″ E	S.F. Chen, Q.L. Liu and F.F. Liu	OK253101	OK253221	OK253365	OK253457	OK253725	OK253929
	CSF9936	AAAADG	Soil (*Eucalyptus* plantation)	Xinluo, Longyan, Fujian, China	25°07′08.597″ N, 116°44′42.257″ E	S.F. Chen, Q.L. Liu and F.F. Liu	OK253102	OK253222	OK253366	OK253458	OK253726	OK253930
	CSF10020	AAAAEA	Soil (*Eucalyptus* plantation)	Yongan, Sanming, Fujian, China	25°55′10.860″ N, 117°16′39.591″ E	S.F. Chen, Q.L. Liu and F.F. Liu	OK253103	OK253223	OK253367	OK253459	OK253727	OK253931
	CSF10021	AAAAEA	Soil (*Eucalyptus* plantation)	Yongan, Sanming, Fujian, China	25°55′10.860″ N, 117°16′39.591″ E	S.F. Chen, Q.L. Liu and F.F. Liu	OK253104	OK253224	OK253368	OK253460	OK253728	OK253932
	CSF10009	AAABBH	Soil (*Eucalyptus* plantation)	Liancheng, Longyan, Fujian, China	25°33′06.994″ N, 116°41′42.328″ E	S.F. Chen, Q.L. Liu and F.F. Liu	OK253105	OK253225	OK253369	OK253461	OK253729	OK253933
	CSF10010	AAABBH	Soil (*Eucalyptus* plantation)	Liancheng, Longyan, Fujian, China	25°33′06.994″ N, 116°41′42.328″ E	S.F. Chen, Q.L. Liu and F.F. Liu	OK253106	OK253226	OK253370	OK253462	OK253730	OK253934

Table 2. *Cont.*

Species [a]	Isolate No. [b,c]	Genotype [d]	Substrate	Sampling Site	GPS Coordinate	Collector	GenBank Accession No. [e]					
							act	*cmdA*	*his3*	*rpb2*	*tef1*	*tub2*
	CSF9997	AABAAB	Soil (*Eucalyptus* plantation)	Liancheng, Longyan, Fujian, China	25°33′06.994″ N, 116°41′42.328″ E	S.F. Chen, Q.L. Liu and F.F. Liu	OK253107	OK253227	OK253371	OK253463	OK253731	OK253935
	CSF9969	AABACB	Soil (natural forest area)	Liancheng, Longyan, Fujian, China	25°26′14.348″ N, 116°38′42.400″ E	S.F. Chen, Q.L. Liu and F.F. Liu	OK253108	OK253228	OK253372	OK253464	OK253732	OK253936
	CSF9972	AABACB	Soil (natural forest area)	Liancheng, Longyan, Fujian, China	25°26′14.348″ N, 116°38′42.400″ E	S.F. Chen, Q.L. Liu and F.F. Liu	OK253109	OK253229	OK253373	OK253465	OK253733	OK253937
	CSF9973	AABACB	Soil (natural forest area)	Liancheng, Longyan, Fujian, China	25°26′14.348″ N, 116°38′42.400″ E	S.F. Chen, Q.L. Liu and F.F. Liu	OK253110	OK253230	OK253374	OK253466	OK253734	OK253938
	CSF10126	AACAAA	Soil (*Eucalyptus* plantation)	Minhou, Fuzhou, Fujian, China	26°15′04.285″ N, 119°02′38.917″ E	S.F. Chen, Q.L. Liu and F.F. Liu	OK253111	OK253231	OK253375	OK253467	OK253735	OK253939
	CSF9962	AACAAD	Soil (natural forest area)	Liancheng, Longyan, Fujian, China	25°26′14.348″ N, 116°38′42.400″ E	S.F. Chen, Q.L. Liu and F.F. Liu	OK253112	OK253232	OK253376	OK253468	OK253736	OK253940
	CSF10019	AADABB	Soil (*Eucalyptus* plantation)	Yongan, Sanming, Fujian, China	25°55′10.860″ N, 117°16′39.591″ E	S.F. Chen, Q.L. Liu and F.F. Liu	OK253113	OK253233	OK253377	OK253469	OK253737	OK253941
	CSF10022	AADABB	Soil (*Eucalyptus* plantation)	Yongan, Sanming, Fujian, China	25°55′10.860″ N, 117°16′39.591″ E	S.F. Chen, Q.L. Liu and F.F. Liu	OK253114	OK253234	OK253378	OK253470	OK253738	OK253942
	CSF10023	AADABB	Soil (*Eucalyptus* plantation)	Yongan, Sanming, Fujian, China	25°55′10.860″ N, 117°16′39.591″ E	S.F. Chen, Q.L. Liu and F.F. Liu	OK253115	OK253235	OK253379	OK253471	OK253739	OK253943
	CSF10045	ABAAAB	Soil (*Cunning-hamia lanceolata*)	Qingliu, Sanming, Fujian, China	26°07′23.497″ N, 116°53′00.762″ E	S.F. Chen, Q.L. Liu and F.F. Liu	OK253116	OK253236	OK253380	OK253472	OK253740	OK253944
	CSF10047	ABAAAB	Soil (*Cunning-hamia lanceolata*)	Qingliu, Sanming, Fujian, China	26°07′23.497″ N, 116°53′00.762″ E	S.F. Chen, Q.L. Liu and F.F. Liu	OK253117	OK253237	OK253381	OK253473	OK253741	OK253945

Table 2. *Cont.*

Species [a]	Isolate No. [b,c]	Genotype [d]	Substrate	Sampling Site	GPS Coordinate	Collector	GenBank Accession No. [e]					
							act	*cmdA*	*his3*	*rpb2*	*tef1*	*tub2*
	CSF9824	ACBAAC	Soil (*Eucalyptus* plantation)	Hua'an, Zhangzhou, Fujian, China	24°53′49.369″ N, 117°32′45.070″ E	S.F. Chen, Q.L. Liu and F.F. Liu	OK253118	OK253238	OK253382	OK253474	OK253742	OK253946
	CSF10004	ADAACB	Soil (*Eucalyptus* plantation)	Liancheng, Longyan, Fujian, China	25°33′06.994″ N, 116°41′42.328″ E	S.F. Chen, Q.L. Liu and F.F. Liu	OK253119	OK253239	OK253383	OK253475	OK253743	OK253947
	CSF10005	ADAACB	Soil (*Eucalyptus* plantation)	Liancheng, Longyan, Fujian, China	25°33′06.994″ N, 116°41′42.328″ E	S.F. Chen, Q.L. Liu and F.F. Liu	OK253120	OK253240	OK253384	OK253476	OK253744	OK253948
Ca. minensis sp. nov.	CSF9941 [g–i]; CGMCC3.18877	AAAAAA	Soil (*Eucalyptus* plantation)	Xinluo, Longyan, Fujian, China	25°07′08.597″ N, 116°44′42.257″ E	S.F. Chen, Q.L. Liu and F.F. Liu	OK253121	OK253259	OK253403	OK253477	OK253814	OK253967
	CSF9974	AAAAAA	Soil (natural forest area)	Liancheng, Longyan, Fujian, China	25°26′14.348″ N, 116°38′42.400″ E	S.F. Chen, Q.L. Liu and F.F. Liu	OK253122	OK253260	OK253404	OK253478	OK253815	OK253968
	CSF9975 [g,h]; CGMCC3.18881	AAAAAA	Soil (natural forest area)	Liancheng, Longyan, Fujian, China	25°26′14.348″ N, 116°38′42.400″ E	S.F. Chen, Q.L. Liu and F.F. Liu	OK253123	OK253261	OK253405	OK253479	OK253816	OK253969
	CSF9976	AAAAAA	Soil (natural forest area)	Liancheng, Longyan, Fujian, China	25°26′14.348″ N, 116°38′42.400″ E	S.F. Chen, Q.L. Liu and F.F. Liu	OK253124	OK253262	OK253406	OK253480	OK253817	OK253970
	CSF9977	AAAAAA	Soil (natural forest area)	Liancheng, Longyan, Fujian, China	25°26′14.348″ N, 116°38′42.400″ E	S.F. Chen, Q.L. Liu and F.F. Liu	OK253125	OK253263	OK253407	OK253481	OK253818	OK253971
	CSF9978	AAAAAA	Soil (natural forest area)	Liancheng, Longyan, Fujian, China	25°26′14.348″ N, 116°38′42.400″ E	S.F. Chen, Q.L. Liu and F.F. Liu	OK253126	OK253264	OK253408	OK253482	OK253819	OK253972
	CSF9933 [g,h]; CGMCC3.18875	ABBABB	Soil (*Eucalyptus* plantation)	Xinluo, Longyan, Fujian, China	25°07′08.597″ N, 116°44′42.257″ E	S.F. Chen, Q.L. Liu and F.F. Liu	OK253127	OK253265	OK253409	OK253483	OK253820	OK253973
	CSF9934	ABBABB	Soil (*Eucalyptus* plantation)	Xinluo, Longyan, Fujian, China	25°07′08.597″ N, 116°44′42.257″ E	S.F. Chen, Q.L. Liu and F.F. Liu	OK253128	OK253266	OK253410	OK253484	OK253821	OK253974

Table 2. *Cont.*

Species [a]	Isolate No. [b,c]	Genotype [d]	Substrate	Sampling Site	GPS Coordinate	Collector	GenBank Accession No. [e]					
							act	*cmdA*	*his3*	*rpb2*	*tef1*	*tub2*
Ca. pacifica	CSF10024	AAAAAA	Soil (*Eucalyptus* plantation)	Yongan, Sanming, Fujian, China	25°55′10.860″ N, 117°16′39.591″ E	S.F. Chen, Q.L. Liu and F.F. Liu	OK253129	OK253267	OK253411	OK253485	OK253822	OK253975
	CSF10129	BAAAAA	Soil (*Eucalyptus* plantation)	Minhou, Fuzhou, Fujian, China	26°15′04.285″ N, 119°02′38.917″ E	S.F. Chen, Q.L. Liu and F.F. Liu	OK253130	OK253268	OK253412	OK253486	OK253823	OK253976
	CSF10070	CABAAA	Soil (natural forest area)	Yanping, Nanping, Fujian, China	26°42′26.672″ N, 118°07′58.317″ E	S.F. Chen, Q.L. Liu and F.F. Liu	OK253131	OK253269	OK253413	OK253487	OK253824	OK253977
	CSF10077	CABAAA	Soil (natural forest area)	Yanping, Nanping, Fujian, China	26°42′26.672″ N, 118°07′58.317″ E	S.F. Chen, Q.L. Liu and F.F. Liu	OK253132	OK253270	OK253414	OK253488	OK253825	OK253978
Ca. pseu-doreteaudii	CSF10059	AAAAAA	Soil (*Eucalyptus* plantation)	Yanping, Nanping, Fujian, China	26°46′19.651″ N, 117°57′37.233″ E	S.F. Chen, Q.L. Liu and F.F. Liu	OK253133	OK253274	OK253418	OK253489	OK253839	OK253982
	CSF10060 [g,h]	AAAAAA	Soil (*Eucalyptus* plantation)	Yanping, Nanping, Fujian, China	26°46′19.651″ N, 117°57′37.233″ E	S.F. Chen, Q.L. Liu and F.F. Liu	OK253134	OK253275	OK253419	OK253490	OK253840	OK253983

[a] New species described in this study are indicated in bold. [b] *CSF* = Culture Collection from Southern Forests (CSF), ZhanJiang, Guangdong Province, China; *CGMCC* = China General Microbiological Culture Collection Center, Beijing, China. [c] Isolates used in phylogenetic analyses. [d] Genotype within each identified species, determined by sequences of *act*, *cmdA*, *his3*, *rpb2*, *tef1* and *tub2* regions; '-' means not available. [e] *act* = actin; *cmdA* = calmodulin; *his3* = histone H3; *rpb2* = the second largest subunit of RNA polymerase; *tef1* = translation elongation factor 1-alpha; *tub2* = β-tubulin. [f] *N/A* represents sequences that are not available. [g] Isolates used in morphological and culture growth studies. [h] Isolates used for mating studies. [i] Isolates that represent ex-type cultures are indicated in bold.

Table 3. Isolates from other studies and used in the phylogenetic analyses.

Code B [a]	Species	Isolates No. [b,c]	Other Collection Number [b]	Substrate	Area of Occurrence	Collector	GenBank accession No. [d] act; cmdA; his3; rpb2; tef1; tub2	References
B1	*Calonectria acaciicola*	CMW 47173[T]	CBS 143557	Soil (*Acacia auriculiformis* plantation)	Do Luong, Nghe An, Vietnam	N.Q. Pham and T.Q. Pham	MT334933; MT335160; MT335399; MT412474; MT412690; MT412930	[16,18]
		CMW 47174	CBS 143558	Soil (*A. auriculiformis* plantation)	Do Luong, Nghe An, Vietnam	N.Q. Pham and T.Q. Pham	MT334934; MT335161; MT335400; MT412475; MT412691; MT412931	[16,18]
B2	*Ca. acicola*	CMW 30996[T]	–	*Phoenix canariensis*	Northland, New Zealand	H. Pearson	MT334935; MT335162; MT335401; MT412476; MT412692; MT412932	[18,36,37]
		CBS 114812	CMW 51216	*P. canariensis*	Northland, New Zealand	H. Pearson	MT334936; MT335163; MT335402; MT412477; MT412693; MT412933	[18,36,37]
B3	*Ca. aciculata*	CERC 5342[T]	CBS 142883; CMW 47645	*Eucalyptus urophylla* × *E. grandis*	Yunnan, China	S.F. Chen and J.Q. Li	MT334937; MT335164; MT335403; MT412478; MT412694; MT412934	[4,18]
B4	*Ca. aconidialis*	CMW 35174[T]	CBS 136086; CERC 1850	Soil (*Eucalyptus* plantation)	Hainan, China	X. Mou and S.F. Chen	MT334938; MT335165; MT335404; MT412479; MT412695; N/A [e]	[11,18]
		CMW 35384	CBS 136091; CERC 1886	Soil (*Eucalyptus* plantation)	Hainan, China	X. Mou and S.F. Chen	MT334939; MT335166; MT335405; N/A; MT412696; N/A	[11,18]
B5	*Ca. aeknauliensis*	CMW 48253[T]	CBS 143559	Soil (*Eucalyptus* plantation)	Aek Nauli, North Sumatra, Indonesia	M.J. Wingfield	MT334953; MT335180; MT335419; MT412486; MT412710; N/A	[16,18]
		CMW 48254	CBS 143560	Soil (*Eucalyptus* plantation)	Aek Nauli, North Sumatra, Indonesia	M.J. Wingfield	MT334954; MT335181; MT335420; MT412487; MT412711; N/A	[16,18]
B8	*Ca. asiatica*	CBS 114073[T]	CMW 23782; CPC 3900	Debris (leaf litter)	Prathet Thai, Thailand	N.L. Hywel-Jones	GQ280428; AY725741; AY725658; N/A; AY725705; AY725616	[23,37]
B10	*Ca. australiensis*	CMW 23669[T]	CBS 112954; CPC 4714	*Ficus pleurocarpa*	Queensland, Australia	C. Pearce and B. Paulus	MT334965; MT335192; MT335432; MT412496; MT412723; MT412946	[18,37,38]
B17	*Ca. brassicicola*	CBS 112841[T]	CMW 51206; CPC 4552	Soil at *Brassica* sp.	Indonesia	M.J. Wingfield	N/A; KX784561; N/A; N/A; KX784689; KX784619	[39]
B19	*Ca. bumicola*	CMW 48257[T]	CBS 143575	Soil (*Eucalyptus* plantation)	Aek Nauli, North Sumatra, Indonesia	M.J. Wingfield	MT334975; MT335205; MT335445; MT412509; MT412736; N/A	[16,18]
B20	*Ca. canadiana*	CMW 23673[T]	CBS 110817; STE-U 499	*Picea* sp.	Canada	S. Greifenhagen	MT334976; MT335206; MT335446; MT412510; MT412737; MT412958	[17,18,40,41]
		CERC 8952	–	Soil	Henan, China	S.F. Chen	MT335058; MT335290; MT335530; MT412587; MT412821; MT413035	[18,25]

Table 3. *Cont.*

Code B [a]	Species	Isolates No. [b,c]	Other Collection Number [b]	Substrate	Area of Occurrence	Collector	GenBank accession No. [d] *act; cmdA; his3; rpb2; tef1; tub2*	References
B23	*Ca. chinensis*	CMW 23674[T]	CBS 114827; CPC 4101	Soil	Hong Kong, China	E.C.Y. Liew	MT334990; MT335220; MT335460; MT412524; MT412751; MT412972	[18,23,37]
		CMW 30986	CBS 112744; CPC 4104	Soil	Hong Kong, China	E.C.Y. Liew	MT334991; MT335221; MT335461; MT412525; MT412752; MT412973	[18,23,37]
B26	*Ca. cochinchinensis*	CMW 49915[T]	CBS 143567	Soil (*Hevea brasiliensis* plantation)	Duong Minh Chau, Tay Ninh, Vietnam	N.Q. Pham, Q.N. Dang and T.Q. Pham	MT334995; MT335225; MT335465; MT412529; MT412756; MT412977	[16,18]
		CMW 47186	CBS 143568	Soil (*A. auriculiformis* plantation)	Song May, Dong Nai, Vietnam	N.Q. Pham and T.Q. Pham	MT334996; MT335226; MT335466; MT412530; MT412757; MT412978	[16,18]
B27	*Ca. colhounii*	CBS 293.79[T]	CMW 30999	*Camellia sinensis*	Mauritius	A. Peerally	GQ280443; GQ267373; DQ190639; KY653376; GQ267301; DQ190564	[17,37,38,42]
B29	*Ca. colombiensis*	CMW 23676[T]	CBS 112220; CPC 723	Soil (*E. grandis* trees)	La Selva, Colombia	M.J. Wingfield	MT334998; MT335228; MT335468; MT412532; MT412759; MT412980	[18,23]
		CMW 30985	CBS 112221; CPC 724	Soil (*E. grandis* trees)	La Selva, Colombia	M.J. Wingfield	MT334999; MT335229; MT335469; MT412533; MT412760; MT412981	[18,23]
B30	*Ca. crousiana*	CMW 27249[T]	CBS 127198	*E. grandis*	Fujian, China	M.J. Wingfield	MT335000; MT335230; MT335470; MT412534; MT412761; MT412982	[7,18]
		CMW 27253	CBS 127199	*E. grandis*	Fujian, China	M.J. Wingfield	MT335001; MT335231; MT335471; MT412535; MT412762; MT412983	[7,18]
B31	*Ca. curvispora*	CMW 23693[T]	CBS 116159; CPC 765	Soil	Tamatave, Madagascar	P.W. Crous	MT335002; MT335232; MT335472; MT412536; MT412763; N/A	[11,17,18,37, 43]
		CMW 48245	CBS 143565	Soil (*Eucalyptus* plantation)	Aek Nauli, North Sumatra, Indonesia	M.J. Wingfield	MT335003; MT335233; MT335473; MT412537; MT412764; N/A	[16,18]
B36	*Ca. eucalypti*	CMW 18444[T]	CBS 125275	*E. grandis*	Aek Nauli, Sumatra Utara, Indonesia	M.J. Wingfield	MT335013; MT335243; MT335483; MT412545; MT412774; MT412992	[18,37]
		CMW 18445	CBS 125276	*E. grandis*	Aek Nauli, Sumatra Utara, Indonesia	M.J. Wingfield	MT335014; MT335244; MT335484; MT412546; MT412775; MT412993	[18,37]
B39	*Ca. fujianensis*	CMW 27257[T]	CBS 127201	*E. grandis*	Fujian, China	M.J. Wingfield	MT335019; MT335249; MT335489; MT412551; MT412780; MT412998	[7,18]
		CMW 27254	CBS 127200	*E. grandis*	Fujian, China	M.J. Wingfield	MT335020; MT335250; MT335490; MT412552; MT412781; MT412999	[7,18]

Table 3. *Cont.*

Code B [a]	Species	Isolates No. [b,c]	Other Collection Number [b]	Substrate	Area of Occurrence	Collector	GenBank accession No. [d] *act; cmdA; his3; rpb2; tef1; tub2*	References
B46	*Ca. heveicola*	CMW 49913[T]	CBS 143570	Soil (*H. brasiliensis* plantation)	Bau Bang, Binh Duong, Vietnam	N.Q. Pham, Q.N. Dang and T.Q. Pham	MT335025; MT335255; MT335495; N/A; MT412786; MT413004	[16,18]
		CMW 49928	CBS 143571	Soil	Bu Gia Map National Park, Binh Phuoc, Vietnam	N.Q. Pham, Q.N. Dang and T.Q. Pham	MT335048; MT335280; MT335520; MT412577; MT412811; MT413025	[16,18]
B47	*Ca. honghensis*	CERC 5572[T]	CBS 142885; CMW 47669	Soil (*Eucalyptus* plantation)	Honghe, Yunnan, China	S.F. Chen and J.Q. Li	MT335026; MT335256; MT335496; MT412557; MT412787; MT413005	[4,18]
		CERC 5571	CBS 142884; CMW 47668	Soil (*Eucalyptus* plantation)	Honghe, Yunnan, China	S.F. Chen and J.Q. Li	MT335027; MT335257; MT335497; MT412558; MT412788; MT413006	[4,18]
B48	*Ca. hongkongensis*	CBS 114828[T]	CMW 51217; CPC 4670	Soil	Hong Kong, China	M.J. Wingfield	MT335028; MT335258; MT335498; MT412559; MT412789; MT413007	[18,23]
		CERC 3570	CMW 47271	Soil (*Eucalyptus* plantation)	Beihai, Guangxi, China	S.F. Chen, J.Q. Li and G.Q. Li	MT335030; MT335260; MT335500; MT412561; MT412791; MT413009	[4,18]
B51	*Ca. ilicicola*	CMW 30998[T]	CBS 190.50; IMI 299389; STE-U 2482	*Solanum tuberosum*	Bogor, Java, Indonesia	K.B. Boedijn and J. Reitsma	MT335036; MT335266; MT335506; MT412564; MT412797; N/A	[17,18,37,44]
B52	*Ca. indonesiae*	CMW 23683[T]	CBS 112823; CPC 4508	*Syzygium aromaticum*	Warambunga, Indonesia	M.J. Wingfield	MT335037; MT335267; MT335507; MT412565; MT412798; MT413015	[18,23]
		CBS 112840	CMW 51205; CPC 4554	*S. aromaticum*	Warambunga, Indonesia	M.J. Wingfield	MT335038; MT335268; MT335508; MT412566; MT412799; MT413016	[18,23]
B53	*Ca. indusiata*	CBS 144.36[T]	CMW 23699	*Camellia sinensis*	Sri lanka	N/A	GQ280536; GQ267453; GQ267262; KY653396; GQ267332; GQ267239	[17,37,39,45]
		CBS 114684	CMW 51213; CPC 2446; UFV16	*Rhododendron* sp.	Florida, USA	N.E. El-Gholl	GQ280537; GQ267454; DQ190653; N/A; GQ267333; AF232862	[17,38,46]
B55	*Ca. kyotensis*	CBS 114525[T]	ATCC 18834; CMW 51824; CPC 2367	*Robinia pseudoacacia*	Japan	T. Terashita	MT335039; MT335271; MT335511; MT412569; MT412802; MT413019	[17,18,39,47]
		CBS 114550	CMW 51825; CPC 2351	Soil	China	M.J. Wingfield	MT335016; MT335246; MT335486; MT412548; MT412777; MT412995	[18,39]
B57	*Ca. lantauensis*	CERC 3302[T]	CBS 142888; CMW 47252	Soil	LiDao, Hong Kong, China	M.J. Wingfield and S.F. Chen	MT335040; MT335272; MT335512; MT412570; MT412803; N/A	[4,18]
		CERC 3301	CBS 142887; CMW 47251	Soil	LiDao, Hong Kong, China	M.J. Wingfield and S.F. Chen	MT335041; MT335273; MT335513; N/A; MT412804; N/A	[4,18]

Table 3. *Cont.*

Code B [a]	Species	Isolates No. [b,c]	Other Collection Number [b]	Substrate	Area of Occurrence	Collector	GenBank accession No. [d] act; cmdA; his3; rpb2; tef1; tub2	References
B58	*Ca. lateralis*	CMW 31412[T]	CBS 136629	Soil (*Eucalyptus* plantation)	Guangxi, China	X. Zhou, G. Zhao and F. Han	MT335042; MT335274; MT335514; MT412571; MT412805; MT413020	[11,18]
B62	*Ca. lichi*	CERC 8866[T]	–	Soil	Henan, China	S.F. Chen	MT335046; MT335278; MT335518; MT412575; MT412809; MT413023	[18,25]
		CERC 8850	–	Soil	Henan, China	S.F. Chen	MT335047; MT335279; MT335519; MT412576; MT412810; MT413024	[18,25]
B63	*Ca. lombardiana*	CMW 30602[T]	CBS 112634; CPC 4233; Lynfield 417	*Xanthorrhoea australis*	Victoria, Australia	T. Baigent	MT335156; MT335395; MT335635; MT412686; MT412926; MT413133	[17,18,24,38]
B64	*Ca. macroconidialis*	CBS 114880[T]	CMW 51219; CPC 307; PPRI 4000	*E. grandis*	Sabie, Mpumalanga, South Africa	P.W. Crous	MT335050; MT335282; MT335522; MT412579; MT412813; MT413027	[17,18,37,48]
B65	*Ca. madagascariensis*	CMW 23686[T]	CBS 114572; CPC 2252	Soil	Rona, Madagascar	J.E. Taylor	MT335052; MT335284; MT335524; MT412581; MT412815; MT413029	[17,18,37,38]
		CMW 30993	CBS 114571; CPC 2253	Soil	Rona, Madagascar	J.E. Taylor	MT335053; MT335285; MT335525; MT412582; MT412816; MT413030	[17,18,37,38]
B66	*Ca. malesiana*	CMW 23687[T]	CBS 112752; CPC 4223	Soil	Northern Sumatra, Indonesia	M.J. Wingfield	MT335054; MT335286; MT335526; MT412583; MT412817; MT413031	[18,23]
		CBS 112710	CMW 51199; CPC 3899	Leaf litter	Prathet, Thailand	N.L. Hywel-Jones	MT335055; MT335287; MT335527; MT412584; MT412818; MT413032	[18,23]
B70	*Ca. monticola*	CBS 140645[T]	CPC 28835	Soil	Chiang Mai, Thailand	P.W. Crous	N/A; KT964771; N/A; N/A; KT964773; KT964769	[49]
		CPC 28836	–	Soil	Chiang Mai, Thailand	P.W. Crous	N/A; KT964772; N/A; N/A; KT964774; KT964770	[49]
B74	*Ca. multiseptata*	CMW 23692[T]	CBS 112682; CPC 1589	*E. grandis*	North Sumatra, Indonesia	M.J. Wingfield	MT335067; MT335299; MT335539; MT412596; MT412830; MT413044	[17,18,38,50]
B80	*Ca. pacifica*	CMW 16726[T]	A1568; CBS 109063; IMI 354528; STE-U 2534	*Araucaria heterophylla*	Hawaii, USA	M. Aragaki	MT335079; MT335311; MT335551; MT412604; MT412842; N/A	[17,18,23,40]
		CMW 30988	CBS 114038	*Ipomoea aquatica*	Auckland, New Zealand	C.F. Hill	MT335080; MT335312; MT335552; MT412605; MT412843; N/A	[17,18,23,37]
B81	*Ca. paracolhounii*	CBS 114679[T]	CMW 51212; CPC 2445	N/A	USA	A.Y. Rossman	N/A; KX784582; N/A; KY653423; KX784714; KX784644	[39,45]
		CBS 114705	CMW 51215; CPC 2423	Fruit of *Annona reticulata*	Australia	D. Hutton	N/A; N/A; N/A; KY653424; KX784715; KX784645	[39,45]

Table 3. *Cont.*

Code B [a]	Species	Isolates No. [b,c]	Other Collection Number [b]	Substrate	Area of Occurrence	Collector	GenBank accession No. [d] *act; cmdA; his3; rpb2; tef1; tub2*	References
B86	*Ca. penicilloides*	CMW 23696[T]	CBS 174.55; STE-U 2388	*Prunus* sp.	Hatizyo Island, Japan	M. Ookubu	MT335106; MT335338; MT335578; MT412631; MT412869; MT413081	[17,18,51]
B97	*Ca. pseudoreteaudii*	CMW 25310[T]	CBS 123694	*E. urophylla* × *E. grandis*	Guangdong, China	M.J. Wingfield and X.D. Zhou	MT335119; MT335354; MT335594; MT412647; MT412885; MT413096	[18,24]
		CMW 25292	CBS 123696	*E. urophylla* × *E. grandis*	Guangdong, China	M.J. Wingfield and X.D. Zhou	MT335120; MT335355; MT335595; MT412648; MT412886; MT413097	[18,24]
B104	*Ca. queenslandica*	CMW 30604[T]	CBS 112146; CPC 3213	*E. urophylla*	Lannercost, Queensland, Australia	B. Brown	MT335132; MT335367; MT335607; MT412660; MT412898; MT413108	[18,24]
		CMW 30603	CBS 112155; CPC 3210	*E. pellita*	Lannercost, Queensland, Australia	P.Q Thu and K.M. Old	MT335133; MT335368; MT335608; MT412661; MT412899; MT413109	[18,24]
B106	*Ca. reteaudii*	CMW 30984[T]	CBS 112144; CPC 3201	*E. camaldulensis*	Chon Thanh, Binh Phuoc, Vietnam	M.J. Dudzinski and P.Q. Thu	MT335135; MT335370; MT335610; MT412663; MT412901; MT413111	[17,18,38,52]
		CMW 16738	CBS 112143; CPC 3200	*Eucalyptus* leaves	Binh Phuoc, Vietnam	M.J. Dudzinski and P.Q. Thu	MT335136; MT335371; MT335611; MT412664; MT412902; MT413112	[17,18,38,52]
B112	*Ca. sumatrensis*	CMW 23698[T]	CBS 112829; CPC 4518	Soil	Northern Sumatra, Indonesia	M.J. Wingfield	MT335145; MT335382; MT335622; MT412674; MT412913; N/A	[18,23]
		CMW 30987	CBS 112934; CPC 4516	Soil	Northern Sumatra, Indonesia	M.J. Wingfield	MT335146; MT335383; MT335623; MT412675; MT412914; N/A	[18,23]
B113	*Ca. syzygiicola*	CBS 112831[T]	CMW 51204; CPC 4511	*S. aromaticum*	Sumatra, Indonesia	M.J. Wingfield	N/A; N/A; N/A; N/A; KX784736; KX784663	[39]
B116	*Ca. uniseptata*	CBS 413.67[T]	CMW 23678; CPC 2391; IMI 299577	*Paphiopedilum callosum*	Celle, Germany	W. Gerlach	GQ280451; GQ267379; GQ267248; N/A; GQ267307; GQ267208	[39]
B123	*Ca. xianrensis*	CSF12909[T]	CGMCC3.19584	Soil (near *Eucalyptus* plantation)	Dacheng Town, Gaozhou County, Maoming Region, Guangdong, China	S.F. Chen, Q.C. Wang and W. Wang	N/A; MK962845; MK962857; N/A; MK962869; MK962833	[21]
		CSF12908	CGMCC3.19518	Soil (near *Eucalyptus* plantation)	Dacheng Town, Gaozhou County, Maoming Region, Guangdong, China	S.F. Chen, Q.C. Wang and W. Wang	N/A; MK962844; MK962856; N/A; MK962868; MK962832	[21]

Table 3. *Cont.*

Code B [a]	Species	Isolates No. [b,c]	Other Collection Number [b]	Substrate	Area of Occurrence	Collector	GenBank accession No. [d] act; cmdA; his3; rpb2; tef1; tub2	References
B120	*Ca. yunnanensis*	CERC 5339[T]	CBS 142897; CMW 47644	Soil (*Eucalyptus* plantation)	Yunnan, China	S.F. Chen and J.Q. Li	MT335157; MT335396; MT335636; MT412687; MT412927; MT413134	[4,18]
		CERC 5337	CBS 142895; CMW 47642	Soil (*Eucalyptus* plantation)	Yunnan, China	S.F. Chen and J.Q. Li	MT335158; MT335397; MT335637; MT412688; MT412928; MT413135	[4,18]
	Curvicladiella cignea	CBS 109167[T]	CPC 1595; MUCL 40269	Decaying leaf	French Guiana	C. Decock	KM231122; KM231287; KM231461; KM232311; KM231867; KM232002	[11,38,53]
		CBS 109168	CPC 1594; MUCL 40268	Decaying seed	French Guiana	C. Decock	KM231121; KM231286; KM231460; KM232312; KM231868; KM232003	[11,38,53]

[a] Codes (B1 to B120) of the 120 accepted *Calonectria* species resulting from Liu and co-authors [18]. [b] *ATCC* = American Type Culture Collection, Virginia, USA; *CBS* = Westerdijk Fungal Biodiversity Institute, Utrecht, The Netherlands; *CERC* = China Eucalypt Research Centre, ZhanJiang, Guangdong Province, China; *CGMCC* = China General Microbiological Culture Collection Center, Beijing, China; *CMW* = Culture collection of the Forestry and Agricultural Biotechnology Institute (FABI), University of Pretoria, Pretoria, South Africa; *CPC* = Pedro Crous working collection housed at Westerdijk Fungal Biodiversity Institute; *CSF* = Culture Collection from Southern Forests (CSF), ZhanJiang, Guangdong Province, China; *IMI* = International Mycological Institute, CABI Bioscience, Egham, Bakeham Lane, UK; *MUCL* = Mycotheque, Laboratoire de Mycologie Systematique st Appliqee, I'Universite, Louvian-la-Neuve, Belgium; *PPRI* = Plant Protection Research Institute, Pretoria, South Africa; *STE-U* = Department of Plant Pathology, University of Stellenbosch, South Africa; '–' represent no other collection number. [c] *T* = ex-type isolates of the species. [d] *act* = actin; *cmdA* = calmodulin; *his3* = histone H3; *rpb2* = the second largest subunit of RNA polymerase; *tef1* = translation elongation factor 1-alpha; *tub2* = β-tubulin. [e] *N/A* represents information not available.

Table 4. Statistics resulting from phylogenetic analyses in this study.

Dataset	No. of Taxa	No. of bp [a]	Maximum Parsimony						
			PIC [b]	No. of Trees	Tree Length	CI [c]	RI [d]	RC [e]	HI [f]
act	147	278	111	4	258	0.636	0.968	0.615	0.364
cmdA	147	672	291	433	677	0.647	0.968	0.626	0.353
his3	143	464	183	1000	830	0.475	0.928	0.440	0.525
rpb2	134	863	269	1000	683	0.530	0.959	0.508	0.470
tef1	149	532	267	1000	758	0.637	0.963	0.613	0.363
tub2	135	597	286	1000	826	0.609	0.958	0.584	0.391
act/cmdA/his3/rpb2/tef1/tub2	149	3406	1407	3000	4408	0.532	0.949	0.504	0.468

Dataset	Maximum Likelihood							
	Subst. Mode [g]	NST [h]	Rate Matrix					Rates
act	TPM2 + G	6	0.5990	4.0516	0.5990	1.0000	4.0516	Gamma
cmdA	TrN + G	6	1.0000	4.1556	1.0000	1.0000	7.1231	Gamma
his3	TPM2uf + I + G	6	1.2442	6.0957	1.2442	1.0000	6.0957	Gamma
rpb2	TrNef + I + G	6	1.0000	9.0443	1.0000	1.0000	13.4319	Gamma
tef1	GTR + G	6	0.9651	1.7160	1.1302	0.5271	3.1484	Gamma
tub2	TPM3uf + I + G	6	1.4044	4.4908	1.0000	1.4044	4.4908	Gamma
act/cmdA/his3/rpb2/tef1/tub2	GTR + I + G	6	1.4593	4.5939	1.1370	0.9972	6.3874	Gamma

[a] *bp* = Base pairs. [b] *PIC* = Number of parsimony informative characters. [c] *CI* = Consistency index. [d] *RI* = Retention index. [e] *RC* = Rescaled consistency index. [f] *HI* = Homoplasy index. [g] *Subst. model* = best fit substitution model. [h] *NST* = Number of substitution rate categories.

2.4. Sexual Compatibility

The mating system as either homothallic or heterothallic was determined for the novel species identified in this study. Representative isolates of this species were crossed with each other in all possible combinations. These crosses were made on minimum salt agar (MSA) [54] with autoclaved toothpicks randomly placed on the agar surface. Petri dishes were then incubated at 25 °C for 2–8 wk, and they were observed regularly for the appearance of perithecia. When perithecia extruding ascospores emerged, germination tests were conducted to determine if the spores were viable. Production of viable ascospores was accepted as an indication of successful mating.

2.5. Morphology

Representative isolates of the novel species identified in this study were selected for morphological characterisation. Synthetic nutrient-poor agar (SNA) [55] was used to induce the asexual morphs. Agar plugs from axenic cultures were transferred to SNA and incubated at 25 °C for seven days. Fungal structures were lifted from the plates using a sterile needle and transferred to a drop of 85% lactic acid on microscope slides. Microscopic structures were examined under a Zeiss Axio Imager A1 microscope (Carl Zeiss Ltd., Jena, Germany).

In the case of sexual structures, the perithecia were transferred to Jung tissue freezing medium (Leica Biosystems, Wetzlar, Germany), which was frozen at −20 °C for ten minutes. Vertical sections (10 μm thick) were cut through the perithecia on a HM550 cryostat microtome (Microme International GmbH, Termo Fisher Scientifc, Walldorf, Germany) at −20 °C and examined under an Axio Imager A1 microscope.

For cultures selected as the ex-type isolates, 50 replicate measurements were made for each taxonomically characteristic structure. For other isolates, 30 replicate measurements were made. Minimum, maximum and average (mean) measurements were recorded as (minimum–) (average–standard deviation)–(average + standard deviation) (–maximum).

Optimal growth temperatures for the novel species were determined on MEA. Agar plugs were removed from the actively growing edges of 7-day-old cultures with a 5 mm diam. cork borer and transferred to the centres of 90 mm Petri dishes containing MEA. Cultures were grown at seven different temperatures ranging from 5 °C to 35 °C, at 5 °C intervals with five replicates per isolate. Colony diameters were measured after seven days. Colony colours were described using the colour charts of Rayner [56] using seven-day-old cultures on MEA incubated at 25 °C. All descriptions were deposited in MycoBank (www.mycobank.org, accessed on 3 October 2021).

3. Results

3.1. Sample Collection and Fungal Isolation

A total of 209 soil samples were collected and 353 isolates having a morphology typical of *Calonectria* were isolated from 79 of these samples (Table 1, Appendix A Table A1). Of these, 121 soil samples were from seven *Eucalyptus* plantations, of which 57 samples yielded 253 *Calonectria* isolates. Forty-three soil samples were collected from four natural forests, of which 14 samples yielded 61 *Calonectria* isolates; 21 soil samples were collected from two *C. lanceolata* plantations, two of which yielded nine *Calonectria* isolates; and 14 soil samples collected from a single *P. heterocycle* plantation, of which five samples yielded 25 *Calonectria* isolates. In addition, ten soil samples were collected from the *Pi. massoniana* plantation, only one of which yielded five *Calonectria* isolates (Table 1).

3.2. Phylogenetic Analyses

The *tef1* fragment was amplified for all of the 353 isolates (Appendix A Table A1), and based on sequence differences for this region and the sampling sites, 144 isolates were selected to amplify the *cmdA*, *his3* and *tub2* gene regions. Subsequently, based on the 37 genotypes revealed by these four gene regions, 71 representative isolates were chosen to amplify the *act* and *rpb2* gene regions (Appendix A Table A1). All of the 71 isolates, representing the 40 genotypes determined from the sequence data for the six gene regions, were used for phylogenetic inference (Table 2). Amplicons generated for the *act*, *cmdA*, *his3*, *rpb2*, *tef1*, and *tub2* gene regions were approximately 300, 700, 500, 860, 550, and 600 bp, respectively.

Sequence data for 46 *Calonectria* species closely related to those collected in this study were downloaded from GenBank and a total of 78 sequences (for ex-type and other strains) from previous studies were included in the phylogenetic analyses (Table 3). Phylogenetic analyses based on the six individual gene regions and the concatenated dataset for those regions were conducted using both MP and ML methods. The results showed that the overall topologies generated from the MP analyses were essentially similar to those from the ML analyses, and consequently, only the ML trees are presented (Figure 3, Appendix B Figures A1–A6).

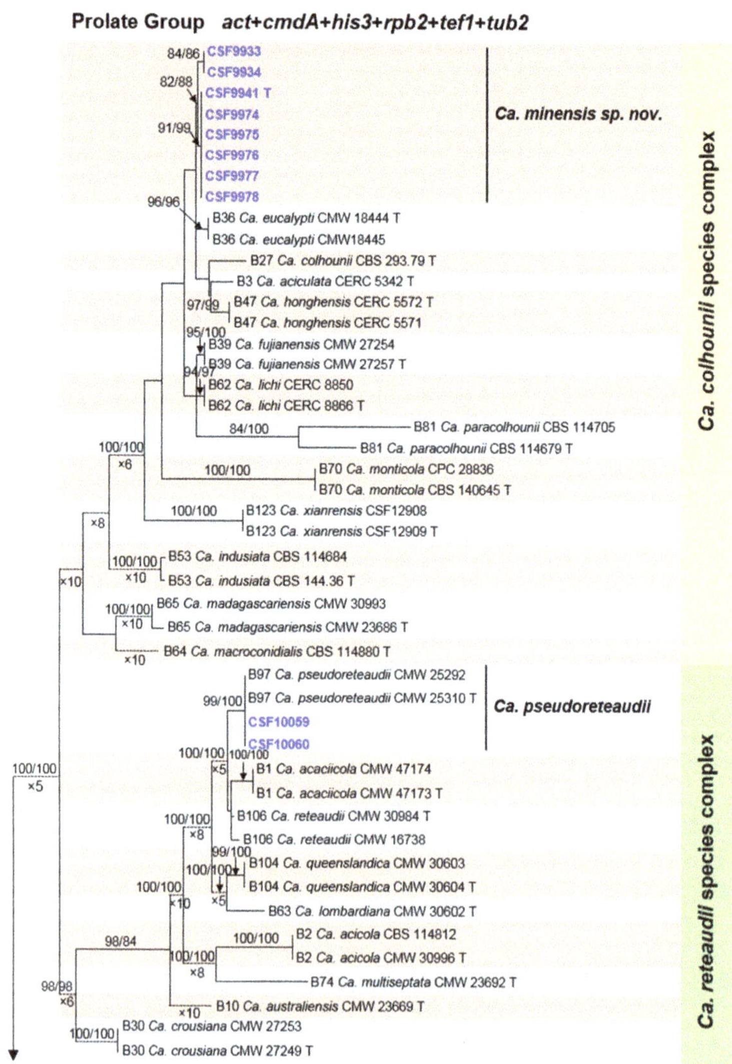

Figure 3. *Cont.*

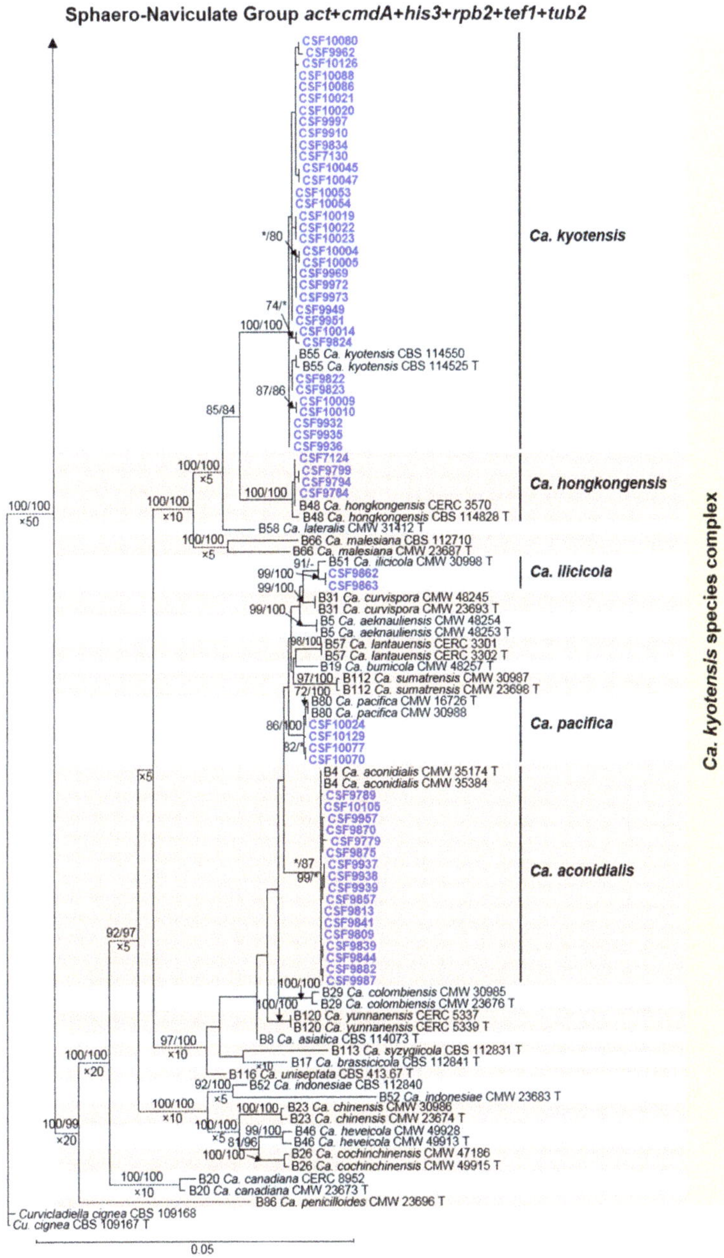

Figure 3. Phylogenetic tree of *Calonectria* species based on maximum likelihood (ML) analyses of combined DNA dataset of *act*, *cmdA*, *his3*, *rpb2*, *tef1*, and *tub2* gene sequences. Bootstrap value ≥70% for ML and MP analyses are presented above the branches. Bootstrap values lower than 70% are marked with "*", and absent analyses values are marked with "-". Ex-type isolates are marked with "T". Isolates sequenced in this study are highlighted in blue and bold type. The "B" species codes are consistent with the recently published results in Liu and co-authors [18]. The tree was rooted to *Curvicladiella cignea* (CBS 109167 and CBS 109168).

The partition homogeneity test carried out on the datasets, for the combined six gene regions, generated *p* values of 0.001. This showed that the accuracy of the combined data did not suffer relative to the individual partitions [57]. Sequence data for the six gene regions were thus combined for analyses. The sequence alignments based on the individual six gene regions and the combination of these were deposited in TreeBASE (No. S28845). Statistics and important parameters emerging from the phylogenetic analyses are presented in Table 4.

Based on the six-gene combined phylogenetic tree (Figure 3), for the 71 isolates used in the phylogenetic analyses, eight isolates resided in the *Ca. colhounii* species complex, two isolates in the *Ca. reteaudii* species complex and 61 isolates in the *Ca. kyotensis* species complex.

3.3. *Species in the Calonectria colhounii Species Complex*

Six isolates (CSF9941, CSF9974, CSF9975, CSF9976, CSF9977 and CSF9978), representing one genotype, formed a distinct lineage in the *cmdA* and *tub2* analyses as well as in the six-gene combined phylogenetic tree (Figure 3, Appendix B Figures A2 and A6). The total number of SNP differences between the six isolates and other phylogenetically closely related species [*Ca. aciculata* (ex-type isolate CERC 5342), *Ca. colhounii* (ex-type isolate CBS 293.79), *Ca. eucalypti* (ex-type isolate CMW 18444) and *Ca. honghensis* (ex-type isolate CERC 5572)] for six gene regions combined, varied between 13 and 31. Thus, this fungus can be regarded as a novel species. Two isolates (CSF9933 and CSF9934) formed an independent clade and were phylogenetically most closely related to the six isolates in the six-gene phylogenetic tree (Figure 3). These two isolates were consequently considered as the same species as the six isolates CSF9941, CSF9974, CSF9975, CSF9976, CSF9977 and CSF9978 and were identified as the novel species.

3.4. *Species in the Calonectria reteaudii Species Complex*

Two isolates (CSF10059 and CSF10060) were phylogenetically closely related to *Ca. pseudoreteaudii* and various other species based on *act* and *cmdA* trees (Appendix B Figures A1 and A2), and clustered with *Ca. pseudoreteaudii* based on *his3*, *rpb2*, *tef1*, *tub2* and the six-gene combined trees (Figure 3, Appendix B Figures A3–A6). In comparisons of DNA sequences for these six gene regions, all the sequences for the two isolates (CSF10059 and CSF10060) were 100% identical to the ex-type isolate (CMW 25310) of *Ca. pseudoreteaudii*. Consequently, they were identified as *Ca. pseudoreteaudii* (Figure 3).

3.5. *Species in the Calonectria kyotensis Species Complex*

Thirty-four isolates representing 20 genotypes were phylogenetically closest to *Ca. kyotensis* in each of the *cmdA*, *his3*, *rpb2* (sequence data for the *rpb2* were not available for isolate CSF9834), *tef1*, *tub2* and the six-gene combined trees (Figure 3, Appendix B Figures A2–A6), and clustered with *Ca. kyotensis* based on the *act* tree (Appendix B Figure A1). Some isolates formed distinct clades based on the six-gene combined trees (Figure 3), while the total number of SNP differences between the 34 isolates and the ex-type isolate of *Ca. kyotensis* (CBS 114525) for six gene regions combined varied between 2 and 8. Based on the phylogenetic analyses, these 34 isolates were identified as *Ca. kyotensis*.

Four isolates (CSF7124, CSF9784, CSF9794 and CSF9799), representing two genotypes, were phylogenetically closest to *Ca. hongkongensis* in each of the *cmdA*, *tub2* and six-gene combined tree (Figure 3, Appendix B Figures A2 and A6), and clustered with *Ca. hongkongensis* based on *act*, *his3*, *rpb2* and *tef1* trees (Appendix B Figures A1 and A3–A5). There were only three or four SNP differences between these four isolates and the ex-type isolate of *Ca. hongkongensis* (CBS 114828) when sequences for six gene regions were combined. Thus, these four isolates were identified as *Ca. hongkongensis*.

Two isolates (CSF9862 and CSF9863), representing one genotype clustered with *Ca. ilicicola* in the *his3* tree (Appendix B Figure A3), formed independent clades but closely related to *Ca. ilicicola* in the *act*, *cmdA*, *rpb2*, *tef1* and six-gene combined trees (Figure 3, Appendix B Figures A1, A2 and A4–A6). There were only six SNP differences between the two isolates and the ex-type isolate of *Ca. ilicicola* (CMW 30998) for five gene regions (*tub2* sequence data were not available for *Ca. ilicicola*) combined. Consequently, these isolates were regarded as *Ca. ilicicola*.

Four isolates (CSF10024, CSF10070, CSF10077 and CSF10129), representing three genotypes, were phylogenetically related to *Ca. pacifica* and various other closely related species based on *act* and *tef1* trees (Appendix B Figures A1 and A5). They were, however, phylogenetically closest to *Ca. pacifica* based on *his3* and six-gene combined trees (Appendix B Figure A3), and clustered with *Ca. pacifica* based on *cmdA* and *rpb2* trees (Appendix B Figures A2 and A4). There were only one or three SNP difference(s) between the four isolates and the ex-type isolate of *Ca. pacifica* (CMW 16726) for five gene regions (*tub2* sequence data were not available for *Ca. pacifica*) combined. These four isolates were thus identified as *Ca. pacifica*.

Seventeen isolates representing 11 genotypes were phylogenetically closest to *Ca. aconidialis* based on *cmdA*, *his3*, *tef1* and six-gene combined trees (Figure 3, Appendix B Figures A2, A3 and A5), and clustered with *Ca. aconidialis* based on *act* and *rpb2* (*rpb2* sequence data were not available for CSF9779 and CSF9875) trees (Appendix B Figures A1 and A4). Some isolates formed distinct clades based on the six-gene combined trees (Figure 3), while the total number of SNP differences between the 17 isolates and the ex-type isolate of *Ca. aconidialis* (CMW 35174) for five gene regions (sequence data for the *tub2* region were not available for *Ca. aconidialis*) combined varied between 0 and 4. Therefore, the 17 isolates were identified as *Ca. aconidialis*.

Seventy-one of the 353 isolates collected in this study were identified based on the DNA sequence of the six gene regions. According to the species identification results, we further identified the remaining 282 isolates based on the DNA sequences for two or four gene regions (Appendix A Table A1). Consequently, for the entire collection of 353 isolates, these were identified as *Ca. aconidialis* (178), *Ca. kyotensis* (103), *Ca. hongkongensis* (37), *Ca. pacifica* (17), *Ca. ilicicola* (five), *Ca. pseudoreteaudii* (five) and a novel species (eight), respectively.

3.6. Sexual Compatibility

Three isolates (CSF9933, CSF9941 and CSF9975) of the novel species were used in the mating tests (Table 2). All of these isolates formed protoperithecia readily within two weeks, and perithecia with viable ascospores were produced within four weeks. This was irrespective of whether they were crossed with each other or with themselves. The species was thus shown to be homothallic.

3.7. Morphology and Taxonomy

Based on multi-gene phylogenetic analyses (Figure 3, Appendix B Figures A1–A6) and morphological characteristics, seven *Calonectria* species were identified in this study, including six described species, i.e., *Ca. aconidialis*, *Ca. kyotensis*, *Ca. hongkongensis*, *Ca. pacifica*, *Ca. ilicicola*, *Ca. pseudoreteaudii* and one novel species. To facilitate future studies, complete morphological descriptions and illustrations have been made for the known species and these are presented in Appendix C (Figures A7–A12). The novel species can be distinguished from the phylogenetically most closely related species (*Ca. aciculata*, *Ca. colhounii*, *Ca. eucalypti* and *Ca. honghensis*) by the dimensions of its macroconidia and ascospores (Table 5). This species is described as follows:

Table 5. Morphological comparisons of *Calonectria* species obtained in this study and other phylogenetically closely related species.

Species	References or Source of Data	Ascospores (L × W) [a,b,c]	Ascospores Average (L × W) [a,b]	Ascospores Septation	Macroconidia (L × W) [a,b,c]	Macroconidia Average (L × W) [a,b]	Macroconidia Septation	Vesicle (Min.–Max.) [a]
Calonectria minensis	this study	(38.5–)46.5–64.5(–80.5) × (6–)6.5–8(–8.5)	55.5 × 7	3	(51–)55–66(–79) × (4.5–)5–6(–7.5)	60.5 × 5.5	(1–)3	3–5
Ca. aciculata	[4]	N/A [d]	N/A	N/A	(53–)62–76(–86) × (4.5–)5–6(–7)6	69 × 5.5	3	(2–)2.5–3.5 (–5)
Ca. colhounii	[17]	(30–)50–65(–75) × (4–)5–6(–8)	55 × 6	(1–)3	(45–)60–70(–80) × (4–)5–(–6)	65 × 5	(1–)3	3–4
Ca. eucalypti	[37]	(25–)30–36(–56) × (3–)5–6(–8)	33 × 6	(1–)3	(66–)69–75(–80) × (5–)6	72 × 6	3	4–6
Ca. honghensis	[4]	(35–)43–55(–65) × (4.5–)5.5–6.5(–7.5)	49 × 6	3	(43–)49–59(–66) × (4.5–)5–5.5(–6)	54 × 5.5	3	(2.5–)3–4.5 (–5.5)

[a] All measurements are in µm. [b] L × W = length × width. [c] Measurements are presented in the format [(minimum–) (average–standard deviation)–(average + standard deviation) (–maximum)]. [d] N/A represents data that is not available.

Taxonomy

Calonectria minensis Q.L. Liu and S.F. Chen, sp. nov.

MycoBank MB841412. (Figure 4).

Etymology: Name refers to the short name of Fujian Province in Chinese "Min", where this fungus was isolated.

Diagnosis: *Calonectria minensis* can be distinguished from the phylogenetically closely related species *Ca. aciculata*, *Ca. colhounii*, *Ca. eucalypti* and *Ca. honghensis* by its distinct ascospore and macroconidia dimensions.

Type: China: Fujian Province, Longyan Region, Xinluo District (25°07′08.597″ N, 116°44′42.257″ E), from soil collected in a *Eucalyptus* plantation, 6 November 2016, *S.F. Chen, Q.L. Liu* and *F.F. Liu* (HMAS249935—holotype, CSF9941 = CGMCC3.18877—ex-type culture).

Description: *Ascomata* perithecial, solitary or in groups of four, bright yellow, becoming orange with age; in section, apex and body yellow, base red-brown, sub-globose to ovoid, 258–395 µm high, 227–330 µm diam, body turning dark yellow, and base dark red-brown in 3% KOH+; ascomatal wall rough, consisting of two thick-walled layers; outer layer of *textura globulosa*, 22–66 µm thick, cells becoming more compressed towards the inner layer of *textura angularis*, 9–21 µm thick, cells becoming thin-walled and hyaline towards the centre; outermost cells 16–31 × 8–16 µm, cells of inner layer 8–33 × 2–8 µm; ascomatal base up to 196 µm wide, consisting of dark red, angular cells, merging with an erumpent stroma; cells of the outer wall layer continuous with the pseudoparenchymatous cells of the erumpent stroma. *Asci* 4-spored, clavate, 80–163 × 11–27 µm, tapering into a long thin stalk. *Ascospores* aggregated in the upper third of the ascus, hyaline, guttulate, fusoid with rounded ends, straight to slightly curved, (1–)3-septate, constricted at the septum, (38.5–)46.5–64.5(–80.5) × (6–)6.5–8(–8.5) µm (av. = 55.5 × 7 µm). *Macroconidiophores* consisting of a stipe, a suite of penicillately arranged fertile branches, a stipe extension, and a terminal vesicle; stipe septate, hyaline, smooth, 33–144 × 4–9 µm, stipe extension septate, straight to flexuous 63–240 µm long, 2–3 µm wide at the apical septum, terminating in a clavate vesicle, 3–5 µm diam; lateral stipe extensions (90° to main axis) absent. *Conidiogenous apparatus* 28–97 µm wide, and 35–83 µm long; primary branches aseptate, 13–40 × 3–7 µm; secondary branches aseptate, 9–31 × 3–6 µm; tertiary branches aseptate, 8–14 × 3–5 µm, quaternary branches aseptate, 7−12 × 3–5 µm, each terminal branch producing 2–4 phialides; phialides allantoid to elongate doliiform to reniform, hyaline, aseptate, 4–14 × 2–7 µm, apex with minute periclinal thickening and inconspicuous collarette. *Macroconidia* cylindrical, rounded at both ends, straight, (51–)55–66(–79) × (4.5–)5–6(–7.5) µm (av. = 60.5 × 5.5 µm), (1–)3-septate, lacking a visible abscission scar, held in parallel cylindrical clusters by colourless slime. Mega- and microconidia not observed.

Figure 4. *Calonectria minensis*. (**a**) Perithecium; (**b**) vertical section through a perithecium; (**c**) cells around ostiolar region of perithecium; (**d**) section through lateral perithecial wall; (**e**,**f**) asci; (**g**,**h**) ascospores; (**i**,**j**) macroconidiophore; (**k**,**m**) clavate vesicles; (**n**,**o**) conidiogenous apparatus with conidiophore branches and elongate doliiform to reniform phialides; (**p**,**q**) macroconidia.—Scale bars: a = 200 μm; b = 100 μm; c, d and f = 20 μm; e and i, j = 50 μm; g, h and n–q = 10 μm; k, m = 5 μm.

Culture characteristics: Colonies forming abundant woolly white to sienna (8) aerial mycelium at 25 °C on MEA, profuse sporulation; surface rust-coloured (39); reverse sienna (8) to rust-coloured (39) after 7 d. Chlamydospores extensive throughout the medium forming microsclerotia. Optimal growth temperature 25 °C, no growth at 5 °C and 35 °C, after 7 d, colonies at 10 °C, 15 °C, 20 °C, 25 °C and 30 °C reached 18.1 mm, 27.0 mm, 58.2 mm, 69.5 mm and 42.4 mm, respectively.

Additional specimens examined: China: Fujian Province, Longyan Region, Xinluo District (25°07′08.597″ N, 116°44′42.257″ E), from soil collected in a *Eucalyptus* plantation, 6 November 2016, S.F. Chen, Q.L. Liu and F.F. Liu (HMAS249936, culture CSF9933 = CGMCC3.18875); Fujian Province, Longyan Region, Liancheng County (25°26′14.348″ N, 116°38′42.400″ E), from soil under a natural forest, 6 November 2016, S.F. Chen, Q.L. Liu and F.F. Liu (HMAS249937, culture CSF9975 = CGMCC3.18881).

Notes: *Calonectria minensis* is a new species in the *Ca. colhounii* species complex. It is closely related to *Ca. aciculata*, *Ca. colhounii*, *Ca. eucalypti*, and *Ca. honghensis*, and can be distinguished from those species by the dimensions of its ascospores and macroconidia. The ascospores of *Ca. minensis* (av. = 55.5 × 7 μm) are larger than those of *Ca. eucalypti* (av. = 33 × 6 μm) [37] and *Ca. honghensis* (av. = 49 × 6 μm) [4]. The macroconidia of *Ca. minensis* (av. = 60.5 × 5.5 μm) are shorter than those of *Ca. aciculata* (av. = 69 × 5.5 μm) [4], *Ca. colhounii* (av. = 65 × 5 μm) [17] and *Ca. eucalypti* (av. = 72 × 6 μm) [37], but longer than those of *Ca. honghensis* (av. = 54 × 5.5 μm) [4]. The total number of SNP differences between the ex-type isolate of *Ca. minensis* (CSF9941), and the ex-type isolates of *Ca. aciculata* (CERC 5342), *Ca. colhounii* (CBS 293.79), *Ca. eucalypti* (CMW 18444) and *Ca. honghensis* (CERC 5572) for six gene regions combined, varied between 13 and 31.

3.8. Distribution of Calonectria Species in Fujian Province

Of the seven *Calonectria* species identified, *Ca. aconidialis* accounted for 50.4% of all the isolates. This was followed in order of occurrence by *Ca. kyotensis* (29.2%), *Ca. hongkongensis* (10.5%), *Ca. pacifica* (4.8%), *Ca. minensis* (2.3%), *Ca. ilicicola* (1.4%) and *Ca. pseudoreteaudii* (1.4%) (Figure 5). *Calonectria aconidialis* and *Ca. kyotensis* can be regarded as the most prevalent species (Figure 5).

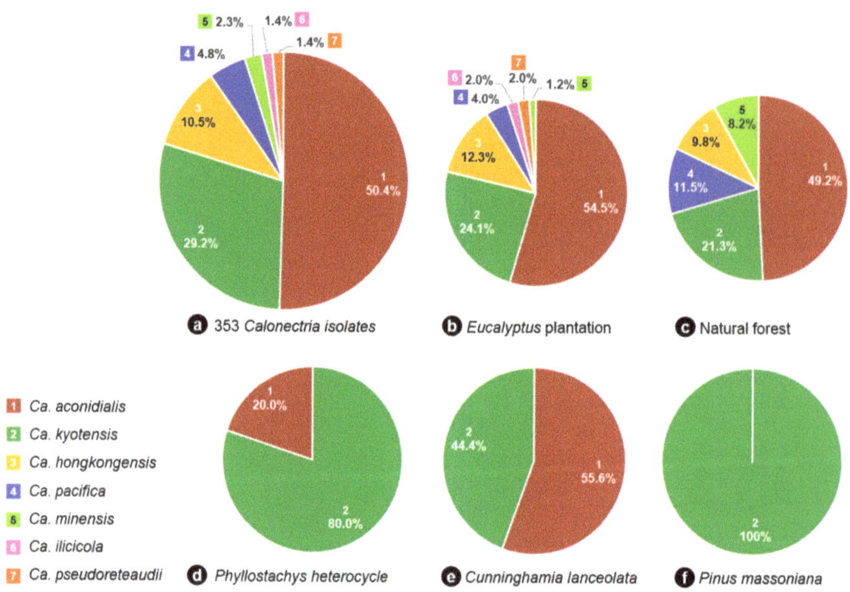

Figure 5. *Calonectria* species collected from soils of five different types of forests in Fujian Province. (**a**). the percentage of each *Calonectria* species accounted for all of the species isolated in this study. Different species are indicated by numbers with different colours; (**b–f**). the percentage of each *Calonectria* species obtained from five different types of forests.

Between two and four *Calonectria* species were isolated from soils sampled at each of the nine Counties or Districts (Figure 2). *Calonectria aconidialis* was found at all sites other than Cangshan District, *Ca. kyotensis* was found at all sites other than Yanping District and Zhangping County, and the remaining five species were found at between one and three sampling sites (Figure 2).

All seven species were isolated from soils collected in *Eucalyptus* plantations. Five of the species were isolated from soils in natural forests, the exception being *Ca. ilicicola* and *Ca. pesudoreteaudii*. Only *Ca. aconidialis* and *Ca. kyotensis* were isolated from soils in *P. heterocycle* and *C. lanceolata* plantations, and only *Ca. kyotensis* was collected from soils in the *Pi. massoniana* plantation (Figure 5). Based on the percentage of soil samples that obtained *Calonectria* from each of the five forest types, the results showed that *Calonectria* was widely distributed in *Eucalyptus* plantation soils (47.1%, 57 of 121 sampled soils), followed by *P. heterocycle* (35.7%, 5 of 14 sampled soils) and natural forests (32.6%, 14 of 43 sampled soils), only 10% of soil samples obtained *Calonectria* from *C. lanceolata* (2 of 21 sampled soils) or *Pi. massoniana* (1 of 10 sampled soils).

Calonectria kyotensis was detected in soils in all of the soil types sampled, while *Ca. aconidialis* was isolated from soils in all forest types other than *Pi. massoniana*. *Calonectria hongkongensis*, *Ca. pacifica* and *Ca. minensis* were found both in *Eucalyptus* plantations and natural forests and the remaining two species were found only in *Eucalyptus* plantations (Figure 5).

4. Discussion

A total of 353 *Calonectria* isolates were collected from soils in *Eucalyptus* plantations and adjacent plantations of other species or natural forests in Fujian Province. Multilocus phylogenetic inference and morphological characteristics revealed seven *Calonectria* species including *Ca. aconidialis*, *Ca. hongkongensis*, *Ca. ilicicola*, *Ca. kyotensis*, *Ca. pacifica* and *Ca. pseudoreteaudii*, and a novel species described here as *Ca. minensis*.

Results in this study showed that *Ca. aconidialis* and *Ca. kyotensis* were the most prevalent species in the soils sampled. *Calonectria aconidialis* accounted for 50.4% of all the isolates, which was found in eight of the nine sampled sites and soils of all forest types other than those of *Pi. massoniana*. The next most common species was *Ca. kyotensis*, accounting for 29.2% of the isolates, which was isolated from seven sites and soils of all five different forest types. The remaining five species were less common, and isolated only from one to three sites, either from *Eucalyptus* plantations or natural forests, or from both of these forest types.

Among the identified species, *Ca. aconidialis* is newly reported in Fujian Province and *Ca. pacifica* represents a first record for China. Eight *Calonectria* species were previously known in Fujian Province. These include *Ca. crousiana*, *Ca. eucalypti*, *Ca. fujianensis*, *Ca. pauciramosa* and *Ca. pseudoreteaudii* collected from diseased *Eucalyptus* leaves [7,8], *Ca. hongkongensis* and *Ca. kyotensis* isolated from soils in unknown forest types [4,18] and *Ca. ilicicola* collected from diseased peanuts (*Arachis hypogaea*) in Longyan Region [58].

The *Calonectria* species diversity in soils was clearly dependent on the forest types sampled. Of the seven species detected, all were obtained from *Eucalyptus* plantations, five were obtained from natural forests and only one or two species were from other forest types. While these observations are convincing in terms of broad patterns, they must be tempered by the fact that the greatest number of soil samples were from *Eucalyptus* plantations and natural forests, which could have influenced the results.

The newly described *Ca. minensis* isolated from soils both in *Eucalyptus* plantations and natural forest, adds a new species to the *Ca. colhounii* species complex. As a consequence, 13 species are now accommodated in this complex [4,7,17,18,21,25,37,39,46,49]. With the exception of *Ca. macroconidialis* [46], *Ca. madagascariensis* [17] and *Ca. paracolhounii* [39], all of the other 10 species have been recorded in southeastern Asia [4,7,17,21,25]. Species in this complex include some important causal agents of CLB on *Eucalyptus* spp. including *Ca. aciculata*, *Ca. eucalypti* and *Ca. fujianensis*, which have all been reported from diseased *Eucalyptus* trees in China plantations [4,7].

Five species residing in the *Ca. kyotensis* species complex were identified in the present study. Of these, *Ca. aconidialis* accounted for more than half of all the isolates collected, and has previously been shown to be widely distributed in soils of *Eucalyptus* plantation in many regions of southern China, including Guangdong [11,18], Guangxi [4,10,11] and Hainan Provinces [11]. In the present study, *Ca. aconidialis* was collected from soils of four types of forests and in eight of the nine sampling sites in Fujian Province (Figure 2), providing new geographic records for this pathogen in China. This species has previously been shown to infect inoculated *Eucalyptus* seedlings [10] and could pose a threat to *Eucalyptus* plantation forestry. *Calonectria pacifica* was isolated from soils both in the *Eucalyptus* plantations

(Minhou and Yongan Counties) and natural forests (Yanping District) in this study. This species was originally described on *Araucaria heterophylla* from Hawaii, USA [40], and this is the first report of the fungus in China.

This study elucidated the diversity and distribution characteristics of *Calonectria* species in soils collected from plantations and natural forests in Fujian Province. Broad patterns of occurrence were clear with *Eucalyptus* soils yielding the largest number of species. The conifer forests had the lowest number of species, which is consistent with the fact that most *Calonectria* spp. are known from Angiosperm hosts or from soils associated with these plants. The results of the present study bring the number of *Calonectria* species recorded in Fujian to 11. Most of these species have also been shown to be pathogenic to *Eucalyptus* in previous studies [7,9,10]. The surprisingly high species diversity in this region suggests that *Calonectria* species will pose long-term challenges for the development of *Eucalyptus* forestry in southern China.

Author Contributions: Conceptualization, Q.L. and S.C.; methodology, Q.L. and S.C.; software, Q.L.; validation, Q.L., M.J.W., T.A.D., B.D.W. and S.C.; formal analysis, Q.L.; investigation, Q.L. and S.C.; resources, Q.L. and S.C.; data curation, Q.L. and S.C; writing—original draft preparation, Q.L.; writing—review and editing, Q.L., M.J.W., T.A.D., B.D.W. and S.C.; visualization, Q.L.; supervision, M.J.W., T.A.D., B.D.W. and S.C.; project administration, S.C.; funding acquisition, S.C. All authors have read and agreed to the published version of the manuscript.

Funding: This study was initiated through the bilateral agreement between the Governments of South Africa and China and supported by The National Key R&D Program of China (China-South Africa Forestry Joint Research Centre Project; project No. 2018YFE0120900), the special fund for basic scientific research of State Key Laboratory of Tree Genetics and Breeding (SKLTGB) of China (project No. TGB2017001), the National Ten-thousand Talents Program (Project No. W03070115) and the Guangdong Top Young Talents Program in China (Project No. 20171172).

Institutional Review Board Statement: Not applicable.

Informed Consent Statement: Not applicable.

Data Availability Statement: The sequences from the current study were submitted to the NCBI database (https://www.ncbi.nlm.nih.gov/, accessed on 24 July 2022) and the accession numbers were listed in Table 2.

Acknowledgments: We thank FeiFei Liu for her assistance in collecting soil samples.

Conflicts of Interest: The authors declare no conflict of interest.

Appendix A. Species of *Calonectria* Collected in This Study

Table A1. Species of *Calonectria* collected in this study.

Species [a]	Isolate No. [b]	Genotype [c]	Substrate	Sampling Site	GPS Coordinate	Collector	GenBank Accession No. [d]					
							act	*cmdA*	*his3*	*rpb2*	*tef1*	*tub2*
Calonectria aconidialis	CSF9779 [e]	AAA-AA	Soil (*Eucalyptus* plantation)	Hua'an, Zhangzhou, Fujian, China	24°53′49.369″ N, 117°32′45.070″ E	S.F. Chen, Q.L. Liu and F.F. Liu	OK253064	OK253135	OK253279	N/A [f]	OK253491	OK253844
	CSF9857 [e]	AAAAAA	Soil (*Eucalyptus* plantation)	Zhangping, Longyan, Fujian, China	25°17′10.882″ N, 117°27′33.635″ E	S.F. Chen, Q.L. Liu and F.F. Liu	OK253065	OK253136	OK253280	OK253423	OK253492	OK253845
	CSF9937 [e]	AAAABA	Soil (*Eucalyptus* plantation)	Xinluo, Longyan, Fujian, China	25°07′08.597″ N, 116°44′42.257″ E	S.F. Chen, Q.L. Liu and F.F. Liu	OK253066	OK253137	OK253281	OK253424	OK253493	OK253846
	CSF9938 [e]	AAAABA	Soil (*Eucalyptus* plantation)	Xinluo, Longyan, Fujian, China	25°07′08.597″ N, 116°44′42.257″ E	S.F. Chen, Q.L. Liu and F.F. Liu	OK253067	OK253138	OK253282	OK253425	OK253494	OK253847
	CSF9939 [e]	AAAABA	Soil (*Eucalyptus* plantation)	Xinluo, Longyan, Fujian, China	25°07′08.597″ N, 116°44′42.257″ E	S.F. Chen, Q.L. Liu and F.F. Liu	OK253068	OK253139	OK253283	OK253426	OK253495	OK253848
	CSF9809 [e]	ABAAAA	Soil (*Eucalyptus* plantation)	Hua'an, Zhangzhou, Fujian, China	24°53′49.369″ N, 117°32′45.070″ E	S.F. Chen, Q.L. Liu and F.F. Liu	OK253069	OK253140	OK253284	OK253427	OK253496	OK253849
	CSF10105 [e]	ABAAAA	Soil (*Eucalyptus* plantation)	Minhou, Fuzhou, Fujian, China	26°15′04.285″ N, 119°02′38.917″ E	S.F. Chen, Q.L. Liu and F.F. Liu	OK253070	OK253141	OK253285	OK253428	OK253497	OK253850
	CSF9789 [e]	ABAAAB	Soil (*Eucalyptus* plantation)	Hua'an, Zhangzhou, Fujian, China	24°53′49.369″ N, 117°32′45.070″ E	S.F. Chen, Q.L. Liu and F.F. Liu	OK253071	OK253142	OK253286	OK253429	OK253498	OK253851
	CSF9839 [e]	ABAAAC	Soil (*Eucalyptus* plantation)	Zhangping, Longyan, Fujian, China	25°17′10.882″ N, 117°27′33.635″ E	S.F. Chen, Q.L. Liu and F.F. Liu	OK253072	OK253143	OK253287	OK253430	OK253499	OK253852
	CSF9844 [e]	ABAAAC	Soil (*Eucalyptus* plantation)	Zhangping, Longyan, Fujian, China	25°17′10.882″ N, 117°27′33.635″ E	S.F. Chen, Q.L. Liu and F.F. Liu	OK253073	OK253144	OK253288	OK253431	OK253500	OK253853

Table A1. *Cont.*

Species [a]	Isolate No. [b]	Genotype [c]	Substrate	Sampling Site	GPS Coordinate	Collector	GenBank Accession No. [d]					
							act	*cmdA*	*his3*	*rpb2*	*tef1*	*tub2*
	CSF9882 [e]	ABAAAD	Soil (*Eucalyptus* plantation)	Zhangping, Longyan, Fujian, China	25°17′10.882″ N, 117°27′33.635″ E	S.F. Chen, Q.L. Liu and F.F. Liu	OK253074	OK253145	OK253289	OK253432	OK253501	OK253854
	CSF9987 [e]	ABAAAD	Soil (natural forest area)	Liancheng, Longyan, Fujian, China	25°26′14.348″ N, 116°38′42.400″ E	S.F. Chen, Q.L. Liu and F.F. Liu	OK253075	OK253146	OK253290	OK253433	OK253502	OK253855
	CSF9813 [e]	ABAACA	Soil (*Eucalyptus* plantation)	Hua'an, Zhangzhou, Fujian, China	24°53′49.369″ N, 117°32′45.070″ E	S.F. Chen, Q.L. Liu and F.F. Liu	OK253076	OK253147	OK253291	OK253434	OK253503	OK253856
	CSF9841 [e]	ABAACA	Soil (*Eucalyptus* plantation)	Zhangping, Longyan, Fujian, China	25°17′10.882″ N, 117°27′33.635″ E	S.F. Chen, Q.L. Liu and F.F. Liu	OK253077	OK253148	OK253292	OK253435	OK253504	OK253857
	CSF9870 [e]	ABBAAA	Soil (*Eucalyptus* plantation)	Zhangping, Longyan, Fujian, China	25°17′10.882″ N, 117°27′33.635″ E	S.F. Chen, Q.L. Liu and F.F. Liu	OK253078	OK253149	OK253293	OK253436	OK253505	OK253858
	CSF9875 [e]	ABB-AA	Soil (*Eucalyptus* plantation)	Zhangping, Longyan, Fujian, China	25°17′10.882″ N, 117°27′33.635″ E	S.F. Chen, Q.L. Liu and F.F. Liu	OK253079	OK253150	OK253294	N/A	OK253506	OK253859
	CSF9957 [e]	ACBAAA	Soil (natural forest area)	Liancheng, Longyan, Fujian, China	25°26′14.348″ N, 116°38′42.400″ E	S.F. Chen, Q.L. Liu and F.F. Liu	OK253080	OK253151	OK253295	OK253437	OK253507	OK253860
	CSF9851	-AA-AA	Soil (*Eucalyptus* plantation)	Zhangping, Longyan, Fujian, China	25°17′10.882″ N, 117°27′33.635″ E	S.F. Chen, Q.L. Liu and F.F. Liu	– [g]	OK253152	OK253296	–	OK253508	OK253861
	CSF9943	-AA-AA	Soil (*Eucalyptus* plantation)	Xinluo, Longyan, Fujian, China	25°07′08.597″ N, 116°44′42.257″ E	S.F. Chen, Q.L. Liu and F.F. Liu	–	OK253153	OK253297	–	OK253509	OK253862
	CSF9812	-BA-AA	Soil (*Eucalyptus* plantation)	Hua'an, Zhangzhou, Fujian, China	24°53′49.369″ N, 117°32′45.070″ E	S.F. Chen, Q.L. Liu and F.F. Liu	–	OK253154	OK253298	–	OK253510	OK253863
	CSF9831	-BA-AA	Soil (*Eucalyptus* plantation)	Hua'an, Zhangzhou, Fujian, China	24°58′22.263″ N, 117°31′09.708″ E	S.F. Chen, Q.L. Liu and F.F. Liu	–	OK253155	OK253299	–	OK253511	OK253864

Table A1. *Cont.*

Species [a]	Isolate No. [b]	Genotype [c]	Substrate	Sampling Site	GPS Coordinate	Collector	GenBank Accession No. [d]					
							act	*cmdA*	*his3*	*rpb2*	*tef1*	*tub2*
	CSF9846	-BA-AA	Soil (*Eucalyptus* plantation)	Zhangping, Longyan, Fujian, China	25°17′10.882″ N, 117°27′33.635″ E	S.F. Chen, Q.L. Liu and F.F. Liu	–	OK253156	OK253300	–	OK253512	OK253865
	CSF9880	-BA-AA	Soil (*Eucalyptus* plantation)	Zhangping, Longyan, Fujian, China	25°17′10.882″ N, 117°27′33.635″ E	S.F. Chen, Q.L. Liu and F.F. Liu	–	OK253157	OK253301	–	OK253513	OK253866
	CSF9895	-BA-AA	Soil (*Eucalyptus* plantation)	Zhangping, Longyan, Fujian, China	25°17′10.882″ N, 117°27′33.635″ E	S.F. Chen, Q.L. Liu and F.F. Liu	–	OK253158	OK253302	–	OK253514	OK253867
	CSF9905	-BA-AA	Soil (*Eucalyptus* plantation)	Zhangping, Longyan, Fujian, China	25°17′10.882″ N, 117°27′33.635″ E	S.F. Chen, Q.L. Liu and F.F. Liu	–	OK253159	OK253303	–	OK253515	OK253868
	CSF9930	-BA-AA	Soil (*Phyllostachys heterocycla*)	Xinluo, Longyan, Fujian, China	25°07′31.133″ N, 116°51′37.485″ E	S.F. Chen, Q.L. Liu and F.F. Liu	–	OK253160	OK253304	–	OK253516	OK253869
	CSF9964	-BA-AA	Soil (natural forest area)	Liancheng, Longyan, Fujian, China	25°26′14.348″ N, 116°38′42.400″ E	S.F. Chen, Q.L. Liu and F.F. Liu	–	OK253161	OK253305	–	OK253517	OK253870
	CSF9970	-BA-AA	Soil (natural forest area)	Liancheng, Longyan, Fujian, China	25°26′14.348″ N, 116°38′42.400″ E	S.F. Chen, Q.L. Liu and F.F. Liu	–	OK253162	OK253306	–	OK253518	OK253871
	CSF9982	-BA-AA	Soil (natural forest area)	Liancheng, Longyan, Fujian, China	25°26′14.348″ N, 116°38′42.400″ E	S.F. Chen, Q.L. Liu and F.F. Liu	–	OK253163	OK253307	–	OK253519	OK253872
	CSF9989	-BA-AA	Soil (natural forest area)	Liancheng, Longyan, Fujian, China	25°26′14.348″ N, 116°38′42.400″ E	S.F. Chen, Q.L. Liu and F.F. Liu	–	OK253164	OK253308	–	OK253520	OK253873
	CSF10032	-BA-AA	Soil (*Eucalyptus* plantation)	Yongan, Sanming, Fujian, China	25°55′10.860″ N, 117°16′39.591″ E	S.F. Chen, Q.L. Liu and F.F. Liu	–	OK253165	OK253309	–	OK253521	OK253874
	CSF10034	-BA-AA	Soil (*Eucalyptus* plantation)	Yongan, Sanming, Fujian, China	25°55′10.860″ N, 117°16′39.591″ E	S.F. Chen, Q.L. Liu and F.F. Liu	–	OK253166	OK253310	–	OK253522	OK253875

Table A1. *Cont.*

Species [a]	Isolate No. [b]	Genotype [c]	Substrate	Sampling Site	GPS Coordinate	Collector	GenBank Accession No. [d]					
							act	*cmdA*	*his3*	*rpb2*	*tef1*	*tub2*
	CSF10041	-BA-AA	Soil (*Eucalyptus* plantation)	Yongan, Sanming, Fujian, China	25°55′10.860″ N, 117°16′39.591″ E	S.F. Chen, Q.L. Liu and F.F. Liu	–	OK253167	OK253311	–	OK253523	OK253876
	CSF10050	-BA-AA	Soil (*Cunning-hamia lanceolata*)	Qingliu, Sanming, Fujian, China	26°07′23.497″ N, 116°53′00.762″ E	S.F. Chen, Q.L. Liu and F.F. Liu	–	OK253168	OK253312	–	OK253524	OK253877
	CSF10064	-BA-AA	Soil (*Eucalyptus* plantation)	Yanping, Nanping, Fujian, China	26°46′19.651″ N, 117°57′37.233″ E	S.F. Chen, Q.L. Liu and F.F. Liu	–	OK253169	OK253313	–	OK253525	OK253878
	CSF10068	-BA-AA	Soil (*Eucalyptus* plantation)	Yanping, Nanping, Fujian, China	26°46′19.651″ N, 117°57′37.233″ E	S.F. Chen, Q.L. Liu and F.F. Liu	–	OK253170	OK253314	–	OK253526	OK253879
	CSF10073	-BA-AA	Soil (natural forest area)	Yanping, Nanping, Fujian, China	26°42′26.672″ N, 118°07′58.317″ E	S.F. Chen, Q.L. Liu and F.F. Liu	–	OK253171	OK253315	–	OK253527	OK253880
	CSF10075	-BA-AA	Soil (natural forest area)	Yanping, Nanping, Fujian, China	26°42′26.672″ N, 118°07′58.317″ E	S.F. Chen, Q.L. Liu and F.F. Liu	–	OK253172	OK253316	–	OK253528	OK253881
	CSF10081	-BA-AA	Soil (*Eucalyptus* plantation)	Minhou, Fuzhou, Fujian, China	26°15′04.285″ N, 119°02′38.917″ E	S.F. Chen, Q.L. Liu and F.F. Liu	–	OK253173	OK253317	–	OK253529	OK253882
	CSF10082	-BA-AA	Soil (*Eucalyptus* plantation)	Minhou, Fuzhou, Fujian, China	26°15′04.285″ N, 119°02′38.917″ E	S.F. Chen, Q.L. Liu and F.F. Liu	–	OK253174	OK253318	–	OK253530	OK253883
	CSF10097	-BA-AA	Soil (*Eucalyptus* plantation)	Minhou, Fuzhou, Fujian, China	26°15′04.285″ N, 119°02′38.917″ E	S.F. Chen, Q.L. Liu and F.F. Liu	–	OK253175	OK253319	–	OK253531	OK253884
	CSF10098	-BA-AA	Soil (*Eucalyptus* plantation)	Minhou, Fuzhou, Fujian, China	26°15′04.285″ N, 119°02′38.917″ E	S.F. Chen, Q.L. Liu and F.F. Liu	–	OK253176	OK253320	–	OK253532	OK253885
	CSF10110	-BA-A-	Soil (*Eucalyptus* plantation)	Minhou, Fuzhou, Fujian, China	26°15′04.285″ N, 119°02′38.917″ E	S.F. Chen, Q.L. Liu and F.F. Liu	–	OK253177	OK253321	–	OK253533	N/A

Table A1. *Cont.*

Species [a]	Isolate No. [b]	Genotype [c]	Substrate	Sampling Site	GPS Coordinate	Collector	GenBank Accession No. [d]					
							act	*cmdA*	*his3*	*rpb2*	*tef1*	*tub2*
	CSF10112	-BA-AA	Soil (*Eucalyptus* plantation)	Minhou, Fuzhou, Fujian, China	26°15′04.285″ N, 119°02′38.917″ E	S.F. Chen, Q.L. Liu and F.F. Liu	–	OK253178	OK253322	–	OK253534	OK253886
	CSF10119	-BA-AA	Soil (*Eucalyptus* plantation)	Minhou, Fuzhou, Fujian, China	26°15′04.285″ N, 119°02′38.917″ E	S.F. Chen, Q.L. Liu and F.F. Liu	–	OK253179	OK253323	–	OK253535	OK253887
	CSF10125	-BA-AA	Soil (*Eucalyptus* plantation)	Minhou, Fuzhou, Fujian, China	26°15′04.285″ N, 119°02′38.917″ E	S.F. Chen, Q.L. Liu and F.F. Liu	–	OK253180	OK253324	–	OK253536	OK253888
	CSF9856	-BA-AC	Soil (*Eucalyptus* plantation)	Zhangping, Longyan, Fujian, China	25°17′10.882″ N, 117°27′33.635″ E	S.F. Chen, Q.L. Liu and F.F. Liu	–	OK253181	OK253325	–	OK253537	OK253889
	CSF9897	-BA-AC	Soil (*Eucalyptus* plantation)	Zhangping, Longyan, Fujian, China	25°17′10.882″ N, 117°27′33.635″ E	S.F. Chen, Q.L. Liu and F.F. Liu	–	OK253182	OK253326	–	OK253538	OK253890
	CSF9814	-BA-CA	Soil (*Eucalyptus* plantation)	Hua'an, Zhangzhou, Fujian, China	24°53′49.369″ N, 117°32′45.070″ E	S.F. Chen, Q.L. Liu and F.F. Liu	–	OK253183	OK253327	–	OK253539	OK253891
	CSF9815	-BA-CA	Soil (*Eucalyptus* plantation)	Hua'an, Zhangzhou, Fujian, China	24°53′49.369″ N, 117°32′45.070″ E	S.F. Chen, Q.L. Liu and F.F. Liu	–	OK253184	OK253328	–	OK253540	OK253892
	CSF9842	-BA-CA	Soil (*Eucalyptus* plantation)	Zhangping, Longyan, Fujian, China	25°17′10.882″ N, 117°27′33.635″ E	S.F. Chen, Q.L. Liu and F.F. Liu	–	OK253185	OK253329	–	OK253541	OK253893
	CSF9843	-BA-CA	Soil (*Eucalyptus* plantation)	Zhangping, Longyan, Fujian, China	25°17′10.882″ N, 117°27′33.635″ E	S.F. Chen, Q.L. Liu and F.F. Liu	–	OK253186	OK253330	–	OK253542	OK253894
	CSF9887	-BA-CA	Soil (*Eucalyptus* plantation)	Zhangping, Longyan, Fujian, China	25°17′10.882″ N, 117°27′33.635″ E	S.F. Chen, Q.L. Liu and F.F. Liu	–	OK253187	OK253331	–	OK253543	OK253895
	CSF9888	-BA-CA	Soil (*Eucalyptus* plantation)	Zhangping, Longyan, Fujian, China	25°17′10.882″ N, 117°27′33.635″ E	S.F. Chen, Q.L. Liu and F.F. Liu	–	OK253188	OK253332	–	OK253544	OK253896

Table A1. *Cont.*

Species [a]	Isolate No. [b]	Genotype [c]	Substrate	Sampling Site	GPS Coordinate	Collector	GenBank Accession No. [d]					
							act	*cmdA*	*his3*	*rpb2*	*tef1*	*tub2*
	CSF9889	-BA-CA	Soil (*Eucalyptus* plantation)	Zhangping, Longyan, Fujian, China	25°17′10.882″ N, 117°27′33.635″ E	S.F. Chen, Q.L. Liu and F.F. Liu	–	OK253189	OK253333	–	OK253545	OK253897
	CSF9890	-BA-CA	Soil (*Eucalyptus* plantation)	Zhangping, Longyan, Fujian, China	25°17′10.882″ N, 117°27′33.635″ E	S.F. Chen, Q.L. Liu and F.F. Liu	–	OK253190	OK253334	–	OK253546	OK253898
	CSF9891	-BA-CA	Soil (*Eucalyptus* plantation)	Zhangping, Longyan, Fujian, China	25°17′10.882″ N, 117°27′33.635″ E	S.F. Chen, Q.L. Liu and F.F. Liu	–	OK253191	OK253335	–	OK253547	OK253899
	CSF9776	—-A-	Soil (*Eucalyptus* plantation)	Hua'an, Zhangzhou, Fujian, China	24°53′49.369″ N, 117°32′45.070″ E	S.F. Chen, Q.L. Liu and F.F. Liu	–	–	–	–	OK253548	–
	CSF9777	—-A-	Soil (*Eucalyptus* plantation)	Hua'an, Zhangzhou, Fujian, China	24°53′49.369″ N, 117°32′45.070″ E	S.F. Chen, Q.L. Liu and F.F. Liu	–	–	–	–	OK253549	–
	CSF9778	—-A-	Soil (*Eucalyptus* plantation)	Hua'an, Zhangzhou, Fujian, China	24°53′49.369″ N, 117°32′45.070″ E	S.F. Chen, Q.L. Liu and F.F. Liu	–	–	–	–	OK253550	–
	CSF9780	—-A-	Soil (*Eucalyptus* plantation)	Hua'an, Zhangzhou, Fujian, China	24°53′49.369″ N, 117°32′45.070″ E	S.F. Chen, Q.L. Liu and F.F. Liu	–	–	–	–	OK253551	–
	CSF9787	—-A-	Soil (*Eucalyptus* plantation)	Hua'an, Zhangzhou, Fujian, China	24°53′49.369″ N, 117°32′45.070″ E	S.F. Chen, Q.L. Liu and F.F. Liu	–	–	–	–	OK253552	–
	CSF9788	—-A-	Soil (*Eucalyptus* plantation)	Hua'an, Zhangzhou, Fujian, China	24°53′49.369″ N, 117°32′45.070″ E	S.F. Chen, Q.L. Liu and F.F. Liu	–	–	–	–	OK253553	–
	CSF9790	—-A-	Soil (*Eucalyptus* plantation)	Hua'an, Zhangzhou, Fujian, China	24°53′49.369″ N, 117°32′45.070″ E	S.F. Chen, Q.L. Liu and F.F. Liu	–	–	–	–	OK253554	–
	CSF9806	—-A-	Soil (*Eucalyptus* plantation)	Hua'an, Zhangzhou, Fujian, China	24°53′49.369″ N, 117°32′45.070″ E	S.F. Chen, Q.L. Liu and F.F. Liu	–	–	–	–	OK253555	–

Table A1. *Cont.*

Species [a]	Isolate No. [b]	Genotype [c]	Substrate	Sampling Site	GPS Coordinate	Collector	GenBank Accession No. [d]					
							act	*cmdA*	*his3*	*rpb2*	*tef1*	*tub2*
	CSF9807	—-A-	Soil (*Eucalyptus* plantation)	Hua'an, Zhangzhou, Fujian, China	24°53′49.369″ N, 117°32′45.070″ E	S.F. Chen, Q.L. Liu and F.F. Liu	–	–	–	–	OK253556	–
	CSF9808	—-A-	Soil (*Eucalyptus* plantation)	Hua'an, Zhangzhou, Fujian, China	24°53′49.369″ N, 117°32′45.070″ E	S.F. Chen, Q.L. Liu and F.F. Liu	–	–	–	–	OK253557	–
	CSF9810	—-A-	Soil (*Eucalyptus* plantation)	Hua'an, Zhangzhou, Fujian, China	24°53′49.369″ N, 117°32′45.070″ E	S.F. Chen, Q.L. Liu and F.F. Liu	–	–	–	–	OK253558	–
	CSF9836	—-A-	Soil (*Eucalyptus* plantation)	Zhangping, Longyan, Fujian, China	25°17′10.882″ N, 117°27′33.635″ E	S.F. Chen, Q.L. Liu and F.F. Liu	–	–	–	–	OK253559	–
	CSF9837	—-A-	Soil (*Eucalyptus* plantation)	Zhangping, Longyan, Fujian, China	25°17′10.882″ N, 117°27′33.635″ E	S.F. Chen, Q.L. Liu and F.F. Liu	–	–	–	–	OK253560	–
	CSF9838	—-A-	Soil (*Eucalyptus* plantation)	Zhangping, Longyan, Fujian, China	25°17′10.882″ N, 117°27′33.635″ E	S.F. Chen, Q.L. Liu and F.F. Liu	–	–	–	–	OK253561	–
	CSF9840	—-A-	Soil (*Eucalyptus* plantation)	Zhangping, Longyan, Fujian, China	25°17′10.882″ N, 117°27′33.635″ E	S.F. Chen, Q.L. Liu and F.F. Liu	–	–	–	–	OK253562	–
	CSF9845	—-A-	Soil (*Eucalyptus* plantation)	Zhangping, Longyan, Fujian, China	25°17′10.882″ N, 117°27′33.635″ E	S.F. Chen, Q.L. Liu and F.F. Liu	–	–	–	–	OK253563	–
	CSF9847	—-A-	Soil (*Eucalyptus* plantation)	Zhangping, Longyan, Fujian, China	25°17′10.882″ N, 117°27′33.635″ E	S.F. Chen, Q.L. Liu and F.F. Liu	–	–	–	–	OK253564	–
	CSF9848	—-A-	Soil (*Eucalyptus* plantation)	Zhangping, Longyan, Fujian, China	25°17′10.882″ N, 117°27′33.635″ E	S.F. Chen, Q.L. Liu and F.F. Liu	–	–	–	–	OK253565	–
	CSF9849	—-A-	Soil (*Eucalyptus* plantation)	Zhangping, Longyan, Fujian, China	25°17′10.882″ N, 117°27′33.635″ E	S.F. Chen, Q.L. Liu and F.F. Liu	–	–	–	–	OK253566	–

Table A1. *Cont.*

Species [a]	Isolate No. [b]	Genotype [c]	Substrate	Sampling Site	GPS Coordinate	Collector	GenBank Accession No. [d]					
							act	*cmdA*	*his3*	*rpb2*	*tef1*	*tub2*
	CSF9850	—-A-	Soil (*Eucalyptus* plantation)	Zhangping, Longyan, Fujian, China	25°17′10.882″ N, 117°27′33.635″ E	S.F. Chen, Q.L. Liu and F.F. Liu	–	–	–	–	OK253567	–
	CSF9852	—-A-	Soil (*Eucalyptus* plantation)	Zhangping, Longyan, Fujian, China	25°17′10.882″ N, 117°27′33.635″ E	S.F. Chen, Q.L. Liu and F.F. Liu	–	–	–	–	OK253568	–
	CSF9853	—-A-	Soil (*Eucalyptus* plantation)	Zhangping, Longyan, Fujian, China	25°17′10.882″ N, 117°27′33.635″ E	S.F. Chen, Q.L. Liu and F.F. Liu	–	–	–	–	OK253569	–
	CSF9854	—-A-	Soil (*Eucalyptus* plantation)	Zhangping, Longyan, Fujian, China	25°17′10.882″ N, 117°27′33.635″ E	S.F. Chen, Q.L. Liu and F.F. Liu	–	–	–	–	OK253570	–
	CSF9855	—-A-	Soil (*Eucalyptus* plantation)	Zhangping, Longyan, Fujian, China	25°17′10.882″ N, 117°27′33.635″ E	S.F. Chen, Q.L. Liu and F.F. Liu	–	–	–	–	OK253571	–
	CSF9858	—-A-	Soil (*Eucalyptus* plantation)	Zhangping, Longyan, Fujian, China	25°17′10.882″ N, 117°27′33.635″ E	S.F. Chen, Q.L. Liu and F.F. Liu	–	–	–	–	OK253572	–
	CSF9859	—-A-	Soil (*Eucalyptus* plantation)	Zhangping, Longyan, Fujian, China	25°17′10.882″ N, 117°27′33.635″ E	S.F. Chen, Q.L. Liu and F.F. Liu	–	–	–	–	OK253573	–
	CSF9860	—-A-	Soil (*Eucalyptus* plantation)	Zhangping, Longyan, Fujian, China	25°17′10.882″ N, 117°27′33.635″ E	S.F. Chen, Q.L. Liu and F.F. Liu	–	–	–	–	OK253574	–
	CSF9861	—-A-	Soil (*Eucalyptus* plantation)	Zhangping, Longyan, Fujian, China	25°17′10.882″ N, 117°27′33.635″ E	S.F. Chen, Q.L. Liu and F.F. Liu	–	–	–	–	OK253575	–
	CSF9867	—-A-	Soil (*Eucalyptus* plantation)	Zhangping, Longyan, Fujian, China	25°17′10.882″ N, 117°27′33.635″ E	S.F. Chen, Q.L. Liu and F.F. Liu	–	–	–	–	OK253576	–
	CSF9868	—-A-	Soil (*Eucalyptus* plantation)	Zhangping, Longyan, Fujian, China	25°17′10.882″ N, 117°27′33.635″ E	S.F. Chen, Q.L. Liu and F.F. Liu	–	–	–	–	OK253577	–

Table A1. *Cont.*

Species [a]	Isolate No. [b]	Genotype [c]	Substrate	Sampling Site	GPS Coordinate	Collector	GenBank Accession No. [d]					
							act	*cmdA*	*his3*	*rpb2*	*tef1*	*tub2*
	CSF9869	—-A-	Soil (*Eucalyptus* plantation)	Zhangping, Longyan, Fujian, China	25°17′10.882″ N, 117°27′33.635″ E	S.F. Chen, Q.L. Liu and F.F. Liu	–	–	–	–	OK253578	–
	CSF9871	—-A-	Soil (*Eucalyptus* plantation)	Zhangping, Longyan, Fujian, China	25°17′10.882″ N, 117°27′33.635″ E	S.F. Chen, Q.L. Liu and F.F. Liu	–	–	–	–	OK253579	–
	CSF9872	—-A-	Soil (*Eucalyptus* plantation)	Zhangping, Longyan, Fujian, China	25°17′10.882″ N, 117°27′33.635″ E	S.F. Chen, Q.L. Liu and F.F. Liu	–	–	–	–	OK253580	–
	CSF9873	—-A-	Soil (*Eucalyptus* plantation)	Zhangping, Longyan, Fujian, China	25°17′10.882″ N, 117°27′33.635″ E	S.F. Chen, Q.L. Liu and F.F. Liu	–	–	–	–	OK253581	–
	CSF9874	—-A-	Soil (*Eucalyptus* plantation)	Zhangping, Longyan, Fujian, China	25°17′10.882″ N, 117°27′33.635″ E	S.F. Chen, Q.L. Liu and F.F. Liu	–	–	–	–	OK253582	–
	CSF9876	—-A-	Soil (*Eucalyptus* plantation)	Zhangping, Longyan, Fujian, China	25°17′10.882″ N, 117°27′33.635″ E	S.F. Chen, Q.L. Liu and F.F. Liu	–	–	–	–	OK253583	–
	CSF9877	—-A-	Soil (*Eucalyptus* plantation)	Zhangping, Longyan, Fujian, China	25°17′10.882″ N, 117°27′33.635″ E	S.F. Chen, Q.L. Liu and F.F. Liu	–	–	–	–	OK253584	–
	CSF9878	—-A-	Soil (*Eucalyptus* plantation)	Zhangping, Longyan, Fujian, China	25°17′10.882″ N, 117°27′33.635″ E	S.F. Chen, Q.L. Liu and F.F. Liu	–	–	–	–	OK253585	–
	CSF9879	—-A-	Soil (*Eucalyptus* plantation)	Zhangping, Longyan, Fujian, China	25°17′10.882″ N, 117°27′33.635″ E	S.F. Chen, Q.L. Liu and F.F. Liu	–	–	–	–	OK253586	–
	CSF9881	—-A-	Soil (*Eucalyptus* plantation)	Zhangping, Longyan, Fujian, China	25°17′10.882″ N, 117°27′33.635″ E	S.F. Chen, Q.L. Liu and F.F. Liu	–	–	–	–	OK253587	–
	CSF9883	—-A-	Soil (*Eucalyptus* plantation)	Zhangping, Longyan, Fujian, China	25°17′10.882″ N, 117°27′33.635″ E	S.F. Chen, Q.L. Liu and F.F. Liu	–	–	–	–	OK253588	–

Table A1. *Cont.*

Species [a]	Isolate No. [b]	Genotype [c]	Substrate	Sampling Site	GPS Coordinate	Collector	GenBank Accession No. [d]					
							act	*cmdA*	*his3*	*rpb2*	*tef1*	*tub2*
	CSF9884	—-A-	Soil (*Eucalyptus* plantation)	Zhangping, Longyan, Fujian, China	25°17′10.882″ N, 117°27′33.635″ E	S.F. Chen, Q.L. Liu and F.F. Liu	–	–	–	–	OK253589	–
	CSF9885	—-A-	Soil (*Eucalyptus* plantation)	Zhangping, Longyan, Fujian, China	25°17′10.882″ N, 117°27′33.635″ E	S.F. Chen, Q.L. Liu and F.F. Liu	–	–	–	–	OK253590	–
	CSF9886	—-A-	Soil (*Eucalyptus* plantation)	Zhangping, Longyan, Fujian, China	25°17′10.882″ N, 117°27′33.635″ E	S.F. Chen, Q.L. Liu and F.F. Liu	–	–	–	–	OK253591	–
	CSF9892	—-A-	Soil (*Eucalyptus* plantation)	Zhangping, Longyan, Fujian, China	25°17′10.882″ N, 117°27′33.635″ E	S.F. Chen, Q.L. Liu and F.F. Liu	–	–	–	–	OK253592	–
	CSF9893	—-A-	Soil (*Eucalyptus* plantation)	Zhangping, Longyan, Fujian, China	25°17′10.882″ N, 117°27′33.635″ E	S.F. Chen, Q.L. Liu and F.F. Liu	–	–	–	–	OK253593	–
	CSF9894	—-A-	Soil (*Eucalyptus* plantation)	Zhangping, Longyan, Fujian, China	25°17′10.882″ N, 117°27′33.635″ E	S.F. Chen, Q.L. Liu and F.F. Liu	–	–	–	–	OK253594	–
	CSF9896	—-A-	Soil (*Eucalyptus* plantation)	Zhangping, Longyan, Fujian, China	25°17′10.882″ N, 117°27′33.635″ E	S.F. Chen, Q.L. Liu and F.F. Liu	–	–	–	–	OK253595	–
	CSF9898	—-A-	Soil (*Eucalyptus* plantation)	Zhangping, Longyan, Fujian, China	25°17′10.882″ N, 117°27′33.635″ E	S.F. Chen, Q.L. Liu and F.F. Liu	–	–	–	–	OK253596	–
	CSF9899	—-A-	Soil (*Eucalyptus* plantation)	Zhangping, Longyan, Fujian, China	25°17′10.882″ N, 117°27′33.635″ E	S.F. Chen, Q.L. Liu and F.F. Liu	–	–	–	–	OK253597	–
	CSF9900	—-A-	Soil (*Eucalyptus* plantation)	Zhangping, Longyan, Fujian, China	25°17′10.882″ N, 117°27′33.635″ E	S.F. Chen, Q.L. Liu and F.F. Liu	–	–	–	–	OK253598	–
	CSF9901	—-A-	Soil (*Eucalyptus* plantation)	Zhangping, Longyan, Fujian, China	25°17′10.882″ N, 117°27′33.635″ E	S.F. Chen, Q.L. Liu and F.F. Liu	–	–	–	–	OK253599	–

Table A1. *Cont.*

Species [a]	Isolate No. [b]	Genotype [c]	Substrate	Sampling Site	GPS Coordinate	Collector	GenBank Accession No. [d]					
							act	*cmdA*	*his3*	*rpb2*	*tef1*	*tub2*
	CSF9902	—-A-	Soil (*Eucalyptus* plantation)	Zhangping, Longyan, Fujian, China	25°17′10.882″ N, 117°27′33.635″ E	S.F. Chen, Q.L. Liu and F.F. Liu	–	–	–	–	OK253600	–
	CSF9903	—-A-	Soil (*Eucalyptus* plantation)	Zhangping, Longyan, Fujian, China	25°17′10.882″ N, 117°27′33.635″ E	S.F. Chen, Q.L. Liu and F.F. Liu	–	–	–	–	OK253601	–
	CSF9904	—-A-	Soil (*Eucalyptus* plantation)	Zhangping, Longyan, Fujian, China	25°17′10.882″ N, 117°27′33.635″ E	S.F. Chen, Q.L. Liu and F.F. Liu	–	–	–	–	OK253602	–
	CSF9906	—-A-	Soil (*Eucalyptus* plantation)	Zhangping, Longyan, Fujian, China	25°17′10.882″ N, 117°27′33.635″ E	S.F. Chen, Q.L. Liu and F.F. Liu	–	–	–	–	OK253603	–
	CSF9927	—-A-	Soil (*Phyllostachys heterocycla*)	Xinluo, Longyan, Fujian, China	25°07′31.133″ N, 116°51′37.485″ E	S.F. Chen, Q.L. Liu and F.F. Liu	–	–	–	–	OK253604	–
	CSF9928	—-A-	Soil (*Phyllostachys heterocycla*)	Xinluo, Longyan, Fujian, China	25°07′31.133″ N, 116°51′37.485″ E	S.F. Chen, Q.L. Liu and F.F. Liu	–	–	–	–	OK253605	–
	CSF9929	—-A-	Soil (*Phyllostachys heterocycla*)	Xinluo, Longyan, Fujian, China	25°07′31.133″ N, 116°51′37.485″ E	S.F. Chen, Q.L. Liu and F.F. Liu	–	–	–	–	OK253606	–
	CSF9931	—-A-	Soil (*Phyllostachys heterocycla*)	Xinluo, Longyan, Fujian, China	25°07′31.133″ N, 116°51′37.485″ E	S.F. Chen, Q.L. Liu and F.F. Liu	–	–	–	–	OK253607	–
	CSF9940	—-A-	Soil (*Eucalyptus* plantation)	Xinluo, Longyan, Fujian, China	25°07′08.597″ N, 116°44′42.257″ E	S.F. Chen, Q.L. Liu and F.F. Liu	–	–	–	–	OK253608	–
	CSF9944	—-A-	Soil (*Eucalyptus* plantation)	Xinluo, Longyan, Fujian, China	25°07′08.597″ N, 116°44′42.257″ E	S.F. Chen, Q.L. Liu and F.F. Liu	–	–	–	–	OK253609	–
	CSF9954	—-A-	Soil (natural forest area)	Liancheng, Longyan, Fujian, China	25°26′14.348″ N, 116°38′42.400″ E	S.F. Chen, Q.L. Liu and F.F. Liu	–	–	–	–	OK253610	–

Table A1. *Cont.*

Species [a]	Isolate No. [b]	Genotype [c]	Substrate	Sampling Site	GPS Coordinate	Collector	GenBank Accession No. [d]					
							act	*cmdA*	*his3*	*rpb2*	*tef1*	*tub2*
	CSF9955	—-A-	Soil (natural forest area)	Liancheng, Longyan, Fujian, China	25°26′14.348″ N, 116°38′42.400″ E	S.F. Chen, Q.L. Liu and F.F. Liu	–	–	–	–	OK253611	–
	CSF9956	—-A-	Soil (natural forest area)	Liancheng, Longyan, Fujian, China	25°26′14.348″ N, 116°38′42.400″ E	S.F. Chen, Q.L. Liu and F.F. Liu	–	–	–	–	OK253612	–
	CSF9958	—-A-	Soil (natural forest area)	Liancheng, Longyan, Fujian, China	25°26′14.348″ N, 116°38′42.400″ E	S.F. Chen, Q.L. Liu and F.F. Liu	–	–	–	–	OK253613	–
	CSF9965	—-A-	Soil (natural forest area)	Liancheng, Longyan, Fujian, China	25°26′14.348″ N, 116°38′42.400″ E	S.F. Chen, Q.L. Liu and F.F. Liu	–	–	–	–	OK253614	–
	CSF9966	—-A-	Soil (natural forest area)	Liancheng, Longyan, Fujian, China	25°26′14.348″ N, 116°38′42.400″ E	S.F. Chen, Q.L. Liu and F.F. Liu	–	–	–	–	OK253615	–
	CSF9967	—-A-	Soil (natural forest area)	Liancheng, Longyan, Fujian, China	25°26′14.348″ N, 116°38′42.400″ E	S.F. Chen, Q.L. Liu and F.F. Liu	–	–	–	–	OK253616	–
	CSF9968	—-A-	Soil (natural forest area)	Liancheng, Longyan, Fujian, China	25°26′14.348″ N, 116°38′42.400″ E	S.F. Chen, Q.L. Liu and F.F. Liu	–	–	–	–	OK253617	–
	CSF9971	—-A-	Soil (natural forest area)	Liancheng, Longyan, Fujian, China	25°26′14.348″ N, 116°38′42.400″ E	S.F. Chen, Q.L. Liu and F.F. Liu	–	–	–	–	OK253618	–
	CSF9979	—-A-	Soil (natural forest area)	Liancheng, Longyan, Fujian, China	25°26′14.348″ N, 116°38′42.400″ E	S.F. Chen, Q.L. Liu and F.F. Liu	–	–	–	–	OK253619	–
	CSF9980	—-A-	Soil (natural forest area)	Liancheng, Longyan, Fujian, China	25°26′14.348″ N, 116°38′42.400″ E	S.F. Chen, Q.L. Liu and F.F. Liu	–	–	–	–	OK253620	–
	CSF9981	—-A-	Soil (natural forest area)	Liancheng, Longyan, Fujian, China	25°26′14.348″ N, 116°38′42.400″ E	S.F. Chen, Q.L. Liu and F.F. Liu	–	–	–	–	OK253621	–

Table A1. *Cont.*

Species [a]	Isolate No. [b]	Genotype [c]	Substrate	Sampling Site	GPS Coordinate	Collector	GenBank Accession No. [d]					
							act	*cmdA*	*his3*	*rpb2*	*tef1*	*tub2*
	CSF9983	—-A-	Soil (natural forest area)	Liancheng, Longyan, Fujian, China	25°26′14.348″ N, 116°38′42.400″ E	S.F. Chen, Q.L. Liu and F.F. Liu	–	–	–	–	OK253622	–
	CSF9984	—-A-	Soil (natural forest area)	Liancheng, Longyan, Fujian, China	25°26′14.348″ N, 116°38′42.400″ E	S.F. Chen, Q.L. Liu and F.F. Liu	–	–	–	–	OK253623	–
	CSF9985	—-A-	Soil (natural forest area)	Liancheng, Longyan, Fujian, China	25°26′14.348″ N, 116°38′42.400″ E	S.F. Chen, Q.L. Liu and F.F. Liu	–	–	–	–	OK253624	–
	CSF9986	—-A-	Soil (natural forest area)	Liancheng, Longyan, Fujian, China	25°26′14.348″ N, 116°38′42.400″ E	S.F. Chen, Q.L. Liu and F.F. Liu	–	–	–	–	OK253625	–
	CSF9988	—-A-	Soil (natural forest area)	Liancheng, Longyan, Fujian, China	25°26′14.348″ N, 116°38′42.400″ E	S.F. Chen, Q.L. Liu and F.F. Liu	–	–	–	–	OK253626	–
	CSF9990	—-A-	Soil (natural forest area)	Liancheng, Longyan, Fujian, China	25°26′14.348″ N, 116°38′42.400″ E	S.F. Chen, Q.L. Liu and F.F. Liu	–	–	–	–	OK253627	–
	CSF9991	—-A-	Soil (natural forest area)	Liancheng, Longyan, Fujian, China	25°26′14.348″ N, 116°38′42.400″ E	S.F. Chen, Q.L. Liu and F.F. Liu	–	–	–	–	OK253628	–
	CSF9992	—-A-	Soil (natural forest area)	Liancheng, Longyan, Fujian, China	25°26′14.348″ N, 116°38′42.400″ E	S.F. Chen, Q.L. Liu and F.F. Liu	–	–	–	–	OK253629	–
	CSF9993	—-A-	Soil (natural forest area)	Liancheng, Longyan, Fujian, China	25°26′14.348″ N, 116°38′42.400″ E	S.F. Chen, Q.L. Liu and F.F. Liu	–	–	–	–	OK253630	–
	CSF10029	—-A-	Soil (*Eucalyptus* plantation)	Yongan, Sanming, Fujian, China	25°55′10.860″ N, 117°16′39.591″ E	S.F. Chen, Q.L. Liu and F.F. Liu	–	–	–	–	OK253631	–
	CSF10030	—-A-	Soil (*Eucalyptus* plantation)	Yongan, Sanming, Fujian, China	25°55′10.860″ N, 117°16′39.591″ E	S.F. Chen, Q.L. Liu and F.F. Liu	–	–	–	–	OK253632	–

Table A1. *Cont.*

Species [a]	Isolate No. [b]	Genotype [c]	Substrate	Sampling Site	GPS Coordinate	Collector	GenBank Accession No. [d]					
							act	*cmdA*	*his3*	*rpb2*	*tef1*	*tub2*
	CSF10031	—-A-	Soil (*Eucalyptus* plantation)	Yongan, Sanming, Fujian, China	25°55′10.860″ N, 117°16′39.591″ E	S.F. Chen, Q.L. Liu and F.F. Liu	–	–	–	–	OK253633	–
	CSF10033	—-A-	Soil (*Eucalyptus* plantation)	Yongan, Sanming, Fujian, China	25°55′10.860″ N, 117°16′39.591″ E	S.F. Chen, Q.L. Liu and F.F. Liu	–	–	–	–	OK253634	–
	CSF10035	—-A-	Soil (*Eucalyptus* plantation)	Yongan, Sanming, Fujian, China	25°55′10.860″ N, 117°16′39.591″ E	S.F. Chen, Q.L. Liu and F.F. Liu	–	–	–	–	OK253635	–
	CSF10036	—-A-	Soil (*Eucalyptus* plantation)	Yongan, Sanming, Fujian, China	25°55′10.860″ N, 117°16′39.591″ E	S.F. Chen, Q.L. Liu and F.F. Liu	–	–	–	–	OK253636	–
	CSF10037	—-A-	Soil (*Eucalyptus* plantation)	Yongan, Sanming, Fujian, China	25°55′10.860″ N, 117°16′39.591″ E	S.F. Chen, Q.L. Liu and F.F. Liu	–	–	–	–	OK253637	–
	CSF10042	—-A-	Soil (*Eucalyptus* plantation)	Yongan, Sanming, Fujian, China	25°55′10.860″ N, 117°16′39.591″ E	S.F. Chen, Q.L. Liu and F.F. Liu	–	–	–	–	OK253638	–
	CSF10043	—-A-	Soil (*Eucalyptus* plantation)	Yongan, Sanming, Fujian, China	25°55′10.860″ N, 117°16′39.591″ E	S.F. Chen, Q.L. Liu and F.F. Liu	–	–	–	–	OK253639	–
	CSF10048	—-A-	Soil (*Cunning-hamia lanceolata*)	Qingliu, Sanming, Fujian, China	26°07′23.497″ N, 116°53′00.762″ E	S.F. Chen, Q.L. Liu and F.F. Liu	–	–	–	–	OK253640	–
	CSF10049	—-A-	Soil (*Cunning-hamia lanceolata*)	Qingliu, Sanming, Fujian, China	26°07′23.497″ N, 116°53′00.762″ E	S.F. Chen, Q.L. Liu and F.F. Liu	–	–	–	–	OK253641	–
	CSF10051	—-A-	Soil (*Cunning-hamia lanceolata*)	Qingliu, Sanming, Fujian, China	26°07′23.497″ N, 116°53′00.762″ E	S.F. Chen, Q.L. Liu and F.F. Liu	–	–	–	–	OK253642	–

Table A1. *Cont.*

Species [a]	Isolate No. [b]	Genotype [c]	Substrate	Sampling Site	GPS Coordinate	Collector	GenBank Accession No. [d]					
							act	*cmdA*	*his3*	*rpb2*	*tef1*	*tub2*
	CSF10052	—-A-	Soil (*Cunninghamia lanceolata*)	Qingliu, Sanming, Fujian, China	26°07′23.497″ N, 116°53′00.762″ E	S.F. Chen, Q.L. Liu and F.F. Liu	–	–	–	–	OK253643	–
	CSF10063	—-A-	Soil (*Eucalyptus* plantation)	Yanping, Nanping, Fujian, China	26°46′19.651″ N, 117°57′37.233″ E	S.F. Chen, Q.L. Liu and F.F. Liu	–	–	–	–	OK253644	–
	CSF10065	—-A-	Soil (*Eucalyptus* plantation)	Yanping, Nanping, Fujian, China	26°46′19.651″ N, 117°57′37.233″ E	S.F. Chen, Q.L. Liu and F.F. Liu	–	–	–	–	OK253645	–
	CSF10066	—-A-	Soil (*Eucalyptus* plantation)	Yanping, Nanping, Fujian, China	26°46′19.651″ N, 117°57′37.233″ E	S.F. Chen, Q.L. Liu and F.F. Liu	–	–	–	–	OK253646	–
	CSF10067	—-A-	Soil (*Eucalyptus* plantation)	Yanping, Nanping, Fujian, China	26°46′19.651″ N, 117°57′37.233″ E	S.F. Chen, Q.L. Liu and F.F. Liu	–	–	–	–	OK253647	–
	CSF10069	—-A-	Soil (*Eucalyptus* plantation)	Yanping, Nanping, Fujian, China	26°46′19.651″ N, 117°57′37.233″ E	S.F. Chen, Q.L. Liu and F.F. Liu	–	–	–	–	OK253648	–
	CSF10074	—-A-	Soil (natural forest area)	Yanping, Nanping, Fujian, China	26°42′26.672″ N, 118°07′58.317″ E	S.F. Chen, Q.L. Liu and F.F. Liu	–	–	–	–	OK253649	–
	CSF10083	—-A-	Soil (*Eucalyptus* plantation)	Minhou, Fuzhou, Fujian, China	26°15′04.285″ N, 119°02′38.917″ E	S.F. Chen, Q.L. Liu and F.F. Liu	–	–	–	–	OK253650	–
	CSF10099	—-A-	Soil (*Eucalyptus* plantation)	Minhou, Fuzhou, Fujian, China	26°15′04.285″ N, 119°02′38.917″ E	S.F. Chen, Q.L. Liu and F.F. Liu	–	–	–	–	OK253651	–
	CSF10100	—-A-	Soil (*Eucalyptus* plantation)	Minhou, Fuzhou, Fujian, China	26°15′04.285″ N, 119°02′38.917″ E	S.F. Chen, Q.L. Liu and F.F. Liu	–	–	–	–	OK253652	–
	CSF10101	—-A-	Soil (*Eucalyptus* plantation)	Minhou, Fuzhou, Fujian, China	26°15′04.285″ N, 119°02′38.917″ E	S.F. Chen, Q.L. Liu and F.F. Liu	–	–	–	–	OK253653	–

Table A1. *Cont.*

Species [a]	Isolate No. [b]	Genotype [c]	Substrate	Sampling Site	GPS Coordinate	Collector	GenBank Accession No. [d]					
							act	*cmdA*	*his3*	*rpb2*	*tef1*	*tub2*
	CSF10102	—-A-	Soil (*Eucalyptus* plantation)	Minhou, Fuzhou, Fujian, China	26°15′04.285″ N, 119°02′38.917″ E	S.F. Chen, Q.L. Liu and F.F. Liu	–	–	–	–	OK253654	–
	CSF10103	—-A-	Soil (*Eucalyptus* plantation)	Minhou, Fuzhou, Fujian, China	26°15′04.285″ N, 119°02′38.917″ E	S.F. Chen, Q.L. Liu and F.F. Liu	–	–	–	–	OK253655	–
	CSF10104	—-A-	Soil (*Eucalyptus* plantation)	Minhou, Fuzhou, Fujian, China	26°15′04.285″ N, 119°02′38.917″ E	S.F. Chen, Q.L. Liu and F.F. Liu	–	–	–	–	OK253656	–
	CSF10106	—-A-	Soil (*Eucalyptus* plantation)	Minhou, Fuzhou, Fujian, China	26°15′04.285″ N, 119°02′38.917″ E	S.F. Chen, Q.L. Liu and F.F. Liu	–	–	–	–	OK253657	–
	CSF10107	—-A-	Soil (*Eucalyptus* plantation)	Minhou, Fuzhou, Fujian, China	26°15′04.285″ N, 119°02′38.917″ E	S.F. Chen, Q.L. Liu and F.F. Liu	–	–	–	–	OK253658	–
	CSF10108	—-A-	Soil (*Eucalyptus* plantation)	Minhou, Fuzhou, Fujian, China	26°15′04.285″ N, 119°02′38.917″ E	S.F. Chen, Q.L. Liu and F.F. Liu	–	–	–	–	OK253659	–
	CSF10109	—-A-	Soil (*Eucalyptus* plantation)	Minhou, Fuzhou, Fujian, China	26°15′04.285″ N, 119°02′38.917″ E	S.F. Chen, Q.L. Liu and F.F. Liu	–	–	–	–	OK253660	–
	CSF10111	—-A-	Soil (*Eucalyptus* plantation)	Minhou, Fuzhou, Fujian, China	26°15′04.285″ N, 119°02′38.917″ E	S.F. Chen, Q.L. Liu and F.F. Liu	–	–	–	–	OK253661	–
	CSF10113	—-A-	Soil (*Eucalyptus* plantation)	Minhou, Fuzhou, Fujian, China	26°15′04.285″ N, 119°02′38.917″ E	S.F. Chen, Q.L. Liu and F.F. Liu	–	–	–	–	OK253662	–
	CSF10114	—-A-	Soil (*Eucalyptus* plantation)	Minhou, Fuzhou, Fujian, China	26°15′04.285″ N, 119°02′38.917″ E	S.F. Chen, Q.L. Liu and F.F. Liu	–	–	–	–	OK253663	–
	CSF10115	—-A-	Soil (*Eucalyptus* plantation)	Minhou, Fuzhou, Fujian, China	26°15′04.285″ N, 119°02′38.917″ E	S.F. Chen, Q.L. Liu and F.F. Liu	–	–	–	–	OK253664	–

Table A1. *Cont.*

Species [a]	Isolate No. [b]	Genotype [c]	Substrate	Sampling Site	GPS Coordinate	Collector	GenBank Accession No. [d]					
							act	*cmdA*	*his3*	*rpb2*	*tef1*	*tub2*
	CSF10116	—-A-	Soil (*Eucalyptus* plantation)	Minhou, Fuzhou, Fujian, China	26°15′04.285″ N, 119°02′38.917″ E	S.F. Chen, Q.L. Liu and F.F. Liu	–	–	–	–	OK253665	–
	CSF10117	—-A-	Soil (*Eucalyptus* plantation)	Minhou, Fuzhou, Fujian, China	26°15′04.285″ N, 119°02′38.917″ E	S.F. Chen, Q.L. Liu and F.F. Liu	–	–	–	–	OK253666	–
	CSF10118	—-A-	Soil (*Eucalyptus* plantation)	Minhou, Fuzhou, Fujian, China	26°15′04.285″ N, 119°02′38.917″ E	S.F. Chen, Q.L. Liu and F.F. Liu	–	–	–	–	OK253667	–
	CSF10120	—-A-	Soil (*Eucalyptus* plantation)	Minhou, Fuzhou, Fujian, China	26°15′04.285″ N, 119°02′38.917″ E	S.F. Chen, Q.L. Liu and F.F. Liu	–	–	–	–	OK253668	–
Ca. hongkongensis	CSF7124 [e]	AAAAAA	Soil (natural forest area)	Cangshan, Fuzhou, Fujian, China	26°5′16.2″ N, 119°14′19.8″ E	S.F. Chen, Q.L. Liu and F.F. Liu	OK253081	OK253192	OK253336	OK253438	OK253669	OK253900
	CSF9784 [e]	AAAAAA	Soil (*Eucalyptus* plantation)	Hua'an, Zhangzhou, Fujian, China	24°53′49.369″ N, 117°32′45.070″ E	S.F. Chen, Q.L. Liu and F.F. Liu	OK253082	OK253193	OK253337	OK253439	OK253670	OK253901
	CSF9794 [e]	ABAAAA	Soil (*Eucalyptus* plantation)	Hua'an, Zhangzhou, Fujian, China	24°53′49.369″ N, 117°32′45.070″ E	S.F. Chen, Q.L. Liu and F.F. Liu	OK253083	OK253194	OK253338	OK253440	OK253671	OK253902
	CSF9799 [e]	ABAAAA	Soil (*Eucalyptus* plantation)	Hua'an, Zhangzhou, Fujian, China	24°53′49.369″ N, 117°32′45.070″ E	S.F. Chen, Q.L. Liu and F.F. Liu	OK253084	OK253195	OK253339	OK253441	OK253672	OK253903
	CSF7139	-AA-AA	Soil (natural forest area)	Cangshan, Fuzhou, Fujian, China	26°5′16.2″ N, 119°14′19.8″ E	S.F. Chen, Q.L. Liu and F.F. Liu	–	OK253196	OK253340	–	OK253673	OK253904
	CSF9804	-AA-AA	Soil (*Eucalyptus* plantation)	Hua'an, Zhangzhou, Fujian, China	24°53′49.369″ N, 117°32′45.070″ E	S.F. Chen, Q.L. Liu and F.F. Liu	–	OK253197	OK253341	–	OK253674	OK253905
	CSF9819	-AA-AA	Soil (*Eucalyptus* plantation)	Hua'an, Zhangzhou, Fujian, China	24°53′49.369″ N, 117°32′45.070″ E	S.F. Chen, Q.L. Liu and F.F. Liu	–	OK253198	OK253342	–	OK253675	OK253906

Table A1. *Cont.*

Species [a]	Isolate No. [b]	Genotype [c]	Substrate	Sampling Site	GPS Coordinate	Collector	GenBank Accession No. [d]					
							act	*cmdA*	*his3*	*rpb2*	*tef1*	*tub2*
	CSF10093	-AA-AA	Soil (*Eucalyptus* plantation)	Minhou, Fuzhou, Fujian, China	26°15′04.285″ N, 119°02′38.917″ E	S.F. Chen, Q.L. Liu and F.F. Liu	–	OK253199	OK253343	–	OK253676	OK253907
	CSF10096	-AA-AA	Soil (*Eucalyptus* plantation)	Minhou, Fuzhou, Fujian, China	26°15′04.285″ N, 119°02′38.917″ E	S.F. Chen, Q.L. Liu and F.F. Liu	–	OK253200	OK253344	–	OK253677	OK253908
	CSF9829	-BA-AA	Soil (*Eucalyptus* plantation)	Hua'an, Zhangzhou, Fujian, China	24°58′22.263″ N, 117°31′09.708″ E	S.F. Chen, Q.L. Liu and F.F. Liu	–	OK253201	OK253345	–	OK253678	OK253909
	CSF7134	––-A-	Soil (natural forest area)	Cangshan, Fuzhou, Fujian, China	26°5′16.2″ N, 119°14′19.8″ E	S.F. Chen, Q.L. Liu and F.F. Liu	–	–	–	–	OK253679	–
	CSF7135	––-A-	Soil (natural forest area)	Cangshan, Fuzhou, Fujian, China	26°5′16.2″ N, 119°14′19.8″ E	S.F. Chen, Q.L. Liu and F.F. Liu	–	–	–	–	OK253680	–
	CSF7136	––-A-	Soil (natural forest area)	Cangshan, Fuzhou, Fujian, China	26°5′16.2″ N, 119°14′19.8″ E	S.F. Chen, Q.L. Liu and F.F. Liu	–	–	–	–	OK253681	–
	CSF7138	––-A-	Soil (natural forest area)	Cangshan, Fuzhou, Fujian, China	26°5′16.2″ N, 119°14′19.8″ E	S.F. Chen, Q.L. Liu and F.F. Liu	–	–	–	–	OK253682	–
	CSF9781	––-A-	Soil (*Eucalyptus* plantation)	Hua'an, Zhangzhou, Fujian, China	24°53′49.369″ N, 117°32′45.070″ E	S.F. Chen, Q.L. Liu and F.F. Liu	–	–	–	–	OK253683	–
	CSF9782	––-A-	Soil (*Eucalyptus* plantation)	Hua'an, Zhangzhou, Fujian, China	24°53′49.369″ N, 117°32′45.070″ E	S.F. Chen, Q.L. Liu and F.F. Liu	–	–	–	–	OK253684	–
	CSF9783	––-A-	Soil (*Eucalyptus* plantation)	Hua'an, Zhangzhou, Fujian, China	24°53′49.369″ N, 117°32′45.070″ E	S.F. Chen, Q.L. Liu and F.F. Liu	–	–	–	–	OK253685	–
	CSF9785	––-A-	Soil (*Eucalyptus* plantation)	Hua'an, Zhangzhou, Fujian, China	24°53′49.369″ N, 117°32′45.070″ E	S.F. Chen, Q.L. Liu and F.F. Liu	–	–	–	–	OK253686	–

Table A1. *Cont.*

Species [a]	Isolate No. [b]	Genotype [c]	Substrate	Sampling Site	GPS Coordinate	Collector	GenBank Accession No. [d]					
							act	*cmdA*	*his3*	*rpb2*	*tef1*	*tub2*
	CSF9791	—-A-	Soil (*Eucalyptus* plantation)	Hua'an, Zhangzhou, Fujian, China	24°53′49.369″ N, 117°32′45.070″ E	S.F. Chen, Q.L. Liu and F.F. Liu	–	–	–	–	OK253687	–
	CSF9792	—-A-	Soil (*Eucalyptus* plantation)	Hua'an, Zhangzhou, Fujian, China	24°53′49.369″ N, 117°32′45.070″ E	S.F. Chen, Q.L. Liu and F.F. Liu	–	–	–	–	OK253688	–
	CSF9793	—-A-	Soil (*Eucalyptus* plantation)	Hua'an, Zhangzhou, Fujian, China	24°53′49.369″ N, 117°32′45.070″ E	S.F. Chen, Q.L. Liu and F.F. Liu	–	–	–	–	OK253689	–
	CSF9795	—-A-	Soil (*Eucalyptus* plantation)	Hua'an, Zhangzhou, Fujian, China	24°53′49.369″ N, 117°32′45.070″ E	S.F. Chen, Q.L. Liu and F.F. Liu	–	–	–	–	OK253690	–
	CSF9796	—-A-	Soil (*Eucalyptus* plantation)	Hua'an, Zhangzhou, Fujian, China	24°53′49.369″ N, 117°32′45.070″ E	S.F. Chen, Q.L. Liu and F.F. Liu	–	–	–	–	OK253691	–
	CSF9800	—-A-	Soil (*Eucalyptus* plantation)	Hua'an, Zhangzhou, Fujian, China	24°53′49.369″ N, 117°32′45.070″ E	S.F. Chen, Q.L. Liu and F.F. Liu	–	–	–	–	OK253692	–
	CSF9801	—-A-	Soil (*Eucalyptus* plantation)	Hua'an, Zhangzhou, Fujian, China	24°53′49.369″ N, 117°32′45.070″ E	S.F. Chen, Q.L. Liu and F.F. Liu	–	–	–	–	OK253693	–
	CSF9802	—-A-	Soil (*Eucalyptus* plantation)	Hua'an, Zhangzhou, Fujian, China	24°53′49.369″ N, 117°32′45.070″ E	S.F. Chen, Q.L. Liu and F.F. Liu	–	–	–	–	OK253694	–
	CSF9803	—-A-	Soil (*Eucalyptus* plantation)	Hua'an, Zhangzhou, Fujian, China	24°53′49.369″ N, 117°32′45.070″ E	S.F. Chen, Q.L. Liu and F.F. Liu	–	–	–	–	OK253695	–
	CSF9816	—-A-	Soil (*Eucalyptus* plantation)	Hua'an, Zhangzhou, Fujian, China	24°53′49.369″ N, 117°32′45.070″ E	S.F. Chen, Q.L. Liu and F.F. Liu	–	–	–	–	OK253696	–
	CSF9817	—-A-	Soil (*Eucalyptus* plantation)	Hua'an, Zhangzhou, Fujian, China	24°53′49.369″ N, 117°32′45.070″ E	S.F. Chen, Q.L. Liu and F.F. Liu	–	–	–	–	OK253697	–

Table A1. *Cont.*

Species [a]	Isolate No. [b]	Genotype [c]	Substrate	Sampling Site	GPS Coordinate	Collector	GenBank Accession No. [d]					
							act	*cmdA*	*his3*	*rpb2*	*tef1*	*tub2*
	CSF9818	—-A-	Soil (*Eucalyptus* plantation)	Hua'an, Zhangzhou, Fujian, China	24°53′49.369″ N, 117°32′45.070″ E	S.F. Chen, Q.L. Liu and F.F. Liu	–	–	–	–	OK253698	–
	CSF9820	—-A-	Soil (*Eucalyptus* plantation)	Hua'an, Zhangzhou, Fujian, China	24°53′49.369″ N, 117°32′45.070″ E	S.F. Chen, Q.L. Liu and F.F. Liu	–	–	–	–	OK253699	–
	CSF9826	—-A-	Soil (*Eucalyptus* plantation)	Hua'an, Zhangzhou, Fujian, China	24°58′22.263″ N, 117°31′09.708″ E	S.F. Chen, Q.L. Liu and F.F. Liu	–	–	–	–	OK253700	–
	CSF9827	—-A-	Soil (*Eucalyptus* plantation)	Hua'an, Zhangzhou, Fujian, China	24°58′22.263″ N, 117°31′09.708″ E	S.F. Chen, Q.L. Liu and F.F. Liu	–	–	–	–	OK253701	–
	CSF9828	—-A-	Soil (*Eucalyptus* plantation)	Hua'an, Zhangzhou, Fujian, China	24°58′22.263″ N, 117°31′09.708″ E	S.F. Chen, Q.L. Liu and F.F. Liu	–	–	–	–	OK253702	–
	CSF9830	—-A-	Soil (*Eucalyptus* plantation)	Hua'an, Zhangzhou, Fujian, China	24°58′22.263″ N, 117°31′09.708″ E	S.F. Chen, Q.L. Liu and F.F. Liu	–	–	–	–	OK253703	–
	CSF10094	—-A-	Soil (*Eucalyptus* plantation)	Minhou, Fuzhou, Fujian, China	26°15′04.285″ N, 119°02′38.917″ E	S.F. Chen, Q.L. Liu and F.F. Liu	–	–	–	–	OK253704	–
	CSF10095	—-A-	Soil (*Eucalyptus* plantation)	Minhou, Fuzhou, Fujian, China	26°15′04.285″ N, 119°02′38.917″ E	S.F. Chen, Q.L. Liu and F.F. Liu	–	–	–	–	OK253705	–
Ca. ilicicola	CSF9862 [e]	AAAAAA	Soil (*Eucalyptus* plantation)	Zhangping, Longyan, Fujian, China	25°17′10.882″ N, 117°27′33.635″ E	S.F. Chen, Q.L. Liu and F.F. Liu	OK253085	OK253202	OK253346	OK253442	OK253706	OK253910
	CSF9863 [e]	AAAAAA	Soil (*Eucalyptus* plantation)	Zhangping, Longyan, Fujian, China	25°17′10.882″ N, 117°27′33.635″ E	S.F. Chen, Q.L. Liu and F.F. Liu	OK253086	OK253203	OK253347	OK253443	OK253707	OK253911
	CSF9864	-AA-AA	Soil (*Eucalyptus* plantation)	Zhangping, Longyan, Fujian, China	25°17′10.882″ N, 117°27′33.635″ E	S.F. Chen, Q.L. Liu and F.F. Liu	–	OK253204	OK253348	–	OK253708	OK253912

Table A1. *Cont.*

Species [a]	Isolate No. [b]	Genotype [c]	Substrate	Sampling Site	GPS Coordinate	Collector	GenBank Accession No. [d]					
							act	*cmdA*	*his3*	*rpb2*	*tef1*	*tub2*
	CSF9865	-AA-AA	Soil (*Eucalyptus* plantation)	Zhangping, Longyan, Fujian, China	25°17′10.882″ N, 117°27′33.635″ E	S.F. Chen, Q.L. Liu and F.F. Liu	–	OK253205	OK253349	–	OK253709	OK253913
	CSF9866	-AA-AA	Soil (*Eucalyptus* plantation)	Zhangping, Longyan, Fujian, China	25°17′10.882″ N, 117°27′33.635″ E	S.F. Chen, Q.L. Liu and F.F. Liu	–	OK253206	OK253350	–	OK253710	OK253914
Ca. kyotensis	CSF7130 [e]	AAAAAA	Soil (natural forest area)	Cangshan, Fuzhou, Fujian, China	26°5′16.2″ N, 119°14′19.8″ E	S.F. Chen, Q.L. Liu and F.F. Liu	OK253087	OK253207	OK253351	OK253444	OK253711	OK253915
	CSF10088 [e]	AAAAAA	Soil (*Eucalyptus* plantation)	Minhou, Fuzhou, Fujian, China	26°15′04.285″ N, 119°02′38.917″ E	S.F. Chen, Q.L. Liu and F.F. Liu	OK253088	OK253208	OK253352	OK253445	OK253712	OK253916
	CSF9834 [e]	AAA-AB	Soil (*Eucalyptus* plantation)	Hua'an, Zhangzhou, Fujian, China	24°58′22.263″ N, 117°31′09.708″ E	S.F. Chen, Q.L. Liu and F.F. Liu	OK253089	OK253209	OK253353	N/A	OK253713	OK253917
	CSF9910 [e]	AAAAAB	Soil (*Phyllostachys heterocycla*)	Xinluo, Longyan, Fujian, China	25°07′31.133″ N, 116°51′37.485″ E	S.F. Chen, Q.L. Liu and F.F. Liu	OK253090	OK253210	OK253354	OK253446	OK253714	OK253918
	CSF10014 [e]	AAAAAC	Soil (*Eucalyptus* plantation)	Yongan, Sanming, Fujian, China	25°55′10.860″ N, 117°16′39.591″ E	S.F. Chen, Q.L. Liu and F.F. Liu	OK253091	OK253211	OK253355	OK253447	OK253715	OK253919
	CSF10080 [e]	AAAAAD	Soil (*Eucalyptus* plantation)	Minhou, Fuzhou, Fujian, China	26°15′04.285″ N, 119°02′38.917″ E	S.F. Chen, Q.L. Liu and F.F. Liu	OK253092	OK253212	OK253356	OK253448	OK253716	OK253920
	CSF10086 [e]	AAAAAE	Soil (*Eucalyptus* plantation)	Minhou, Fuzhou, Fujian, China	26°15′04.285″ N, 119°02′38.917″ E	S.F. Chen, Q.L. Liu and F.F. Liu	OK253093	OK253213	OK253357	OK253449	OK253717	OK253921
	CSF10053 [e]	AAAABB	Soil (*Pinus massoniana*)	Qingliu, Sanming, Fujian, China	26°10′54.311″ N, 116°52′50.901″ E	S.F. Chen, Q.L. Liu and F.F. Liu	OK253094	OK253214	OK253358	OK253450	OK253718	OK253922
	CSF10054 [e]	AAAABB	Soil (*Pinus massoniana*)	Qingliu, Sanming, Fujian, China	26°10′54.311″ N, 116°52′50.901″ E	S.F. Chen, Q.L. Liu and F.F. Liu	OK253095	OK253215	OK253359	OK253451	OK253719	OK253923

Table A1. *Cont.*

Species [a]	Isolate No. [b]	Genotype [c]	Substrate	Sampling Site	GPS Coordinate	Collector	GenBank Accession No. [d]					
							act	*cmdA*	*his3*	*rpb2*	*tef1*	*tub2*
	CSF9922 [e]	AAAABF	Soil (*Phyllostachys heterocycla*)	Xinluo, Longyan, Fujian, China	25°07′31.133″ N, 116°51′37.485″ E	S.F. Chen, Q.L. Liu and F.F. Liu	OK253096	OK253216	OK253360	OK253452	OK253720	OK253924
	CSF9923 [e]	AAAABF	Soil (*Phyllostachys heterocycla*)	Xinluo, Longyan, Fujian, China	25°07′31.133″ N, 116°51′37.485″ E	S.F. Chen, Q.L. Liu and F.F. Liu	OK253097	OK253217	OK253361	OK253453	OK253721	OK253925
	CSF9949 [e]	AAAADB	Soil (*Eucalyptus* plantation)	Xinluo, Longyan, Fujian, China	25°07′08.597″ N, 116°44′42.257″ E	S.F. Chen, Q.L. Liu and F.F. Liu	OK253098	OK253218	OK253362	OK253454	OK253722	OK253926
	CSF9951 [e]	AAAADB	Soil (*Eucalyptus* plantation)	Xinluo, Longyan, Fujian, China	25°07′08.597″ N, 116°44′42.257″ E	S.F. Chen, Q.L. Liu and F.F. Liu	OK253099	OK253219	OK253363	OK253455	OK253723	OK253927
	CSF9932 [e]	AAAADG	Soil (*Eucalyptus* plantation)	Xinluo, Longyan, Fujian, China	25°07′08.597″ N, 116°44′42.257″ E	S.F. Chen, Q.L. Liu and F.F. Liu	OK253100	OK253220	OK253364	OK253456	OK253724	OK253928
	CSF9935 [e]	AAAADG	Soil (*Eucalyptus* plantation)	Xinluo, Longyan, Fujian, China	25°07′08.597″ N, 116°44′42.257″ E	S.F. Chen, Q.L. Liu and F.F. Liu	OK253101	OK253221	OK253365	OK253457	OK253725	OK253929
	CSF9936 [e]	AAAADG	Soil (*Eucalyptus* plantation)	Xinluo, Longyan, Fujian, China	25°07′08.597″ N, 116°44′42.257″ E	S.F. Chen, Q.L. Liu and F.F. Liu	OK253102	OK253222	OK253366	OK253458	OK253726	OK253930
	CSF10020 [e]	AAAAEA	Soil (*Eucalyptus* plantation)	Yongan, Sanming, Fujian, China	25°55′10.860″ N, 117°16′39.591″ E	S.F. Chen, Q.L. Liu and F.F. Liu	OK253103	OK253223	OK253367	OK253459	OK253727	OK253931
	CSF10021 [e]	AAAAEA	Soil (*Eucalyptus* plantation)	Yongan, Sanming, Fujian, China	25°55′10.860″ N, 117°16′39.591″ E	S.F. Chen, Q.L. Liu and F.F. Liu	OK253104	OK253224	OK253368	OK253460	OK253728	OK253932
	CSF10009 [e]	AAABBH	Soil (*Eucalyptus* plantation)	Liancheng, Longyan, Fujian, China	25°33′06.994″ N, 116°41′42.328″ E	S.F. Chen, Q.L. Liu and F.F. Liu	OK253105	OK253225	OK253369	OK253461	OK253729	OK253933
	CSF10010 [e]	AAABBH	Soil (*Eucalyptus* plantation)	Liancheng, Longyan, Fujian, China	25°33′06.994″ N, 116°41′42.328″ E	S.F. Chen, Q.L. Liu and F.F. Liu	OK253106	OK253226	OK253370	OK253462	OK253730	OK253934

Table A1. *Cont.*

Species [a]	Isolate No. [b]	Genotype [c]	Substrate	Sampling Site	GPS Coordinate	Collector	GenBank Accession No. [d]					
							act	*cmdA*	*his3*	*rpb2*	*tef1*	*tub2*
	CSF9997 [e]	AABAAB	Soil (*Eucalyptus* plantation)	Liancheng, Longyan, Fujian, China	25°33′06.994″ N, 116°41′42.328″ E	S.F. Chen, Q.L. Liu and F.F. Liu	OK253107	OK253227	OK253371	OK253463	OK253731	OK253935
	CSF9969 [e]	AABACB	Soil (natural forest area)	Liancheng, Longyan, Fujian, China	25°26′14.348″ N, 116°38′42.400″ E	S.F. Chen, Q.L. Liu and F.F. Liu	OK253108	OK253228	OK253372	OK253464	OK253732	OK253936
	CSF9972 [e]	AABACB	Soil (natural forest area)	Liancheng, Longyan, Fujian, China	25°26′14.348″ N, 116°38′42.400″ E	S.F. Chen, Q.L. Liu and F.F. Liu	OK253109	OK253229	OK253373	OK253465	OK253733	OK253937
	CSF9973 [e]	AABACB	Soil (natural forest area)	Liancheng, Longyan, Fujian, China	25°26′14.348″ N, 116°38′42.400″ E	S.F. Chen, Q.L. Liu and F.F. Liu	OK253110	OK253230	OK253374	OK253466	OK253734	OK253938
	CSF10126 [e]	AACAAA	Soil (*Eucalyptus* plantation)	Minhou, Fuzhou, Fujian, China	26°15′04.285″ N, 119°02′38.917″ E	S.F. Chen, Q.L. Liu and F.F. Liu	OK253111	OK253231	OK253375	OK253467	OK253735	OK253939
	CSF9962 [e]	AACAAD	Soil (natural forest area)	Liancheng, Longyan, Fujian, China	25°26′14.348″ N, 116°38′42.400″ E	S.F. Chen, Q.L. Liu and F.F. Liu	OK253112	OK253232	OK253376	OK253468	OK253736	OK253940
	CSF10019 [e]	AADABB	Soil (*Eucalyptus* plantation)	Yongan, Sanming, Fujian, China	25°55′10.860″ N, 117°16′39.591″ E	S.F. Chen, Q.L. Liu and F.F. Liu	OK253113	OK253233	OK253377	OK253469	OK253737	OK253941
	CSF10022 [e]	AADABB	Soil (*Eucalyptus* plantation)	Yongan, Sanming, Fujian, China	25°55′10.860″ N, 117°16′39.591″ E	S.F. Chen, Q.L. Liu and F.F. Liu	OK253114	OK253234	OK253378	OK253470	OK253738	OK253942
	CSF10023 [e]	AADABB	Soil (*Eucalyptus* plantation)	Yongan, Sanming, Fujian, China	25°55′10.860″ N, 117°16′39.591″ E	S.F. Chen, Q.L. Liu and F.F. Liu	OK253115	OK253235	OK253379	OK253471	OK253739	OK253943
	CSF10045 [e]	ABAAAB	Soil (*Cunninghamia lanceolata*)	Qingliu, Sanming, Fujian, China	26°07′23.497″ N, 116°53′00.762″ E	S.F. Chen, Q.L. Liu and F.F. Liu	OK253116	OK253236	OK253380	OK253472	OK253740	OK253944
	CSF10047 [e]	ABAAAB	Soil (*Cunninghamia lanceolata*)	Qingliu, Sanming, Fujian, China	26°07′23.497″ N, 116°53′00.762″ E	S.F. Chen, Q.L. Liu and F.F. Liu	OK253117	OK253237	OK253381	OK253473	OK253741	OK253945

Table A1. *Cont.*

Species [a]	Isolate No. [b]	Genotype [c]	Substrate	Sampling Site	GPS Coordinate	Collector	GenBank Accession No. [d]					
							act	*cmdA*	*his3*	*rpb2*	*tef1*	*tub2*
	CSF9824 [e]	ACBAAC	Soil (*Eucalyptus* plantation)	Hua'an, Zhangzhou, Fujian, China	24°53′49.369″ N, 117°32′45.070″ E	S.F. Chen, Q.L. Liu and F.F. Liu	OK253118	OK253238	OK253382	OK253474	OK253742	OK253946
	CSF10004 [e]	ADAACB	Soil (*Eucalyptus* plantation)	Liancheng, Longyan, Fujian, China	25°33′06.994″ N, 116°41′42.328″ E	S.F. Chen, Q.L. Liu and F.F. Liu	OK253119	OK253239	OK253383	OK253475	OK253743	OK253947
	CSF10005 [e]	ADAACB	Soil (*Eucalyptus* plantation)	Liancheng, Longyan, Fujian, China	25°33′06.994″ N, 116°41′42.328″ E	S.F. Chen, Q.L. Liu and F.F. Liu	OK253120	OK253240	OK253384	OK253476	OK253744	OK253948
	CSF7123	-AA-AB	Soil (natural forest area)	Cangshan, Fuzhou, Fujian, China	26°5′16.2″ N, 119°14′19.8″ E	S.F. Chen, Q.L. Liu and F.F. Liu	–	OK253241	OK253385	–	OK253745	OK253949
	CSF9915	-AA-AB	Soil (*Phyllostachys heterocycla*)	Xinluo, Longyan, Fujian, China	25°07′31.133″ N, 116°51′37.485″ E	S.F. Chen, Q.L. Liu and F.F. Liu	–	OK253242	OK253386	–	OK253746	OK253950
	CSF9999	-AA-AB	Soil (*Eucalyptus* plantation)	Liancheng, Longyan, Fujian, China	25°33′06.994″ N, 116°41′42.328″ E	S.F. Chen, Q.L. Liu and F.F. Liu	–	OK253243	OK253387	–	OK253747	OK253951
	CSF10124	-AA-AB	Soil (*Eucalyptus* plantation)	Minhou, Fuzhou, Fujian, China	26°15′04.285″ N, 119°02′38.917″ E	S.F. Chen, Q.L. Liu and F.F. Liu	–	OK253244	OK253388	–	OK253748	OK253952
	CSF10055	-AA-BB	Soil (*Pinus massoniana*)	Qingliu, Sanming, Fujian, China	26°10′54.311″ N, 116°52′50.901″ E	S.F. Chen, Q.L. Liu and F.F. Liu	–	OK253245	OK253389	–	OK253749	OK253953
	CSF10056	-AA-BB	Soil (*Pinus massoniana*)	Qingliu, Sanming, Fujian, China	26°10′54.311″ N, 116°52′50.901″ E	S.F. Chen, Q.L. Liu and F.F. Liu	–	OK253246	OK253390	–	OK253750	OK253954
	CSF10057	-AA-BB	Soil (*Pinus massoniana*)	Qingliu, Sanming, Fujian, China	26°10′54.311″ N, 116°52′50.901″ E	S.F. Chen, Q.L. Liu and F.F. Liu	–	OK253247	OK253391	–	OK253751	OK253955
	CSF9952	-AA-DB	Soil (*Eucalyptus* plantation)	Xinluo, Longyan, Fujian, China	25°07′08.597″ N, 116°44′42.257″ E	S.F. Chen, Q.L. Liu and F.F. Liu	–	OK253248	OK253392	–	OK253752	OK253956

Table A1. *Cont.*

Species [a]	Isolate No. [b]	Genotype [c]	Substrate	Sampling Site	GPS Coordinate	Collector	GenBank Accession No. [d]					
							act	*cmdA*	*his3*	*rpb2*	*tef1*	*tub2*
	CSF9953	-AA-DB	Soil (*Eucalyptus* plantation)	Xinluo, Longyan, Fujian, China	25°07′08.597″ N, 116°44′42.257″ E	S.F. Chen, Q.L. Liu and F.F. Liu	–	OK253249	OK253393	–	OK253753	OK253957
	CSF9924	-AA-BF	Soil (*Phyllostachys heterocycla*)	Xinluo, Longyan, Fujian, China	25°07′31.133″ N, 116°51′37.485″ E	S.F. Chen, Q.L. Liu and F.F. Liu	–	OK253250	OK253394	–	OK253754	OK253958
	CSF9925	-AA-BF	Soil (*Phyllostachys heterocycla*)	Xinluo, Longyan, Fujian, China	25°07′31.133″ N, 116°51′37.485″ E	S.F. Chen, Q.L. Liu and F.F. Liu	–	OK253251	OK253395	–	OK253755	OK253959
	CSF9926	-AA-BF	Soil (*Phyllostachys heterocycla*)	Xinluo, Longyan, Fujian, China	25°07′31.133″ N, 116°51′37.485″ E	S.F. Chen, Q.L. Liu and F.F. Liu	–	OK253252	OK253396	–	OK253756	OK253960
	CSF10011	-AA-BH	Soil (*Eucalyptus* plantation)	Liancheng, Longyan, Fujian, China	25°33′06.994″ N, 116°41′42.328″ E	S.F. Chen, Q.L. Liu and F.F. Liu	–	OK253253	OK253397	–	OK253757	OK253961
	CSF10012	-AA-BH	Soil (*Eucalyptus* plantation)	Liancheng, Longyan, Fujian, China	25°33′06.994″ N, 116°41′42.328″ E	S.F. Chen, Q.L. Liu and F.F. Liu	–	OK253254	OK253398	–	OK253758	OK253962
	CSF10013	-AA-BH	Soil (*Eucalyptus* plantation)	Liancheng, Longyan, Fujian, China	25°33′06.994″ N, 116°41′42.328″ E	S.F. Chen, Q.L. Liu and F.F. Liu	–	OK253255	OK253399	–	OK253759	OK253963
	CSF10006	-DA-CB	Soil (*Eucalyptus* plantation)	Liancheng, Longyan, Fujian, China	25°33′06.994″ N, 116°41′42.328″ E	S.F. Chen, Q.L. Liu and F.F. Liu	–	OK253256	OK253400	–	OK253760	OK253964
	CSF10007	-DA-CB	Soil (*Eucalyptus* plantation)	Liancheng, Longyan, Fujian, China	25°33′06.994″ N, 116°41′42.328″ E	S.F. Chen, Q.L. Liu and F.F. Liu	–	OK253257	OK253401	–	OK253761	OK253965
	CSF10008	-DA-CB	Soil (*Eucalyptus* plantation)	Liancheng, Longyan, Fujian, China	25°33′06.994″ N, 116°41′42.328″ E	S.F. Chen, Q.L. Liu and F.F. Liu	–	OK253258	OK253402	–	OK253762	OK253966
	CSF9821	—-A-	Soil (*Eucalyptus* plantation)	Hua'an, Zhangzhou, Fujian, China	24°53′49.369″ N, 117°32′45.070″ E	S.F. Chen, Q.L. Liu and F.F. Liu	–	–	–	–	OK253763	–

Table A1. *Cont.*

Species [a]	Isolate No. [b]	Genotype [c]	Substrate	Sampling Site	GPS Coordinate	Collector	GenBank Accession No. [d]					
							act	*cmdA*	*his3*	*rpb2*	*tef1*	*tub2*
	CSF9822	—-A-	Soil (*Eucalyptus* plantation)	Hua'an, Zhangzhou, Fujian, China	24°53′49.369″ N, 117°32′45.070″ E	S.F. Chen, Q.L. Liu and F.F. Liu	–	–	–	–	OK253764	–
	CSF9823	—-A-	Soil (*Eucalyptus* plantation)	Hua'an, Zhangzhou, Fujian, China	24°53′49.369″ N, 117°32′45.070″ E	S.F. Chen, Q.L. Liu and F.F. Liu	–	–	–	–	OK253765	–
	CSF9825	—-A-	Soil (*Eucalyptus* plantation)	Hua'an, Zhangzhou, Fujian, China	24°53′49.369″ N, 117°32′45.070″ E	S.F. Chen, Q.L. Liu and F.F. Liu	–	–	–	–	OK253766	–
	CSF9832	—-A-	Soil (*Eucalyptus* plantation)	Hua'an, Zhangzhou, Fujian, China	24°58′22.263″ N, 117°31′09.708″ E	S.F. Chen, Q.L. Liu and F.F. Liu	–	–	–	–	OK253767	–
	CSF9833	—-A-	Soil (*Eucalyptus* plantation)	Hua'an, Zhangzhou, Fujian, China	24°58′22.263″ N, 117°31′09.708″ E	S.F. Chen, Q.L. Liu and F.F. Liu	–	–	–	–	OK253768	–
	CSF9835	—-A-	Soil (*Eucalyptus* plantation)	Hua'an, Zhangzhou, Fujian, China	24°58′22.263″ N, 117°31′09.708″ E	S.F. Chen, Q.L. Liu and F.F. Liu	–	–	–	–	OK253769	–
	CSF9907	—-A-	Soil (*Phyllostachys heterocycla*)	Xinluo, Longyan, Fujian, China	25°07′31.133″ N, 116°51′37.485″ E	S.F. Chen, Q.L. Liu and F.F. Liu	–	–	–	–	OK253770	–
	CSF9908	—-A-	Soil (*Phyllostachys heterocycla*)	Xinluo, Longyan, Fujian, China	25°07′31.133″ N, 116°51′37.485″ E	S.F. Chen, Q.L. Liu and F.F. Liu	–	–	–	–	OK253771	–
	CSF9909	—-A-	Soil (*Phyllostachys heterocycla*)	Xinluo, Longyan, Fujian, China	25°07′31.133″ N, 116°51′37.485″ E	S.F. Chen, Q.L. Liu and F.F. Liu	–	–	–	–	OK253772	–
	CSF9911	—-A-	Soil (*Phyllostachys heterocycla*)	Xinluo, Longyan, Fujian, China	25°07′31.133″ N, 116°51′37.485″ E	S.F. Chen, Q.L. Liu and F.F. Liu	–	–	–	–	OK253773	–
	CSF9912	—-A-	Soil (*Phyllostachys heterocycla*)	Xinluo, Longyan, Fujian, China	25°07′31.133″ N, 116°51′37.485″ E	S.F. Chen, Q.L. Liu and F.F. Liu	–	–	–	–	OK253774	–

Table A1. *Cont.*

Species [a]	Isolate No. [b]	Genotype [c]	Substrate	Sampling Site	GPS Coordinate	Collector	GenBank Accession No. [d]					
							act	*cmdA*	*his3*	*rpb2*	*tef1*	*tub2*
	CSF9913	—-A-	Soil (*Phyllostachys heterocycla*)	Xinluo, Longyan, Fujian, China	25°07′31.133″ N, 116°51′37.485″ E	S.F. Chen, Q.L. Liu and F.F. Liu	–	–	–	–	OK253775	–
	CSF9914	—-A-	Soil (*Phyllostachys heterocycla*)	Xinluo, Longyan, Fujian, China	25°07′31.133″ N, 116°51′37.485″ E	S.F. Chen, Q.L. Liu and F.F. Liu	–	–	–	–	OK253776	–
	CSF9916	—-A-	Soil (*Phyllostachys heterocycla*)	Xinluo, Longyan, Fujian, China	25°07′31.133″ N, 116°51′37.485″ E	S.F. Chen, Q.L. Liu and F.F. Liu	–	–	–	–	OK253777	–
	CSF9917	—-A-	Soil (*Phyllostachys heterocycla*)	Xinluo, Longyan, Fujian, China	25°07′31.133″ N, 116°51′37.485″ E	S.F. Chen, Q.L. Liu and F.F. Liu	–	–	–	–	OK253778	–
	CSF9918	—-A-	Soil (*Phyllostachys heterocycla*)	Xinluo, Longyan, Fujian, China	25°07′31.133″ N, 116°51′37.485″ E	S.F. Chen, Q.L. Liu and F.F. Liu	–	–	–	–	OK253779	–
	CSF9919	—-A-	Soil (*Phyllostachys heterocycla*)	Xinluo, Longyan, Fujian, China	25°07′31.133″ N, 116°51′37.485″ E	S.F. Chen, Q.L. Liu and F.F. Liu	–	–	–	–	OK253780	–
	CSF9920	—-A-	Soil (*Phyllostachys heterocycla*)	Xinluo, Longyan, Fujian, China	25°07′31.133″ N, 116°51′37.485″ E	S.F. Chen, Q.L. Liu and F.F. Liu	–	–	–	–	OK253781	–
	CSF9921	—-A-	Soil (*Phyllostachys heterocycla*)	Xinluo, Longyan, Fujian, China	25°07′31.133″ N, 116°51′37.485″ E	S.F. Chen, Q.L. Liu and F.F. Liu	–	–	–	–	OK253782	–
	CSF9959	—-A-	Soil (natural forest area)	Liancheng, Longyan, Fujian, China	25°26′14.348″ N, 116°38′42.400″ E	S.F. Chen, Q.L. Liu and F.F. Liu	–	–	–	–	OK253783	–
	CSF9960	—-A-	Soil (natural forest area)	Liancheng, Longyan, Fujian, China	25°26′14.348″ N, 116°38′42.400″ E	S.F. Chen, Q.L. Liu and F.F. Liu	–	–	–	–	OK253784	–
	CSF9961	—-A-	Soil (natural forest area)	Liancheng, Longyan, Fujian, China	25°26′14.348″ N, 116°38′42.400″ E	S.F. Chen, Q.L. Liu and F.F. Liu	–	–	–	–	OK253785	–

Table A1. *Cont.*

Species [a]	Isolate No. [b]	Genotype [c]	Substrate	Sampling Site	GPS Coordinate	Collector	GenBank Accession No. [d]					
							act	*cmdA*	*his3*	*rpb2*	*tef1*	*tub2*
	CSF9963	—-A-	Soil (natural forest area)	Liancheng, Longyan, Fujian, China	25°26′14.348″ N, 116°38′42.400″ E	S.F. Chen, Q.L. Liu and F.F. Liu	–	–	–	–	OK253786	–
	CSF9994	—-A-	Soil (*Eucalyptus* plantation)	Liancheng, Longyan, Fujian, China	25°33′06.994″ N, 116°41′42.328″ E	S.F. Chen, Q.L. Liu and F.F. Liu	–	–	–	–	OK253787	–
	CSF9995	—-A-	Soil (*Eucalyptus* plantation)	Liancheng, Longyan, Fujian, China	25°33′06.994″ N, 116°41′42.328″ E	S.F. Chen, Q.L. Liu and F.F. Liu	–	–	–	–	OK253788	–
	CSF9996	—-A-	Soil (*Eucalyptus* plantation)	Liancheng, Longyan, Fujian, China	25°33′06.994″ N, 116°41′42.328″ E	S.F. Chen, Q.L. Liu and F.F. Liu	–	–	–	–	OK253789	–
	CSF9998	—-A-	Soil (*Eucalyptus* plantation)	Liancheng, Longyan, Fujian, China	25°33′06.994″ N, 116°41′42.328″ E	S.F. Chen, Q.L. Liu and F.F. Liu	–	–	–	–	OK253790	–
	CSF10000	—-A-	Soil (*Eucalyptus* plantation)	Liancheng, Longyan, Fujian, China	25°33′06.994″ N, 116°41′42.328″ E	S.F. Chen, Q.L. Liu and F.F. Liu	–	–	–	–	OK253791	–
	CSF10001	—-A-	Soil (*Eucalyptus* plantation)	Liancheng, Longyan, Fujian, China	25°33′06.994″ N, 116°41′42.328″ E	S.F. Chen, Q.L. Liu and F.F. Liu	–	–	–	–	OK253792	–
	CSF10002	—-A-	Soil (*Eucalyptus* plantation)	Liancheng, Longyan, Fujian, China	25°33′06.994″ N, 116°41′42.328″ E	S.F. Chen, Q.L. Liu and F.F. Liu	–	–	–	–	OK253793	–
	CSF10003	—-A-	Soil (*Eucalyptus* plantation)	Liancheng, Longyan, Fujian, China	25°33′06.994″ N, 116°41′42.328″ E	S.F. Chen, Q.L. Liu and F.F. Liu	–	–	–	–	OK253794	–
	CSF10015	—-A-	Soil (*Eucalyptus* plantation)	Yongan, Sanming, Fujian, China	25°55′10.860″ N, 117°16′39.591″ E	S.F. Chen, Q.L. Liu and F.F. Liu	–	–	–	–	OK253795	–
	CSF10016	—-A-	Soil (*Eucalyptus* plantation)	Yongan, Sanming, Fujian, China	25°55′10.860″ N, 117°16′39.591″ E	S.F. Chen, Q.L. Liu and F.F. Liu	–	–	–	–	OK253796	–

Table A1. *Cont.*

Species [a]	Isolate No. [b]	Genotype [c]	Substrate	Sampling Site	GPS Coordinate	Collector	GenBank Accession No. [d]					
							act	*cmdA*	*his3*	*rpb2*	*tef1*	*tub2*
	CSF10044	—-A-	Soil (*Cunning-hamia lanceolata*)	Qingliu, Sanming, Fujian, China	26°07′23.497″ N, 116°53′00.762″ E	S.F. Chen, Q.L. Liu and F.F. Liu	–	–	–	–	OK253797	–
	CSF10046	—-A-	Soil (*Cunning-hamia lanceolata*)	Qingliu, Sanming, Fujian, China	26°07′23.497″ N, 116°53′00.762″ E	S.F. Chen, Q.L. Liu and F.F. Liu	–	–	–	–	OK253798	–
	CSF10084	—-A-	Soil (*Eucalyptus* plantation)	Minhou, Fuzhou, Fujian, China	26°15′04.285″ N, 119°02′38.917″ E	S.F. Chen, Q.L. Liu and F.F. Liu	–	–	–	–	OK253799	–
	CSF10085	—-A-	Soil (*Eucalyptus* plantation)	Minhou, Fuzhou, Fujian, China	26°15′04.285″ N, 119°02′38.917″ E	S.F. Chen, Q.L. Liu and F.F. Liu	–	–	–	–	OK253800	–
	CSF10087	—-A-	Soil (*Eucalyptus* plantation)	Minhou, Fuzhou, Fujian, China	26°15′04.285″ N, 119°02′38.917″ E	S.F. Chen, Q.L. Liu and F.F. Liu	–	–	–	–	OK253801	–
	CSF10089	—-A-	Soil (*Eucalyptus* plantation)	Minhou, Fuzhou, Fujian, China	26°15′04.285″ N, 119°02′38.917″ E	S.F. Chen, Q.L. Liu and F.F. Liu	–	–	–	–	OK253802	–
	CSF10090	—-A-	Soil (*Eucalyptus* plantation)	Minhou, Fuzhou, Fujian, China	26°15′04.285″ N, 119°02′38.917″ E	S.F. Chen, Q.L. Liu and F.F. Liu	–	–	–	–	OK253803	–
	CSF10091	—-A-	Soil (*Eucalyptus* plantation)	Minhou, Fuzhou, Fujian, China	26°15′04.285″ N, 119°02′38.917″ E	S.F. Chen, Q.L. Liu and F.F. Liu	–	–	–	–	OK253804	–
	CSF10092	—-A-	Soil (*Eucalyptus* plantation)	Minhou, Fuzhou, Fujian, China	26°15′04.285″ N, 119°02′38.917″ E	S.F. Chen, Q.L. Liu and F.F. Liu	–	–	–	–	OK253805	–
	CSF10121	—-A-	Soil (*Eucalyptus* plantation)	Minhou, Fuzhou, Fujian, China	26°15′04.285″ N, 119°02′38.917″ E	S.F. Chen, Q.L. Liu and F.F. Liu	–	–	–	–	OK253806	–
	CSF10122	—-A-	Soil (*Eucalyptus* plantation)	Minhou, Fuzhou, Fujian, China	26°15′04.285″ N, 119°02′38.917″ E	S.F. Chen, Q.L. Liu and F.F. Liu	–	–	–	–	OK253807	–

Table A1. *Cont.*

Species [a]	Isolate No. [b]	Genotype [c]	Substrate	Sampling Site	GPS Coordinate	Collector	GenBank Accession No. [d]					
							act	*cmdA*	*his3*	*rpb2*	*tef1*	*tub2*
	CSF10123	—-A-	Soil (*Eucalyptus* plantation)	Minhou, Fuzhou, Fujian, China	26°15′04.285″ N, 119°02′38.917″ E	S.F. Chen, Q.L. Liu and F.F. Liu	–	–	–	–	OK253808	–
	CSF10127	—-A-	Soil (*Eucalyptus* plantation)	Minhou, Fuzhou, Fujian, China	26°15′04.285″ N, 119°02′38.917″ E	S.F. Chen, Q.L. Liu and F.F. Liu	–	–	–	–	OK253809	–
	CSF10128	—-A-	Soil (*Eucalyptus* plantation)	Minhou, Fuzhou, Fujian, China	26°15′04.285″ N, 119°02′38.917″ E	S.F. Chen, Q.L. Liu and F.F. Liu	–	–	–	–	OK253810	–
	CSF7122	—-A-	Soil (natural forest area)	Cangshan, Fuzhou, Fujian, China	26°5′16.2″ N, 119°14′19.8″ E	S.F. Chen, Q.L. Liu and F.F. Liu	–	–	–	–	OK253811	–
	CSF7128	—-A-	Soil (natural forest area)	Cangshan, Fuzhou, Fujian, China	26°5′16.2″ N, 119°14′19.8″ E	S.F. Chen, Q.L. Liu and F.F. Liu	–	–	–	–	OK253812	–
	CSF7129	—-A-	Soil (natural forest area)	Cangshan, Fuzhou, Fujian, China	26°5′16.2″ N, 119°14′19.8″ E	S.F. Chen, Q.L. Liu and F.F. Liu	–	–	–	–	OK253813	–
Ca. minensis sp. nov.	CSF9941 [e,h–j]; CGMCC3.18877	AAAAAA	Soil (*Eucalyptus* plantation)	Xinluo, Longyan, Fujian, China	25°07′08.597″ N, 116°44′42.257″ E	S.F. Chen, Q.L. Liu and F.F. Liu	OK253121	OK253259	OK253403	OK253477	OK253814	OK253967
	CSF9974 [e]	AAAAAA	Soil (natural forest area)	Liancheng, Longyan, Fujian, China	25°26′14.348″ N, 116°38′42.400″ E	S.F. Chen, Q.L. Liu and F.F. Liu	OK253122	OK253260	OK253404	OK253478	OK253815	OK253968
	CSF9975 [e,h,i]; CGMCC3.18881	AAAAAA	Soil (natural forest area)	Liancheng, Longyan, Fujian, China	25°26′14.348″ N, 116°38′42.400″ E	S.F. Chen, Q.L. Liu and F.F. Liu	OK253123	OK253261	OK253405	OK253479	OK253816	OK253969
	CSF9976 [e]	AAAAAA	Soil (natural forest area)	Liancheng, Longyan, Fujian, China	25°26′14.348″ N, 116°38′42.400″ E	S.F. Chen, Q.L. Liu and F.F. Liu	OK253124	OK253262	OK253406	OK253480	OK253817	OK253970
	CSF9977 [e]	AAAAAA	Soil (natural forest area)	Liancheng, Longyan, Fujian, China	25°26′14.348″ N, 116°38′42.400″ E	S.F. Chen, Q.L. Liu and F.F. Liu	OK253125	OK253263	OK253407	OK253481	OK253818	OK253971

Table A1. *Cont.*

Species [a]	Isolate No. [b]	Genotype [c]	Substrate	Sampling Site	GPS Coordinate	Collector	GenBank Accession No. [d]					
							act	*cmdA*	*his3*	*rpb2*	*tef1*	*tub2*
	CSF9978 [e]	AAAAAA	Soil (natural forest area)	Liancheng, Longyan, Fujian, China	25°26′14.348″ N, 116°38′42.400″ E	S.F. Chen, Q.L. Liu and F.F. Liu	OK253126	OK253264	OK253408	OK253482	OK253819	OK253972
	CSF9933 [e,h,i]; CGMCC3.18875	ABBABB	Soil (*Eucalyptus* plantation)	Xinluo, Longyan, Fujian, China	25°07′08.597″ N, 116°44′42.257″ E	S.F. Chen, Q.L. Liu and F.F. Liu	OK253127	OK253265	OK253409	OK253483	OK253820	OK253973
	CSF9934 [e]	ABBABB	Soil (*Eucalyptus* plantation)	Xinluo, Longyan, Fujian, China	25°07′08.597″ N, 116°44′42.257″ E	S.F. Chen, Q.L. Liu and F.F. Liu	OK253128	OK253266	OK253410	OK253484	OK253821	OK253974
Ca. pacifica	CSF10024 [e]	AAAAAA	Soil (*Eucalyptus* plantation)	Yongan, Sanming, Fujian, China	25°55′10.860″ N, 117°16′39.591″ E	S.F. Chen, Q.L. Liu and F.F. Liu	OK253129	OK253267	OK253411	OK253485	OK253822	OK253975
	CSF10129 [e]	BAAAAA	Soil (*Eucalyptus* plantation)	Minhou, Fuzhou, Fujian, China	26°15′04.285″ N, 119°02′38.917″ E	S.F. Chen, Q.L. Liu and F.F. Liu	OK253130	OK253268	OK253412	OK253486	OK253823	OK253976
	CSF10070 [e]	CABAAA	Soil (natural forest area)	Yanping, Nanping, Fujian, China	26°42′26.672″ N, 118°07′58.317″ E	S.F. Chen, Q.L. Liu and F.F. Liu	OK253131	OK253269	OK253413	OK253487	OK253824	OK253977
	CSF10077 [e]	CABAAA	Soil (natural forest area)	Yanping, Nanping, Fujian, China	26°42′26.672″ N, 118°07′58.317″ E	S.F. Chen, Q.L. Liu and F.F. Liu	OK253132	OK253270	OK253414	OK253488	OK253825	OK253978
	CSF10027	-AA-AA	Soil (*Eucalyptus* plantation)	Yongan, Sanming, Fujian, China	25°55′10.860″ N, 117°16′39.591″ E	S.F. Chen, Q.L. Liu and F.F. Liu	–	OK253271	OK253415	–	OK253826	OK253979
	CSF10039	-AA-AA	Soil (*Eucalyptus* plantation)	Yongan, Sanming, Fujian, China	25°55′10.860″ N, 117°16′39.591″ E	S.F. Chen, Q.L. Liu and F.F. Liu	–	OK253272	OK253416	–	OK253827	OK253980
	CSF10130	-AA-AA	Soil (*Eucalyptus* plantation)	Minhou, Fuzhou, Fujian, China	26°15′04.285″ N, 119°02′38.917″ E	S.F. Chen, Q.L. Liu and F.F. Liu	–	OK253273	OK253417	–	OK253828	OK253981
	CSF10025	—-A-	Soil (*Eucalyptus* plantation)	Yongan, Sanming, Fujian, China	25°55′10.860″ N, 117°16′39.591″ E	S.F. Chen, Q.L. Liu and F.F. Liu	–	–	–	–	OK253829	–

Table A1. *Cont.*

Species [a]	Isolate No. [b]	Genotype [c]	Substrate	Sampling Site	GPS Coordinate	Collector	GenBank Accession No. [d]					
							act	*cmdA*	*his3*	*rpb2*	*tef1*	*tub2*
	CSF10026	—-A-	Soil (*Eucalyptus* plantation)	Yongan, Sanming, Fujian, China	25°55′10.860″ N, 117°16′39.591″ E	S.F. Chen, Q.L. Liu and F.F. Liu	–	–	–	–	OK253830	–
	CSF10028	—-A-	Soil (*Eucalyptus* plantation)	Yongan, Sanming, Fujian, China	25°55′10.860″ N, 117°16′39.591″ E	S.F. Chen, Q.L. Liu and F.F. Liu	–	–	–	–	OK253831	–
	CSF10038	—-A-	Soil (*Eucalyptus* plantation)	Yongan, Sanming, Fujian, China	25°55′10.860″ N, 117°16′39.591″ E	S.F. Chen, Q.L. Liu and F.F. Liu	–	–	–	–	OK253832	–
	CSF10040	—-A-	Soil (*Eucalyptus* plantation)	Yongan, Sanming, Fujian, China	25°55′10.860″ N, 117°16′39.591″ E	S.F. Chen, Q.L. Liu and F.F. Liu	–	–	–	–	OK253833	–
	CSF10071	—-A-	Soil (natural forest area)	Yanping, Nanping, Fujian, China	26°42′26.672″ N, 118°07′58.317″ E	S.F. Chen, Q.L. Liu and F.F. Liu	–	–	–	–	OK253834	–
	CSF10072	—-A-	Soil (natural forest area)	Yanping, Nanping, Fujian, China	26°42′26.672″ N, 118°07′58.317″ E	S.F. Chen, Q.L. Liu and F.F. Liu	–	–	–	–	OK253835	–
	CSF10076	—-A-	Soil (natural forest area)	Yanping, Nanping, Fujian, China	26°42′26.672″ N, 118°07′58.317″ E	S.F. Chen, Q.L. Liu and F.F. Liu	–	–	–	–	OK253836	–
	CSF10078	—-A-	Soil (natural forest area)	Yanping, Nanping, Fujian, China	26°42′26.672″ N, 118°07′58.317″ E	S.F. Chen, Q.L. Liu and F.F. Liu	–	–	–	–	OK253837	–
	CSF10079	—-A-	Soil (natural forest area)	Yanping, Nanping, Fujian, China	26°42′26.672″ N, 118°07′58.317″ E	S.F. Chen, Q.L. Liu and F.F. Liu	–	–	–	–	OK253838	–
Ca. pseudoreteaudii	CSF10059 [e]	AAAAAA	Soil (*Eucalyptus* plantation)	Yanping, Nanping, Fujian, China	26°46′19.651″ N, 117°57′37.233″ E	S.F. Chen, Q.L. Liu and F.F. Liu	OK253133	OK253274	OK253418	OK253489	OK253839	OK253982
	CSF10060 [e]	AAAAAA	Soil (*Eucalyptus* plantation)	Yanping, Nanping, Fujian, China	26°46′19.651″ N, 117°57′37.233″ E	S.F. Chen, Q.L. Liu and F.F. Liu	OK253134	OK253275	OK253419	OK253490	OK253840	OK253983

Table A1. *Cont.*

Species [a]	Isolate No. [b]	Genotype [c]	Substrate	Sampling Site	GPS Coordinate	Collector	GenBank Accession No. [d]					
							act	*cmdA*	*his3*	*rpb2*	*tef1*	*tub2*
	CSF10058	-AA-AA	Soil (*Eucalyptus* plantation)	Yanping, Nanping, Fujian, China	26°46′19.651″ N, 117°57′37.233″ E	S.F. Chen, Q.L. Liu and F.F. Liu	–	OK253276	OK253420	–	OK253841	OK253984
	CSF10061	-AA-AA	Soil (*Eucalyptus* plantation)	Yanping, Nanping, Fujian, China	26°46′19.651″ N, 117°57′37.233″ E	S.F. Chen, Q.L. Liu and F.F. Liu	–	OK253277	OK253421	–	OK253842	OK253985
	CSF10062	-AA-AA	Soil (*Eucalyptus* plantation)	Yanping, Nanping, Fujian, China	26°46′19.651″ N, 117°57′37.233″ E	S.F. Chen, Q.L. Liu and F.F. Liu	–	OK253278	OK253422	–	OK253843	OK253986

[a] New species described in this study are indicated in bold. [b] *CSF* = Culture Collection at the Research Institute of Fast-growing Trees (RIFT)/China Eucalypt Research Centre (CERC), Chinese Academy of Forestry (CAF), ZhanJiang, Guangdong Province, China; *CGMCC* = China General Microbiological Culture Collection Center, Beijing, China. [c] Genotype within each identified species, determined by sequences of *act*, *cmdA*, *his3*, *rpb2*, *tef1* and *tub2* regions; '-' means not available. [d] *act* = actin; *cmdA* = calmodulin; *his3* = histone H3; *rpb2* = the second largest subunit of RNA polymerase; *tef1* = translation elongation factor 1-alpha; *tub2* = β-tubulin. [e] Isolates used in phylogenetic analyses. [f] *N/A* represents the relative locus was not successfully amplified in the current study. [g] '–' represents the relative locus was not amplified in the current study. [h] Isolates used in morphological and culture growth studies. [i] Isolates used for mating studies. [j] Isolates that represent ex-type cultures are indicated in bold.

Appendix B. Phylogenetic Tree of *Calonectria* Species Based on Maximum Likelihood (ML) Analyses of *act, cmdA, his3, rpb2, tef1* and *tub2* Gene Sequences

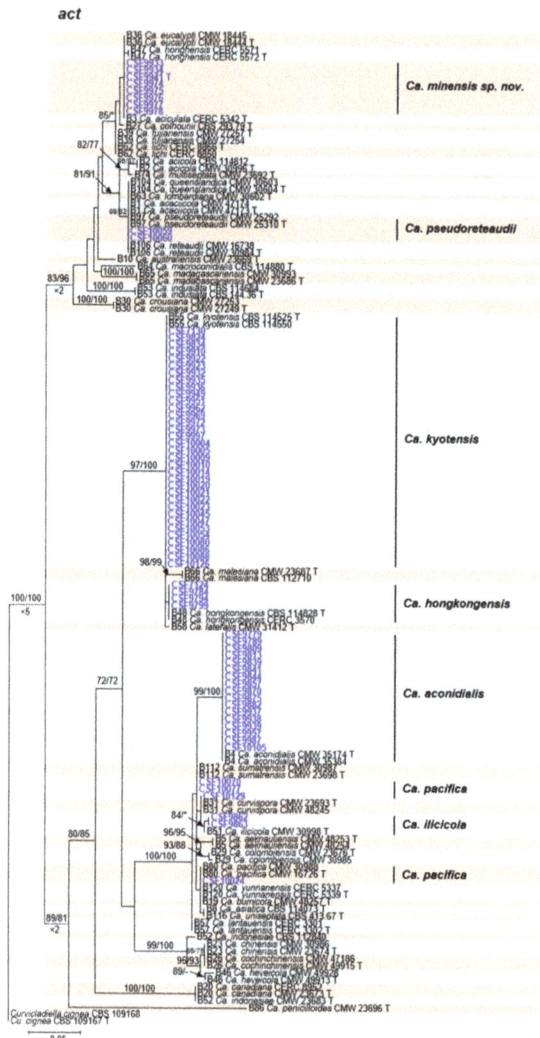

Figure A1. Phylogenetic tree of *Calonectria* species based on maximum likelihood (ML) analyses of *act* gene sequences. Bootstrap value ≥70% for ML and MP analyses are presented at the branches. Bootstrap values lower than 70% are marked with "*", and absent analyses values are marked with "-". Ex-type isolates are marked with "T". Isolates sequenced in this study are highlighted in blue and bold type. The "B" species codes are consistent with the recently published results in Liu and co-authors [18]. The tree was rooted to *Curvicladiella cignea* (CBS 109167 and CBS 109168).

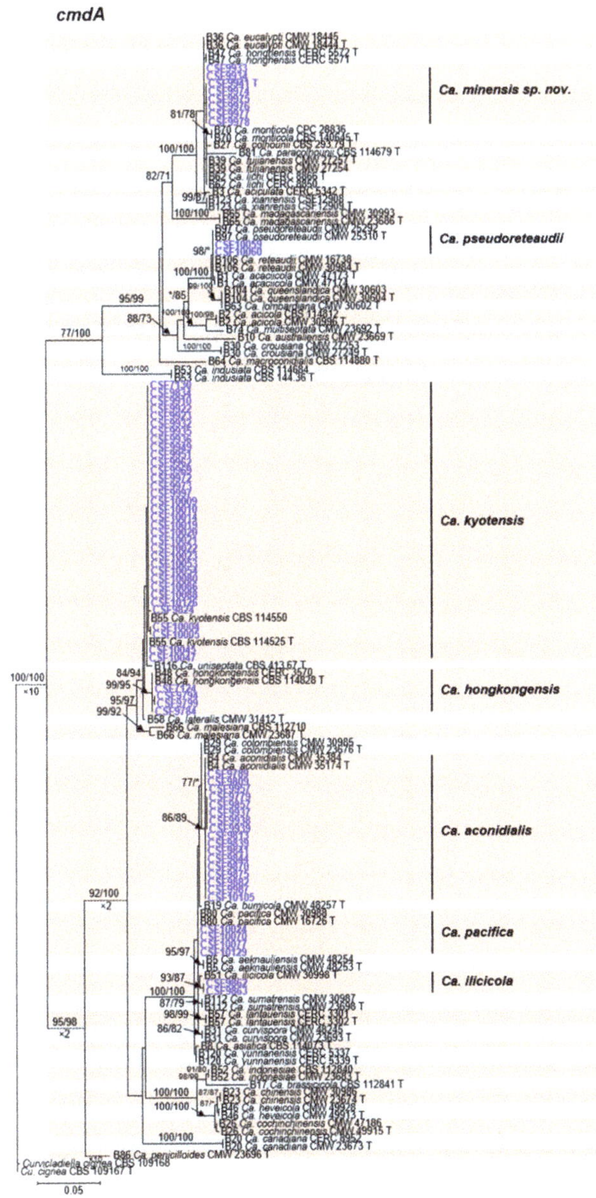

Figure A2. Phylogenetic tree of *Calonectria* species based on maximum likelihood (ML) analyses of *cmdA* gene sequences. Bootstrap value ≥70% for ML and MP analyses are presented at the branches. Bootstrap values lower than 70% are marked with "*", and absent analyses values are marked with "-". Ex-type isolates are marked with "T". Isolates sequenced in this study are highlighted in blue and bold type. The "B" species codes are consistent with the recently published results in Liu and co-authors [18]. The tree was rooted to *Curvicladiella cignea* (CBS 109167 and CBS 109168).

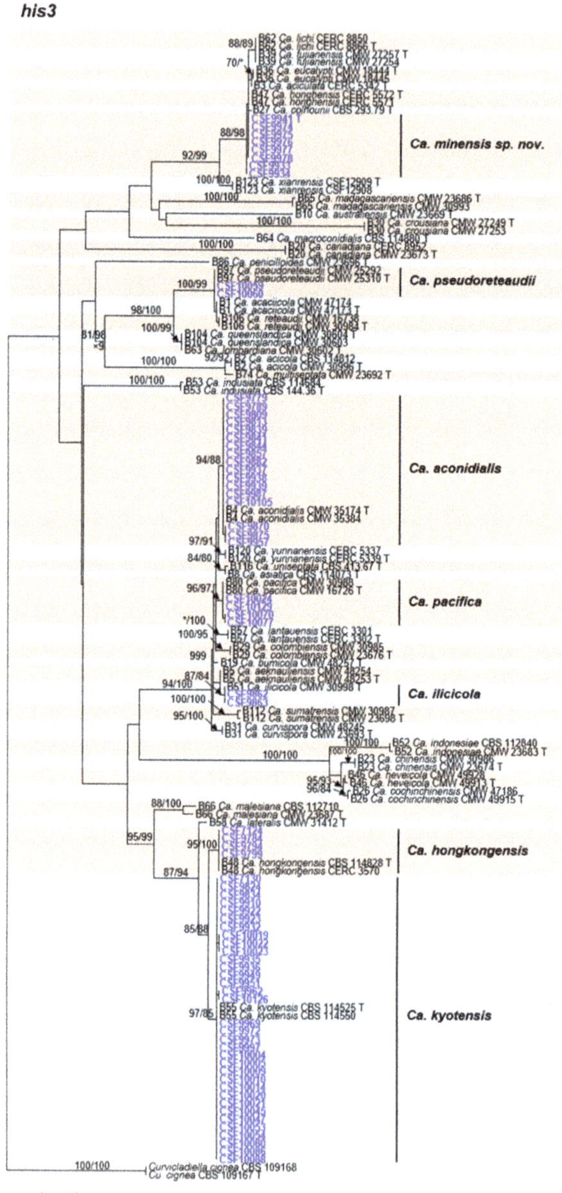

Figure A3. Phylogenetic tree of *Calonectria* species based on maximum likelihood (ML) analyses of *his3* gene sequences. Bootstrap value ≥70% for ML and MP analyses are presented at the branches. Bootstrap values lower than 70% are marked with "*", and absent analyses values are marked with "-". Ex-type isolates are marked with "T". Isolates sequenced in this study are highlighted in blue and bold type. The "B" species codes are consistent with the recently published results in Liu and co-authors [18]. The tree was rooted to *Curvicladiella cignea* (CBS 109167 and CBS 109168).

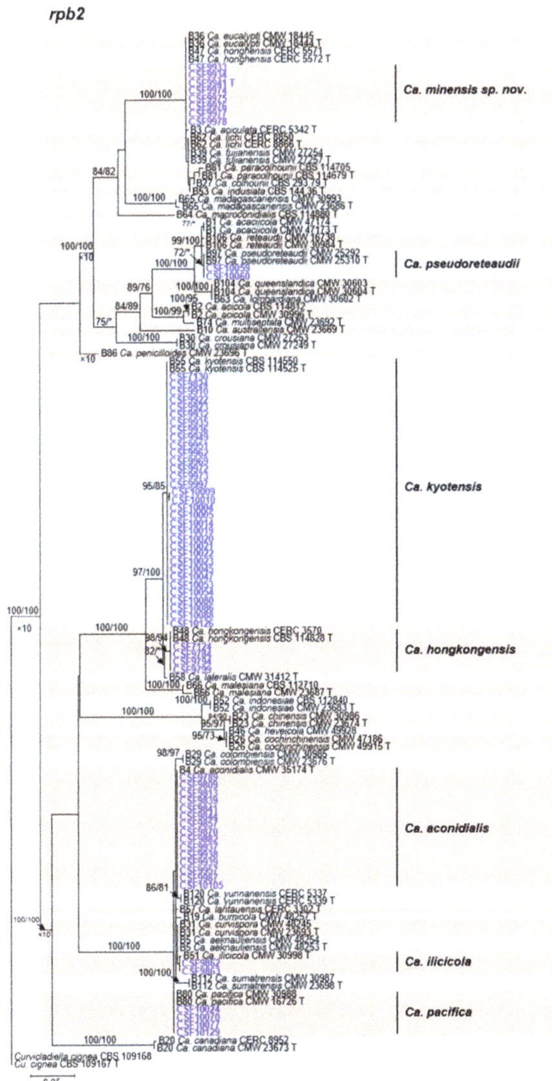

Figure A4. Phylogenetic tree of *Calonectria* species based on maximum likelihood (ML) analyses of *rpb2* gene sequences. Bootstrap value ≥70% for ML and MP analyses are presented at the branches. Bootstrap values lower than 70% are marked with "*", and absent analyses values are marked with "-". Ex-type isolates are marked with "T". Isolates sequenced in this study are highlighted in blue and bold type. The "B" species codes are consistent with the recently published results in Liu and co-authors [18]. The tree was rooted to *Curvicladiella cignea* (CBS 109167 and CBS 109168).

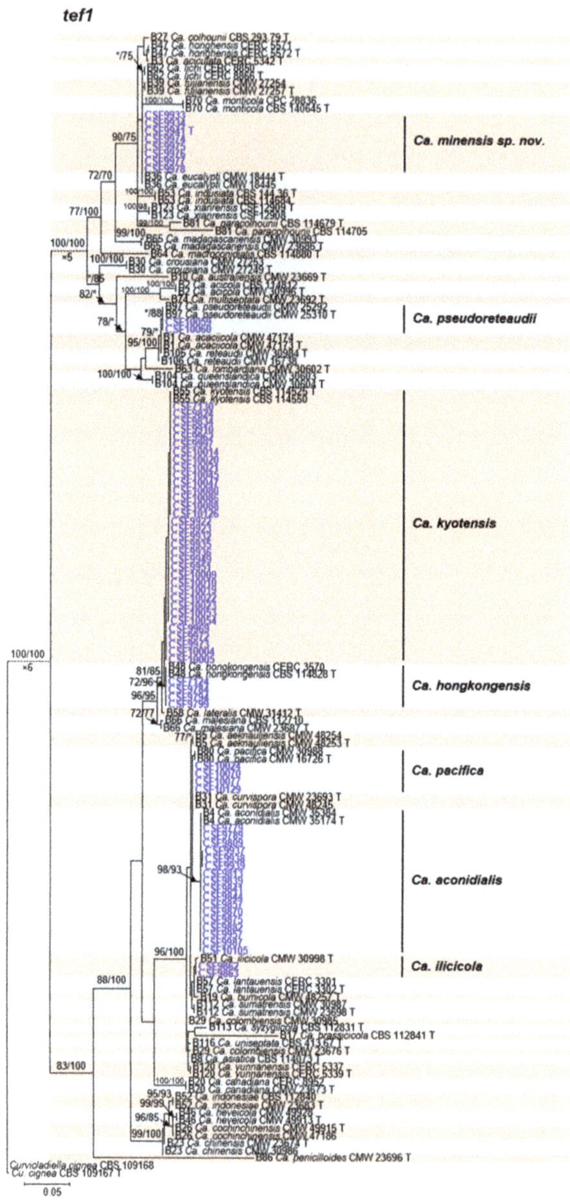

Figure A5. Phylogenetic tree of *Calonectria* species based on maximum likelihood (ML) analyses of *tef1* gene sequences. Bootstrap value ≥70% for ML and MP analyses are presented at the branches. Bootstrap values lower than 70% are marked with "*", and absent analyses values are marked with "-". Ex-type isolates are marked with "T". Isolates sequenced in this study are highlighted in blue and bold type. The "B" species codes are consistent with the recently published results in Liu and co-authors [18]. The tree was rooted to *Curvicladiella cignea* (CBS 109167 and CBS 109168).

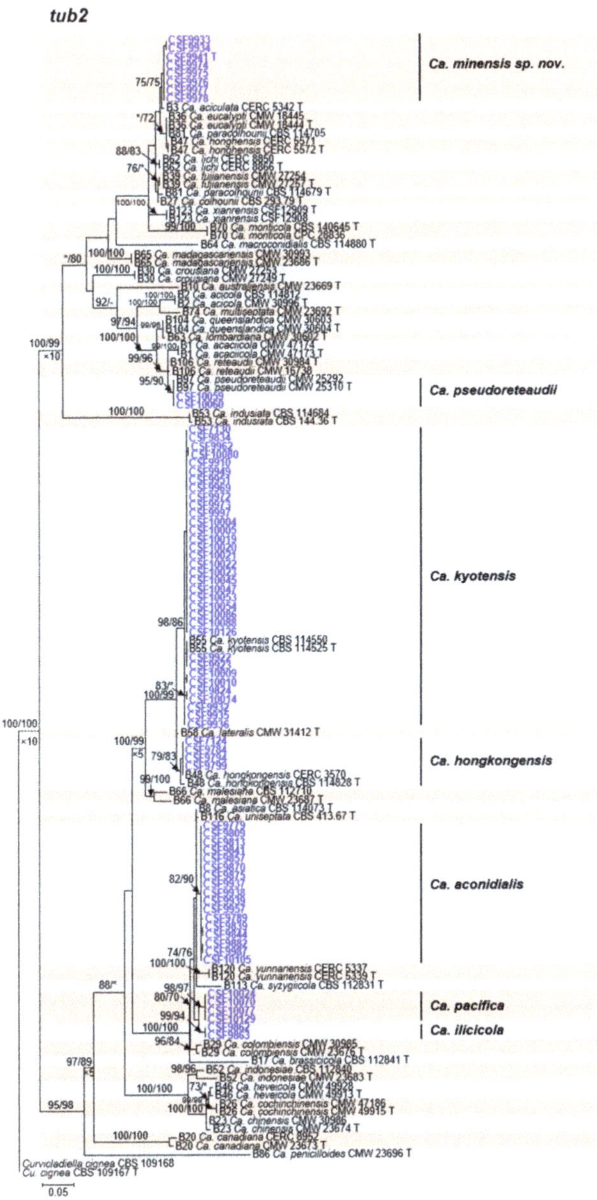

Figure A6. Phylogenetic tree of *Calonectria* species based on maximum likelihood (ML) analyses of *tub2* gene sequences. Bootstrap value ≥70% for ML and MP analyses are presented at the branches. Bootstrap values lower than 70% are marked with "*", and absent analyses values are marked with "-". Ex-type isolates are marked with "T". Isolates sequenced in this study are highlighted in blue and bold type. The "B" species codes are consistent with the recently published results in Liu and co-authors [18]. The tree was rooted to *Curvicladiella cignea* (CBS 109167 and CBS 109168).

Appendix C. Morphology of Six Previously Described *Calonectria* Species Collected in This Study

Calonectria aconidialis

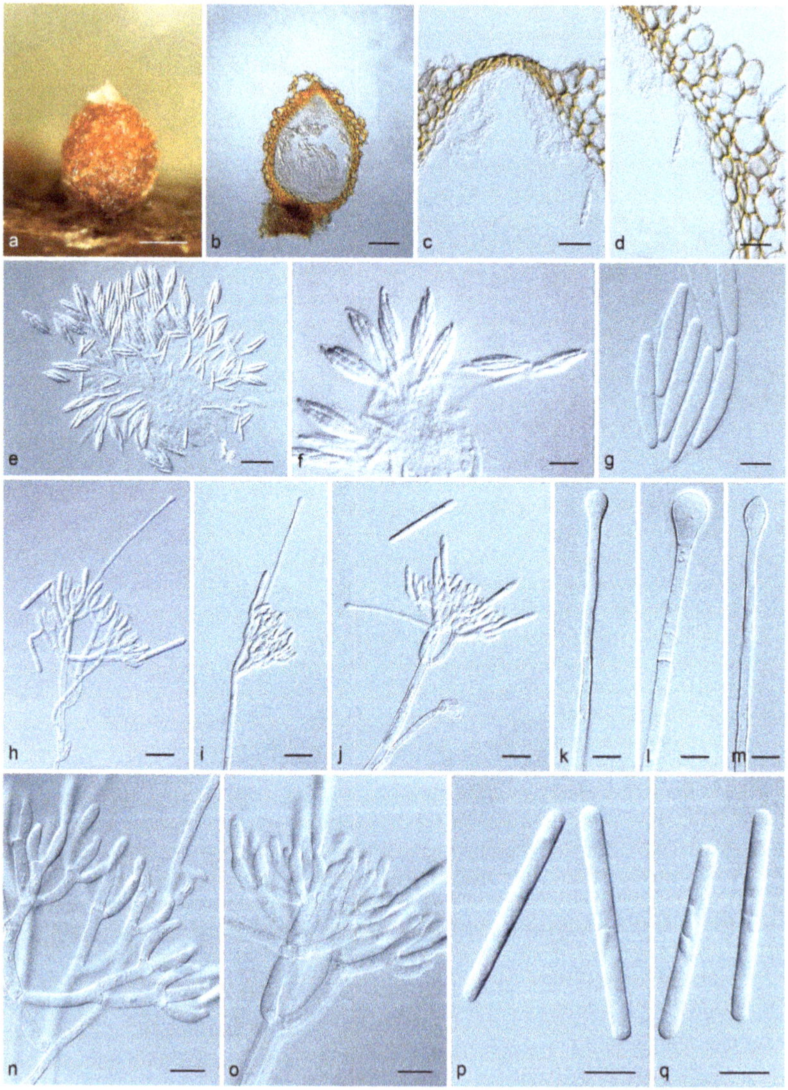

Figure A7. *Calonectria aconidialis*. (**a**). Perithecium; (**b**). vertical section through a perithecium; (**c**). cells around ostiolar region of perithecium; (**d**). section through lateral perithecial wall; (**e**,**f**). asci; (**g**). ascospores; (**h**–**j**). macroconidiophore; (**k**–**m**). obpyriform to sphaeropedunculate vesicles; (**n**,**o**). conidiogenous apparatus with conidiophore branches and elongate doliiform to reniform phialides; (**p**,**q**). macroconidia.—Scale bars: a = 200 μm; b = 100 μm; c, d, f and h–j = 20 μm; e = 50 μm; g and n–q = 10 μm; k–m = 5 μm.

Description: *Ascomata* perithecial, solitary or in groups of two, orange, becoming orange-brown with age; in section, apex and body orange, base red-brown, subglobose to ovoid, 368–491 μm high, 335–455 μm diam, body turning dark orange to red, and base dark red-brown in 3% KOH+; ascomatal wall rough, consisting of two thick-walled layers; outer layer of *textura globulosa*, 23–82 μm thick, cells becoming more compressed towards the inner layer of *textura angularis*, 8–21 μm thick, cells becoming thin-walled and hyaline towards the centre; outermost cells 21–35 × 7–21 μm, cells of inner layer 9–34 × 2–9 μm; ascomatal base up to 201 μm wide, consisting of dark red, angular cells, merging with an erumpent stroma; cells of the outer wall layer continuous with the pseudoparenchymatous cells of the erumpent stroma. *Asci* 8-spored, clavate, 68–143 × 10–22 μm, tapering into a long thin stalk. *Ascospores* aggregated in the upper third of the ascus, hyaline, guttulate, fusoid with rounded ends, straight to slightly curved, 1-septate, constricted at the septum, (24.5–)30.5–37.5(–42.5) × (4–)4.5–5.5(–7) μm (av. = 34 × 5 μm). *Macroconidiophores* consisting of a stipe, a suite of penicillate arranged fertile branches, a stipe extension, and a terminal vesicle; stipe septate, hyaline, smooth, 27–134 × 4–6 μm, stipe extension septate, straight to flexuous 64–129 μm long, 2–4 μm wide at the apical septum, terminating in a obpyriform to sphaeropedunculate vesicle, 3–7 μm diam; lateral stipe extensions (90° to main axis) moderate, 31–88 μm long, 1.5–3 μm wide at the apical septum, terminating in obpyriform vesicles, 2–5 μm. *Conidiogenous apparatus* 37–134 μm wide, and 41–128 μm long; primary branches aseptate, 15–27 × 3–5 μm; secondary branches aseptate, 12–20 × 3–4.5 μm; tertiary branches aseptate, 11–15 × 3–4 μm; quaternary branches aseptate, 8–16 × 3–5 μm, each terminal branch producing 2–6 phialides; phialides elongate doliiform to reniform, hyaline, aseptate, 9–18 × 2–5 μm, apex with minute periclinal thickening and inconspicuous collarette. *Macroconidia* cylindrical, rounded at both ends, straight, (40–)46–54.5(–63.5) × (3.5–)4.5–5(–6) μm (av. = 50 × 5 μm), 1-septate, lacking a visible abscission scar, held in parallel cylindrical clusters by colourless slime. Mega- and microconidia not observed.

Culture characteristics: Colonies producing abundant white to cinnamon (62) aerial mycelium at 25 °C on MEA, moderate sporulation on the medium surface; reverse sienna (8) to umber (9) after 7 d; chlamydospores extensive throughout the medium forming microsclerotia. Optimal growth temperature 25 °C, no growth at 5 °C and 35 °C, after 7 d, colonies at 10 °C, 15 °C, 20 °C, 25 °C and 30 °C reached 21.5 mm, 31.2 mm, 57.1 mm, 81.3 mm and 55.2 mm, respectively.

Specimens examined: China: Fujian Province, Longyan Region, Xinluo District (25°07′08.597″ N, 116°44′42.257″ E), from soil collected in a *Eucalyptus* plantation, 6 November 2016, S.F. Chen, Q.L. Liu and F.F. Liu (HMAS249929, culture CSF9937); Fujian Province, Longyan Region, Liancheng County (25°26′14.348″ N, 116°38′42.400″ E), from soil under a natural forest, 6 November 2016, S.F. Chen, Q.L. Liu and F.F. Liu (HMAS249930, culture CSF9957).

Notes: Isolates CSF9937, CSF9938 and CSF9957 were crossed with each other in all possible combinations on MSA to which autoclaved toothpicks had been placed, randomly distributed on the agar surface. Isolates CSF9937 and CSF9938 readily formed protoperithecia within two weeks, and perithecia with viable ascospores were produced within four weeks, when they crossed with themselves. After eight weeks of incubation, isolate CSF9957 failed to form sexual structures in any combination. *Calonectria aconidialis* is a species in the *Ca. kyotensis* species complex. The ascospores of *Ca. aconidialis* obtained in this study (av. = 34 × 5 μm) were smaller than those of the originally described *Ca. aconidialis* (av. = 36 × 6 μm) [11].

Calonectria hongkongensis

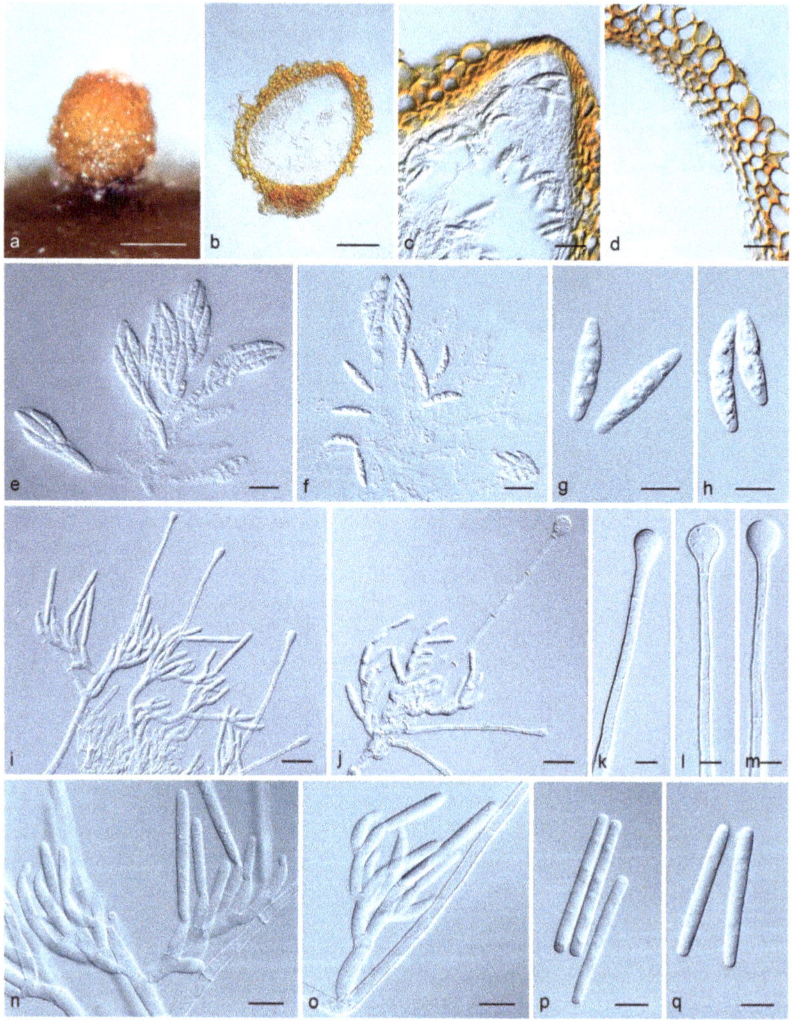

Figure A8. *Calonectria hongkongensis*. (**a**). Perithecium; (**b**). vertical section through a perithecium; (**c**). cells around ostiolar region of perithecium; (**d**). section through lateral perithecial wall; (**e**,**f**). asci; (**g**,**h**). ascospores; (**i**,**j**). macroconidiophore; (**k**–**m**). sphaeropedunculate vesicles; (**n**,**o**). conidiogenous apparatus with conidiophore branches and elongate doliiform to reniform phialides; (**p**,**q**). macroconidia.—Scale bars: a = 200 μm; b = 100 μm; c–f and i,j = 20 μm; g,h and n–q = 10 μm; k–m = 5 μm.

Description: *Ascomata* perithecial, solitary or in groups of up to three, orange, becoming red-brown with age; in section, apex and body orange, base dark red-brown, subglobose to ovoid, 243–376 μm high, 219–355 μm diam, body turning red, and base dark red-brown in 3% KOH+; ascomatal wall rough, consisting of two thick-walled layers; outer layer of *textura globulosa*, 31–54 μm thick, cells becoming more compressed towards the inner layer of *textura angularis*, 10–28 μm thick, cells becoming thin-walled and hyaline towards the centre; outermost cells 10–25 × 9–23 μm, cells of inner layer 6–24 × 2–6 μm; ascomatal base up to 168 µm wide, consisting of dark red, angular cells, merging with an erumpent stroma; cells of the outer wall layer continuous with the pseudoparenchymatous cells of the erumpent stroma. *Asci* 8-spored, clavate, 82–148 × 12–32 µm, tapering into a long thin stalk. *Ascospores* aggregated in the upper third of the ascus, hyaline, guttulate, fusoid with rounded ends, straight to slightly curved, 1-septate, constricted at the septum, (23–)25–30(–34) × (4–)5–7(–8) µm (av. = 28 × 6 µm). *Macroconidiophores* consisting of a stipe, a suite of penicillate arranged fertile branches, a stipe extension, and a terminal vesicle; stipe septate, hyaline, smooth, 47–117 × 4–8 µm, stipe extension septate, straight to flexuous 68–198 µm long, 1–4 µm wide at the apical septum, terminating in a sphaeropedunculate vesicle, 4–10 µm diam; lateral stipe extensions (90° to main axis) abundant, 42–111 µm long, 1–3 µm wide at the apical septum, terminating in obpyriform vesicles, 2–6 µm. *Conidiogenous apparatus* 37–146 µm wide, and 41–111 µm long; primary branches aseptate, 12–28 × 3–5.5 µm; secondary branches aseptate, 9.5–19 × 3–6 µm; tertiary branches aseptate, 9–13 × 3–5 µm, additional branches –5, aseptate, 8–15 × 2–4.5 µm, each terminal branch producing 2–4 phialides; phialides elongate doliiform to reniform, hyaline, aseptate, 8–14 × 2–5 µm, apex with minute periclinal thickening and inconspicuous collarette. *Macroconidia* cylindrical, rounded at both ends, straight, (34–)37–41(–44) × (3–)3.5–4(–5) µm (av. = 39 × 4 µm), 1-septate, lacking a visible abscission scar, held in parallel cylindrical clusters by colourless slime. Mega- and microconidia not observed.

Culture characteristics: Colonies forming abundant white to sienna (8) aerial mycelium at 25 °C on MEA, with irregular margins, abundant sporulation; surface rust-coloured (39); reverse sienna (8) to umber (9) after 7 d. Chlamydospores extensive throughout the medium forming microsclerotia. Optimal growth temperature 25 °C, no growth at 5 °C and 35 °C, after 7 d, colonies at 10 °C, 15 °C, 20 °C, 25 °C and 30 °C reached 21.2 mm, 26.1 mm, 46.3 mm, 69.1 mm and 64.1 mm, respectively.

Specimens examined: China: Fujian Province, Zhangzhou Region, Hua'an county (24°53′49.369″ N, 117°32′45.070″ E), from soil collected in a *Eucalyptus* plantation, 5 November 2016, S.F. Chen, Q.L. Liu and F.F. Liu (HMAS249931, culture CSF9784).

Notes: Isolates CSF7124, CSF9784 and CSF9794 were crossed with each other in all possible combinations on MSA. Isolates CSF7124 and CSF9784 readily formed protoperithecia within two weeks, and perithecia with viable ascospores were produced within four weeks, when they crossed with themselves. After eight weeks of incubation, isolate CSF9794 failed to form sexual structures in any combination. *Calonectria hongkongensis* is a species in the *Ca. kyotensis* species complex. The ascospores and macroconidia of *Ca. hongkongensis* obtained in this study (ascospores: av. = 28 × 6 µm; macroconidia: av. = 39 × 4 µm) were shorter than those of the originally described *Ca. hongkongensis* (ascospores: av. = 31 × 6 µm; macroconidia: av. = 46.5 × 4 µm) [23]. The vesicle of *Ca. hongkongensis* obtained in this study (4–10 µm) was narrower than those of the originally described *Ca. hongkongensis* (8–14 µm) [23].

Calonectria ilicicola

Figure A9. *Calonectria ilicicola.* (**a**). Perithecium; (**b**). vertical section through a perithecium; (**c**). cells around ostiolar region of perithecium; (**d**). section through lateral perithecial wall; (**e**,**f**). asci; (**g**). ascospores; (**h**,**i**). macroconidiophore; (**j**,**k**). ovoid to sphaeropedunculate vesicles; (**l**,**m**). conidiogenous apparatus with conidiophore branches and elongate doliiform to reniform phialides; (**n**,**o**). macroconidia.—Scale bars: a = 200 μm; b = 100 μm; c, d, f and i = 20 μm; e and h = 50 μm; g and l–o = 10 μm; j, k = 5 μm.

Description: *Ascomata* perithecial, solitary or in groups of two, orange to red, becoming red-brown with age; in section, apex and body red-brown, base dark red-brown, subglobose to ovoid, 375–509 μm high, 363–474 μm diam, body turning dark red, and base dark red-brown in 3% KOH+; ascomatal wall rough, consisting of two thick-walled layers; outer layer of *textura globulosa*, 47–75 μm thick, cells becoming more compressed towards the inner layer of *textura angularis*, 14–30 μm thick, cells becoming thin-walled and hyaline towards the centre; outermost cells 9–40 × 8–36 μm, cells of inner layer 10–23 × 2–7 μm; ascomatal base up to 208 μm wide, consisting of dark red, angular cells, merging with an erumpent stroma; cells of the outer wall layer continuous with the pseudoparenchymatous cells of the erumpent stroma. *Asci* 8-spored, clavate, 70–137 × 12–34 μm, tapering into a long thin stalk. *Ascospores* aggregated in the upper third of the ascus, hyaline, guttulate, fusoid with rounded ends, straight to slightly curved, 1-septate, not or slightly constricted at the septum, (30–)37–46.5(–58) × (4–)5–6(–8) μm (av. = 42 × 5 μm). *Macroconidiophores* consisting of a stipe, a suite of penicillate arranged fertile branches, a stipe extension, and a terminal vesicle; stipe septate, hyaline, smooth, 12–98 × 4–7 μm, stipe extension septate, straight to flexuous 111–216 μm long, 2–4.5 μm wide at the apical septum, terminating in an ovoid to sphaeropedunculate vesicle, 6–13 μm diam; lateral stipe extensions (90° to main axis) absent. *Conidiogenous apparatus* 32–94 μm wide, and 49–106 μm long; primary branches aseptate, 12–34 × 4–6 μm; secondary branches aseptate, 4–21 × 3.5–6 μm; tertiary branches aseptate, 9–17 × 4–6 μm, each terminal branch producing 2–4 phialides; phialides elongate doliiform to reniform, hyaline, aseptate, 8–15 × 3–5 μm, apex with minute periclinal thickening and inconspicuous collarette. *Macroconidia* cylindrical, rounded at both ends, straight, (58–)63–70(–76) × 6–7(–8) μm (av. = 67 × 7 μm), (1–)3-septate, lacking a visible abscission scar, held in parallel cylindrical clusters by colourless slime. Mega- and microconidia not observed.

Culture characteristics: Colonies forming abundant white to cinnamon (62) aerial mycelium at 25 °C on MEA, with irregular margins, profuse sporulation; reverse with cinnamon (62) outer margin, and rust (39) inner region after 7 d. Chlamydospores extensive throughout the medium forming microsclerotia. Optimal growth temperature 25 °C, no growth at 5 °C and 35 °C, after 7 d, colonies at 10 °C, 15 °C, 20 °C, 25 °C and 30 °C reached 16.1 mm, 24.9 mm, 54.8 mm, 74.3 mm and 66.4 mm, respectively.

Specimens examined: China: Fujian Province, Longyan Region, Zhangping County (25°17′10.882″ N, 117°27′33.635″ E), from soil collected in a *Eucalyptus* plantation, 6 November 2016, S.F. Chen, Q.L. Liu and F.F. Liu (HMAS249932, culture CSF9862).

Notes: Isolates CSF9862 and CSF9863 were crossed with each other on MSA and they were readily formed protoperithecia within two weeks, and perithecia with viable ascospores were produced within four weeks, when they crossed with themselves. *Calonectria ilicicola* is a species in the *Ca. kyotensis* species. The ascospores of *Ca. ilicicola* (av. = 42 × 5.5 μm) obtained in this study were smaller than those of the originally described *Ca. ilicicola* (av. = 45 × 6 μm) [17], and the macroconidia of *Ca. ilicicola* (av. = 67 × 7 μm) were larger than those of the originally described *Ca. ilicicola* (av. = 62 × 6 μm) [17], and they share similar vesicle dimensions.

Calonectria kyotensis

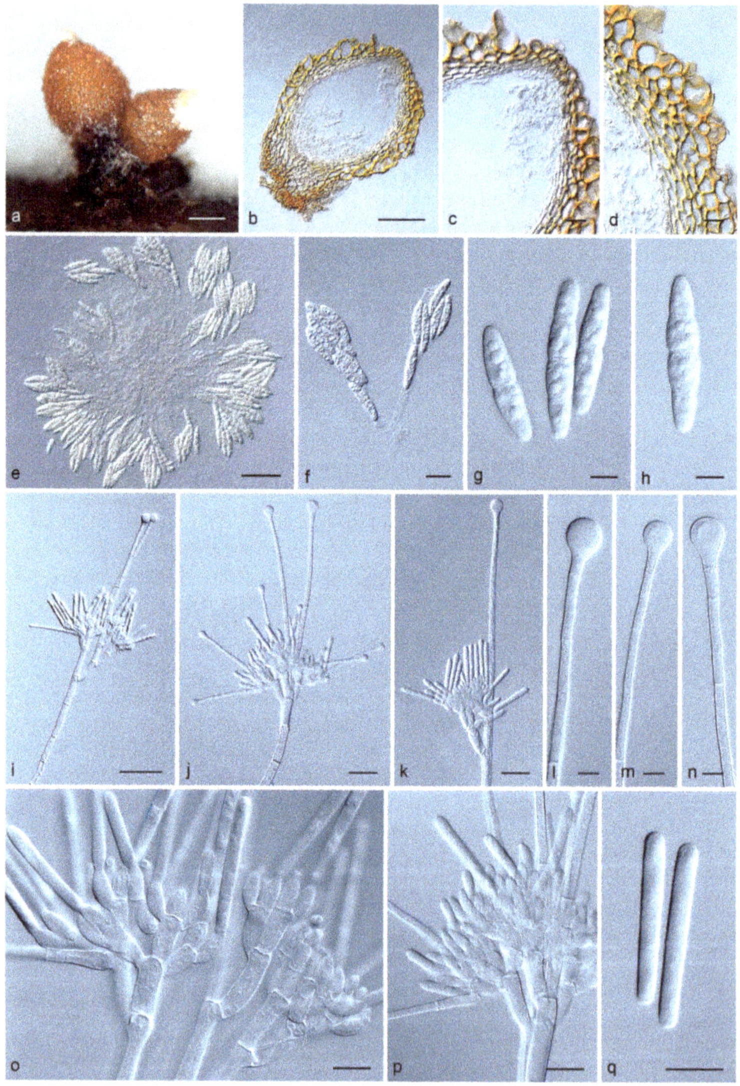

Figure A10. *Calonectria kyotensis*. (**a**). Perithecium; (**b**). vertical section through a perithecium; (**c**). cells around ostiolar region of perithecium; (**d**). section through lateral perithecial wall; (**e**,**f**). asci; (**g**,**h**). ascospores; (**i**–**k**). macroconidiophore; (**l**–**n**). sphaeropedunculate vesicles; (**o**,**p**). conidiogenous apparatus with conidiophore branches and elongate doliiform to reniform phialides; (**q**). macroconidia.—Scale bars: a = 200 μm; b = 100 μm; c, d, f, j and k = 20 μm; e and i = 50 μm; g, h and o–q = 10 μm; l–n = 5 μm.

Description: *Ascomata* perithecial, solitary or in groups of up to four, orange, becoming red-brown with age; in section, apex and body orange, base dark red-brown, subglobose to ovoid, 322–482 μm high, 296–432 μm diam, body turning red, and base dark red-brown in 3% KOH+; ascomatal wall rough, consisting of two thick-walled layers; outer layer of *textura globulosa*, 8–24 μm thick, cells becoming more compressed towards the inner layer of *textura angularis*, 25–59 μm thick, cells becoming thin-walled and hyaline towards the centre; outermost cells 14–25 × 8–13 μm, cells of inner layer 10–30 × 2–6 μm; ascomatal base up to 234 μm wide, consisting of dark red, angular cells, merging with an erumpent stroma; cells of the outer wall layer continuous with the pseudoparenchymatous cells of the erumpent stroma. *Asci* 8-spored, clavate, 73–125 × 15–29 μm, tapering into a long thin stalk. *Ascospores* aggregated in the upper third of the ascus, hyaline, guttulate, fusoid with rounded ends, straight to slightly curved, 1(–3)-septate, constricted at the septum, (26–)31–38.5(–43.5) × (5–)5.5–7.5(–9.5) μm (av. = 34.5 × 6.5 μm). *Macroconidiophores* consisting of a stipe, a suite of penicillate arranged fertile branches, a stipe extension, and a terminal vesicle; stipe septate, hyaline, smooth, 36–135 × 4–9 μm, stipe extension septate, straight to flexuous 69.5–222 μm long, 2–4 μm wide at the apical septum, terminating in a sphaeropedunculate vesicle, 4–10 μm diam; lateral stipe extensions (90° to main axis) abundant, 41–108 μm long, 1–3 μm wide at the apical septum, terminating in sphaeropedunculate vesicles, 3−7 μm. *Conidiogenous apparatus* 40–110 μm wide, and 36–108 μm long; primary branches aseptate, 14–31 × 4–6 μm; secondary branches aseptate, 9–22 × 3–5 μm; tertiary branches aseptate, 7–16 × 3–5 μm, quaternary branches aseptate, 8−11 × 3–5 μm, each terminal branch producing 2–4 phialides; phialides doliiform to reniform, hyaline, aseptate, 6–10 × 2–4 μm, apex with minute periclinal thickening and inconspicuous collarette. *Macroconidia* cylindrical, rounded at both ends, straight, (28–)32–35.5(–39.5) × (2.5–)3–4(–4.5) μm (av. = 33.5 × 3.5 μm), 1-septate, lacking a visible abscission scar, held in parallel cylindrical clusters by colourless slime. Mega- and microconidia not observed.

Culture characteristics: Colonies forming abundant white to sienna (8) aerial mycelium at 25 °C on MEA, with feather, irregular margins, profuse sporulation; reverse sienna (8) to umber (9) after 7 d. Chlamydospores extensive throughout the medium forming microsclerotia. Optimal growth temperature 25 °C, no growth at 5 °C and 35 °C, after 7 d, colonies at 10 °C, 15 °C, 20 °C, 25 °C and 30 °C reached 16.2 mm, 23.2 mm, 52.1 mm, 66.3 mm and 61.5 mm, respectively.

Specimens examined: China: Fujian Province, Zhangzhou Region, Hua'an county (24°53′49.369″ N, 117°32′45.070″ E), from soil collected in a *Eucalyptus* plantation, 5 November 2016, S.F. Chen, Q.L. Liu and F.F. Liu (HMAS249933, culture CSF9824); Fujian Province, Longyan Region, Liancheng county (25°33′06.994″ N, 116°41′42.328″ E), from soil collected in a *Eucalyptus* plantation, 6 November 2016, S.F. Chen, Q.L. Liu and F.F. Liu (HMAS249934, culture CSF10004).

Notes: Isolates CSF7130, CSF9824 and CSF10004 were crossed with each other in all possible combinations on MSA. Isolates CSF9824 readily formed protoperithecia within two weeks. After eight weeks of incubation, isolates CSF7130 and CSF10004 failed to form sexual structures in any combination. *Calonectria kyotensis* is a species in the *Ca. kyotensis* species. The ascospores of *Ca. kyotensis* (av. = 34.5 × 6.5 μm) obtained in this study were longer than those of the originally described *Ca. kyotensis* (av. = 29 × 6 μm) [47], while the macroconidia of *Ca. kyotensis* (av. = 33.5 × 3.5 μm) in this study were shorter than those of the originally described *Ca. kyotensis* (av. = 41 × 4 μm) [47], and the vesicle in this study (4–10 μm) was narrower than those of originally described *Ca. kyotensis* (8.8–19 μm) [47].

Calonectria pacifica

Figure A11. *Calonectria pacifica*. (**a–c**). Macroconidiophore; (**d–g**). ovoid to sphaeropedunculate vesicles; (**h–j**). conidiogenous apparatus with conidiophore branches and doliiform to reniform phialides; (**k**,**l**). macroconidia.—Scale bars: a–c = 20 μm; d–g = 5 μm; h–l = 10 μm.

Description: *Sexual morph* unknown. *Macroconidiophores* consisting of a stipe, a suite of penicillate arranged fertile branches, a stipe extension, and a terminal vesicle; stipe septate, hyaline, smooth, 44–115 × 4–7 μm; stipe extensions septate, straight to flexuous 73.5–171 μm long, 2–3.5 μm wide, at the apical septum, terminating in an ovoid to sphaeropedunculate vesicle, 4–10 μm diam; lateral stipe extensions (90° to main axis) abundant, 36–98 μm long, 1.5–2.5 μm wide at the apical septum, terminating in an ovoid vesicles, 3–5 μm diam. *Conidiogenous apparatus* 45–105 μm wide, and 35–81 μm long; primary branches aseptate, 12.5–23 × 4–6 μm; secondary branches aseptate, 10–20 × 3–6 μm; tertiary branches aseptate, 10–15 × 3–5 μm, each terminal branch producing 2–4 phialides; phialides doliiform to reniform, hyaline, aseptate, 6–15 × 3–5 μm, apex with minute periclinal thickening and inconspicuous collarette. *Macroconidia* cylindrical, rounded at both ends, straight, (36–)40–46(–48) × (3.5–)4–5(–6) μm, (av. = 43 × 5 μm), 1-septate, lacking a visible abscission scar, held in parallel cylindrical clusters by colourless slime. Mega- and microconidia not observed.

Culture characteristics: Colonies forming sparse white to sienna (8) aerial mycelium at 25 °C on MEA, with feathery, irregular margins at the edges, abundant sporulation; reverse sienna (8) to umber (9) after 7 d. Optimal growth temperature 25 °C, no growth at 5 °C and

35 °C, after 7 d, colonies at 10 °C, 15 °C, 20 °C, 25 °C and 30 °C reached 15.1 mm, 21.4 mm, 45.1 mm, 58.2 mm and 42.1 mm, respectively.

Specimens examined: China: Fujian Province, Nanping Region, Yanping District (26°42′26.672″ N, 118°07′58.317″ E), from soil under a natural forest, 08 November 2016, S.F. Chen, Q.L. Liu and F.F. Liu (HMAS249938, culture CSF10070); Fujian Province: Nanping Region, Yanping District (26°42′26.672″ N, 118°07′58.317″ E), from soil under a natural forest, 08 November 2016, S.F. Chen, Q.L. Liu and F.F. Liu (HMAS249939, culture CSF10077).

Notes: Isolates CSF10024, CSF10070 and CSF10077 were crossed with each other in all possible combinations on MSA and failed to form sexual structures in any combination. *Calonectria pacifica* is a species in the *Ca. kyotensis* species complex. The macroconidia of *Ca. pacifica* (av. = 43 × 5 μm) obtained in this study were shorter than those of the originally described *Ca. pacifica* (av. = 55 × 4.5 μm) [17], and the vesicles were narrower than those of originally described strains of *Ca. pacifica* (7–15 μm) [17].

Calonectria pseudoreteaudii

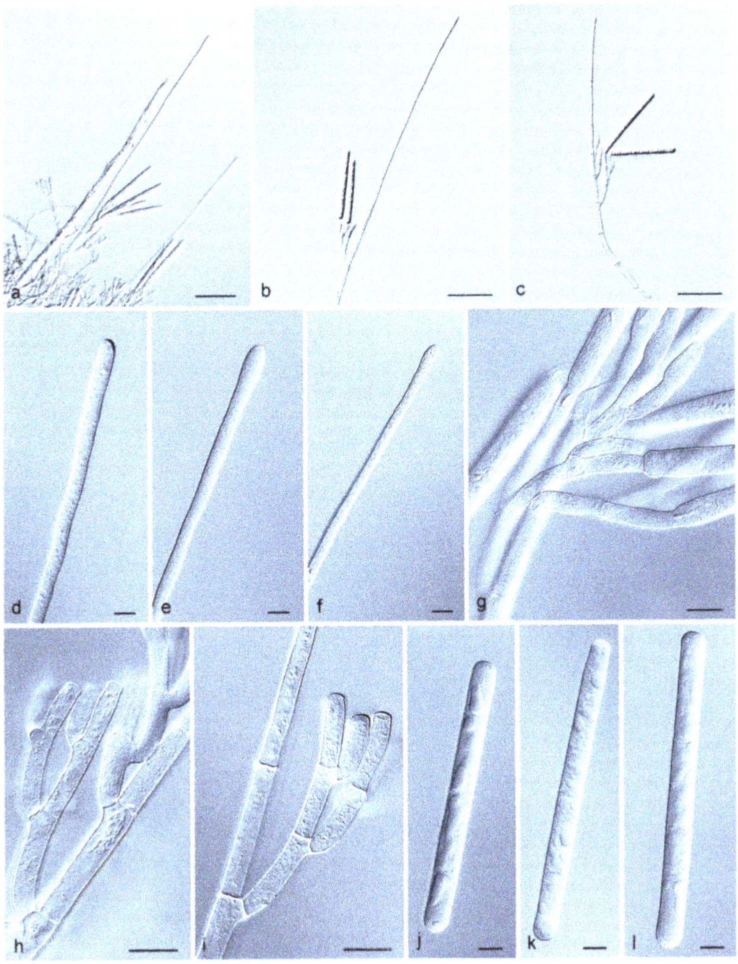

Figure A12. *Calonectria pseudoreteaudii*. (**a**–**c**). Macroconidiophore; (**d**–**f**). clavate to narrowly clavate vesicle; (**g**–**i**) conidiogenous apparatus with conidiophore branches and cylindrical to allantoid phialides; (**j**–**l**). macroconidia.—Scale bars: a–c = 20 μm; d–f = 5 μm; g–l = 10 μm.

Description: *Sexual morph* unknown. *Macroconidiophores* consisting of a stipe, a suite of penicillate arranged fertile branches, a stipe extension, and a terminal vesicle; stipe septate, hyaline, smooth, 81–145 × 3–8 µm; stipe extensions septate, straight to flexuous 150–268 µm long, 5–7 µm wide, at the apical septum, terminating in a narrowly clavate vesicle, 3–5 µm diam. *Conidiogenous apparatus* 68–140 µm long, and 30–92 µm wide; primary branches aseptate or 1-septate, 19–34 × 4–6 µm; secondary branches aseptate, 16–25 × 4–5 µm; tertiary branches aseptate, 13–22 × 3–5 µm, each terminal branch producing 1–3 phialides; phialides cylindrical to allantoid, hyaline, aseptate, 10−18 × 3–5 µm, apex with minute periclinal thickening and inconspicuous collarette. *Macroconidia* cylindrical, rounded at the apex, flattened at the base, straight, (54.5–)73–88.5(–96) × (6–)6.5–8(–9) µm, (av. = 81 × 7.5 µm), 5-septate, lacking a visible abscission scar, held in parallel cylindrical clusters by colourless slime. Mega- and microconidia not observed.

Culture characteristics: Colonies forming white to sienna (8) aerial mycelium at 25 °C on MEA, with feathery, regular margins at the edges, abundant sporulation; reverse sienna (8) to chestnut (40) after 7 d; chlamydospores extensive throughout the medium, forming microsclerotia. Optimal growth temperature 25 °C, no growth at 5 °C and 35 °C, after 7 d, colonies at 10 °C, 15 °C, 20 °C, 25 °C and 30 °C reached 19.3 mm, 25.1 mm, 49.2 mm, 59.1 mm and 47.1 mm, respectively.

Specimens examined: China: Fujian Province, Nanping Region, Yanping District (26°46′19.651″ N, 117°57′37.233″ E), from soil collected in a *Eucalyptus* plantation, 08 November 2016, S.F. Chen, Q.L. Liu and F.F. Liu (HMAS249940, culture CSF10059); Fujian Province: Nanping Region, Yanping District (26°46′19.651″ N, 117°57′37.233″ E), from soil collected in a *Eucalyptus* plantation, 08 November 2016, S.F. Chen, Q.L. Liu and F.F. Liu (HMAS249941, culture CSF10060).

Notes: Isolates CSF10059 and CSF10060 were crossed with each other on MSA and failed to form sexual structures in any combination. *Calonectria pseudoreteaudii* is a species in the *Ca. reteaudii* species complex. The macroconidia of isolates obtained in this study (av. = 81 × 7.5 µm) were much shorter than those of the originally described strains of *Ca. pseudoreteaudii* (av. = 104 × 8 µm) [24].

References

1. Liu, T.; Xie, Y.J. Studies on the causes of rapid development of *Eucalyptus* plantations in China. *Eucalyptus Sci. Technol.* **2020**, *27*, 38–47. (In Chinese)
2. Zhou, X.D.; Wingfield, M.J. Eucalypt diseases and their management in China. *Australas. Plant Pathol.* **2011**, *40*, 339–345. [CrossRef]
3. Carstensen, G.D.; Venter, S.N.; Wingfield, M.J.; Coutinho, T.A. Two *Ralstonia* species associated with bacterial wilt of *Eucalyptus*. *Plant Pathol.* **2017**, *66*, 393–403. [CrossRef]
4. Li, J.Q.; Wingfield, M.J.; Liu, Q.L.; Barnes, I.; Roux, J.; Lombard, L.; Crous, P.W.; Chen, S.F. *Calonectria* species isolated from *Eucalyptus* plantations and nurseries in South China. *IMA Fungus* **2017**, *8*, 259–294. [CrossRef]
5. Li, G.Q.; Liu, F.F.; Li, J.Q.; Liu, Q.L.; Chen, S.F. Botryosphaeriaceae from *Eucalyptus* plantations and adjacent plants in China. *Persoonia* **2018**, *40*, 63–95. [CrossRef]
6. Wang, W.; Li, G.Q.; Liu, Q.L.; Chen, S.F. *Cryphonectriaceae* on *Myrtales* in China: Phylogeny, host range, and pathogenicity. *Persoonia* **2020**, *45*, 101–131. [CrossRef]
7. Chen, S.F.; Lombard, L.; Roux, J.; Xie, Y.J.; Wingfield, M.J.; Zhou, X.D. Novel species of *Calonectria* associated with *Eucalyptus* leaf blight in Southeast China. *Persoonia* **2011**, *26*, 1–12. [CrossRef]
8. Ye, X.Z.; Zhong, Z.H.; Liu, H.Y.; Lin, L.Y.; Guo, M.M.; Guo, W.S.; Wang, Z.H.; Zhang, Q.H.; Feng, L.Z.; Lu, G.D.; et al. Whole genome and transcriptome analyses reveal adaptive strategies and pathogenesis of *Calonectria pseudoreteaudii* to *Eucalyptus*. *BMC Genom.* **2018**, *19*, 358. [CrossRef]
9. Wang, Q.C.; Chen, S.F. *Calonectria pentaseptata* causes severe leaf disease on cultivated *Eucalyptus* in Leizhou Peninsula of southern China. *Plant Dis.* **2020**, *104*, 493–509. [CrossRef]
10. Wu, W.X.; Chen, S.F. Species diversity, mating strategy and pathogenicity of *Calonectria* species from diseased leaves and soils in the *Eucalyptus* plantation in Southern China. *J. Fungi* **2021**, *7*, 73. [CrossRef]
11. Lombard, L.; Chen, S.F.; Mou, X.; Zhou, X.D.; Crous, P.W.; Wingfield, M.J. New species, hyper-diversity and potential importance of *Calonectria* spp. from *Eucalyptus* in South China. *Stud. Mycol.* **2015**, *80*, 151–188. [CrossRef]

12. Booth, T.H.; Jovanovic, T.; Old, K.M.; Dudzinski, M.J. Climatic mapping to identify high-risk areas for *Cylindrocladium quinqueseptatum* leaf blight on eucalypts in mainland South East Asia and around the world. *Environ. Pollut.* **2000**, *108*, 365–372. [CrossRef]
13. Park, R.F.; Keane, P.J.; Wingfield, M.J.; Crous, P.W.; Kile, G.A.; Podger, F.D.; Brown, B.N. Fungal disease of eucalypt foliage. In *Disease and Pathogens of Eucalypts*; Keane, P.J., Kile, G.A., Podger, F.D., Brown, B.N., Eds.; CSIRO Publishing: Clayton, Australia, 2000; pp. 153–239.
14. Alfenas, R.F.; Lombard, L.; Pereira, O.L.; Alfenas, A.C.; Crous, P.W. Diversity and potential impact of *Calonectria* species in *Eucalyptus* plantations in Brazil. *Stud. Mycol.* **2015**, *80*, 89–130. [CrossRef]
15. Jessadarom, H.; Phetruang, W.; Haitook, S.; Cheewangkoon, R. Isolation of *Calonectria sulawesiensis* from soil in Thailand and its pathogenicity against *Eucalyptus camaldulensis*. *Plant Pathol. Quar.* **2018**, *8*, 1–8. [CrossRef]
16. Pham, N.; Barnes, I.; Chen, S.F.; Liu, F.F.; Dang, Q.N.; Pham, T.Q.; Lombard, L.; Crous, P.W.; Wingfield, M.J. Ten new species of *Calonectria* from Indonesia and Vietnam. *Mycologia* **2019**, *111*, 78–102. [CrossRef]
17. Crous, P.W. *Taxonomy and Pathology of Cylindrocladium (Calonectria) and Allied Genera*; APS Press: St. Paul, MN, USA, 2002.
18. Liu, Q.L.; Li, J.Q.; Wingfield, M.J.; Duong, T.A.; Wingfield, B.D.; Crous, P.W.; Chen, S.F. Reconsideration of species boundaries and proposed DNA barcodes for *Calonectria*. *Stud. Mycol.* **2020**, *97*, 100106. [CrossRef]
19. Crous, P.W.; Luangsa-Ard, J.J.; Wingfield, M.J.; Carnegie, A.J.; Hernández-Restrepo, M.; Lombard, L.; Roux, J.; Barreto, R.W.; Baseia, I.G.; Cano-Lira, J.F.; et al. Fungal Planet description sheets: 785–867. *Persoonia* **2018**, *41*, 238–417. [CrossRef]
20. Crous, P.W.; Carnegie, A.J.; Wingfield, M.J.; Sharma, R.; Mughini, G.; Noordeloos, M.E.; Santini, A.; Shouche, Y.S.; Bezerra, J.P.D.; Dima, B.; et al. Fungal Planet description sheets: 868–950. *Persoonia* **2019**, *42*, 291–473. [CrossRef]
21. Wang, Q.C.; Liu, Q.L.; Chen, S.F. Novel species of *Calonectria* isolated from soil near *Eucalyptus* plantations in southern China. *Mycologia* **2019**, *111*, 1028–1040. [CrossRef]
22. Crous, P.W.; Hernández-Restrepo, M.; Schumacher, R.K.; Cowan, D.A.; Maggs-Kölling, G.; Marais, E.; Wingfield, M.J.; Yilmaz, N.; Adan, O.C.G.; Akulov, A.; et al. New and Interesting Fungi. 4. *Fungal Syst. Evol.* **2021**, *7*, 255–343. [CrossRef]
23. Crous, P.W.; Groenewald, J.Z.; Risède, J.M.; Simoneau, P.; Hywel-Jones, N.L. *Calonectria* species and their *Cylindrocladium* anamorphs: Species with sphaeropedunculate vesicles. *Stud. Mycol.* **2004**, *50*, 415–430.
24. Lombard, L.; Zhou, X.D.; Crous, P.W.; Wingfield, B.D.; Wingfield, M.J. *Calonectria* species associated with cutting rot of *Eucalyptus*. *Persoonia* **2010**, *24*, 1–11. [CrossRef]
25. Liu, Q.L.; Chen, S.F. Two novel species of *Calonectria* isolated from soil in a natural forest in China. *MycoKeys* **2017**, *26*, 25–60. [CrossRef]
26. Liu, L.L.; Wu, W.X.; Chen, S.F. Species diversity and distribution characteristics of *Calonectria* in five soil layers in a *Eucalyptus* Plantation. *J. Fungi* **2021**, *7*, 857. [CrossRef]
27. Van Burik, J.A.H.; Schreckhise, R.W.; White, T.C.; Bownen, R.A. Comparison of six extraction techniques for isolation of DNA from filamentous fungi. *Med. Mycol.* **1998**, *36*, 299–303. [CrossRef]
28. Kearse, M.; Moir, R.; Wilson, A.; Stones-Havas, S.; Cheung, M.; Sturrock, S.; Buxton, S.; Cooper, A.; Markowitz, S.; Duran, C.; et al. Geneious Basic: An integrated and extendable desktop software platform for the organization and analyses of sequence data. *Bioinformatics* **2012**, *28*, 1647–1649. [CrossRef]
29. Katoh, K.; Standley, D.M. MAFFT multiple sequence alignment software version 7: Improvements in performance and usability. *Mol. Biol. Evol.* **2013**, *30*, 772–780. [CrossRef]
30. Tamura, K.; Stecher, G.; Peterson, D.; Filipski, A.; Kumar, S. MEGA6: Molecular evolutionary genetics analyses version 6.0. *Mol. Biol. Evol.* **2013**, *30*, 2725–2729. [CrossRef]
31. Swofford, D.L. *Phylogenetic Analyses Using Parsimony (*and Other Methods)*; V. 4.0b10; Sinauer Associates: Sunderland, MA, USA, 2003.
32. Guindon, S.; Gascuel, O. A simple, Fast, and Accurate Algorithm to Estimate Large Phylogenies by Maximum Likelihood. *Syst. Biol.* **2003**, *52*, 696–704. [CrossRef]
33. Farris, J.S.; Källersjö, M.; Kluge, A.G.; Bult, C. Testing significance of incongruence. *Cladistics* **1994**, *10*, 315–319. [CrossRef]
34. Felsenstein, J. Confidence intervals on phylogenetics: An approach using bootstrap. *Evolution* **1985**, *39*, 783–791. [CrossRef] [PubMed]
35. Posada, D. jModelTest: Phylogenetic model averaging. *Mol. Biol. Evol.* **2008**, *25*, 1253–1256. [CrossRef] [PubMed]
36. Gadgil, P.D.; Dick, M.A. Fungi silvicolae novazelandae: 5. *N. Z. J. For. Sci.* **2004**, *34*, 316–323.
37. Lombard, L.; Crous, P.W.; Wingfield, B.D.; Wingfield, M.J. Phylogeny and systematics of the genus *Calonectria*. *Stud. Mycol.* **2010**, *66*, 31–69. [CrossRef]
38. Crous, P.W.; Groenewald, J.Z.; Risède, J.M.; Simoneau, P.; Hyde, K.D. *Calonectria* species and their *Cylindrocladium* anamorphs: Species with clavate vesicles. *Stud. Mycol.* **2006**, *55*, 213–226. [CrossRef]
39. Lombard, L.; Wingfield, M.J.; Alfenas, A.C.; Crous, P.W. The forgotten *Calonectria* collection: Pouring old wine into new bags. *Stud. Mycol.* **2016**, *85*, 159–198. [CrossRef]
40. Kang, J.C.; Crous, P.W.; Schoch, C.L. Species concepts in the *Cylindrocladium floridanum* and *Cy. spathiphylli* complexes (*Hypocreaceae*) based on multi-allelic sequence data, sexual compatibility and morphology. *Syst. Appl. Microbiol.* **2001**, *24*, 206–217. [CrossRef]

41. Lechat, C.; Crous, P.W.; Groenewald, J.Z. The enigma of *Calonectria* species occurring on leaves of *Ilex aquifolium* in Europe. *IMA Fungus* **2010**, *1*, 101–108. [CrossRef]
42. Peerally, M.A. *Calonectria colhounii* sp. nov., a common parasite of tea in Mauritius. *Trans. Br. Mycol. Soc.* **1973**, *61*, 89–93. [CrossRef]
43. Victor, D.; Crous, P.W.; Janse, B.J.H.; Wingfield, M.J. Genetic variation in *Cylindrocladium floridanum* and other morphologically similar *Cylindrocladium* species. *Syst. Appl. Microbiol.* **1997**, *20*, 268–285. [CrossRef]
44. Boedijn, K.B.; Reitsma, J. Notes on the genus *Cylindrocladium* (Fungi: Mucedineae). *Reinwardtia* **1950**, *1*, 51–60.
45. Marin-Felix, Y.; Groenewald, J.Z.; Cai, L.; Chen, Q.; Marincowitz, S.; Barnes, I.; Bensch, K.; Braun, U.; Camporesi, E.; Damm, U.; et al. Genera of phytopathogenic fungi: GOPHY 1. *Stud. Mycol.* **2017**, *86*, 99–216. [CrossRef]
46. Crous, P.W.; Kang, J.C.; Schoch, C.L.; Mchau, G.R.A. Phylogenetic relationships of *Cylindrocladium pseudogracile* and *Cylindrocladium rumohrae* with morphologically similar taxa, based on morphology and DNA sequences of internal transcribed spacers and β-tubulin. *Can. J. Bot.* **1999**, *77*, 1813–1820. [CrossRef]
47. Terashita, T. A new species of *Calonectria* and its conidial state. *Trans. Mycol. Soc. Jpn.* **1968**, *8*, 124–129.
48. Crous, P.W.; Wingfield, M.J.; Alfenas, A.C. Additions to *Calonectria*. *Mycotaxon* **1993**, *46*, 217–234.
49. Crous, P.W.; Wingfield, M.J.; Le Roux, J.J.; Richardson, D.M.; Strasberg, D.; Shivas, R.G.; Alvarado, P.; Edwards, J.; Moreno, G.; Sharma, R.; et al. Fungal Planet description sheets: 371–399. *Persoonia* **2015**, *35*, 264–327. [CrossRef]
50. Crous, P.W.; Wingfield, M.J.; Mohammed, C.; Yuan, Z.Q. New foliar pathogens of *Eucalyptus* from Australia and Indonesia. *Mycol. Res.* **1998**, *102*, 527–532. [CrossRef]
51. Tubaki, K. Studies on the Japanese *Hyphomycetes*. V. Leaf & stem group with a discussion of the classification of *Hyphomycetes* and their perfect stages. *J. Hattori Bot. Lab.* **1958**, *20*, 142–244.
52. Kang, J.C.; Crous, P.W.; Old, K.M.; Dudzinski, M.J. Non-conspecificity of *Cylindrocladium quinqueseptatum* and *Calonectria quinqueseptata* based on a β-tubulin gene phylogeny and morphology. *Can. J. Bot.* **2001**, *79*, 1241–1247. [CrossRef]
53. Decock, C.; Crous, P.W. *Curvicladium gen. nov.*, a new hyphomycete genus from French Guiana. *Mycologia* **1998**, *90*, 276–281. [CrossRef]
54. Guerber, J.C.; Correll, J.C. Characterization of *Glomerella acutata*, the teleomorph of *Colletotrichum acutatum*. *Mycologia* **2001**, *93*, 216–229. [CrossRef]
55. Nirenburg, H.I. A simplified method for identifying *Fusarium* spp. occurring on wheat. *Can. J. Bot.* **1981**, *59*, 1599–1609. [CrossRef]
56. Rayner, R.W. *A Mycological Colour Chart*; Commonwealth Mycological Institute: Kew, UK, 1970.
57. Cunningham, C.W. Can three incongruence tests predict when data should be combined? *Mol. Biol. Evol.* **1997**, *14*, 733–740. [CrossRef]
58. Pan, R.; Deng, Q.; Xu, D.; Meng, M.; Chen, W. First Report of Peanut Cylindrocladium Black Rot Caused by *Cylindrocladium parasiticum* in Fujian Province, Eastern China. *Plant Dis.* **2012**, *99*, 890. [CrossRef] [PubMed]

Journal of *Fungi*

MDPI

Article

The Early Terrestrial Fungal Lineage of *Conidiobolus*—Transition from Saprotroph to Parasitic Lifestyle

Andrii P. Gryganskyi [1,*], Yong Nie [2], Ann E. Hajek [3], Kathie T. Hodge [4], Xiao-Yong Liu [5], Kelsey Aadland [6], Kerstin Voigt [7], Iryna M. Anishchenko [8], Vira B. Kutovenko [9], Liudmyla Kava [9], Antonina Vuek [9], Rytas Vilgalys [10], Bo Huang [11,*] and Jason E. Stajich [6]

1 UES Inc., Dayton, OH 45432, USA
2 School of Civil Engineering and Architecture, Anhui University of Technology, Ma'anshan 243002, China; nieyong@ahut.edu.cn
3 Department of Entomology, Cornell University, Ithaca, NY 14853, USA; aeh4@cornell.edu
4 Section of Plant Pathology & Plant-Microbe Biology, School of Integrative Plant Science, Cornell University, Ithaca, NY 14853, USA; kh11@cornell.edu
5 College of Life Sciences, Shandong Normal University, Jinan 250014, China; liuxiaoyong@im.ac.cn
6 Department of Microbiology and Plant Pathology, University of California, Riverside, CA 92521, USA; kelseya@ucr.edu (K.A.); jason.stajich@ucr.edu (J.E.S.)
7 Department of Molecular and Applied Microbiology, Leibniz Institute for Natural Product Research and Infection Biology, 07745 Jena, Germany; kerstin.voigt@hki-jena.de
8 M.G. Kholodny Institute of Botany, National Academy of Sciences of Ukraine, 02000 Kyiv, Ukraine; ira_anishchenko@hotmail.com
9 Agrobiological Department, National University of Life and Environmental Sciences of Ukraine, 03041 Kyiv, Ukraine; virakutovenko@gmail.com (V.B.K.); kavalyuda@ukr.net (L.K.); antoninavuek@mail.ru (A.V.)
10 Department of Biology, Duke University, Durham, NC 27708, USA; fungi@duke.edu
11 Anhui Provincial Key Laboratory for Microbial Pest Control, Anhui Agricultural University, Hefei 230036, China
* Correspondence: andrii.gryganskyi@gmail.com (A.P.G.); bhuang@ahau.edu.cn (B.H.)

Citation: Gryganskyi, A.P.; Nie, Y.; Hajek, A.E.; Hodge, K.T.; Liu, X.-Y.; Aadland, K.; Voigt, K.; Anishchenko, I.M.; Kutovenko, V.B.; Kava, L.; et al. The Early Terrestrial Fungal Lineage of *Conidiobolus*—Transition from Saprotroph to Parasitic Lifestyle. *J. Fungi* **2022**, *8*, 789. https://doi.org/10.3390/jof8080789

Academic Editors: Lei Cai and Cheng Gao

Received: 2 June 2022
Accepted: 26 July 2022
Published: 28 July 2022

Abstract: Fungi of the *Conidiobolus* group belong to the family Ancylistaceae (Entomophthorales, Entomophthoromycotina, Zoopagomycota) and include over 70 predominantly saprotrophic species in four similar and closely related genera, that were separated phylogenetically recently. Entomopathogenic fungi of the genus *Batkoa* are very close morphologically to the *Conidiobolus* species. Their thalli share similar morphology, and they produce ballistic conidia like closely related entomopathogenic Entomophthoraceae. Ballistic conidia are traditionally considered as an efficient tool in the pathogenic process and an important adaptation to the parasitic lifestyle. Our study aims to reconstruct the phylogeny of this fungal group using molecular and genomic data, ancestral lifestyle and morphological features of the conidiobolus-like group and the direction of their evolution. Based on phylogenetic analysis, some species previously in the family Conidiobolaceae are placed in the new families Capillidiaceae and Neoconidiobolaceae, which each include one genus, and the Conidiobolaceae now includes three genera. Intermediate between the conidiobolus-like groups and Entomophthoraceae, species in the distinct *Batkoa* clade now belong in the family Batkoaceae. Parasitism evolved several times in the *Conidiobolus* group and Ancestral State Reconstruction suggests that the evolution of ballistic conidia preceded the evolution of the parasitic lifestyle.

Keywords: ballistic conidia; entomopathogenicity; evolution; ancestral state reconstruction

1. Introduction

The subphylum Entomophthoromycotina Humber is comprised of over 300 species which occupy various ecological niches, from saprotrophs to pathogens of insects [1]. Furthermore, some species parasitize a wide range of other hosts from different kingdoms of life: mushroom fruit bodies, fern gametophytes, protists, nematodes, millipedes, spiders, reptiles, and other animals, including humans [2]. Most of the entomopathogenic species are highly specialized to their hosts, and the efficiency of the infection even in closely related arthropod species is very low [3]. However, there are several species of pathogen-generalists that are able to infect the hosts from a wide range of insect families, such as *Batkoa major* [4]. The hallmarks of most fungi of this group are ballistic conidia, which are dispersed from the cadavers of deceased insects and infect new victims. Some of the entomophthoralean fungi can efficiently control the populations of harmful insects in natural and agricultural ecosystems and are thought to be useful as biocontrol agents [5].

Species of *Conidiobolus* sensu lato are saprobic or pathogenic fungi with forcibly discharged globose conidia, simple phototropic conidiophores, and secondary conidia similar to the primary, of a different shape, or absent. They have been considered a basal group in the subphylum Entomophthoromycotina. It is clear that their simple morphology has masked their phylogenetic diversity. Humber [6] moved several species to *Batkoa* based on their micromorphology and karyology. Recently, Nie et al. [7] reduced *Conidiobolus* sensu lato from 80 to 37 species and assigned the remaining species to three newly described genera: *Capillidium* B. Huang & Y. Nie, *Microconidiobolus* B. Huang & Y. Nie, and *Neoconidiobolus* B. Huang & Y. Nie. Cai et al. [8] described an allied genus *Azygosporus* B. Huang & Y. Nie as a close relative of *Conidiobolus* sensu stricto. This group of morphologically similar fungi belonging to *Conidiobolus* sensu lato may exceed one hundred species, and discovery of new species continues [9].

Most conidiobolus-like fungi are saprotrophs that dwell in organic detritus, and therefore grow well on artificial media under laboratory conditions. They are ubiquitous soil inhabitants, are easy to isolate from a variety of different soils [10], and are common in fungal culture collections across the world. Some of the species have been recorded infecting various insect species, and these are often known as having broad host ranges (ARSEF, ATCC and CBS web pages). Two species are nematode parasites, five can infect other fungi, lichens, and mosses, and nearly a dozen are found in or on dead or living arthropods, mostly insects (Table S1). Some *Conidiobolus* species can infect warm-blooded animals, including humans, and these animal diseases are called conidiobolomycoses [11,12]. Together with morphologically similar species of *Batkoa*, which include insect pathogens of wide and narrow host ranges, species of *Conidiobolus* sensu lato might serve as good examples of the evolutionary transition from a saprotrophic lifestyle to highly specialized parasitism on arthropods, as exhibited in the family Entomophthoraceae which originated from conidiobolus-like ancestors. We see the evolutionary trajectory of entomophthoralean fungi as follows: organic detritus, litter, and soil –> dead insects –> living insects: wide host range, weak pathogens -> living insects: narrow range, strong pathogens [8,9].

Entomophthoralean fungi might acquire this ability to parasitize insects through the process of their evolution as primarily soil inhabitants, first growing on insect cadavers in litter and soil and later infecting living insects. Infecting nematodes, other fungi and lichens, and plants might evolve as secondary adaptations, and therefore are not major evolutionary paths. Sporulation of this fungal group plays an essential role in dissemination and pathogenicity processes. It is diverse with the major spore type being ballistic conidia, with several ejection types (Figure 1).

The goal of our study was to describe the lifestyle of the ancestors of conidiobolus-like fungi, and thus the econiche occupied by the ancestors of Entomophthoromycotina. Our hypothesis is that these ancestors were soil saprotrophs, which over evolutionary time acquired the ability to use dead and later living insects as the substrate for their growth, development, and dissemination. A second goal of our study was to evaluate whether a hallmark feature of Entomophthoromycotina, forcibly discharged conidia, was an adaptation to a parasitic lifestyle, or whether it originated earlier evolutionarily.

We also removed species of *Conidiobolus* sensu lato from the family Ancylistaceae which has an atypical type genus *Ancylistes*—pathogens of algae. Relationships of this genus will be possible to resolve based on genome sequences of its species (T. James, personal communication). Based on the data obtained through the phylogenetic reconstruction, we elevated the ranks of the genera *Batkoa*, *Capillidium*, *Conidiobolus* sensu stricto, and *Neoconidiobolus* to the family level. Our study aims to organize these morphologically similar yet genetically diverse fungal groups based on revealing their evolutionary trajectories.

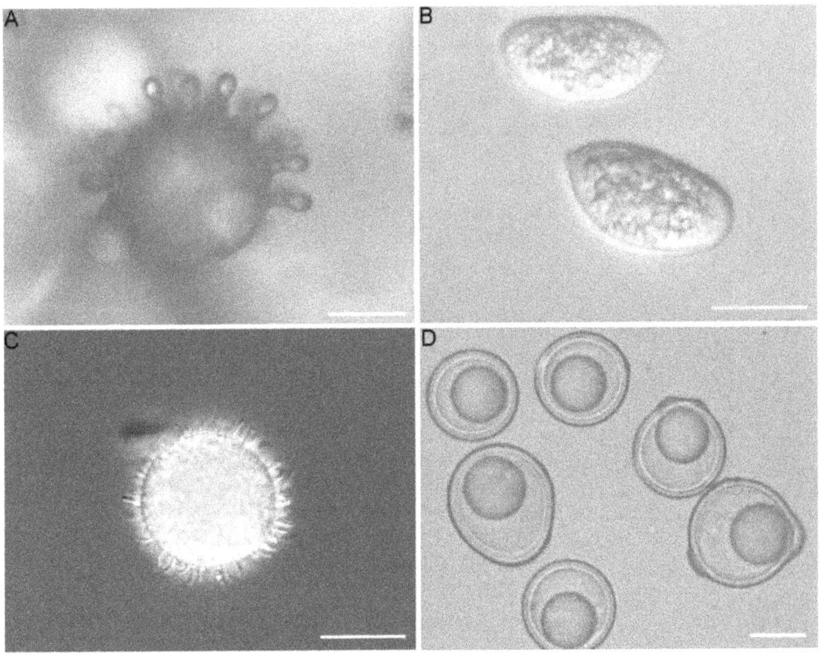

Figure 1. (**A**). *Conidiobolus coronatus* (microconidia); (**B**). *Capillidium macrocapilliconidium* (capilliconidia) [13]; (**C**). *Conidiobolus coronatus* (villose conidia); (**D**). *Conidiobolus brefeldianus* (zygospores); Scale bars: (**A–D**) = 20 μm.

2. Materials and Methods

2.1. Phylogenetic Reconstruction

We used molecular data (single copy genes and 17 available genomic/transcriptomic datasets) of 41 species of the genera *Batkoa*, *Capillidium*, *Conidiobolus*, *Microconidiobolus*, and *Neoconidiobolus* in our study. All of these species morphologically resemble *Conidiobolus*, and often have similar lifestyles. We considered the size of the genera as small if they had less than five taxa, and large for the genera with more than ten species. We also included the molecular data of species in the family Entomophthoraceae, including eight genomic and transcriptomic datasets. As outgroups we used sequences of *Schizangiella serpentis* and genomic data of *Syncephalis pseudoplumigaleata* S71, *Smittium culicis* ARSEF 9010, and *Coemansia reversa* NRRL 1564 [14].

The alignment with proteins and DNA was combined with partitions in NieGenomeComb.mfa. The partitions file is from Nie et al. [7] NieGenomeComb.partitions.nex and the IQTREE best merged partitions were used to create combined alignment NieGenomeComb.partitions.nex.best_scheme.nex, submitted to TreeBase (S29885).

Phylogenetic reconstruction was done by combining single-gene trees [7], which were rebuilt using a partitioned subset of overlapping taxa, and genome-scale coalescent-based species tree estimation using ASTRAL [15] to compare the multipartitioned concatenated topology. The resulting combined coalescent trees contained 400 protein-coding genes. Trees were tested for congruence and compared to the tree built with just genome and transcriptome data. A bootstrap analysis was done using 1000 repetitions. The phylogenetic tree was visualized using FigTree v 1.4.4 [16], saved as a PDF file and adjusted in Microsoft Office PowerPoint.

2.2. Ancestral States Reconstruction (ACR)

Maximum likelihood ancestral character states were reconstructed in R [17] using the rerooting method in phytools [18] under the equal rates, single parameter ER model. Two-character states were reconstructed—lifestyle and ballistic conidia, with each having four conditions. Conditions for lifestyle (only evaluating entomopathogenicity) were (0) non-pathogen, (1) pathogen of wide host range, (2) pathogen of moderate range, and (3) pathogen of narrow range. Conditions for ballistic conidia were (0) not forcible or absent, (1) forcible small and round (≤20 μm), (2) forcible large and round (>20 μm), and (3) forcible other than round. Additionally, ASR was performed using Mesquite [19] for the same two characters. Conditions for lifestyle (only evaluating entomopathogenicity) were (0) non-pathogen and (1) pathogen. Conditions for ballistic conidia were (0) not forcible or absent, (1) forcible. For each ancestral node on both the lifestyle and ballistic conidia trees, the log-likelihood was plotted as a pie graph of colors corresponding to the two or four conditions for each character.

2.3. Light Microscopy

Microscopic structures were observed under a BX51 light microscope and imaged with a DP25 microscope-camera system (Olympus Corporation, Tokyo, Japan) to obtain the photoplates. Images used for figures were processed with Adobe Photoshop CS3 Extended version 10.0 software (Adobe Systems, San Jose, CA, USA).

3. Results

3.1. Phylogenetic Reconstruction

Phylogenetic reconstruction based on the combined 4-gene (18S, 28S, EF-1α and mtSSU) and genome dataset revealed the polyphyletic composition of *Conidiobolus* sensu lato (Figure 2). It consists of several large clades, and is well supported, with bootstrap values between 90 and 100. This topology is also preserved when only genome data were used (Figure S1).

For most clades we have assigned new families. Basal to all sampled Entomophthoromycotina is a new family Conidiobolaceae fam. nov., described below. This is the group with the most molecular data: both single genes and genome data are available for 17 species. Adjacent and intermediate to it is a small group well supported by bootstrap values which includes representatives of the genus *Microconidiobolus*. It is represented only by single gene data and therefore it would be meaningful to collect more data to determine the affinities of this genus. Another big new family, Capillidiaceae fam. nov., consists of ten species, as shown below.

Another large group of *Conibiobolus* sensu lato, represented by 18 sets of molecular data for 11 species forms the Neoconidiobolaceae fam. nov., which in our phylogenetic reconstruction is located next to the entomopathogenic species of two families: Batkoaceae fam. nov., and Entomophthoraceae.

The Batkoaceae fam. nov. arises basal to the entomopathogenic Entomophthoraceae, on a branch that is well separated with strong bootstrap support. It includes several *Conidiobolus*-like taxa *C. obscurus* (=*Batkoa gigantea* or *B. obscura*), *C. pseudapiculatus*, and *Entomophaga conglomerata* (=*B. apiculata*).

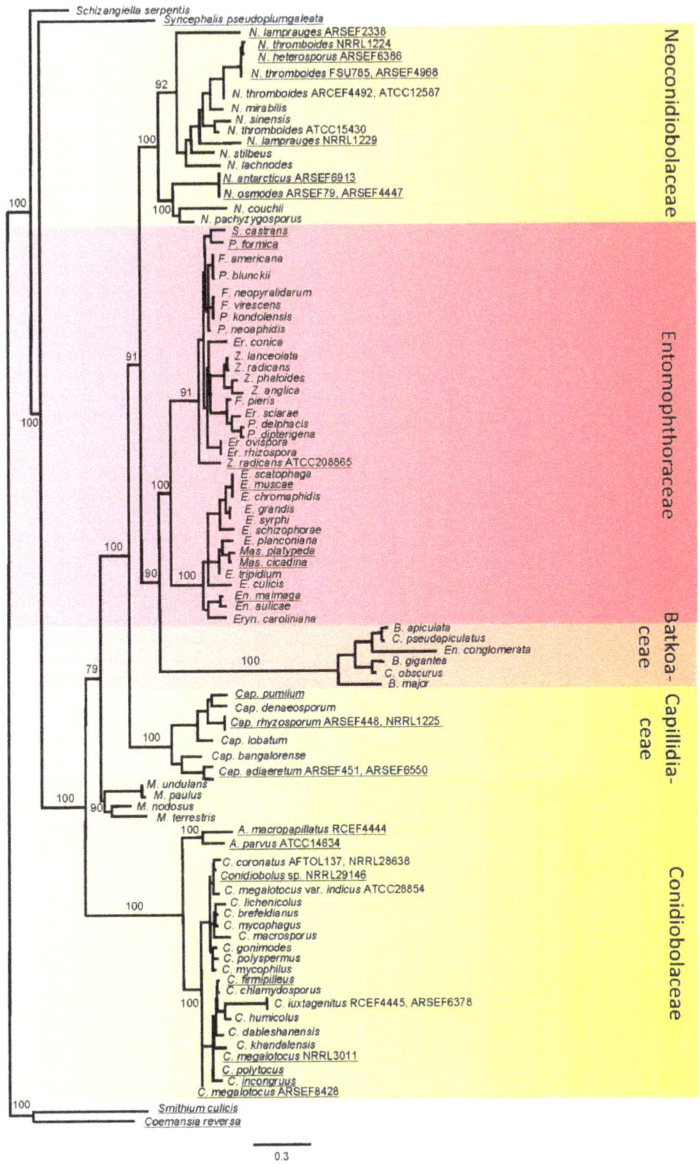

Figure 2. Maximum likelihood phylogenetic tree of conidiobolus-like fungi and related groups. Bootstrap values shown only at the basal nodes. In the genus *Batkoa,* the old taxonomic names of *B. apiculata* (*Conidiobolus pseudapiculatus* and *Entomophaga conglomerata*) and *B. gigantea* (*C. obscurus*) are shown to demonstrate the difficulties of species identification using only morphological features. Underlined are the taxa with existing NGS data. Strain names are shown for the taxa where more than one strain was used. Background colors are various shades of yellow for the mostly saprotrophic families of *Conidiobolus*-like fungi, orange for culturable entomopathogenic Batkoaceae, and red for strict entomopathogens of Entomophthoraceae. A.—*Azygosporus,* B.—*Batkoa,* C.—*Conidiobolus,* Cap.—*Capillidium,* E.—*Entomophthora,* En.—*Entomophaga,* Er.—*Erynia,* Eryn.—*Eryniopsis,* F.—*Furia,* M.—*Microconidiobolus,* Mas.—*Massospora,* N.—*Neoconodiobolus,* P.—*Pandora,* S.—*Strongwellsea,* Z.—*Zoophthora.*

3.2. Taxonomy

3.2.1. **Batkoaceae** Gryganskyi, A.E. Hajek & Stajich, fam. nov. [MB 844345]

Type genus: Batkoa Humber, Mycotaxon 34 (2): 446 (1989) (Humber 1989), [MB 25280].

Type species: Batkoa apiculata (Thaxter) Humber, Mycotaxon 34 (2): 446 (1989), [MB 135576].

=*Entomophthora apiculata* (Thaxt.) M.A. Gust., Kungliga Landbruks-Höngskolans Annaler 31: 131 (1965) [MB 330591].

=*Conidiobolus apiculatus* (Thaxt.) Remaud. & S. Keller Mycotaxon 11 (1): 330 (1980) [MB 118560].

Description: Mycelia of hyaline, septate, branching hyphae. Hyphal bodies hypha- or ameboid-like, subglobose to elongate, multinucleate, nuclei staining with aceto-orcein. Primary conidiophores simple, positively phototropic, bearing a single apical primary conidium. Primary conidia forcibly discharged, single-celled, multinucleate, globose, with prominent conical papilla. Replicative conidia similar and smaller than primary conidia. Chlamydospores globose, hyaline. Zygospores globose, hyaline or yellowish. Rhizoids present or absent. Obligate pathogens of insects.

Notes: Members of the Batkoaceae are entomopathogenic, infecting insects from various orders, ballistospore-forming fungi with a broad global distribution. Their cultures are easily to isolate on simple culture media.

Accepted species:

Batkoa amrascae S. Keller & Villac., Philippine Entomologist 11 (1): 81 (1997) [MB 313160].

Batkoa apiculata (Thaxt.) Humber, Mycotaxon 34 (2): 446 (1989) [MB 135576].

Batkoa cercopidis (S. Keller) B. Huang, Humber & K.T. Hodge, Mycotaxon 100: 231 (2007) [MB 510686].

Batkoa dysderci (Viegas) Humber, Mycotaxon 34 (2): 446 (1989) [MB 135577].

Batkoa gigantea (S. Keller) Humber, ibid. [MB 135578].

Batkoa hydrophila S. Keller, Sydowia 59 (1): 77 (2007) [MB 529508].

Batkoa limoniae (S. Keller) S. Keller, Nova Hedwigia 73 (1–2): 171 (2001) [MB 484564]

Batkoa major (Thaxter) Humber, Mycotaxon 34 (2): 446 (1989) [MB 135579].

Batkoa obscura (Hall & Dunn) Gryganskyi, comb. nov. [MB 844349].

Basionym: Entomophthora obscura I.M. Hall & P.H. Dunn, Hilgardia 27: 162 (1957) [MB 297265] = *Conidiobolus obscurus* (I.M. Hall & P.H. Dunn) Remaud.; S. Keller, Mycotaxon 11 (1): 331 (1980) [MB 118567].

Batkoa papillata (Thaxter) Humber, Mycotaxon 34 (2): 446 (1989) [MB 135580].

Batkoa pseudapiculata (S. Keller) B. Huang, Humber & K.T. Hodge, Mycotaxon 100: 231 (2007) [MB 510687].

3.2.2. **Capillidiaceae** Y. Nie, Stajich & K.T. Hodge, fam. nov. [MB 844346]

Type genus: Capillidium B. Huang & Y. Nie, MycoKeys 66: 62 (2020) [2] [MB 831596]

Type species: Capillidium heterosporum (Drechsler) B. Huang & Y. Nie, MycoKeys 66: 62 (2020) [MB 831601] = *Conidiobolus heterosporus* Drechsler, Am. J. Botany 40: 107 (1953) [MB 295472]

Description: Mycelia hyaline. Primary conidiophores simple, bearing single primary conidia. Primary conidia forcibly discharged, multinucleate, hyaline, globose, pyriform to obovoid. Two kinds of replicative conidia, the first similar and smaller than primary conidia, the second (capilliconidia) arise singly, off-axis at the top of slender, elongate conidiophores, and are not forcibly discharged. Zygospores present or absent, formed in axial alignment with conjugating segments, globose to subglobose, often smooth, sometimes rough, hyaline or yellowish.

Accepted species:

Capillidium adiaeretum (Drechsler) B. Huang & Y. Nie, MycoKeys 66: 61 (2020) [MB 831602]

Capillidium bangalorense (Sriniv. & Thirum.) B. Huang & Y. Nie, ibid. [MB 831607]

Capillidium denaeosporum (Drechsler) B. Huang & Y. Nie, ibid. [MB 831608]

Capillidium heterosporum (Drechsler) B. Huang & Y. Nie, ibid. [MB 831601]

Capillidium jiangsuense B. Huang & Y. Nie, MycoKeys 89: 146 (2022) [MB 842228]

Capillidium lobatum (Sriniv. & Thirum.) B. Huang & Y. Nie, MycoKeys 66: 62 (2020) [MB 831609]

Capillidium macrocapilliconidium B. Huang & Y. Nie, MycoKeys 89: 143 (2022) [MB 842227]

Capillidium pumilum (Drechsler) B. Huang & Y. Nie, MycoKeys 66: 61 (2020) [MB 831610]

Capillidium rhysosporum (Drechsler) B. Huang & Y. Nie, ibid. [MB 831611]

Capillidium rugosum (Drechsler) B. Huang & Y. Nie, ibid. [MB 842229].

3.2.3. **Conidiobolaceae** B. Huang, Stajich & K.T. Hodge, fam. nov. [MB 844347]

Type genus: Conidiobolus Bref., Untersuchungen aus dem Gesamtgebiete der Mykologie 4: 35 (1884) [20] [MB 20144]

Type species: Conidiobolus utriculosus Bref., Untersuchungen aus dem Gesamtgebiete der Mykologie 4: 35 (1884) [MB 144259]

Description: Mycelium of hyaline, septate, branching hyphae. Hyphae multinucleate, with small nuclei that do not stain with aceto-orcein. Simple hyphal conidiogenous cells each develop one apical conidium, a ballistospore that is forcibly discharged by rapid circumcissile rupture and papillar eversion. Detached conidia are hyaline, single-celled, more or less globose, with an everted blunt conical papilla where they were once attached to the conidiophore. Sexual zygospores, present or absent, formed in axial alignment with conjugating segments, globose to subglobose.

Notes: Members of the Conidiobolaceae are saprobic, ballistospore-forming fungi with a broad global distribution. They grow readily on simple culture media, typically as white to cream colonies, often forming satellite colonies derived from the asexual ballistospores. The primary ballistospores may germinate in three ways: to form a smaller ballistospore of similar form as the parent spore, to form a passively dispersed capilliconidium atop a long, attenuated stalk, or they may form a hyphal germ tube. Some species are able to opportunistically infect humans and other animals.

Conidiobolaceae B. Huang, Stajich & K.T. Hodge

Included genera (3) and species:

Azygosporus B. Huang & Y. Nie, MycoKeys 85: 165 (2021) [MB 840849]

Accepted species:

Azygosporus macropappilatus B. Huang & Y. Nie, ibid. [MB 840848]

Azygosporus parvus (Drechsler) B. Huang & Y. Nie, ibid. [MB 840850]

Conidiobolus sensu stricto according to B. Huang & Y. Nie 2020 [MB 20144]

Accepted species:

Conidiobolus coronatus (Costantin) Batko, Entomophaga 2: 129 (1964) [MB 283037]

Conidiobolus bifurcatus B. Huang & Y. Nie, MycoKeys 73: 137 (2021) [MB 831599]

Conidiobolus brefeldianus Couch, American Journal of Botany 26: 119 (1939) [MB 258852]

Conidiobolus dabieshanensis Y. Nie & B. Huang, Mycosphere 8 (7): 811 (2017) [MB 552756]

Conidiobolus gonimodes Drechsler, Mycologia 53: 292 (1961) [MB 328751]

Conidiobolus iuxtagenitus S.D. Waters & Callaghan, Mycological Research 93 (2): 223 (1989) [MB 135617]

Conidiobolus khandalensis Sriniv. & Thirum., Mycologia 54 (6): 692 (1963) [MB 328754]

Conidiobolus lichenicola Sriniv. & Thirum., Mycopathologia et Mycologia Applicata 36: 344 (1968) [MB 328755]

Conidiobolus lunulus D. Goffre, R.A. Humber & P.J. Folgarait, Mycologia: 131 (1): 56 (2020) [MB 834443] should be in this group by morphology.

Conidiobolus macrosporus Sriniv. & Thirum., Mycologia 59: 702 (1967) [MB 328757]

Conidiobolus mycophagus Sriniv. & Thirum., Sydowia 19 (1–6): 88 (1965) [MB 328759]

Conidiobolus mycophilus Sriniv. & Thirum., ibid. [MB 328760]

Conidiobolus polyspermus Drechsler, Mycologia 53: 279 (1961) [MB 328763]

Conidiobolus polytocus Drechsler, American Journal of Botany 42: 793 (1955) [MB 295480]

Conidiobolus taihushanensis B. Huang & Y. Nie, MycoKeys 73: 140 (2021) [MB 835124]

Conidiobolus utriculosus Bref., Untersuchungen aus dem Gesamtgebiete der Mykologie 4: 35 (1884) [MB 144259] HOLOTYPE SPECIES

Conidiobolus variabilis B. Huang & Y. Nie, MycoKeys 73: 142 (2021) [MB 835125]

Additional species to be considered: *C. chlamydosporus*, *C. firmipileus*, *C. humicolus*, *C. incongruous* and *C. megalotocus*.

Microconidiobolus B. Huang & Y. Nie, MycoKeys 66: 72 (2020) [MB 831597]

Accepted species:

Microconidiobolus nodosus (Sriniv. & Thirum.) B. Huang & Y. Nie, ibid. [MB 831624]

Microconidiobolus paulus (Drechsler) B. Huang & Y. Nie, ibid. [MB 831605]

Microconidiobolus terrestris (Sriniv. & Thirum.) B. Huang & Y. Nie, ibid. [MB 831625]

Additional species to be considered: *Microconidiobolus undulatus*.

3.2.4. **Neoconidiobolaceae** X.Y. Liu, Stajich & K.T. Hodge, fam. nov. [MB 844348]

Type genus: Neoconidiobolus B. Huang & Y. Nie, MycoKeys 66: 70 (2020) [MB 831598]

Type species: Neoconidiobolus thromboides (Drechsler) B. Huang & Y. Nie, MycoKeys 66: 70 (2020) [MB 831606] = *Conidiobolus thromboides* Drechsler, J. Washington Acad. Sci. 43: 38 (1953) [MB 295484]

Description: Mycelia hyaline. Primary conidiophores are simple or sometimes branched, positively phototropic, bearing a single apical primary conidium. Primary conidia forcibly discharged, multinucleate, hyaline, globose, pyriform to obovoid. Replicative conidia similar and smaller than primary conidia. Chlamydospores globose, formed terminally on hyphae or from globose cells by thickening of the wall. Zygospores formed in axial alignment with two conjugating segments, globose to ellipsoidal, smooth, hyaline, rarely pale yellowish.

Notes: Species of the family Neoconidiobolaceae resemble those of the Conidiobolaceae in lacking both microconidia and capilliconidia. All members in the clade of Neoconidiobolus share the following characteristics: forcibly discharged, hyaline, globose, pyriform to obovoid primary conidia. Two kinds of replicative conidia are produced: One is discharged, similar to and smaller than primary conidia; the other is elongate and forcibly discharged. Two types of resting spores are produced: zygospores and chlamydospores.

Accepted species:

Neoconidiobolus couchii (Sriniv. & Thirum.) B. Huang & Y. Nie, MycoKeys 66: 73 (2020) [MB 831626]

Neoconidiobolus kunyushanensis B. Huang & Y. Nie, Mycological Progress 20: 1233 (2021) [MB 831600]

Neoconidiobolus lachnodes (Drechsler) B. Huang & Y. Nie, MycoKeys 66: 73 (2020) [MB 831627]

Neoconidiobolus mirabilis (Y. Nie & B. Huang) B. Huang & Y. Nie, ibid. [MB 831628]

Neoconidiobolus osmodes (Drechsler) B. Huang & Y. Nie, ibid. [MB 831629]

Neoconidiobolus pachyzygosporus (Y. Nie & B. Huang) B. Huang & Y. Nie, ibid. [MB 831630]

Neoconidiobolus sinensis (Y. Nie, X.Y. Liu & B. Huang) B. Huang & Y. Nie, ibid. [MB 831631]

Neoconidiobolus stilbeus (Y. Nie & B. Huang) B. Huang & Y. Nie, ibid. [MB 831632]

Neoconidiobolus stromoideus (Sriniv. & Thirum.) B. Huang & Y. Nie, ibid. [MB 831633]

Neoconidiobolus thromboides (Drechsler) B. Huang & Y. Nie, ibid. [MB 831606]

Neoconidiobolus vermicola (J.S. McCulloch) B. Huang & Y. Nie, ibid. [MB 831634]

4. Ancestral State Reconstruction

The ancestors of entomophthoralean fungi were with high probability saprotrophic, as with most of their extant basal lineages *Azygosporus*, *Conidiobolus*, *Capillidium*, *Microconidiobolus* and *Neoconidiobolus*. The ability to infect insects was developed in various groups and multiple times during the evolution of these fungi. Of the 18 species included in the Conidiobolaceae, *C. macrosporus* is host specific pathogen. Of the 11 species included in *Neoconidiobolus*, two are pathogens with broader host ranges (*Neoconidiobolus osmodes* and *Neoconidiobolus thromboides*). None of the species included in Capillidiaceae (three *Microconidiobolus* and six *Capillidium*) are pathogens. Also, ancestors of the entomophthoralean fungi became entomopathogenic and didn't lose this ability further on; all extant members of this group (families Batkoaceae and Entomophthoraceae) are entomopathogenic (Table S1).

A similar evolutionary trajectory was reconstructed for another character—ballistic conidia. This character evolved very early in Entomophthoromycotina, in the ancestors with saprotrophic lifestyles, possibly as an adaptation for spore dissemination (Figure 1), and this character is seen throughout the phylogenetic tree for these groups. Ballistic conidia preceded the entomopathogenic lifestyle since this ability is attributed to the earlier (lower) basal nodes on the phylogenetic tree (Figure 3 and Figure S2). This character was inherited by the early entomopathogens and served as an efficient tool for insect infection as taxa evolved further.

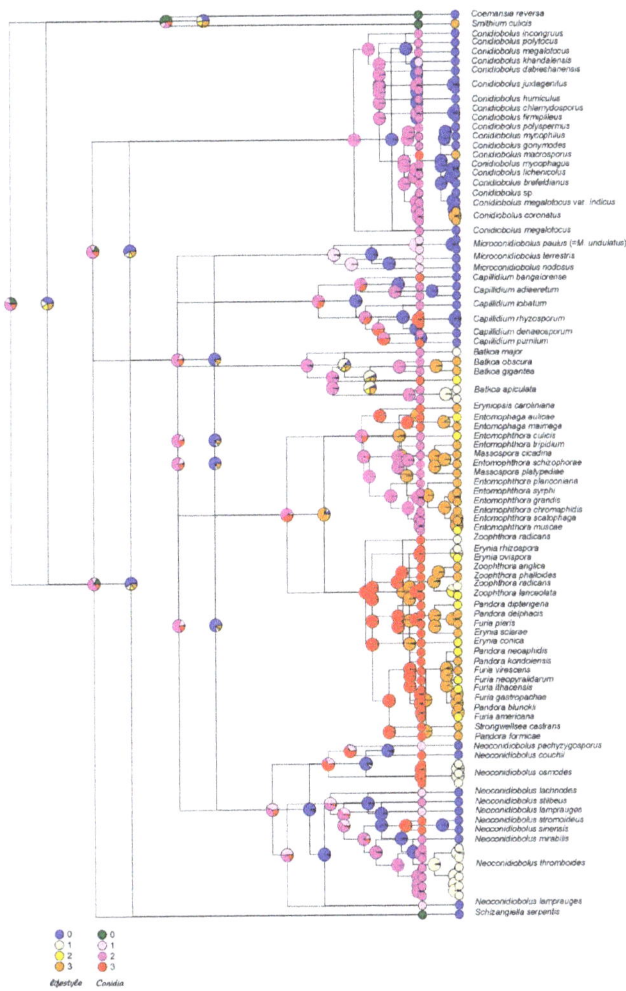

Figure 3. Ancestral State Reconstruction suggests ballistic conidia evolutionarily precedes the parasitic lifestyle. Reconstruction is done for four states. Lifestyle: non-pathogenic to insects, grow well on artificial media (0, blue), insect pathogen of wide range, growing well on artificial media (1, beige), insect pathogen of moderate range (2, yellow), insect pathogen of narrow range (3, orange). Ballistic conidia: not forcible, absent, or unknown (0, green), forcible small and round ≤20 μm (1, pink), forcible large and round (>20 μm) in diameter (2, purple), presence of forcible conidia other than round (3, red).

5. Discussion

Phylogenetic reconstruction suggests polyphyletic origins of conidiobolus-like fungi, and not a single origin. The polyphyletic origin of this composite fungal group was already suggested by our previous works [7,21–24], and one of the tasks of this study was to determine the corresponding taxonomic levels of the main branches. Alternatively, the whole group should be treated as a single family (the Entomophthoraceae), and all genera could be subsumed into it. However, we believe that the polyphyletic origin of most branches at the base of Entomophthorales also suggests independent origins of the newly described families.

There are some differences in the placement of certain groups and taxa. In the study of Nie et al. [7], the genera *Conidiobolus* and *Microconidiobolus* were grouped together, while on our tree they are located separately. Therefore, we haven't assigned a corresponding unique taxonomic level to *Microconidiobolus* that is higher than genus, as we do not yet have genomic or transcriptomic data. Species of *Microconidiobolus* differ morphologically in producing microconidia and this is not characteristic of other members of the Conidiobolaceae. Taxonomic level needs more taxa and more genomes involved for phylogenetic reconstruction.

Also, there are some discrepancies inside the family Entomophthoraceae. The issue with having *Massospora* and two species of *Entomophthora* together in one clade is caused by the lack of the genes for both *Entomophthora* species. Their missing data might cause some interference with better gene sampling for *Massospora*. In any case, *Massospora* is next to *Entomophthora*, as in most previous phylogenetic reconstructions.

Neoconidiobolus thromboides appears in our phylogenetic reconstruction not as a single clade. This might be an indication that this taxon represents a species complex or group of species, and reflects the diversity of this fungal group, which is similar to another complex species, *Conidiobolus coronatus*. The resolution of these two species complexes possibly containing several cryptic species needs better genome sampling. The presence of *Neoconidiobolus heterosporus* inside it can provide a hint. In any case, the spore sizes and especially hosts and substrates of *Neoconidiobolus thromboides* are definitely worth studying in more detail with support from genome references.

The division of Entomophthoromycotina reflects the gradual evolutionary switch from saprotrophy to the parasitic lifestyle, with multiple origins. While all Entomophthoraceae are insect pathogens, conidiobolus-like fungi occupy more diverse ecological niches. They developed the adaptation of infecting insects several times during their evolution. The family Batkoaceae presents intermediate placement between these two groups. While all known species in the Batkoaceae are insect pathogens, they are easy to culture on artificial nutrient media like *Azygosporus*, *Capillidium*, *Micro-*, *Neo-* and *Conidiobolus* isolates. Only a few species of Entomophthoraceae grow well under laboratory conditions in pure culture, mostly requiring special nutritive media, and the majority of species in this fungal lineage are not yet culturable [6]. Typical entomopathogenic fungi of the family Entomophthoraceae have rather atypically large genomes for the Zoopagomycota, being over 600 Mb. Sizes of known genomes from species in the *Conidiobolus* group range from 25 to 90 Mb and this is rather more typical for the saprotrophs and insect symbionts in terrestrial fungal lineages [25].

Similarly to the parasitism of insects, parasitism of other groups of living organisms like lichens, nematodes, and mushrooms developed independently several times in different branches among saprotrophic conidiobolus-like fungi over their evolutionary trajectories. Ancestral state reconstruction for lifestyle and ballistic conidia appearance suggests that these two features coexisted for a long-time during evolution, and ballistic conidia were present in the early ancestors of the whole group already, before pathogenic lifestyles were adopted. Ballistic conidiospores were already present at the very beginning of Entomophthorales evolution in saprotrophic lineages. Although ballistic conidia were described as a hallmark of entomopathogens, they could not be an adaptation to the parasitic lifestyle, namely an adaptation to infect insects. However, the ability of these fungi to eject spores for significant distances, possibly along with light sensing mechanisms, and involving attachment to the substrate due to conidiophore content [26], essentially increased their chances of finding new habitats or infecting hosts, when they had switched to pathogenic lifestyles. Ballistic conidia evolved earlier and was successfully adopted by newly evolved entomopathogens.

Our study aims to add some structure to the very unclear taxonomic positioning of conidiobolus-like fungi. Despite many morphological similarities, they have polyphyletic origins and occupy various econiches ranging from saprotrophy to entomopathogenicity. We hope that our study will help the researchers of this fungal group to assess relatedness between these families and with the entomopathogenic Entomophthoraceae, and also predict the ecological niches of new species of these families, as further discoveries continue.

Supplementary Materials: The following supporting information can be downloaded at: https://www.mdpi.com/article/10.3390/jof8080789/s1, Figure S1: Phylogeny based on 27 genomes of Conidiobolus-like fungi; Figure S2: Ancestral state reconstruction of insect-pathogenic lifestyle and ballistic conidia (gray), which evolutionary precedes parasitic lifestyle (black) for two states: present (black) and absent (white); Table S1: Taxa used for phylogenetic and ancestral state reconstructions, their pores and lifestyle.

Author Contributions: A.E.H., A.P.G., B.H., J.E.S., K.T.H., K.V. and R.V., Conceptualization; A.E.H., A.P.G., J.E.S. and K.T.H., methodology; A.P.G., J.E.S., K.A. and Y.N., software; A.P.G., B.H., I.M.A., K.T.H., L.K., V.B.K. and Y.N., validation; A.E.H., A.P.G., B.H., I.M.A., K.T.H., L.K. and V.B.K., formal analysis; A.P.G., A.V., B.H., I.M.A., J.E.S., L.K., K.A., V.B.K., X.-Y.L. and Y.N., investigation; B.H., J.E.S., K.V., X.-Y.L. and Y.N., resources; A.P.G., B.H., I.M.A., J.E.S., L.K., V.B.K., X.-Y.L. and Y.N., data curation; A.E.H., A.P.G., J.E.S., K.A., K.T.H. and Y.N., writing; A.P.G., A.V., I.M.A., L.K., V.B.K. and Y.N., visualization; A.P.G., B.H., J.E.S., K.V., R.V., X.-Y.L. and Y.N., supervision; A.P.G., B.H., J.E.S., R.V., X.-Y.L. and Y.N., project administration; B.H., J.E.S., R.V., X.-Y.L. and Y.N., funding acquisition. All authors have read and agreed to the published version of the manuscript.

Funding: This material is based upon work supported by the grants of National Science Foundation DEB-1441715 to K.A. and J.E.S., and National Natural Science Foundation of China 31900008, 30770008 and 31970009 to Y.N., B.H. and X.L.

Institutional Review Board Statement: Not applicable.

Informed Consent Statement: Not applicable.

Data Availability Statement: All data underlying the presented study will be made available at an online data repository.

Conflicts of Interest: The authors declare that they have no conflicts of interest.

References

1. Humber, R.A. Evolution of entomopathogenicity in rungi. *J. Invertebr. Pathol.* **2008**, *98*, 262–266. [CrossRef]
2. Möckel, L.; Meusemann, K.; Misof, B.; Schwartze, V.U.; De Fine Licht, H.H.; Voigt, K.; Stielow, B.; de Hoog, S.; Beutel, R.G.; Buellesbach, J. Phylogenetic revision and patterns of host specificity in the fungal Subphylum Entomophthoromycotina. *Microorganisms* **2022**, *10*, 256. [CrossRef] [PubMed]
3. Mullens, B.A. Cross-transmission of *Entomophthora muscae* (Zygomycetes: Entomophthoraceae) among naturally infected muscoid fly (Diptera: Muscidae) hosts. *J. Invertebr. Pathol.* **1989**, *53*, 272–275. [CrossRef]
4. Gryganskyi, A.P.; Golan, J.; Hajek, A.E. Season-long infection of diverse hosts by the entomopathogenic fungus *Batkoa Major*. *PLoS ONE* **2022**, *17*, e0261912. [CrossRef]
5. Jaronski, S.T. Chapter 11—Mass production of entomopathogenic fungi: State of the art. In *Mass Production of Beneficial Organisms*; Morales-Ramos, J.A., Rojas, M.G., Shapiro-Ilan, D.I., Eds.; Academic Press: San Diego, CA, USA, 2014; pp. 357–413. ISBN 978-0-12-391453-8.
6. Humber, R.A. Synopsis of a revised classification for the Entomophthorales (Zygomycotina). *Mycotaxon* **1989**, *34*, 441–460. [CrossRef]
7. Nie, Y.; Yu, D.S.; Wang, C.F.; Liu, X.Y.; Huang, B. A taxonomic revision of the genus *Conidiobolus* (*Ancylistaceae*, Entomophthorales): Four clades including three new genera. *Mycokeys* **2020**, *66*, 55–81. [CrossRef] [PubMed]
8. Cai, Y.; Nie, Y.; Zhao, H.; Wang, Z.M.; Zhou, Z.Y.; Liu, X.Y.; Huang, B. *Azygosporus* gen. nov., a synapomorphic clade in the family *Ancylistaceae*. *MycoKeys* **2021**, *85*, 161–172. [CrossRef]

9. Goffre, D.; Jensen, A.B.; Lopez Lastra, C.C.; Humber, R.A.; Folgarait, P.J. *Conidiobolus lunulus*, a new entomophthoralean species isolated from leafcutter ants. *Mycologia* **2021**, *113*, 56–64. [CrossRef]
10. Stajich, J.E. *Andrii Shows How to Culture Conidiobolus from Leaf Litter*; ZyGoLife: Gainesville, FL, USA, 2015; Available online: https://www.youtube.com/watch?v=Ea6RoU9kbvg (accessed on 25 July 2022).
11. Kimura, M.; Yaguchi, T.; Sutton, D.A.; Fothergill, A.W.; Thompson, E.H.; Wickes, B.L. Disseminated human conidiobolomycosis due to *Conidiobolus lamprauges*. *J. Clin. Microbiol.* **2011**, *49*, 752–756. [CrossRef]
12. Vilela, R.; Mendoza, L. Human pathogenic Entomophthorales. *Clin. Microbiol. Rev.* **2018**, *31*, e00014-18. [CrossRef]
13. Nie, Y.; Zhao, H.; Wang, Z.M.; Zhou, Z.Y.; Liu, X.Y.; Huang, B. Two new species in *Capillidium* (*Ancylistaceae*, Entomophthorales) from China, with a proposal for a new combination. *MycoKeys* **2022**, *89*, 139–153. [CrossRef]
14. Grigoriev, I.V.; Nordberg, H.; Shabalov, I.; Aerts, A.; Cantor, M.; Goodstein, D.; Kuo, A.; Minovitsky, S.; Nikitin, R.; Ohm, R.A.; et al. The Genome Portal of the Department of Energy Joint Genome Institute. *Nucleic Acids Res.* **2012**, *40*, D26–D32. [CrossRef] [PubMed]
15. Mirarab, S.; Reaz, R.; Bayzid, M.D.S.; Zimmermann, T.; Swenson, M.S.; Warnow, T. ASTRAL: Genome-scale coalescent-based species tree estimation. *Bioinformatics* **2014**, *30*, 541–548. [CrossRef] [PubMed]
16. Rambaut, A. *Title of the Software, FigTree v1.3.1*; Institute of Evolutionary Biology, University of Edinburgh: Edinburgh, UK, 2010; Available online: http://tree.bio.ed.ac.uk/software/figtree/ (accessed on 25 July 2022).
17. R Core Team. R: A Language and Environment for Statistical Computing. 2021. Available online: http://www.R-project.org/ (accessed on 25 July 2022).
18. Revell, L.J. Phytools: An R package for phylogenetic comparative biology (and other things). *Methods Ecol. Evol.* **2012**, *3*, 217–223. [CrossRef]
19. Madisson, W.P.; Madisson, D.R. Mesquite: A Modular System for Evolutionary Analysis. Version 3.04. 2011. Available online: https://www.mesquiteproject.org/ (accessed on 25 July 2022).
20. Brefeld, O. *Conidiobolus utriculosus* and *minor*. *Unters. Gesammtgebiete Mykol.* **1884**, *6*, 35–78.
21. Gryganskyi, A.P.; Humber, R.A.; Smith, M.E.; Miadlikowska, J.; Wu, S.; Voigt, K.; Walther, G.; Anishchenko, I.M.; Vilgalys, R. Molecular phylogeny of the Entomophthoromycota. *Mol. Phylogenet. Evol.* **2012**, *65*, 682–694. [CrossRef]
22. Gryganskyi, A.P.; Humber, R.A.; Smith, M.E.; Hodge, K.; Huang, B.; Voigt, K.; Vilgalys, R. Phylogenetic lineages in Entomophthoromycota. *Persoonia* **2013**, *30*, 94–105. [CrossRef]
23. Nie, Y.; Wang, L.; Cai, Y.; Tao, W.; Zhang, Y.J.; Huang, B. Mitochondrial genome of the entomophthoroid fungus *Conidiobolus heterosporus* provides insights into evolution of basal fungi. *Appl. Microbiol. Biotechnol.* **2019**, *103*, 1379–1391. [CrossRef] [PubMed]
24. Nie, Y.; Zhao, H.; Wang, Z.M.; Zhou, Z.Y.; Liu, X.Y.; Huang, B. The gene rearrangement, loss, transfer and deep intronic variation in mitochondrial genomes of *Conidiobolus*. *Front. Microbiol.* **2021**, *12*, 765733. [CrossRef]
25. Gryganskyi, A.P.; Golan, J.; Muszewska, A.; Idnurm, A.; Dolatabadi, S.; Kutovenko, V.B.; Kutovenko, V.O.; Gajdeczka, M.T.; Anishchenko, I.M.; Pawlowska, J. Sequencing the genomes of the terrestrial fungal lineages, What Have We Learned? *Microorganisms* **2022**, *unpublished*.
26. Elya, C.; De Fine Licht, H.H. The genus *Entomophthora*: Bringing the insect destroyers into the twenty-first century. *IMA Fungus* **2021**, *12*, 34. [CrossRef]

Journal of *Fungi*

MDPI

Article

Colletotrichum Species Associated with Anthracnose Disease of Watermelon (*Citrullus lanatus*) in China

Zhen Guo [1,2], Chao-Xi Luo [2], Hui-Jie Wu [1], Bin Peng [1], Bao-Shan Kang [1], Li-Ming Liu [1], Meng Zhang [3] and Qin-Sheng Gu [1,*]

[1] Key Laboratory of Fruit and Cucurbit Biology, Zhengzhou Fruit Research Institute, Chinese Academy of Agricultural Sciences, Zhengzhou 450009, China; guozhenguorui@163.com (Z.G.); wuhuijie@caas.cn (H.-J.W.); pengbin@caas.cn (B.P.); kangbaoshan@caas.cn (B.-S.K.); liuliming@caas.cn (L.-M.L.)

[2] Key Laboratory of Horticultural Plant Biology, Ministry of Education, College of Plant Science and Technology, Huazhong Agricultural University, Wuhan 430070, China; cxluo@mail.hzau.edu.cn

[3] Department of Plant Pathology, Henan Agricultural University, Zhengzhou 450002, China; zm2006@126.com

* Correspondence: guqinsheng@caas.cn

Abstract: *Colletotrichum* species are important plant pathogens, causing anthracnose in virtually every crop grown throughout the world. However, little is known about the species that infect watermelon. A total of 526 strains were isolated from diseased watermelon samples of eight major watermelon growing provinces in China. Phylogenetic analyses using seven loci (ITS, *gadph*, *chs-1*, *his3*, *act*, *tub2*, and *gs*) coupled with morphology of 146 representative isolates showed that they belonged to 12 known species of *Colletotrichum*, including *C. aenigma*, *C. chlorophyti*, *C. fructicola*, *C. jiangxiense*, *C. karstii*, *C. magnum*, *C. nymphaeae*, *C. nigrum*, *C. orbiculare*, *C. plurivorum*, *C. sojae*, and *C. truncatum* and three new species, here described as *C. citrulli*, *C. kaifengense*, and *C. qilinense*. *Colletotrichum orbiculare* was the dominant species. Pathogenicity tests revealed that all isolates of the species described above were pathogenic, with *C. magnum* and *C. kaifengense* being the most aggressive to leaves and fruits, respectively. This is the first report of *C. aenigma*, *C. chlorophyti*, *C. fructicola*, *C. jiangxiense*, *C. nymphaeae*, *C. nigrum*, *C. plurivorum*, and *C. sojae* on watermelon. These findings shed light on the *Colletotrichum* spp. involved in watermelon anthracnose and provide useful information for implementing effective control of watermelon anthracnose in China.

Keywords: multi-locus phylogeny; pathogenicity; plant pathogen; taxonomy

Citation: Guo, Z.; Luo, C.-X.; Wu, H.-J.; Peng, B.; Kang, B.-S.; Liu, L.-M.; Zhang, M.; Gu, Q.-S. *Colletotrichum* Species Associated with Anthracnose Disease of Watermelon (*Citrullus lanatus*) in China. *J. Fungi* **2022**, *8*, 790. https://doi.org/10.3390/jof8080790

Academic Editors: Cheng Gao and Lei Cai

Received: 8 July 2022
Accepted: 25 July 2022
Published: 28 July 2022

1. Introduction

Colletotrichum is the most common and important genus of plant pathogenic fungi, saprobes, and endophytes [1–3]. Species of *Colletotrichum* spp. infect numerous plant crops worldwide, e.g., apple (*Malus pumila*), chili (*Capsicum* spp.), coffee (*Coffea* spp.), grape (*Vitis vinifera*), longan (*Dimocarpus longan*), mango (*Mangifera indica*), olive (*Canarium album*), orange (*Citrus* spp.), pear (*Pyrus* spp.), peach (*Prunus persica*), strawberry (*Fragaria ananassa*), and tea (*Camellia* spp.) [4–21].

During the first half of the 20th century, many species of the plant-pathogenic fungal genus *Colletotrichum* were defined, relying on the hosts from which they were originally isolated [22]. Based on his revision of the genus primarily on morphological characteristics in culture, Von Arx reduced approximately 750 species to only 11 [23]. However, identifying *Colletotrichum* species is challenging due to the instability of their morphological characteristics, which are affected by experimental methods and conditions [1,24,25]. Along with the increased availability of DNA sequencing technology, a very large volume of DNA sequence data has been generated, allowing fungal taxonomy through phylogenetics, including genealogical concordance [26,27]. Following the adoption of multi-locus phylogenetic analysis together with morphological characteristics, the classification and species concepts in *Colletotrichum* taxa worldwide have changed significantly [1,28–33]. Nearly all acknowledged species studied were grouped into 16 *Colletotrichum* species complexes, and more than 16 singleton species have been identified [9,30–43].

Watermelon (*Citrullus lanatus* (Thunb.) Matsum. et Nakai), belongs to the xerophytic genus *Citrullus* Schrad. ex Eckl. et Zeyh of the botanical family *Cucurbitaceae* and is one of the most important commercial crops worldwide [44]. It is the third-most widespread fruit crop after apple and orange in China (China Agriculture Research System Statistical Data 2018). In 2019, the total area harvested in China was 1,471,581 ha, with a yield of 60,861,241 t, accounting for 60.6% of the world's production of watermelon [45]. However, anthracnose diseases are one of the important factors, limiting the commercial production of watermelon [26]. Anthracnose displays spots and blights on the aerial parts of plant in the fields. Owing to anthracnose, watermelon production can be reduced by 5–20%, and even causes no harvest [46]. Moreover, it is an important post-harvest pathogen and becomes active after the fruit has been stored or appears on the market shelf. Owing to *Colletotrichum* disease, up to 100% of the fruit can be lost during storage and transportation [2].

Cucurbitaceae crops, especially watermelon (*Citrullus lanatus*), melon (*Cucumis melo*), and cucumber (*Cucumis sativus*), are important plant hosts of *Colletotrichum* spp. [47]. Seven *Colletotrichum* species have been reported in watermelon, belonging to *C. gloeosporioides*, *C. gloeosporioides* f.sp. cucurbitae, *C. karstii*, *C. magnum*, *C. scovillei*, *C. orbiculare*, and *C. truncatum* [32,37,38,48–51]. First described from cucumber, the species *C. gloeosporioides* f.sp. cucurbitae is widely regarded as a synonym for *C. orbiculare* [52]. However, *Colletotrichum* species associated with watermelon remain largely unresolved, with only six individual species identified. Moreover, these studies used a limited number of samples and areas, therefore species diversity might have been underestimated.

The objectives of the present study were as follows: (i) to identify the *Colletotrichum* species causing anthracnose in watermelon in the major production provinces in China based on multi-locus phylogenetic analyses combined with morphological characterization; (ii) to evaluate the pathogenicity of the different *Colletotrichum* species.

2. Materials and Methods

2.1. Collection and Isolates

The survey was conducted from 2018 to 2020 in watermelon fields in various geographical areas of China. Samples were collected from eight provinces (Henan, Jiangsu, Zhejiang, Jilin, Liaoning, Hebei, Jiangxi, and Hainan) (Table 1). Anthracnose symptoms on watermelon leaves were small circular or irregular spots, pale brown in the center with medium to dark brown margins, while rows of fusiform brown sunken spots were formed on stems (Figure 1a–g). Three symptom types were observed on fruits: (1) small round sunken spots (Figure 1k,l) and (2) large brown or black sunken rot lesions, forming orange conidia under humid conditions (Figure 1h–j,m,n).

Table 1. A list of *Colletotrichum* isolates collected from *Citrullus lanatus* in China.

Location	Host Tissue	Year	Latitude and Longitude	Number of Isolates
Jiyang, Hainan	leaf and stem	2018	109.58° E, 18.28° N	20
Tongxu, Henan	fruit	2019	114.47° E, 34.48° N	8
Weishi, Henan	fruit	2019	114.29° E, 34.35° N	16
Xiangfu, Henan	fruit	2019	114.35° E, 34.77° N	6
Jinming, Henan	fruit	2019	114.31° E, 34.87° N	19
	fruit	2020		26
Zhongmou, Henan	leaf and stem	2019	113.98° E, 34.71° N	25
	leaf and stem	2020		9

Table 1. *Cont.*

Location	Host Tissue	Year	Latitude and Longitude	Number of Isolates
Xinzheng, Henan	fruit	2019	113.74° E, 34.39° N	8
Fengcheng, Liaoning	leaf and stem	2019	124.06° E, 40.45° N	24
Ningbo, Zhejiang	leaf and stem	2019	121.54° E, 29.87° N	18
Taikang, Hennan	fruit	2019	114.62° E, 34.11° N	44
Ganzhou, Jingxi	leaf and stem	2020	115.78° E, 25.60° N	48
Xinxiang, Henan	leaf, stem and fruit	2020	113.85° E, 35.30° N	91
Enshi, Hubei	leaf and stem	2020	109.47° E, 30.29° N	52
Fugou, Henan	fruit	2020	114.39° E, 34.05° N	31
Yancheng, Jiangsu	fruit	2020	120.15° E, 33.34° N	22
Gaoan, Jiangxi	leaf and stem	2020	115.37° E, 28.41° N	35
Tongyu, Jilin	leaf	2020	123.08° E, 44.80° N	24
Total				526

Figure 1. Symptoms of watermelon anthracnose on leaves, stems, and fruits in the field. (**a–d**) Symptoms on leaves of *Citrullus lanatus*; (**e–g**) symptoms on stems of *Citrullus lanatus*; (**h–n**) symptoms on fruits of *Citrullus lanatus*.

Samples were surface-sterilized by dipping in 75% ethanol for 10 s and rinsing three times with sterile distilled water. Tissue pieces (5 mm) were excised from areas neighboring the diseased tissue. Excised tissue was placed onto potato dextrose agar (PDA, 20% diced potato, 2% dextrose, and 1.5% agar, and distilled water) after surface sterilization (70% ethanol for 15 s, 5% NaOCl for 5 min, washed three times in sterile water and dried on sterilized filter paper) [53]. The plates were incubated at 27 °C in the dark for 5 d. Six single colonies of each strain were isolated using the single-spore isolation method [54]. All isolates were stored in 25% glycerol at −80 °C. Type specimens of the new species from this study were deposited in the Mycological Herbarium, Institute of Microbiology, Chinese Academy of Sciences, Beijing, China (HMAS), and ex-type living cultures were deposited in the China General Microbiological Culture Collection Center (CGMCC), Beijing, China.

2.2. DNA Extraction, PCR Amplification, and Sequencing

Isolates were transferred to fresh PDA plates and incubated at 27 °C for 7–19 days. Genomic DNA of 146 representative isolates was extracted using the Ezup Column Fungi Genomic DNA Purification Kit (Sangon Biotech, Shanghai, China). Seven loci, including the internal transcribed spacer regions and intervening 5.8S nrRNA gene (ITS) and partial sequences of the glyceraldehyde-3-phosphate dehydrogenase (*gadph*), chitin synthase 1 (*chs-1*), actin (*act*), histone3 (*his3*), beta-tubulin (*tub2*), and glutamine synthetase (*gs*) genes were amplified and sequenced using the primer pairs ITS-1F [55] + ITS-4 [56], GDF1 + GDR1 [57], CHS-354R + CHS-79F [58], ACT-512 F + ACT-783R [58], CYLH3F + CYLH3R [59], T1 [60] +Bt2b [61] and GSF1 + GSR1 [57], respectively. PCR was performed in a total volume of 25 μL. The PCR mixture contained 1 μL 20× diluted genomic DNA, 12.5 μL 2 × Rapid Taq Master Mix (Vazyme, Nanjing, China), and 0.2 μM of each primer. The PCR reactions were mostly as described by Woudenberg et al. (2009) [62], but were modified by using an annealing temperature of 58 °C for *act*, *gadph*, and *chs-1*, 56 °C for *tub2*, *his3*, and *gs*. PCR amplifications were performed in Mastercycler X50 (Eppendorf, Hamburg, Germany). PCR amplicons were purified and sequenced at the Sangon Biotech (Shanghai, China) Company, Ltd. Forward and reverse primers were assembled to obtain consensus sequences using DNAMAN (v. 9.0; Lynnon Biosoft, San Ramon, USA). Sequences generated in this study were deposited in GenBank (Supplementary Table S1).

2.3. Phylogenetic Analyses

The DNA sequences were aligned using MAFFT v. 7 [63]. and manually adjusted using MEGA (version 11.0.10, Mega Limited, Auckland, New Zealand) where necessary. Multiple sequences of ITS, *gadph*, *act*, *tub2*, *chs-1*, and *gs* were concatenated using Sequence-Matrix 1.8 [64]. A Markov Chain Monte Carlo (MCMC) algorithm was used to generate phylogenetic trees with Bayesian inference (BI) using MrBayes v. 3.2.6 [65] for the combined sequence datasets. Best-fit models of nucleotide substitution for each gene partition were determined by Akaike information criterion (AIC) of MrModeltest v. 2.3 [66] (Table 2). Two analyses of four MCMC chains were conducted from random trees with 6×10^6 generations for *C. orbiculare* and *C. gloeosporioides* species complexes, and 2×10^6 generations for the *C. boninense*, *C. orchidearum*, *C. magnum* species complexes, and *C. acutatum*, *C. truncatum* species complexes and the related reference species involved in the same phylogenetic tree. The analyses were sampled every 1000 generations.

Alignment gaps were treated as missing and data all characters were unordered and of equal weight. The first 25% of trees were discarded as the burn-in phase of each analysis and posterior probabilities were determined from the remaining trees. Convergence of all parameters (i.e., effective sample sizes >200) was visually confirmed using Tracer 1.7.1 [67]. The maximum parsimony analyses (MP) were performed on the multi-locus alignment using PAUP (Phylogenetic Analysis Using Parsimony) v. 4.0a169 [68]. Phylogenetic trees were generated using the heuristic search option with 1000 random sequence additions and tree bisection and reconstruction (TBR) as the branch-swapping algorithm. Maxtrees were set up to 5000, branches of zero length were collapsed, and all multiple parsimonious trees were saved. Clade stability was assessed by bootstrap analysis with 1000 replicates. Afterward, tree length (TL), consistency index (CI), retention index (RI), rescaled consistency index (RC), and homoplasy index (HI) were calculated for the resulting tree. Furthermore, maximum likelihood (ML) analyses were implemented on the multi-locus alignments using the IQ-TREE 2 [69]. Clade stability was assessed using a bootstrap analysis with 1000 replicates. Phylogenetic trees were visualized in FigTree v. 1.4.4 [70]. The alignments and phylogenetic trees were deposited in TreeBASE (submission number: 29157).

The phylogenetically related ambiguous species were analyzed using the Genealogical Concordance Phylogenetic Species Recognition (GCPSR) model with a pairwise homoplasy index (PHI) test as described by Quaedvlieg et al. (2014) [26]. The PHI test was performed in SplitsTree4 [71–73] to determine the recombination level within phylogenetically closely related species using a six-locus concatenated dataset (ITS, *gadph*, *chs-1*, *his3*, *act*, and *tub2*). Pairwise homoplasy index results below 0.05 indicated significant recombination in the dataset. The relationship between closely related species was visualized by constructing a splits graph.

Table 2. Nucleotide substitution models used in the phylogenetic analyses.

Gene	ITS	*gadph*	*chs-1*	*his3*	*act*	*tub2*	*gs*
Gloeosporioides clade	GTR + I	HKY + I	K80 + G	GTR + G	GTR + I	SYM + G	–
Boninense clade	SYM + I + G	HKY + G	K80 + G	GTR + I+G	GTR + G	K80 + I	–
Orbiculare clade	GTR	SYM	SYM + I	GTR + G	HKY + G	HKY	GTR + G
Orchidearum clade	GTR + I	HKY	GTR + G	HKY + G	HKY	HKY + I	–
Magnum clade	GTR + G	HKY + G	GTR + I	HKY + I	HKY + G	GTR + G	–
Acutatum clade and other taxa	GTR + I + G	HKY + I + G	GTR + I+G	–	HKY + I + G	HKY + I + G	–

2.4. Morphological Analysis

Mycelial plugs (5-mm diam) from the margin of actively growing cultures were transferred on PDA, oatmeal agar (OA) [53], and synthetic nutrient-poor agar medium (SNA) [74] and incubated at 27 °C in the dark. Colony characteristics were noted after 9 d, and colony diameters were measured daily for 3 d, 5 d, and 7 d to calculate their mycelial average growth rates (mm/d). Additionally, the shape, color, and size of acervali, conidia, conidiophores, asci, ascospores, and seta were observed using light microscopy (Nikon SMZ-1500 and Olympus BX51, Japan or Leica TCS SP5, Germany). Moreover, to determine their sizes, 30 conidia, seta, or ascospores were measured. The formation of conidial appressoria was induced by dropping a conidial suspension in 1% glucose (10^7 conidia/mL; 50 μL) on a concavity slide, placed inside plates containing moistened filter papers with sterile distilled water, and then incubated at 27 °C in darkness. After 48 h, the sizes of 30 conidial appressoria were measured.

2.5. Prevalence

To determine the prevalence of *Colletotrichum* species in the sampled provinces, *Colletotrichum* species were isolated from infected watermelon organs (leafs, stems, or fruits). The isolation rate (IR) for each species was calculated using the formula, IR% = (Ns/Nt) × 100, where Ns is the number of isolates from the same species and Nt is the total number of isolates from each sample-collected province [11,14].

2.6. Pathogenicity Tests

Sixteen representative isolates of each *Colletotrichum* species were used in the pathogenicity tests on detached leaves of *Citrullus lanatus* cv. Hongheping (5–6 true leaves). Healthy leaves were collected from plants growing in pots in a greenhouse. The leaves were washed with tap water, then submerged in 75% ethanol for 30 s, washed three times with sterile water, and finally air-dried on sterilized filter paper. The leaves were inoculated using the wound/drop and non-wound/drop inoculation methods [75,76]. A drop of 10 μL spore suspension (10^6 conidia/mL) was individually placed onto the left side of the leaf after wounding once by pin-pricking the upper surface with a sterilized needle (insect pin, 0.5 mm diam). A 10 μL drop of sterile water was placed on the right side of the same leaf as a control. The process was repeated with unwounded leaves. The leaves were incubated at 27 °C and 100% humidity in the dark. The infection rate was calculated 4 d post-inoculation (dpi) using the formula PI (%) = Nf/Np × 100, where Nf is the number of infected points and Np is the total number of inoculated points [19]. All isolates were tested in triplicate.

Pathogenicity tests were also performed on detached mature watermelon fruits of *Citrullus lanatus* cv. Motong in triplicate. Healthy-looking fruits were collected from plants growing in the greenhouse. The fruits were washed with tap water, then submerged in 75% ethanol for 30 s and washed three times with sterile water, finally air-dried on sterilized filter paper. Wound/drop and non-wound/drop inoculation methods were used [28,75,76]. For the wound inoculation, a drop of 10 μL spore suspension (10^6 conidia/mL) was placed directly onto fruit surfaces after wounding once by pin-pricking with a sterilized needle (5 mm deep). In the non-wound inoculation, the same spore suspension was placed directly on the unwounded watermelon fruit skin. Sterile water was used as the control. The fruits were incubated at 27 °C and at 100% humidity in the dark. The infection rate was calculated at 6 dpi using the formula PI (%) = Nf/Np × 100, where Nf is the number of infected points and Np is the total number of inoculated points [19].

3. Results

3.1. Collection of Watermelon Anthracnose Samples and Strain Isolation

Watermelon anthracnose was common in all the fields and provinces surveyed. A total of 224 diseased watermelon samples (146 leaves, 57 stems, and 21 fruits) were collected from eight provinces in China. From them, 526 *Colletotrichum* strains were isolated based on the colony morphologies on PDA and ITS sequence data (Table 1). Based on megablast searches in GenBank using ITS sequences, the colony morphological characteristics on PDA, and morphological characteristics conidia isolates were primarily divided into eight groups. Of those, 146 representative isolates were selected for further analysis.

3.2. Multi-Locus Phylogenetic Analyses

The 146 representative isolates and related *Colletotrichum* species including the outgroup were subjected to multi-locus phylogenetic analyses (Supplementary Tables S1 and S2) with 7-locus phylogenetic analyses (ITS, *gadph*, *chs-1*, *his3*, *act*, *tub2* and *gs*) for *C. orbiculare* species complexes, 6-locus phylogenetic analyses (ITS, *gadph*, *chs-1*, *his3*, *act*, *and tub2*) for those belonging to the *C. gloeosporioides*, *C. boninense*, *C. magnum*, and *C. orchidearum* species complexes, or with 5-locus phylogenetic analyses (ITS, *gadph*, *chs-1*, *his3*, *act*, and *tub2*) for other species complexes of without available *his3* sequences. The consensus tree obtained from maximum parsimony and maximum likelihood analyses confirmed the tree topology obtained with Bayesian inference (Figures 2–7).

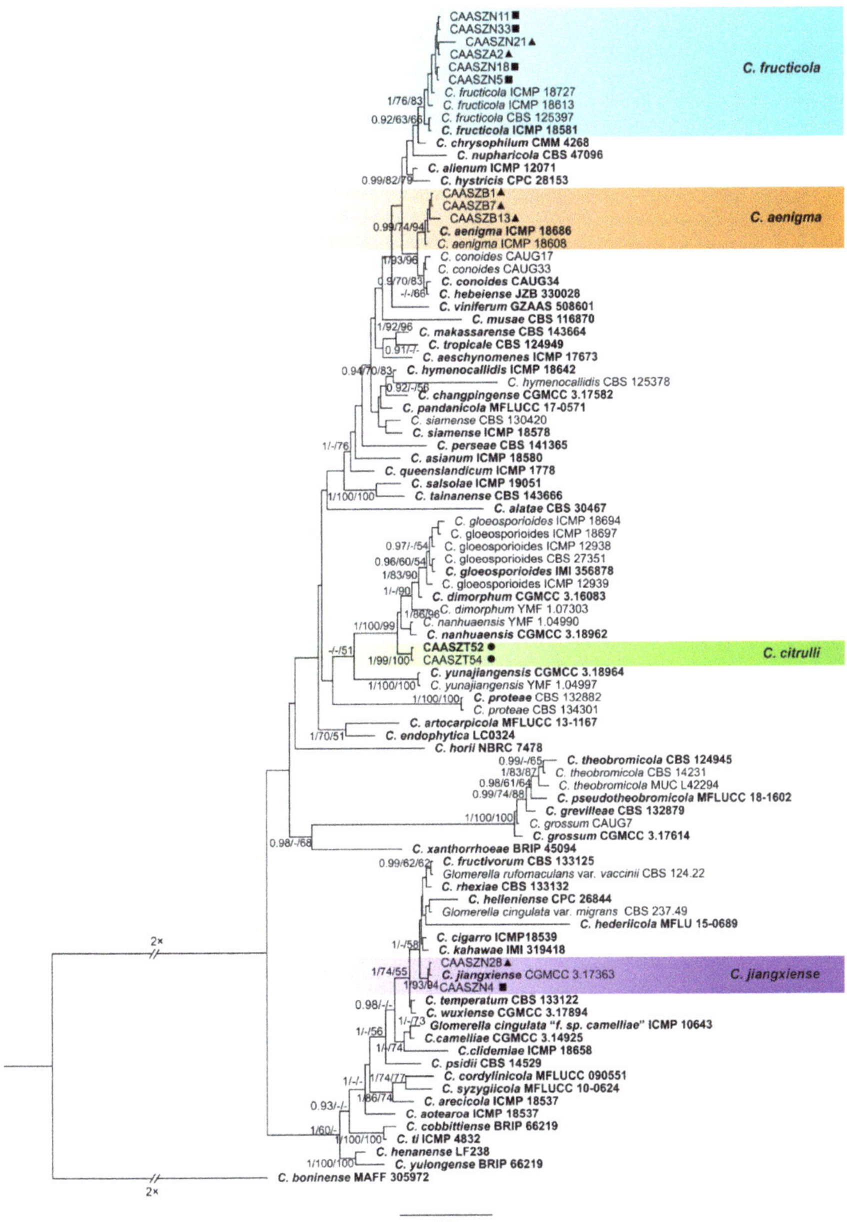

Figure 2. A Bayesian inference phylogenetic tree of 92 isolates in the *C. gloeosporioides* species complex. The species *C. boninense* (MAFF 305972) was selected as an outgroup. The tree was built using concatenated sequences of the ITS, *gadph*, *chs-1*, *his3*, *act*, and *tub2* genes. Bayesian posterior probability (PP $\geq$ 0.90), MP bootstrap support values (MP $\geq$ 60%), and iqtree bootstrap support values (ML $\geq$ 50%) were shown at the nodes (PP/MP/ML). Ex-type isolates are in bold. Colored blocks indicate clades containing isolates from *Citrullus lanatus* in this study; triangles indicate strains isolated from leaves, rectangle indicate strains isolated from stems, circles indicate strains isolated from fruits. The scale bar indicates 0.02 expected changes per site.

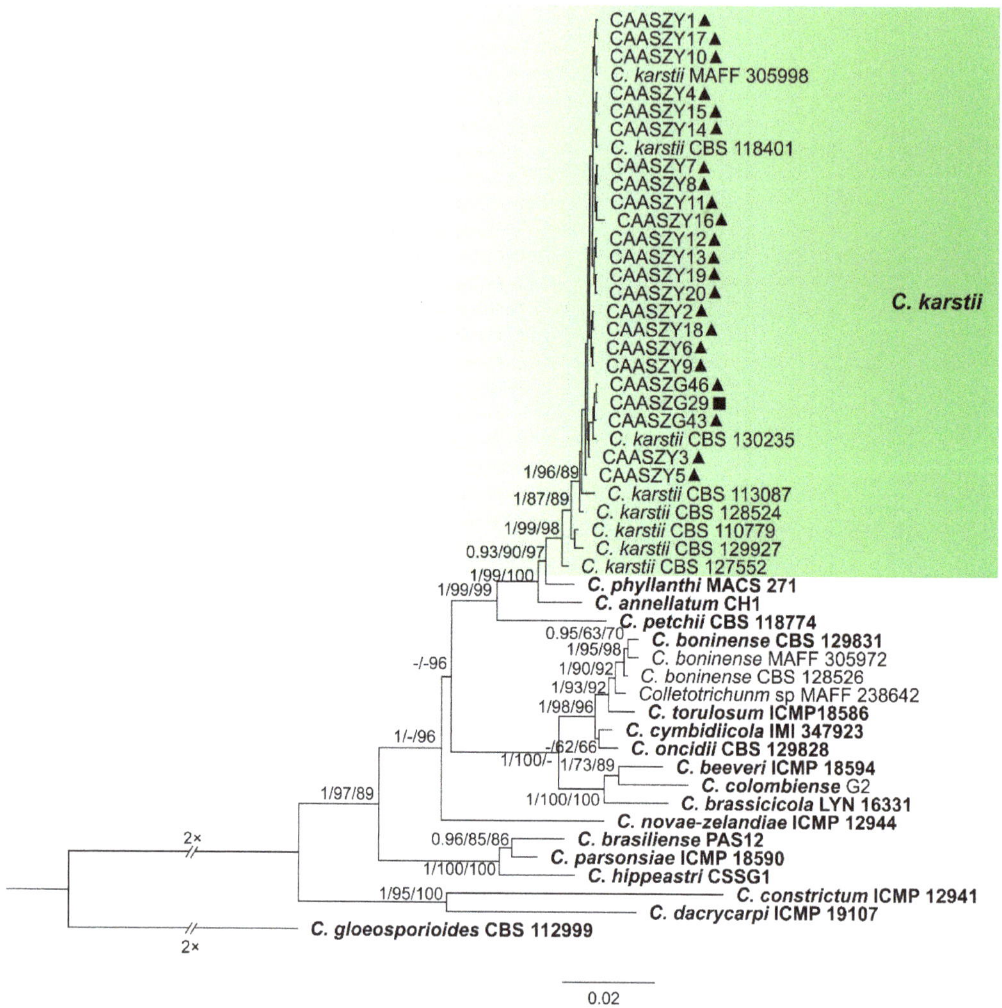

Figure 3. A Bayesian inference phylogenetic tree of 50 isolates in the *C. boninense* species complex. The species *C. gloeosporioides* (CBS 112999) was selected as an outgroup. The tree was built using concatenated sequences of the ITS, *gadph*, *chs-1*, *his3*, *act*, and *tub2* genes. Bayesian posterior probability (PP ≥ 0.90), MP bootstrap support values (MP ≥ 60%), and iqtree bootstrap support values (ML ≥ 50%) were shown at the nodes (PP/MP/ML). Ex-type isolates are in bold. Colored blocks indicate clades containing isolates from *Citrullus lanatus* in this study; triangles indicate strains isolated from leaves, rectangle indicate strains isolated from stems. The scale bar indicates 0.02 expected changes per site.

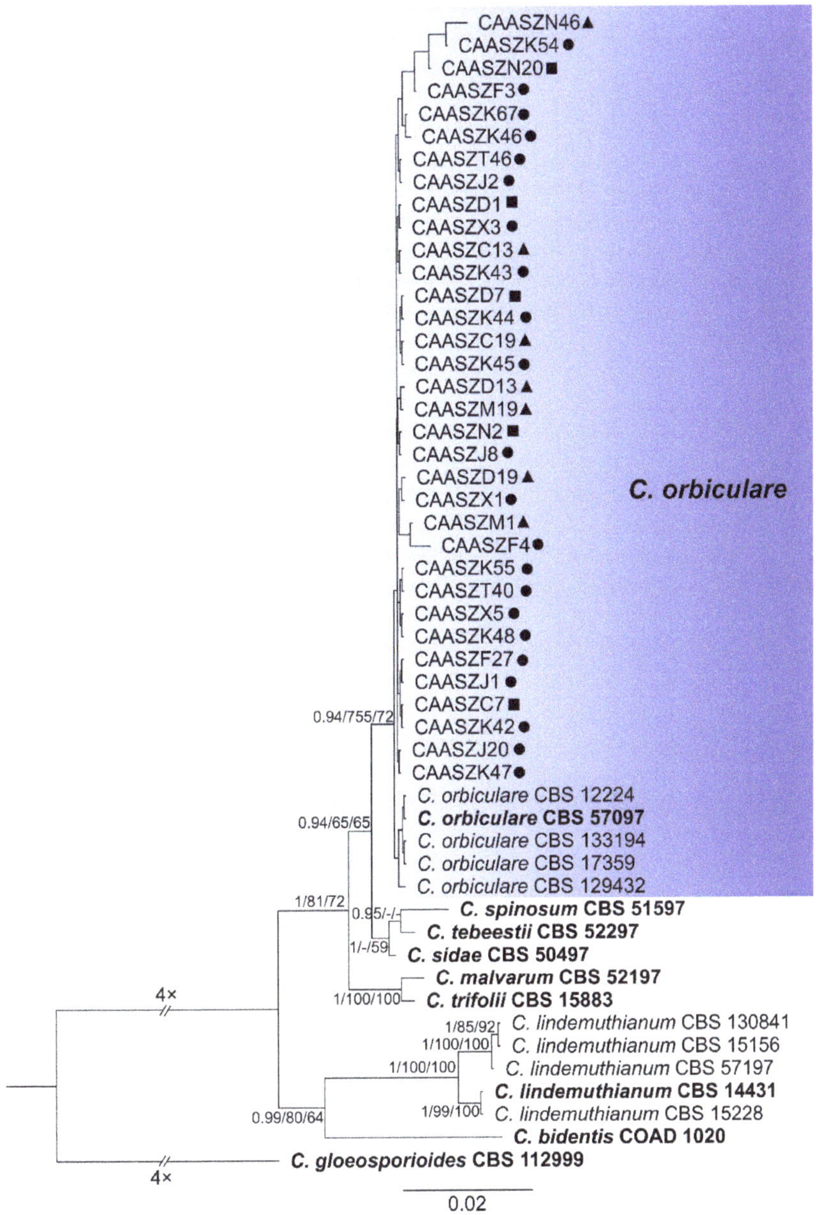

Figure 4. A Bayesian inference phylogenetic tree of 50 isolates in the *C. orbiculare* species complex. The species *C. gloeosporioides* (CBS 112999) was selected as an outgroup. The tree was built using concatenated sequences of the ITS, *gadph*, *chs-1*, *his3*, *act*, *tub2*, and *GS* genes. Bayesian posterior probability (PP $\geq$ 0.90), MP bootstrap support values (MP $\geq$ 60%), and iqtree bootstrap support values (ML $\geq$ 50%) were shown at the nodes (PP/MP/ML). Ex-type isolates are in bold. Colored blocks indicate clades containing isolates from *Citrullus lanatus* in this study; triangles indicate strains isolated from leaves, rectangle indicate strains isolated from stems, circles indicate strains isolated from fruits. The scale bar indicates 0.02 expected changes per site.

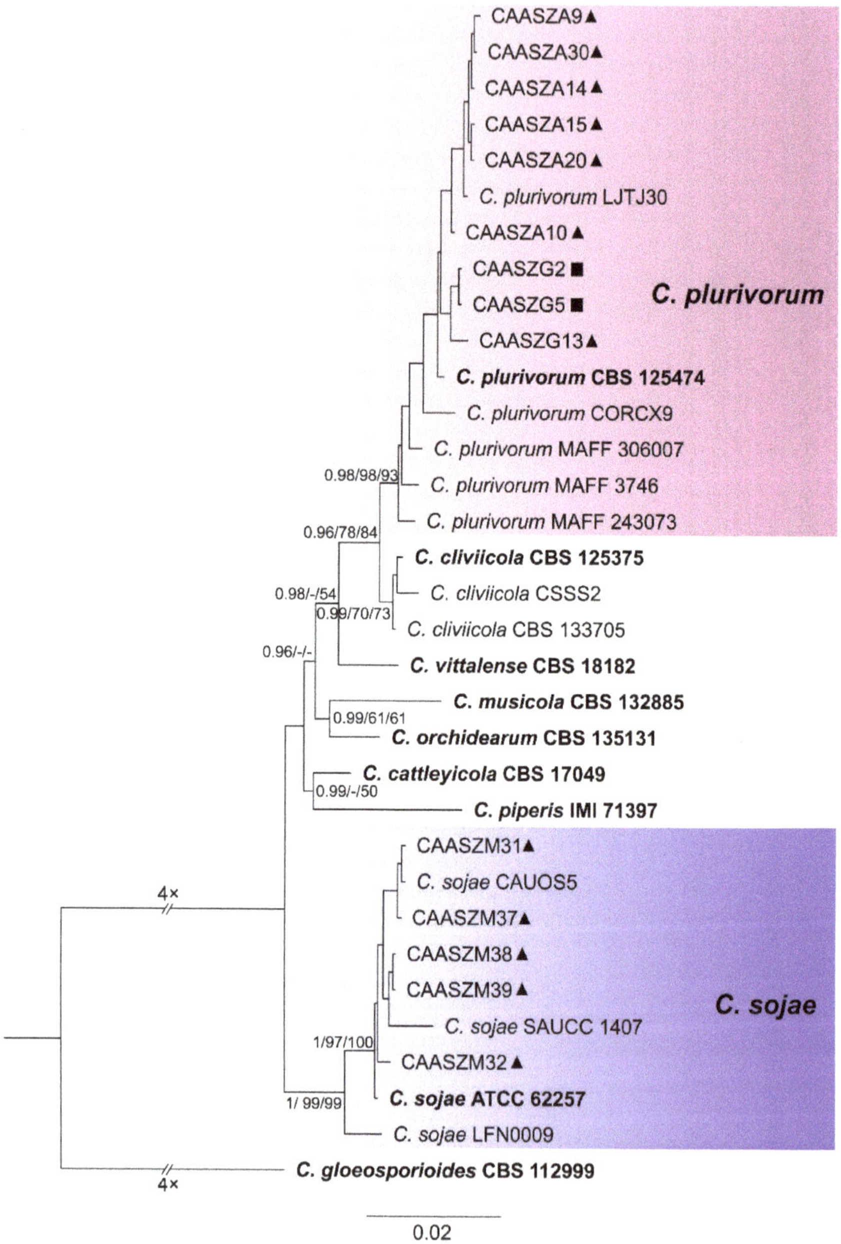

Figure 5. A Bayesian inference phylogenetic tree of 32 isolates in the *C. orchidearum* species complex. The species *C. gloeosporioides* (CBS 112999) was selected as an outgroup. The tree was built using concatenated sequences of the ITS, *gadph*, *chs-1*, *his3*, *act*, and *tub2* genes. Bayesian posterior probability (PP $\geq$ 0.90), MP bootstrap support values (MP $\geq$ 60%), and iqtree bootstrap support values (ML $\geq$ 50%) were shown at the nodes (PP/MP/ML). Ex-type isolates are in bold. Colored blocks indicate clades containing isolates from *Citrullus lanatus* in this study; triangles indicate strains isolated from leaves, rectangle indicate strains isolated from stems. The scale bar indicates 0.02 expected changes per site.

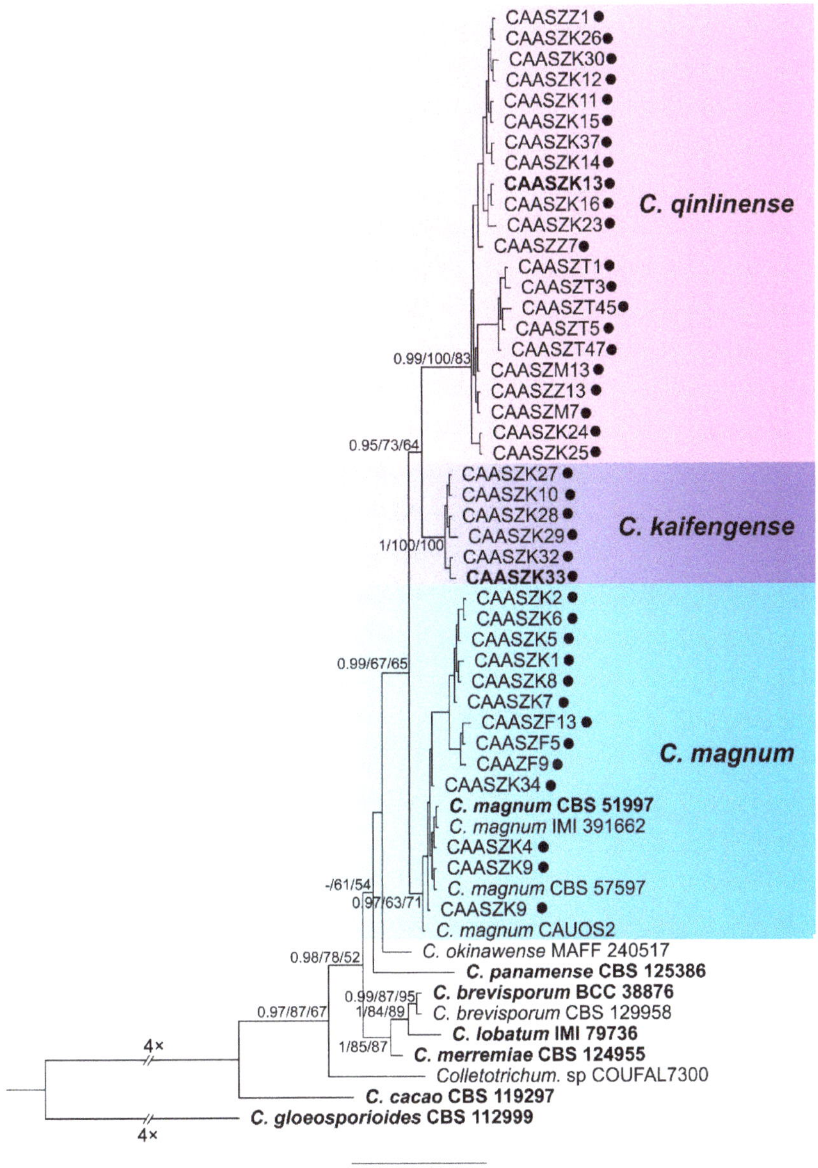

Figure 6. A Bayesian inference phylogenetic tree of 52 isolates in the *C. magnum* species complex. The species *C. gloeosporioides* (CBS 112999) was selected as an outgroup. The tree was built using concatenated sequences of the ITS, *gadph*, *chs-1*, *his3*, *act*, and *tub2* genes. Bayesian posterior probability (PP $\geq$ 0.90), MP bootstrap support values (MP $\geq$ 60%), and iqtree bootstrap support values (ML $\geq$ 50%) were shown at the nodes (PP/MP/ML). Ex-type isolates are in bold. Colored blocks indicate clades containing isolates from *Citrullus lanatus* in this study; circles indicate strains isolated from fruits. The scale bar indicates 0.02 expected changes per site.

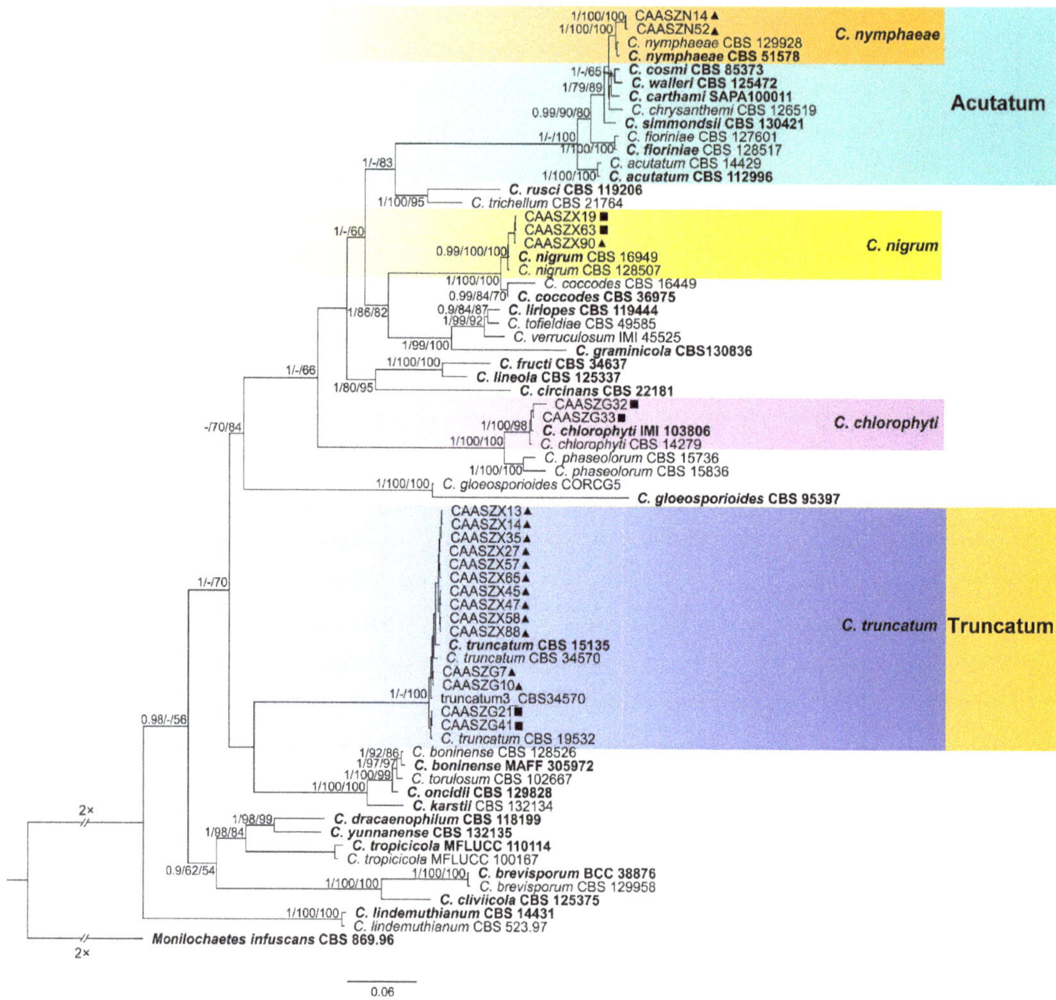

Figure 7. Phylogenetic tree generated by Bayesian inference based on concatenated sequences of the ITS, *gadph*, *chs-1*, *his3*, *act* and *tub2* genes. *Monilochaetes infuscans* (CBS 869.96) was selected as an outgroup. Bayesian posterior probability (PP ≥ 0.90), MP bootstrap support values (MP ≥ 60%), and iqtree bootstrap support values (ML ≥ 50%) were shown at the nodes (PP/MP/ML). Ex-type isolates are in bold. Colored blocks indicate clades containing isolates from *Citrullus lanatus* in this study; triangles indicate strains isolated from leaves, rectangle indicate strains isolated from stems. The scale bar indicates 0.06 expected changes per site.

The phylogenetic tree was constructed for the isolates of the *C. gloeosporioides* species complex, in which six isolates were clustered with *C. fructicola*, six with *C. aenigma*, and two with *C. jiangxiense*. In addition, the two isolates formed a distinct clade (Bayesian posterior probabilities value 1/PAUP bootstrap support value 99/iqtree bootstrap support value 100) as a sister group to *C. nanhuaensis*, which clustered distantly from any known species in the complex (Figure 2). For isolates in the *C. boninense* species complex, 23 isolates were clustered with *C. karstii* (Figure 3). For the isolates in the *C. orbiculare* species complex, 34 isolates were clustered with *C. orbiculare* (Figure 4). For isolates of *C. orchidearum* species complex, 14 isolates were clustered in two clades, 9 with *C. plurivorum*, and 5 with *C. sojae*

(Figure 5). In the *C. magnum* species complex, 14 isolates were clustered with *C. magnum*. The remaining 29 isolates were clustered in two distinct clades (1/99/100, 1/99/96) as sisters to *C. magnum*, and distanted from any known species in the complex (Figure 6). The remaining 21 isolates were clustered in four clades corresponding to *C. nymphaeae* (2 isolates), *C. nigrum* (3 isolates), *C. chlorophyti* (2 isolates), and *C. truncatum* (14 isolates) (Figure 7). The PHI test indicated no significant recombination events between *C. citrulli* and *C.nanhuaensis* (Figure 8a). In addition, the results showed that there were no significant recombination events between *C. qilinense* and *C. magnum* (Figure 8b), or between *C. kaifengense* and *C. magnum* (Figure 8c). The results revealed no significant recombination events between *C. kaifengense*, a sister group of *C. qilinense*, and *C. magnum* (Figure 8b,c).

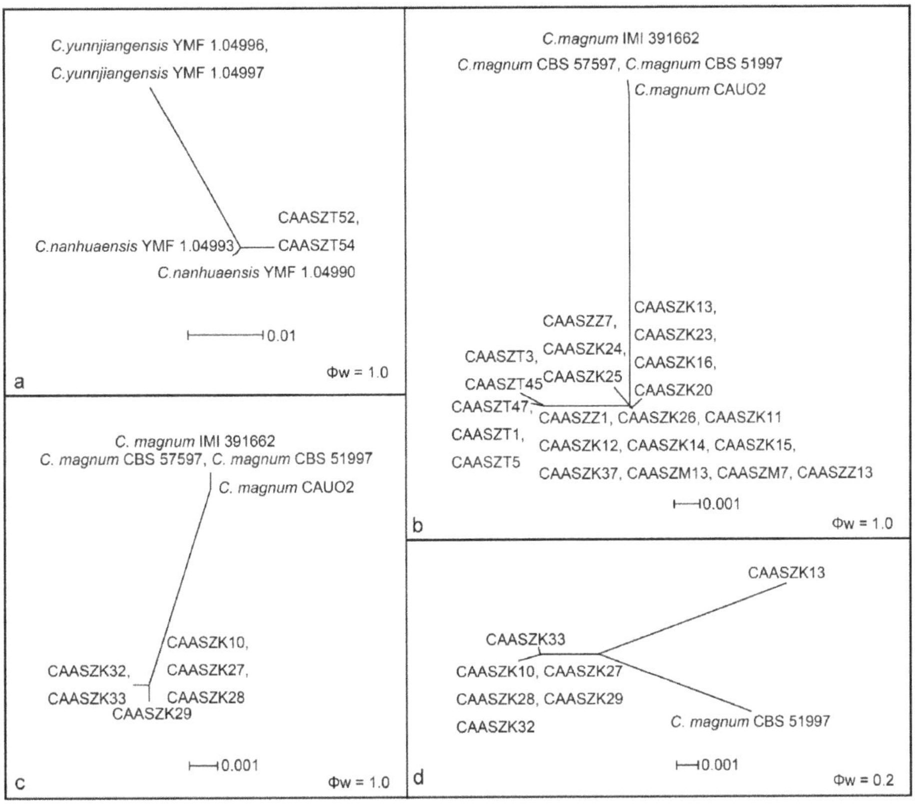

Figure 8. The result of the pairwise homoplasy index (PHI) tests of closely related species using both LogDet transformation and Splits decomposition. The PHI of *C. citrulli* (**a**) or *C. kaifengense* (**b**) or *C. qilinense* (**c**) or *C. kaifengense* and *C. qilinense* (**d**) and their phylogenetically related isolates or species, respectively. PHI test value (Φw) < 0.05 indicate significant recombination within the dataset.

3.3. Multi-Locus Phylogenetic Analyses

Analyses of the prevalence of the *Colletotrichum* species associated with watermelon in China showed that *C. orbiculare* isolates (160 isolates, 28.9% of the total isolates) were predominantly isolated from five provinces (Henan, Hubei, Liaoning, Jilin, and Jiangsu), followed by *C. truncatum* (67 isolates, 8.8%, isolated from Henan and Jiangxi), *C. qilinense* (44 isolates, 8.4%, isolated from Henan), *C. plurivorum* (41 isolates, 8.4%, isolated from Jiangxi), *C. magnum* (56 isolates, 8.4%, isolated from Henan), *C. nigrum* (37 isolates, 6.8%, isolated from Henan), *C. karstii* (32 isolates, 6.1%, isolated from Jiangxi and Hainan), *C.*

fructicola (23 isolates, 4.4%, isolated from Jiangxi and Hubei), and *C. aenigma* (18 isolates, 3.4%, isolated from Zhejiang). The remaining six species account for 9.3% of total isolates (Figure 9a). These results suggested that *C. orbiculare* is the most prevalent species of *Colletotrichum* sp. on watermelon in China (Figure 9a).

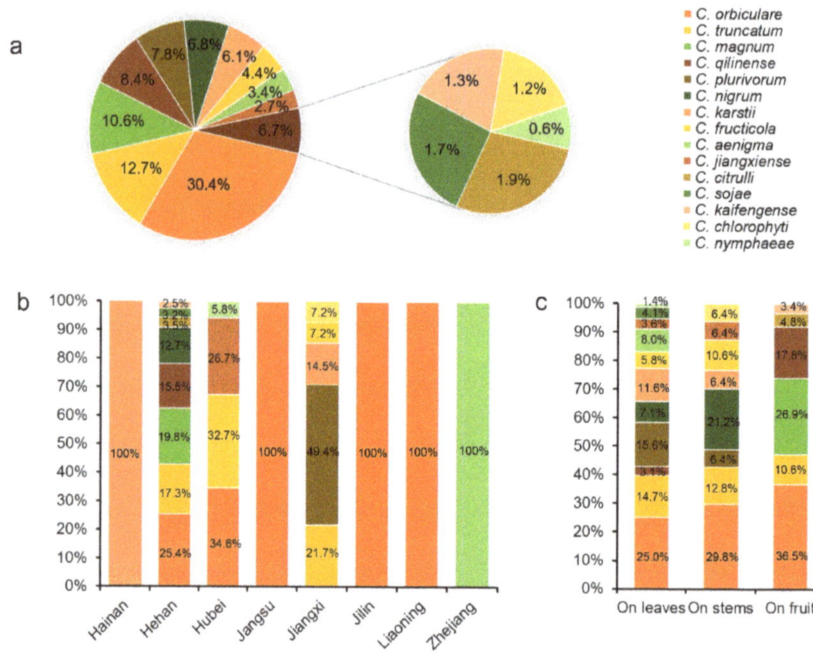

Figure 9. The prevalence of *Colletotrichum* species isolated from watermelon. (**a**) Overall isolation rate (%) of *Colletotrichum* species; (**b**,**c**) isolation rate (%) of *Colletotrichum* species from part of sampled province (**b**) and watermelon organs (**c**), respectively.

Analyses of the isolation rate of 15 *Colletotrichum* species in each sampled province revealed that *C. orbiculare* was commonly isolated in Henan and Hubei provinces, accounting for 25.4% and 34.6% of the obtained isolates, respectively (Figure 9b). In the Jiangxi province, *C. plurivorum* strains were dominant, accounting for 49.4% of total isolates from this province (Figure 9b). Moreover, all the *C. orbiculare* isolates were from Liaoning, Jilin, and Jiangsu provinces, and *C. aenigma* isolates were isolated from Zhejiang province (Figure 9b). Analyses of the *Colletotrichum* species isolated from three organs of watermelon revealed that 11 *Colletotrichum* species (*C. orbiculare*, *C. truncatum*, *C. qilinense*, *C. plurivorum*, *C. nigrum*, *C. karstii*, *C. fructicola*, *C. aenigma*, *C. jiangxiense*, *C. sojae*, and *C. nymphaeae*) were isolated from leaves, 8 (*C. orbiculare*, *C. karstii*, *C. truncatum*, *C. plurivorum*, *C. fructicola*, *C. nigrum*, *C. jiangxiense*, and *C. chlorophyti*) were isolated from stems, and 5 (*C. orbiculare*, *C. citrulli*, *C. magnum*, *C. qilinnense*, and *C. kaifengense*) were isolated from the fruits. Of those, *C. orbiculare* strains were dominant and accounted for 25.0%, 29.8%, and 36.5% of total isolates from leaves, stems, and fruits, respectively (Figure 9c).

3.4. Pathogenicity Assay

Fifteen representative *Colletotrichum* isolates were selected to prove Koch's postulates. At 4 dpi, all the *Colletotrichum* isolates developed brown to black lesions symptoms of anthracnose on detached wounded leaves of *Citrullus lanatus* cv. Hongheping inoculated by a spore suspension, and *C. fructicola*, *C. chlorophyte*, and *C. nymphaeae* were unable to infect non-wounded leaves (Figure 10(a2,a10,a15), Table 3). Under wounded conditions, pathogenicity tests demonstrated that *C. kaifengense* (mean ± SD = 32.33 ± 6.35 mm), *C. citrulli* (mean ± SD = 27.89 ± 4.71 mm), *C. qilinense* (mean ± SD = 26.84 ± 9.45 mm), and *C. magnum* (mean ± SD = 24.42 ± 0.17 mm) were highly aggressive on watermelon leaves (Figure 11). However, *C. fructicola, C. karstii,* and *C. nymphaeae* showed only weak moderate aggression on watermelon leaves (Figure 12). Under unwounded conditions, *C. qilinense* (isolate CAASZK13) had the highest infection rate at 77.8% (Table 3). No lesions were induced in the control leaves inoculated with sterile water.

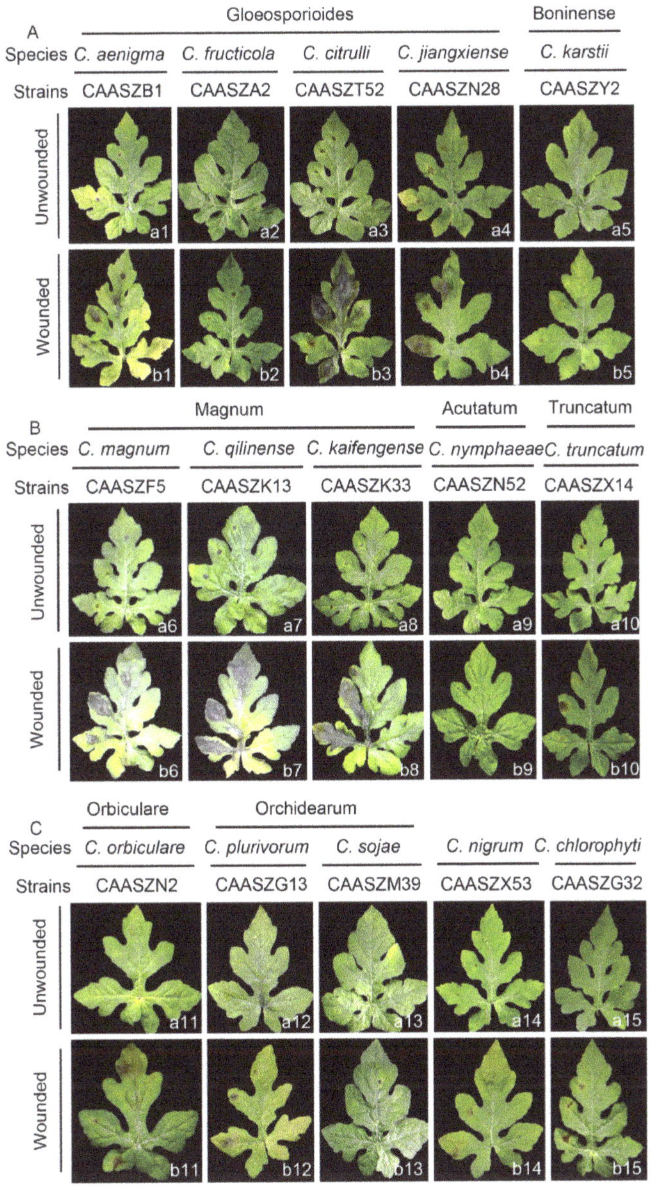

Figure 10. Symptoms of watermelon leaves (*Citrullus lanatus* cv. Hongheping) induced by inoculation of spore suspensions of 15 *Colletotrichum* spp. under unwounded and wounded conditions. The symptoms

caused by these species were photographed at 4 dpi. (**a1–a15**) Under wounded condi-tions, symptoms induced by the isolates/species belonging to the *C. gloeosporioides, C. boninense, C. magnum, C. acutatum, C. truncatum, C. orbiculare,* and *C. orchidearum* complexe species or singleton species, respectively. (**b1–b15**) Under unwounded conditions, symptoms induced by the iso-lates/species belonging to the *C. gloeosporioides, C. boninense, C. magnum, C. acutatum, C. truncatum, C. orbiculare, C. orchidearum* complexe species or singleton species, respectively. The symptoms induced by the isolates/species belonging to the *C. gloeosporioides* complex and the *C. boninense* complex (**A**), the *C. magnum* complex (**B**), and other complexes or singleton species (**C**), respectively. The inoculation was conducted by dropping 1×10^7 spores (conidia or ascospores) per mL on about 6–8 detached true leaves of *Citrullus lanatus* cv. Hongheping in three replicates after wounded by pin-pricking each leaf for one time with a sterilized needle (wounded) or kept unwounded (unwounded).

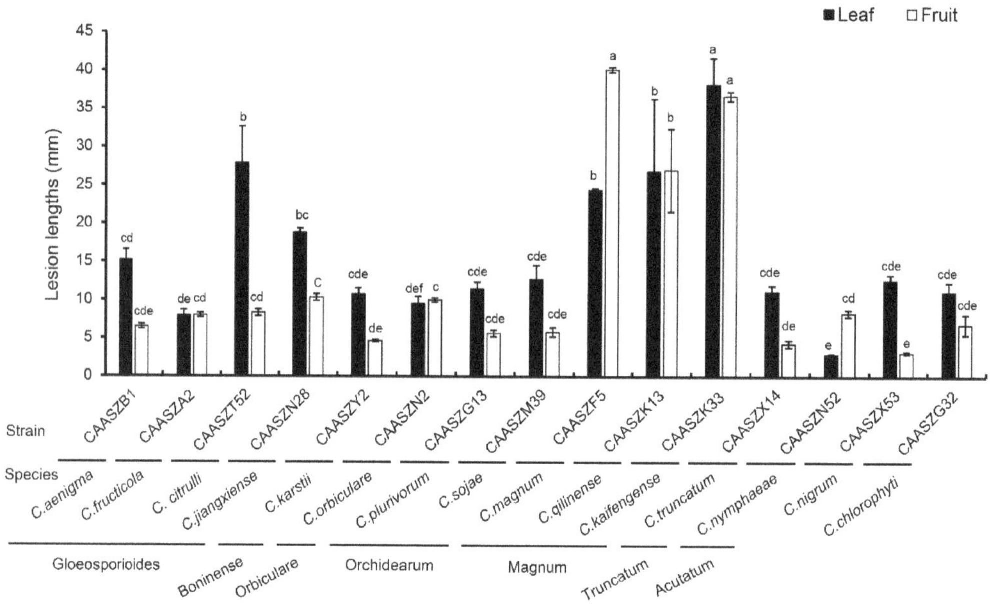

Figure 11. Lesion lengths on wounded watermelon leaves (*Citrullus. lanatus* cv. Hongheping) and fruits (*Citrullus. lanatus* cv. Motong) at 4 dpi and 6 dpi, respectively, induced by conidial suspensions of 16 representative isolates of 16 *Colletotrichum* spp. The involved isolates and their belonging are indicated at the bottom of the bars. Data were analyzed with SPSS Statistics 19.0 (WinWrap Basic; http://www.winwrap.com (accessed on 18 December 2021)) by one-way analysis of variance, and means were compared using Duncan's test at a significance level of $p = 0.05$. Letters over the error bars indicate the significant difference at the $p = 0.05$ level.

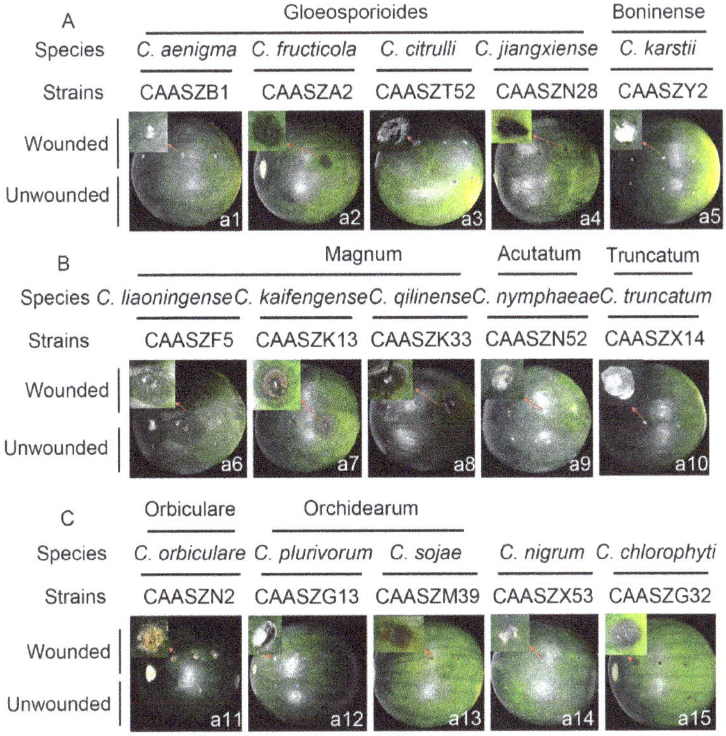

Figure 12. Symptoms of watermelon fruits (*Citrullus. lanatus* cv. Motong) induced by inoculation of spore suspensions of 15 Colletotrichum spp. under unwounded and wounded conditions. The symptoms caused by these species were photographed at 6 dpi. (**a1–a15**) Under wounded and unwounded conditions, s symptoms induced by the isolates/species belonging to the *C. gloeosporioides*, *C. boninense*, *C. magnum*, *C. acutatum*, *C. truncatum*, *C. orbiculare*, and *C. orchidearum* complexe species or singleton species, respectively. The symptoms induced by the isolates/species belonging to the *C. gloeosporioides* complex and the *C. boninense* complex (**A**), the *C. magnum* complex (**B**), and other complexes or singleton species (**C**), respectively. The inoculation was conducted by dropping 1×10^7 spores (conidia or ascospores) per mL on detached fruits of *Citrullus. lanatus* cv. Motong in three replicates after wounded by pin-pricking each leaf for one time with a sterilized needle (wounded) or kept unwounded (unwounded).

Pathogenicity was also accessed on detached watermelon fruits of *Citrullus lanatus* cv. Motong. Under unwounded conditions, *C. aenigma*, *C. fructicola*, *C. nigrum*, and *C. truncatum* did not infect the non-wounded fruits (Table 3). *C. karstii*, *C. orbiculare*, *C. magnum*, and *C. sojae* had the highest infection incidence at 100% (Table 3). Under wounded conditions, there was clear variation in aggression among species. Importantly, *C. qilinense* induced large brown or dark brown lesions and formed concentric rings of conidia at 6 dpi (Figure 12). The isolates of the *C. magnum* species complex, including *C. magnum* (mean ± SD = 40.2 ± 0.3 mm), *C. qilinense* (mean ± SD = 36.7 ± 0.6 mm), and *C. kaifengense* (mean ± SD = 27.0 ± 5.4 mm) induced significantly longer lesions than others (Figure 12).

Table 3. Incidence of infection on leaves of *Citrullus lanatus* cv. Hongheping and fruit of *Citrullus lanatus* cv. Motong by *Colletotrichum* species.

Species	Strain	Infection Incidence %	
		Leaf Bioassay	Fruit Bioassay
C.aenigma	CAASZB1	44.4	0.0
C.fructicola	CAASZA2	0.0	0.0
C. citrullus	CAASZT52	66.7	66.7
C.jiangxiense	CAASZN28	44.4	33.3
C.karstii	CAASZY2	44.4	100.0
C.orbiculare	CAASZN2	44.4	100.0
C.plurivorum	CAASZG13	55.6	66.7
C.sojae	CAASZM39	33.3	100.0
C.magnum	CAASZF5	66.7	100.0
C.qilinense	CAASZK13	77.8	66.7
C.kaifengense	CAASZK33	66.7	66.7
C.nymphaeae	CAASZN52	0.0	66.7
C.nigrum	CAASZX53	44.4	0.0
C.chlorophyti	CAASZG32	0.0	33.3
C.truncatum	CAASZX14	33.3	0.0

3.5. Taxonomy

Based on DNA sequence data and the morphological characteristics, the 146 isolates were assigned to 15 species, including 8 species reported from watermelon for the first time, and 3 species proved to represent new taxa. The three new species in culture are characterized below.

Colletotrichum citrulli—Z. Guo & Q.S. Gu, sp. nov. (Figure 13).

MycoBank Number: MB842245.

Etymology: The species epithet is derived from the host plant, *Citrullus lanatus*.

Holotype: China, Henan Province, Zhoukou City, Taikang County, on fruits of *Citrullus lanatus*, June 2020, Z. Guo. Holotype HMAS 351572, Ex-type culture CGMCC 3.20769 = CAASZT52.

Sexual morph not observed. *Asexual* morph developed on OA. *Chlamydospores* not observed. *Conidiomata acervular* formed on a cushion of angular cells. *Conidiophores* hyaline to pale brown, aseptate, unbranched. *Conidiogenous cells* hyaline, cylindrical, 10.5–17 × 3–6 μm, opening 1–2 μm. *Conidia* hyaline, smooth-walled, aseptate, apex subacute or obtuse, contents with 1–2 guttules, 15–18.5 × 5–7 μm, mean ± SD = 16.2 ± 0.9 × 5.6 ± 0.5 μm, L/W ratio = 2.9; *Appressoria* single or in groups, medium to dark brown, variable in shape, often navicular to bullet-shaped, circular, clavate, smooth-walled to undulate, 8–12 × 6–10 μm, mean ± SD = 9.9 ± 0.2 × 7.3 ± 0.8 μm, L/W ratio = 1.4.

Asexual morph developed on SNA. *Setae* medium to dark brown, smooth to finely verruculose close to the tip, the tip rounded, 1–3- aseptate, 42–79 μm long, base inflated, usually paler, 3–6.5 μm diam. *Conidia* hyaline, smooth-walled, aseptate, apex subacute or obtuse, contents with 1–2 guttules, 14–17.5 × 4–6 μm, mean ± SD = 15.9 ± 1.0 × 5.3 ± 0.4 μm, L/W ratio = 3.0.

Culture characteristics: Colonies on PDA flat with entire margin, aerial mycelium dense, cottony, surface pale cinnamon, covered with irregular grey to black conidiomata, reverse same colors. Colony diam 69.0–70.0 mm in 5 d. Colonies on OA flat with entire margin, aerial mycelium lacking, partly covered with numerous orange conidia, surface white, reverse pale grey, 73.0–73.5 mm in 5 d. Colonies on SNA flat with entire margin, aerial mycelium sparse, surface pale grey, reverse same colors, 66.5–68.0 mm in 5 d.

Notes: isolates of *C. citrulli* are phylogenetically most closely related to *C. nanhuaensis* (Figure 2). They were distinguished by ITS (with 99.15% sequence identity) and *gadph* (97.63%). Furthermore, the PHI test (Φw = 1) did not detect recombination events between *C. citrulli* and *C. nanhuaensis* (Figure 8a). In morphology, *C. citrulli* differs from *C. nanhuaensis* by having longer conidia (14.65–18.4 5 × 4.75–6.6 5 μm vs. 10.5–16 × 4.5–6 μm) [33].

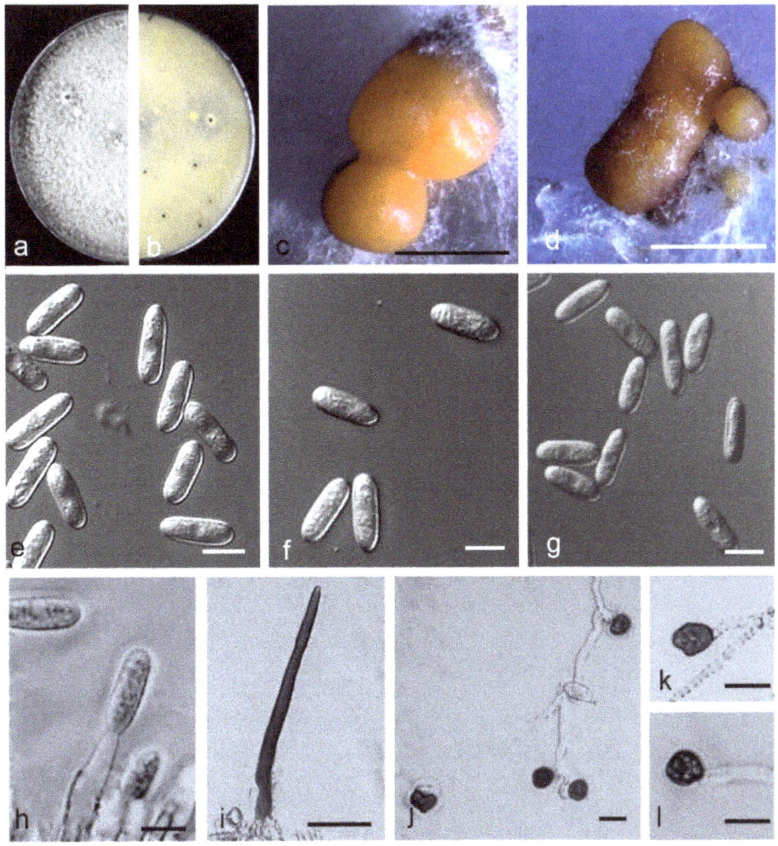

Figure 13. *Colletotrichum citrulli*. (**a**,**b**) Front and back view, respectively, of 9-d-old PDA culture; (**c**,**d**) conidiomata; (**e–g**) conidia; (**h**) conidiophores; (**i**) seta; (**j–l**) appressoria; (**a–q**) isolate CAASZT52; (**a**,**b**) produced on PDA agar medium; (**c**,**e–h**) produced on OA agar medium; (**d**,**i–l**) produced on SNA agar medium. Scale bars: (**c**,**d**) = 1000 μm; (**i**) = 20 μm; (**h**,**j–l**) = 10 μm.

Colletotrichum kaifengense Z. Guo & Q. S. Gu, sp. nov. (Figure 14).

MycoBank Number: MB842244.

Etymology: Named after Kaifeng, the city in Henan Province, China, where the species was collected.

Holotype: China, Henan Province, Kaifeng City, Jinming County, on fruits of *Citrullus lanatus*, August 2019, Z. Guo. Holotype HMAS 351574, Ex-type culture CGMCC 3.20768 = CAASZK33.

Sexual morph not observed. *Asexual* morph developed on PDA. Chlamydospores not observed. *Conidiomata acervular*, conidiophores, and setae formed on a cushion of angular cells. *Conidiophores* hyaline, pale-to medium brown, smooth-walled, aseptate, unbranched. *Conidiogenous cells* hyaline, cylindrical, 11–21 × 4–5 μm. *Setae* medium to dark brown, smooth-walled, the tip acuted, aseptate, 70–153 μm. *Conidia* hyaline, smooth-walled, aseptate, cylindrical, both ends bluntly rounded, 13.5–20 × 4–6.5 μm, mean ± SD = 17.4 ± 1.7× 4.9 ± 0.5 μm, L/W ratio = 3.5. *Appressoria* single or in loose groups, pale to medium brown, bullet-shaped, circular, smooth-walled, spathulate or irregular, 8–14.5 × 5.5–11 μm, mean ± SD = 11.1 ± 1.9× 7.9 ± 1.4μm, L/W ratio = 1.5.

Asexual morph developed on OA. *Setae* medium to dark brown, smooth-walled, aseptate, 130.0–200 μm, the tip acute. *Conidia* hyaline, smooth-walled, aseptate, cylindrical, both ends bluntly rounded, 16–20.5 × 4–6.5 μm, mean ± SD = 18.2 ± 1.1× 5.3 ± 0.5 μm, L/W ratio = 3.5.

Asexual morph developed on SNA. *Conidiophores,* and Setae formed on hyphae. *Setae* medium to dark brown, smooth-walled, aseptate, 82.5–177.5 μm, the tip acute. *Conidia* hyaline, smooth-walled, aseptate, cylindrical, both ends bluntly rounded, 16–22.5 × 4–5.5 μm, mean ± SD = 20.0 ± 1.2× 4.8 ± 0.4 μm, L/W ratio = 4.2.

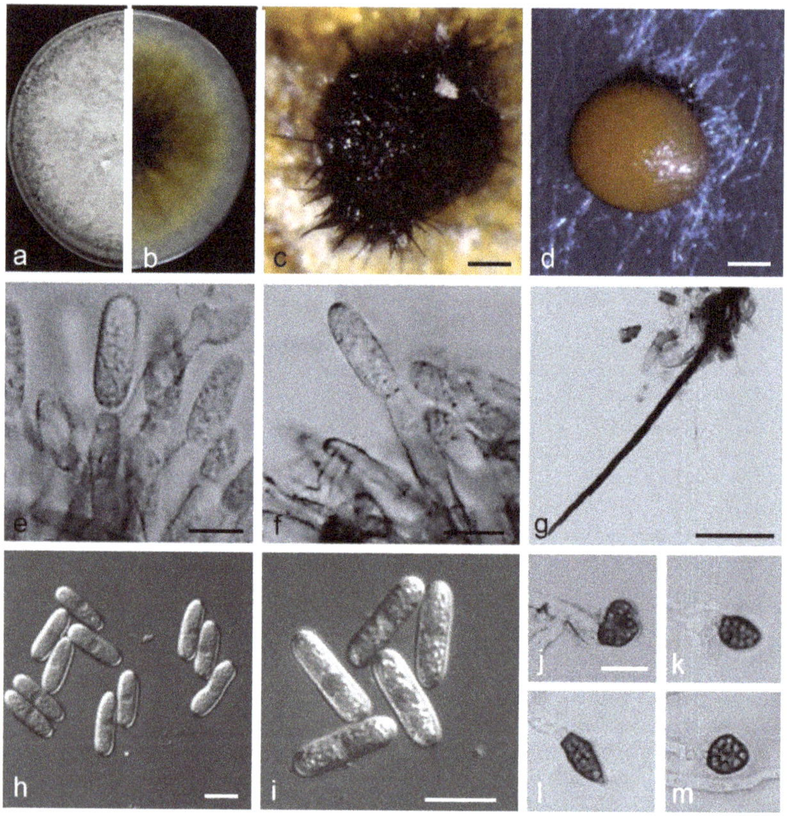

Figure 14. *Colletotrichum kaifengense*. (**a**,**b**) Front and back view, respectively, of 9-d-old PDA culture; (**c**,**d**) conidiomata; (**e**,**f**) conidiophores; (**g**) seta; (**h**,**i**) conidia; (**j**–**m**) appressoria; (**a**–**m**) isolate CAASZK33; (**a**–**c**,**g**) produced on PDA agar medium; (**d**–**f**,**h**–**m**) produced on SNA agar medium). Scale bars: (**c**,**d**) = 100 μm; (**g**) = 50 μm; (**e**,**f**,**h**–**m**) = 10 μm.

Culture characteristics: Colonies on PDA flat with entire margin, aerial mycelium dense, floccose, surface pale grey to white, reverse grey in the center with cinnamon margin, 75.0–76.0 mm in 7 d. Colonies on OA flat with entire margin, aerial mycelium lacking, surface white, covered with small black conidiomata with orange conidial masses, reverse white, covered with irregular black conidiomata, 50.0–52.0 mm in 7 d. Colonies on SNA flat with entire margin, short felty whitish aerial mycelium, surface white, reverse same colors, 50.5–51.5 mm in 7 d.

Notes: isolates of *C. kaifengense* are phylogenetically closely related to *C. magnum* (Figure 6). They were distinguished by *gadph* (93.24%) and *his3* (98.55%). Furthermore, the PHI test (Φw = 1) did not detect recombination events between *C. kaifengense* and *C. magnum* (Figure 8c). They were distinguished by *gadph* (88.14%) and *chs-1* (98.65%).

Furthermore, the PHI test (Φw = 1) did not detect recombination events between *C. qilinense* and *C. kaifengense* (Figure 8d).

Colletotrichum qilinense Z. Guo & Q.S. Gu, sp. nov. (Figure 15).

MycoBank Number: MB842247;

Etymology. Referring to the host variety (*Citrullus lanatus* cv. qilinwang) from which the fungus was collected.

Holotype: China, Henan Province, Kaifeng City, Weishi County, on fruits of *Citrullus lanatus*, 20 August 2019, Z. Guo. Holotype HMAS 351573, Ex-type culture CGMCC 3.20767 = CAASZK13).

Sexual morph not observed. *Asexual* morph developed on PDA. *Chlamydospores* not observed. *Conidiomata acervular*, conidiophores, and setae formed on a cushion of angular cells. *Conidiophores* hyaline, medium to dark brown, smooth-walled, aseptate, unbranched. *Conidiogenous cells* hyaline, cylindrical, 9–23 × 4–6 μm. *Setae* medium to dark brown, smooth-walled, the tip acuted, 75.5–188 μm. *Conidia* hyaline, smooth-walled, cylindrical, the apex rounded, the base rounded to truncate, 17.5–23 × 5–6 μm, mean ± SD = 19.7 ± 1.4 × 5.4 ± 0.4 μm, L/W ratio = 3.7. *Appressoria* single or in loose groups, medium to dark brown, bullet-shaped, smooth-walled, aseptate, mostly ovoid or ellipsoidal to irregular in outline, 8–16 × 5–11 μm, mean ± SD = 10.6 ± 2.1 × 7.7 ± 1.5 μm, L/W ratio = 1.4.

Asexual morph developed on OA. *Setae* medium to dark brown, smooth-walled, the tip acuted, base inflated or not, 96.7–178.4 μm. *Conidia* hyaline, smooth-walled, aseptate or aseptate, cylindrical, the apex rounded, the base rounded to truncate, 14–20 × 5.5–6.5 μm, mean ± SD = 17.6 ± 1.5 × 5.3 ± 0.5 μm, L/W ratio = 3.3.

Asexual morph developed on SNA. Conidiophores are directly formed on hyphae. *Setae* medium to dark brown, smooth-walled, the tip acuted, aseptate, 83–203 μm. *Conidia* hyaline, smooth-walled, aseptate, cylindrical, both ends bluntly rounded, 16–20 × 4.5–6 μm, mean ± SD = 17.8 ± 1.0 × 5.1 ± 0.4μm, L/W ratio = 3.5.

Culture characteristics: colonies (CAASZK13) on PDA flat with entire margin, aerial mycelium lacking, floccose, surface dark grey to black, reverse black in center with white margin, 75.50–76.0 mm in 7 d. Colonies on OA flat with entire margin, surface white, partly covered with short felty whitish aerial mycelium, reverse white, covered with irregular black conidiomata, 53.0–54.0 mm in 7 d. Colonies on SNA flat with entire margin, short felty whitish aerial mycelium, surface white, reverse same colors, 51.0–52.0 mm in 7 d.

Notes: isolates of *C. qilinense* are phylogenetically closely related to *C. magnum* (Figure 6). They were distinguished by *gadph* (87.08%), *chs-1* (99.71%), and *act* (96.57%). Furthermore, the PHI test (Φw = 1) did not detect recombination events between *C. qilinense* and *C. magnum* (Figure 8d).

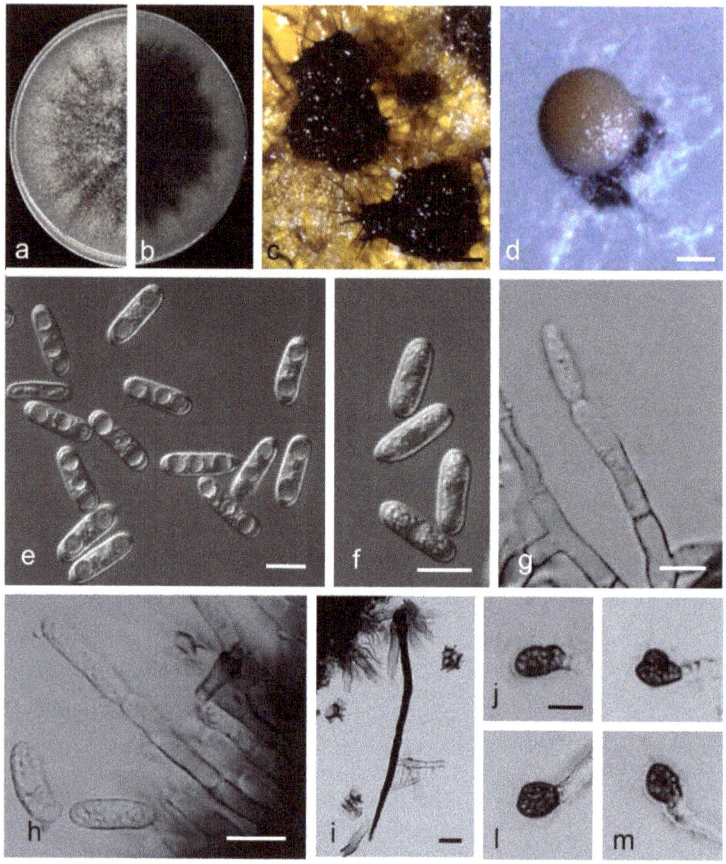

Figure 15. *Colletotrichum qilinense*. (**a**,**b**) Front and back view, respectively, of 9-d-old PDA culture; (**c**,**d**) conidiomata; (**e**,**f**) conidia; (**g**,**h**) conidiophores; (**i**) seta; (**j–m**) appressoria; (**a–e**,**h–m**) isolate CAASZK13; (**f**,**g**) isolate CAASZZ7; (**a–c**) produced on PDA agar medium; (**d–m**) produced on OA agar medium). Scale bars: (**c**,**d**) = 100 μm; (**i**) = 20 μm; (**e–h**,**i–m**) = 10 μm; j applies to (**j–m**).

4. Discussion

The main objective of this study was to identify the *Colletotrichum* species associated with watermelon in China. We obtained 526 single spore strains from watermelon stems, leaves, and fruits displaying anthracnose symptoms. Based on multi-locus data combined with morphological characteristics we revealed 15 species allocated in 7 species complexes, including gloeosporioides (*C. aenigma*, *C. fructicola*, *C. jiangxiense*, and *C. citrulli*), boninense (*C. karstii*), orbiculare (*C. orbiculare*), orchidearum (*C. plurivorum* and *C. sojae*), magnum (*C. liaoningense*, *C. magnum*, *C. qilinnense*, and *C. kaifengense*), Truncatum (*C. truncatum*), and acutatum (*C. nymphaeae*) and two singleton species (*C. nigrum*, and *C. chlorophyti*). It is the first report of *C. aenigma*, *C. fructicola*, *C. jiangxiense*, *C. plurivorum*, *C. sojae*, *C. magnum*, *C. nymphaeae*, *C. nigrum*, and *C. chlorophyti* causing anthracnose of watermelon. Importantly, this study differentiated three new species, namely *C. citrulli*, *C. qilinnense*, and *C. kaifengense*.

The species complexes of the *Colletotrichum* include several taxa [30–32,36–39,43]. Thus, the complex species related to the 15 *Colletotrichum* species recovered in this study were selected and included in the analyses. In previous studies, although morphological characters delimited some *Colletotrichum* species, some species were difficult to describe due to their morphological variability [25,77]. Thus, in recent studies, the multi-locus phylogenetic analyses combined with morphological characteristics including measurements of setae, conidia, and appressoria, colony characters, and growth rates were able to resolve the complexes and species within *Colletotrichum*. spp [12,16–18,78,79]. In this study, most of the *Colletotrichum* species were determined by multi-locus DNA sequence analyses, and some species also exhibited specific morphological characteristics. For example, the growth rate of *C. orbiculare* was significantly lower than that of the other 14 *Colletotrichum* species (Supplementary Figure S1). It is worth noting that the *Colletotrichum* species associated with watermelon differ in their ability to form sexual morphs and setae. Importantly, only four species, *C. aenigma*, *C. fructicola*, *C. karstii*, and *C. plurivorum*, produced ascospores on OA plates (Supplementary Table S3). Additionally, setae of *C. jiangxiense*, *C. nigrum*, and *C. nymphaeae* were not observed on PDA, OA, and SNA plates (Supplementary Table S3).

The prevalence of a *Colletotrichum* species associated with watermelon anthracnose is closely related to the sampling area and plant tissue. For example, some of *Colletotrichum* species, including *C. jiangxiense*, *C. citrulli*, *C. sojae*, *C. magnum*, *C. qilinnense*, *C. kaifengense*, *C. nymphaeae*, *C. nigrum*, and *C. chlorophyti* associated with watermelon showed restricted distribution. Notably, *C. orbiculare* was isolated from five provinces (Liaoning, Jilin, Henan, Hubei, and Jiangsu), and was dominant in our investigation of the major watermelon-growing regions in China (Figure 9a). It is worth noting that geographical preference was found for *C. aenigma*, in agreement with recent results in China [18].

Some *Colletotrichum* species are host-organ-specific and, for example, predominantly infect leaves of apple and pear [80]. This study revealed that more *Colletotrichum* species infect watermelon leaves than stems or fruits. Three *Colletotrichum* species including *C. aenigma*, *C. sojae*, and *C. nymphaeae* were only isolated from the leaves; *C. chlorophyti* was only isolated from the stems; and five *Colletotrichum* species, including *C. citrulli*, *C. magnum*, *C. qilinnense*, and *C. kaifengense*, were only isolated from fruits (Figure 9c).

Pathogenicity of 15 *Colletotrichum* species isolated from watermelon samples in China was tested on watermelon fruits and leaves. It is worth noting that all *Colletotrichum* species could infect wounded fruits and leaves. However, five *Colletotrichum* species, including *C. aenigma*, *C. fructicola*, *C. magnum*, *C. nigrum*, and *C. truncatum*, did not cause disease symptoms on the non-wounded watermelon fruits. Three *Colletotrichum* species (*C. fructicola*, *C. nymphaeae*, and *C. chlorophyti*) did not cause lesions on the non-wounded leaves (Table 3). This phenomenon is related to inoculation methods and conditions, quiescent infection, and structural defense of the host tissue. In this study, pathogenicity experiments were conducted on wounded or unwounded tissues under extreme conditions of artificial inoculation. It remains to be seen how the symptoms and lesion development are displayed under natural conditions. The quiescent infection is an important feature of *Colletotrichum* spp., and it mainly occurs prior to harvest and eventually leads to fruit rot [81–83]. A previous study indicated that the cuticle and epidermis were important as resistance barriers to the initial infection. Wounding can break the quiescent infection and enhance the infectiousness [84].

Different *Colletotrichum* species have varying aggressiveness on the host [17,18,85]. In this study, most isolates belonging to the *C. magnum* species complex showed higher aggressiveness than those of the *C. gloeosporioides*, *C. boninense*, *C. orbiculare*, *C. orchidearum*, *C. truncatum*, and *C. acutatum* species complexes (Figure 11). Studies suggest that *C. magnum* caused watermelon anthracnose in USA and chili anthracnose in China [16,32,43]. However, it was never reported on watermelon in China. Thus, this study here represents the first report of *C. magnum* on watermelon in China. *Colletotrichum qilinense* and *C. kaifengense* were found to cause anthracnose on watermelon fruits and were described as new species in this study. Noteworthy, among the 15 representative *Colletotrichum* species in inoculated wounded watermelon leaves, *C. kaifengense* was the most aggressive one (Figure 11).

Previous studies revealed that 15 *Colletotrichum* species were identified from chili fruits with anthracnose symptoms in China, including *C. fioriniae*, *C. fructicola*, *C. gloeosporioides*, *C. scovillei*, *C. truncatum*, *C. conoides*, *C. grossum*, *C. magnum*, *C. aenigma*, *C. cliviae*, *C. endophytica*,

C. hymenocallidis, *C. incanum*, *C. karstii*, and *C. viniferum* [16,43]. Furthermore, it is interesting that five *Colletotrichum* species, including *C. fructicola*, *C. truncatum*, *C. magnum*, *C. aenigma*, and *C. karstii*, were also isolated from watermelon in this study. *Colletotrichum fructicola* was reported for the first time, associated with anthracnose symptoms from strawberry in USA and Canada [30]. This is the first report of *C. fructicola* to induce anthracnose of watermelon (Figure 2). *Colletotrichum truncatum* was the dominant species on chili in China [16], and it was also found to cause anthracnose on watermelon leaves, stems, and fruits. *Colletotrichum aenigma* was first reported on chili in China, where it was found to cause anthracnose on fruits [16], but this is the first report of *C. aenigma* on *Citrullus lanatus*, where it was found to cause anthracnose on watermelon leaves. *Colletotrichum karstii* was reported causing fruit spots on chili in China [16] and bitter rot on strawberry in the Australia [86]. This is the first report of *C. karstii* in China, which caused watermelon leaf spot.

This study provides useful information about ecology and pathogenicity of 15 *Colletotrichum* spp. involved in watermelon anthracnose. This is of great significance for the control of anthracnose disease in different regions in China.

5. Conclusions

This study provides the first summary of 15 *Colletotrichum* spp. associated with watermelon in China. Furthermore, this study presents the first reports of *C. aenigma*, *C. fructicola*, *C. jiangxiense*, *C. plurivorum*, *C. sojae*, *C. nymphaeae*, *C. nigrum*, and *C. chlorophyti*, and three new species (*C. citrulli*, *C. qilinnense*, and *C. kaifengense*), causing anthracnose of watermelon in China.

Supplementary Materials: The following supporting information can be downloaded at: https://www.mdpi.com/article/10.3390/jof8080790/s1, Table S1. List of 148 representative isolates of 15 *Colletotrichum* spp. collected from watermelon in China, with details host and location, and Genbank No; Table S2. Strains of the *Colletotrichum* species in this paper with details about host and location, and Genbank No; Table S3. The sizes of seta, conidia, appresoria, and ascospores of the representative isolates of Colletotrichum spp. obtained in this study; Figure S1. The growth rate of the representative isolates of *Colletotrichum* spp. obtained in this study. Data were analyzed with SPSS Statistics 19.0 (WinWrap Basic; http://www.winwrap.com (accessed on 18 December 2021)) by one-way analysis of variance, and means were compared using Duncan's test at a significance level of $p = 0.05$. Letters over the error bars indicate the significant difference at the $p = 0.05$ level.

Author Contributions: Conceptualization, Z.G. and Q.-S.G.; methodology, Z.G.; software, Z.G. and H.-J.W.; validation, Z.G. and Q.-S.G.; formal analysis, Z.G.; investigation, Z.G., B.P. and Q.-S.G.; data curation, Z.G. and Q.-S.G.; writing—original draft preparation, Z.G. and Q.-S.G.; writing—review and editing, H.-J.W., M.Z., C.-X.L. and Q.-S.G.; visualization, Z.G.; supervision, B.-S.K., L.-M.L. and Q.-S.G.; project administration and funding acquisition, Q.-S.G. All authors have read and agreed to the published version of the manuscript.

Funding: This work was financially supported Agricultural Science and Technology Innovation Program (CAAS-ASTIP-2022-ZFRI-09), Central Public-interest Scientific Institution Basal Research Fund (No. 1610192022105), and National Natural Science Foundation of China (No. U21A20229).

Institutional Review Board Statement: Not applicable.

Informed Consent Statement: Not applicable.

Data Availability Statement: Alignments generated during the current study are available from TreeBASE (study 29157). All sequence data are available in the NCBI GenBank, following the accession numbers in the manuscript.

Acknowledgments: We sincerely thank the reviewers for their contributions during the revision process.

Conflicts of Interest: The authors declare no conflict of interest. The funders had no role in the design of the study; in the collection, analyses, or interpretation of data; in the writing of the manuscript; or in the decision to publish the results.

References

1. Cannon, P.F.; Damm, U.; Johnston, P.R.; Weir, B.S. *Colletotrichum*-current status and future directions. *Stud. Mycol.* **2012**, *73*, 181–213. [CrossRef] [PubMed]
2. Dean, R.; Van Kan, J.A.; Pretorius, Z.A.; Hammond-Kosack, K.E.; Di Pietro, A.; Spanu, P.D.; Rudd, J.J.; Dickman, M.; Kahmann, R.; Ellis, J.; et al. The Top 10 fungal pathogens in molecular plant pathology. *Mol. Plant Pathol.* **2012**, *13*, 414–430. [CrossRef] [PubMed]
3. Zheng, H.; Yu, Z.; Jiang, X.; Fang, L.; Qiao, M. Endophytic *Colletotrichum* Species from Aquatic Plants in Southwest China. *J. Fungi* **2022**, *8*, 87. [CrossRef]
4. Denoyes-Rothan, B.; Guerin, G.; Delye, C.; Smith, B.; Minz, D.; Maymon, M.; Freeman, S. Genetic diversity and pathogenic variability among isolates of *Colletotrichum* species from strawberry. *Phytopathology* **2003**, *93*, 219–228. [CrossRef]
5. Prihastuti, H.; McKenzie, E.; Hyde, K.; Cai, L.; Mckenzie, H.; Hyde, E. Characterization of *Colletotrichum* species associated with coffee berries in northern Thailand. *Fungal Divers.* **2009**, *39*, 89–109.
6. Phoulivong, S.; McKenzie, E.H.C.; Hyde, K.D. Cross infection of *Colletotrichum* species; a case study with tropical fruits. *Curr. Res. Environ. Appl. Mycol.* **2012**, *2*, 99–111. [CrossRef]
7. Huang, F.; Chen, G.Q.; Hou, X.; Fu, Y.S.; Cai, L.; Hyde, K.D.; Li, H.Y. *Colletotrichum* species associated with cultivated citrus in China. *Fungal Divers.* **2013**, *61*, 61–74. [CrossRef]
8. Lima, N.B.; Batista, M.V.D.A.; De Morais, M.A.; Barbosa, M.A.G.; Michereff, S.J.; Hyde, K.D.; Câmara, M.P.S. Five *Colletotrichum* species are responsible for mango anthracnose in northeastern Brazil. *Fungal Divers.* **2013**, *61*, 75–88. [CrossRef]
9. Liu, F.; Damm, U.; Cai, L.; Crous, P.W. Species of the *Colletotrichum gloeosporioides* complex associated with anthracnose diseases of Proteaceae. *Fungal Divers.* **2013**, *61*, 89–105. [CrossRef]
10. Schena, L.; Mosca, S.; Cacciola, S.O.; Faedda, R.; Sanzani, S.M.; Agosteo, G.E.; Sergeeva, V.; di San Lio, G.M. Species of the *Colletotrichum gloeosporioides* and *C. boninense* complexes associated with olive anthracnose. *Plant Pathol.* **2014**, *63*, 437–446. [CrossRef]
11. Vieira, W.A.S.; Michereff, S.J.; de Morais, M.A.; Hyde, K.D.; Câmara, M.P.S. Endophytic species of *Colletotrichum* associated with mango in northeastern Brazil. *Fungal Divers.* **2014**, *67*, 181–202. [CrossRef]
12. Yan, J.-Y.; Jayawardena, M.M.R.S.; Goonasekara, I.D.; Wang, Y.; Zhang, W.; Liu, M.; Huang, J.B.; Wang, Z.Y.; Shang, J.J.; Peng, Y.L.; et al. Diverse species of *Colletotrichum* associated with grapevine anthracnose in China. *Fungal Divers.* **2014**, *71*, 233–246. [CrossRef]
13. La Hoz, C.J.P.-D.; Calderón, C.; Rincón, A.M.; Cárdenas, M.; Danies, G.; López-Kleine, L.; Restrepo, S.; Jiménez, P. Species from the *Colletotrichum acutatum*, *Colletotrichum boninense* and *Colletotrichum gloeosporioidesspecies* complexes associated with tree tomato and mango crops in Colombia. *Plant Pathol.* **2016**, *65*, 227–237. [CrossRef]
14. Wang, Y.C.; Hao, X.Y.; Wang, L.; Bin, X.; Wang, X.C.; Yang, Y.J. Diverse *Colletotrichum* species cause anthracnose of tea plants (*Camellia sinensis* (L.) O. Kuntze) in China. *Sci. Rep.* **2016**, *6*, 35287. [CrossRef]
15. De Silva, D.D.; Ades, P.K.; Crous, P.W.; Taylor, P.W.J. *Colletotrichum* species associated with chili anthracnose in Australia. *Plant Pathol.* **2017**, *66*, 254–267. [CrossRef]
16. Diao, Y.Z.; Zhang, C.; Liu, F.; Wang, W.Z.; Liu, L.; Cai, L.; Liu, X.L. *Colletotrichum* species causing anthracnose disease of chili in China. *Persoonia* **2017**, *38*, 20–37. [CrossRef] [PubMed]
17. Guarnaccia, V.; Groenewald, J.Z.; Polizzi, G.; Crous, P.W. High species diversity in *Colletotrichum associated* with citrus diseases in Europe. *Persoonia* **2017**, *39*, 32–50. [CrossRef]
18. Fu, M.; Crous, P.W.; Bai, Q.; Zhang, P.F.; Xiang, J.; Guo, Y.S.; Zhao, F.F.; Yang, M.M.; Hong, N.; Xu, W.X.; et al. *Colletotrichum* species associated with anthracnose of *Pyrus* spp. in China. *Persoonia* **2019**, *42*, 1–35. [CrossRef]
19. Wang, Y.; Chen, J.Y.; Xu, X.; Cheng, J.; Zheng, L.; Huang, J.; Li, D.W. Identification and Characterization of *Colletotrichum* Species Associated with Anthracnose Disease of Camellia oleifera in China. *Plant Dis.* **2020**, *104*, 474–482. [CrossRef]
20. Wang, W.; de Silva, D.D.; Moslemi, A.; Edwards, J.; Ades, P.K.; Crous, P.W.; Taylor, P.W.J. *Colletotrichum* Species Causing Anthracnose of Citrus in Australia. *J. Fungi* **2021**, *7*, 47. [CrossRef]
21. Tan, Q.; Schnabel, G.; Chaisiri, C.; Yin, L.F.; Yin, W.X.; Luo, C.X. *Colletotrichum* Species Associated with Peaches in China. *J. Fungi* **2022**, *8*, 313. [CrossRef] [PubMed]
22. Sutton, B.C. *The Coelomycetes: Fungi Imperfecti with Pycnidia Acervuli and Stromata*; Commonwelth Mycological Institute: Kew, UK, 1980.
23. Von Arx, J.A. Die Arten der Gattung *Colletotrichum* Cda. *Phytopathol. Z.* **1957**, *29*, 413–468.
24. Hyde, K.D.; Cai, L.; McKenzie, E.; Yang, Y.L.; Zhang, J.Z.; Prihastuti, H. *Colletotrichum*: A catalogue of confusion. *Fungal Divers.* **2009**, *39*, 1–17.
25. Hyde, K.D.; Chomnunti, P.; Crous, P.W.; Groenewald, J.Z.; Damm, U.; Ko, T.W.K.; Shivas, R.G.; Summerell, B.A.; Tan, Y.P. A case for re-inventory of Australia's plant pathogens. *Persoonia* **2010**, *25*, 50–60. [CrossRef] [PubMed]
26. Quaedvlieg, W.; Binder, M.; Groenewald, J.Z.; Summerell, B.A.; Carnegie, A.J.; Burgess, T.I.; Crous, P.W. Introducing the Consolidated Species Concept to resolve species in the *Teratosphaeriaceae*. *Persoonia* **2014**, *33*, 1–40. [CrossRef] [PubMed]
27. Tsang, C.C.; Tang, J.Y.M.; Lau, S.K.P.; Woo, P.C.Y. Taxonomy and evolution of *Aspergillus*, *Penicillium* and *Talaromyces* in the omics era-Past, present and future. *Comput. Struct. Biotechnol. J.* **2018**, *16*, 197–210. [CrossRef]

28. Cai, L.; Hyde, K.; Taylor, P.; Weir, B.; Waller, J.; Abang, M.; Zhang, J.Z.; Yang, Y.L.; Phoulivong, S.; Liu, Z.Y. A polyphasic approach for studying *Colletotrichum*. *Fungal Divers.* **2009**, *39*, 183–204.
29. Yang, Y.L.; Liu, Z.Y.; Cai, L.; Hyde, K.; Yu, Z.N.; McKenzie, E. *Colletotrichum* anthracnose of Amaryllidaceae. *Fungal Divers.* **2009**, *39*, 123–146.
30. Weir, B.S.; Johnston, P.R.; Damm, U. The *Colletotrichum gloeosporioides* species complex. *Stud. Mycol.* **2012**, *73*, 115–180. [CrossRef]
31. Liu, F.; Cai, L.; Crous, P.W.; Damm, U. The *Colletotrichum gigasporum* species complex. *Persoonia* **2014**, *33*, 83–97. [CrossRef]
32. Damm, U.; Sato, T.; Alizadeh, A.; Groenewald, J.Z.; Crous, P.W. The *Colletotrichum dracaenophilum*, *C. magnum* and *C. orchidearum* species complexes. *Stud. Mycol.* **2019**, *92*, 1–46. [CrossRef] [PubMed]
33. Yu, Z.; Jiang, X.; Zheng, H.; Zhang, H.; Qiao, M. Fourteen New Species of Foliar *Colletotrichum* Associated with the Invasive Plant *Ageratina adenophora* and Surrounding Crops. *J. Fungi* **2022**, *8*, 185. [CrossRef] [PubMed]
34. Damm, U.; Woudenberg, J.H.C.; Cannon, P.; Crous, P.W. *Colletotrichum* species with curved conidia from herbaceous hosts. *Fungal Divers.* **2009**, *39*, 45.
35. Shivas, R.; Tan, Y.P. A taxonomic re-assessment of *Colletotrichum acutatum*, introducing *C. fioriniae* comb. et stat. nov. and *C. simmondsii* sp. nov. *Fungal Divers.* **2009**, *39*, 111–122.
36. Damm, U.; Cannon, P.F.; Woudenberg, J.H.; Crous, P.W. The *Colletotrichum acutatum* species complex. *Stud. Mycol.* **2012**, *73*, 37–113. [CrossRef] [PubMed]
37. Damm, U.; Cannon, P.F.; Woudenberg, J.H.; Johnston, P.R.; Weir, B.S.; Tan, Y.P.; Shivas, R.G.; Crous, P.W. The *Colletotrichum boninense* species complex. *Stud. Mycol.* **2012**, *73*, 1–36. [CrossRef] [PubMed]
38. Damm, U.; Cannon, P.F.; Liu, F.; Barreto, R.W.; Guatimosim, E.; Crous, P.W. The *Colletotrichum orbiculare* species complex: Important pathogens of field crops and weeds. *Fungal Divers.* **2013**, *61*, 29–59. [CrossRef]
39. Damm, U.; O'Connell, R.J.; Groenewald, J.Z.; Crous, P.W. The *Colletotrichum destructivum* species complex-hemibiotrophic pathogens of forage and field crops. *Stud. Mycol.* **2014**, *79*, 49–84. [CrossRef] [PubMed]
40. Marin-Felix, Y.; Groenewald, J.Z.; Cai, L.; Chen, Q.; Marincowitz, S.; Barnes, I.; Bensch, K.; Braun, U.; Camporesi, E.; Damm, U.; et al. Genera of phytopathogenic fungi: GOPHY 1. *Stud. Mycol.* **2017**, *86*, 99–216. [CrossRef]
41. Jayawardena, R.S.; Hyde, K.D.; Chen, Y.J.; Papp, V.; Palla, B.; Papp, D.; Bhunjun, C.S.; Hurdeal, V.G.; Senwanna, C.; Manawasinghe, I.S.; et al. One stop shop IV: Taxonomic update with molecular phylogeny for important phytopathogenic genera: 76–100 (2020). *Fungal Divers.* **2020**, *103*, 87–218. [CrossRef]
42. Bhunjun, C.S.; Phukhamsakda, C.; Jayawardena, R.S.; Jeewon, R.; Promputtha, I.; Hyde, K.D. Investigating species boundaries in *Colletotrichum*. *Fungal Divers.* **2021**, *107*, 107–127. [CrossRef]
43. Liu, F.; Ma, Z.Y.; Hou, L.W.; Diao, Y.Z.; Wu, W.P.; Damm, U.; Song, S.; Cai, L. Updating species diversity of *Colletotrichum*, with a phylogenomic overview. *Stud. Mycol.* **2022**, *101*, 1–56. [CrossRef]
44. Guo, S.; Zhang, J.; Sun, H.; Salse, J.; Lucas, W.J.; Zhang, H.; Zheng, Y.; Mao, L.; Ren, Y.; Wang, Z.; et al. The draft genome of watermelon (*Citrullus lanatus*) and resequencing of 20 diverse accessions. *Nat. Genet.* **2013**, *45*, 51–58. [CrossRef]
45. FAO. Food and Agricultural Organization of the United Nations (FAOSTAT) Website. 2020. Available online: https://www.fao.org/faostat/zh/#data/QCL (accessed on 7 January 2022).
46. Prusky, D. Pathogen quiescence in postharvest diseases. *Annul. Rev. Phytopathol.* **1996**, *34*, 413–434. [CrossRef] [PubMed]
47. Farr, D.F.; Aime, M.C.; Rossman, A.Y.; Palm, M.E. Species of *Colletotrichum* on agavaceae. *Mycol. Res.* **2006**, *110*, 1395–1408. [CrossRef] [PubMed]
48. Du, M.Z.; Schardl, C.L.; Nuckles, E.M.; Vaillancourt, L.J. Using mating-type gene sequences for improved phylogenetic resolution of *Colletotrichum* species complexes. *Mycologia* **2005**, *97*, 641–658. [CrossRef]
49. Zivkovic, S.T.; Stosic, S.S.; Stevanovic, M.L.; Gasic, K.M.; Aleksic, G.A.; Vucurovic, I.B.; Ristic, D.T. *Colletotrichum orbiculare* on watermelon: Identification and in vitro inhibition by antagonistic fungi. *Zb. Matice Srp. Za Prir. Nauk.* **2017**, 331–343. [CrossRef]
50. UASD (U.S. Department of Agriculture). U.S. National Fungus Collections Statistics. 2022. Available online: https://www.fao.org/faostat/zh/#data/QCL (accessed on 20 April 2022).
51. Goh, K.S.; Balasubramaniam, J.; Sani, S.F.; Alam, M.W.; Ismail, N.A.; Gleason, M.L.; Rosli, H. First Report of *Colletotrichum scovillei* Causing Anthracnose on Watermelon (*Citrullus lanatus*) in Malaysia. *Plant Dis.* 2022; *Advance online publication.* [CrossRef]
52. Da Silva, V.N.; Guzzo, S.D.; Lucon, C.M.M.; Harakava, R. Promoção de crescimento e indução de resistencia a antracnose por *Trichoderma* spp. em pepineiro. *Pesqui. Agropecu. Bras.* **2011**, *46*, 1609–1618. [CrossRef]
53. Fang, Z. *Plant Disease Research Methods*, 3rd ed.; China Agriculture Press: Beijing, China, 1998; pp. 122–145.
54. Chomnunti, P.; Hongsanan, S.; Aguirre-Hudson, B.; Tian, Q.; Peršoh, D.; Dhami, M.K.; Alias, A.S.; Xu, J.; Liu, X.; Stadler, M.; et al. The sooty moulds. *Fungal Divers.* **2014**, *66*, 1–36. [CrossRef]
55. Gardes, M.; Bruns, T.D. ITS primers with enhanced specificity for basidiomycetes-application to the identification of mycorrhizae and rusts. *Mol. Ecol.* **1993**, *2*, 113–118. [CrossRef]
56. White, T.; Bruns, T.; Lee, S.; Taylor, J.; Innis, M.; Gelfand, D.; Sninsky, J. Amplification and Direct Sequencing of Fungal Ribosomal RNA Genes for Phylogenetics. In *PCR Protocols*; Academic Press: San Diego, CA, USA, 1990; Volume 31, pp. 315–322.
57. Guerber, J.C.; Liu, B.; Correll, J.C.; Johnston, P.R. Characterization of Diversity in *Colletotrichum acutatum* sensu lato by Sequence Analysis of Two Gene Introns, mtDNA and Intron RFLPs, and Mating Compatibility. *Mycologia* **2003**, *95*, 872–895. [CrossRef] [PubMed]

58. Carbone, I.; Kohn, L.M. A method for designing primer sets for speciation studies in filamentous *ascomycetes*. *Mycologia* **1999**, *91*, 553–556. [CrossRef]
59. Crous, P.W.; Groenewald, J.Z.; Risede, J.-M.; Simoneau, P.; Hywel-Jones, N. *Calonectria* species and their *Cylindrocladium* anamorphs: Species with sphaeropedunculate vesicles. *Stud. Mycol.* **2004**, *50*, 415–430.
60. O'Donnell, K.; Cigelnik, E. Two Divergent Intragenomic rDNA ITS2 Types within a Monophyletic Lineage of the Fungus *Fusarium* Are Nonorthologous. *Mol. Phyl. Evol.* **1997**, *7*, 103–116. [CrossRef]
61. Glass, L.N.; Donaldson, G.C. Development of primer sets designed for use with the PCR to amplify conserved genes from Filamentous *Ascomycetes*. *Am. Soc. Microbiol.* **1995**, *61*, 1320–1330. [CrossRef]
62. Woudenberg, J.H.; Aveskamp, M.M.; de Gruyter, J.; Spiers, A.G.; Crous, P.W. Multiple *Didymella* teleomorphs are linked to the *Phoma clematidina* morphotype. *Persoonia* **2009**, *22*, 56–62. [CrossRef] [PubMed]
63. Katoh, K.; Standley, D.M. MAFFT multiple sequence alignment software version 7: Improvements in performance and usability. *Mol. Biol. Evol.* **2013**, *30*, 772–780. [CrossRef] [PubMed]
64. Vaidya, G.; Lohman, D.J.; Meier, R. SequenceMatrix: Concatenation software for the fast assembly of multi-gene datasets with character set and codon information. *Cladistics* **2011**, *27*, 171–180. [CrossRef]
65. Ronquist, F.; Huelsenbeck, J.P. MrBayes 3: Bayesian phylogenetic inference under mixed models. *Bioinformatics* **2003**, *19*, 1572–1574. [CrossRef]
66. Nylander, J. MrModeltest V2. Program Distributed by the Author. *Bioinformatics* **2004**, *24*, 581–583. [CrossRef]
67. Rambaut, A.; Drummond, A.J.; Xie, D.; Baele, G.; Suchard, M.A. Posterior Summarization in Bayesian Phylogenetics Using Tracer 1.7. *Syst. Biol.* **2018**, *67*, 901–904. [CrossRef] [PubMed]
68. Swofford, L.D.; Bell, D.C. PAUP 4.0 b10. In *Phylogenetic Analysis Using Parsimony (* and Other Methods)*; Sinauer Associates: Sunderland, MA, USA, 2002.
69. Minh, B.Q.; Schmidt, H.A.; Chernomor, O.; Schrempf, D.; Woodhams, M.D.; von Haeseler, A.; Lanfear, R. IQ-TREE 2: New Models and Efficient Methods for Phylogenetic Inference in the Genomic Era. *Mol. Biol. Evol.* **2020**, *37*, 1530–1534. [CrossRef] [PubMed]
70. Rambaut, A. FigTree v. 1.4.4. Institute of Evolutionary Biology, University of Edinburgh. 2018. Available online: https://tree.bio.ed.ac.uk/software/figtree/ (accessed on 1 March 2022).
71. Huson, D.H. SplitsTree: Analyzing and visualizing evolutionary data. *Bioinformatics* **1998**, *14*, 68–73. [CrossRef] [PubMed]
72. Huson, D.H.; Kloepper, T.H. Computing recombination networks from binary sequences. *Bioinformatics* **2005**, *21* (Suppl. S2), ii159–ii165. [CrossRef] [PubMed]
73. Huson, D.H.; Bryant, D. Application of phylogenetic networks in evolutionary studies. *Mol. Biol. Evol.* **2006**, *23*, 254–267. [CrossRef] [PubMed]
74. Nirenberg, H. Untersuchungen über die Morphologische und Biologische Differenzierung in der Fusarium-Sektion Liseola. Mitteilungen aus der Biologischen Bundesanstalt für Landund Forstwirtschaft (Berlin-Dahlem). 1976. No. 169. Available online: http://pascal-francis.inist.fr/vibad/index.php?action=getRecordDetail&idt=PASCAL7710018408 (accessed on 1 March 2022).
75. Lin, Q.; Kanchana-udomkarn, C.; Jaunet, T.; Mongkolporn, O. Genetic analysis of resistance to pepper anthracnose caused by *Colletotrichum capsici*. *Thail. J. Agric. Sci.* **2004**, *35*, 259–264.
76. Than, P.P.; Jeewon, R.; Hyde, K.D.; Pongsupasamit, S.; Mongkolporn, O.; Taylor, P.W.J. Characterization and pathogenicity of *Colletotrichum* species associated with anthracnose on chilli (*Capsicum* spp.) in Thailand. *Plant Pathol.* **2008**, *57*, 562–572. [CrossRef]
77. Crouch, J.A.; Clarke, B.; Hillman, B. Unraveling Evolutionary Relationships Among the Divergent Lineages of *Colletotrichum* Causing Anthracnose Disease in Turfgrass and Corn. *Phytopathology* **2006**, *96*, 46–60. [CrossRef]
78. Liu, F.; Weir, B.S.; Damm, U.; Crous, P.W.; Wang, Y.; Liu, B.; Wang, M.; Zhang, M.; Cai, L. Unravelling *Colletotrichum* species associated with Camellia: Employing ApMat and GS loci to resolve species in the *C. gloeosporioides* complex. *Persoonia* **2015**, *35*, 63–86. [CrossRef]
79. Zhao, Q.; Chen, X.; Ren, G.; Wang, J.; Liu, L.; Qian, W.; Wang, J. First report of *Colletotrichum chlorophyti* causing peanut anthracnose in China. *Plant Dis.* **2020**, *105*, 226. [CrossRef]
80. Tashiro, N.; Manabe, K.; Ide, Y. Emergence and frequency of highly benzimidazole-resistant *Colletotrichum gloeosporioides*, pathogen of Japanese pear anthracnose, after discontinued use of benzimidazole. *J. Gen. Plant Pathol.* **2012**, *78*, 221–226. [CrossRef]
81. Peres, N.A.; Timmer, L.W.; Adaskaveg, J.E.; Correll, J.C. Lifestyles of *Colletotrichum acutatum*. *Plant Dis.* **2005**, *89*, 784–796. [CrossRef] [PubMed]
82. Alkan, N.; Friedlander, G.; Ment, D.; Prusky, D.; Fluhr, R. Simultaneous transcriptome analysis of *Colletotrichum gloeosporioides* and tomato fruit pathosystem reveals novel fungal pathogenicity and fruit defense strategies. *New Phytol.* **2015**, *205*, 801–815. [CrossRef] [PubMed]
83. De Silva, D.D.; Crous, P.W.; Ades, P.K.; Hyde, K.D.; Taylor, P.W.J. Life styles of *Colletotrichum* species and implications for plant biosecurity. *Fungal Biol. Rev.* **2017**, *31*, 155–168. [CrossRef]
84. Jiang, J.; Zhai, H.; Li, H.; Wang, Z.; Chen, Y.; Hong, N.; Wang, G.; Chofong, G.N.; Xu, W. Identification and characterization of *Colletotrichum fructicola* causing black spots on young fruits related to bitter rot of pear (*Pyrus bretschneideri* Rehd.) in China. *Crop Prot.* **2014**, *58*, 41–48. [CrossRef]

85. De Silva, D.D.; Groenewald, J.Z.; Crous, P.W.; Ades, P.K.; Nasruddin, A.; Mongkolporn, O.; Taylor, P.W.J. Identification, prevalence and pathogenicity of *Colletotrichum* species causing anthracnose of *Capsicum annuum* in Asia. *IMA Fungus* **2019**, *10*, 8. [CrossRef]
86. Shivas, R.G.; Tan, Y.P.; Edwards, J.; Dinh, Q.; Maxwell, A.; Andjic, V.; Liberato, J.R.; Anderson, C.; Beasley, D.R.; Bransgrove, K.; et al. *Colletotrichum* species in Australia. *Australas. Plant Pathol.* **2016**, *45*, 447–464. [CrossRef]

Journal of *Fungi*

MDPI

Article

Hidden Species Diversity was Explored in Two Genera of Catapyrenioid Lichens (Verrucariaceae, Ascomycota) from the Deserts of China

Tingting Zhang [1,2], Xin Zhang [1,3], Qiuxia Yang [1] and Xinli Wei [1,2,*]

[1] State Key Laboratory of Mycology, Institute of Microbiology, Chinese Academy of Sciences, Beijing 100101, China; tingtingzhang813@163.com (T.Z.); monetax@163.com (X.Z.); yangqx@im.ac.cn (Q.Y.)
[2] College of Life Sciences, University of Chinese Academy of Sciences, Beijing 100049, China
[3] College of Plant Sciences, Tibet Agricultural and Animal Husbandry University, Linzhi 860000, China
* Correspondence: weixl@im.ac.cn

Abstract: Verrucariaceae is the third-largest lichen family with high species diversity. However, this diversity has not been well-explored in China. We carried out a wide-scale field investigation in the arid and semi-arid regions of Northwest China from 2017 to 2021. A large number of lichen groups, especially those commonly distributed in deserts, were collected. Based on molecular phylogeny using ITS and nuLSU sequences by Bayesian and maximum likelihood analyses, combining morphological characters, seven taxa of catapyrenioid lichens in Verricariaceae were found in this study, including one genus (*Clavascidium*) and one species (*Clavascidium lacinulatum*) new to China; one genus (*Placidium*) new to the mainland of China; and four species (*Clavascidium sinense*, *Placidium nitidulum*, *Placidium nigrum*, and *Placidium varium*) new to science. It enriched our understanding of the high species diversity in Verrucariaceae and the lichen flora of Chinese arid and semi-arid deserts.

Keywords: catapyrenioid lichens; *Clavascidium*; new species; *Placidium*; taxonomy; Verrucariaceae

Citation: Zhang, T.; Zhang, X.; Yang, Q.; Wei, X. Hidden Species Diversity was Explored in Two Genera of Catapyrenioid Lichens (Verrucariaceae, Ascomycota) from the Deserts of China. *J. Fungi* **2022**, *8*, 729. https://doi.org/10.3390/jof8070729

Academic Editor: Pradeep K. Divakar

Received: 14 June 2022
Accepted: 11 July 2022
Published: 13 July 2022

1. Introduction

The lichen family Verrucariaceae Eschw. is affiliated with Verrucariales, Eurotiomycetes, and Ascomycota, including 43 genera and 943 species [1]. Members of this family can colonize on various substrates, such as rock, soil, wood or bark, moss, and even other lichens [2]. Many species can tolerate harsh environments and participate in forming biological soil crusts (BSCs) in arid and semi-arid regions, such as catapyrenioid lichens [3].

The original catapyrenioid lichens are characterized by squamulose thalli, including *Catapyrenium* s.l. and *Endocarpon* Hedw. [4], between which *Catapyrenium* s.l. can be distinguished by simple ascospores and the absence of hymenial algae [5]. *Catapyrenium* s.l. generally referred to eight genera [6], i.e., *Anthracocarpon* Breuss, *Catapyrenium* Flot., *Clavascidium* Breuss, *Heteroplacidium* Breuss, *Involucropyrenium* Breuss, *Neocatapyrenium* H. Harada, *Placidium* A. Massal., and *Scleropyrenium* H. Harada; however, later research showed that the eight genera did not cluster into the same lineage [2] but scattered in two lineages and at least three groups, i.e., the *Endocarpon* group, *Placidium* group, and *Staurothele* group [2,4,7], among which only the *Placidium* group still represented the originally defined catapyrenioid lichens that have squamulose thalli, simple ascospores, and the absence of hymenial algae, including *Clavascidium*, *Heteroplacidium*, and *Placidium*, although sometimes, in *Heteroplacidium*, areolate to squamulose-areolate thallus also exist besides the smaller squamulose thallus [2]. *Placidium* refers to the members with squamulose thalli, usually well-developed medulla and a lower cortex, cylindrical or clavate asci, and laminal or marginal *Dermatocarpon*-type pycnidia [8]. Based on the thallus structure and asci type, *Clavascidium* and *Placidium* are delimited into different genera [9]. The characters such as medulla, asci shape, pycnidia position, and the presence or absence of rhizines are crucial taxonomic criteria, especially the medulla type, which is a significant character in the delimitation of the genera *Clavascidium*, *Placidium*, and *Heteroplacidium* [10]. The kind of medulla in Verrucariaceae is highly variable, which contains three types: prosoplectenchymatous, paraplectenchymatous, and mixed-type. The prosoplectenchymatous type is characterized by loosely interlaced hyphae with elongated cells [11], the paraplectenchymatous type is composed of tightly arranged and rounded cells, and the mixed-type medulla ("Mischtyp") is composed of both rounded and elongated cells [12]. However, the variable medulla types generally occurred in *Clavascidium* and *Placidium*, not in *Heteroplacidium*, which only has the paraplectenchymatous medulla type.

In China, catapyrenioid lichens have not been well-studied; up to now, only two species of *Placidium* were reported, including *Placidium pilosellum* (Breuss) Breuss and *Placidium squamulosum* (Ach.) Breuss [13]. There is still a large space to explore the unknown species diversity in catapyrenioid lichens. Especially, those lichens are often the main components of biological soil crust in arid and semi-arid desert regions with important ecological functions, such as sand fixation by rhizines or lower surface and carbon fixations by photosynthesis [14,15]. Therefore, it has the important significance of exploring the catapyrenioid lichen species diversity for both taxonomy and ecology.

2. Materials and Methods

2.1. Taxon Sampling and Morphological Examination

About 3000 specimens were collected (Figure 1) from Northwest China in 2017–2021 and are preserved in the Lichen Section of Herbarium Mycologicum Academiae Sinicae, Beijing, China (HMAS-L), among which, 46 are *Clavascidium* and *Placidium* samples attracting our attention, because the two taxa have rarely been studied in China. The morphology and anatomy were examined using a MOTIC SMZ-168 stereomicroscope, a LEICA M125 dissecting microscope equipped with a Leica DFC450 camera, and a Zeiss Axio Imager A2-M2 equipped with Zeiss AxioCam MRC5 camera. The internal anatomy of the thallus was studied in sections of 10–15-μm thick cuts by a Leica CM1950 freezing microtome or by hand. Measurements were described as (a) b ± c (d), a = minimum value, b = mean value, c = standard deviation, and d = maximum value. Spot tests were performed using 10% KOH aqueous solution (K test) and Lugol's iodine solution. Lichen secondary metabolites were examined using standardized thin-layer chromatography (TLC, solvent C) [16].

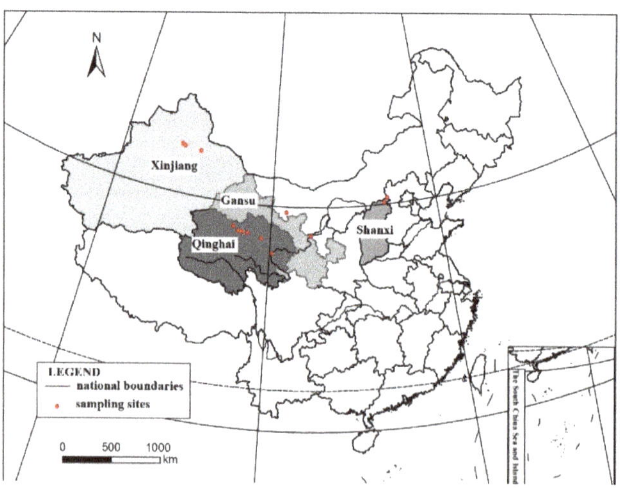

Figure 1. Collection sites. The detailed collection areas are marked in a solid red circle, and the corresponding four provinces involved are marked with varying degrees of a gray color.

2.2. DNA Extraction, PCR, and Sequencing

The forty-six *Clavascidium* and *Placidium* specimens were extracted from DNA, followed by a modified CTAB method [17]. The PCR amplification of nrDNA ITS (internal transcribed spacer) was performed using the primers ITS4 and ITS5 [18] and nuLSU (large subunit) using the primers PRI1 and PRI2 [7]. Amplifications were performed in 25-μL volumes containing 12.5 μL 2 × Taq PCR MasterMix (Nanjing Vazyme Co., Ltd., Nanjing, China), 1 μL of each primer solution (10 μM), 9.5 μL ddH_2O, and 1 μL dilutions (1:10) of genomic DNA. Amplifications were carried out in an ETC-811 plus thermal cycler (Beijing Eastwin Co., Ltd., Beijing, China), following conditions: an initial heating step for 5 min at 95 °C, followed by 35 cycles of 30 s at 94 °C, 30 s at 52 °C, and 1 min 30 s at 72 °C, with a final extension step of 10 min at 72 °C. The target PCR products were checked by electrophoresis on 1% agarose gels and then sequenced in SinoGenoMax Co., Ltd. (Beijing, China).

2.3. Sequence Alignment and Phylogenetic Analysis

A total of 160 DNA sequences including 61 new sequences (44 ITS, 17 nuLSU) were used in this study (Table S1) [2,4,7,19]. The new sequences generated for this study were deposited in GenBank. Twenty-nine accepted species previously sequenced, including 3 species of *Clavascidium*, accounting for 37.5% of the total 8 *Clavascidium* species, 19 species of *Placidium*, accounting for 57.6% of the total 33 *Placidium* species, and 7 species of *Heteroplacidium*, accounting for 58.3% the total 12 *Heteroplacidium* species, were included. *Placopyrenium* spp. were used as the outgroup. Raw sequences were assembled and edited with SeqMan [20] and then aligned using MAFFT v.7 [21]. The program Gblocks V0.19b [22,23] was used to remove ambiguously aligned sites. PAUP* v. 4.0 [24] was used for homogeneity testing ($p > 0.05$) before combining the two loci (ITS and nuLSU). All Maximum Likelihood and Bayesian analyses were performed using the GTR + I + G model selected by jModelTest 2 [25]. The randomized accelerated maximum likelihood (RAxML) analysis involving 1000 pseudoreplicates with RaxML v. 8.2.6 [26] was run on the Cipres Science Gateway (http://www.phylo.org, (accessed on 18 July 2011)). The Bayesian analysis performed using MrBayes v3.2.7 [27,28] with two parallel Markov chain Monte Carlo (MCMC), each using 5 million generations and sampling every 1000 steps, generating a 50% majority rule consensus tree after discarding the first 25% as the burn-in. TRACER v.1.7.2 [29] determined the burn-in value with effective sample sizes (ESS) higher than 200. All tree files were visualized with FigTree v.1.4.3 (http://tree.bio.ed.ac.uk/software/figtree, (accessed on 28 August 2014)). The clades with bootstrap (BP) values above 75 or posterior probability (PP) values above 0.95 were considered highly supported.

3. Results

3.1. Phylogenetic Analysis

A total of 61 sequences newly obtained in this study were used in the single-locus phylogenetic tree construction; among which, 30 sequences of the ITS and nuLSU generated from 15 specimens representing 5 species (7 *Clavascidium* specimens and 8 *Placidium* specimens) were combined for constructing the concatenated tree (Figure 2). The concatenated matrix included 1204 variable positions (448 ITS and 756 nuLSU) after excluding ambiguous regions. The aligned matrix contained 1204 nucleotide position characteristics for the complete data set of 67 members. The concatenated BI phylogenetic tree and two single-gene-locus RAxML trees including 44 and 16 samples are shown in Figures S1–S3. *Placidium*, *Clavascidium*, and *Heteroplacidium* all formed monophyletic branches. Both the ML and BI phylogenetic trees produced similar topologies. There are four new species: *Clavascidium sinense* sp. nov., *Placidium nitidulum* sp. nov., *Placidium nigrum* sp. nov., and *Placidium varium* sp. nov. are well-supported.

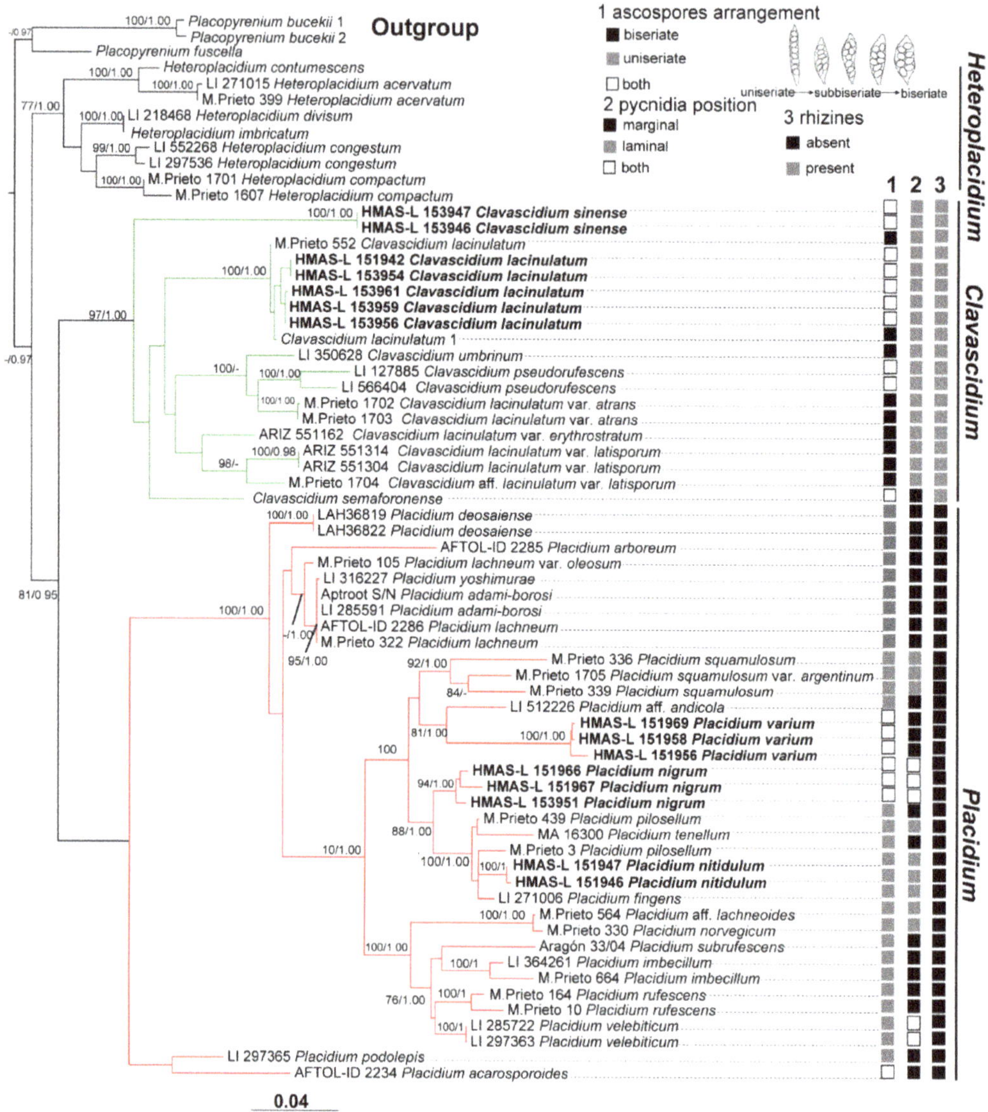

Figure 2. The RAxML tree is based on the concatenated ITS + nuLSU data set representing both ML and BI trees. The number in each node represents bootstrap support (BS) and posterior probability (PP) values. BS values ≥ 75 and PP values ≥ 0.95 were plotted on the branches. The taxa in bold indicate that these sequences were newly generated for this study. Green and red clades show *Clavascidium* and *Placidium*, respectively. The character status of 1 ascospores arrangement, 2 pycnidia position, and 3 rhizines are listed at the right of the tree, corresponding to each sample of *Clavascidium* and *Placidium*. Scale in 0.04 substitution per site.

3.2. Taxonomy

A key to the species of Clavascidium and Placidium is listed in Table 1.

Table 1. Key to the species of *Clavascidium* and *Placidium* in China.

1. Presence of rhizines	2
1′. Absence of rhizines	3
2. Upper cortex thinner 50–70 μm; medulla paraplectenchymatous	*Cl. lacinulatum*
2′. Upper cortex thicker 60–100 μm; medulla mixed type	*Cl. sinense*
3. Pycnidia both marginal and laminal	*Pl.nigrum*
3′. Pycnidia either marginal or laminal	4
4. Pycnidia laminal	5
4′. Pycnidia marginal	6
5. Squamules tiny 0.7–2 mm; asci slender 45–65 × 5–8 μm; medulla paraplectenchymatous	*Pl. nitidulum*
5′. Squamules larger 2–7 mm; asci 70–90 × 10–15 μm; medulla mixed type	*Pl. squamulosum*
6. Ascospores more elliptical (12–17 × 5.5–7 μm); squamules margin hairy	*Pl. pilosellum*
6′. Acospores are more rounded (10–13.5 × 7.5–9.5 μm) with central guttule; squamules margin smooth	*Pl. varium*

Clavascidium **Breuss**, Annln naturh. Mus. Wien, Ser. B, Bot. Zool. 98 (Suppl.): 41, 1996
The genus *Clavascidium* is characterized by squamulose thallus, the presence of rhizines, and clavate to (sub) cylindrical asci containing biseriate ascospores. Gueidan [2,9] introduced *Clavascidium* as a sister genus of *Placidium* based on morphological characters and a phylogenetic analysis, between which *Clavascidium* has clavate asci with biseriate ascospores, whereas *Placidium* has cylindrical asci with uniseriate ascospores. However, uniseriate or both biseriate and uniseriate ascospores were also reported [30].
Clavascidium lacinulatum (Ach.) M. Prieto, in Prieto et al., Am. J. Bot. 99 (1): 28, 2012 (Figure 3)
≡ *Endocarpon hepaticum* var. *Lacinulatum* Ach Lich. univ.: 299, 1810.
Description: Thallus squamulose, terricolous, 300 μm thick; lobes 1–4 mm wide, roundish to deeply lobed, contiguous, rarely overlapping; upper surface dark brown, dull; lower surface pale ± rhizines; epinecral layer transparent, up to 15 μm thick; upper cortex paraplectenchymous, 50–70 μm thick, cells 4 × 10 μm in diam.; uppmost layer brown, 25–30 μm thick; photobiont layer 50–100 μm thick, algal cells 6–10 μm in diam; medullar tissue paraplectenchymous; lower cortex not delimited from the medulla, 40 μm thick, paraplectenchymous, composed of irregularly arranged roundish cells up to 7.5 μm in diam. Rhizohyphae hyaline, 4–6 μm wide. Pycnidia laminal, *Dermatocarpon*–type, immersed, subglobose, light brown, conidia oblong–ellipsoid to bacilliform, 1–1.3 × 3–3.7 μm in size. Perithecia immersed, broadly pyriform (150 × 180 μm) to subglobose (up to 230 μm) wide; perithecia wall bright, 30–35 μm thick; hymenium bright, 35–45 μm thick; involucrellum brown, 42–44 μm thick; pyriphyses 35–40 × 2.5–3 μm; asci cylindrical to clavate, 14–18 × 40–52 μm, ascospores 8 per ascus, uniseriate to biseriate, ellipsoid to fusiform to ovoid, 5–7 × 10–12 μm.
Chemistry: All the spot tests were negative, and no substances were detected by TLC.
Habitat and distribution: These species grow on the surface of sandy soil in the semi-arid and arid region of Northwest China, located in the open areas with sun exposure. The surrounding environment is characterized by interlace of meadows, fixed undulating sand dunes under shrubs, and the Gobi Desert, with the elevations greatly varying from 923 to 3175 m. It distributes worldwide [30] and is new to China.
Specimens examined: CHINA. SHANXI: Datong City, Yanggao County. 40.98° N 113.88° E, 1264 m alt., on the sand, 14 April 2021, X. Qian & T.T. Zhang 20210273 (HMAS–L 153955), 20210274 (HMAS–L 153956), 20210283 (HMAS–L 153957); 40°25′51″N 113°44′19″E, 923 m alt., on the sand, 15 April 2021, X. Qian & T.T. Zhang 20210299 (HMAS–L 153962), 20210300 (HMAS–L 153958), 20210302 (HMAS–L 153959), 20210306 (HMAS–L 153960), 20210307 (HMAS–L 153961). GANSU: Baiyin City, Jingtai County. 37°19′41″ N 104°35′37″ E, 1524 m alt., on the sand, 21 October 2020, X. Qian et al. 20201529 (HMAS–L 153781). QING-

HAI: Haixi Mongolian and Tibetan Autonomous Prefecture, Delhi City. 37°23′02″ N 97°08′44″ E, 3041 m alt., on the sand, 24 April 2021, X.L. Wei & T.T. Zhang 20210436 (HMAS–L 151942); 37°24′29″ N 96°30′43″ E, 3175 m alt., on the sand, 24 April 2021, X.L. Wei & T.T. Zhang 20210499 (HMAS–L 153954); Hainan Tibetan Autonomous Prefecture, Guinan County. 35°40′54″ N 100°15′08″ E, 1606 m alt., on the sand, 16 October 2020, X. Qian et al. QX20200026 (HMAS–L 153953).

Notes: This species is widespread in the desert regions of China as a crust (Figure 1).Our specimens are well-clustered with *Cl. lacinulatum* in the phylogeny [4] (Figures 2 and S1–S3) and, also, with a high consistency of the phenotype described by Nash et al. [30]. Although this species is variable in the external morphology of thallus and squamules, several important taxonomic characters such as laminal pycnidia, oblong-ellipsoidal to subcylindrical conidia, and rhizines are constantly present. Breuss described uniseriate ascospores in this species [31]; however, we found this character is variable (Figure 3).

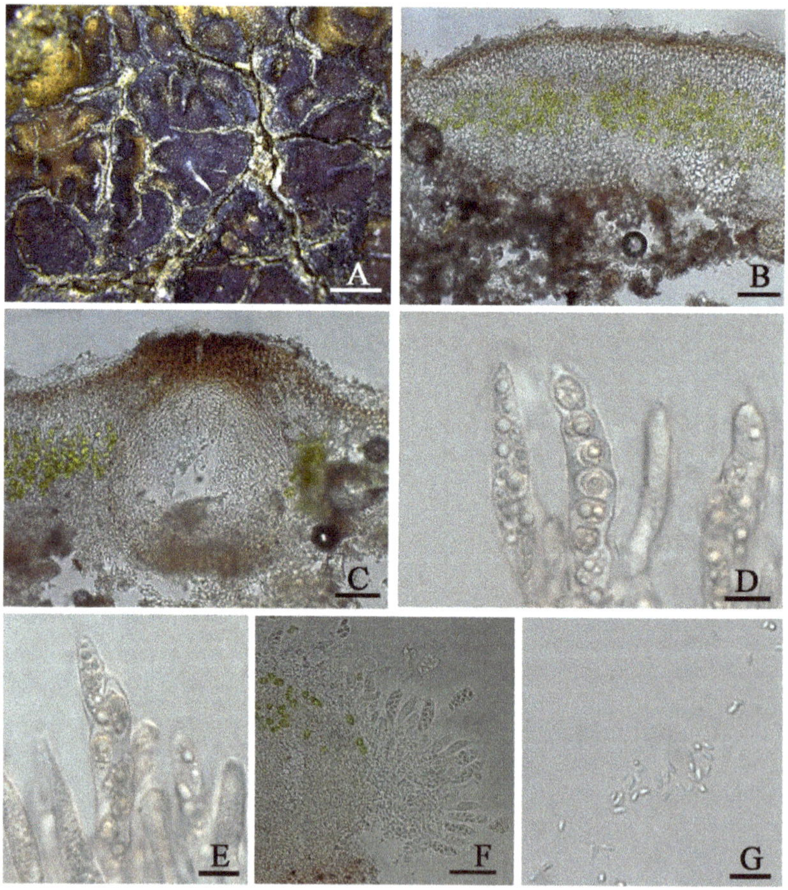

Figure 3. The thallus habit and the anatomic structure of *Clavascidium lacinulatum* (HMAS-L 153962). (**A**). Squamulose thallus. (**B**). Transversal section of thallus. (**C**). Immersed perithecium. (**D**,**E**). Uniseriate ascospores in cylindrical asci. (**F**). Biseriate ascospores in clavate asci. (**G**). Conidia. Bars: (**A**) =2 mm, (**B**,**C**) =50 μm, (**D**,**E**) =10 μm, (**F**) =50 μm, and (**G**) =10 μm.

Clavascidium sinense T.T. Zhang & X.L. Wei, **sp. nov.** (Figure 4)
Fungal Names No.: FN571022
Etymology: The epithet '*sinense*' refers to the Chinese distribution of this species.
Typus: CHINA. SHANXI: Datong City, Yanggao County. 40°25′28″ N 113°46′07″ E, 1177 m alt., on soil, 14 April 2021, X. Qian & T.T. Zhang 20210245 (holotype, HMAS–L 153946, ITS ON712842, nuLSU ON712829).
Diagnosis: It is characterized by the co-existence of both uniseriate and biseriate ascospores in asci, laminal pycnidia, and mixed-type medulla.
Description: Thallus squamulose, terricolous; lobes 1.5–4 mm wide, (194) 255 ± 32 (314) μm thick, deeply lobed, contiguous, rarely overlapping, full appressed to the substrate or with raised margins free from the substrate; upper surface medium to dark brown, dull; lower surface pale ± rhizines; epinecral layer often absent, if present up to 10 μm thick; upper cortex paraplectenchymatous, (57) 81 ± 19 (109) μm thick, cells (7.8) 9.9 ± 1.7 (12.6) μm in diam., uppmost layer bright brown, 15–25 μm thick; algal layer (70) 107 ± 24 (157) μm thick, composed of rounded cells (7.4) 9.5 ± 1.6 (12) μm in diam., globose to sub-globose; medulla mixed type, 60–100 μm thick; lower cortex not clearly delimited from the medulla, transparent, outer layer of lower cortex bright brown, (37) 51 ± 8.5 (59) μm thick, composed of rounded cells (7.2) 8.2 ± 0.9 (10.2) μm diam. Rhizohyphae 4–5 μm thick, colorless. Pycnidia laminal, *Dermatocarpon*-type, subglobose, light brown, conidia oblong-ellipsoid to bacilliform, (3.0) 3.4 ± 0.29 (3.7) × (1.26) 1.4 ± 0.18 (1.75) μm.
Perithecia laminal, immersed, subglobular, up to 400 μm in diam.; perithecia wall 40–55 μm thick; hymenium pale brown, 65–100 μm thick; involucrellum brown, 38–50 μm thick; pyriphyses 30–40 μm long; asci (sub)cylindrical to clavate, 64–84 × 12–15 μm, 8 spored, ascospores uniseriate to biseriate, hyaline, narrow ellipsoid, (10.2) 13 ± 1.2 (13.9) × (4.8) 6.3 ± 0.7 (7.3) μm.
Chemistry: All the spot tests were negative, and no substances were detected by TLC.
Habitat and distribution: On the surface of a broken great wall built by clay in the semi-arid region of Northwest China. It has been known only in China up to now.
Additional specimens examined: CHINA. SHANXI: Datong City, Yanggao County. 40°25′28″ N 113°46′07″ E, 1177 m alt., on soil, 14 April 2021, X. Qian & T.T. Zhang 20210246 (HMAS-L 153947).
Notes: Biseriate and uniseriate ascospores in asci are generally diagnostic characters of *Clavascidium* and *Placidium*, respectively [2,9]. However, uniseriate or both biseriate and uniseriate ascospores in asci were also found in *Clavascidium* [30], including *Cl. sinense*, indicating ascospores arrangement is a continuously changing character between these two closely related genera in phylogeny. The phylogenetic analysis clearly supported *Cl. sinense* obviously separated from other *Clavascidium* species with both types of ascospores arrangements, such as *Cl. lacinulatum*, *Cl. pseudorufescens* (Breuss) M. Prieto, and *Cl. semaforonense* (Breuss) M. Prieto. However, *Cl. sinense* has mixed-type medulla, distinct from *Cl. lacinulatum* with paraplectenchymous medulla and *Cl. pseudorufescens* with filamentous-hyphae medulla [30]. *Cl. sinense* has laminal pycnidia, but *Cl. semaforoense* has marginal pycnidia, which is an exception within the genus *Clavascidium* [8]. The phylogenetic trees (Figures 2 and S1–S3) provided further support, indicating *Cl. semaforoense* formed a separate clade, far from all the other species of *Clavascidium*. However, the new species *Cl. sinense* situated in the outermost clade, and *Cl. semaforoense* situated in the secondary outermost, indicating that laminal or marginal pycnidia is not an absolutely unchanging character within this genus but can be used in delimitation species. There are three *Clavascidium* species absence of DNA sequences: *Clavascidium antillarum* (Breuss) Breuss, *Clavascidium imitans* (Breuss) M. Prieto, and *Clavascidium krylovianum* (Tomin) M. Prieto. However, *Cl. sinense* has distinct traits compared with these three species, for example, *Cl. antillarum* is characterized by dark brown to black lower surface and rhizines [5], while *Cl. sinense* is characterized by a pale lower surface and rhizines. *Cl. imitans* and *Cl. krylovianum* can be distinguished by the morphology of medulla; the former medullary hyphae divide into many short and swollen cells, but the latter medullary hyphae are filamentous [5]; in comparison, *Cl. sinense* has a mixed-type medulla.

Figure 4. The thallus habit and the anatomic structure of *Clavascidium sinense* sp. nov (holotype HMAS-L 153946). (**A**). Squamulose thallus and immersed apothecia with the appearance of black spots. (**B**). Transversal section of thallus. (**C**). Perithecium. (**D**). Pycnidium. (**E**). Thallus section. (**F**). Uniseriate ascospores in cylindrical asci. (**G**). Biseriate ascospores in clavate ascus. (**H**,**I**). Ascospores. Bars: (**A**) =2 mm, (**B**) =100 μm, (**C**) =50 μm, (**D**) =100 μm, (**E**) =50 μm, (**F**,**G**) =10 μm, and (**H**,**I**) =5 μm.

Placidium A. Massal. Symmict. Lich.: 75, 1855

The genus *Placidium* has squamulose thallus, laminal or marginal *Dermatocarpon*-type pycnidia, cylindrical asci, uniseriate ascospores, and rhizohyphae [8]. Two *Placidium* species have been reported from Taiwan by Dr. Aptroot [13]. This genus is firstly reported from Mainland China in this study, and we found some additional morphological characters in *Placidium* such as a glossy upper surface and aggregated pycnidia, which can be used to distinguish some species.

Placidium nitidulum T.T. Zhang & X.L. Wei **sp. nov.** (Figure 5)

Fungal Names No.: FN571023

Etymology: The epithet '*nitidulum*' refers to the glossy upper surface of the thallus in this species.

Typus: CHINA. QINGHAI: Haixi Mongolian and Tibetan Autonomous Prefecture, Wulan County. 36°57′31′′ N 98°54′03′′ E, 3314 m alt., on the sand, 25 April 2021, X.L. Wei & T.T. Zhang 20210560 (holotype, HMAS–L 151947, ITS ON712844, nuLSU ON712835).

Diagnosis: It is characterized by a glossy upper surface, tiny lobes, and a thick algal layer.
Description: Thallus squamulose; terricolous; lobes roundish, tiny, 0.7–2 mm wide, (356) 442 ± 55 (523) μm thick; contiguous to densely aggregated, tightly adnate to the substrate; upper surface medium brown, glossy, brown rimmed; lower surface pale to pale yellow; epinecral layer transparent, (33.4) 35.4 ± 1.6 (37.5) μm thick; upper cortex (41) 58 ± 15 (78) μm thick, paraplectenchymatous, cells 8–10 μm in diam.; uppermost layer pale yellow to light brown, 33–36 μm thick; algal layer (104) 159 ± 34 (195) μm thick, algal cells globular, (7.3) 7.8 ± 0.5 (8.7) μm in diam. dispersed over the whole medulla; medulla zone not clear; lower cortex not delimited from the medulla, (42) 63 ± 16 (88) μm thick, composed of densely aggregated globular cells, cells (8.4) 10.6 ± 1.5 (12.6) μm in diam. Rhizohyphae (4.0) 5.2 ± 0.7 (5.9) μm thick, colorless. Pycnidia laminal, immersed, *Dermatocarpon*-type, immature, subglobular, 250 μm in dim., conidia not seen.
Perithecia laminal, immersed, occasionally aggregated, narrowly pyriform, up to 300 μm wide; perithecia wall grey, 28–42 μm thick; hymenium pale yellow, 45–75 μm thick; involucrellum brown or absence, 68–90 μm thick; paraphyses branched, 35–55 × 1.5–2.5 μm; asci cylindrical, 45–65 × 5–8 μm, 8 spored, uniseriate, ascospores hyaline, narrow ellipsoid, (8.4) 8.7 ± 1.5 (11.5) × (5.1) 6.1 ± 0.5 (7.0) μm; paraphyses well-developed.
Chemistry: All the spot tests were negative, and no substances were detected by TLC.
Ecology and distribution: On the surface of sandy soil in high altitude areas of the Qinghai–Tibet plateau. It has been known only in China up to now.
Additional specimens examined: CHINA. QINGHAI: Haixi Mongolian and Tibetan Autonomous Prefecture, Delhi City. 37°23′02″ N 97°08′44″ E, 3041 m alt., on the sand, 24 April 2021, X.L. Wei & T.T. Zhang 20210421 (HMAS-L 151962); Wulan County. 36°57′31″ N 98°54′03″ E, 3314 m alt., on the sand, 25 April 2021, X.L. Wei & T.T. Zhang 20210552 (HMAS–L 151946).
Notes: This new species can be easily recognized by its glossy appearance of the upper surface and tiny lobes, which is very distinctive and different from all the other known *Placidium* species. The medulla zone is obscure due to being fully covered by the thick algal layer, so the lower cortex is difficult to delimit from the medulla, similar to *Placidium tenellum* (Breuss) Breuss in this character. However, the upper surface of *Pl. tenellum* is matt, the algal layer is thinner ((40) 93 ± 24 (155) μm), and the pycnidia is much broader (up to 500 μm). The distribution is more frequent in coastal areas [8]. Based on the phylogenetic trees (Figure 2 and Figures S1–S3), the new species is close to *Placidium fingens* (Breuss) Breuss, *Pl. pilosellum*, and *Pl. tenellum*, among which the first two are more intimate than the last one to the new species in phylogeny; however, the lobe widths of *Pl. fingens* and *Pl. pilosellum* (up to 6 mm) are nearly three times the new species (0.7–2 mm). *Pl. nitidulum* is closer to *Pl. fingens*, both of which have laminal pycnidia, while the species *Pl. pilosellum* within this subclade with a little far distance has marginal pycnidia. *Pl. nitidulum* has more slender asci (45–65 × 5–8 μm) than the species *Pl. pilosellum* (70–90 × 10–15 μm). Additionally, *Pl. nitidulum* has squamules with a smooth margin, but *Pl. pilosellum* has squamules with hairy margins [8].

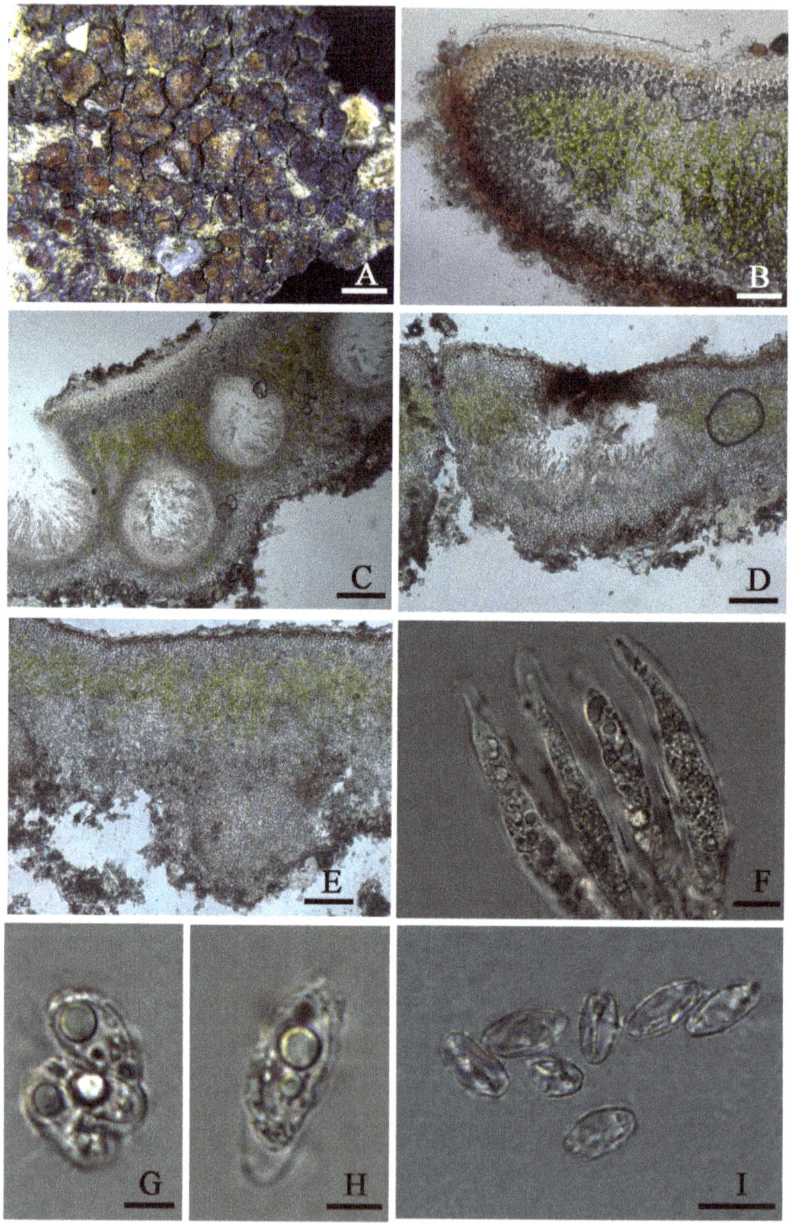

Figure 5. The thallus habit and the anatomic structure of *Placidium nitidulum* sp. nov (holotype HMAS-L 151947). (**A**). Squamulose thallus. (**B**). Transversal section of thallus. (**C**). Immersed perithecia. (**D**). Aggregated pyrithecia. (**E**). Pycnidium. (**F**). Cylindrical asci with uniseriate ascospores. (**G–I**). Ascospores. Bars: (**A**) =2 mm, (**B**) =50 μm, (**C–E**) =100 μm, (**F**) =10 μm, (**G,H**) =5 μm, and (**I**) =10 μm.

Placidiumnigrum T.T. Zhang & X.L. Wei, **sp. nov.** (Figure 6)
Fungal Names No.: FN571024
Etymology: The epithet '*nigrum*' refers to the surrounding black area due to the aggregation of abundant pycnidia of this species.
Typus: CHINA. QINGHAI: Haixi Mongolian and Tibetan Autonomous Prefecture, Dachaidan Region. 37°52′11′′ N 95°15′54′′ E, 3515 m alt., on the sand, 25 April 2021, X.L. Wei & T.T. Zhang 20210516 (holotype, HMAS–L 151944, ITS ON712851).
Diagnosis: It is characterized by both laminal and marginal pycnidia, with abundant aggregation forming into the surrounding black area.
Description: Thallus squamulose, terricolous; lobes 1–3 mm wide, (211) 336 ± 81 (462) μm thick, tumid roundish to lobate, scattered or contiguous to densely aggregated, tightly adnate to the substrate or with slightly raised margins; upper surface pale to medium brown, dull, pruinose; epinecral layer transparent, (12) 20.4 ± 7.2 (33) μm thick; upper cortex paraplectenchymatous, (31) 52 ± 11 (70) μm thick, cells (8.5) 10.5 ± 1.9 (13.8) μm in diam., uppermost layer brown, 8–12 μm thick; algal layer (37) 67 ± 14 (85) μm thick, composed of spherical cells, cells (7.4) 13.3 ± 2.9 (17.6) μm in diam., globose to subglobose, in clusters of 1–3 cells; medulla mixed type, (37) 69 ± 17 (102) μm thick; lower cortex delimited from the medulla, hyaline, (32) 44 ± 9.8 (73) μm thick, composed of spherical cells, cells (8.1) 9.8 ± 1.4 (12.4) μm in diam. Rhizohyphae (3.3) 4.4 ± 0.77 (5.7) μm thick, colorless. Pycnidia laminal and marginal, abundant, aggregated into all-round black area, immersed, *Dermatocarpon*-type, subglobular to irregular, up to (230) 274 ± 33 (317) μm wide; conidia bacilliform, (2.7) 3 ± 0.39 (3.5) × (0.96) 1.3 ± 0.29 (1.89) μm.
Perithecia laminal, immersed, subglobular, 320 × 350 μm; perithecia wall bright, 28–50 μm thick; hymenium 70–85 μm thick; involucrellum brown, 43–55 μm thick; periphyses 40–55 μm long; asci cylindrical to clavate, 45–65 × 9–18 μm, 8 spored, uniseriate to (sub)biseriate, ascospores hyaline, ellipsoid, (8.4) 9.9 ± 1 (11.5) × (4.0) 5.7 ± 0.9 (8.0).
Chemistry: All the spot tests were negative, and no substances were detected by TLC.
Habitat and distribution: On the surface of sandy soil in the semi-arid and arid region of Northwest China and the Qinghai–Tibet plateau. The distribution range is relatively wide, from low altitude to high altitude (438–3515 m alt.). The environment in which they grow is also more varied, including naturally formed bush bottoms and artificial vegetation fix sand forests (i.e., *Ammopiptanthus mongolicus* (Maxim. ex Kom.) Cheng f. and *Populus alba* var. *pyramidalis* Bge.), and the Gobi Desert, common in dry and open habitats. It has been known only in China up to now.
Additional specimens examined: CHINA. QINGHAI: Haixi Mongolian and Tibetan Autonomous Prefecture, Dachaidan Region. 37°52′11′′ N 95°15′54′′ E, 3515 m alt., on sand, 25 April 2021, X.L. Wei & T.T. Zhang 20210535 (HMAS–L 153951), 20210522 (HMAS–L 151945), 20210523 (HMAS–L 151965), 20210524 (HMAS–L 151966), 20210534 (HMAS–L 153950); Delhi City. 37°23′02′′ N 97°08′44′′ E, 3041 m alt., on the sand, 24 April 2021, X.L. Wei & T.T. Zhang 20210461 (HMAS–L 151963); Hainan Tibetan Autonomous Prefecture, Guinan County. 35°40′54′′ N 100°15′08′′ E, 1606 m alt., on the sand, 16 October 2020, X. Qian et al. 20201384 (HMAS–L 153732). GANSU: Baiyin City, Jingtai County. 37°25′06′′ N 104°34′56′′ E, 1591 m alt., on the sand, 21 October 2020, X. Qian et al. 20201486 (HMAS–L 153786). XINJIANG: Changji Hui Autonomous Prefecture, Fukang County. 44°22′59′′ N 87°52′34′′ E, 438 m alt., on the sand, 10 May 2021, X. Qian & T.T. Zhang 20210666 (HMAS–L 151952); Qitai County. 44°13′32′′ N 90°02′32′′ E, 723 m alt., on the sand, 10 May 2021, X. Qian & T.T. Zhang 20210746 (HMAS–L 151941); Wujiaqu City. 44°29′54′′ N 87°28′21′′ E, 408 m alt., on the sand, 9 May 2021, X. Qian & T.T. Zhang 20210606 (HMAS–L 151967). INNER MONGOLIA: Alxa Right Banner, 39°28′24′′ N 101°04′03′′ E, 1564 m alt., on sand, 22 July 2017, D.L. Liu & R.D. Liu XL2017267 (HMAS–L 140940); 39°28′22′′ N 101°04′04′′ E, 1563 m alt., on sand, 5 June 2018, D.L. Liu et al. ALS2018022 (HMAS–L 143912); 39°32′30″ N 101°06′34′′ E, 1478 m alt., on sand, 5 June 2018, D.L. Liu et al. ALS2018040 (HMAS–L 153952).

Notes: This species is distinctive by having both uniseriate and (sub) biseriate ascospores arrangement, both laminal and marginal pycnidia, and a surrounding black area due to the aggregation of abundant pycnidia, which well-separate it from all the other *Placidium* species. *Placidium* is generally known as the only genus comprising species with laminal or marginal pycnidia [31]. Within this genus, very few species have both laminal and marginal pycnidia such as *Pl. velebiticum* (Zahlbr. ex Zschacke) Breuss [4,8]; however, *Pl. velebiticum* only has a uniseriate ascospores arrangement, which is different from *Pl. nigrum*.

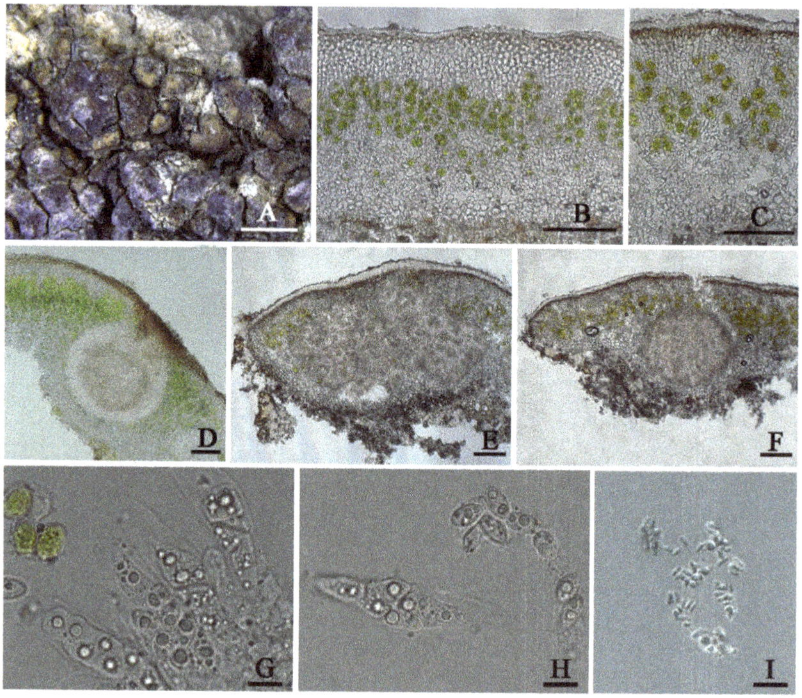

Figure 6. The thallus habit and the anatomic structure of *Placidium nigrum* sp. nov (HMAS-L 153950). (**A**). Squamulose thallus. (**B**,**C**). Thallus section. (**D**). Immersed perithecium. (**E**,**F**). *Dermatocarpon*-type pycnidia. (**G**,**H**). Asci with ascospores. (**I**). Conidia. Bars: (**A**) =2 mm, (**B**,**C**) =100 μm, (**D**) =200 μm, (**E**,**F**) =100 μm, and (**G**–**I**) =10 μm.

Placidium varium T.T. Zhang & X.L. Wei, **sp. nov.** (Figure 7)
Fungal Names No.: FN571025
Etymology: The epithet '*varium*' refers to the morphology and arrangement of the variable ascospores in this species.
Typus: CHINA. XINJIANG: Changji Hui Autonomous Prefecture, Qitai County. 44°13′32″ N 90°02′32″ E, 723 m alt., on the sand, 10 May 2021, X. Qian & T.T. Zhang 20210751 (holotype, HMAS–L 151970, nuLSU ON712841).
Diagnosis: This new species is characterized chiefly by both (sub)biseriate and uniseriate ascospores arrangement and having macro guttule in the center of each ascospore.
Description: Thallus squamulose, terricolous; lobes up to 4 mm wide, (183) 202 ± 10 (224) μm thick, deeply lobate, scattered or contiguous to densely aggregated, sometimes overlapped or imbricated, adpressed to the substrate or with raised margins free from the substrate, sometimes with a black margin; upper surface red-brown to dark brown, pruinose, dull; lower surface greyish brown; epinecral layer transparent, very thin, up to 20 μm thick; upper cortex (31) 46 ± 9 (65) μm thick, paraplectenchymatous cells (7.8) 8.2 ± 0.3 (8.6) μm diam, uppermost layer light brown, 10–13 μm thick; algal layer (73)

93 ± 14 (112) μm thick, algal cells globular, (7.2) 8.5 ± 0.9 (10) μm in diam., in clusters of 1–3 cells; medulla mixed type, (46) 54 ± 5(62) μm thick; lower cortex not delimited from the medulla, hyaline, (28) 33 ± 5.9 (43) μm thick, composed of spherical cells, (5.9) 7.7 ± 1.5 (10) μm in diam. Rhizohyphae (3.2) 4.4 ± 0.78 (5.7) μm thick, colorless. Pycnidia marginal, immersed, *Dermatocarpon*-type, subglobular to pyriform, up to 256–320 × 220–288 μm in size; conidia bacilliform to oblong-ellipsoid, (3.1) 3.5 ± 0.21 (3.9) × (1.0) 1.5 ± 0.2 (1.9) μm. Perithecia immersed, subglobular to broad pyriform, up to 260 μm wide; perithecia wall bright, 19–25 μm thick; hymenium bright, 22–46 μm thick; involucrellum abscence; periphyses 18–26 μm long; asci cylindrical to clavate, 55–73 × 8–21 μm, 8 spored, uniseriate to (sub)biseriate, ascospores hyaline, ellipsoid to ovoid, (8.5) 11.8 ± 1.8 (15.3) × (7.6) 8.6 ± 1.0 (11.6) μm in size, central guttule (6.6) 8.0 ± 1.1 (10.8) × (4.9) 5.8 ± 0.68 (7.0) μm.

Chemistry: All the spot tests were negative, and no substances were detected by TLC.

Ecology and distribution: On the surface of sandy soil in the arid region of Xinjiang Autonomous, Northwest China. The specimens were collected at the bottom of a fixed sand area of cultivated shrubs. The habitat is a low-altitude open area with long daylight exposure, low precipitation and high evaporation. It has been known only in China up to now.

Additional specimens examined: CHINA. XINJIANG: Changji Hui Autonomous Prefecture, Fukang County. 44°22′59″ N 87°52′34″ E, 438 m alt., on the sand, 10 May 2021, X. Qian & T.T. Zhang 20210639 (HMAS–L 153949), 20210676 (HMAS–L 151953), 20210689 (HMAS–L 151955), 20210692 (HMAS–L 1519561), 20210698 (HMAS–L 151958), 20210696 (HMAS–L 151957); Wujiaqu City. 44°29′54″ N 87°28′21″ E, 408 m alt., on the sand, 9 May 2021, X. Qian & T.T. Zhang 20210576 (HMAS–L 153948), 20210599 (HMAS–L 151948), 20210608 (HMAS–L 151949), 20210611 (HMAS–L 151940), 20210623 (HMAS–L 151951); Qitai County. 44°13′32″ N 90°02′32″ E, 723 m alt., on the sand, 10 May 2021, X. Qian & T.T. Zhang 20210725 (HMAS–L 151969), 20210716 (HMAS–L 152826).

Notes: The ascospores arrangement is variable in this new species. Compared with other *Placidium* species with small ascospores, such as *Placidium arboretum* (Schwein. ex E. Michener) Lendemer (syn. *Placidium tuckermanii* (Rav. ex Mont.) Breuss), *Placidium andicola* (Breuss) Breuss, *Placidium chilense* (Breuss) Breuss, *Placidium corticola* (Räsänen) Breuss, and *Placidium ruiz-lealii* (Räsänen) Breuss, characterized by oval and uniseriate ascospores [30,32], the new species has more variable ascospores in shape and arrangement. Moreover, several species are not terricolous like this new species but corticolous in *Pl. corticola* and *Pl. arboretum* or saxicolous in *Pl. ruiz-lealii* [30]. Although *Pl. andicola* and *Pl. chilense* are terricolous, they are different from the new species in morphology besides the different shapes of ascospores; for example, *Pl. andicola* has brown but not red-brown thallus and only uniseriate ascospores in asci [30]. *Pl. chilense* has much more extensive (4–10 (−20) mm wide) and thicker (600 μm thick) lobes and much larger perithecia (up to 600 μm wide) than *Pl. varium* [30]. In the phylogenetic analyses, due to the absence of DNA sequences, among the species mentioned earlier related to *Pl. varium*, only *Pl. andicola* and *Pl. arboretum* can be further compared. It can be seen from the phylogenetic trees (Figures 2 and S1–S3) *Pl. varium* clustered more closely to *Pl. andicola* than other species, although it is also pronounced there is a much distant relationship between these two species, and *Pl. varium* formed a well-supported separate clade, indicating it is a new species. As comparison, there is farther phylogenetic relationship between terricolous *Pl. varium* and corticolous *Pl. arboretum*, indicating the substrate type contributes to the species delimitation in *Placidium*.

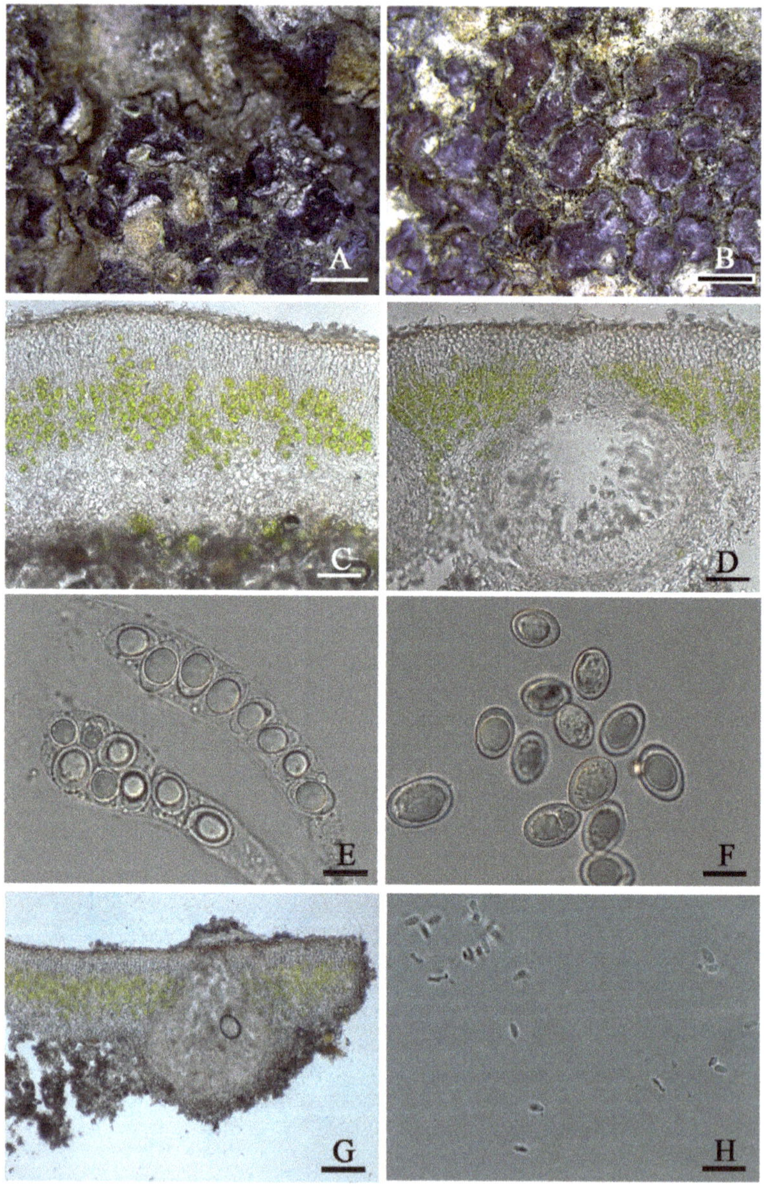

Figure 7. The thallus habit and the anatomic structure of *Placicium varium* sp. nov (holotype, HMAS–L 151970). (**A**). Squamulose thallus (HMAS–L 153948). (**B**). Squamulose thallus (HMAS–L 151967). (**C**). Thallus section. (**D**). Immersed perithecium. (**E**). Asci with biseriate and uniseriate ascospores. (**F**). Ascospores. (**G**). Pycnidium. (**H**). Conidia. Bars: (**A**,**B**) =2 mm, (**C**,**D**) =50 μm, (**E**,**F**) =10 μm, (**G**) =100 μm, and (**H**) =10 μm.

4. Discussion

Nowadays, dry lands cover about 41% of Earth's land surface and influence more than 38% of the global population [33]. Curbing the spread of land desertification in arid and semi-arid areas has become an urgent problem that needs to be focused on. Biological soil crusts (BSCs), as a ubiquitous phenomenon in these regions, play an essential role in soil nutrient cycling, sand stability, and hydrological processes [34]. When the BSCs developed to the lichen crust stage, they would contribute to the greater compressive strength and carbon and nitrogen fixation [35,36]. Lichens existing as the crust in the desert regions are very popular, such as *Endocarpon* and *catapyrenioid* lichens, which have started to be explored as a species diversity in China [37,38], but due to the small size of the lichen thallus and diverse, continuously changing, and transitional phenotypes, sometimes there will inevitably bring difficulties in correctly recognizing and choosing some related genera and species in the field survey, which then affects the sampling for the further study in the lab especially when we also should consider small sampling for not destroying too much the BSCs, all of which could lead to the missing findings for new taxa and the hidden species diversity in deserts.

Our phylogenetic analysis (Figure 2) well-supported *Clavascidium*, *Heteroplacidium*, and *Placidium* formed separate monophyletic clades, especially the genus *Heteroplacidium* situated in the outermost position, consistent with its only-exiting paraplectenchymatous medulla, while variable medulla types exited in *Clavascidium* and *Placidium*. *Clavascidium* and *Placidium* can be clearly distinguished by presence or absence of rhizines. Within the genus *Placidium*, it is generally known only uniseriate ascospores arrangement existed [8,9], but sometimes, both uniseriate and biseriate ascospores arrangements co-existed in some species such as *Placidium acarosporoides* (Zahlbr.) Breuss [30] and two new species *Pl. nigrum* and *Pl. varium*, indicating uniseriate and biseriate ascospores arrangements are continuously changing character between *Clavascidium* and *Placidium*, and within each of these two genera; nevertheless, a single biseriate ascospores arrangement was not seen in *Placidium*. As comparison, within the genus *Clavascidium*, it is generally known only biseriate ascospores arrangement existed [6,8,9], but sometimes, both uniseriate and biseriate ascospores arrangements also coexisted in some species such as *Cl. pseudorufescens* and *Cl. semaforonense* [30] and *Cl. lacinulatum*; however, single a uniseriate ascospores arrangement was not seen in *Clavascidium*. Laminal pycnidia are almost an exclusive character in most genera of catapyrenioid lichens, except *Placidium* with laminal or marginal pycnidia; however, marginal pycnidia were also found in *Cl. semaforonense* [30] and new species *Cl. sinensel*; further supporting pycnidia position is also a continuously changing character between *Clavascidium* and *Placidium*. The only difference in the pycnidia position between *Clavascidium* and *Placidium* is both laminal and marginally coexisted in some species of *Placidium* but not in *Clavascidium*. Besides, some new characters have been put forward to define species in *Placidium*, well-supported by the phylogenetic analysis, such as aggregated pycnidia and a glossy upper surface.

This study found that *Clavascidium* and *Placidium* are distributed in Northwest China, where are harsh environments such as drought and oligotrophic. The previous studies showed that ecological environments could influence the lichen distribution, similar to *Placidium* [8,39]; therefore, the relationship between the adaptative characters and distribution should be paid more attention. The finding of the new species *Cl. sinense* provides a further clue to reconsider the generic relationship between *Clavascidium* and *Placidium*. Moreover, the coexistence of uniseriate and biseriate ascospores arrangement may promote the chance of ascospores discharge for better reproduction. *Pl. nitidulum* only grows at high altitudes above 3000 m alt., and the adaptive phenotype may have produced, for example, tiny lobes and thick algal layer may increase photosynthetic efficiency, and the developed paraphyses in the perithecia would be helpful in squeezing the asci for a better discharge of ascospores through swelling caused by absorbing water. *Pl. nigrum* has a relatively wide ecological niche, ranging from low altitude to high altitude (400–3500 m alt.), with the climate type from arid to wet (90–523 mm annual precipitation) [40]; we hypothesize it is so abundant, pycnidia could help itself to propagate in the variable environments through its asexual reproduction. *Pl. varium* grows in the Xinjiang Autonomous Region dry area with annual precipitation of 180–270 mm; small size and macro guttule of the ascospores may decrease their weight and correspondingly increase their propagative velocity for colonizing a farther and broader living space. However, whether these phenotypic characters are significantly related to the adaptive mechanism need to be further studied.

Therefore, to better recognize the species diversity, understand the adaptive phenotype and other mechanism, and explore more potential species resources that could be applied to restrain sustainable desertification; more taxonomic studies should be continuously carried out on the BSCs-related taxa, such as catapyrenioid lichens, and genomic-scale adaptive evolution studies also need to be explored in the near future.

Supplementary Materials: The following supporting information can be downloaded at https://www.mdpi.com/article/10.3390/jof8070729/s1: Table S1: Details of specimens (species name, voucher information, and GenBank numbers) used in this study. Figure S1: The Bayesian tree based on the concatenated ITS + nuLSU (two genes) data set. Figure S2: The maximum-likelihood tree based on the ITS data set. Figure S3: The maximum-likelihood tree based on the nuLSU data set.

Author Contributions: X.W. conceived and designed the study. X.W. and T.Z. collected the specimens from China. T.Z. and X.Z. generated the DNA sequence data, T.Z., Q.Y. and X.W. performed the phenotypic analysis, T.Z. and X.W. analyzed the DNA data. T.Z. and X.W. checked the issues related to the nomenclatural articles. T.Z. wrote the manuscript draft. T.Z. and X.W. revised the draft. All authors have read and agreed to the published version of the manuscript.

Funding: This research was funded by the National Natural Science Foundation of China (32070096).

Institutional Review Board Statement: Not applicable.

Data Availability Statement: Publicly available datasets were analyzed in this study. All newly generated sequences were deposited in GenBank (https://www.ncbi.nlm.nih.gov/genbank/; Table S1, accessed on 4 June 2022). All new taxa were deposited in Fungal names (https://www.fungalinfo.im.ac.cn, accessed on 8 June 2022).

Acknowledgments: We thank X. Qian and Reyim Mamut for their help during the field collections.

Conflicts of Interest: The authors declare no conflict of interest.

References

1. Lücking, R.; Hodkinson, B.P.; Leavitt, S.D. The 2016 classification of lichenized fungi in the Ascomycota and Basidiomycota–Approaching one thousand genera. *Bryologist* **2017**, *119*, 361–416. [CrossRef]
2. Gueidan, C.; Roux, C.; Lutzoni, F. Using a multigene phylogenetic analysis to assess generic delineation and character evolution in Verrucariaceae (Verrucariales, Ascomycota). *Mycol. Res.* **2007**, *111*, 1145–1168. [CrossRef]
3. Belnap, J.; Lange, O.L. Biological Soil Crusts Characteristics and Distribution. In *Biological Soil Crusts: Structure, Function, and Management*; Springer: Berlin/Heidelberg, Germany, 2003; pp. 3–30.
4. Prieto, M.; Martinez, I.; Aragón, G.; Gueidan, C.; Lutzoni, F. Molecular phylogeny of *Heteroplacidium*, *Placidium*, and related catapyrenioid genera (Verrucariaceae, lichen-forming Ascomycota). *Am. J. Bot.* **2012**, *99*, 23–35. [CrossRef] [PubMed]
5. Breuss, O. An Updated World-Wide Key to the Catapyrenioid Lichens (Verrucariaceae). *Herzogia* **2010**, *23*, 205–216. [CrossRef]
6. Breuss, O. Ein verfeinertes Gliederungskonzept für *Catapyrenium* (lichenisierte Ascomyceten, Verrucariaceae). *Ann. Nat. Mus. Wien* **1996**, *98*, 35–50.
7. Prieto, M.; Martinez, I.; Aragón, G.; Otalora, M.A.; Lutzoni, F. Phylogenetic study of *Catapyrenium* s. str. (Verrucariaceae, lichen-forming Ascomycota) and related genus Placidiopsis. *Mycologia* **2010**, *102*, 291–304. [CrossRef]
8. Prieto, M.; Aragón, G.; Martínez, I. The genus *Catapyrenium* s. lat. (Verrucariaceae) in the Iberian Peninsula and the Balearic Islands. *Lichenologist* **2010**, *42*, 637–684. [CrossRef]
9. Gueidan, C.; Tibell, S.; Thüs, H.; Roux, C.; Keller, C.; Tibell, L.; Prieto, M.; Heiðmarsson, S.; Breuss, O.; Orange, A.; et al. Generic classification of the Verrucariaceae (Ascomycota) based on molecular and morphological evidence: Recent progress and remaining challenges. *Taxon* **2009**, *58*, 184–208. [CrossRef]
10. Arnold, A.E.; Miadlikowska, J.; Higgins, K.L.; Sarvate, S.D.; Gugger, P.; Way, A.; Hofstetter, V.; Kauff, F.; Lutzoni, F. A phylogenetic estimation of trophic transition networks for ascomycetous fungi: Are lichens cradles of symbiotrophic fungal diversification? *Syst. Biol.* **2009**, *58*, 283–297. [CrossRef]
11. Hanneman, B. Anhangsorgane der Flechten, ihre Strukturen und ihre systematische Verteilung. *Bibl. Lichenol.* **1973**, *1*, 1–123.
12. Breuss, O. Die Flechtengattung *Catapyrenium* (Verrucariaceae) in Europa. *Stapfia* **1990**, 1–174.
13. Aptroot, A. Pyrenocarpous lichens and related non-lichenized ascomycets from Taiwan. *J. Hattori Bot. Lab.* **2003**, *93*, 155–173.
14. Wu, L.; Lan, S.; Zhang, D.; Hu, C. Small-scale vertical distribution of algae and structure of lichen soil crusts. *Microb. Ecol.* **2011**, *62*, 715–724. [CrossRef]

15. Ding, L.; Zhou, Q.; Wei, J. Estimation of *Endocarpon pusillum* Hedwig carbon budget in the Tengger Desert based on its photosynthetic rate. *Sci. China Life Sci.* **2013**, *56*, 848–855. [CrossRef] [PubMed]
16. Brodo, I. Microchemical Methods for the Identification of Lichens. *Bryologist* **2003**, *106*, 345. [CrossRef]
17. Rogers, S.O.; Bendich, A.J. Extraction of DNA from plant tissues. In *Plant Molecular Biology Manual A6*; Gelvin, S.B., Schilperoort, R.A., Eds.; Kluwer Academic Publishers: Boston, MA, USA, 1988; pp. 1–10.
18. White, T.; Bruns, T.; Lee, S.; Taylor, J.; Innis, M.; Gelfand, D.; Sninsky, J. Amplification and Direct Sequencing of Fungal Ribosomal RNA Genes for Phylogenetics. *PCR Protoc. A Guide Methods Appl.* **1990**, *31*, 315–322.
19. Savic, S.; Tibell, L.; Gueidan, C.; Lutzoni, F. Molecular phylogeny and systematics of *Polyblastia* (Verrucariaceae, Eurotiomycetes) and allied genera. *Mycol. Res.* **2008**, *112*, 1307–1318. [CrossRef]
20. Swindell, S.R.; Plasterer, T.N. SEQMAN. In *Sequence Data Analysis Guidebook*; Swindell, S.R., Ed.; Springer New York: Totowa, NJ, USA, 1997; pp. 75–89.
21. Katoh, K.; Asimenos, G.; Toh, H. Multiple alignment of DNA sequences with MAFFT. *Methods Mol. Biol.* **2009**, *537*, 39–64. [CrossRef]
22. Castresana, J. Selection of conserved blocks from multiple alignments for their use in phylogenetic analysis. *Mol. Biol. Evol.* **2000**, *17*, 540–552. [CrossRef]
23. Talavera, G.; Castresana, J. Improvement of phylogenies after removing divergent and ambiguously aligned blocks from protein sequence alignments. *Syst. Biol.* **2007**, *56*, 564–577. [CrossRef]
24. Swofford, D. *PAUP*. Phylogenetic Analysis Using Parsimony (*and Other Methods). Version 4.0b10*; Sinauer Associates Inc.: Sunderland, MA, USA, 2002.
25. Darriba, D.; Taboada, G.L.; Doallo, R.; Posada, D. jModelTest 2: More models, new heuristics and parallel computing. *Nat. Methods* **2012**, *9*, 772. [CrossRef] [PubMed]
26. Stamatakis, A. RAxML Version 8: A Tool for Phylogenetic Analysis and Post-Analysis of Large Phylogenies. *Bioinformatics* **2014**, *30*, 1312–1313. [CrossRef] [PubMed]
27. Ronquist, F.; Teslenko, M.; van der Mark, P.; Ayres, D.L.; Darling, A.; Höhna, S.; Larget, B.; Liu, L.; Suchard, M.A.; Huelsenbeck, J.P. MrBayes 3.2: Efficient Bayesian phylogenetic inference and model choice across a large model space. *Syst. Biol.* **2012**, *61*, 539–542. [CrossRef] [PubMed]
28. Huelsenbeck, J.P.; Ronquist, F. MRBAYES: Bayesian inference of phylogenetic trees. *Bioinformatics* **2001**, *17*, 754–755. [CrossRef]
29. Rambaut, A.; Drummond, A.J.; Xie, D.; Baele, G.; Suchard, M.A. Posterior Summarization in Bayesian Phylogenetics Using Tracer 1.7. *Syst. Biol.* **2018**, *67*, 901–904. [CrossRef]
30. Nash, T.H.; Ryan, B.D.; Diederich, P.; Gries, C.; Bungartz, F. *Placidium, In Lichen Flora of the Greater Sonoran Desert Region Volume I*; Lichens Unlimited: Tempe, AZ, USA, 2002; pp. 384–393.
31. Breuss, O. *Catapyrenium* (Verrucariaceae) species from South America. *Plant Syst. Evol.* **1993**, *185*, 17–33. [CrossRef]
32. Lendemer, J.C.; Yahr, R. Changes and additions to the Checklist of North American Lichens-II. *Mycotaxon* **2004**, *90*, 319–322.
33. Reynolds, J.F.; Smith, D.M.; Lambin, E.F.; Turner, B.L., 2nd; Mortimore, M.; Batterbury, S.P.; Downing, T.E.; Dowlatabadi, H.; Fernandez, R.J.; Herrick, J.E.; et al. Global desertification: Building a science for dryland development. *Science* **2007**, *316*, 847–851. [CrossRef]
34. Li, X.-R.; He, M.-Z.; Zerbe, S.; Li, X.-J.; Liu, L.-C. Micro-geomorphology determines community structure of biological soil crusts at small scales. *Earth Surf. Processes Landf.* **2010**, *35*, 932–940. [CrossRef]
35. Hu, C.; Liu, Y.; Zhang, D. Effect of desert soil algae on the stabilization of fine sands. *J. Appl. Phycol.* **2002**, *14*, 281–292. [CrossRef]
36. Belnap, J. Nitrogen fixation in biological soil crusts from southeast Utah, USA. *Biol. Fertil. Soils* **2002**, *35*, 128–135. [CrossRef]
37. Zhang, T.; Liu, M.; Wang, Y.Y.; Wang, Z.J.; Wei, X.L.; Wei, J.C. Two new species of *Endocarpon* (Verrucariaceae, Ascomycota) from China. *Sci. Rep.* **2017**, *7*, 7193. [CrossRef] [PubMed]
38. Cheng, X.-M.; Liu, D.-L.; Wei, X.-L.; Wei, J.-C. *Heteroplacidium compactum* reported as a genus new to China. *Mycotaxon* **2019**, *134*, 369–376. [CrossRef]
39. Usman, M.; Dyer, P.S.; Khalid, A.N. A novel arctic-alpine lichen from Deosai National Park, Gilgit Baltistan, Pakistan. *Bryologist* **2021**, *124*, 484–493. [CrossRef]
40. Qinghai Provincial Bureau of Statistics. *Precipitation Statistics of Main Areas in Qinghai Province (2001–2020)*; National Tibetan Plateau Data Center: Beijing, China, 2021.

Journal of *Fungi*

MDPI

Article

Identification and Characterization of *Calonectria* Species Associated with Plant Diseases in Southern China

Yunxia Zhang [1,2,*], Cantian Chen [1,2], Chao Chen [1,2], Jingwen Chen [1,2], Meimei Xiang [1,2], Dhanushka N. Wanasinghe [3], Tom Hsiang [4], Kevin D. Hyde [1,5] and Ishara S. Manawasinghe [1,2,*]

1 Innovative Institute for Plant Health, Zhongkai University of Agriculture and Engineering, Guangzhou 510225, China; 18825187974@163.com (C.C.); chenchao_cc98@163.com (C.C.); jingwen075@163.com (J.C.); mm_xiang@163.com (M.X.); kdhyde3@gmail.com (K.D.H.)

2 Key Laboratory of Green Prevention and Control on Fruits and Vegetables in South China, Ministry of Agriculture and Rural Affairs, Guangzhou 510225, China

3 Center for Mountain Futures, Kunming Institute of Botany, Chinese Academy of Sciences, Honghe 654400, China; dnadeeshan@gmail.com

4 School of Environmental Sciences, University of Guelph, Guelph, ON N1G 2W1, Canada; thsiang@uoguelph.ca

5 Center of Excellence in Fungal Research, Mae Fah Luang University, Chiang Rai 57100, Thailand

* Correspondence: yx_zhang08@163.com (Y.Z.); ishara9017@gmail.com (I.S.M.); Tel.: +86-20-8900-3192 (Y.Z.); +86-20-8900-3192 (I.S.M.)

Abstract: *Calonectria* species are important plant pathogens on a wide range of hosts, causing significant losses to plant production worldwide. During our survey on phytopathogenic fungi from 2019 to 2021, diseased samples were collected from various hosts in Guangdong Province, China. In total, 16 *Calonectria* isolates were obtained from leaf spots, stem blights and root rots of species of *Arachis*, *Cassia*, *Callistemon*, *Eucalyptus*, *Heliconia*, *Melaleuca* and *Strelitzia* plants. Isolates were identified morphologically, and a multigene phylogenetic analysis of combined partial sequences of calmodulin (*cmd*A), translation elongation factor 1-alpha (*tef*1-α) and beta-tubulin (*β-tubulin*) was performed. These sixteen isolates were further identified as nine *Calonectria* species, with five new species: *Ca. cassiae*, *Ca. guangdongensis*, *Ca. melaleucae*, *Ca. shaoguanensis* and *Ca. strelitziae*, as well as four new records: Ca. *aconidialis* from *Arachis hypogaea*, Ca. *auriculiformis* from *Eucalyptus* sp., *Ca. eucalypti* from *Callistemon rigidus*, and *Ca. hongkongensis* from *Eucalyptus gunnii*. Moreover, we provide updated phylogenetic trees for four *Calonectria* species complexes viz. *Ca. colhounii*, *Ca. cylindrospora*, *Ca. kyotensis* and *Ca. reteaudii*. Our study is the first comprehensive study on *Calonectria* species associated with various hosts from subtropical regions in China. Results from the present study will be an addition to the biodiversity of microfungi in South China.

Keywords: new taxa; Nectriaceae; pairwise homoplasy index; polyphasic approaches; species complexes

Citation: Zhang, Y.; Chen, C.; Chen, C.; Chen, J.; Xiang, M.; Wanasinghe, D.N.; Hsiang, T.; Hyde, K.D.; Manawasinghe, I.S. Identification and Characterization of *Calonectria* Species Associated with Plant Diseases in Southern China. *J. Fungi* **2022**, *8*, 719. https://doi.org/10.3390/jof8070719

Academic Editors: Cheng Gao and Lei Cai

Received: 23 June 2022
Accepted: 5 July 2022
Published: 9 July 2022

1. Introduction

Calonectria De Not. (Nectriaceae, Ascomycota) was introduced and typified by *Ca. daldiniana* De Not. which was later changed to *Ca. pyrochroa* (Desm.) Sacc. [1]. *Calonectria* species are characterized by producing a red perithecium, a vesicle with a long stipe produced from conidiophores, and cylindrical multi-septate conidia [2]. There are 126 species accepted in *Calonectria* [3–6] and 423 species epithets are listed in the Index Fungorum [7]. *Calonectria* species are widely distributed in tropical and subtropical regions [8]. They are important phytopathogens causing leaf spots, stem blights and root rots, leading to plant death on a wide range of hosts [9]. It has been reported that 335 plant species belonging to around 100 families including forest trees, crops and ornamental plants can be infected by *Calonectria* species [9]. Some prominent diseases caused by *Calonectria* species such as *Eucalyptus* leaf blight, *Calonectria* black rot on peanuts, and box blight can cause serious threats to plant production worldwide [10–13]. *Calonectria* species are regarded as the most important causal agents of leaf blight on *Eucalyptus* spp., which is the most devastating disease in Southeast Asia and South America [11,14,15]. Around 40 *Calonectria* species have been identified from *Eucalyptus* plantations and nurseries in Brazil and China [11,16]. *Calonectria ilicicola* Boedijn & Reitsma has been reported as the peanut black rot pathogen (CBR) in Africa, Asia, Australia and North America [13]. This disease incidence has reached 50% in some peanut plantations [10]. Box blight is caused by *Ca. pseudonaviculata* (Crous, J.Z. Groenew. & C.F. Hill) L. Lombard, M.J. Wingf. & Crous and *Ca. henricotiae* Gehesquière, Heungens & J.A. Crouch, and the disease is now found in over 20 countries throughout temperate regions and can cause severe losses [12].

Even though there are over 100 *Calonectria* species that have been identified, taxonomic studies on *Calonectria* species in China are comparatively few. There are 25 *Calonectria* species recorded from China, and 17 of them are associated with *Eucalyptus* [14–17]. In the tropics, the warm and humid climate is suitable for fungal infection and growth [9]. Thus, isolation and characterisation of tropical microfungi in China have a great significance to fungal biodiversity in this country. The objectives of this study were to isolate and identify *Calonectria* species associated with various plant diseases in Guangdong Province, China. In total, 11 isolates were collected and identified based on their morphology and the multi-gene molecular approaches. Seven species belonging to four species complexes were identified and characterised. Complete species descriptions and illustrations are provided for identified taxa.

2. Materials and Methods

2.1. Sample Collection

Infected plant materials were collected from 2019 to 2021 in Guangdong, China from commercially grown plantations and nurseries. These samples included plants showing typical diseased symptoms of leaf spots, stem blights and root rots (Figure 1). In total, over 30 samples were collected from eight plant species, namely, *Arachis hypogaea* L., *Callistemon rigidus* R. Br., *Cassia surattensis* Burm., *Eucalyptus gunnii* Hook. f., *Eucalyptus* sp., *Heliconia metallica* Planch. et Linden ex Hook. f., *Melaleuca bracteata* F. Muell. and *Strelitzia reginae* Aiton. Photographs were taken and symptoms were recorded. Samples were placed loosely in bags with valves and kept cool and then brought back to the lab for further study.

Figure 1. Field symptoms of diseased plants were collected from this study. (**a**) Leaf of *Callistemon rigidus*; (**b**) Leaf of *Eucalyptus* sp.; (**c**) Leaf of *Strelitzia reginae*; (**d**) Fruits of *Arachis hypogaea*; (**e**) Stems of *Cassia surattensis*; Scale bars: (**a–e**) = 1 cm.

2.2. Fungal Isolation

Diseased plant tissues were cut into small pieces (5 × 5 mm) which contained both healthy and diseased tissue. Then the surface was disinfected with 75% ethanol for 15–25 s and 2.5% NaClO for 40 s and rinsed in sterile water three times. After that, tissue pieces were dried on sterilized filter paper inside a laminar flow hood, and placed on potato dextrose agar (PDA) and incubated in the dark at 25 °C. Pure cultures were obtained after three serial transfers of hyphal tips. All cultures obtained in this study were deposited in the culture collection of Zhongkai University of Agriculture and Engineering (ZHKUCC). Herbarium materials (as dry cultures) were deposited in the herbarium of Zhongkai University of Agriculture and Engineering (ZHKU).

2.3. DNA Extraction and PCR Amplification

Genomic DNA was extracted using a DNA rapid Extraction Kit (Aidlab Biotechnologies Co., Ltd., Beijing, China) on 5-day-old cultures grown on PDA. Three loci, the calmodulin (*cmdA*), the translation elongation factor 1-alpha (*tef1-α*) and the beta-tubulin (*β-tubulin*) were amplified and sequenced using primers that were previously designed [18–21] (Table 1). PCR reaction mixtures consisted of 12.5 μL of 2 × Easy Taq PCR SuperMix (TransGen Biotech, Beijing, China), 2 μL DNA, 1 μL of each of the paired 5 μM primers, and ddH2O (8.5 μL). PCR reactions were conducted with an initial step of 95 °C for 3 min, followed by 35 cycles consisting of denaturation at 95 °C for 30 s, annealing from 53 to 55 °C (Table 1) for 30 s and extension at 72 °C for 1 min, and a final extension at 72 °C for 10 min. The reactions were performed in a C1000 TouchTM thermal cycler (Guangzhou Hongtu instrument Co., Ltd., Guangzhou, China). Amplified fragments were sequenced in both directions with forward and reverse primers by Guangzhou Tianyi Science and Technology Co., Ltd. (Guangzhou, China) and consensus sequences derived using BioEdit v.7.0.5.2 [22]. All sequence data generated in this study were submitted to NCBI GenBank (Supplementary Table S1).

Table 1. Primers information in this study.

Genes	Primers	Sequence 5′-3′	Annealing Temperature	References
*cmd*A	CAL-228F	GAGTTCAAGGAGGCCTTCTCCC	55 °C	[18]
	CAL-737R	CATCTTTCTGGCCATCATGG		
*tef*1-α	EF1-728F	CATCGAGAAGTTCGAGAAGG	54 °C	[18]
	EF2	GGARGTACCAGTSATCATGTT		[19]
β-tubulin	T1	AACATGCGTGAGATTGTAAGT	53 °C	[20]
	CYLTUB1R	AGTTGTCGGGACGGAAGAG		[21]

2.4. Phylogenetic Analyses

For all isolates, the genus level was confirmed using the BLASTn tool (Basic Local Alignment Search Tool; https://blast.ncbi.nlm.nih.gov/Blast.cgi) at the National Center for Biotechnology Information (NCBI). To conduct phylogenetic analysis, sequences of *Calonectria* and other related species were obtained following Liu et al. [3]. Downloaded sequences were aligned with newly generated sequences using MAFFT v. 7 (https://mafft.cbrc.jp/alignment/server/). Sequences were improved manually when necessary, using BioEdit 7.0.5.2 [22]. Phylogenetic analyses were performed using concatenated datasets of *cmd*A, *tef*1-α and *β-tubulin* sequence data.

Phylogenetic relationships were inferred using maximum likelihood (ML) in RAxML [23] and Bayesian posterior probability analysis (BYPP) in MrBayes (v3.0b4) [24]. Maximum likelihood analyses and Bayesian analyses were accomplished on the CIPRES science gateway platform (http://www.phylo.org). The GTR + I + G evolution model was used with 1000 non-parametric bootstrapping iterations. The ML analysis was done with RAxML–HPC2 on XSEDE (8.2.8) [25,26] in the CIPRES Science Gateway platform [27]. For each phylogenetic tree, 1000 nonparametric bootstrapping iterations were used. Bayesian analyses were based on 2,000,000 generations, sampling every 100 generations, with four simultaneous Markov chains. Bayesian posterior probabilities were calculated after discarding a burn-in phase. The stability of the trees was evaluated by 1000 bootstrap replications. Descriptive statistics were calculated for the resulting trees.

2.5. Pairwise Homoplasy Index (PHI)

To confirm the species novelties, the pairwise homoplasy index (PHI index) was calculated. Here, the PHI index was calculated to determine species boundaries for the taxa with low tree values and significant evolution length. The PHI test was performed using SplitsTree4 v. 4 [28]. The concatenated three-locus dataset (*cmd*A + *tef*1-α + *β-tubulin*) was used for the analyses. The relationships between isolates belonging to this study and closely related taxa were visualized in split graphs with both the Log-Det transformation and split decomposition options.

2.6. Morphological Characterisation

Representative isolates incubated on carnation leaf agar (CLA) at 25 °C were used for morphological characterization [9]. Teleomorphic structures such as perithecia, asci and ascospores, and anamorphic structures such as the conidiophores, vesicles and conidia were photographed, and measurements were taken. The Cnoptec SZ650 (Chongqing Optec Instrument Co., Ltd., Chongqing, China) series stereomicroscope was used to observe macro-morphological characteristics. Micromorphological characters were observed using Nikon Eclipse 80i (Nikon, Tokyo, Japan). Morphological features including conidial length, width, and size were measured (at least 40 per isolate) using NISElements BR 3.2.

To observe culture characteristics, representative isolates were grown on malt extract agar (MEA) at 25 °C in the dark [9]. Colony diameters were examined after seven days with three replicate plates. Colony colors were recorded following the Rayner [29] color chart, and textures were observed daily until colonies covered the whole plate.

3. Results

3.1. Phylogenetic Analyses

In total 16 *Calonectria* isolates were obtained. Phylogenetic analyses for *Calonectria* species were done using the concatenated dataset of *cmd*A, *tef* 1-α and *β-tubulin*. In total, sequences from 137 *Calonectria* strains including 16 strains from this study were used. *Curvicladiella cignea* (CBS 109167 and CBS 109168) was used as the outgroup. Tree topologies derived from the ML and BI analyse were congruent with each other; only the best scoring RAxML tree is presented (Figure 2). The best scoring RAxML tree had -14467.831159 as a final likelihood. The matrix had 998 distinct alignment patterns, with 14.59% of undetermined characters or gaps. Estimated base frequencies were as follows: A = 0.221391, C = 0.308101, G = 0.226807, T = 0.243702; substitution rates AC = 1.449245, AG = 3.672733, AT = 1.291386, CG = 0.870745, CT = 4.692338, GT = 1.000000; gamma distribution shape parameter α = 0.883863. The Bayesian analyses generated 4002 trees (average standard deviation of split frequencies: 0.015016) from which 3002 were sampled after 25% of the trees were discarded as burn-ins. The alignment contained a total of 988 unique site patterns.

In the phylogenetic trees, 16 isolates from this study were clustered in four species complexes in *Calonectria*: *Ca. colhounii*, *Ca. cylindrospora*, *Ca. kyotensis* and *Ca reteaudii*. These species formed nine distinct groups, from which four were grouped with already known species and five were novel species.

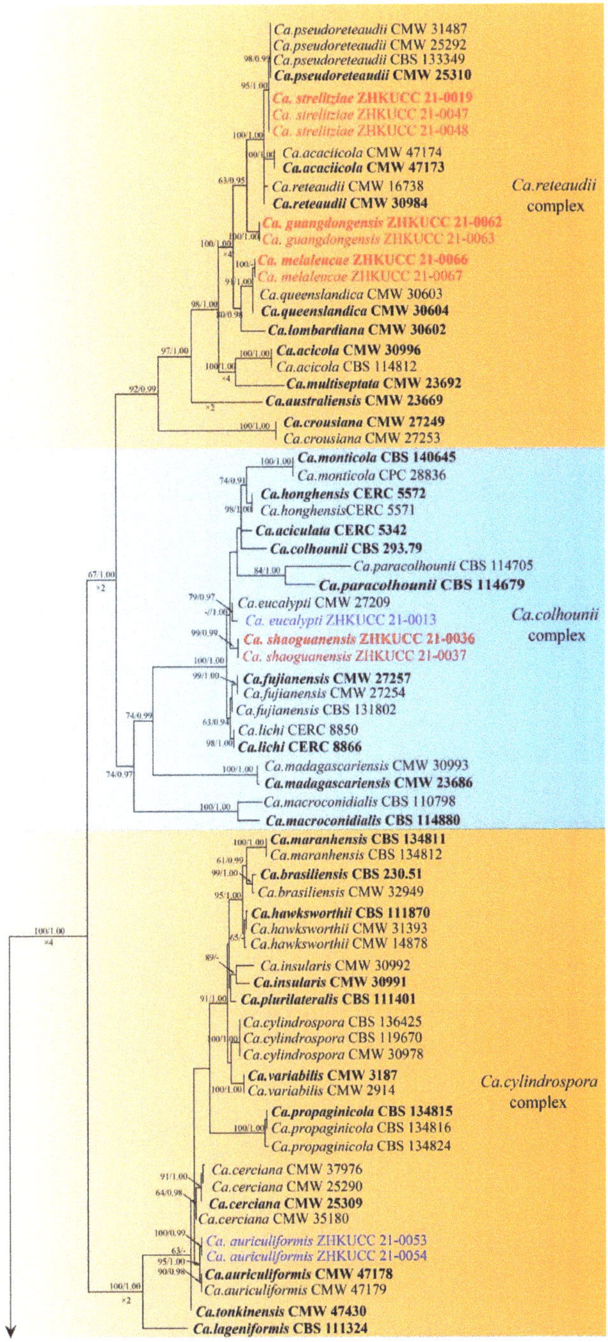

Figure 2. *Cont.*

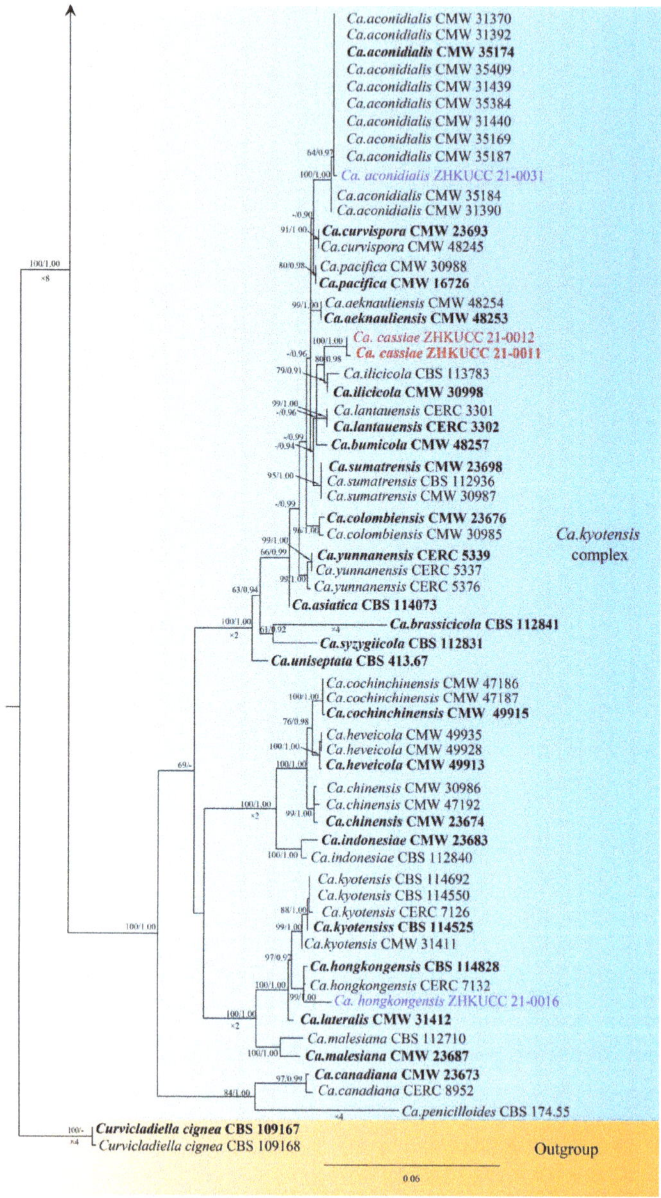

Figure 2. Phylogenetic tree of *Calonectria* species based on Maximum Likelihood (ML) analyses of the combined DNA dataset of *cmd*A, *tef*1-α and *β-tubulin* gene sequences. The ML bootstrap support values ≥60% and BYPP higher than 0.90 are indicated above the nodes and branches, and the ×2, ×4, and ×8 below the branches indicate that their lengths are compressed two-, four-, or eight-fold, respectively. The scale bar indicates 0.06 changes per site. Ex-type isolates of *Calonectria* species are marked in bold. Isolates for already known species in this study are in blue and novel taxa in this study are in red.

3.2. PHI Analyses

To confirm the species novelties, the PHI index was calculated. The PHI analysis of five new species (*Ca. cassiae, Ca. guangdongensis, Ca. melaleucae, Ca. shaoguanensis* and *Ca. strelitziae*) and closely related taxa did not show significant recombination (the P-value was 1.0, 1.0, 0.06, 1.0 and 1.0, respectively) (Figure 3). This evidence provides support that the new taxa and closely species were different from each other. These results confirmed that these five taxa were different from the already known species of *Calonectria*.

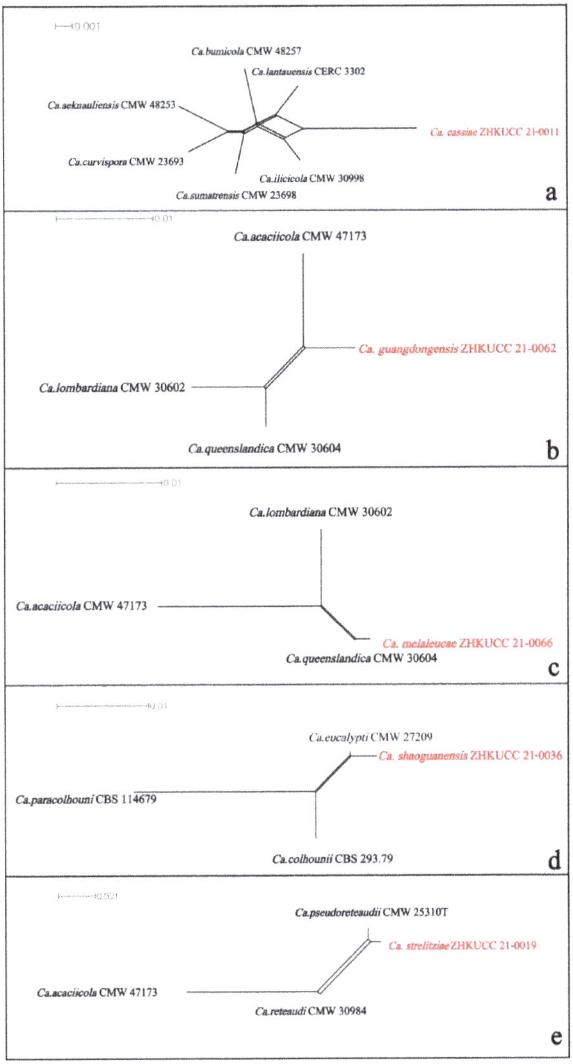

Figure 3. The results of the pairwise homoplasy index (PHI) test of five *Calonectria* new species and closely related species using both LogDet transformation and split decomposition. (**a**) *Ca. cassiae*; (**b**) *Ca. guangdongensis*; (**c**) *Ca. melaleucae*; (**d**) *Ca. shaoguanensis*; (**e**) *Ca. strelitziae*.

3.3. Taxonomy

Calonectria aconidialis L. Lombard, Crous & S.F. Chen bis, Stud. Mycol. 80: 162 (2015) Figure 4.

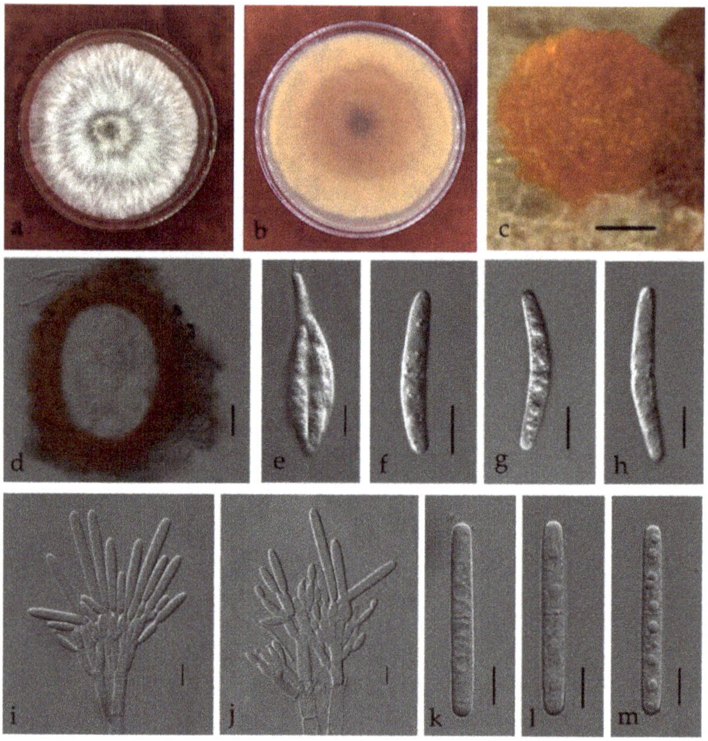

Figure 4. *Calonectria aconidialis* (ZHKUCC 21-0031: New host record) (**a**,**b**) Culture characteristics on MEA after 7 days ((**a**), upper view; (**b**), reverse view); (**c**) Ascomata; (**d**) Vertical section through an ascoma; (**e**) An ascus; (**f–h**) Ascospores; (**i**,**j**) Conidiogenous apparatus; (**k–m**) Macroconidia; Scale bars: c = 100 μm; d = 50 μm; (**e–m**) = 10 μm.

Index Fungorum number: IF 809043.

Associated with stem rot of *Arachis hypogaea.* Telemorph: *Perithecia* 290–600 × 240– 480 μm, solitary, orange–red, subglobose, perithecial walls rough. *Asci* 110–140 × 15–20 μm, fusiform, eight-spored. *Ascospores* (32–)37–40(–46) × (5–)6–8(–9) μm ($\bar{x}$ = 38.5 × 6.5 μm, *n* = 50), hyaline, fusoid, straight to curved, 1-septate, sometimes constricted at septum. Anamorph: *Macroconidiophores* 15–25 × 4–6 μm, septate, hyaline. Primary branches of conidiogenous apparatus; secondary branches aseptate, 10–20 × 4–6 μm; tertiary branches 10–15 × 3–6 μm; each terminal branch producing two to four phialides, 10–15 × 3–6 μm. *Macroconidia* (44–)48–50(–55) × 4–7(–8) μm ($\bar{x}$ = 49.5 × 7 μm, *n* = 50), cylindrical, straight, 1-septate. Megaconidia and microconidia not observed.

Culture characteristics—Colonies on MEA fast growing at 25 °C growth rate 11 mm/d (*n* = 5), circular, producing abundant white aerial mycelium; reverse orange.

Material examined—China, Guangdong Province, Shaoguan City, *Arachis hypogaea* Linn., (*Fabaceae*), 27 September 2019, C.T. Chen, dried culture (ZHKU 21-0028), and living culture (ZHKUCC 21-0031).

Notes—One isolate from this study clustered with *Ca. aconidialis* in the multigene phylogeny with 64% ML and 0.97 BYPP values (Figure 2). Morphologically our isolate is similar to *Ca. aconidialis* as described by Lombard et al. [16] (Table 2). However, Mega, macro and microconidia were not observed in Lombard et al. [16], while our isolate produces macroconidia. To our knowledge, this is the first report of *Ca. aconidialis* from *Arachis hypogaea* [30].

Calonectria auriculiformis N.Q. Pham, T.Q. Pham & M.J. Wingf., Mycologia 111(1): 85 (2019) Figure 5.

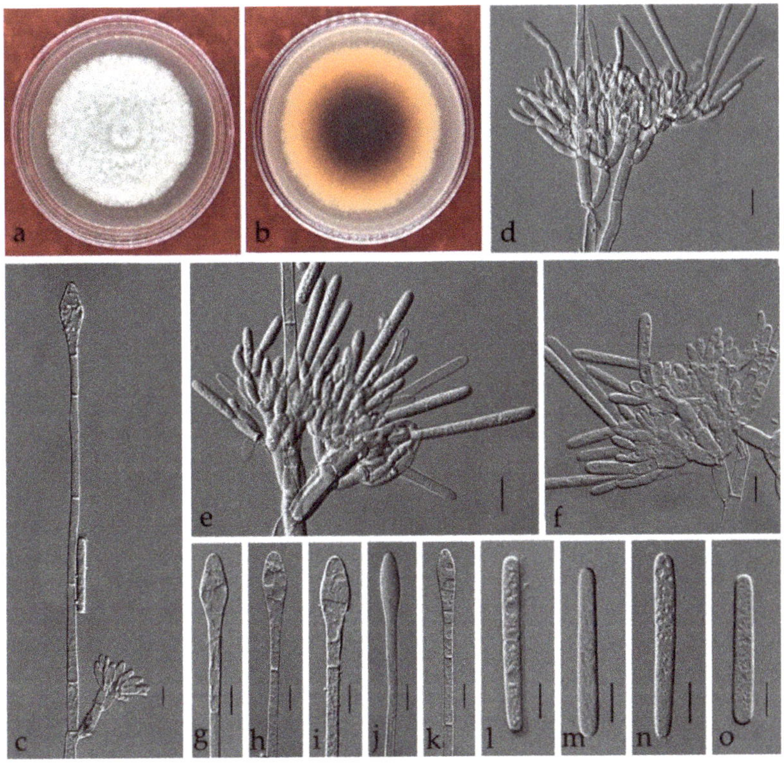

Figure 5. *Calonectria auriculiformis* (ZHKUCC 21-0053: New host record) (**a**,**b**) Culture characteristics on MEA after 7 days (**a**) upper view; (**b**) reverse view); (**c**) Macroconidiophores; (**d**–**f**) Conidiogenous apparatus; (**g**–**k**) Vesicles; (**l**–**o**) Macroconidia; Scale bars: (**c**–**o**) = 10 μm.

Index Fungorum number: IF 825527.

Associated with leaf spot of *Eucalyptus* sp. Telemorph: not observed. Anamorph: *Macroconidiophores* septate, hyaline. Primary branches of conidiogenous apparatus 10–30 × 3–10 μm; secondary branches aseptate, 10–25 × 3–8 μm; third branches 10–16 × 3–8 μm; tertiary branches 10–15 × 3–8 μm; each terminal branch producing two to four phialides, 10–20 × 3–6 μm. *Vesicles* 5–13 μm diamater, ellipsoidal to obpyriform. *Macroconidia* (38–)40–43(–47) × (3–)4–6 (–7) μm ($\bar{x}$ = 41.5 × 5 μm, *n* = 50), cylindrical, straight, 1-septate. Megaconidia and microconidia not observed.

Culture characteristics—Colonies on MEA fast growing at 25 °C, growth rate 9.5 mm/d (*n* = 5), circular, with regular margin, producing white aerial mycelium; reverse umber.

Material examined—China, Guangdong Province, Guangzhou City, *Eucalyptus* sp. L.Herit, (*Myrtaceae*), 11 July 2020, C.T. Chen, dried cultures (ZHKU 21-0050), and living culture (ZHKUCC 21-0053).

Notes—In the phylogenetic tree, two isolates from our study were closed to *Ca. auriculiformis* with 95% in ML, and 1.00 in BYPP support (Figure 2). Morphologically, our isolates are similar to *Ca. auriculiformis* described by Pham et al. [31] (Table 2).

Calonectria cassiae Y. X. Zhang, C. T. Chen, Manawas., & M. M. Xiang, sp. nov. Figure 6.

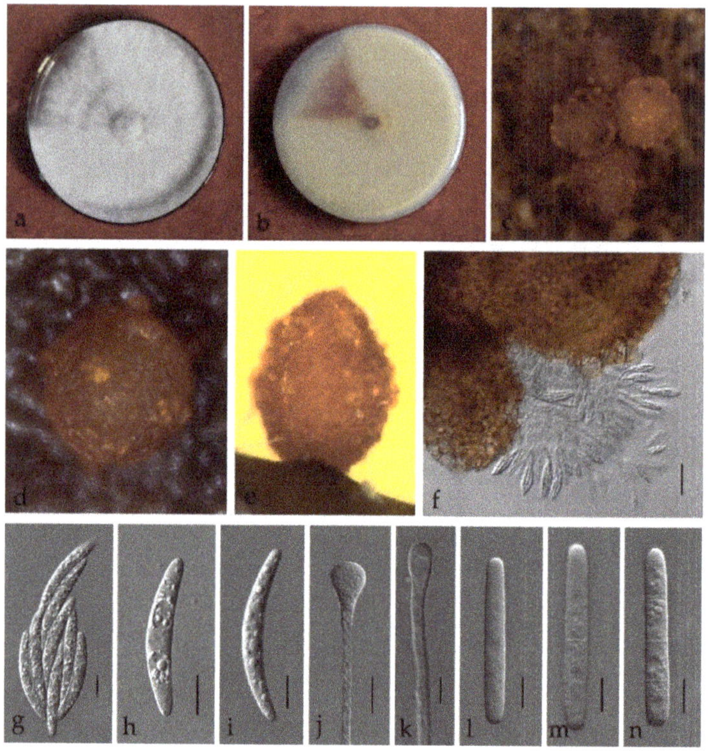

Figure 6. *Calonectria cassiae* (ZHKUCC 21-0011: Ex-type) (**a**,**b**) Culture characteristics on MEA after 7 days ((**a**), upper view; (**b**), reverse view); (**c**–**e**) Ascomata; (**f**,**g**) Asci; (**h**,**i**) Ascospores; (**j**,**k**) Vesicles; (**l**–**n**) Macroconidia; Scale bars: (**c**–**e**) = 100 μm; f = 50 μm; (**g**–**n**) = 10 μm.

Index Fungorum number: IF 553370.

Associated with stem rot of *Cassia surattensis*. Telemorph: *Perithecia* 400–700 × 350–550 μm, solitary, red–brown, subglobose to ovoid, perithecial walls rough. *Asci* fusiform with eight-spored. *Ascospores* 40–50 × 4–8 μm ($\overline{x}$ = 45 × 6 μm, n = 50), hyaline, fusoid, straight to slightly curved, 1-septate, constricted at septum. Anamorph: *Macroconidiophores* septate, hyaline. Primary branches of conidiogenous apparatus 15–25 × 4–6 μm; secondary branches aseptate, 10–20 × 4–6 μm; tertiary branches 10–15 × 3–6 μm; each terminal branch producing two to four phialides, 10–15 × 3–6 μm. *Vesicles* 7–12 μm diameter, sphaeropedunculate. *Macroconidia* 40–65 × 4–8 μm ($\overline{x}$ = 54 × 6 μm, n = 50), cylindrical, straight, 1–3-septate. Megaconidia and microconidia not observed.

Culture characteristics—Colonies on MEA fast growing at 25 °C, growth rate 11.2 mm/d (n = 5), circular, producing abundant white aerial mycelium; reverse red brown.

Material examined—China, Guangdong Province, Guangzhou City, *Cassia surattensi* Burm. F., (*Fabaceae*), 30 March 2019, Y.X. Zhang, dried culture (ZHKU 21-0008, holotype), and living culture (ZHKUCC 21-0011, ex-type).

Notes—Two isolates obtained in this study developed a distinct sister clade to the *Ca. ilicicola* clade with 80% in ML and 0.98 in BYPP supports in phylogenetic analyses (Figure 2). Morphologically, the macroconidia of our isolates ($\overline{x}$ = 54 × 6 μm) are shorter than those of *Ca. ilicicola* ($\overline{x}$ = 62 × 6 μm) described by Lombard et al. [32] (Table 2), and microconidia were not observed in this study. In the PHI analysis of closely related taxa, there is no significant evidence of recombination among our isolates and other related species (p = 1.0). Therefore, we introduce *Ca. cassiae* from *Cassia surattensis* as a novel species based on phylogenetic analyses, morphological analyses and recombination analysis.

Calonectria eucalypti L. Lombard, M.J. Wingf. & Crous, Stud. Mycol. 66: 47 (2010) Figure 7.

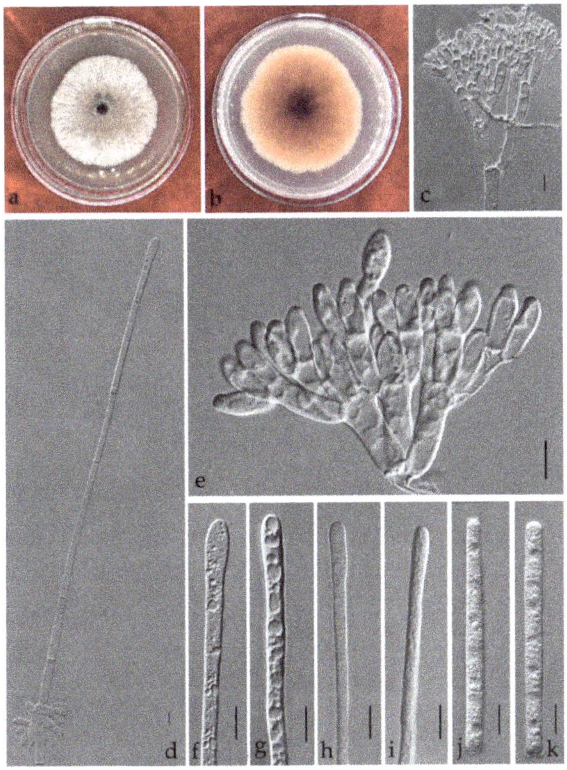

Figure 7. *Calonectria eucalypti* (ZHKUCC 21-0013: New host record) (**a**,**b**) Culture characteristics on MEA after 7 days (**a**) upper view; (**b**) reverse view); (**d**) Macroconidiophores; (**c**,**e**) Conidiogenous apparatus; (**f–i**) Vesicles; (**j**,**k**) Macroconidia; Scale bars: (**c–k**) = 10 μm.

Index Fungorum number: IF 515530.

Associated with leaf spot of *Callistemon rigidus*. Telemorph: not observed. Anamorph: *Macroconidiophores* 15–45 × 4–8 μm, septate, hyaline. Primary branches of conidiogenous apparatus; secondary branches aseptate, 10–25 × 4–8 μm; tertiary branches 10–20 × 3–8 μm; each terminal branch producing two to six phialides, 8–20 × 3–6 μm. *Vesicles* 4–8 (−10) μm diameter, clavate to broadly clavate. *Macroconidia*(65–)70–80(–87) × (5–)6 μm ($\bar{x}$ = 75 × 6.5 μm, n = 50), cylindrical, straight, 1–3-septate. Megaconidia and microconidia not observed.

Culture characteristics—Colonies on MEA at 25 °C, growth rate 8.7 mm/d (n = 5), circular, producing abundant white aerial mycelium; reverse brown.

Material examined—China, Guangdong Province, Guangzhou City, *Callistemon rigidus* R. Br., (*Myrtaceae*), 10 November 2020, dried culture (ZHKU 21-0010), and living culture (ZHKUCC 21-0013).

Notes—In the phylogenetic tree, our isolates were grouped with *Ca. eucalypti* with 79% in ML and 0.97 in BYPP support (Figure 2). The macroconidia of the isolate belonging to this study ($\bar{x}$ = 75 × 6.5 μm) are similar to the size of the *Ca. eucalypti* ($\bar{x}$ = 72 × 6 μm) described by Lombard et al. [32]. In addition, the vesicle shape and the dimension of our isolate are similar to *Ca. eucalypti* [32] (Table 2). We introduce our isolates from *C. rigidus* as an anamorph of *Ca. eucalypti* based on phylogenetic analyses and asexual morphological characteristics. To our knowledge, this is the first report of *Ca. eucalypti* from *Callistemon rigidus* [30].

Calonectria guangdongensis Y. X. Zhang, C. T. Chen, Manawas., & M. M. Xiang, sp. nov. Figure 8

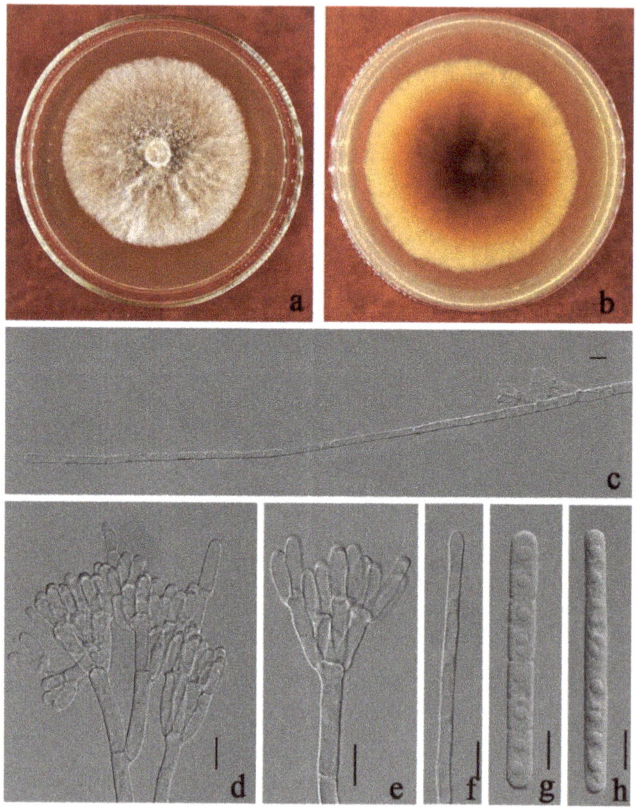

Figure 8. *Calonectria guangdongensis* (ZHKUCC 21-0062: Ex–type) (**a**,**b**) Culture characteristics on MEA after 7 days (**a**), upper view; (**b**), reverse view); (**c**) Macroconidiophores; (**d**,**e**) Conidiogenous apparatus; (**f**) Vesicles; (**g**,**h**) Macroconidia; Scale bars: (**c**–**h**) = 10 μm.

Index Fungorum number: IF 553380.

Etymology—Epithet refers to the Guangdong Province from where the type was collected.

Holotype—ZHKU 21-0059.

Associated with leaf spot of *Heliconia metallica*. Telemorph: not observed. Anamorph: *Macroconidiophores* septate, hyaline. Primary branches of conidiogenous apparatus 15–30 × 3–8 μm; secondary branches aseptate, 10–20 × 3–7 μm; tertiary branches aseptate, 10–15 × 3–6 μm; each terminal branch producing two to four phialides, 7–15 × 3–6 μm. *Vesicles* 3–7 μm diameter, narrowly clavate. *Macroconidia* 55–70 × 5–7(–9) μm ($\overline{x}$ = 64 × 6 μm, *n* = 50), cylindrical, straight, 1–3-septate. Megaconidia and microconidia not observed.

Culture characteristics—Colonies fast growing at 25 °C on MEA, growth rate 9.2 mm/d (*n* = 5), circular, producing abundant white aerial mycelium, reverse red brown.

Material examined—China, Guangdong Province, *Heliconia metallica* Planch. et Linden ex Hook. f., (*Musaceae*). 26 July 2020, Y.X. Zhang, dried cultures (ZHKU 21-0059, holotype), and living culture (ZHKUCC 21-0062, ex-type).

Notes—Our isolates from *Heliconia metallica* formed a distinct clade from the closely related taxa with 63% ML and 0.95 BYPP support. Morphologically our isolate differs from the other four closely related species by the size of the macroconidia. Our isolate (64 × 6 μm) developed macroconidia smaller than *Ca. acaciicola* N.Q. Pham, T.Q. Pham & M.J. Wingf. (94 × 7 μm), *Ca. pseudoreteaudii* L. Lombard, M.J. Wingf. & Crous (104 × 8 μm), *Ca. reteaudii* C. Booth (84 × 6.5 μm) and *Ca. strelitziae* (87 × 8 μm) [9,14,31] (Table 2). In the PHI analysis of closely related taxa, there is no significant evidence of recombination among our isolate and other related species (p = 1.0). Based on these polyphasic approaches, we introduce *Ca. guangdongensis* as a novel species from *H. metallica.*

Calonectria hongkongensis Crous, Stud. Mycol. 50(2): 422 (2004) Figure 9.

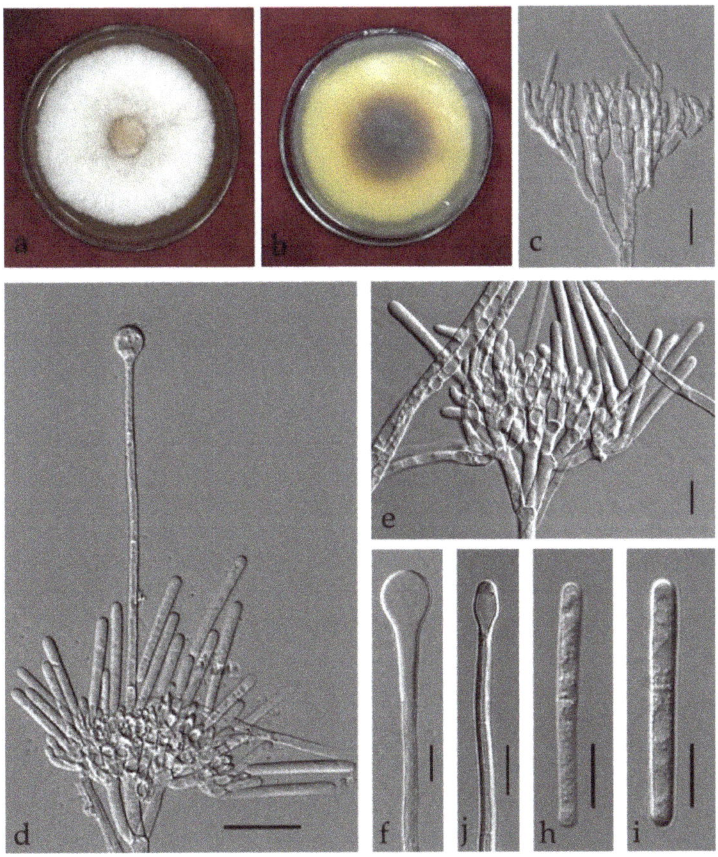

Figure 9. *Calonectria hongkongensis* (ZHKUCC 21-0016: New host record) (**a**,**b**) Culture characteristics on MEA after 7 days ((**a**) upper view; (**b**) reverse view); (**d**) Macroconidiophores; (**c**,**e**) Conidiogenous apparatus; (**f**–**j**) Vesicles; (**h**,**i**) Macroconidia; Scale bars: (**c**–**i**) = 10 μm.

Index Fungorum number: IF 500107.

Associated with leaf spot of *Eucalyptus gunnii*. Telemorph: not observed. Anamorph: *Macroconidiophores* septate, hyaline. Primary branches of conidiogenous apparatus 10–20 × 2–4 μm; secondary branches aseptate, 7–15 × 2–4 μm; tertiary branches 6–13 × 2–4 μm; each terminal branch producing two to three phialides, 5–10 × 1–3 μm. *Vesicles* 8–10 μm diameter, sphaeropedunculate to obpyriform. *Macroconidia* (38–)40–43(–46) × 4–6 μm ($\bar{x}$ = 42 × 5 μm, n = 50), cylindrical, straight, 1-septate. Megaconidia and microconidia not observed.

Culture characteristics—Colonies on MEA fast growing at 25 °C, growth rate 10.7 mm/d (n = 5), circular, with regular margin, producing abundant white aerial mycelium; reverse light-yellow to dark-brown.

Material examined—China, Guangdong Province, Guangzhou City, *Eucalyptus gunnii* Hook, (*Myrtaceae*), 11 July 2019, C.T. Chen and X. Sun, dried culture (ZHKU 21-0013), and living culture (ZHKUCC 21-0016).

Notes—In the phylogenetic analysis, a single isolate was obtained from *Eucalyptus gunnii* clustered with *Ca. hongkongensis* with 99% ML and 1.00 BYPP support (Figure 2). The anamorph of the isolate from this study is similar to *Ca. hongkongensis* described by Crous et al. [21] (Table 2). The telemorph was not observed in this study. To our knowledge, this is the first report of *Ca. hongkongensis* from *E. gunnii* [30].

Calonectria melaleucae Y. X. Zhang, C. T. Chen, Manawas., & M. M. Xiang, sp. nov. Figure 10.

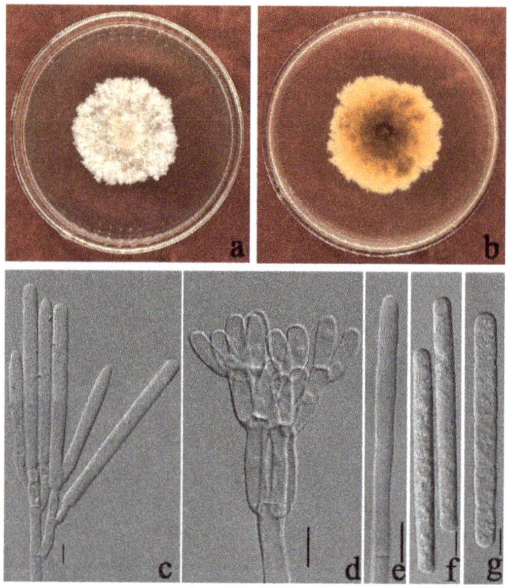

Figure 10. *Calonectria melaleucae* (ZHKUCC 21-0066: Ex–type) (**a**,**b**) Culture characteristics on MEA after 7 days (**a**) upper view; (**b**) reverse view); (**c**,**d**) Conidiogenous apparatus; (**e**) Vesicle; (**f**,**g**) Macroconidia; Scale bars: (**c**–**g**) = 10 μm.

Index Fungorum number: IF 553393.

Etymology—Epithet refers to the host (*Melaleuca bracteata*) from which the type was collected.

Holotype—ZHKU 21-0063.

Associated with leaf spot of *Melaleuca bracteata*. Telemorph: not observed. Anamorph: *Macroconidiophores* septate, hyaline. Primary branches of conidiogenous apparatus 15–35 × 3–7 μm; secondary branches aseptate, 15–25 × 3–6 μm; tertiary branches aseptate, 10–20 × 3–5 μm; each terminal branch producing two to four phialides, 10–15 × 3–5 μm. *Vesicles* 3–7 μm diameter, narrowly clavate. *Macroconidia* 80–95(–100) × (5–)7–10 μm ($\bar{x}$ = 88 × 8 μm, n = 50), cylindrical, straight, 3–5-septate. Megaconidia and microconidia not observed.

Culture characteristics—Colonies fast growing at 25 °C on MEA, growth rate 7.2 mm/d (n = 5), circular, producing abundant white aerial mycelium, reverse lightly red brown.

Material examined—China, Guangdong Province, Guangzhou City, *Melaleuca bracteata* F. Muell., (*Myrtaceae*). 26 July 2020, Y.X. Zhang, dried cultures (ZHKU 21-0063 holotype), and living culture (ZHKUCC 21-0066, ex–type).

Notes—Our isolates from *Melaleuca bracteata* formed a distinct clade sister to *Ca. queenslandica* L. Lombard, M.J. Wingf. & Crous with 91% in ML and 1.00 in BYPP support. Our isolate differs from *Ca. queenslandica* by the size of the macroconidia [14]. Our isolate (88 × 8 μm) developed macroconidia larger than *Ca. queenslandica* (69 × 6 μm). In addition, our isolates are different from *Ca. queenslandica* in the number of macroconidia septa [3–5 vs. 4–6], as well as the size of vesicles (3–7 vs. 3–4 μm diameter) [14] (Table 2). In the PHI analysis of closely related taxa, there is no significant evidence of recombination among our isolate and other related species (p = 0.06). Based on these polyphasic approaches, we introduce *Ca. melaleucae* as a novel species from *M. bracteata*.

Calonectria shaoguanensis Y. X. Zhang, C. T. Chen, Manawas., & M. M. Xiang, sp. nov. Figure 11.

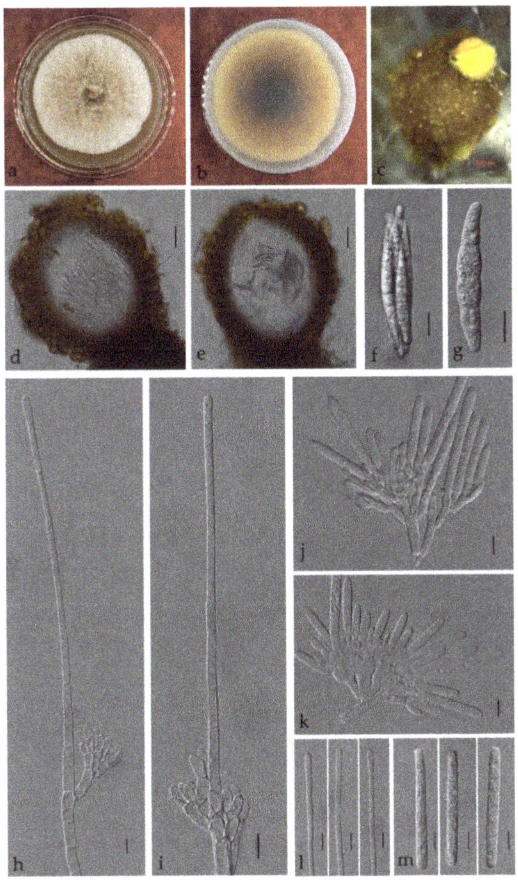

Figure 11. *Calonectria shaoguanensis* (ZHKUCC 21-0036: Ex-type) (**a**,**b**) Culture characteristics on MEA after 7 days (**a**) upper view; (**b**) reverse view); (**c**) Ascomata; (**d**,**e**) Vertical section through ascomata; (**f**) An ascus; (**g**) A ascospore; (**h**,**i**) Macroconidiophores; (**j**,**k**) Conidiogenous apparatus; (**l**) Vesicles; (**m**) Macroconidia; Scale bars: (**c**) = 100 μm; (**d**,**e**) = 50 μm; (**f**–**m**) = 10 μm.

Index Fungorum number: IF 555217.

Etymology—Epithet refers to the Shaoguan city from where the type was collected.

Holotype—ZHKU 21-0033.

Associated with leaf spot of *Callistemon rigidus*. Telemorph: *Perithecia* 250–500 × 200–380 μm, solitary, yellow, subglobose to ovoid, and rough perithecial walls. *Asci* fusiform, eight-spored. *Ascospores* (45–)50–70 × (3–)4–8(–9) μm ($\bar{x}$ = 56.5 × 6.5 μm, *n* = 50), hyaline, fusiform, straight to curved, 1-septate, sometimes constricted at septum. Anamorph: *Macroconidiophores* septate, hyaline. Primary branches of conidiogenous apparatus 10–25 × 3–8 μm; secondary branches aseptate, 10–18 × 3–7 μm; tertiary branches aseptate, 8–15 × 3–6 μm; each terminal branch producing two to four phialides, 7–12 × 2–5 μm. *Vesicles* (2–)4–7 μm diameter, narrowly clavate. *Macroconidia* (55–)60–70(–75) × (4–)5–8 μm ($\bar{x}$ = 65 × 6.5 μm, *n* = 50), cylindrical, straight, 1–3-septate. Megaconidia and microconidia not observed.

Culture characteristics—Colonies fast growing at 25 °C on MEA, growth rate 10 mm/d (*n* = 5), circular, producing abundant white aerial mycelium, reverse red brown.

Material examined—China, Guangdong Province, Shaoguan City, *Callistemon rigidus* R. Br., (*Myrtaceae*). 21 June 2020, Y.X. Zhang, dried cultures (ZHKU 21-0033 holotype), and living culture (ZHKUCC 21-0036 ex-type).

Notes—Our isolates from *Callistemon rigidus* formed a distinct clade from the other six closely related species with less than 60% ML and 1.00 BYPP support (Figure 2). Morphologically our isolate differs from *Ca. eucalypti* by the size of the macroconidia [32]. Our isolate (65 × 6.5 μm) developed macroconidia shorter than the two isolates of *Ca. eucalypti* described by Lombard et al. (72 × 6 μm) [32] and in this study (75 × 6.5 μm). In the PHI analysis of closely related taxa, there is no significant evidence of recombination among our isolate and other related species (*p* = 1.0). Based on these polyphasic approaches we introduce *Ca. shaoguanensis* as a novel species from *C. rigidus*.

Calonectria strelitziae Y. X. Zhang, C. T. Chen, Manawas., & M. M. Xiang, sp. nov. Figure 12.

Index Fungorum number: IF 553395.

Etymology—Epithet refers to the host (*Strelitzia reginae*) from which the type was collected.

Holotype—ZHKU 21-0016.

Associated with leaf spot of *Strelitzia reginae*. Telemorph: not observed. Anamorph: *Macroconidiophores* septate, hyaline. Primary branches of conidiogenous apparatus 15–30 × 3–9 μm; secondary branches aseptate, 10–25 × 3–9 μm; tertiary branches aseptate, 10–25 × 4–8 μm; each terminal branch producing two to four phialides, 10–20 × 3–8 μm. *Vesicles* 3–8 μm diameter, narrowly clavate. *Macroconidia* (65–)80–95(–115) × (4–)6–10(–12) μm ($\bar{x}$ = 87 × 8 μm, *n* = 50), cylindrical, straight, 3–5-septate. Microconidia 45–55 × 4.7–5.5 ($\bar{x}$ = 49 × 5.1 μm, *n* = 9), cylindrical, 1-septate. Megaconidia not observed.

Culture characteristics—Colonies fast growing at 25 °C on MEA, growth rate 5.4 mm/d (*n* = 5), circular, with irregular edge, producing white aerial mycelium, reverse red brown.

Material examined—China, Guangdong Province, Guangzhou City, *Strelitzia reginae* Aiton, (*Musaceae*). 11 July 2020, Y.X. Zhang and C.T. Chen, dried cultures (ZHKU 21-0016, holotype), and living culture (ZHKUCC 21-0019, ex-type).

Notes—Our isolates from *Strelitzia reginae* formed a single lineage sister to *Ca. pseudoreteaudii* with 95% in ML, and 1.00 in BYPP support (Figure 2). Morphologically our isolate differs from *Ca. pseudoreteaudii* by the shorter macroconidia (87 × 8 μm vs. 104 × 8 μm) and smaller microconidia (49 × 5 μm vs. 44 × 4 μm) [14] (Table 2). In the PHI analysis of closely related taxa, there is no significant evidence of recombination among our isolate and other related species (*p* = 1.0). Based on these polyphasic approaches, we introduce *Ca. strelitziae* as a novel species from *S. reginae*.

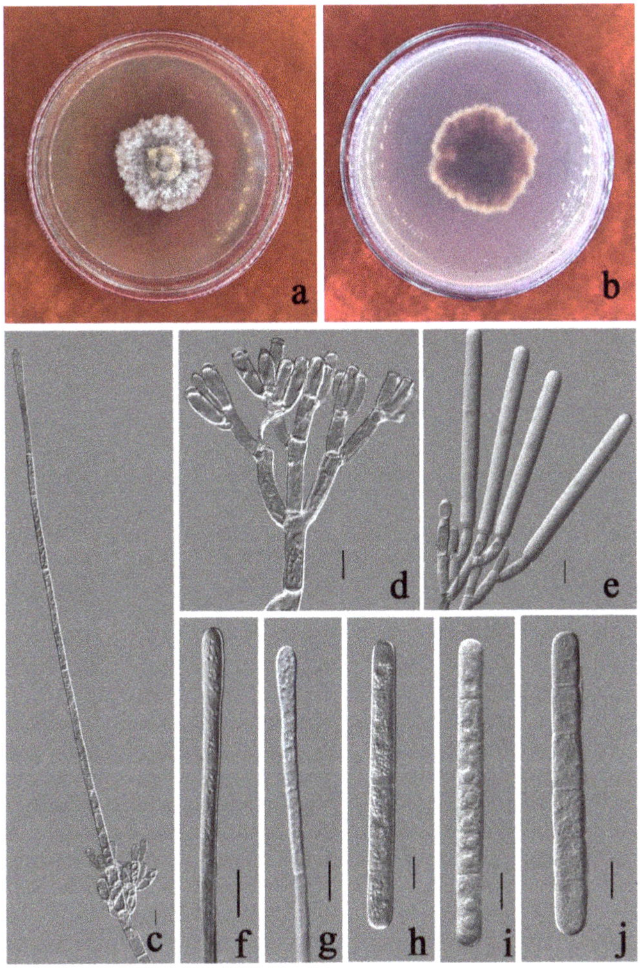

Figure 12. *Calonectria strelitziae* (ZHKUCC 21-0019: Ex-type) (**a**,**b**) Culture characteristics on MEA after 7 days (**a**) upper view; (**b**) reverse view); (**c**) Macroconidiophores; (**d**,**e**) Conidiogenous apparatus; (**f**,**g**) Vesicles; (**h**–**j**) Macroconidia; Scale bars: (**c**–**j**) = 10 μm.

Table 2. Morphological comparison of *Calonectria* species obtained in this study with their relatives.

Complex	Species	Ascospores		Macroconidia		Vesicle		References
		Size/μm	Septa	Size/μm	Septa	Shape	Diam/μm	
Ca. colhounii	*Ca. eucalypti*	(25–)30–36(–56) × (3–)5–6(–8) x̄ = 33 × 6	(1–)3	(66–)69–75(–80) × (5–)6 x̄ = 72 × 6	3	broad clavate	4–6	[32]
	Ca. eucalypti	/	/	(65–)70–80(–87) × (5–)6 x̄ = 75 × 6.5	1–3	Clavate to broad clavate	4–10	(This study)
	Ca. shaoguanensis	(45–)50–70 × (3–)4–8(–9) x̄ = 56.5 × 6.5	1	(55–)60–70(–75) × (4–)5–8 x̄ = 65 × 6.5	1–3	narrowly clavate	(2–)4–7	(This study)
Ca. kyotensis	*Ca. ilicicola*	x̄ = 45 × 6	1	x̄ = 62 × 6	1(–3)	sphaeropedunculate	8–12	[32]

Table 2. *Cont.*

Complex	Species	Ascospores		Macroconidia		Vesicle		References
		Size/µm	Septa	Size/µm	Septa	Shape	Diam/µm	
Ca. reteaudii	*Ca. cassiae*	40–50 × 4–8 $\bar{x}$ = 45 × 6	1	40–65 × 4–8 $\bar{x}$ = 54 × 6	1–3	sphaeropedunculate	8–12	(This study)
	Ca. strelitziae	/	/	(65–)80–95(–115) × (4–)6–10(–12) $\bar{x}$ = 87 × 8	3–5	narrowly clavate to broad clavate	3–8	(This study)
	Ca. pseudoreteaudii	/	/	(88–)96–112(–119) × 7–9(–10) $\bar{x}$ = 104 × 8	5–8	narrowly clavate	3–5	[14]
	Ca. acaciicola	/	/	(85–)90–98(–105) × (6–)6.5–7.5 $\bar{x}$ = 94 × 7	5	narrowly clavate	4–7	[31]
	Ca. reteaudii	/	/	(50–)75–95(–120) × (5–)6–7 $\bar{x}$ = 84 × 6.5	(1–) 5(–6)	clavate	3–5(–6)	[9]
	Ca. guangdongensis	/	/	55–70 × 5–7(–9) $\bar{x}$ = 64 × 6	1–3	narrowly clavate	3–7	(This study)
	Ca. melaleucae	/	/	80–95(–100) × (5–)7–10 $\bar{x}$ = 88 × 8	3–5	narrowly clavate	3–7	(This study)
	Ca. queenslandica	/	/	(61–)65–73(–78) × (4–)5–6(–7) $\bar{x}$ = 69 × 6	4–6	narrowly clavate	3–4	[14]
	Ca. lombardiana	/	/	(64–)74–86(–98) × (5–)5.5–6.5(–7.5) $\bar{x}$ = 80 × 6	5	narrowly clavate	2–4	[3]

4. Discussion

Species identification is the most important step in plant pathology to develop control measures [33,34]. Thus, proper species identification is necessary following polyphasic approaches [33–36]. *Calonectria* represents one of the most important phytopathogenic genera in Nectriaceae [37]. In the present study, we isolated and identified seven *Calonectria* species associated with various leaf and root diseases from eight different hosts in China. These species were identified as five novel species: *Ca. cassiae*, *Ca. guangdongensis*, *Ca. melaleucae*, *Ca. shaoguanensis* and *Ca. strelitziae*, as well as four new host records: *Ca. aconidialis* from *Arachis hypogaea*, *Ca. auriculiformis* from *Eucalyptus* sp., *Ca. eucalypti* from *Callistemon rigidus*, and *Ca. hongkongensis* from *Eucalyptus gunnii*. *Calonectria* species belong to 11 species complexes [3] revealing that the diversity among these species is considerably high. Thus, the use of either morphology or phylogeny alone is difficult to provide strong support to delineate the species [11,16]. Therefore, in the present study, we used polyphasic approaches including morphology, phylogeny and recombination analysis to introduce novel species.

In this study, four *Calonectria* species were identified as new host records. *Calonectria eucalypti* were isolated from *Eucalyptus grandis* leaf from Sumatera Utara [32]. Here *Ca. eucalypti* is reported to be associated with leaf spots on *Callistemon rigidus*. Previous studies have mentioned that *Ca. colhounii* Peerally, *Ca. kyotensis* Terash. and *Ca. pteridis* Crous, M.J. Wingf. & Alfenas are associated with *C. rigidus* [9], and our collection, *Ca. eucalypti* will be a new addition. *Calonectria aconidialis* was introduced as a soil fungus from *Eucalyptus* plantation [16]. In the present study, *Ca. aconidialis* was isolated from rotted *A. hypogaea* pods. There are no host records for *Ca. aconidialis* worldwide [30]. Therefore, our study provides a novel ecological niche for *Ca. aconidialis*. However, further studies are required to understand the host shifting and pathogenicity mechanisms of this fungus.

Calonectria hongkongensis was described by Crous et al. from the soil in Hong Kong [21]. Up to now, 15 *Calonectria* species, including *Ca. hongkongensis*, have been isolated from *Eucalyptus* plants or plantation soils in China [17]. This species was only isolated from *Eucalyptus* plantation soil and was not identified as a pathogen on *Eucalyptus* leaves [16,17,38]. Therefore, *Ca. hongkongensis* is considered a soil-borne species that inhabits the soil [17]. This is the first report of *Ca. hongkongensis* being isolated from diseased *E. gunnii* roots, reflecting that this species can be a potential soil-borne pathogen in *Eucalyptus* plantations. *Calonectria auriculiformis* was described by Pham et al. from *Acacia auriculiformis* plantation soil in Vietnam [31]. Later it was isolated from soil in *E. urophylla* hybrid plantations in China [17]. Similar to *Ca. hongkongensis, Ca. auriculiformis* has not been reported as a plant pathogen in previous studies. In the present study, we isolated *Ca. auriculiformis* from diseased leaves of *Eucalyptus* in Guangdong, China. However, further assays including controlled inoculation studies are required to confirm the pathogenicity of these isolated taxa.

Calonectria shaoguanensis was introduced as a new species, while adding one more species to the *Ca. colhounii* complex. This species complex includes 11 species: *Ca. aciculata* Jie Q. Li, Q.L. Liu & S.F. Chen, *Ca. colhounii*, *Ca. eucalypti*, *Ca. fujianensis* S.F. Chen, L. Lombard, M.J. Wingf. & X.D. Zhou, *Ca. honghensis* Jie Q. Li, Q.L. Liu & S.F. Chen, *Ca. indusiata* (Seaver) Crous, *Ca. lichi* Q.L. Liu & S.F. Chen, *Ca. macroconidialis* (Crous, M.J. Wingf. & Alfenas) Crous, *Ca. madagascariensis* Crous, *Ca. monticola* L. Lombard & Crous and *Ca. paracolhounii* L. Lombard & Crous [3]. *Calonectria shaoguanensis* can be distinguished from its closely related species by its macroconidial dimensions and by the shape of vesicles as well as by the number of ascosporous septa and macroconidial septa [32]. This species was isolated from *Callistemon rigidus* leaf spot. There have been three *Calonectria* species found on *C. rigidus* previously [30].

Three novel species, *Ca. guangdongensis, Ca. melaleucae* and *Ca. strelitziae* were added into the *Ca. reteaudii* complex, which accommodates nine species [3]. Species in the *Ca. reteaudii* complex have narrowly clavate vesicles, orange to red–brown perithecia and generally >3-septate macroconidia [3]. These three new species differ from closely related taxa by the size of macroconidia, *Ca. guangdongensis* and *Ca. strelitziae* developed shorter macroconidia, *Ca. melaleucae* formed larger macroconidia [3,9,14,31]. In addition, the macroconidia septa number of *Ca. guangdongensis* was one to three, similar to that of *Ca. crousiana* and *Ca. australiensis*, while distinct from most species having over 3-septate macroconidia in the *Ca. reteaudii* complex [3]. *Calonectria cassiae* was introduced as a new species while adding one more species to the *Ca. kyotensis* complex, which is a larger complex with twenty-four species [3]. The novel species differ from closely related species *Ca. ilicicola* by producing shorter macroconidia [32]. Until now, there are fifteen fungal species associated with *Cassia surattensis*, while there are no previous records from *Calonectria* species [30]. This is the first report of *Calonectria* species being associated with *C. surattensis* stem and root rots.

Overall, in the present study, novel *Calonectria* species and new host records were identified. Here, these taxa were isolated from diseased plant tissues including leaf spots, stem and root rots. However, the pathogenicity of these isolates was not confirmed in this study. The identification and characterization of novel taxa from southern China contributes to the knowledge of the biodiversity resources in tropical regions. Moreover, this study adds information on the taxonomy and diversity of *Calonectria* species in China. Future studies are required to confirm the pathogenicity of these isolated species on different plant hosts, and to examine their biology and ecology.

Supplementary Materials: The following supporting information can be downloaded at: https://www.mdpi.com/article/10.3390/jof8070719/s1, Table S1. GeneBank accession numbers of Calonectria strains used in phylogenetic analysis.

Author Contributions: Conceptualization, Y.Z. and I.S.M.; methodology, Y.Z.; software, C.C. (Cantian Chen), C.C. (Chao Chen) and J.C.; validation, Y.Z. and C.C. (Cantian Chen); formal analysis, Y.Z., C.C. (Cantian Chen) and C.C. (Chao Chen); investigation, Y.Z. and C.C. (Cantian Chen); resources, Y.Z. and C.C. (Cantian Chen); data curation, Y.Z. and C.C. (Cantian Chen); writing—original draft preparation, Y.Z.; writing—review and editing, Y.Z., I.S.M., D.N.W., T.H. and K.D.H.; visualization, C.C. (Chao Chen) and C.C. (Cantian Chen); supervision, I.S.M. and M.X.; project administration, Y.Z.; funding acquisition, Y.Z. All authors have read and agreed to the published version of the manuscript.

Funding: This research was funded by the National Natural Science Foundation of China (grant no. 31600019), Modern Agricultural Industry Technology System Flower Innovation Team of Guangdong Province (grant no. 2021KJ121 and 2022KJ121), the Project of Educational Commission of Guangdong Province of China (grant no. 2021KTSCX045) and Key Realm R&D Program of Guangdong Province (grant no. 2018B020205003). Ishara S. Manawasinghe would like to thank Zhongkai University of Agriculture and Engineering, Guangzhou 510225, P.R China for research funding (grant no. KA210319288).

Institutional Review Board Statement: Not applicable.

Informed Consent Statement: Not applicable.

Data Availability Statement: All sequence data generated in this study is deposited in NCBI GenBank and accession numbers are given in the Supplementary Materials.

Acknowledgments: We would like to thank Shaun Pennycook, Nomenclature Editor, Mycotaxon, for his guidance on the species names. We also would like to thank Paul Kirk, Royal Botanic Gardens Kew, for his help with new species name registration in Index Fungorum.

Conflicts of Interest: The authors declare no conflict of interest.

References

1. Rossman, A.Y. Calonectria and its type species, Calonectria daldiniana, a later synonym of Calonectria pyrochroa. *Mycotaxon* **1979**, *8*, 321–328.
2. Crous, P.W.; Wingfield, M.J. A monograph of Cylindrocladium, including anamorphs of Calonectria. *Mycotaxon* **1994**, *51*, 393–404.
3. Liu, Q.L.; Li, J.; Wingfield, M.J.; Duong, T.A.; Wingfield, B.D.; Crous, P.W.; Chen, S.F. Reconsideration of species boundaries and proposed DNA barcodes for Calonectria. *Stud. Mycol.* **2020**, *97*, 1–73. [CrossRef]
4. Crous, P.W.; Hernández-Restrepo, M.; Schumacher, R.K.; Cowan, D.A.; Maggs-Kölling, G.; Marais, E.; Wingfield, M.J.; Yilmaz, N.; Adan, O.C.G.; Akulov, A.; et al. New and interesting fungi. 4. *Fungal Syst. Evol.* **2021**, *7*, 255–343. [CrossRef] [PubMed]
5. Mohali, S.R.; Stewart, J.E. Calonectria vigiensis sp. nov. (Hypocreales, Nectriaceae) associated with dieback and sudden-death symptoms of Theobroma cacao from Mérida state, Venezuela. *Botany* **2021**, *99*, 683–693. [CrossRef]
6. Pham, N.Q.; Marincowitz, S.; Chen, S.; Yaparudin, Y.; Wingfield, M.J. Calonectria species, including four novel taxa, associated with Eucalyptus in Malaysia. *Mycol. Prog.* **2022**, *21*, 181–197. [CrossRef]
7. Index Fungorum. 2022. Available online: http://www.indexfungorum.org/Names/Names.asp (accessed on 11 March 2022).
8. Crous, P.W.; Phillips, A.J.L.; Wingfield, M.J. The genera Cylindrocladium and Cylindrocladiella in South Africa, with special reference to forest nurseries. *S. Afr. For. J.* **1991**, *157*, 69–85.
9. Crous, P.W. *Taxonomy and Pathology of Cylindrocladium (Calonectria) and Allied Genera*; American Phytopathological Society (APS Press.): Saint Paul, MI, USA, 2002.
10. Pan, R.; Guan, M.; Xu, D.; Gao, X.; Yan, X.; Liao, H. Cylindrocladium black rot caused by Cylindrocludium parasiticum newly reported on peanut in China. *Plant Pathol.* **2009**, *58*, 1176. [CrossRef]
11. Alfenas, R.F.; Lombard, L.; Pereira, O.L.; Alfenas, A.C.; Crous, P.W. Diversity and potential impact of Calonectria species in Eucalyptus plantations in Brazil. *Stud. Mycol.* **2015**, *80*, 89–130. [CrossRef]
12. Gehesquière, B.; Crouch, J.A.; Marra, R.E.; Van Poucke, K.; Rys, F.; Maes, M.; Heungens, K. Characterization and taxonomic reassessment of the box blight pathogen Calonectria pseudonaviculata, introducing Calonectria henricotiae sp. nov. *Plant Pathol.* **2016**, *65*, 37–52. [CrossRef]
13. Gai, Y.; Deng, Q.; Chen, X.; Guan, M.; Xiao, X.; Xu, D.; Pan, R. Phylogenetic diversity of Calonectria ilicicola causing Cylindrocladium black rot of peanut and red crown rot of soybean in southern China. *J. Gen. Plant Pathol.* **2017**, *83*, 273–282. [CrossRef]
14. Lombard, L.; Zhou, X.D.; Crous, P.W.; Wingfield, B.D.; Wingfield, M.J. Calonectria species associated with cutting rot of Eucalyptus. *Persoonia* **2010**, *24*, 1–11. [CrossRef]
15. Chen, S.; Lombard, L.; Roux, J.; Xie, Y.; Wingfield, M.J.; Zhou, X.D. Novel species of Calonectria associated with Eucalyptus leaf blight in Southeast China. *Persoonia* **2011**, *26*, 1–12. [CrossRef]
16. Lombard, L.; Chen, S.F.; Mou, X.; Zhou, X.D.; Crous, P.W.; Wingfield, M.J. New species, hyper-diversity and potential importance of Calonectria spp. from Eucalyptus in South China. *Stud. Mycol.* **2015**, *80*, 151–188. [CrossRef]
17. Wu, W.; Chen, S. Species diversity, mating strategy and pathogenicity of Calonectria species from diseased leaves and soils in the Eucalyptus plantation in southern China. *J. Fungi* **2021**, *7*, 73. [CrossRef]
18. Carbone, I.; Kohn, L.M. A method for designing primer sets for speciation studies in filamentous ascomycetes. *Mycologia* **1999**, *91*, 553–556. [CrossRef]
19. O'Donnell, K.; Kistler, H.C.; Cigelnik, E.; Ploetz, R.C. Multiple evolutionary origins of the fungus causing Panama disease of banana: Concordant evidence from nuclear and mitochondrial gene genealogies. *Proc. Natl. Acad. Sci. USA* **1998**, *95*, 2044–2049. [CrossRef]

20. Donnell, K.O.; Cigelnik, E. Two divergent intragenomic rDNA ITS2 types within a monophyletic lineage of the fungus Fusarium are nonorthologous. *Mol. Phylogenet. Evol.* **1997**, *7*, 103–116. [CrossRef]
21. Crous, P.W.; Groenewald, J.Z.; Risède, J.M.; Simoneau, P.; Hywel-Jones, N.L. Calonectria species and their Cylindrocladium anamorphs: Species with sphaeropedunculate vesicles. *Stud. Mycol.* **2004**, *50*, 415–430.
22. Hall, T.A. BioEdit: A user-friendly biological sequence alignment editor and analysis program for Windows 95/98/NT. *Nucleic Acid Symp. Ser.* **1999**, *41*, 95–98.
23. Silvestro, D.; Michalak, I. RaxmlGUI: A graphical front-end for RAxML. *Org. Divers. Evol.* **2012**, *12*, 335–337. [CrossRef]
24. Ronquist, F.; Huelsenbeck, J.P. MrBayes 3: Bayesian phylogenetic inference under mixed models. *Bioinformatics* **2003**, *19*, 1572–1574. [CrossRef]
25. Stamatakis, A.; Hoover, P.; Rougemont, J. A rapid bootstrap algorithm for the RAxML web servers. *Syst. Biol.* **2008**, *57*, 758–771. [CrossRef]
26. Stamatakis, A. RAxML version 8: A tool for phylogenetic analysis and post-analysis of large phylogenies. *Bioinformatics* **2014**, *30*, 1312–1313. [CrossRef]
27. Miller, M.A.; Pfeiffer, W.T.; Schwartz, T. Creating the CIPRES Science Gateway for Inference of Large Phylogenetic Trees. In *Gateway Computing Environments Workshop (GCE)*; Institute of Electrical and Electronics, Engineers: New Orleans, LA, USA, 2010.
28. Huson, D.H.; Bryant, D. Application of phylogenetic networks in evolutionary studies. *Mol. Biol. Evol.* **2006**, *23*, 254–267. [CrossRef]
29. Rayner, R.W. *A Mycological Colour Chart*; Commonwealth Mycological Institute and British Mycological Society: Kew, UK, 1970.
30. Farr, D.F.; Rossman, A.Y. Fungal Databases, U.S. National Fungus Collections, ARS, USDA. Available online: https://nt.ars-grin.gov/fungaldatabases/ (accessed on 11 March 2022).
31. Pham, N.Q.; Barnes, I.; Chen, S.; Liu, F.; Dang, Q.N.; Pham, T.Q.; Wingfield, M.J. Ten new species of Calonectria from Indonesia and Vietnam. *Mycologia* **2019**, *111*, 78–102. [CrossRef]
32. Lombard, L.; Crous, P.W.; Wingfield, B.D.; Wingfield, M.J. Phylogeny and systematics of the genus Calonectria. *Stud. Mycol.* **2010**, *66*, 31–69. [CrossRef]
33. Manawasinghe, I.S.; Phillips, A.J.L.; Xu, J.; Balasuriya, A.; Hyde, K.D.; Tępień, L.; Harischandra, D.L.; Karunarathna, A.; Yan, J.; Weerasinghe, J.; et al. Defining a species in fungal plant pathology: Beyond the species level. *Fungal Divers.* **2021**, *109*, 267–282. [CrossRef]
34. Chethana, K.W.; Manawasinghe, I.S.; Hurdeal, V.G.; Bhunjun, C.S.; Appadoo, M.A.; Gentekaki, E.; Raspé, O.; Promputtha, I.; Hyde, K.D. What are fungal species and how to delineate them? *Fungal. Divers.* **2021**, *109*, 1–25. [CrossRef]
35. Jayawardena, R.S.; Hyde, K.D.; de Farias, A.R.G.; Bhunjun, C.S.; Ferdinandez, H.S.; Manamgoda, D.S.; Thines, M. What is a species in fungal plant pathogens? *Fungal Divers.* **2021**, *109*, 239–266. [CrossRef]
36. Maharachchikumbura, S.S.; Chen, Y.; Ariyawansa, H.A.; Hyde, K.D.; Haelewaters, D.; Perera, R.H.; Samarakoon, M.C.; Wanasinghe, D.N.; Bustamante, D.E.; Liu, J.K. Integrative approaches for species delimitation in Ascomycota. *Fungal Divers.* **2021**, *109*, 155–179. [CrossRef]
37. Jayawardena, R.S.; Hyde, K.D.; Jeewon, R.; Ghobad-Nejhad, M.; Wanasinghe, D.N.; Liu, N.; Phillips, A.J.L.; Oliveira-Filho, J.R.C.; da Silva, G.A.; Gibertoni, T.B.; et al. One stop shop ii: Taxonomic update with molecular phylogeny for important phytopathogenic genera: 26–50. *Fungal Divers.* **2019**, *94*, 41–129. [CrossRef]
38. Li, J.; Wingfield, M.J.; Liu, Q.; Barnes, I.; Roux, J.; Lombard, L.; Crous, P.W.; Chen, S.F. Calonectria species isolated from Eucalyptus plantations and nurseries in South China. *IMA fungus* **2017**, *8*, 259–286. [CrossRef]

Journal of *Fungi*

MDPI

Article

New Species of Large-Spored *Alternaria* in Section *Porri* Associated with Compositae Plants in China

Lin Zhao [1], Huan Luo [2], Hong Cheng [1], Ya-Nan Gou [1], Zhi-He Yu [3] and Jian-Xin Deng [1,*]

[1] Department of Plant Protection, College of Agriculture, Yangtze University, Jingzhou 434025, China; zhaolin999@hotmail.com (L.Z.); hongcheng7777@hotmail.com (H.C.); gynan024@hotmail.com (Y.-N.G.)
[2] Department of Applied Biology, Chungnam National University, Daejeon 34134, Korea; luohuan_0813@163.com
[3] Department of Biology, College of Life Sciences, Yangtze University, Jingzhou 434025, China; zhiheyu@hotmail.com
* Correspondence: djxin555@yangtzeu.edu.cn

Abstract: *Alternaria* is a ubiquitous fungal genus including saprobic, endophytic, and pathogenic species associated with a wide variety of substrates. It has been separated into 29 sections and seven monotypic lineages based on molecular and morphological data. *Alternaria* sect. *Porri* is the largest section, containing the majority of large-spored *Alternaria* species, most of which are important plant pathogens. Since 2015, of the investigations for large-spored *Alternaria* species in China, 13 species were found associated with Compositae plants based on morphological comparisons and phylogenetic analyses. There were eight known species and five new species (*A. anhuiensis* sp. nov., *A. coreopsidis* sp. nov., *A. nanningensis* sp. nov., *A. neimengguensis* sp. nov., and *A. sulphureus* sp. nov.) distributed in the four sections of *Helianthiinficientes*, *Porri*, *Sonchi*, and *Teretispora*, and one monotypic lineage (*A. argyranthemi*). The multi-locus sequence analyses encompassing the internal transcribed spacer region of rDNA (ITS), glyceraldehydes-3-phosphate dehydrogenase (GAPDH), *Alternaria* major allergen gene (Alt a 1), translation elongation factor 1-alpha (TEF1), and RNA polymerase second largest subunit (RPB2), revealed that the new species fell into sect. *Porri*. Morphologically, the new species were illustrated and compared with other relevant large-spored *Alternaria* species in the study. Furthermore, *A. calendulae*, *A. leucanthemi*, and *A. tagetica* were firstly detected in *Brachyactis ciliate*, *Carthamus tinctorius*, and *Calendula officinalis* in China, respectively.

Keywords: *Alternaria*; compositae; morphology; multi-locus sequence analyses; taxonomy

Citation: Zhao, L.; Luo, H.; Cheng, H.; Gou, Y.-N.; Yu, Z.-H.; Deng, J.-X. New Species of Large-Spored *Alternaria* in Section *Porri* Associated with Compositae Plants in China. *J. Fungi* **2022**, *8*, 607. https://doi.org/10.3390/jof8060607

Academic Editors: Lei Cai and Cheng Gao

Received: 17 May 2022
Accepted: 3 June 2022
Published: 6 June 2022

1. Introduction

Alternaria is a cosmopolitan and widely distributed fungal genus described originally by Nees (1816), which is characterized by the dark-coloured phaeodictyospores in chains and a beak of tapering apical cells [1]. It is also associated with nearly every environmental substrate including animal, plant, agricultural product, soil, and the atmosphere. Species of *Alternaria* are known as serious plant pathogens, causing enormous losses on many crops [1,2]. The taxonomy is mainly based on sporulation patterns and their conidial shape, size, and septation [2,3]. Around 280 species are summarised and recognised on the basis of morphology [2], comprising two groups, large-spored (60–100 μm long conidial body) and small-spored (below 60 μm conidial body) [4–6].

Since the 20th century, molecular approaches, especially multi-locus phylogenetic analyses, have been used to identify *Alternaria* species [7–10]. Over ten gene regions are used in the classification, such as the internal transcribed spacer region of rDNA (ITS), large subunit ribosomal DNA (LSU), mitochondrial small subunit (mtSSU), glyceraldehydes-3-phosphate dehydrogenase (GAPDH), *Alternaria* major allergen gene (Alt a 1), translation elongation factor 1-alpha (TEF1), RNA polymerase second largest subunit (RPB2), and plasma membrane ATPase [1,4,7,9,11–18]. *Alternaria* has been separated into 29 sections and seven monotypic lineages [19–21]. The introduction of a molecular phylogenetic approach has helped to clarify their taxonomy, combining many allied genera into one large genus of *Alternaria* complex [1].

Due to the effects of *Alternaria* on humans and their surroundings, the identification is particularly important to agriculture, medicine, and science. The Compositae plants serve as food plant, oil seed, seed plant, ornamental, and sources of medicine and insecticide worldwide [22], of which nearly 3000 species almost 240 genera have been found in China [23]. Most *Alternaria* are commonly plant pathogens leading to substantial economic losses caused by Alternaria leaf spots and defoliation [18,24–26]. Large-spored *Alternaria* species encompassing 148 species are almost phytopathogenic demonstrated [2].

During the investigation of large-spored *Alternaria* in China, five new species were encountered from diseased leave samples of composite plants. The objectives of this study were to identify them on the basis of the cultural and conidial morphology incorporate with multi-loci phylogeny (ITS, GAPDH, Alt a 1, TEF1, and RPB2). The present multi-locus analysis supplemented with cultural and morphological data forms an example for *Alternaria* species recognition. The five new species described in this study add species diversity to large-spored *Alternaria* and provide theoretical and practical basis for the further identification and disease management.

2. Materials and Methods

2.1. Sample Collection and Fungal Isolation

Symptomatic samples of composite plants (14) have been randomly collected from different provinces in China since 2015. For fungal isolation, the samples were put into sterile plastic bags and taken to the laboratory. Small leaf segments (2 mm) with disease lesions were placed into petri dishes with moist filter papers and incubated at 25 °C in dark for conidial sporulation. Single spore of large-spored *Alternaria* was picked by a sterilized glass needle under the stereoscopic microscope and transferred to potato dextrose agar (PDA: Difco, Montreal, Canada). Over ten similar spores were randomly picked from a sample for sub-culturing to obtain the pure cultures, and two to three strains were selected for deposition when exhibiting similar cultural morphology on PDA. A total of 81 strains were kept in test-tube slants and deposited at 4 °C. Living ex-type strains were preserved in the Fungi Herbarium of Yangtze University (YZU), in Jingzhou, Hubei, China.

2.2. Morphological Observations

To determine cultural characteristics including growth rate, color and texture of colonies [27], mycelial plugs (6 mm in diameter) were taken from the edge of colonies grown on PDA. Then, the plugs were put on fresh PDA plates (90 mm) at 25 °C for 7 days in darkness. To observe the conidial morphology (conidial sporulation patterns, shape, size, etc.), mycelia were grown on potato carrot agar (PCA) and V8 juice agar (V8A) inoculated at 22 °C with a light period of 8 h light/16 h dark [2]. After 7 days, conidia and sporulation patterns were observed. Conidiophores and conidia were mounted with lactophenol picric acid solution and photographed with a Nikon ECLIPSE Ni-U microscope (Nikon, Japan). Randomly selected conidia (n = 50) were separately measured for each characterization.

2.3. DNA Extraction and PCR Amplification

Genomic DNA extraction was performed using fresh mycelia collected from colonies grown on PDA [28]. Polymerase chain reaction (PCR) amplifications of the internal transcribed spacer region of rDNA (ITS), glyceraldehydes-3-phosphate dehydrogenase (GAPDH), *Alternaria* major allergen gene (Alt a 1), translation elongation factor 1-alpha (TEF1), and RNA polymerase second largest subunit (RPB2) gene regions were amplified with the primer pairs ITS5/ITS4 [29], EF1-728F/EF1-986R [30], gpd1/gpd2 [31], Alt-for/Alt-rev [12], and RPB2-5F2/RPB2-7cR [32,33], respectively. A 25 μL of the PCR reaction mixture comprising 21 μL of 1.1 × Taq PCR Star Mix (TSINGKE, Beijing, China), 2 μL template DNA and 1 μL of each primer was applied and performed in a BIORAD T100 thermocycler [1].

Successfully amplified PCR products were purified and sequenced by TSINGKE company (Beijing, China).

2.4. Phylogenetic Analyses

The resulted sequences were examined by BioEdit v.7.0.9 [34] and assembled with PHYDIT 3.2 [35]. All newly generated sequences were deposited in GenBank (Table 1). Relevant sequences [4] were retrieved from NCBI database based on the results of BLAST searches (Table 1). The concatenated sequence dataset of multiple loci was aligned using MEGA v.6.0 [36]. Phylogenetic analyses of each alignment were performed using maximum likelihood (ML) and Bayesian inference (BI) methods. ML analysis was conducted using RAxML v.7.2.8 [37]. Bootstrapping with 1000 replicates was performed using the model of nucleotide substitution obtained by MrModeltest. For the BI analysis, it was performed using parameters including 1,000,000 Markov chain Monte Carlo (MCMC) algorithm with Bayesian posterior probabilities [38]. MrModel test v.2.3 used the best-fit model (GTR+I+G) according to the Akaike Information Criterion (AIC). Two MCMC chains were run from random trees for 10^6 generations, and the trees were sampled every 100th generation. After discarding the first 25% of the samples, the 50% majority rule consensus tree and posterior probability values were calculated. Finally, the resulting trees were edited in FigTree v.1.3.1 [39]. Branch support of the groupings (>60%/0.6 for ML bootstrap value-BS/posterior probability-PP) were indicated in the phylogram. *Alternaria gypsophilae* CBS 107.41 in sect. *Gypsophilae* was used as an outgroup.

Table 1. GenBank accession numbers of *Alternaria* species used in phylogenetic analyses.

Section	Species	Strain	Locality	Substrate	ITS	GAPDH	Alt a 1	TEF1	RPB2
Porri	*A. acalyphicola*	CBS 541.94 T	Seychelles	*Acalypha indica*	KJ718097	KJ717952	KJ718617	KJ718446	KJ718271
Porri	*A. agerati*	CBS 117221 R	USA	*Ageratum houstonianum*	KJ718098	KJ717953	KJ718618	KJ718447	KJ718272
Porri	*A. agripestis*	CBS 577.94 T	Canada	*Euphorbia esula*, stem lesion	KJ718099	JQ646356	KJ718619	KJ718448	KJ718273
Porri	*A. allii*	CBS 116701 R	USA	*Allium cepa* var. *viviparum*	KJ718103	KJ717957	KJ718623	KJ718452	KJ718277
Porri	*A. alternariacida*	CBS 105.51 T	UK	*Solanum lycopersicum*, fruit	KJ718105	KJ717959	KJ718625	KJ718454	KJ718279
Porri	*A. anagallidis*	CBS 117129 R	New Zealand	*Anagallis arvensis*, leaf spot	KJ718109	KJ717962	KJ718629	KJ718457	KJ718283
Porri	***A. anhuiensis* sp. nov.**	**YZU 171206 T**	**China**	***Coreopsis basalis*, leaf**	**MK264916**	**MK303949**	**MK303953**	**MK303958**	**MK303960**
Porri	*A. anodae*	PPRI 12376	South Africa	*Anoda cristata*, leaf	KJ718110	KJ717963	KJ718630	KJ718458	KJ718284
Porri	*A. aragakii*	CBS 594.93 T	USA	*Passiflora edulis*	KJ718111	KJ717964	KJ718631	KJ718459	KJ718285
Porri	*A. argyroxiphii*	CBS 117222 T	USA	*Argyroxiphium* sp.	KJ718112	JQ646350	KJ718632	KJ718460	KJ718286
Porri	*A. azadirachtae*	CBS 116444 T	Australia	*Azadirachta indica*, leaf spot	KJ718115	KJ717967	KJ718635	KJ718463	KJ718289
Porri	*A. bataticola*	CBS 531.63 T	Japan	*Ipomoea batatas*	KJ718117	JQ646349	JQ646433	KJ718465	KJ718291
Porri	*A. blumeae*	CBS 117364 T	Thailand	*Blumea aurita*	KJ718126	AY562405	AY563291	KJ718474	KJ718300
Porri	*A. calendulae*	CBS 224.76 T	Germany	*Calendula officinalis*	KJ718127	KJ717977	KJ718648	KJ718475	KJ718301
Porri	*A. calendulae*	CBS 101498	New Zealand	*Calendula officinalis*, leaf	KJ718128	KJ717978	KJ718645	KJ718476	KJ718302
Porri		CBS 116439 T	New Zealand	*Rosa* sp., leaf spot	KJ718129	KJ717979	KJ718646	KJ718477	KJ718303
Porri		CBS 116650 R	Japan	*Calendula officinalis*, leaf spot	KJ718130	KJ717980	KJ718647	KJ718478	KJ718304
Porri	*A. carthami*	CBS 117091 R	USA	*Carthamus tinctorius*, leaf spot	KJ718133	KJ717983	KJ718651	KJ718481	KJ718307
Porri	*A. carthamicola*	CBS 117092 T	Iraq	*Carthamus tinctorius*	KJ718134	KJ717984	KJ718652	KJ718482	KJ718308

Table 1. *Cont.*

Section	Species	Strain	Locality	Substrate	ITS	GAPDH	Alt a 1	TEF1	RPB2
Porri	*A. cassiae*	CBS 116119 T	Malaysia	*Sauropus androgynus*	KJ718136	KJ717986	KJ718654	KJ718484	KJ718310
Porri	*A. catananches*	CBS 137456 T	Netherlands	*Catananche caerulea*	KJ718139	KJ717989	KJ718657	KJ718487	KJ718313
Porri	*A. centaureae*	CBS 116446 T	USA	*Centaurea solstitialis*, leaf spot	KJ718140	KJ717990	KJ718658	KJ718488	KJ718314
Porri	*A. cichorii*	CBS 102.33 T	Cyprus	*Cichorium intybus*, leaf spot	KJ718141	KJ717991	KJ718659	KJ718489	KJ718315
Porri	*A. cirsinoxia*	CBS 113261 T	Canada	*Cirsium arvense*, stem lesion	KJ718143	KJ717993	KJ718661	KJ718491	KJ718317
Porri	*A. citrullicola*	CBS 103.32 T	Cyprus	*Citrullus vulgaris*, fruit	KJ718144	KJ717994	KJ718662	KJ718492	KJ718318
Porri	***A. coreopsidis* sp. nov.**	**YZU 161159**	**China**	***Coreopsis basalis*, leaf**	**MK264914**	**MK303947**	**MK303951**	**MK303955**	**MK303971**
Porri		**YZU 161160 T**	**China**	***Coreopsis basalis*, leaf**	**ON130144**	**ON229924**	**ON229926**	**ON229928**	**ON229930**
Porri	*A. crassa*	CBS 110.38 T	Cyprus	*Datura stramonium*, leaf spot	KJ718147	KJ717997	KJ718665	KJ718495	KJ718320
Porri		CBS 122590 R	USA	*Datura stramonium*, leaf spot	KJ718152	GQ180072	GQ180088	KJ718500	KJ718325
Porri	*A. cucumerina*	CBS 116114 T	USA	*Luffa acutangula*	KJ718153	KJ718000	KJ718668	KJ718501	KJ718326
Porri		CBS 117225 R	USA	*Cucumis melo*, leaf spot	KJ718154	KJ718001	KJ718669	KJ718502	KJ718327
Porri	*A. cyamopsidis*	CBS 117219 R	USA	*Cyamopsis tetragonoloba*, leaf spot	KJ718157	KJ718004	KJ718672	KJ718505	KJ718330
Porri	*A. dauci*	CBS 111.38 T	Italy	*Daucus carota*, seed	KJ718158	KJ718005	KJ718673	KJ718506	KJ718331
Porri	*A. deserticola*	CBS 110799 T	Namibia	desert soil	KJ718249	KJ718077	KJ718755	KJ718595	KJ718424
Porri	*A. dichondrae*	CBS 199.74 T	Italy	*Dichondra repens*, leaf spot	KJ718166	JQ646357	JQ646441	KJ718514	KJ718339
Porri	*A. echinaceae*	CBS 116117 T	New Zealand	*Echinacea* sp., leaf lesion	KJ718170	KJ718015	KJ718684	KJ718518	KJ718343
Porri	*A. grandis*	CBS 116695 R	USA	*Solanum tuberosum*, leaf spot	KJ718241	KJ718070	KJ718748	KJ718587	KJ718416
Porri	*A. ipomoeae*	CBS 219.79 T	Ethiopia	*Ipomoea batatas*, stem and petiole	KJ718175	KJ718020	KJ718689	KJ718523	KJ718348
Porri	*A. jesenskae*	CBS 133855 T	Slovakia	*Fumana procumbens*, seed	KJ718177	KJ718022	KJ718691	KJ718525	KJ718350
Porri	*A. linariae*	CBS 105.41 T	Denmark	*Linaria maroccana*, seedling	KJ718180	KJ718024	KJ718692	KJ718528	KJ718353
Porri	*A. passiflorae*	CBS 630.93 T	USA	*Passiflora edulis*	KJ718210	JQ646352	KJ718718	KJ718556	KJ718383
Porri		CBS 116333 T	New Zealand	*Gaura lindheimeri*, leaf	KJ718211	KJ718046	KJ718719	KJ718557	KJ718384
Porri	*A. pipionipisi*	CBS 116115 T	India	*Cajanus cajan*, seed	KJ718214	KJ718049	KJ718722	KJ718560	KJ718387
Porri	*A. porri*	CBS 116699 T	USA	*Allium cepa*, leaf spot	KJ718218	KJ718053	KJ718727	KJ718564	KJ718391
Porri	*A. protenta*	CBS 116437 T	New Zealand	*Hordeum vulgare*, seed	KJ718220	KJ718055	KJ718729	KJ718566	KJ718393
Porri	*A. pseudorostrata*	CBS 119411 T	USA	*Euphorbia pulcherrima*	JN383483	AY562406	AY563295	KC584680	KC584422
Porri	*A. ranunculi*	CBS 116330 T	Israel	*Ranunculus asiaticus*, seed	KJ718225	KJ718058	KJ718732	KJ718571	KJ718398
Porri	*A. ricini*	CBS 215.31 T	Japan	*Ricinus communis*	KJ718226	KJ718059	KJ718733	KJ718572	KJ718399
Porri	*A. rostellata*	CBS 117366 T	USA	*Euphorbia pulcherrima*, leaf	KJ718229	JQ646332	KJ718736	KJ718575	KJ718402
Porri	*A. scorzonerae*	CBS 478.83 T	Netherlands	*Scorzonera hispanica*, leaf spot	KJ718191	JQ646334	KJ718699	KJ718538	KJ718364
Porri	*A. sennae*	CBS 477.81 T	India	*Senna corymbosa*, leaf	KJ718230	JQ646344	JQ646428	EU130543	KJ718403

Table 1. *Cont.*

Section	Species	Strain	Locality	Substrate	ITS	GAPDH	Alt a 1	TEF1	RPB2
Porri	*A. sesami*	CBS 115264 R	India	*Sesamum indicum*, seedling	JF780939	KJ718061	KJ718738	KJ718577	KJ718405
Porri	*A. sidae*	CBS 117730 T	Kiribati	*Sida fallax*, leaf spot	KJ718232	KJ718062	KJ718739	KJ718578	KJ718406
Porri	*A. silybi*	CBS 134092 T	Russia	*Silybum marianum*, leaf	KJ718233	KJ718063	KJ718740	KJ718579	KJ718407
Porri		CBS 134093	Russia	*Silybum marianum*, leaf	KJ718234	KJ718064	KJ718741	KJ718580	KJ718408
Porri		CBS 134094	Russia	*Silybum marianum*, leaf	KJ718235	KJ718065	KJ718742	KJ718581	KJ718409
Porri	*A. solani*	CBS 109157 R	USA	*Solanum tuberosum*, leaf spot	KJ718238	GQ180080	KJ718746	KJ718585	KJ718413
Porri	*A. solani-nigri*	CBS 117101 R	New Zealand	*Solanum nigrum*, leaf spot	KJ718247	KJ718075	KJ718753	KJ718593	KJ718422
Porri	*A. steviae*	CBS 117362 T	Japan	*Stevia rebaudiana*, leaf spot	KJ718252	KJ718079	KJ718758	KJ718598	KJ718427
Porri	*A. tagetica*	CBS 117217 R	USA	*Tagetes* sp., leaf spot	KJ718256	KJ718083	KJ718763	KJ718602	KJ718431
Porri		CBS 297.79	UK	*Tagetes* sp., seed	KJ718253	KJ718080	KJ718759	KJ718599	KJ718428
Porri		CBS 298.79	UK	*Tagetes* sp., seed	KJ718254	KJ718081	KJ718760	KJ718600	KJ718429
Porri		CBS 479.81 R	UK	*Tagetes erecta*, seed	KC584221	KC584143	KJ718761	KC584692	KC584434
Porri		CBS 480.81 R	USA	*Tagetes* sp., seed	KJ718255	KJ718082	KJ718762	KJ718601	KJ718430
Porri	*A. thunbergiae*	CBS 116331 T	Australia	*Thunbergia alata*, leaf spot	KJ718257	KJ718084	KJ718764	KJ718603	KJ718432
Porri	*A. tillandsiae*	CBS 116116 T	New Zealand	*Tillandsia usneoides*	KJ718260	KJ718087	KJ718767	KJ718606	KJ718435
Porri	*A. tropica*	CBS 631.93 T	USA	*Passiflora edulis*, fruit	KJ718261	KJ718088	KJ718768	KJ718607	KJ718436
Porri	*A. venezuelensis*	CBS 116121 T	Venezuela	*Phaseolus vulgaris*, leaf spot	KJ718263	KJ718263	KJ718770	KJ718609	KJ718438
Porri	*A. zinniae*	CBS 117223 R	New Zealand	*Zinnia elegans*, leaf spot	KJ718270	KJ718096	KJ718777	KJ718616	KJ718445
Porri		CBS 118.44	Hungary	*Callistephus chinensis*, seed	KJ718264	JQ646361	KJ718771	KJ718610	KJ718439
Porri		CBS 117.59	Italy	*Zinnia elegans*	KJ718266	KJ718092	KJ718773	KJ718612	KJ718441
Porri		CBS 299.79	UK	*Zinnia* sp., seed	KJ718268	KJ718094	KJ718775	KJ718614	KJ718443
Gypsophilae	*A. gypsophilae*	CBS 107.41 T	Netherlands	*Gypsophila elegans*, seed	KC584199	KC584118	KJ718688	KC584660	KC584401

Note: The bold indicate the newly generated sequences. T, ex-type strain; R, representative strain.

3. Results

In the present study, large-spored *Alternaria* species associated with Compositae leaf spot in China since a survey from 2015 are summarized based on the phylogenetic analysis of GAPDH and RPB2 gene fragments (Figure S1 and Table S1). A total of 13 species including the present five new taxa revealed in four sections of *Helianthiinficientes* (*A. helianthiinficiens*), *Porri* (*A. calendulae*, *A. tagetica* and *A. zinniae*), *Sonchi* (*A. cinerariae* and *A. sonchi*), and *Teretispora* (*A. leucanthemi*), and one monotypic lineage (*A. argyranthemi*) (Figure S1). Meanwhile, a comprehensive description of the five new species in sect. *Porri* are described as *A. anhuiensis* sp. nov., *A. coreopsidis* sp. nov., *A. nanningensis* sp. nov., *A. neimengguensis* sp. nov., and *A. sulphureus* sp. nov..

3.1. Phylogenetic Analysis

The multi-gene phylogeny was constructed to determine the accurate positions of the new *Alternaria* based on five sequence loci (ITS + GAPDH + Alt a 1 + TEF1 + RPB2) (Table 1). The analysis comprised sequences of the ITS (504 characters), GAPDH (526 characters), Alt a 1 (457 characters), TEF1 (342 characters), and RPB2 (672 characters) gene regions with a total length of 2501 characters. The tree topologies (Figure 1) computed from the ML and BI analyses, were similar to each other, resulting in identical species-clades and the ML topology was presented as basal tree. The present strains fell into five separate branches in sect. *Porri* of *Alternaria*. Strain YZU 171206 was sister to *A. alternariacida* supported with a PP value of 1.0, which close to *A. silybi* with low BS and PP values surpport. Strains YZU 161159 and YZU 161160 formed an independent clade (BS/PP = 100%/1.0). Strain YZU 171523 fell into an individual branch close to *A. obtecta* and *A. tillandsiae* well-supported by 97%/0.99 (BS/PP). Strain YZU 171784 was clustered with *A. cirsinoxia*, *A. centaureae*, *A. cichorii*, and *A. cantannaches* supported by values of 79%/1.0 (BS/PP). Strain YZU 191448 was out group of strain YZU 171206, *A. silybi* and *A. alternariacida* with BS and PP values below 60% and 0.6. The results indicated that the five branches represent five new species from three different hosts (*Coreopsis basalis*, *Cosmos sulphureus*, and *Lactuca seriola*).

3.2. Taxonomy

Alternaria anhuiensis H. Luo and J.X. Deng, sp. nov. (Figure 2).

MycoBank No: 844033.

Etymology: Named after the collecting locality, Anhui Province.

Typification: China, Anhui Province, Hefei City, from leaf spot of *Coreopsis basalis*. June, 2017, J.X Deng, ex-type culture YZU 171206.

Description: Colonies on PDA circular, buff in the centre, flocculent with brown halo at the edge; reverse crimson pigment at centers, light yellow at margins, 59–60 mm in diam, at 25 °C for 7 days. On V8A, conidiophores arising from substrate or lateral of aerial hyphae with geniculate conidiogenous loci at or near apex, straight or curved, smooth-walled, septate, pale to dark brown, (40–) 60–145 (–203) × (4.5–) 5–7.5 (–8) μm; conidia solitary, long-narrow ovoid or ellipsoid body, apex rounded, base narrow, smooth-walled, single to double beak, dark brown, 61–100 (–111.5) × (11.5–) 13–19.5 μm, 6–11 transverse septa, 0–1 (–2) longitudinal septa; beak long-narrowed filiform, 1-beak, (32–) 58–133 (–150.5) × 2.5–4 (–4.5) μm; 2-beak, (22–) 60.5–90.5 (–116.5) × 2.5–3.5 μm. On PCA, conidiophores straight or curved, smooth-walled, septate, (42.5–) 50–140 × 4.5–6.5 (–9) μm; conidia solitary, long-narrow ovoid or ellipsoid body, single to double beak, triple or quadruple beaks not common, black brown, (55–) 66–105 × 11–16 μm, 5–10 (–11) transverse septa, 0–1 longitudinal septum; beak long-narrowed filiform, 1-beak, 100–180 (–202) × 2.5–4 μm; 2-beak, 95–217 (–236) × 2.5–4 (–5.5) μm; 3-beak, 60–140 × 2.5–3.5 μm; 4-beak (n = 1), 82 × 3 μm.

Notes: Phylogenetic analysis of the species based on a combined dataset of ITS, GAPDH, Alt a 1, TEF1, and RPB2 gene fragments falls in an individual clade close to *A. alternariacida* and *A. silybi* in sect. *Porri* (Figure 1). Morphologically, its primary conidiophores can generate geniculate conidiogenous loci at or near apex which differed from those two species (Figure 2, Table 2). It can be easily distinguished from *A. alternariacida* by producing more transverse septa and shorter beaks. Moreover, its conidia are solitary while *A. alternariacida* forms solitary or in unbranched chains of 2 (–3) conidia.

Alternaria coreopsidis H. Luo and J.X. Deng, sp. nov. (Figure 3).

MycoBank No: 844034.

Etymology: Named after the host genus name, *Coreopsis*.

Typification: China, Shaanxi Province, Xian City, from leaf spot of *Coreopsis basalis*. June, 2016, J.X Deng, ex-type culture YZU 161160.

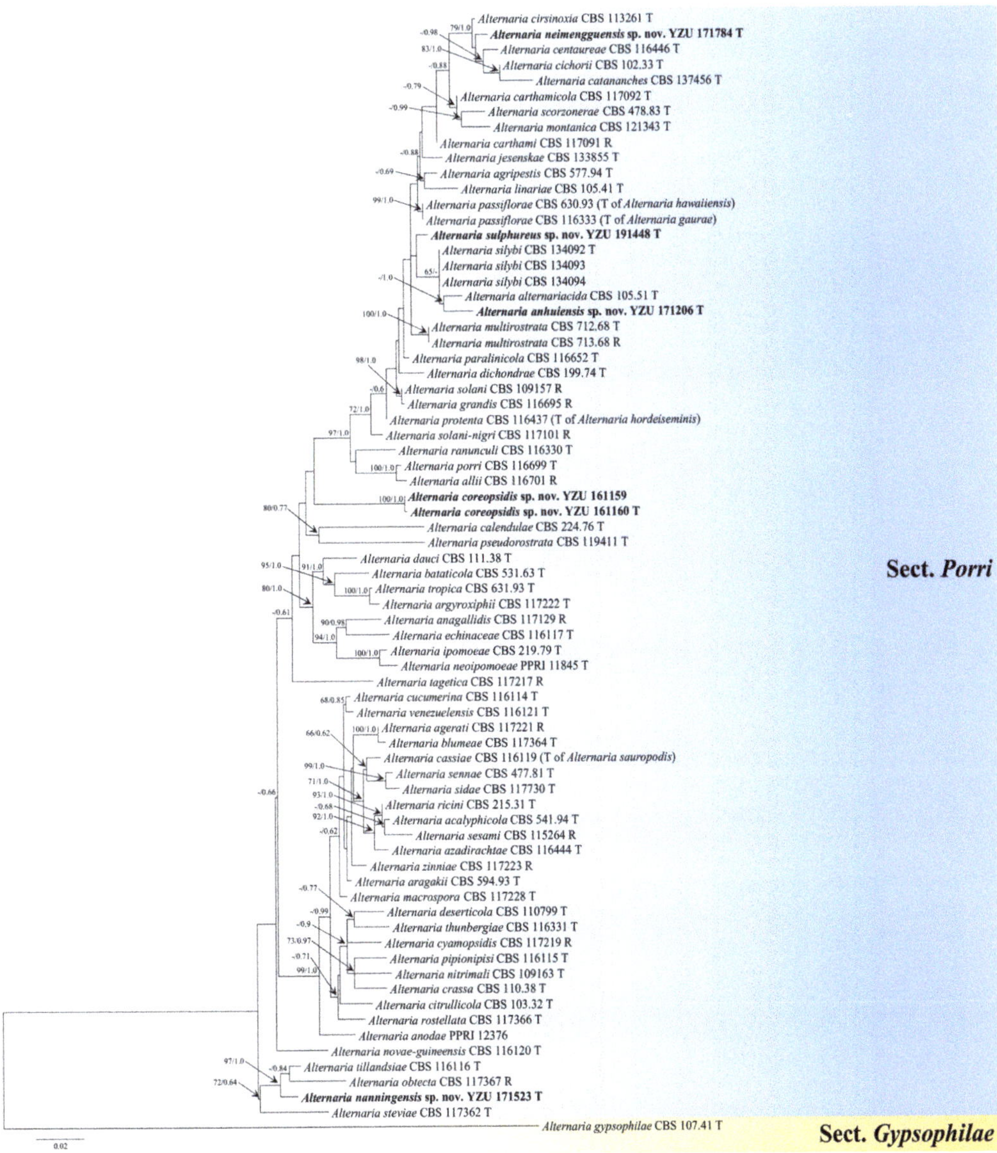

Figure 1. Maximum likelihood (ML) phylogram of new five *Alternaria* species from the Compositae family based on a combined dataset of ITS, GAPDH, Alt a 1, TEF1, and RPB2 gene sequences. The RAxML bootstrap support values >60%(ML) and Bayesian posterior probabilities >0.6 (PP) are given at the nodes (ML/PP). The present strains are in bold.

Description: Colonies on PDA circular, buff halo in the centre, villiform with white at the edge; reverse dark brown at centers, vinaceous buff pigment at margins, 47–48 mm in diam, at 25 °C for 7 days. On V8A, conidiophores arising from substrate or lateral of aerial hyphae, solitary, simple, straight to slightly curved, septate, pale to dark brown, apical conidiogenous locus, pale brown, (34–) 50–86 (–115.5) × 5–7 (–9) µm; conidia solitary or in unbranched chains of 2 conidia, long-narrow ovoid or ellipsoid body, smooth-walled, single beak, yellow or brown, (48.5–) 55–80 (–85) × (9–) 10–15 µm, 6–9 transverse septa,

0–1 longitudinal septa; beak filamentous, 1-beak, (20–) 30–140 (–206) × (2–) 2.5–4 µm; normally, false beak swollen at the apex, around 8–10.5 (–14) × 4.5–5 (–6) µm. On PCA, conidiophores straight or curved, smooth-walled, septate, (24–) 50–90 (–135) × 5–7.5 (–9) µm; conidia long-narrow ovoid or ellipsoid body, apex rounded, single beak, pale brown, (40–) 45–70 × 9–13 µm, (5–) 6–8 (–9) transverse septa, 0–1 longitudinal septa; beak filamentous, 1-beak, (0–) 15–100 (–175) × (0–) 2–4 µm; swollen apex of false beak commonly 10–13 (–16.5) × 5–6 (–6.5) µm.

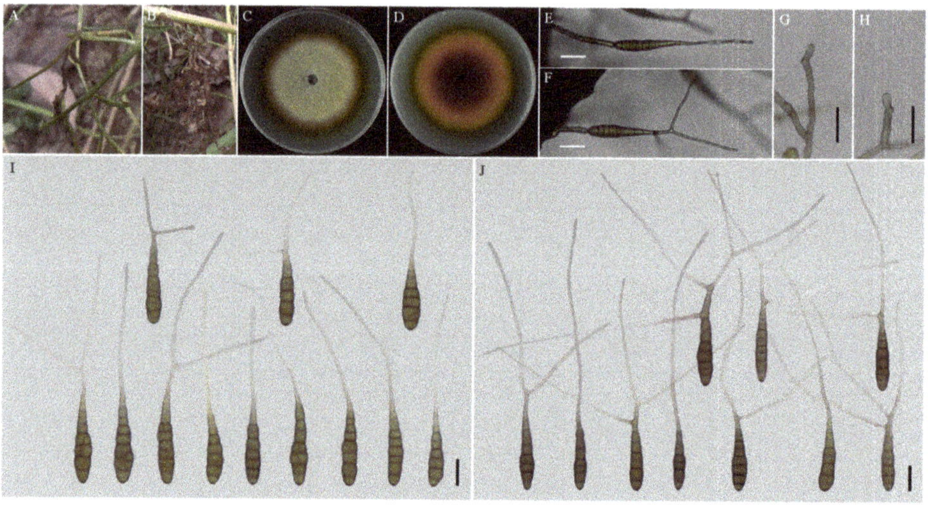

Figure 2. Morphology of *Alternaria anhuiensis* sp. nov. (**A**,**B**) Natural symptoms of *Coreopsis basalis*; (**C**,**D**) Colony phenotypes (on PDA for 7 days at 25 °C); (**E**,**F**) Sporulation patterns (on V8A at 22 °C); (**G**,**H**) Conidiophores (on V8A at 22 °C); (**I**) Conidia (on V8A at 22 °C); (**J**) Conidia (on PCA at 22 °C). Bars: (**E**–**J**) = 25 µm.

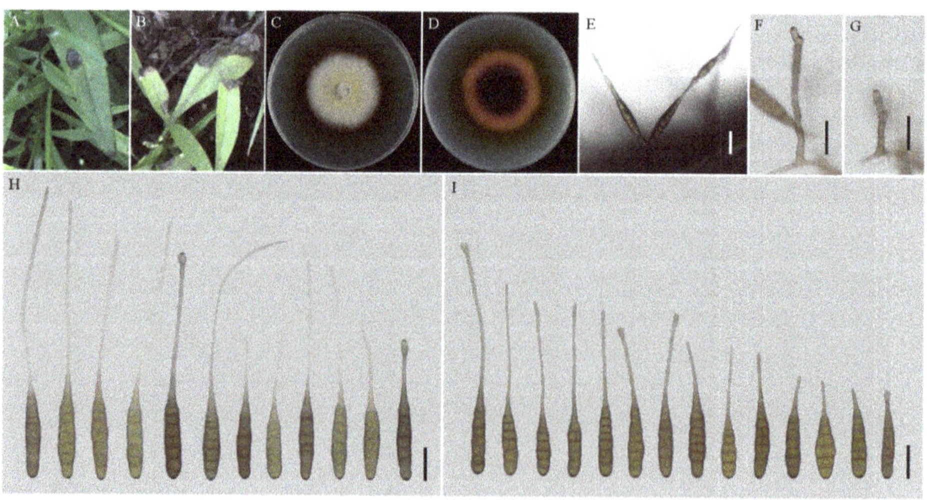

Figure 3. Morphology of *Alternaria coreopsidis* sp. nov. (**A**,**B**) Natural symptoms of *Coreopsis basalis*; (**C**,**D**) Colony phenotypes (on PDA for 7 days at 25 °C); (**E**) Sporulation patterns (on V8A at 22 °C); (**F**,**G**) Conidiophores (on V8A at 22 °C); (**H**) Conidia (on V8A at 22 °C); (**I**) Conidia (on PCA at 22 °C). Bars: (**E**–**I**) = 25 µm.

Table 2. Morphological comparisons of the five new *Alternaria* species and their closely related species.

Species	Strain	Conidia				Sporulation Pattern	Medium	Reference
		Shape	Size (μm)	Transversesepta	Beak (μm)			
A. alternariacida	CBS 105.51	Smooth-walled, narrowly ovoid; smooth-walled	(85–) 99–111 (–121) × (6–) 7–8 (–10)	(3–) 5–6 (–8)	(47–) 129–257 (–610) × 2	Solitary or in unbranched chains of 2 (–3) conidia	SNA	[4]
***A. anhuiensis* sp. nov.**	**YZU 171206**	**Long-narrow ovoid or ellipsoid; smooth-walled**	**61–100 (–111.5) × (11.5–) 13–19.5**	**6–11**	**(22–) 58–133 (–150.5) × 2.5–4 (–4.5)**	**Solitary**	**V8A**	**This study**
A. catananches	CBS 137456	Narrowly ovoid; ornamented in lower half of the conidium	(26–) 37–43 (–57) × (7–) 8–9 (–11)	(2–) 4 (–6)	(77–) 126–160 (–260) × 2	Solitary	SNA	[4]
A. centaureae	CBS 116446	Long narrow-ellipsoid or long-ovoid; ornamentation and punctate to pustulate	75–100 × 15–24	7–10	140–190 × 1.5–6	Solitary	V8A	[2]
A. cichorii	CBS 117218	Narrow-ovoid or narrow-ellipsoid; smooth-walled	60–80 × 14–18	7–12	120–240 × 2.5–7	Terminal clumps of 4–5 conidia	V8A	[2]
A. cirsinoxia	CBS 113261	Long-obclavate, short-ovoid; punctulate-walled	70–90 × 12–22	7–9	80–165 × 2.5–4	Solitary or tufts of 2–7 conidia	V8A	[2]
***A. coreopsidis* sp. nov.**	**YZU 161160**	**Long-narrow ovoid or ellipsoid; smooth-walled**	**(48.5–) 55–80 (–85) × (9–) 10–15**	**6–9**	**(20–) 30–140 (–206) × (2–) 2.5–4**	**Solitary or 2-conidium chains**	**V8A**	**This study**
***A. nanningensis* sp. nov.**	**YZU 171523**	**Ovoid or ellipsoid; smooth-walled**	**(40.5–) 47–79 (–87) × 9–13.5 (–15)**	**6–10 (–11)**	**10–30 × (1–) 1.5–2 (–3)**	**Solitary**	**V8A**	**This study**
***A. neimengguensis* sp. nov.**	**YZU 171784**	**Ovoid or ellipsoid; smooth-walled**	**(70–) 77–130 (–143.5) × (13–) 15–20 (–23)**	**6–11 (–12)**	**(24.5–) 35–65 (–92) × (1.5–) 2–3 (–4)**	**Solitary**	**V8A**	**This study**
A. obtecta	CBS 134278	Long-ovoid or ellipsoid; smooth or punctulate-walled	65–95 × 18–22	7–10	55–150 × 2	Solitary	PCA	[2]
A. porri	CBS 116698	Ovoid, sometimes broad or nearly cylindrical; smooth or punctulate-walled	70–105 × 19–24	8–12	95–160 × 2–6.5	Solitary	V8A	[2]
A. silybi	CBS 134093	Long-ellipsoid, subcylindrical or long-ovoid	50–80 × 15–20 (–22)	(5–) 7–10	70–130 (–190) × 3	Solitary	V4A	[40]
A. steviae	CBS 117362	Long-ovoid, subellipsoid, or obovoid; smooth or punctulate-walled	55–95 × 18–30	7–10	60–120 × 1.5–2.5	Solitary or tiny distal clumps	V8A	[2]
***A. sulphureus* sp. nov.**	**YZU 191448**	**Ovoid, ellipsoid, or obovoid; smooth-walled**	**(64–) 74–116 × (12.5–) 14–20 (–25.5)**	**(5–) 7–11**	**(25.5–) 34–151 (–159.5) × 2.5–4.5 (–5.5)**	**Solitary**	**V8A**	**This study**
A. tillandsiae	CBS 116116	Long-ovoid, ellipsoid, long-obovoid; smooth or a minor punctulate-walled	70–102 × 16–19	8–11	75–120 × 2	Solitary	V8A	[2]

Materials examined: China, Shaanxi Province, Xian City, from leaf spot of *Coreopsis basalis*. June 2016, J.X Deng, living culture YZU 161159.

Notes: Phylogenetically, the species falls into an independent lineage outside of a clade comprising type species of *A. porri* of sect. *Porri* (Figure 1). It can be delimited based on either of GAPDH and RPB2 gene sequences (Figure S1). The species is characterized by producing conidia with false beak swollen at the apex up to 8–13 (–16.5) × 4.5–6.5 μm (Figure 3; Table 2).

Alternaria nanningensis H. Luo and J.X. Deng, sp. nov. (Figure 4).

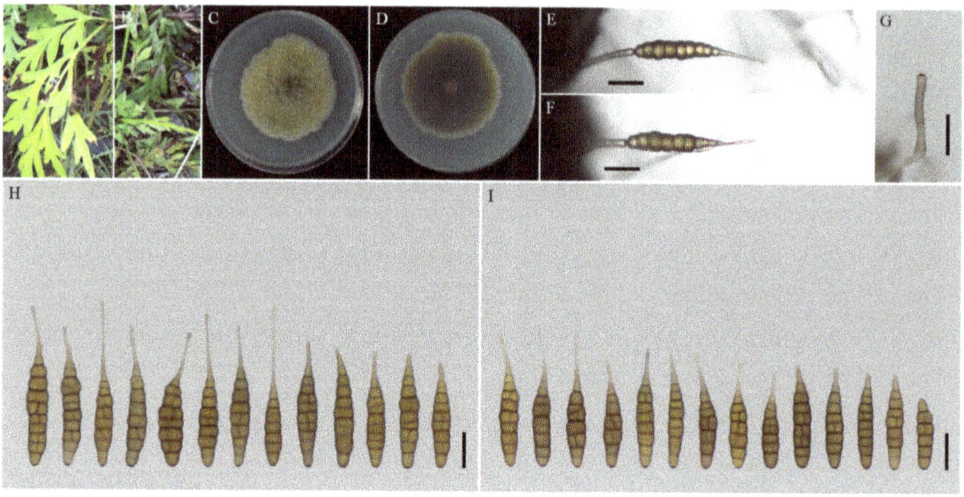

Figure 4. Morphology of *Alternaria nanningensis* sp. nov. (**A**,**B**) Natural symptoms of *Cosmos sulphureus*; (**C**,**D**) Colony phenotypes (on PDA for 7 days at 25 °C); (**E**,**F**) Sporulation patterns (on V8A at 22 °C); (**G**) Conidiophores (on V8A at 22 °C); (**H**) Conidia (on V8A at 22 °C); (**I**) Conidia (on PCA at 22 °C). Bars: (**E–J**) = 25 μm.

MycoBank No: 844035.

Etymology: Named after the collecting locality, Nanning City.

Typification: China, Guangxi Province, Nanning City, from leaf spot of *Cosmos sulphureus*. July 2017, J.X Deng, ex-type culture YZU 171523.

Description: Colonies on PDA irregular, pistac, entire; reverse dark olive green, slightly protuberant with white at margins, 56–57 mm in diam, at 25 °C for 7 days. On V8A, conidiophores arising from substrate or lateral of aerial hyphae with geniculate conidiogenous loci at apex, straight or curved, smooth-walled, septate, pale brown, 38–59 (–64)× 4–5 (–6) μm; conidia solitary, ovoid or ellipsoid body, base narrow, smooth-walled, single beak, pale to yellow brown, (40.5–) 47–79 (–87) × 9–13.5 (–15) μm, 6–10 (–11) transverse septa, 0–1 longitudinal septa; beak long-narrowed filiform, 1-beak, 10–30 × (1–) 1.5–2 (–3) μm. On PCA, conidiophores straight or curved, smooth-walled, septate; 32–70 (–86) × 4–5.5 μm; conidia solitary, ovoid or ellipsoid body, single beak, pale to yellow brown, (49–) 55–77 (–82) × 10.5–13.5 (–15) μm, (5–) 6–9 (–10) transverse septa, 0–1 longitudinal septum; beak long-narrowed filiform, 1-beak, 13–26 (–44) × 1.5–2 (–2.5) μm.

Notes: The species is phylogenetically recognized as a distinct species in sect. *Porri* based on ITS, GAPDH, Alt a 1, TEF1, and RPB2 which displays a close relationship with *A. obtecta*, *A. tillandsiae*, and *A. steviae* (Figure 1). Compared with them, it is quite different by producing smaller conidia with short beaks (Figure 4; Table 2). Furthermore, its conidia are smooth-walled while some conidia of *A. obtecta* and *A. steviae* are minutely punctulate. *Alternaria nanningensis* forms simple conidiophores (solitary). But many conidiophores of *A. steviae* produce geniculate extensions and additional conidia, yielding tiny distal clumps of sporulation.

Alternaria neimengguensis H. Luo and J.X. Deng, sp. nov. (Figure 5).

MycoBank No: 844036.

Etymology: Named after the collecting locality, Inner Mongolia Autonomous Region.

Typification: China, Inner Mongolia Autonomous Region, Inner Mongolia Agricultural University, IMAU, from leaf spot of *Lactuca seriola*. September 2017, J.X Deng, ex-type culture YZU 171784.

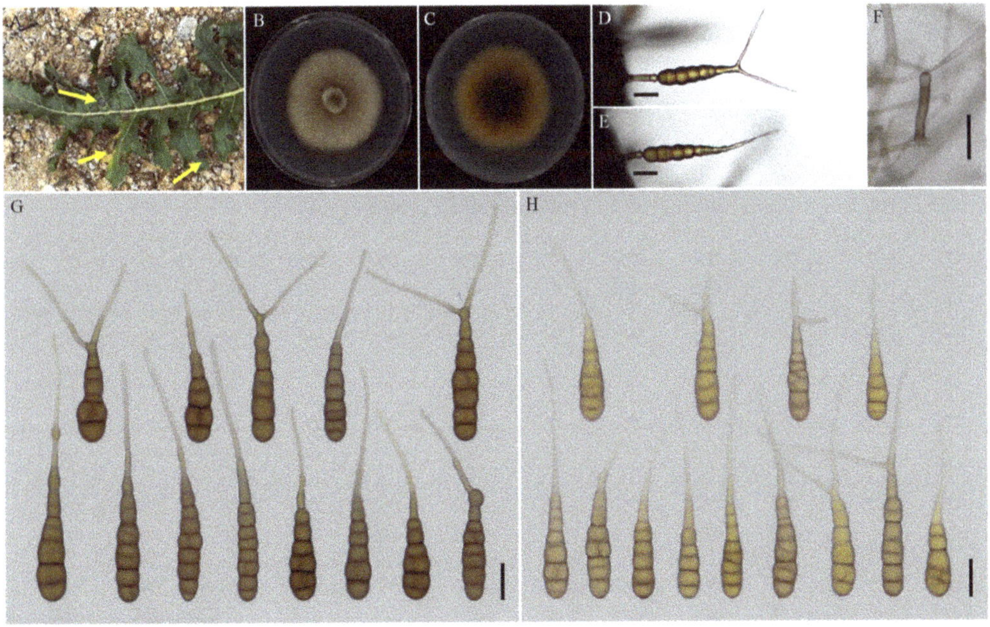

Figure 5. Morphology of *Alternaria neimengguensis* sp. nov. (**A**) Natural symptoms of *Lactuca seriola*; (**B**,**C**) Colony phenotypes (on PDA for 7 days at 25 °C); (**D**,**E**) Sporulation patterns (on V8A at 22 °C); (**F**) Conidiophores (on V8A at 22 °C); (**G**) Conidia (on V8A at 22 °C); (**H**) Conidia (on PCA at 22 °C). Bars: (**D–H**) = 25 μm.

Description: Colonies on PDA circular, pale brown en masse, flocculent, reverse dark olive green at centers, pale brown at margins, 51–54 mm in diam, at 25 °C for 7 days. On V8A, conidiophores arising from substrate or lateral of aerial hyphae, straight or curved, smooth-walled, septate, brown, 26–45 (–51) × 5–7 (–8) μm; conidia solitary, ovoid or ellipsoid body, apex rounded, base wide, smooth-walled, single to double beak, brown, (70–) 77–130 (–143.5) × (13–) 15–20 (–23) μm, 6–11 (–12) transverse septa, 0–1 (–2) longitudinal septa; beak long-narrowed filiform, 1-beak, (24.5–) 35–65 (–76) × (1.5–) 2–3 (–4) μm; 2-beak, (33–) 45–65 (–92) × (2–) 2.5–3 (–3.5) μm. On PCA, conidiophores straight or curved, smooth-walled, septate; 35–70 (–75)× 5–6.5 (–7.5) μm; conidia solitary, ovoid or ellipsoid body, apex rounded, single to double beak, pale to yellow brown, (59–) 66–104 (–120.5) × 13–18 (–20) μm, (5–) 6–10 (–11) transverse septa, 0–1 (–2) longitudinal septa; beak long-narrowed filiform, 1-beak, (13–) 31.5–60 (–93) × 1.5–3 μm; 2-beak, (12–) 26–53 (–80) × 1.5–2.5 (–3) μm.

Notes: In the phylogeny, the species is sister to *A. cirsinoxia*, *A. centaureae*, *A. cichorii*, and *A. catananches* (Figure 1). The conidiophores are distinct to *A. cirsinoxia* whose are 2–3 arm branches near a conidiophore tip and progressively geniculate, yielding tufts of several conidia. They are different from *A. cichorii* whose are frequently branch or proliferate in a geniculate manner near the apex, yielding terminal clumps of 4–5 conidia. In conidial morphology, it is obviously different from those four species by producing larger conidia (Table 2).

Alternaria sulphureus L. Zhao and J.X. Deng, sp. nov. (Figure 6).

Figure 6. Morphology of *Alternaria sulphureus* sp. nov. (**A**,**B**) Natural symptoms of *Cosmos sulphureus*; (**C**,**D**) Colony phenotypes (on PDA for 7 days at 25 °C); (**E**,**F**) Sporulation patterns (on PCA at 22 °C); (**G**) Conidiophores (on PCA at 22 °C); (**H**) Conidia (on V8A at 22 °C); (**I**) Conidia (on PCA at 22 °C). Bars: (**E**–**I**) = 25 μm.

MycoBank No: 844037.

Etymology: Named after the host species name, *Cosmos sulphureus*.

Typification: China, Shanxi Province, from leaf spot of *Cosmos sulphureus*. September 2019, J.X Deng, ex-type culture YZU 191448.

Description: Colonies on PDA circular, light brown in the centre, buff texture velutinous at the edge, reverse black brown at centers, 62–63 mm in diam, at 25 °C for 7 days. On V8A, conidiophores arising from substrate, solitary, simple, straight to slightly curved, septate, apical conidiogenous locus, pale brown; (50–) 63–100 (–108) × 6–8 (–9) μm; conidia solitary, sometimes in chains of two conidia, ovoid, ellipsoid or obovoid body, smooth-walled, pale to yellow, (64–) 74–116 × (12.5–) 14–20 (–25.5) μm, (5–) 7–11 transverse septa, 0–1 (–2) longitudinal septa; beak long-narrowed filiform, 1-beak, (25.5–) 34–151 (–159.5) × 2.5–4.5 (–5.5) μm; 2-beak (n = 1), 129 × 4 μm. On PCA, conidiophores straight or curved, smooth-walled, septate, (34.5–) 40.5–56 (–85) × 5–7.5 μm; conidia ovoid, ellipsoid, or obovoid body, apex rounded, single to double beak, triple beaks not common, pale brown, 80–110 × 16–24 μm, 6–10 transverse septa, 0–1 longitudinal septum; beak long-narrowed filiform, 1-beak, (73–) 110–195 × 3–5 μm; 2-beak, (74–) 96–170 × 3–4 μm; 3-beak (n = 1), 109.5 × 3.5 μm.

Notes: This species is phylogenetically related to *A. silybi*, *A. alternariacida* and *A. anhuiensis* sp. nov. in sect. *Porri* (Figure 1). It could be distinguished from *A. silybi* and *A. alternariacida* by forming larger conidia (Figure 6; Table 2) and is quite different from *A. alternariacida* by producing multiple and shorter beaks.

4. Discussion

Thirteen large-spored *Alternaria* species associated with Compositae leaf spot in China were assigned to four sections and one monotypic lineage in this study. Among theses species, five new species (*A. anhuiensis* sp. nov., *A. coreopsidis* sp. nov., *A. nanningensis* sp. nov., *A. neimengguensis* sp. nov., and *A. sulphureus* sp. nov.) were clearly recognized in section *Porri*. The section is speciose assessing encompassing 117 large-spored *Alternaria* [5]. In 2014, the section is reduced 82 morphospecies in to 63 phylogenetic species [2]. They are commonly pathogenic and could induce typical black necrotic lesions surrounded by chlorotic areas. There are some important famous plant pathogens, such as *A. porri* on *Allium* plants (Liliaceae), *A. solani* for potato (Solanaceae), *A. sesami* for sesame (Pedaliaceae) and *A. dauci* for carrot (Umbelliferae) [2]. Twenty-one species are comprised in sect. *Porri* associated with the Compositae family [4]. This study provides new data supplements for the *Alternaria* taxonomy of sect. *Porri*.

Morphologically, large-spored *Alternaria* species in sect. *Porri* are characterised by broadly ovoid, obclavate, ellipsoid, subcylindrical or obovoid, medium to large conidia containing multiple transverse and longitudinal septa, solitary or in short chains with a simple or branched, long to filamentous beak [4]. Among these characteristics, sporulation patterns, conidial body, transverse septa, and beak type provide useful information for the preliminary separation into sections [2]. Morphology is quite important for new fungal species identification, which can be defined based on unique morphological characters when the molecular data is not well-supported [41]. Morphological comparisons of the present new species and their relevant species in sect. *Porri* were conducted (Table 2). For the sporulation patterns, the conidia of *A. anhuiensis*, *A. nanningensis*, *A. neimengguensis*, and *A. sulphureus* are solitary produced except *A. coreopsidis*, which similar to *A. alternariacida*, *A. cichorii*, *A. cirsinoxia*, and *A. steviae* forming chain of 2 (–3) units [2,4]. In conidial morphology, *A. anhuiensis*, *A. nanningensis*, *A. neimengguensis*, and *A. sulphureus* are distinguishable from their closely related species based on the size of conidial bodies (Table 2) and also the wall ornamentations [2,4]. On the other hand, *A. anhuiensis*, *A. neimengguensis*, and *A. sulphureus* are readily be distinguished by producing multiple beaks. By the way, there are no significant differences on conidial morphology of PCA and V8A medium for all species.

In addition, morphological variation and fundamental pleomorphism complicate the *Alternaria* species recognition, and host plants reflect some evidences for the identification [3]. With the discovery of *Alternaria* species, it has been found from various plants of Compositae [1,4,21,42,43]. *Alternaria calendulae* has been reported from *Calendula officinalis* in Czech Republic [2], Germany [4], Japan [4], and Korea [44]. It also is found on *C. officinalis* in China and firstly on *Brachyactis ciliata* in the study. *Alternaria leucanthemi* has previously been found on *Chrysanthemum maximum* from Netherlands [1] and *Helianthus annuus* from China [45]. It is firstly isolated from *Carthamus tinctorius* in this study. In addition, *A. tagetica* is commonly associated with *Tagetes* plants (*Tagetes erecta* and *Tagetes patula*) [3,4,46–48], which firstly encountered from *Calendula officinalis* in this study. Interestedly, the five new species are isolated from three different composite hosts (*Coreopsis basalis*, *Cosmos sulphureus*, and *Lactuca seriola*) and *A. cinerariae* are found on five different composite plants in China (Figure S1; Table S1). The results suggest that an *Alternaria* species may associated with several host plants.

5. Conclusions

The present data indeed revealed a diversity of large-spored *Alternaria* associated with Compositae plants in China. A total of 13 large-spored *Alternaria* species were obtained and circumscribed as eight known species and five new species belonging to the four sections of *Helianthiinficientes*, *Porri*, *Sonchi*, and *Teretispora*, and one monotypic lineage (*A. argyranthemi*) based on the morphological characteristics and molecular properties of multiple DNA sequences (ITS, GAPDH, Alt a 1, TEF1, and RPB2). *Alternaria calendulae*, *A. leucanthemi*, and *A. tagetica* were firstly isolated from *Brachyactis ciliate*, *Carthamus tinctorius*, and *Calendula officinalis* in China, respectively. Since large-spored *Alternaria* species are almost demonstrated phytopathogens, further study on the pathogenicity is needed to verify in the future.

Supplementary Materials: The following supporting information can be downloaded at: https://www.mdpi.com/article/10.3390/jof8060607/s1, Figure S1: Phylogenetic tree of large-spored *Alternaria* from the Compositae family in China using a maximum likelihood (ML) analysis based on combined GAPDH and RPB2 gene sequences. The RAxML bootstrap support values > 60% (ML) and Bayesian posterior probabilities >0.6 (PP) are given at the nodes (ML/PP); Table S1: The other *Altenraria* species associated with the Compositae plants from China analyzed by phylogeny.

Author Contributions: The contributions of L.Z. and H.L. are consistent. Conceptualization, L.Z. and J.-X.D.; methodology, L.Z. and J.-X.D.; software, H.L.; validation, H.L., H.C. and Y.-N.G.; formal analysis, H.C. and Y.-N.G.; data curation, L.Z. and H.L.; writing—original draft preparation, L.Z. and H.L.; writing—review and editing, L.Z., J.-X.D. and Z.-H.Y.; visualization, L.Z.; supervision, J.-X.D.; project administration, J.-X.D. All authors have read and agreed to the published version of the manuscript.

Funding: The financial support was given by the National Natural Science Foundation of China (No. 31400014 and No. 31570022).

Institutional Review Board Statement: Not applicable.

Informed Consent Statement: Not applicable.

Data Availability Statement: The sequences newly generated in this study have been submitted to the GenBank database.

Acknowledgments: The authors would like to thank Xue-Feng Wei for providing the leaf samples.

Conflicts of Interest: The authors declare no conflict of interest.

References

1. Woudenberg, J.H.; Groenewald, J.Z.; Binder, M.; Crous, P.W. *Alternaria* redefined. *Stud. Mycol.* **2013**, *75*, 171–212. [CrossRef] [PubMed]
2. Simmons, E.G. *Alternaria: An Identification Manual*; CBS Fungal Biodiversity Centre: Utrecht, The Netherlands, 2007.
3. Zhang, T.Y. *Flora Fungorum Sinicorum, Alternaria*; Science Press: Beijing, China, 2003; Volume 16.
4. Woudenberg, J.H.C.; Truter, M.; Groenewald, J.Z.; Crous, P.W. Large-spored *Alternaria* pathogens in section *Porri* disentangled. *Stud. Mycol.* **2014**, *79*, 1–47. [CrossRef]
5. Gannibal, P.B. Distribution of *Alternaria* species among sections. 1. Section *Porri*. *Mycotaxon* **2015**, *130*, 207–213. [CrossRef]
6. Pinto, V.E.F.; Patriarca, A. *Alternaria Species and Their Associated Mycotoxins*; Mycotoxigenic Fungi Humana Press: New York, NY, USA, 2017; pp. 13–32. [CrossRef]
7. Pryor, B.M.; Bigelow, D.M. Molecular characterization of *Embellisia* and *Nimbya* species and their relationship to *Alternaria*, *Ulocladium* and *Stemphylium*. *Mycologia* **2003**, *95*, 1141–1154. [CrossRef]
8. Park, M.S.; Romanoski, C.E.; Pryor, B.M. A re-examination of the phylogenetic relationship between the causal agents of carrot black rot, *Alternaria radicina* and *A. carotiincultae*. *Mycologia* **2008**, *100*, 511–527. [CrossRef]
9. Liu, H.F.; Liao, J.; Chen, X.Y.; Liu, Q.K.; Yu, Z.H.; Deng, J.X. A novel species and a new record of *Alternaria* isolated from two Solanaceae plants in China. *Mycol. Prog.* **2019**, *18*, 1005–1012. [CrossRef]
10. Htun, A.A.; Liu, H.F.; He, L.; Zhou, X.Z.; Aung, S.L.L.; Deng, J.X. New species and new record of *Alternaria* from onion leaf blight in Myanmar. *Mycol. Prog.* **2022**, *21*, 59–69. [CrossRef]
11. Pryor, B.M.; Gilbertson, R.L. Molecular phylogenetic relationships amongst *Alternaria* species and related fungi based upon analysis of nuclear ITS and mt SSU rDNA sequences. *Mycol. Res.* **2000**, *104*, 1312–1321. [CrossRef]
12. Hong, G.S.; Cramer, R.A.; Lawrence, C.B.; Pryor, B.M. Alt a 1 allergen homologs from *Alternaria* and related taxa: Analysis of phylogenetic content and secondary structure. *Fungal Genet. Biol.* **2005**, *42*, 119–129. [CrossRef]
13. Lawrence, D.P.; Park, M.S.; Pryor, B.M. *Nimbya* and *embellisia* revisited, with nov.comb. for *Alternaria celosiae* and *A. perpunctulata*. *Mycol. Prog.* **2012**, *11*, 799–815. [CrossRef]
14. Lawrence, D.P.; Gannibal, P.B.; Peever, T.L.; Pryor, B.M. The sections of *Alternaria*: Formalizing species-group concepts. *Mycologia* **2013**, *105*, 530–546. [CrossRef] [PubMed]
15. Lawrence, D.P.; Gannibal, P.B.; Dugan, F.M.; Pryor, B.M. Characterization of *Alternaria* isolates from the infectoria species-group and a new taxon from *Arrhenatherum*, *Pseudoalternaria arrhenatheria* sp. nov. *Mycol. Prog.* **2014**, *13*, 257–276. [CrossRef]
16. Woudenberg, J.H.C.; Seidl, M.F.; Groenewald, J.Z.; Vries, M.D.; Stielow, J.B.; Thomma, B.P.H.J.; Crous, P.W. *Alternaria* section *Alternaria*: Species, formae speciales or pathotypes? *Stud. Mycol.* **2015**, *82*, 1–21. [CrossRef]
17. Pei, D.F.; Aung, S.L.L.; Liu, H.F.; Liu, Q.K.; Yu, Z.H.; Deng, J.X. *Alternaria hydrangeae* sp. nov. (Ascomycota: Pleosporaceae) from *Hydrangea paniculata* in China. *Phytotaxa* **2019**, *401*, 287–295. [CrossRef]

18. He, L.; Cheng, H.; Htun, A.A.; Ge, H.; Xia, Z.Z.; Guo, J.J.; Deng, J.X.; Du, T. Phylogeny and taxonomy of two new *Alternaria* (Ascomycota: Pleosporaceae) species in section *Gypsophilae* from China. *Mycol. Prog.* **2021**, *20*, 355–363. [CrossRef]
19. Lawrence, D.P.; Rotondo, F.; Gannibal, P.B. Biodiversity and taxonomy of the pleomorphic genus *Alternaria*. *Mycol. Prog.* **2016**, *15*, 3. [CrossRef]
20. Ghafri, A.A.; Maharachchikunbura, S.S.; Hyde, K.D.; Nadiya, A.A.S.; Abdullah, M.A.S. A new section and a new species of *Alternaria* encountered from Oman. *Phytotaxa* **2019**, *405*, 279–289. [CrossRef]
21. Gannibal, P.B.; Orina, A.S.; Gasich, E.L. A new section for *Alternaria helianthiinficiens* found on sunflower and new asteraceous hosts in Russia. *Mycol. Prog.* **2022**, *21*, 34. [CrossRef]
22. Hind, N. Introduction to the compositae, the largest family of flowering plants. *Curtis's Bot. Mag.* **2018**, *35*, 332–338. [CrossRef]
23. Luo, H.; Yu, Z.; Lu, H.; Zhao, M.; Zhou, Y.; Deng, J. Taxonomy of *Alternaria* from Compositae: Research progress. *Chin. Agric. Sci. Bull.* **2018**, *34*, 63–70.
24. Mackinnon, S.L.; Keifer, P.; Ayer, W.A. Components from the phytotoxic extract of *Alternaria brassicicola*, a black spot pathogen of canola. *Phytochemistry* **1999**, *51*, 215–221. [CrossRef]
25. Andersen, B.; Krøger, E.; Roberts, R.G. Chemical and morphological segregation of *Alternaria alternata*, *A. gaisen* and *A. longipes*. *Mycol. Res.* **2001**, *105*, 291–299. [CrossRef]
26. Thomma, B.P. *Alternaria* spp.: From general saprophyte to specific parasite. *Mol. Plant Pathol.* **2003**, *4*, 225–236. [CrossRef] [PubMed]
27. Deng, J.X.; Li, M.J.; Paul, N.C.; Oo, M.M.; Lee, H.B.; Oh, S.K.; Yu, S.H. *Alternaria brassicifolii* sp. nov. isolated from *Brassica rapa* subsp. pekinensis in Korea. *Mycobiology* **2018**, *46*, 172–176. [CrossRef]
28. Cenis, J.L. Rapid extraction of fungal DNA for PCR amplification. *Nucleic Acids Res.* **1992**, *20*, 2380. [CrossRef] [PubMed]
29. White, T.J.; Bruns, T.; Lee, S.; Taylor, J. Amplification and direct sequencing of fungal ribosomal RNA genes for phylogenetics. In *PCR Protocols: A Guide to Methods and Applications*; Innis, M.A., Gelfand, D.H., Sninsky, J.J., Eds.; Academic Press: San Diego, CA, USA, 1990; pp. 315–322. [CrossRef]
30. Carbone, I.; Kohn, L.M. A method for designing primer sets for speciation studies in filamentous ascomycetes. *Mycologia* **1999**, *91*, 553–556. [CrossRef]
31. Berbee, M.L.; Pirseyedi, M.; Hubbard, S. *Cochliobolus* phylogenetics and the origin of known, highly virulent pathogens, inferred from ITS and glyceraldehyde-3-phosphate dehydrogenase gene sequences. *Mycologia* **1999**, *91*, 964–977. [CrossRef]
32. Liu, Y.J.; Whelen, S.; Hall, B.D. Phylogenetic relationships among ascomycetes: Evidence from an RNA polymerse II subunit. *Mol. Biol. Evol.* **1999**, *16*, 1799–1808. [CrossRef]
33. Sung, G.H.; Sung, J.M.; Hywel-Jones, N.L.; Spatafora, J.W. A multi–gene phylogeny of *Clavicipitaceae* (Ascomycota, fungi): Identification of localized incongruence using a combinational bootstrap approach. *Mol. Phylogenetics Evol.* **2007**, *44*, 1204–1223. [CrossRef] [PubMed]
34. Hall, T.A. Bioedit: A user-friendly biological sequence alignment editor and analysis program for Windows 95/98/NT. *Nucleic Acids Symp. Ser.* **1999**, *41*, 95–98.
35. Chun, J. Computer Assisted Classification and Identification of Actinomycetes. Ph.D. Thesis, Unversity of Newcastle, Newcastle upon Tyne, UK, 1995. Available online: http://theses.ncl.ac.uk/jspui/handle/10443/410 (accessed on 1 May 2022).
36. Tamura, K.; Stecher, G.; Peterson, D.; Filipski, A.; Kumar, S. MEGA6: Molecular evolutionary genetics analysis version 6.0. *Mol. Biol. Evol.* **2013**, *30*, 2725–2729. [CrossRef] [PubMed]
37. Stamatakis, A. RAxML-VI-HPC: Maximum likelihood-based phylogenetic analyses with thousands of taxa and mixed models. *Bioinformatics* **2006**, *22*, 2688–2690. [CrossRef] [PubMed]
38. Rannala, B.; Yang, Z. Probability distribution of molecular evolutionary trees: A new method of phylogenetic inference. *J. Mol. Evol.* **1996**, *43*, 304–311. [CrossRef]
39. Rambaut, A.; Drummond, A. *FigTree v.1.3.1. Institute of Evolutionary Biology*; University of Edinburgh: Edinburgh, UK, 2010.
40. Gannibal, P.B. Taxonomic studies of *Alternaria* from Russia: New species on Asteraceae. *Mycotaxon* **2010**, *114*, 109–114. [CrossRef]
41. Jeewon, R.; Hyde, K.D. Establishing species boundaries and new taxa among fungi: Recommendations to resolve taxonomic ambiguities. *Mycosphere* **2016**, *7*, 1669–1677. [CrossRef]
42. Luo, H.; Xia, Z.Z.; Chen, Y.Y.; Zhou, Y.; Deng, J.X. Morphology and Molecular Characterization of *Alternaria argyranthemi* on *Chrysanthemum coronarium* in China. *Mycobiology* **2018**, *46*, 278–282. [CrossRef]
43. He, L.; Liu, H.F.; Htun, A.A.; Ge, H.; Deng, J.X.; Du, T. First Report of *Alternaria cinerariae* Causing Leaf Spot on *Tussilago farfara* in China. *Plant Dis.* **2020**, *104*, 3264. [CrossRef]
44. Lee, Y.H.; Cho, W.D.; Kim, W.K.; Jin, K.S.; Lee, E.J. Report on host-unrecorded diseases identified from economical crops in Korea. *Res. Rep. Rural Developm. Admin.* **1991**, *33*, 15–19.
45. Zhao, G.; Zhang, T.Y.; Cao, A.X.; Wang, H.K. Phylogenetic relationships of *Alternaria* and related genera and taxonomic status of *A. leucanthemi* inferred from ITS rDNA sequence analysis. *Mycosystema* **2006**, *25*, 184–191. [CrossRef]
46. Tomioka, K.; Sato, T.; Koganezawa, H. Marigold leaf spot caused by *Alternaria tagetica* new to Japan. *J. Gen. Plant Pathol.* **2000**, *664*, 294–298. [CrossRef]
47. Li, M.J.; Deng, J.X.; Paul, N.C.; Lee, H.B.; Yu, S.H. Characterization and pathogenicity of *Alternaria vanuatuensis*, a new record from *Allium* plants in Korea and China. *Mycobiology* **2014**, *42*, 412–415. [CrossRef] [PubMed]
48. Paul, N.C.; Deng, J.X.; Lee, H.B.; Yu, S.H. Characterization and pathogenicity of *Alternaria burnsii* from seeds of *Cucurbita maxima* (Cucurbitaceae) in Bangladesh. *Mycobiology* **2015**, *43*, 384–391. [CrossRef]

Journal of *Fungi*

MDPI

Article

Four New Species of Dictyostelids from Soil Systems in Northern Thailand

James C. Cavender [2], Eduardo M. Vadell [3], Allison L. Perrigo [4], John C. Landolt [5], Steven L. Stephenson [6] and Pu Liu [1,*]

1 Engineering Research Center of Edible and Medicinal Fungi, Ministry of Education, Jilin Agricultural University, Changchun 130118, China
2 Departmental of Environmental and Plant Biology, Ohio University, Athens, OH 45701, USA; cavender@ohio.edu
3 Museo de Historia Natural R.S.V.—Microbiology, Suipacha 6612 Pcia. de Buenos Aires, Viamonte 1716 6p, Buenos Aires City, Argentina; eduardo.vadell@gmail.com
4 Gothenburg Global Biodiversity Centre, Box 461, 405 30 Gothenburg, Sweden; allison.l.perrigo@gmail.com
5 Department of Biology, Shepherd University, Shepherdstown, WV 25443, USA; landolt@shepherd.edu
6 Department of Biological Sciences, University of Arkansas, Fayetteville, AR 72701, USA; slsteph@uark.edu
* Correspondence: pul@jlau.edu.cn

Abstract: Dictyostelid cellular slime molds (dictyostelids) are ubiquitous microorganisms found in the uppermost layers of most soils. Reports on the species diversity of dictyostelids in Southeast Asia, particularly Thailand, are few in number. A survey for dictyostelids performed in northern Thailand in 2008 recovered 15 distinctive forms, including several common species and a number of forms morphologically different from anything already described. Five of the latter were formally described as new to science in a previous paper. An additional five isolates appeared to be morphologically distinct, and this was supported by DNA sequence data and phylogenetic analysis. These isolates representing four species are described herein as species new to science. Detailed descriptions and illustrations of these new species are provided.

Keywords: *Cavenderia*; cellular slime molds; species concept; taxonomy; biodiversity

Citation: Cavender, J.C.; Vadell, E.M.; Perrigo, A.L.; Landolt, J.C.; Stephenson, S.L.; Liu, P. Four New Species of Dictyostelids from Soil Systems in Northern Thailand. *J. Fungi* **2022**, *8*, 593. https://doi.org/10.3390/jof8060593

Academic Editors: Cheng Gao and Lei Cai

Received: 8 April 2022
Accepted: 21 May 2022
Published: 31 May 2022

1. Introduction

Dictyostelid cellular slime molds (dictyostelids) are a ubiquitous component of soils, where they feed upon bacteria and other microbes and thus play a major role in maintaining the balance between these organisms in the soil microhabitat [1,2]. Dictyostelids are amoebozoans, a distinct branch of eukaryotes, separate from plants, fungi, and animals. During the past 20 years, as a result of ongoing surveys in areas of the world where dictyostelids remain an understudied group, the number of species has essentially doubled. Molecular studies performed over the past decade have revised the traditional system of classification used for dictyostelids [3], and a SSU-based phylogeny has established that there are four major clades [4].

In 2008, a total of 40 samples for isolation of dictyostelids were collected from four localities in northern Thailand to obtain data on the occurrence and distribution of dictyostelids in this region of the world. One of the localities sampled was located within a tropical cloud forest in Doi Inthanon National Park, Chom Thong District, Chiang Mai Province (18°35′32″ N, 98°29′12″ E). In addition to several common species of dictyostelids, nine isolates that could not be assigned to any described species were also recovered from the sampling. All of these isolates were subjected to a detailed morphological study of subcultures in addition to DNA sequence analyses, and five were described as new to science in a previous publication [5]. Five additional isolates, all obtained from samples collected from the tropical cloud forest (elevation 2500 m) in Doi Inthanon National Park, were also considered be morphologically distinct, albeit by only minor differences, and required additional investigation.

2. Materials and Methods

2.1. Sampling

Samples of soil/litter were collected from a tropical cloud forest in northern Thailand in January 2008. These samples, each approximately 30–50 g, were placed in sterile whirl-pack plastic bags, returned to the laboratory, and processed as soon as possible, as recommended by Cavender and Raper [6].

2.2. Isolation and Cultivation

The samples were processed using the methods described by Cavender and Raper [6]. Each sample was weighed, and enough sterile distilled water was added to obtain an initial soil/water dilution of 1:10. This mixture was shaken to disperse the material and to suspend the cells of the dictyostelids present. A 5.0 mL volume of this initial dilution was added to 7.5 mL of sterile, distilled water to create a 1:25 dilution of sample material. Aliquots (each 0.5 mL) of this suspension were added to each of two or three 95 × 15 mm Petri dishes prepared with hay (leached and dried, mostly Poa sp.) infusion agar [2]. This produced a final dilution of 0.02 g of soil per plate. Approximately 0.4 mL of a heavy suspension of 12–24 h *Escherichia coli* was added to each culture plate, and plates were incubated under diffused light at 20–25 °C. Each plate was examined at least once daily for several days following the appearance of initial aggregations, and the location of each aggregate clone was marked. Aggregations, pseudoplasmodia, and sorocarps appeared in the plates over a period that ranged from 2 d to 3 w. Isolates of interest were subcultured from spores on low-nutrient agar with *E. coli*. Spores were also conserved in tubes of silica gel granules at 4 °C, as described by Raper [2]. Isolates considered to be of interest were studied in more detail, which involved an initial characterization of morphological features and obtaining a first set of images of morphological structures. Later, subcultures of these isolates were sent to Vadell and Liu. All isolates considered in the present study were deposited in the Dicty Stock Center at Northwestern University in the United States and then at Jilin Agricultural University (HMJAU) in China.

2.3. Morphological Observations

The characteristic stages in the life cycle, including cell aggregation and the formation of pseudoplasmodia and sorocarps, were photographed in the Liu laboratory under a dissecting microscope (Axio Zoom V16, Carl Zeiss Microscopy GmbH, Göttingen, Germany) with a 1.5× objective and a 10× ocular. Slides with sorocarps were prepared with water as the mounting medium. Features of spores, sorophores, and sorocarps were observed and measured on the slides by using a light microscope (Axio Imager A2, Carl Zeiss Microscopy GmbH, Göttingen, Germany), with a 10× ocular and 10, 40, and 100× (oil) objectives. Photographs were taken with a Axiocam 506 color microscope camera (Carl Zeiss Microscopy GmbH, Göttingen, Germany).

Detailed examinations of the development and overall morphology of subcultured clones of the five isolates from Doi Inthanon National Park were performed. These isolates were observed at magnifications from 50 to 200×, either in hydric conditions or in xeric media conditions to observe the effects of dehydration and conservation of the hydric contents of the sori over time, migration, stolon formation, and sorogen development. Observations of early aggregations and fruiting bodies were made after 2–30 d incubation under diffuse illumination at 18–26 °C. Optimal temperatures for growth and fruiting body formation, when determined, were measured in an incubator at 18, 20, 24, and 26 (±0.5 °C), under low-diffuse illumination. All major stages of development and behavior (in young and old cultures) of each isolate were observed, described, measured, and drawn by hand in India ink, carefully noting the patterns and shapes of early and late sorogens and behavior [7,8]. The morphological behavior was compared and contrasted with independent observations. Structures were studied through the top, side, and bottom of the Petri dish, making use of variations in light intensity to discriminate the various special perspectives and differences (e.g., basal zones, type and arrangement of cells or the accumulations of slime supporting the base of the sporophore, presence of granulated slime and cushions, relative degree of hydric conditions, and mound shapes) [8], using magnifications provided by the 4 and 12× lenses of a dissecting microscope along with observations made with a compound phase contrast microscope. The general criteria used for the features observed were based on Raper [2], and the drawings of the four species described herein are presented.

2.4. DNA Isolation, PCR Amplification and Sequencing

The spores of all five isolates being studied were collected with a sterile tip and mixed with the lysis buffer of the MiniBEST Universal Genomic DNA Extraction Kit Ver.5.0 (Takara Bio Inc., Kusatsu, Japan) following the manufacturer's protocol. The genomic DNA solution was used directly for the small subunit (SSU) PCR amplification using the primers 18SF–A (AACCTGGTTGATCCTGCCAG) and 18SR–B (TGATCCTTCTGCAGGTTCAC) [9] along with D542F (ACAATTGGAGGGCAAGTCTG3) and D1340R (TCGAGGTCTCGTC-CGTTATC) [4]. PCR products were sent to Sangon Biotech Co., Ltd. (Shanghai, China) for sequencing. Sequences obtained were deposited in the GenBank database. The isolates and the NCBI GenBank accession numbers of SSU DNA sequences considered in present study are listed in Table 1.

Table 1. NCBI GenBank accession information for SSU sequences of all isolates included in the phylogenetic analysis. New sequences are indicated in bold.

Taxon	Isolate No.	Accession No.
Cavenderia amphispora	BM9A	HQ141521.1
C. antarctica	NZ43B	AM168080.1
C. aureostabilis	ALP-2018a	MH745571.1
C.aureostipes	B15A	KF662199.1
C. aureostipes	YA6	AM168083.1
C. aureostipes	OH396	KF662201.1
C. aureostipes var. *helvetia*	HM592	KF662214.1
C. basinodulosa	ALP-2019a	MN338955.1
C. bifurcata	UK5	AM168084.1
C. bhumiboliana	THC11X	HQ141523.1
C. boomerangispora	K26B	HQ141520.1
C. canoespora	ALP-2019b	MN338956.1
C. delicata	TNS-C-226	AM168093.1
C. deminutiva	MexM19A	AM168092.1
C. exigua	TNS-C-199	AM168085.1
C. fasciculata	SH3	AM168087.1
C. fasciculata	SmokOW9A	AM168086.1
C. fasciculoidea	Cavender Puelo 1B	GQ496157.1
C. granulophora	CHII-4	AM168072.1
C. helicoidea	**TH19B**	**OM677255**
C. macrocarpa	MGE2	HQ141519.1
C. medusoides	OH592	AM168088.1
C. mexicana	MexTF4B1	AM168089.1
C. microspora	TNS-C-38	AM168090.1
C. multistipes	UK26b	AM168070.1
C. myxobasis	NT2A	HQ141522.1
C. parvibrachiata	**TH20C**	**OM677256**
C. parvibrachiata	**2019TH20C**	**OM677257**
C. parvispora	OS126	AM168091.1
C. protodigitata	ALP-2018b	MH745572.1
C. protumula	**TH20A**	**OM677258**
C. pseudoaureostipes	TH39A	HQ141518.1

Table 1. *Cont.*

Taxon	Isolate No.	Accession No.
C. sp.	TAS30A	HQ141516.1
C. stellata	SAB7B	AM168081.1
C. subdiscoidea	TH1A	HQ141515.1
C. ungulata	**TH18B**	**OM677259**
Dictyostelium brefeldianum	TNS-C-115	AM168030.1
D. macrocephalum	B33	AM168049.1
D. medium	TNS-C-205	AM168050.1
D. mucoroides	sweden 20	HQ141482.1

2.5. Phylogenetic Analysis

The five newly generated sequences were checked and then submitted to GenBank, as noted above. The SSU sequences were aligned and compared using the program ClustalW Multiple alignment version 2.1 (Institut Pasteur, Paris, France) [10] and then manually adjusted in BioEdit version 7.0.9.0 (Manchester, UK) [11]. Maximum likelihood (ML) analyses were performed using IQ-TREE v.1.6.12 (Insitut Pasteur, Paris, France) [12] with 1000 replicates of ultrafast-likelihood bootstrapping to obtain node support values by the "-bb 1000" option, and further optimized using a hill-climbing nearest-neighbor interchange (NNI) by the "-bnni" option [13]. The "-nt AUTO" option was used to automatically determine the best number of cores given the current data. In the ML analyses of SSU sequences, TVMe+R5 (IQTree, Vienna, Austria) model was chosen as the best-fit model according to BIC by IQ-TREE with the "-m TVMe+R5" option.

2.6. Data Availability

Sequence data are available in GenBank (www.ncbi.nlm.nih.gov/genbank/, accessed on 14 February 2022, Accession Numbers OM677255, OM677256, OM677257, OM677258, and OM677259). The nomenclature of the new species in the present study is available in MycoBank (www.mycobank.org, accessed on 13 February 2022 for MB842981, MB842982 and MB842984, accessed on 14 February 2022 for MB842989). Sequence alignment was uploaded as supplementary material (File S1).

3. Results

Five isolates representing four new species [*Cavenderia helicoidea* (Figures 1 and 2), *C. parvibrachiata* (Figures 3 and 4), *C. protumula* (Figures 5 and 6) and *C. ungulata* (Figures 7 and 8)] of dictyostelids were recovered from samples collected from a tropical cloud forest in northern Thailand. Morphological characteristics and phylogenetic studies of the SSU sequences both support the taxonomic placement in *Cavenderia* of all four of the species (Table 2, Figure 9).

Table 2. Summary data on a comparison of the new species with closely related species.

Species	Sorocarp	Sorophore	Sorophore Cell	Sorocarp Size (mm)	Base	Base Size	Tip	Tip Size	Branch	Sorous	Spore	Polar Granule	Aggregation	Yellow Pigmentation
Cavenderia aureostabilis	prone	slender to curved		L	Disk, round to clavate	S–L	unfinished capitate or small irregular cells	L		M	M–L	consolidated irregular	Radiate	Intense
C. aureostipes	erect	crowded		M	Round-irregular	M		M	>20	M	M	conspicuous	Polysphondylium violaceum type	Strong
C. bhumiboliana	prone	uneven, irregular	one tier	S	clavate, curved or not	L	Flexuous, piliform or round	L	1–4	S	L	prominent and large consolidated	Mounds	Fades
C. helicoidea	prone		one to several tiers	S	clavate to round	S–M	acuminate or piliform	S	a few	S–L	L	irregular consolidated	Radiate or taking the shape of irregular mounds	Fades
C. parvibrachiata	Erect to pros-trate	slender	one or two tiers	S–L	round, clavate or curved hook-shaped	S–M	obtuse, capitate or clavate	S–M	Unbranched or irregular	S–L	M–L	pronounced consolidated polar-subpolar two unequal	Radiate	
C. protodigitata	Erect to prone	uneven	one tier	S	Clavate-digitate	S	piliform filaments or irregularly capitate	S	unbranched or secondary branched	S	S-M	medium to large consolidated, regular	Mounds	Fades
C. protumula	Erect		one to several tiers	S–L	clavate or round	M–L	obutuse, capitate	S–M	0–2	S–L	M–L	+	Radiate	
C. ungulata	erect to prone	very irregular, uneven	one or two tiers	S–M	claw-like clavate	S	variable, obtuse to acuminate or clavate	M	unbranched but sometimes with 1–5	S–L	S-L	large regular (2 μm)	Radiate	

Notes: Spores: S small (most common range: 4.5–5.5 × 2–3 μm), M median (most common range: 5.5–7.5 × 2.5–3.5 μm), L large (most common range: 6.5–9 × 3–5 μm) [3]. Sori: S small (most common range: 20–100 μm), M median (commonest range: 80–150 μm), L large (commonest range: 150–250 μm) [3]. Sorocarp Size: S small (commonest range: 0.2–3 mm), M median (commonest range: 3–5 mm), L large (most common range: 5–8 mm). Tip Size: S small (commonest range: 2–6 μm), M median (commonest range: 6.1–11 μm), L large (most common range: 15–30 μm). Bases Size: S small (commonest range: 8.4–20 μm), M median (commonest range: 20–30 μm), L large (most common range: 30–55 μm).

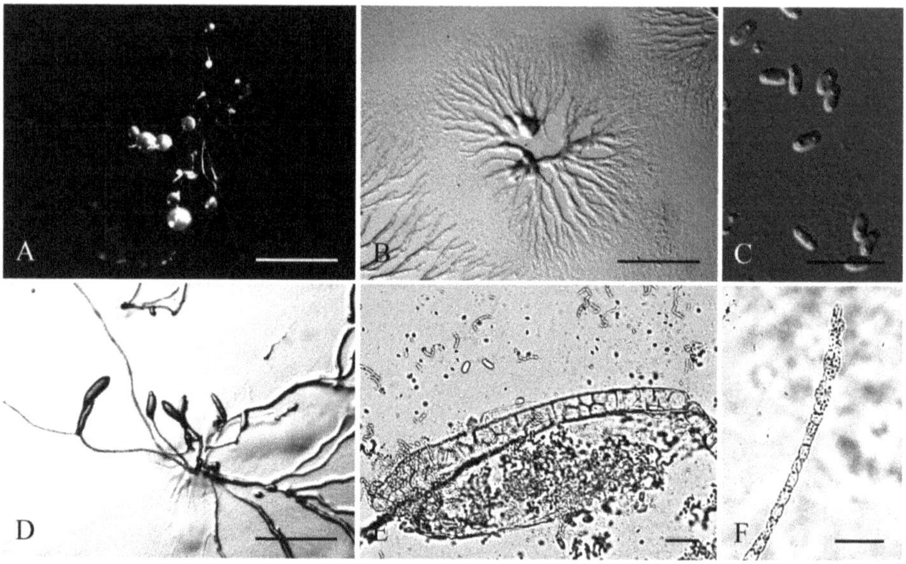

Figure 1. Morphological features of *Cavenderia helicoidea* (TH19B). (**A**), Sorocarps. (**B**), Aggregations. (**C**), Spores. (**D**), Clustered pseudoplasmodia. (**E**), Sorophore base. (**F**), Sorophore tip. Bars: (**A**,**B**) = 1 mm; (**C**) = 20 µm; (**D**) = 1 mm; (**E**,**F**) = 20 µm.

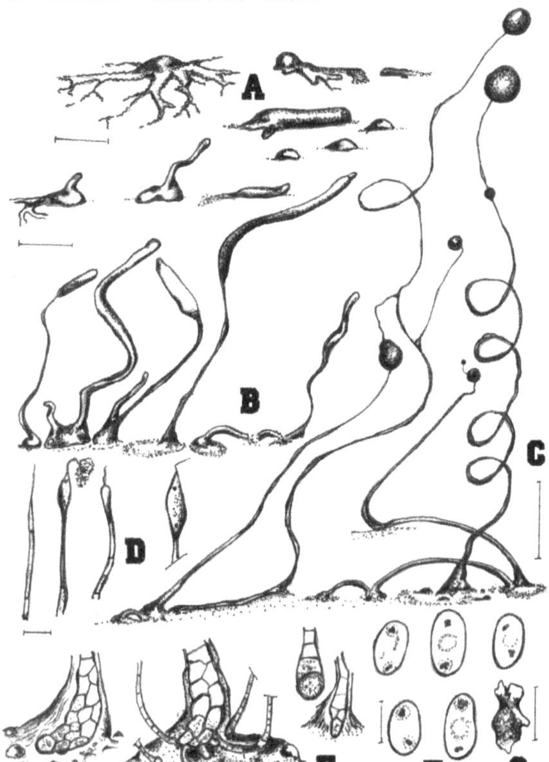

Figure 2. Morphological features of *Cavenderia helicoidea* (TH19B). (**A**), Medium-sized streamed

aggregation, with edges (left) and the most common type consisting of small mounds, some with short blocky streams (right). (**B**), Early central sorogen (above, left); these settle without streams (above, center) and migrate with short stalk formation; small mounds, unstreamed (above, right); Group of solitary (below left and center) and tightly clustered sorogens (below center); the stoloniferous migrating habit of a late sorogen (below, right). (**C**), Clustered sorocarps, one with a prostrate lower stalk (left); two solitary sorocarps, one with a typical stoloniferous habit and the other with its frequent helical architecture; two sori are tangled, one small example refruits, and small mounds are present within the halo at bases (right). (**D**), Four simple tips, mostly piliform, and one ampule-shaped, a mass of dense slime attached to a filament. (**E**), Curved clavate base with flexuous small cells, except the terminal one (left); a curved base with small satellite sorocarps and some digitate cells, within a dense matrix of slime (center, aged culture); a one-celled round base and a clavate base, both larger and flexuose (right); all flexuose. (**F**), Elliptical slightly larger spores with prominent irregular PG. (**G**). Myxamoeba. Bars: (**A**) = 300 μm; (**B**) = 200 μm; (**C**) = 0.5 mm; (**D**) = 10 μm; (**E**) = 20 μm; (**F**) = 10 μm; (**G**) = 5 μm.

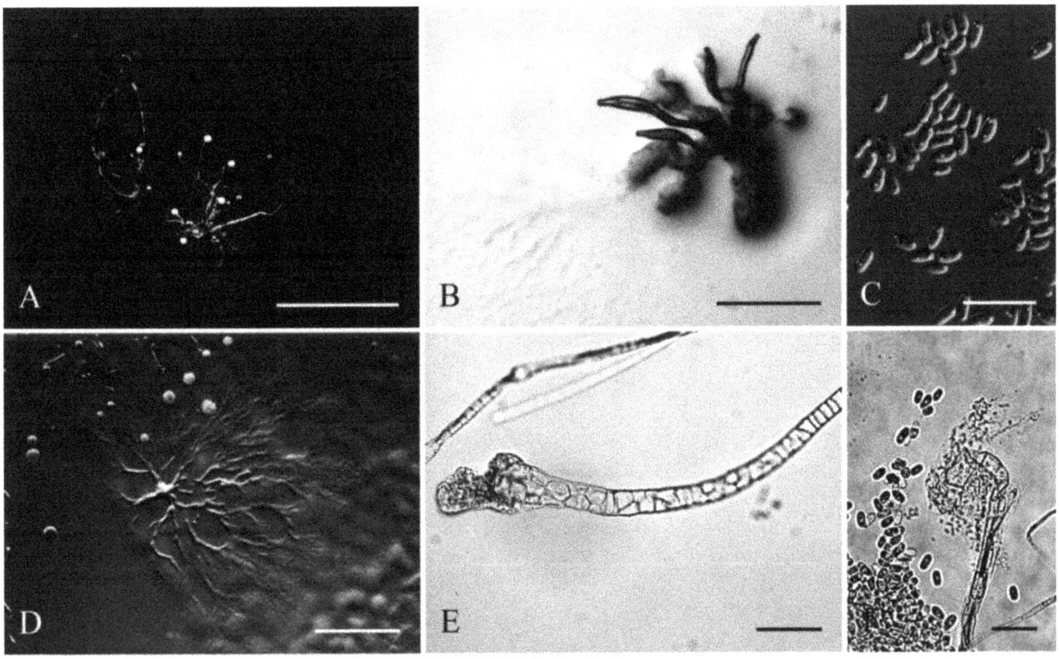

Figure 3. Morphological features of *Cavenderia parvibrachiata* (TH20C). (**A**), Sorocarps. (**B**), Clustered pseudoplasmodia. (**C**), Spores. (**D**), Aggregations. (**E**), Sorophore base. (**F**), Sorophore tip. Bars: (**A**) = 2 mm; (**B**) = 500 μm; (**C**) = 20 μm; (**D**) = 2 mm; (**E**) = 40 μm; (**F**) = 20 μm.

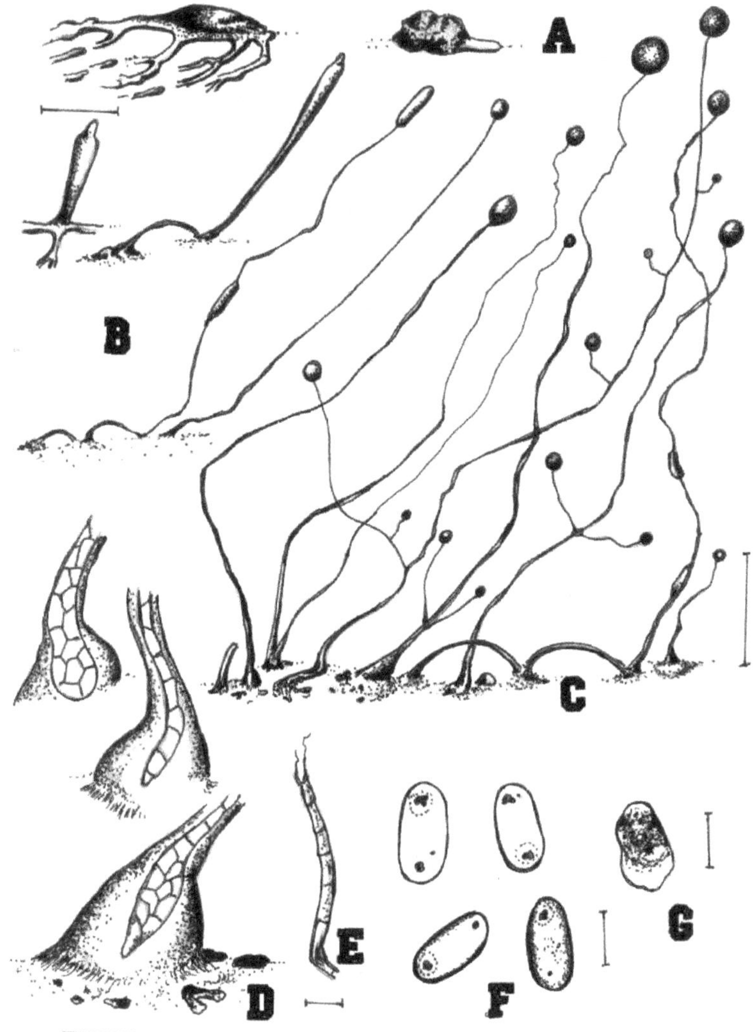

Figure 4. Morphological features of *Cavenderia parvibrachiata* (TH20C). (**A**), Small shortly radiated aggregation with partite streams (left); very short blocky-streamed mound-like aggregation (right). (**B**), Early central sorogen with short streams at base (above left), a stoloniferous early late elongated sorogen (above, right); stoloniferous habit of migrating late sorogen with an ascending pseudoplasmodial mass, stalk broken from early stages (below, left); solitary unbranched early mature slender small sorocarp (below, right). (**C**), Crowded group of sorocarps; tight clustered unbranched sorocarps within numerous smaller curved-broken fruiting bodies and sorogens (left); lower branched sorocarps, branches are short, sometimes coincident at one point (center); a stoloniferous sorocarp with a more evident broken sorophore and its companion small fruiting body (right). (**D**), Round base (left, above) clavate one-celled sorophore base (center); acutely clavate regular base within a hyaline mass of slime, sheath is not profuse and there are small masses of slime at base (below). (**E**), Simple tip frequently with a small cell with two piliform ends, flexuous. (**F**), Short small elliptical spores with irregular consolidated PG, one with halo. (**G**), Myxamoeba. Bars: (**A**,**B**) = 200 μm; (**C**) = 0.5 mm; (**D**) = 20 μm; (**E**) = 10 μm; (**F**) = 5 μm; (**G**) = 10 μm.

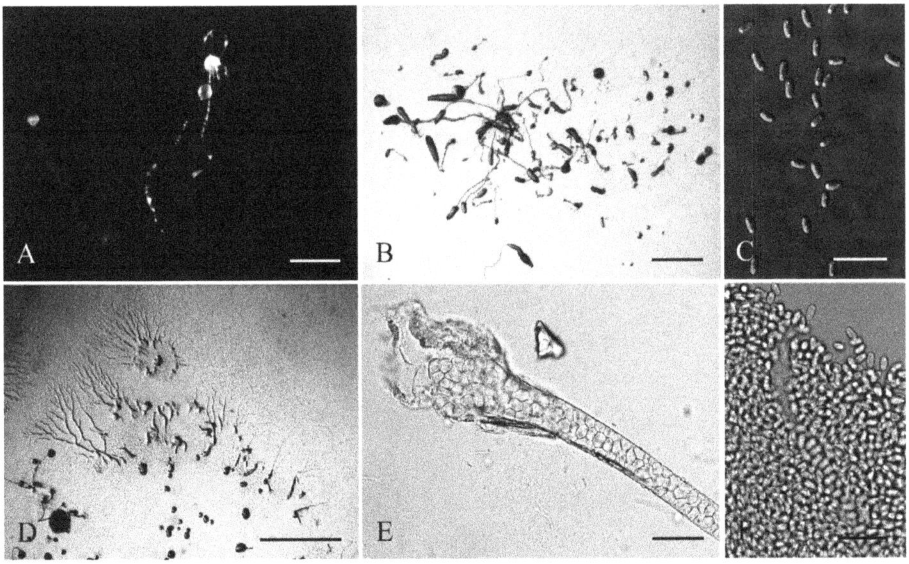

Figure 5. Morphological features of *Cavenderia protumula* (TH20A). (**A**), Sorocarps. (**B**), Clustered pseudoplasmodia. (**C**), Spores. (**D**), Aggregations. (**E**), Sorophore base. (**F**), Sorophore tip. Bars: (**A**) = 200 μm; (**B**) = 1 mm; (**C**) = 20 μm; (**D**) = 2 mm; (**E**) = 40 μm; (**F**) = 20 μm.

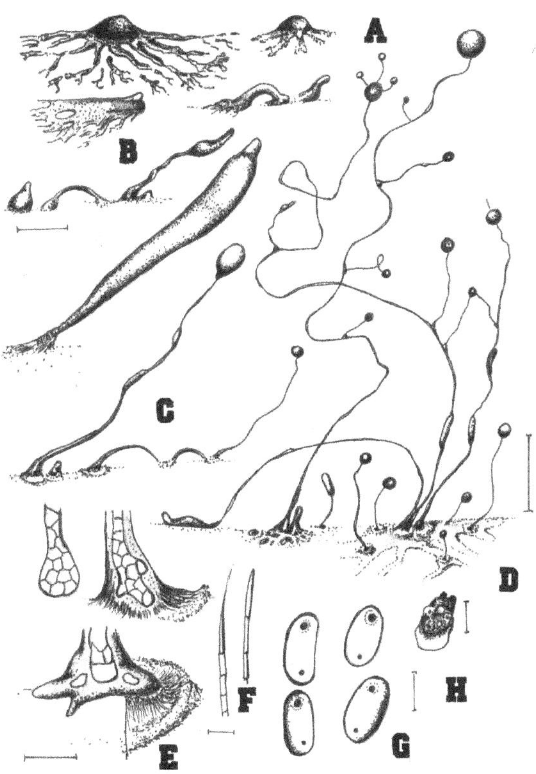

Figure 6. Morphological features of *Cavenderia protumula* (TH20A). (**A**), Radiate aggregation with

partite streams (left); a small aggregation with short blocky streams (right). (**B**), Migrating solitary pseudoplasmodium with flat anastomosed streams (above, left); a stoloniferous migrating early sorogen (above, right); progression of an early-late solitary migrating sorogen (below, left); a single enlarged early late migrating sorogen with stalk formation at its basal portion (out of scale, below left). (**C**), Late solitary sorogen with masses of ascending pseudoplasmodia and a companion small pseudoplasmodium mass close to the base (left); stoloniferous habit of a solitary sorocarp, basal halo present (right). (**D**), Solitary sigmoid low-branched mature sorocarp with its satellite early sorogen at base, surrounded by small pseudoplasmodial mases (left); clustered sorocarps with small satellite unbranched sorocarps, bases with halos and a trace of the early aggregation remains. Sori refruits, collapse soon or slides down and hang together; a frequent stoloniferous habit. (**E**), Round base (above, left); shortly digitate base with abundant mucilage sheath, a basal ring, or halo (above, right); young base terminating in a single larger cell, the slime is densely abundant and the sheath prolonged until the elevated ring or halo (below). (**F**), Piliform (left) and simple narrowed one-celled tip (right). (**G**), Short regular spores with polar to subpolar compound granules, some with halos. (**H**), Myxamoeba. Bars: (**A**,**B**) = 200 μm; (**C**,**D**) = 0.5 mm; (**E**) = 25 μm; (**F**) = 10 μm; (**G**) = 5 μm; (**H**) = 10 μm.

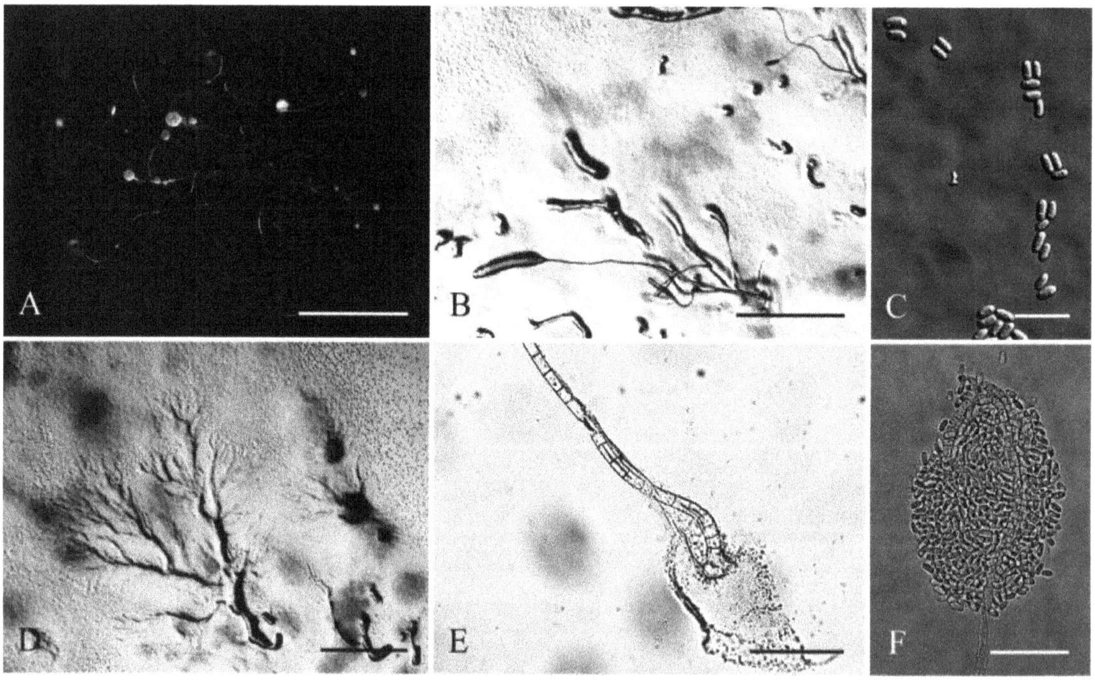

Figure 7. Morphological features of *Cavenderia ungulata* (TH18B). (**A**), Sorocarps. (**B**), Clustered pseudoplasmodia. (**C**), Spores. (**D**), Aggregations. (**E**), Sorophore base. (**F**), Sorophore tip. Bars: (**A**,**B**) = 1 mm; (**C**) = 20 μm; (**D**) = 1 mm; (**E**) = 60 μm; (**F**) = 40 μm.

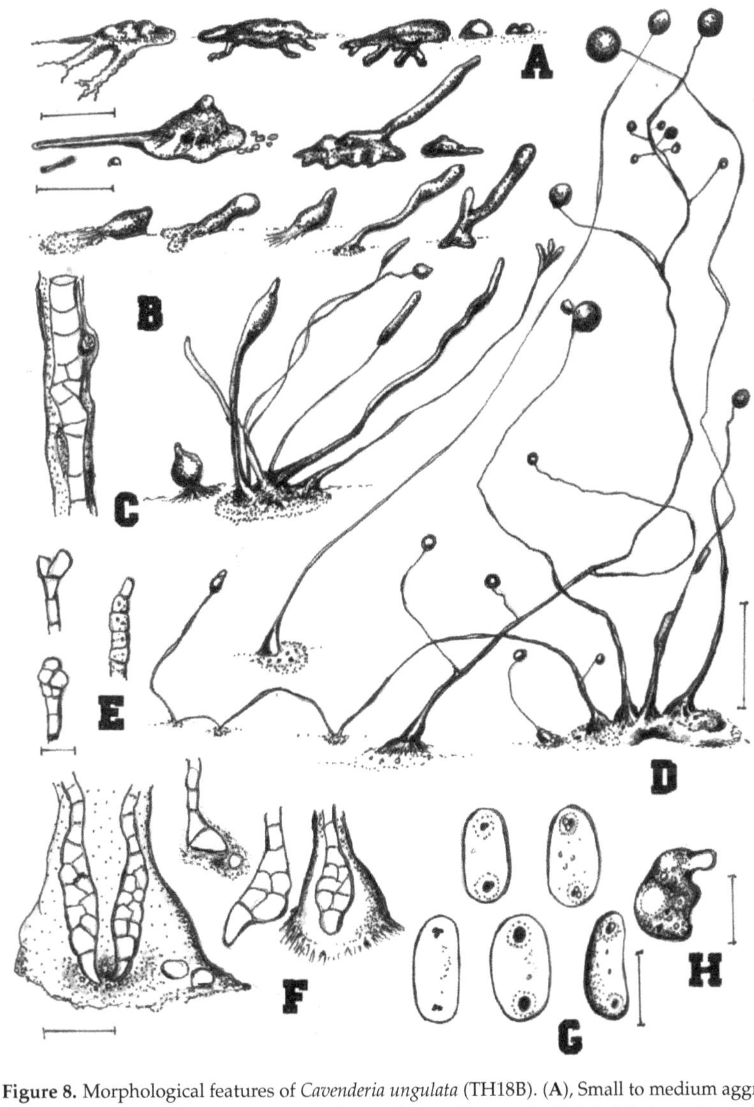

Figure 8. Morphological features of *Cavenderia ungulata* (TH18B). (**A**), Small to medium aggregations with short thick streams (left, center) and irregular small mounds (right). (**B**), An early sorogen rises up from the center, one elongated pseudoplasmodium stream remains (above, left); one early-late sorogen emerges from the side of the mound (above, right); from left to right: solitary migrating early sorogens (center); a cluster of late sorogens (below, right). (**C**), Irregular segment of a sorophore. (**D**), Solitary unbranched mature sorocarp with mucilage rest at the base (left); curved sorocarp with some branches (center); clustered sorocarps with ascending masses of small pseudoplasmodia, on the left margin there is a stoloniferous habit that progress to the left (right). (**E**), Two types of capitates tips (left); a flexuous simple tip (right). (**F**), Two tightly clustered clavate bases with terminal cells as cat claws, the darken matrix of slime is very dense and with large granules of yellow pigment (left); two solitary bases with enlarged terminal cells (center); a clavate regular base with its basal slime and sheath (right). (**G**), Slightly large elliptical spores with protruding PG (regular in shape) generally of the same size in both poles, also with vacuoles and halos. (**H**), Myxamoeba with many small-medium vacuoles. Bars: (**A**) = 300 μm; (**B**) = 200 μm; (**C–E**) = 15 μm; (**D**) = 0.5 mm; (**F**) = 20 μm; (**G**) = 6 μm; (**H**) = 10 μm.

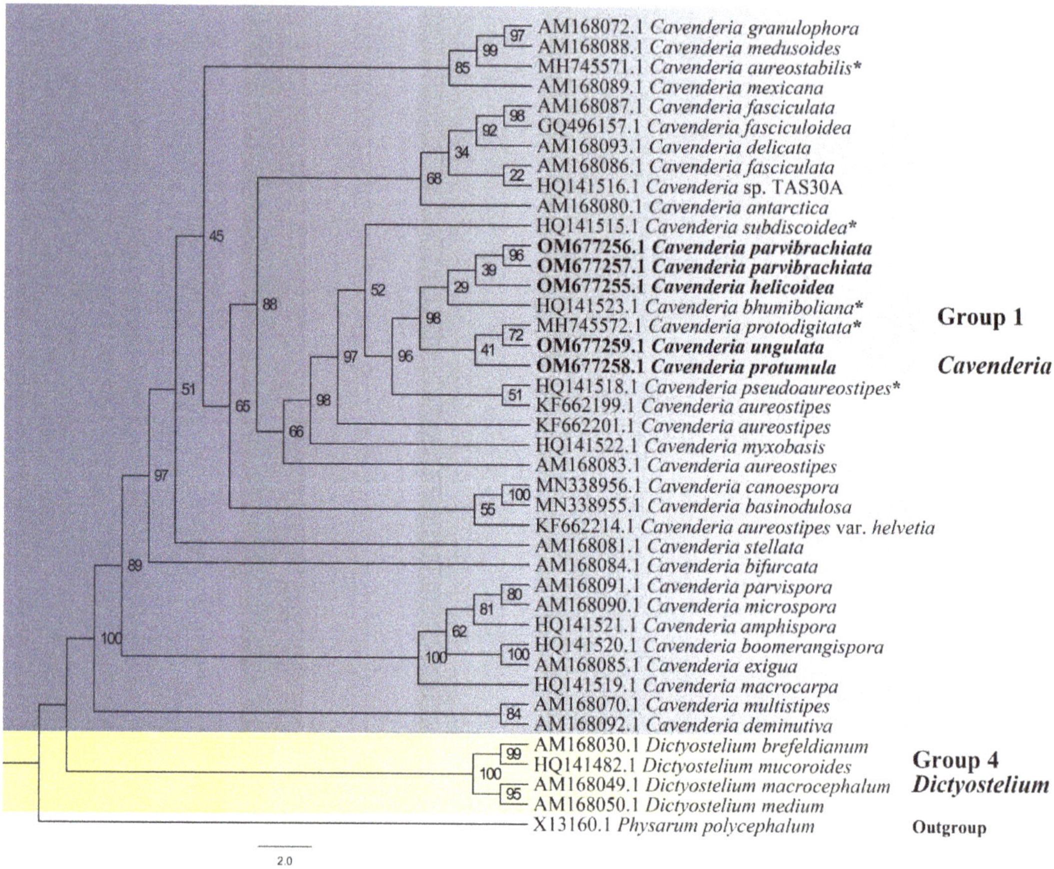

Figure 9. Phylogeny of four new species obtained in this study along with other species of *Cavenderia* based on SSU rRNA. Names in bold are the sequences obtained in this study, names with asterisks are the other five new species that were discovered and reported in this locality previously. The accession numbers are listed to the left of the species names. Complete strain information and GenBank accession numbers for all taxa in this tree can be observed in Table 1. The tree was derived using IQ-TREE with ultrafast bootstrap approximation (UFBoot) by TVMe+R5 model. The phylogeny is rooted according to Sheikh et al. [3].

Taxonomy and Phylogeny

Cavenderia helicoidea Cavender, P. Liu, Vadell, A. L. Perrigo, J. C. Landolt & S. L. Stephenson, sp. Nov.

MycoBank accession number: MB842981; GenBank Accession Number: OM677255 (SSU), Figures 1 and 2.

Etymology: The specific epithet is derived from the ancient Greek word *helix*, meaning twist or turn, and referring to the helical ascending movement during the development of the sorogens and sorocarps.

The culture examined was from: Thailand, Chiang Mai Province, Chom Thong District, Doi Inthanon National Park, 18°35′32″ N 98°29′12″ E, tropical cloud forest, isolated from a sample collected by Stephenson SL in January 2008, Landolt TH19B (Holotype) deposited in dictyostelid collection at Jilin Agricultural University (HMJAU), ex-Landolt TH19B deposited in the Dicty Stock Center at Northwestern University (No. DBS0350782).

Sorocarps were mostly prone, decumbent, helical, solitary, or clustered to gregarious, 0.23–2.85 mm long, zero to five branches (2C). Sorophores were hyaline to white or slightly yellow, with one to several tiers of cells, sometimes with a few branches. Branches 0.4–0.7 mm long. Tips acuminate or piliform with one or two tiers of cells, 4.8–5.6 μm wide (2D). Bases were clavate to round, consisting of one to six cells, with one or more terminal cells protruding, 13.3–23.3 μm wide (2E). Sori were hyaline-white to cream, globose, with a 39–241 μm diam. Spores of fallen sori germinated immediately. Spores were elliptical-oblong, 5.6–10.5 × 2.9–4.4 μm, with irregular consolidated polar granules (PG), occasionally slightly subpolar, germinating rapidly (2F). Typical myxamoeba (2G). Aggregations radiated or took the shape of irregular mounds, as psedoplasmodia with stalks (2A). Sorogens were in tight clusters. Early and late sorogens may migrate, leaving traces of slime and sections of immature sorophores (2B).

Comments: This species appears morphologically most similar to *C. protodigitata*, but the latter is smaller in length. Both share the fast fading of the yellow pigment very early during morphogenesis. This fading of the yellow pigment is a common feature among the smaller members of the Thailand species in group 1 (*sensu* Sheikh et al. [3]) that have been studied, as a relict feature tending to disappear. This isolate is also larger than *C. ungulata* and *C. aureostabilis* [5], having similarly elongated, thin fruiting bodies also common in both species, and its sorocarps are more regular, sigmoid, and broken, and the bases are more variable. Spores are relatively large, and do not germinate immediately.

Cavenderia parvibrachiata Cavender, P. Liu, Vadell, A. L. Perrigo, J. C. Landolt & S. L. Stephenson, sp. nov.

MycoBank accession number: MB842982; GenBank Accession Numbers: OM677256 (SSU), OM677257 (SSU), Figures 3 and 4.

Etymology: The specific epithet *parvibrachiata* refers to the poor and small production of branches and is derived from the Latin words *parvus* (small, low, poor) and *brachium* (branch).

The culture examined was from: Thailand, Chom Thong District, Chiang Mai Province, Doi Inthanon National Park, 18°35′32″ N 98°29′12″ E, tropical cloud forest, isolated from a sample collected by Stephenson SL in January 2008, Landolt TH20C (Holotype) deposited in the dictyostelid collection at Jilin Agricultural University (HMJAU), ex–Landolt TH20C deposited in the Dicty Stock Center at Northwestern University (No. DBS0350791). Landolt 2019TH20C represents the same taxon and was isolated from the same sample that yielded Landlt TH20C.

Sorocarps were solitary or clustered to gregarious, erect to prostrate, 0.3–6.0 mm, mostly unbranched but sometimes irregularly branched, yellowish (4C). Branches were 0.4–1.2 mm. Sorophores were tenuous, slender, first straight and then curved, with one or two tiers of cells. Tips were obtuse, capitate, or clavate with one or two tiers of cells, 3.2–9.7 μm wide (4E). Bases were round, clavate, or curved hook-shaped with one to several tiers of cells, always surrounded by a layer of dense granular slime, 8.4–27.6 μm wide (4D). Sori pearls were white to slightly yellow, subglobose to globose, varying considerably in diam. (43–252 μm). Spores were elliptical-oblong, sticky, 5.4–9.7 × 2.6–4.8 μm, with pronounced consolidated subpolar to polar PG (4F). Aggregation radiated, and pseudoplasmodia was yellowish with stalks (4A). Sorogens were clustered and sometimes bifurcated (4B). Myxamobae were small at first, well-dispersed, and active, soon rounding up, with many vacuoles and a few dark granules (4G).

Comments: *Cavenderia parvibrachiata* belongs to dictyostelid group 1 in an SSU ribosomal DNA (rDNA) phylogeny (Figure 9). It forms a clade together with *C. bhumiboliana* and *C. protodigitata* and several other species. However, it differs morphologically from *C. bhumiboliana* in the width of the sorophores, branches, and spore size. *Cavenderia bhumiboliana* has tips that are 15–25 μm wide at the apex and 30–45 μm wide at the base. In *C. parvibrachiata*, the tips are 3.2–10.5 μm wide at the apex and 8.4–27.6 μm wide at the base. *Cavenderia bhumiboliana* has one to four branches whereas *C. parvibrachiata* has irregular branches. *C. parvibrachiata* differs morphologically from *C. protodigitata* with respect to sorophore tips, branches, sorocarps, and spores. *C. protodigitata* has piliform filaments or irregularly capitate tips. *C. protodigitata* is also unbranched or has secondary branches.

Cavenderia protumula Cavender, P. Liu, Vadell, A. L. Perrigo, J. C. Landolt & S. L. Stephenson, sp. nov.

MycoBank accession number: MB842989; GenBank Accession Number. OM677258 (SSU), Figures 5 and 6.

Etymology: The specific epithet *protumulus* refers to the production of elevated small mounds (*tumulus* in Latin) that surround the bases of sorocarps.

The culture examined was from: Thailand, Chom Thong District, Chiang Mai Province, Doi Inthanon National Park, 18°35′32″ N 98°29′12″ E, tropical cloud forest, isolated from a sample collected by Stephenson SL in January 2008, Landolt TH20A (Holotype) deposited in dictyostelid collection at Jilin Agricultural University (HMJAU), ex-Landolt TH20A was deposited in the Dicty Stock Center at Northwestern University (No. DBS0350786).

Sorocarps were solitary to gregarious, erect or inclined, thin, 0.5–5.6 mm, had zero to two branches, mostly in tight clusters, sometimes colligated and coremiform, and stoloniferous (6A). Smaller solitary unbranched sorocarps, varying greatly in length, occur close to or at the base of the larger ones (6C). Sorophores had one to several tiers of cells. Branches were delicate, sigmoid, and well separated, consisting of one tier of a few cells (one to six but mostly four), arising at a right angle, 0.4–0.6 mm. Tips were obtuse and capitate, with one or two tiers of cells, 5.2–11.9 μm wide (6F). Bases were variable, regular or irregular, clavate or round with several tiers of cells, and 27.5–39.2 μm wide (6E). Sori were white, globose, 42–419 μm wide, sometimes with protruding larger cells, immersed in a bell-shaped dense granular matrix of slime. Spores were elliptical, 5.2–8.9 × 2.5–3.9 μm, with PG (6G). Typical myxamoeba (6H). Aggregation radiated and pseudoplasmodia migrated with stalks (6B). Sorogens in tight clusters (6D).

Comments: This species belongs to dictyostelid group 1 in an SSU rDNA phylogeny (Figure 9). It forms a clade together with *C. bhumiboliana* and *C. protodigitata* and several other species. However, it differs morphologically from *C. bhumiboliana* in the tips of the sorocarps, the bases of sorophores, and the sori. Sorocarps of *C. bhumiboliana* are prostrate, but *C. ungulata* has erect sorocarps. *C. bhumiboliana* has one tier of cells on the sorophores, whereas *C. protumula* has one to several tiers of cells. *Cavenderia bhumiboliana* has tips that are 15–25 μm wide, and *C. protumula* has tips that are 5.2–11.9 μm wide. *C. bhumiboliana* has bases that are 30–45 μm wide, while in *C. protumula,* they are 27.5–39.2 μm wide. *C. bhumiboliana* has sori 20–180 μm in diam, but those of *C. protumula* are 42–419 μm in diam. *Cavenderia protumula* also differs morphologically from *C. protodigitata* with respect to features of the sorocarps, the bases of the branches of the sorophores branches, and the sori. *Cavenderia protodigitata* has prostrate sorocarps, one tier of cells in the sorophores, is unbranched or has secondary branches, and sori are 50–150 μm in diam.

Cavenderia ungulata Cavender, P. Liu, Vadell, A. L. Perrigo, J. C. Landolt & S. L. Stephenson, sp. nov.

MycoBank accession number: MB842984; GenBank Accession Number. OM677259 (SSU), Figures 7 and 8.

Etymology: The specific epithet *ungulata* is derived from the Latin word *unguis*, meaning claw or finger claw, referring to the cat's claw shape of the sharp terminal cells of the bases of the sorocarps.

The culture examined was from: Thailand, Chom Thong District, Chiang Mai Province, Doi Inthanon National Park, 18°35′32″ N 98°29′12″ E, tropical cloud forest, isolated from a sample collected by Stephenson SL in January 2008, Landolt TH18B (Holotype) deposited in dictyostelid collection at Jilin Agricultural University (HMJAU), ex–Landolt TH18B deposited in the Dicty Stock Center at Northwestern University (No. DBS0350791).

Sorocarps were erect to prone, solitary, clustered to gregarious, sinuous, 0.4–5.0 mm, slightly phototrophic, and mostly unbranched but sometimes with one to five branches (8D). Sorophores were extremely delicate, tenuous, uneven, white with one or two tiers of cells, slender and regular when young or commonly very irregular, consisting of one tier of cells (8C). Smaller solitary sorocarps were unbranched and varying considerably in height, sometimes surround the bases of larger sorophores (8B). Branches were 0.5–1.0 mm long. Tips were variable, obtuse to acuminate or clavate, with one to two tiers of cells (mostly one), 6.1–10.0 µm wide (8E). Bases were clavate with one or two tiers of cells, 8.9–17.2 µm wide (8F). Sori were pale white, globose, and 29–182 µm wide. Spores were elliptical-oblong, 5.0–9.8 × 2.3–4.6 µm, with prominent, large, consolidated PG, sticky, and often surrounded by a clear narrow halo (8G). Aggregations radiated, consisting of irregular mounds with short irregular streams (8A). Pseudoplasmodia migrated. Myxamoebae had many median vacuoles (8H).

Comments: *Cavenderia ungulata* is separated from any other known species of dictyostelids by its very irregular sorophores with claw-like clavate bases, sticky regular spores with large regular PG (2 µm), with halos and dispersed small vacuoles in the spore body, a homogeneous content, continuous development and growth, and migration of early sorogens. This species adapts well to different conditions, and aggregations are either those with small irregular radiate streams or tenuous aggregations that develop dendroid ample streams partitioning into small, thick, round, or slightly elongated separate pseudoplasmodia. Some features resemble those of *C. aureostipes*, although this species is very small, crowded, and delicate. This species is much smaller than *C. aureostabilis* [5] and even smaller than *C. helicoidea* with a different spore morphology and aggregation.

4. Discussion

Historically, species of dictyostelids have been described solely on the basis of morphological characters [2] and only recently has it been possible to complement such morphological information with DNA sequence analysis. In general, DNA sequence data have supported morphology-based species determination of dictyostelids. However, DNA sequence information recently has led to an extensive revision of the higher levels of classification within the dictyostelids [3].

In little more than 50 years, the number of known species of dictyostelids has increased from about 40 to more than 160. Surveys for these organisms performed in regions of the world where previous investigations were either limited or completely lacking [14–17] largely account for this dramatic increase. However, more intensive studies in areas where previous records of dictyostelids already existed have yielded unexpected results. For example, Cavender et al. [18] reported 10 new species of dictyostelids from the Great Smoky Mountains National Park in the eastern United States.

In an earlier survey of dictyostelids in Southeast Asia performed in 1970 by Cavender, samples were collected from three localities in Thailand [19]. These were Kao Yai National Park in southern Thailand along with collecting sites near Chiang Dao and Chiang Mai in northern Thailand. All three sites were characterized by tropical semi-deciduous forests, and only Chiang Dao is located at an elevation comparable to that of Doi Inthanon, where the samples considered herein were collected. The data from this survey were summarized in Cavender [19]. A total of nine species were isolated, including a new group 1 species (*Cavenderia bifurcata*) from Chiang Dao.

The sequences obtained for the four species described in the present paper appear to indicate that they are closely related and probably evolved in situ. A similar situation was reported by Cavender et al. [18] for several morphologically similar species collected in a single small collecting site located at higher elevations the Great Smoky Mountains National Park in the United States. As a general observation, the species composition and diversity of the assemblage of dictyostelids in northern Thailand are relatively similar to what has been reported in previous studies of both the dictyostelids of Southeast Asia [19] and the American tropics [15]. In both regions, those species that appear to be endemic are rare. The four new and apparently rare species isolated from the tropical cloud forest at Doi Inthanon appear to be surviving as organisms adapted to a cool environment characterized by high levels of organic matter.

The major contribution of the present study is that it adds four new species of dictyostelids to the known biodiversity of these organisms in Thailand, all of Southeast Asia, and the world as a whole. The sampling effort involved certainly was not adequate to characterize the entire assemblage of dictyostelids present in the general study areas investigated. As such, it represents a potential starting point for additional future studies.

Supplementary Materials: The following supporting information can be downloaded at: https://www.mdpi.com/article/10.3390/jof8060593/s1, File S1: Sequence alignment.

Author Contributions: S.L.S. collected the samples in northern Thailand; E.M.V., J.C.C., J.C.L. and P.L. were responsible for the laboratory isolations and observations; P.L. and A.L.P. performed the molecular work; and J.C.C., S.L.S. and P.L. prepared the manuscript. All authors have read and agreed to the published version of the manuscript.

Funding: The collecting performed in Thailand was supported in part by a grant from the National Science Foundation (to S.L.S, grant No. 0701464.), whereas some of the laboratory research was supported by funding from the National Natural Science Foundation of China (to P.L., grants No. 32070009 and 31870015) and the Science and Technology Development Program of Jilin Province (to P.L., grant No. 20200801068GH).

Institutional Review Board Statement: Not applicable.

Informed Consent Statement: Not applicable.

Data Availability Statement: The datasets generated for this study can be found in the GenBank and MycoBank.

Acknowledgments: We thank Yue Zou and Shu Li from Jilin Agricultural University for their assistance on several aspects of the work performed on the new species of dictyostelids described herein. Appreciation is extended to Thida Win Ko Ko for helping collect samples in Thailand.

Conflicts of Interest: The authors declare no conflict of interest.

References

1. Raper, K.B. Acrasiomycetes. In *The Fungi*; Ainsworth, G.C., Sparrow, F.K., Sussman, A.S., Eds.; Academic Press: Cambridge, MA, USA, 1973; pp. 9–36.
2. Raper, K.B. *The Dictyostelids*; Princeton University Press: Princeton, NJ, USA, 1984; pp. 1–453.
3. Sheikh, S.; Thulin, M.; Cavender, J.C.; Hernández, R.E.; Kawakami, S.I.; Lado, C.; Landolt, J.C.; Nanjundiah, V.; Queller, D.; Shaap, P.; et al. A new classification of the dictyostelids. *Protist* **2018**, *169*, 1–28. [CrossRef] [PubMed]
4. Schaap, P.; Winckler, T.; Nelson, M.; Alvarez-Curto, E.; Elgie, B.; Hagiwara, H.; Cavender, J.; Milano-Curto, A.; Rozen, D.E.; Dingermann, T.; et al. Molecular phylogeny and evolution of morphology in the social amoebas. *Science* **2006**, *314*, 661–663. [CrossRef] [PubMed]
5. Vadell, E.M.; Cavender, J.C.; Landolt, J.C.; Perrigo, A.L.; Liu, P.; Stephenson, S.L. Five new species of dictyostelid social amoebae (Amoebozoa) from Thailand. *BMC Evol. Biol.* **2018**, *18*, 198. [CrossRef] [PubMed]
6. Cavender, J.C.; Raper, K.B. The Acrasieae in nature. I. Isolation. *Am. J. Bot.* **1965**, *52*, 294–296. [CrossRef] [PubMed]
7. Vadell, E.M. Taxonomy, Ecology and Karyotypes of the Cellular Slime Molds of Tikal, Guatemala. Master's Dissertation, Ohio University, Athens, OH, USA, 1993.
8. Vadell, E.M. Contribución a la Sistemática y Ecología de los Dictyostélidos del Parque Nacional Iguazú, Misiones, Argentina. Ph.D. Dissertation, University of Buenos Aires, Buenos Aires, Argentina, 2003.
9. Medlin, L.; Elwood, H.J.; Stickel, S.; Sogin, M.L. The characterization of enzymatically amplified eukaryotic 16S-like rRNA-coding regions. *Gene* **1988**, *71*, 491–499. [CrossRef]
10. Thompson, J.D.; Higgins, D.G.; Gibson, T.J. CLUSTAL W: Improving the sensitivity of progressive multiple sequence alignment through sequence weighting, position specific gap penalties and weight maxtrix choise. *Nucleic Acids Res.* **1994**, *22*, 4673–4680. [CrossRef] [PubMed]
11. Hall, T.A. BioEdit: A user-friendly biological sequence alignment editor and analysis program for Windows 95/98/NT. *Nucleic Acids Symp. Ser.* **1999**, *41*, 95–98.
12. Nguyen, L.T.; Schmidt, H.A.; Haeseler, A.; Minh, B.Q. IQ-TREE: A fast and effective stochastic algorithm for estimating maximum-likelihood phylogenies. *Mol. Biol. Evol.* **2015**, *32*, 268–274. [CrossRef] [PubMed]
13. Hoang, D.T.; Chernomor, O.; von Haeseler, A.; Minh, B.Q.; Vinh, L.S. Ufboot2: Improving the ultrafast bootstrap approximation. *Mol. Biol. Evol.* **2018**, *35*, 518–522. [CrossRef] [PubMed]

14. Landolt, J.C.; Cavender, J.C.; Stephenson, S.L.; Vadell, E.M. New species of dictyostelid cellular slime moulds from Australia. *Aust. Syst. Bot.* **2008**, *21*, 50–66. [CrossRef]
15. Cavender, J.C.; Vadell, E.M.; Landolt, J.C.; Winsett, K.E.; Stephenson, S.L.; Rollins, A.W.; Romeralo, M. New small dictyostelids from seasonal rain forests of Central America. *Mycologia* **2013**, *105*, 610–635. [CrossRef] [PubMed]
16. Cavender, J.C.; Landolt, J.C.; Romeralo, M.; Perrigo, A.; Vadell, E.M.; Stephenson, S.L. New species of *Polysphondylium* from Madagascar. *Mycologia* **2016**, *108*, 80–109. [CrossRef] [PubMed]
17. Liu, P.; Zou, Y.; Li, W.; Li, Y.; Li, X.; Che, S.; Stephenson, S.L. Dictyostelid Cellular Slime Molds from Christmas Island, Indian Ocean. *mSphere* **2019**, *4*, e00133-19. [CrossRef] [PubMed]
18. Cavender, J.C.; Vadell, E.M.; Landolt, J.C.; Stephenson, S.L. New species of small dictyostelids from the Great Smoky Mountains National Park. *Mycologia* **2005**, *97*, 493–512. [CrossRef] [PubMed]
19. Cavender, J.C. Cellular slime molds of Southeast Asia. II. Occurrence and distribution. *Am. J. Bot.* **1976**, *63*, 60–70. [CrossRef]

Journal of
Fungi

MDPI

Article

Three New Species of *Microdochium* (*Sordariomycetes*, *Amphisphaeriales*) on *Miscanthus sinensis* and *Phragmites australis* from Hainan, China

Shubin Liu [1,2], Xiaoyong Liu [1], Zhaoxue Zhang [2], Jiwen Xia [2], Xiuguo Zhang [1,2] and Zhe Meng [1,*]

1 College of Life Sciences, Shandong Normal University, Jinan 250358, China; shubinliu2022@126.com (S.L.); liuxiaoyong@im.ac.cn (X.L.); zhxg@sdau.edu.cn (X.Z.)
2 Shandong Provincial Key Laboratory for Biology of Vegetable Diseases and Insect Pests, College of Plant Protection, Shandong Agricultural University, Taian 271018, China; zhangzhaoxue2022@126.com (Z.Z.); xiajiwen163@163.com (J.X.)
* Correspondence: zmeng@sdnu.edu.cn

Abstract: Species in *Microdochium*, potential agents of biocontrol, have often been reported as plant pathogens, occasionally as endophytes and fungicolous fungi. Combining multiple molecular markers (ITS rDNA, LSU rDNA, TUB2 and RPB2) with morphological characteristics, this study proposes three new species in the genus *Microdochium* represented by seven strains from the plant hosts *Miscanthus sinensis* and *Phragmites australis* in Hainan Island, China. These three species, *Microdochium miscanthi* sp. Nov., *M. sinense* sp. Nov. and *M. hainanense* sp. Nov., are described with MycoBank number, etymology, typification, morphological features and illustrations, as well as placement on molecular phylogenetic trees. Their affinity with morphologically allied and molecularly closely related species are also analyzed. For facilitating identification, an updated key to the species of *Microdochium* is provided herein.

Keywords: *Ascomycota*; *Amphisphaeriaceae*; taxonomy; multigene phylogeny; new taxon

Citation: Liu, S.; Liu, X.; Zhang, Z.; Xia, J.; Zhang, X.; Meng, Z. Three New Species of *Microdochium* (*Sordariomycetes*, *Amphisphaeriales*) on *Miscanthus sinensis* and *Phragmites australis* from Hainan, China. *J. Fungi* **2022**, *8*, 577. https://doi.org/10.3390/jof8060577

Academic Editor: Samantha C. Karunarathna

Received: 30 April 2022
Accepted: 26 May 2022
Published: 27 May 2022

1. Introduction

Microdochium Syd. & P. Syd. is a fungal genus in the family *Amphisphaeriaceae* G. Winter of the order *Amphisphaeriales* D. Hawksw. & O.E. Erikss., which was established by Sydow [1] and typified by *M. phragmitis* Syd. & P. Syd. on living leaves of the plant host *Phragmites australis* (Cav.) Trin. ex Steud. This genus is characterized by spherical and erumpent stromata composed of minute and transparent cells, small papilla, conical sporulation cells and solitary transparent spindle-shaped-to-oval conidia. In recent years, many taxonomists have continuously enriched the known diversity in *Microdochium* [2–8]. Currently, 54 names are listed for this genus in the Index Fungorum [9], but only 37 species are accepted in the Catalogue of Life [10]. They are difficult to cultivate; therefore, just two-fifths have been studied in pure culture [4–7].

Microdochium sensu lato is known to be polyphyletic [2]. While one species, *M. oryzae* (Hashioka & Yokogi) Samuels & I.C. Hallett, was synonymized with *M. albescens* (Thüm.) Hern.-Restr. & Crous [2], and one species, *M. sorghi* (D.C. Bain & Edgerton ex Deighton) U. Braun, was recognized as a synonym of its basionym *Gloeocercospora sorghi* D.C. Bain & Edgerton ex Deighton [9,10], seven species were reclassified to other genera [2,11,12]. In detail, *M. dimerum* (Penz.) Arx, *M. falcatum* B. Sutton & Hodges, *M. fusarioides* D.C. Harris, *M. gracile* Mouch. & Samson, *M. lunatum* (Ellis & Everh.) Arx, *M. tabacinum* (J.F.H. Beyma) Arx, and *M. tripsaci* were transferred to genera *Bisifusarium* L. Lombard, Crous & W. Gams, *Idriella* P.E. Nelson & S. Wilh., *Hyalorbilia* Baral & G. Marson, *Paramicrodochium* Hern-Restr. & Crous, *Bisifusarium*, *Plectosphaerella* Kleb., and *Ephelis* Fr., respectively [2,13–18]. Currently, *Microdochium* sensu stricto is a monophyletic clade. A phylogenetic analysis of translation elongation factor 1-alpha gene (TEF1) showed that the isolates of *M. nivale* (Fr.) Samuels & I.C. Hallett were heterogeneous, and hence the variety *M. nivale* var. *majus* (Wollenw.) Samuels & I.C. Hallett was raised to a species rank as *M. majus* (Wollenw.) Glynn & S.G. Edwards, which was still thought to be sister to *M. nivale* [11].

Microdochium is an important plant pathogen in grasses and cereals. Liang et al. [19] identified *M. poae* J.M. Liang & Lei Cai as pathogen of *Poa pratensis* L. (Kentucky bluegrass) and *Agrostis stolonifera* L. (creeping bentgrass) which are both cold-season turfgrasses and widely grown on golf courses in northern China. In cold temperate regions, *M. nivale* (=*M. nivale* var. *nivale*) and *M. majus* (=*M. nivale* var. *majus*) [11,12,20] cause "Microdochium patch" on wheat and barley, resulting in significant economic losses. Some species of *Microdochium* are Brassicaceae-associated endophytes in low-Pi conditions (2.48 mg/L) and low-pH conditions (3.4–4.4) [21,22], and *M. bolleyi* is found to be endophytically associated with plant shoots and roots [23] and further to be biocontrol-active against *Gaeumannomyces graminis* var. *tritici* Walker, which causes barley's take-all disease [24]. A few species are fungicolous, such as *M. fusarioides* D.C. Harris on the oospore of *Phytophthora syringae* (Kleb.) Kleb [17].

In this study, three new pathogenic species in *Microdochium* were found among samples collected in Hainan Island, China. Two of them were isolated from *Miscanthus sinensis* Anderss, and a third one from *Phragmites australis* (Cav.) Trin. ex Steud. Their morphological characteristics and molecular-sequence data are described and discussed below.

2. Materials and Methods

2.1. Isolation and Morphology

Samples were collected from Hainan Province, China (108°37′–117°50′ E, 3°58′–20°20′ N). The strains of *Microdochium* were isolated from diseased leaves of *Miscanthus sinensis* and *Phragmites australis* using a tissue-isolation method [25]. Tissue fragments (5 × 5 mm) were taken from the margin of leaf lesions and surface-sterilized by immersing consecutively in 75% ethanol solution for 1 min, 5% sodium hypochlorite solution for 30 s, and then rinsing in sterile distilled water for 1 min [26,27]. The sterilized leaf fragments were dried with sterilized paper towels and placed on potato-dextrose agar (PDA) [28]. All the plates were incubated in a biochemical incubator at 25 °C for 3–4 days, after which hyphae were picked out of the periphery of the colonies and transferred onto new PDA plates and oatmeal-agar (OA) [29] plates.

Pure cultures transferred to PDA and OA plates were incubated at 25 °C for 15 days and photographed twice at the 7th and 15th days using a Powershot G7X mark II digital camera. Macro- and micromorphological characteristics were observed using an Olympus SZX10 stereomicroscope and an Olympus BX53 light microscope, respectively. These two microscopes were both fitted with an Olympus DP80 high-definition color digital camera to photo-document fungal structures. All fungal strains were preserved at 4 °C in sterilized 10% glycerin for further studies. Voucher specimens were deposited in the Herbarium Mycologicum Academiae Sinicae, Institute of Microbiology, Chinese Academy of Sciences, Beijing, China (HMAS) and Herbarium of the Department of Plant Pathology, Shandong Agricultural University, Taian, China (HSAUP).

Living cultures were deposited in the Shandong Agricultural University Culture Collection (SAUCC). Taxonomic information on the new taxa was submitted to MycoBank (http://www.mycobank.org/, accessed on 25 April 2022).

2.2. DNA Extraction and Amplification

Genomic DNAs were extracted from fungal mycelia grown on PDA, using a modified cetyltrimethylammonium bromide (CTAB) protocol as described in Guo et al. [30]. Four pairs of primers were adopted to amplify four genetic markers [2]. Partial nuclear ribosomal large subunit (LSU), entire internal transcribed spacer (ITS) of rDNA, partial beta-tubulin gene (TUB2), and partial RNA polymerase II second-largest subunit (RPB2) were amplified and sequenced using primer pairs LR0R/LR5 [31], ITS4/ITS5 [32], Btub526F and Btub1332R [12], and RPB2-5F2/fRPB2-7cR [33,34], respectively.

PCRs were performed using an Eppendorf Master Thermocycler (Hamburg, Germany). Amplification reactions were carried out in a volume of 25 μL, containing 12.5 μL 2 × Green Taq Mix (Vazyme, Nanjing, China), 1 μL of each forward and reverse primers (10 μM) (Biosune, Shanghai, China), 1 μL of template genomic DNA (approximately 10 ng/μL), and 9.5 μL of distilled deionized water.

The PCR program consisted of an initial denaturation at 94 °C for 5 min, 35 cycles × [denaturation at 94 °C for 30 s, annealing at a suitable temperature for 30 s, extension at 72 °C for 1 min] and a final elongation at 72 °C for 10 min. Annealing temperatures were 55 °C for ITS, 51 °C for LSU, 56 °C for RPB2 and 53 °C for TUB2. PCR products were visualized through 1% agarose-gel electrophoresis. Paired-end sequencing was conducted by Biosune Company Limited (Shanghai, China). Sequences were proofread for basic authenticity and reliability according to the five simple guidelines established by Nilsson et al. [35]. Consensus sequences were obtained using MEGA 7.0 [36]. All sequences generated in this study were deposited in GenBank (Table 1).

Table 1. Information of specimens used in this study.

Species	Voucher	Host/Substrate	Country	GenBank Accession Numbers			
				LSU	ITS	BTUB	RPB2
Idriella. lunata	CBS 204.56 *	*Fragaria chiloensis*	USA	KP858981	KP859044	–	–
Microdochium. albescens	CBS 291.79	*Oryza sativa*	Ivory Coast	KP858932	KP858996	KP859059	KP859105
	CBS 243.83	*Oryza sativa*	Unknown	KP858930	KP858994	KP859057	KP859103
M. bolleyi	CBS 540.92	*Hordeum vulgare*	Syria	KP858946	KP859010	KP859073	KP859119
M. chrysanthemoides	CGMCC3.17929 *	Unnamed Karst Cave	China	KU746736	KU746690	–	–
	CGMCC3.17930 *	Unnamed Karst Cave	China	KU746735	KU746689	–	–
M. citrinidiscum	CBS 109067 *	*Eichhornia crassipes*	Peru	KP858939	KP859003	KP859066	KP859112
M. colombiense	CBS 624.94 *	*Musa sapientum*	Colombia	KP858935	KP858999	KP859062	KP859108
M. dawsoniorum	BRIP 65649	*Sporobolus*	Australia	–	MK966337	–	–
M. fisheri	CBS 242.90 *	*Oryza sativa*	UK	KP858951	KP859015	KP859079	KP859124
M. hainanense	SAUCC210781 *	*Phragmites australis*	China	OM959323	OM956295	OM981146	OM981153
	SAUCC210782	*Phragmites australis*	China	OM959324	OM956296	OM981147	OM981154
M. indocalami	SAUCC1016 *	*Indocalamus longiauritus*	China	MT199878	MT199884	MT435653	MT510550
M. lycopodinum	CBS 146.68	Air samples	The Netherlands	KP858929	KP858993	KP859056	KP859102
	CBS 109397	*Phragmites australis*	Germany	KP858940	KP859004	KP859067	KP859113
	CBS 109398	*Phragmites australis*	Germany	KP858941	KP859005	KP859068	KP859114
M. majus	CBS 741.79	*Triticum aestivum*	Germany	KP858937	KP859001	KP859064	KP859110
M. miscanthi	SAUCC211092 *	*Miscanthus sinensis*	China	OM957532	OM956214	OM981141	OM981148
	SAUCC211093	*Miscanthus sinensis*	China	OM957533	OM956215	OM981142	OM981149
	SAUCC211094	*Miscanthus sinensis*	China	OM957534	OM956216	OM981143	OM981150
M. musae	CBS 111018 = CPC 5380	*Musa* cv. *cavendish*	Costa Rica	–	AY293061	–	–
	CBS 143499 = CPC 32809	*Musa* sp.	Malaysia	MH107941	MH107894	–	–
	CBS 143500 * = CPC 32689	*Musa* sp.	Malaysia	MH107942	MH107895	–	MH108003

Table 1. *Cont.*

Species	Voucher	Host/Substrate	Country	GenBank Accession Numbers			
				LSU	ITS	BTUB	RPB2
	CPC 11234	*Musa* sp.	Mauritius	MH107943	MH107896	–	–
	CPC 11240	*Musa* sp.	Mauritius	MH107944	MH107897	–	–
	CPC 16258	*Musa* sp.	Mexico	MH107945	MH107898	–	–
	CPC 32681	*Musa* sp.	Malaysia	MH107946	MH107899	–	–
M. neoqueenslandicum	CBS 445.95	*Juncus effusus*	The Netherlands	KP858933	KP858997	KP859060	KP859106
	CBS 108926 *	*Agrostis* sp.	New Zealand	KP858938	KP859002	KP859065	KP859111
M. nivale	CBS 116205 *	*Triticum aestivum*	UK	KP858944	KP859008	KP859071	KP859117
	CBS 288.50	Unknown	Unknown	MH868135	MH856626	–	–
M. novae-zelandiae	CBS 143847	Turf leaves (*Poaceae*)	New Zealand	–	LT990655	LT990608	LT990641
	CPC 29693	Turf leaves (*Poaceae*)	New Zealand	–	LT990656	LT990609	LT990642
M. paspali	HK-ML-1371	*Paspalum vaginatum*	China	–	KJ569509	KJ569514	–
	QH-BA-48	*Paspalum vaginatum*	China	–	KJ569510	KJ569515	–
	SY-LQG66	*Paspalum vaginatum*	China	–	KJ569511	KJ569516	–
	WC-WC-85	*Paspalum vaginatum*	China	–	KJ569512	KJ569517	–
	WN-BD-452	*Paspalum vaginatum*	China	–	KJ569513	KJ569518	–
M. phragmitis	CBS 285.71 *	*Phragmites australis*	Poland	KP858949	KP859013	KP859077	KP859122
	CBS 423.78	*Phragmites communis*	Germany	KP858948	KP859012	KP859076	KP859121
M. ratticaudae	BRIP 68298	introduced giant rat's tail grasses	Australia	MW481666	MW481661	–	MW626890
M. rhopalostylidis	CPC 34449 = CBS 145125 *	*Rhopalostylis sapida*	New Zealand	MK442532	MK442592	–	MK442667
M. seminicola	KAS3576 = CBS 139951 *	Maize kernels	Switzerland	KP858974	KP859038	KP859101	KP859147
	KAS1516 = CPC 26001	Grain	Canada	KP858961	KP859025	KP859088	KP859134
	KAS3574 = DAOM 250155	*Maize kernels*	Switzerland	KP858973	KP859037	KP859100	KP859146
	KAS3158 = DAOM 250161	*Triticum aestivum*	Canada	KP858970	KP859034	KP859097	KP859143

Table 1. *Cont.*

Species	Voucher	Host/Substrate	Country	GenBank Accession Numbers			
				LSU	ITS	BTUB	RPB2
	KAS1527 = DAOM 250165	Grain	Canada	KP858966	KP859030	KP859093	KP859139
	KAS1473 = DAOM 250176	*Triticum aestivum*	Canada	KP858955	KP859019	KP859082	KP859128
M. sinense	SAUCC211097 *	*Miscanthus sinensis*	China	OM959225	OM956289	OM981144	OM981151
	SAUCC211098	*Miscanthus sinensis*	China	OM959226	OM956290	OM981145	OM981152
M. sorghi	CBS 691.96	*Sorghum halepense*	Cuba	KP858936	KP859000	KP859063	KP859109
M. sp. indet.	SAUCC1017	*Indocalamus longiauritus*	China	MT199879	MT199885	MT435654	–
M. tainanense	CBS 269.76 *	*Saccharum officinarum*	Taiwan	KP858945	KP859009	KP859072	KP859118
	CBS 270.76	*Saccharum officinarum*	Taiwan	KP858931	KP858995	KP859058	KP859104
M. trichocladiopsis	CBS 623.77 *	*Triticum aestivum*	Unknown	KP858934	KP858998	KP859061	KP859107
M. yunnanense	SAUCC1011 *	*Indocalamus longiauritus*	China	MT199875	MT199881	MT435650	MT510547
	SAUCC1012	*Indocalamus longiauritus*	China	MT199876	MT199882	–	MT510548
	SAUCC1015	*Indocalamus longiauritus*	Chima	MT199877	MT199883	MT435652	MT510549
	SAUCC1018	*Indocalamus longiauritus*	Chima	MT199880	MT199886	MT435655	–

Notes: New species established in this study are in bold. Ex-types, ex-epitypes or holotype strains are marked with "*".

2.3. Phylogenetic Analyses

Twenty-eight new sequences were generated in this study, and available reference sequences of *Microdochium* species were retrieved from GenBank [2–7]. Four genetic markers (ITS, LSU, TUB2 and RPB2) were separately aligned using MAFFT v.7.110 (Osaka, Japan) [37]. Phylogenetic analyses were conducted individually for each marker at first and then for a combined dataset of the four genetic markers (Supplementary File S1).

Phylogenetic analyses were conducted with Bayesian inference (BI) and maximum-likelihood (ML) algorithms on the CIPRES Science Gateway portal (https://www.phylo.org/, accessed on 15 April 2022;) [38]. The BI ran with MrBayes on XSEDE v. 3.2.7a (Stockholm, Sweden) [39–41], and the ML ran with RAxML-HPC2 on XSEDE v. 8.2.12 (Heidelberg, Germany) [42]. The best evolutionary model for each partition was determined using MrModelTest v. 2.3 [43]. Default parameters were used for 1000 bootstrap ML analysis. In BI analysis, starting trees were random, and four MCMC chains ran simultaneously for five million generations. Trees were sampled once every 500 generations. These chains stopped when all convergences met and the standard deviation fell below 0.01. The burn-in fraction was set to 0.25 and Posterior Probabilities (PP) were determined from the remaining trees. All resulting trees were plotted using FigTree v. 1.4.4 (http://tree.bio.ed.ac.uk/software/figtree, accessed on 15 April 2022) and the layout of the trees was carried out with Adobe Illustrator CC 2019.

3. Results

3.1. Phylogenetic Analyses

Seven *Microdochium* strains isolated from plant hosts were sequenced. Multilocus data (ITS, LSU, TUB2 and RPB2) were composed of 52 strains of *Microdochium* as ingroup and a strain CBS 204.56 of *Idriella lunata* as outgroup. A total of 2957 characters were fed to the phylogenetic analysis, viz. 1–573 (ITS), 574–1423 (LSU), 1424–2117 (TUB2), and 2118–2957 (RPB2). Of these characters, 2223, 97 and 637 were constant, variable parsimony-uninformative and parsimony-informative, respectively. For the BI and ML analyses, the evolutionary model of GTR+I+G was selected for ITS, TUB2 and RPB2, while SYM+I+G was selected for LSU (Figure 1). The topology of the phylogenetic tree generated by the ML method was highly similar to that by BI, and therefore it was chosen to represent the evolutionary history of *Microdochium*.

The 59 strains are assigned to 29 species clades based on the four-marker phylogeny (Figure 1). The seven strains isolated herein represent three novel species. The new species *M. miscanthi* (SAUCC211092, SAUCC211093 and SAUCC211094) has a sister relationship to another new species, *M. sinense* (SAUCC211097 and SAUCC211098), with robust support values (BIPP 1.00 and MLBV 100%). These two species are closely related to *M. rhopalostylidis* (CBS 145125), *M. phragmitis* (CBS 285.71 and CBS 423.78), *M. lycopodinum* (CBS 146.68, CBS 109397 and CBS 109398), *M. indocalami* (SAUCC1016) and *M. fisheri* (CBS 242.90) with high support values (BIPP 1.00 and MLBV 100%). The last new species, *M. hainanense* (SAUCC210781 and SAUCC210782), forms the sister group of the seven species mentioned above with reasonable support (MLBV 92%).

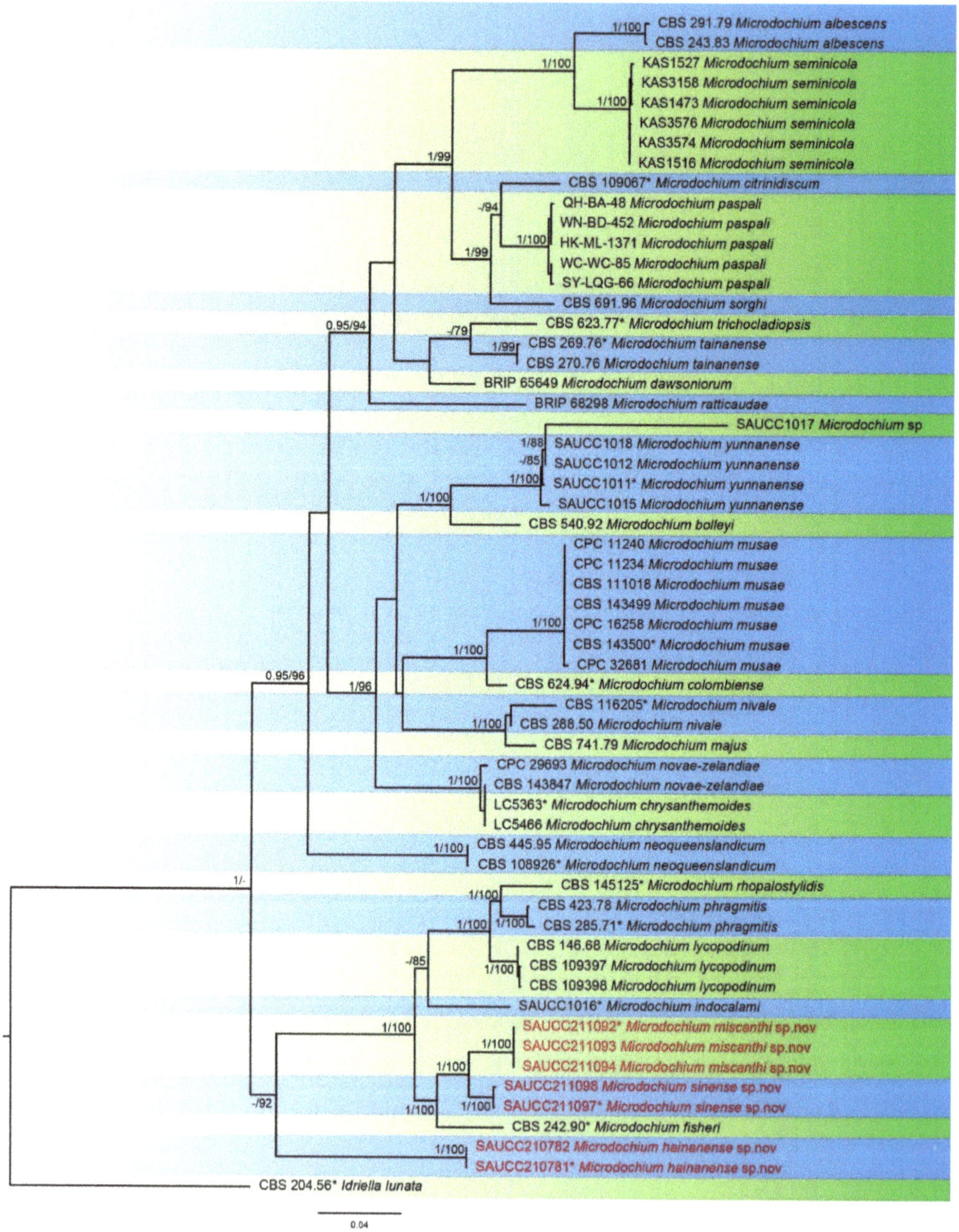

Figure 1. A maximum-likelihood phylogram of *Microdochium* based on combined ITS, LSU, TUB2 and RPB2 sequences with CBS 204.56 of *Idriella lunata* as outgroup. The maximum-likelihood Bootstrap Value (MLBV $\geq$ 75%) and Bayesian Inference Posterior Probability (BIPP $\geq$ 0.95) are shown at the first and second position, respectively. Strains marked with "*" are ex-types, ex-epitypes or holotypes. Strains from the current study are in red. The scale bar at the bottom middle indicates 0.08 substitutions per site.

3.2. *Taxonomy*

Microdochium miscanthi (Figure 2) S.B. Liu, X.Y. Liu, Z. Meng & X.G. Zhang, sp. nov.

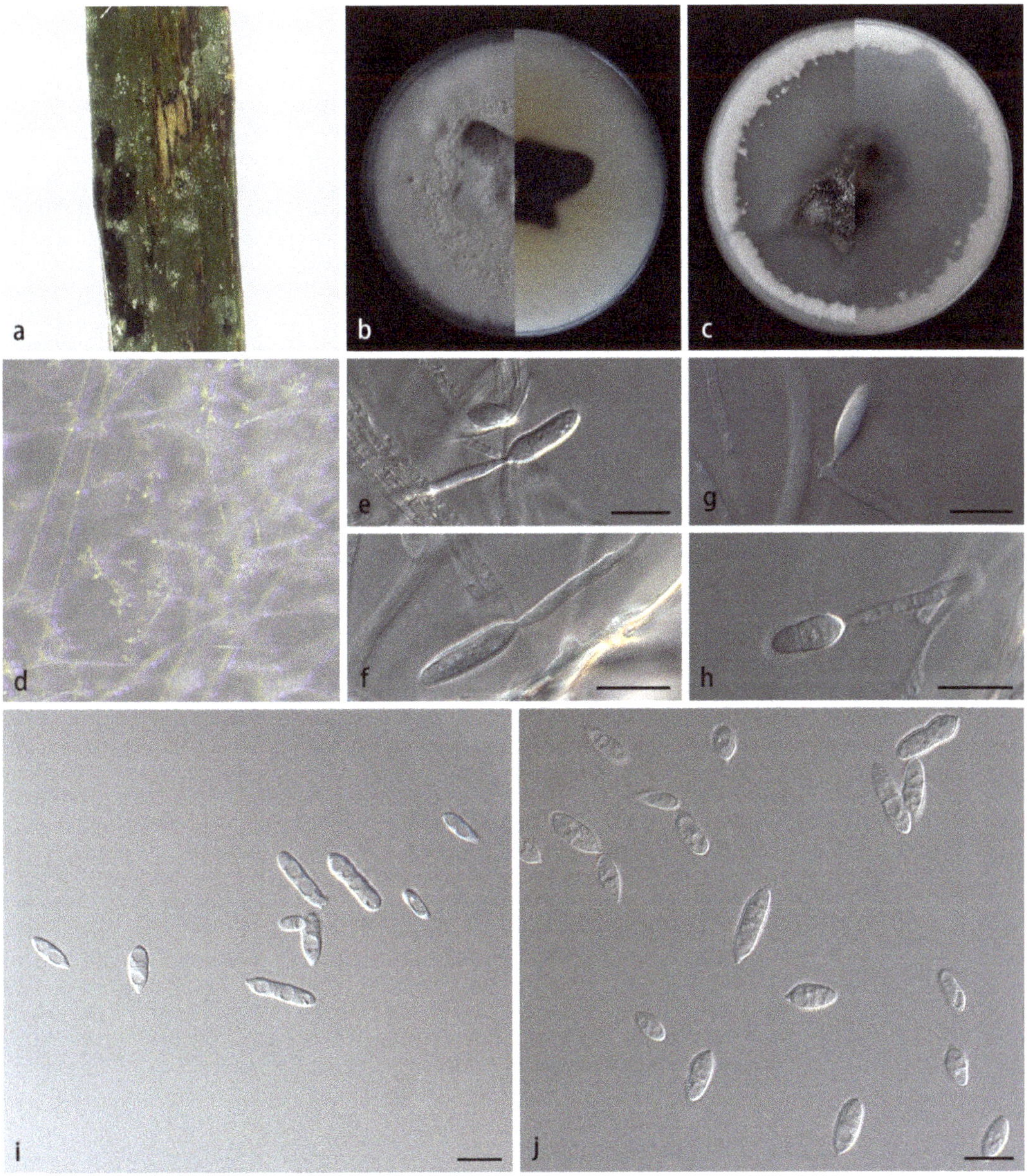

Figure 2. *Microdochium miscanthi* (holotype HMAS352151, ex-holotype SAUCC211092). (**a**) Leaves of host plant; (**b**) inverse and reverse sides of colony after 15 days on PDA; (**c**) inverse and reverse sides of colony after 15 days on OA; (**d**) a colony overview; (**e**–**h**) conidiophores and conidiogenous cells; (**i**,**j**) conidia. Scale bars: (**e**–**j**) 10 μm.

MycoBank No.: 843867

Etymology—The epithet "*miscanthi*" refers to the genus name of the host plant *Miscanthus sinensis*.

Type—China, Hainan Province: Diaoluoshan National Forest Park, on diseased leaves of *Miscanthus sinensis*, 21 May 2021, S.B. Liu, holotype HMAS352151, isotype HSAUP211092, ex-holotype living culture SAUCC211092.

Description—Colonies on PDA at 25 °C for 14 days attain 87.2–89.1 mm in diameter. When young, round in shape, dark green in the center and white at the edge, with some dark green parts covered with continuously growing mycelia. When old, tight, uneven and pale yellow in the center, fluffy, flat, white at the edge. Mycelia are superficial and immersed, 1.5–2.3 μm wide, transparent, branched and diaphragmatic. Conidiophores are straight or slightly curved, produced on aerial mycelia, septate and often reduced to conidiogenous cells borne directly from hyphae. Conidiogenous cells are mono- or polyblastic, terminal, denticulate, transparent, smooth and cylindrical, 9.7–14.5 × 3.6–4.1 μm. Conidia are solitary, transparent, spindle-to-rod-shaped, 0–2-septate, 7.0–16.1 × 2.5–4.7 μm, 0–5 guttulate when mature and sometimes borne directly from hyphae. Chlamydospores were not observed. Sexual morphs unknown.

Culture characteristics—Colonies on OA at 25 °C for 14 days, reach 88.4–89.3 mm in diameter, and are circular, black-green in the center and irregular in shape, covered with a thin layer of white mycelia, dense at the edge and forming a white ring. Substrate hyphae are transparent and smooth. Vegetative hyphae are transparent, smooth, branched and diaphragmatic.

Notes—Strains SAUCC211092, SAUCC211093 and SAUCC211094 are identified as the same new species *Microdochium miscanthi*. They have similar morphological characteristics, including culture characteristics, sporodochia and conidia. They are also the same in DNA sequences, gathering together with robust support values (MLBV 100% and BIPP 1.00, Figure 1). Phylogenetic analyses on a combined dataset of four genetic markers showed that *M. miscanthi*, *M. lycopodinum*, *M. phragmites*, *M. rhopalostylidis*, *M. fisheri* and *M. sinense* formed a clade. *M. miscanthi* and *M. sinense* form sister clades on the phylogenetic trees, but they are different in culture characteristics, conidia and DNA sequences. In *M. miscanthi*, colonies on PDA are overall white, with central dark-green plaque covered by white mycelia; conidiogenous cells are 9.7–14.5 × 3.6–4.1 μm, without diaphragms; conidia are 7.0–16.1 × 2.5–4.7 μm, spindle-to-rod-shaped. In *M. sinense*, colonies are overall pale yellow; conidiogenous cells are 6.3–22.4 × 4.1–5.7 μm, with single or multiple diaphragms; conidia are 11.5–19.34 × 2.8–5.4 μm, spindle-shaped or cylindrical. As for molecular differences between *M. miscanthi* and *M. sinense*, ITS, BTUB, LSU and RPB2 had 10, 21, 2 and 35 bp of dissimilarity, respectively. Therefore, we assign them in two different species. In addition, conidiogenous cells in *M. miscanthi* are terminal or sympodial, denticulate, transparent, smooth and cylindrical, which are similar to the species in this clade. The conidia of *M. miscanthi* (7.0–16.1 × 2.5–4.7 μm) differs in size from those of *M. lycopodinum* (8.0–15.5 × 2.5–4.0 μm), *M. phragmites* (10.0–14.5 × 2.0–3.0 μm), *M. fisheri* (7.0–12.0 × 3.0–4.0 μm) and *M. rhopalostylidis* (16.0–20.0 × 3.0–4.0 μm) [2,5]. Furthermore, mature conidia are guttulate in *M. miscanthi*.

Microdochium sinense (Figure 3) S.B. Liu, X.Y. Liu, Z. Meng & X.G. Zhang, sp. nov.

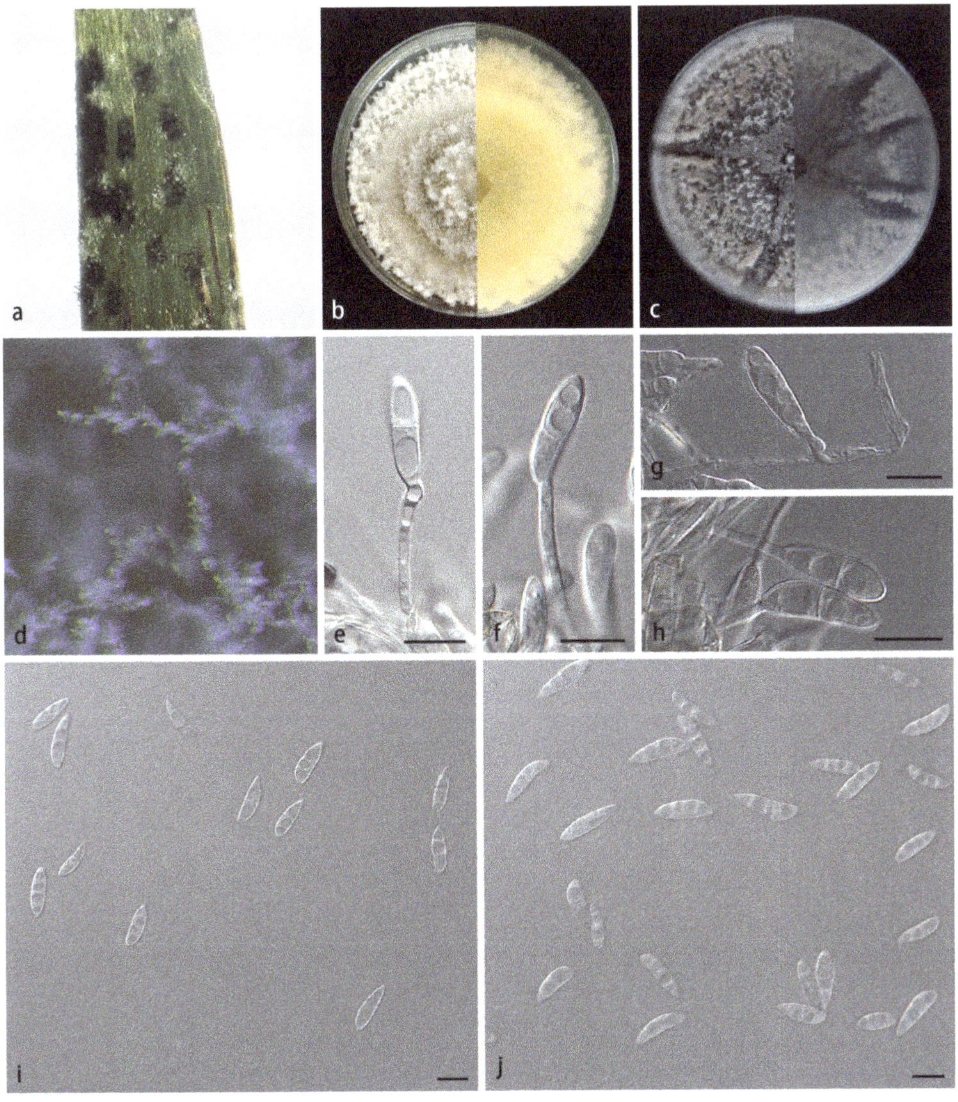

Figure 3. *Microdochium sinense* (holotype HMAS352154, ex-holotype SAUCC211097). (**a**) Leaves of host plant; (**b**) inverse and reverse sides of colony after 15 days on PDA; (**c**) inverse and reverse sides of colony after 15 days on OA; (**d**) colony overview; (**e–h**) conidiophores and conidiogenous cells; (**i**,**j**) conidia. Scale bars: (**e–j**) 10 μm.

MycoBank—No: 843868

Etymology—The epithet "*sinense*" (Lat.) refers to China, where the species was collected.

Type—China, Hainan Province: *Diaoluoshan* National Forest Park, on diseased leaves of *Miscanthus sinensis*, 21 May 2021, S.B. Liu, holotype HMAS352154, isotype HSAUP211097, ex-holotype living culture SAUCC211097.

Description—Colonies on PDA at 25 °C for 14 days attain 87.2–89.3 mm in diameter; when young, they are irregular in shape, dark green in the center and covered by white hyphae; when old, they are dark green overall, covered completely by white, lush, fluffy and beige hyphae. Mycelia are superficial and immersed, 1.3–2.3 μm wide, transparent, branched and diaphragmatic. Conidiophores are straight or slightly curved, produced from aerial hyphae, septate and often reduced to conidiogenous cells borne directly from hyphae. Conidiogenous cells are monoblastic, terminal, hyaline, smooth and cylindrical, 16.3–22.4 × 4.1–5.7 μm. Conidia are solitary, hyaline, spindle-shaped or cylindrical, 1–3-septate, 11.5–19.34 × 2.8–5.4 μm, 2–9 guttulate when mature and sometimes borne directly from hyphae. Chlamydospores were not observed. Sexual morphs unknown.

Culture characteristics—Colonies on OA at 25 °C for 14 days, reach 86.4–88.9 mm in diameter; when young, they are circular gray in the center and wax yellow at the edge; when old, they have ravines, dense, yellow-brown overall and fluffy at the edge. Vegetative hyphae are transparent, branched and diaphragmatic.

Notes—Strains SAUCC211097 and SAUCC211098 are identified to the same species *Microdochium sinense* sp. nov. For details, refer to the notes for *M. miscanthi*.

Microdochium hainanense (Figure 4) S.B. Liu, X.Y. Liu, Z. Meng & X.G. Zhang, sp. nov.

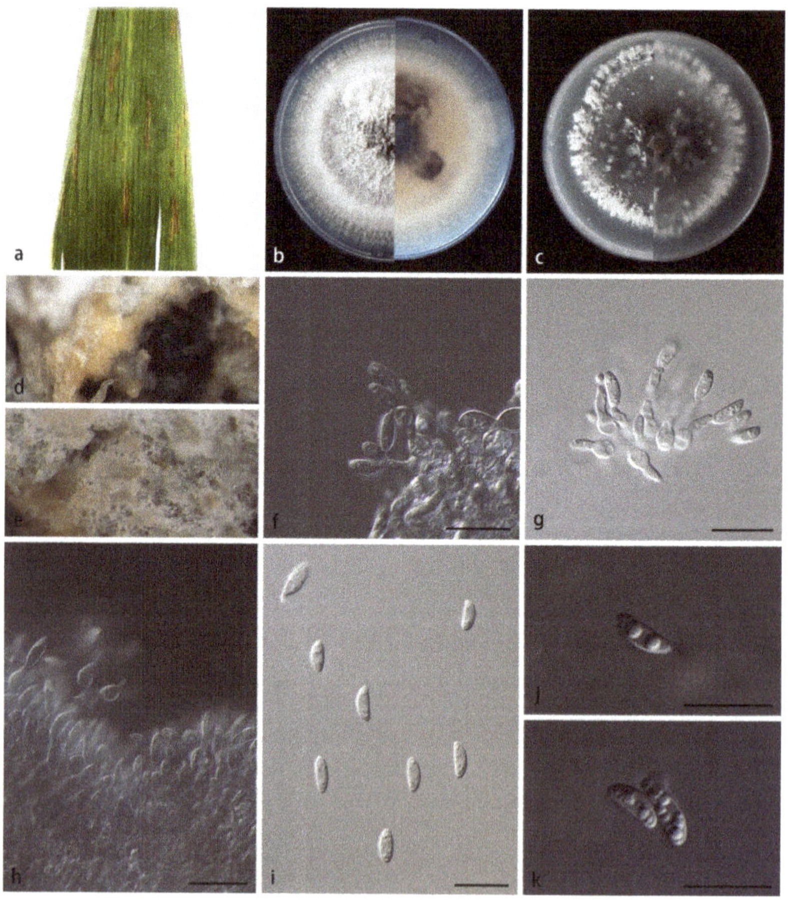

Figure 4. *Microdochium hainanense* (holotype HMAS352156, ex-holotype SAUCC210781). (**a**) leaves of host plant; (**b**) inverse and reverse sides of colony after 15 days on PDA; (**c**) inverse and reverse sides of colony after 15 days on OA; (**d**) sporodochia after removing the surface mycelia; (**e**) the mixture of conidia and secretions on mycelium; (**f**–**h**) conidiophores and conidiogenous cells; (**i**–**k**) conidia. Scale bars: (**f**–**k**) 10 μm.

MycoBank—No:843869

Etymology—The epithet "*hainanense*" is named after Hainan Province, where the fungus was collected.

Type—China, Hainan Province: Diaoluoshan National Forest Park, on diseased leaves of *Phragmites australis*, 21 May 2021, S.B. Liu, holotype HMAS352156, isotype HSAUP210781, ex-holotype living culture SAUCC210781.

Description—Colonies on PDA for 14 days attain 75.4–77.2 mm in diameter; when young, they form a conspicuously concentric circle, brown and dense in the center, white and sparse at the edge; when old, they produce sporodochia in aerial mycelia or on agar surface, slimy, hyaline or orange, colorless-to-brownish in reverse due to secreted soluble pigments. Mycelia are superficial and immersed, width 1.5–3.0 μm, transparent, smooth, branched and diaphragmatic. Conidiophores are reduced to conidiogenous cells. Conidiogenous cells are monoblastic, terminal, hyaline, smooth, ampulliform and lageniform, with percurrent proliferations, 4.8–8.2 × 2.0–2.5 μm. Conidia are solitary, hyaline, aseptate and spindle-to-rod-shaped, 7.0–16.1 × 2.5–4.7 μm, 0–8 guttulate when mature. Chlamydospores were not observed. Sexual morphs unknown.

Culture characteristics—Colonies on OA at 25 °C for 14 days reach 69.7–71.9 mm in diameter; they are circular, with hyphae mostly immersed in agar and occasionally scattered on the agar surface; light black and sparse in the center, white and dense at the edge. Substrate hyphae are transparent and smooth. Vegetative hyphae are transparent, septate and branched.

Notes—Strains SAUCC210781 and SAUCC210782 are identified to the same new species, *M. hainanense*. They share morphological characteristics, including culture characteristics, sporodochia and conidia. They are also identical in DNA sequences, gathering together with robust support values (MLBV 100% and BIPP 1.00, Figure 1). Phylogenetic analysis of the four genetic markers of *M. hainanense* showed that *M. hainanense* formed an independent branch, sister to the group of *M. indocalami*, *M. sinense*, *M. miscanthi*, *M. rhopalostylidis*, *M. phragmites*, *M. fisheri* and *M. lycopodinum* with satisfactory support (MLBV 92%, Figure 1). *Microdochium. hainanense* produces sporodochia, similar to *M. phragmitis* (CBS 423.78) and *M. rhopalostylidis*, but *M. hainanense* produces clear-to-orange soluble pigments, while the conidia of other species are directly produced from hyphae. Conidia are single, ellipsoid or spindle-shaped, similar to all the related species mentioned above. Conidia of *M. hainanense* (5.5–8.1 × 2.2–3.0 μm) differ in size from those of *M. lycopodinum* (8.0–15.5 × 2.5–4.0 μm), *M. phragmites* (10.0–14.5 × 2.0–3.0 μm), *M. rhopalostylidis* (16.0–20.0 × 3.0 –4.0 μm), *M. indocalami* (13.0–15.5 × 3.5–5.5 μm), *M. fisheri* (7.0–12.0 × 3.0–4.0 μm), *M. miscanthi* (7.0–16.1 × 2.5–4.7 μm) and *M. sinense* (11.5–19.34 × 2.8–5.4 μm) [2,5].

3.3. Key to the Species of *Microdochium*

Together with the three new species proposed in this study, we currently accepted a worldwide total of 47 species in the genus *Microdochium*. In order to facilitate identification in the future, a key to the species of *Microdochium* is provided herein, updating the key compiled 46 years ago [15]. Characteristics adopted in the key include perithecia, septa, asci, ascospores, conidiogenous cells, conidia and chlamydospores.

1. Sexual morph known——2
1. Sexual morph unknown——13
2. Perithecia maximum diameter > 200 μm——3
2. Perithecia maximum diameter < 200 μm——9
3. Maximum number of septa of ascospores > 3——4
3. Maximum number of septa of ascospores ≤ 3——5
4. Asci size 90.0–120.0 × 21.0–25.0 μm——*M. consociatum*
4. Asci size 80.0–100.0 × 17.0–22.0 μm——*M. musae*
5. Asci size = 50.0–70.0 × 7.0–9.0 μm——6
5. Asci size ≠ 50.0–70.0 × 7.0–9.0 μm——7
6. Ascospores size 9.5–17.0 × 3.0–4.5 μm——*M. majus*

6. Ascospores size 10.0–17.0 × 3.5–4.5 µm——*M. nivale*
7. Ascospores 1–3 septa——8
7. Ascospores 1–2 septa——*M. stevensonii*
8. Ascospores size 20.0–32.0 × 3.0–3.5 µm——*M. fusariisporum*
8. Ascospores size 15.0–25.0 × 4.0–5.0 µm——*M. passiflorae*
9. Perithecia maximum diameter < 150 µm——10
9. Perithecia maximum diameter > 150 µm——11
10. Ascospores size 20.0–22.0 × 3.5 µm——*M. opuntiae*
10. Ascospores size 12.0–22.0 × 3.0–5.0 µm——*M. seminicola*
11. Chlamydospores known——*M. ratticaudae*
11. Chlamydospores unknown——12
12. Conidia falcate, 11.0–16.0 × 3.5–4.5 µm, 0–3 septa——*M. albescens*
12. Conidia lunate, 8.0–15.0 × 2.5–3.5 µm, 0–1 septa——*M. lycopodinum*
13. Chlamydospores known——14
13. Chlamydospores unknown——16
14. Conidia oblong——*M. trichocladiopsis*
14. Conidia lunate——15
15. Chlamydospores chain or clusters——*M. bolleyi*
15. Chlamydospores rounded or obovoid——*M. poae*
16. Conidia aseptate——17
16. Conidia septate——23
17. Conidiogenous cells two types——*M. yunnanense*
17. Conidiogenous cells one type——18
18. Conidiogenous cells with denticulate——19
18. Conidiogenous cells not denticulate——21
19. Conidiogenous cells ampulliform——20
19. Conidiogenous cells cylindrical——*M. sclerotiorum*
20. Conidia pointed at both ends, no appendages——*M. griseum*
20. Conidia with straight appendages at both ends——*M. queenslandicum*
21. Conidiogenous cells monoblastic——***M. hainanense* sp. nov.**
21. Conidiogenous cells sympodial——22
22. Conidia filiform, 7.0–16.0 × 1.0 µm——*M. palmicola*
22. Conidia lunate, 7.5.0–11.0 × 1.8–2.0 µm——*M. queenslandicum*
23. Conidiogenous cells two types——*M. colombiense*
23. Conidiogenous cells one type——24
24. Conidia relatively narrow, acicular, filiform, falcate or lunate——25
24. Conidia relatively rounded, ellipsoid, fusiform, cylindrical or obovoid——32
25. Conidia with long appendages at both ends——*M. linariae*
25. Conidia without appendages at both ends——26
26. Conidia with conspicuous rhachides——*M. tainanense*
26. Conidia without conspicuous rhachides——27
27. Conidiogenous cells ampulliform——28
27. Conidiogenous cells cylindrical——31
28. Maximum number of septa of conidia = 10——*M. sorghi*
28. Maximum number of septa of conidia < 10——29
29. Conidia lunate——*M. neoqueenslandicum*
29. Conidia falcate——30
30. Conidia size 25.0–30.0 × 1.5–2.0 µm, 0–1 septa——*M. caespitosum*
30. Conidia size 7.0– 20.5 ×2.5–4.5 µm, 0–3 septa——*M. paspali*
31. Conidia size 25.0–75.0 × 1.0–2.0 µm, 0–3 septa——*M. dawsoniorum*
31. Conidia size 5.5–10.0 × 2.0–2.5 µm, 0–1 septa——*M. novae-zelandiae*
32. Conidia with guttulate——33
32. Conidia no guttulate——35
33. Conidiogenous cells solitary——*M. chrysanthemoides*

33. Conidiogenous cells sympodial——34
34. Conidia size 10.0–14.5 × 2.0–3.0 μm, 0–1 septa——*M. phragmitis*
34. Conidia size 13.0–23.0 × 2.5–4.0 μm, 1–3 septa——*M. rhopalostylidis*
35. Conidia cylindrical——36
35. Conidia fusiform——42
36. Conidiogenous cells denticulate——37
36. Conidiogenous cells not denticulate——40
37. Conidiogenous cells blastic-sympodial——*M. cylindricum*
37. Conidiogenous cells mono- or polyblastic——38
38. Conidia spindle-to-rod-shaped——***M. miscanthi* sp. nov.**
38. Conidia clavate to obovoid——39
39. Conidia size 7.0–31.0 × 2.0–3.0 μm, 0–3 septa——*M. citrinidiscum*
39. Conidia size 13.0–15.5 × 3.5–5.5 μm, 1–3 septa——*M. indocalami*
40. Conidiogenous cells ampulliform——*M. maydis*
40. Conidiogenous cells cylindrical——41
41. Conidiogenous cells monoblastic, 16.3–22.4 × 4.1–5.7 μm——***M. sinense* sp. nov.**
41. Conidiogenous cells sympodial, 6.5–15.0 × 2.5–3.5 μm——*M. stoveri*
42. Conidiogenous cells ampulliform——43
42. Conidiogenous cells cylindrical——44
43. Conidiogenous cells solitary——*M. punctum*
43. Conidiogenous cells sympodial——*M. triticicola*
44. Conidiogenous cells mono- or polyblastic——*M. maculosum*
44. Conidiogenous cells sympodial——45
45. Conidiogenous cells not denticulate——*M. panattonianum*
45. Conidiogenous cells denticulate——46
46. Conidia size 7.0–12.0 × 3.0–4.0 μm, 0–1 septa——*M. fisheri*
46. Conidia size 8.0–15.0 × 3.0–4.5 μm, 1–2 septa——*M. intermedium*

4. Discussion

Microdochium was established in 1924, and *Monographella* Petr. also established in 1924 was previously described as a sexual morph of *Microdochium* [16,44–46]. With the application of "one fungus one name" declaration [47], *Microdochium* was retained as the correct genus name because it accommodates more species and is used more frequently [2]. Due to their phylogenetic affinity, *Microdochium*, *Idriella* and *Selenodriella* were introduced into a new family, namely *Microdochiaceae* [2]. This new family is characterized by (1) *Monographella*-like sexual morphs; and (2) asexual morphs of polyblastic, sympodial or annellidic conidiogenous cells with hyaline conidia, but no appendages. As an important basis for classification, conidia of *Microdochium* vary in shape, i.e., cylindrical, fusiform, elliptical, stick-shaped, vertical or curved, with truncate bases and apices mainly rounded.

Since the inception of *Microdochium* in 1924, its delimitation has undergone changes, and currently, 47 species are accepted in the genus. Although the number is small, there are still some problems in the classification. For example, Catalogue of Life accepts the basionym *Gloeocercospora sorghi* rather than the combination *Microdochium sorghi* but without any explanation [10]. It is possibly because *M. sorghi* remains sterile and only produces black sclerotia in culture [2]. However, in this study, the phylogenetic analysis based on the based on four genetic markers showed that *M. sorghi* formed a separated branch closely related to the clade of *M. citrinidiscum* and *M. paspali* with strong support (MLBV: 100% and BIPP: 1.00, Figure 1). Upon this molecular evidence, we accept *M. sorghi* as the correct name for this species.

Microdochium is mainly distributed in warm and humid areas, and most prefer to parasitize on gramineous plants. Our finding of the new species *M. miscanthi*/*M. sinense* on *Miscanthus sinensis* (*Poaceae*) and *M. hainanense* on *Phragmites australis* (*Poaceae*), confirms this phenomenon well. Hainan Province is located in the tropical region of southern China. Its annual average temperature is 22–27 °C, and its annual precipitation is 1000–2600 mm, with a typical tropical rainforest climate. This kind of environment is conducive to the growth of unusual microbial species, resulting in a high species diversity.

In order to accurately identify the species of *Microdochium*, molecular analysis is needed. In this study, the four genetic markers ITS, LSU, RPB2 and TUB2 were selected according to previous molecular studies of *Microdochium*. LSU provides enough information for the generic placement of *Microdochium*. Although any of the genetic markers ITS, TUB2 or RPB2 can be used for phylogenetic analysis at the species level in *Microdochium* (results not shown), TUB2 has more phylogenetic information, with longer distances between species and higher support values. This is consistent with previous studies on other xylariaceous genera [2,48,49].

Supplementary Materials: The following supporting information can be downloaded at: https://www.mdpi.com/article/10.3390/jof8060577/s1, Supplementary File S1: The combined ITS, LSU, TUB2, and RPB2 multiple sequence alignment. Table S1: Specimens and GenBank accession numbers of DNA sequences used in this study.

Author Contributions: Conceptualization, S.L. and X.L.; methodology, Z.M.; software, Z.Z.; validation, J.X.; formal analysis, X.L.; investigation, S.L.; resources, Z.Z.; data curation, S.L.; writing—original draft preparation, S.L.; writing—review and editing, X.L.; visualization, X.L.; supervision, X.Z.; project administration, Z.M.; funding acquisition, X.Z. All authors have read and agreed to the published version of the manuscript.

Funding: This research was funded by National Natural Science Foundation of China (nos. U2002203, 31750001, 31900014).

Institutional Review Board Statement: Not applicable for studies involving humans or animals.

Informed Consent Statement: Not applicable.

Data Availability Statement: The sequences from the present study were submitted to the NCBI database (https://www.ncbi.nlm.nih.gov/, accessed on 25 April 2022) and the accession numbers were listed in Table 1.

Conflicts of Interest: The authors declare no conflict of interest.

References

1. Sydow, H. Mycotheca germanic. Fasc. XLII–XLV (No. 2051–2250). *Annls Mycol.* **1924**, *22*, 257–268.
2. Hernández-Restrepo, M.; Groenewald, J.Z.; Crous, P.W. Taxonomic and phylogenetic re-evaluation of *Microdochium*, *Monographella* and *Idriella*. *Pers. Mol. Phylogeny Evol. Fungi* **2016**, *36*, 57–82. [CrossRef] [PubMed]
3. Zhang, Z.F.; Liu, F.; Zhou, X.; Liu, X.Z.; Liu, S.J.; Cai, L. Culturable Mycobiota from karst caves in China, with descriptions of 20 new species. *Pers. Mol. Phylogeny Evol. Fungi* **2017**, *39*, 1–31. [CrossRef] [PubMed]
4. Crous, P.W.; Schumacher, R.K.; Wingfield, M.J.; Akulov, A.; Denman, S.; Roux, J.; Braun, U.; Burgess, T.I.; Carnegie, A.J.; Váczy, K.Z. New and interesting fungi 1. *Fuse* **2018**, *1*, 169–215. [CrossRef] [PubMed]
5. Crous, P.W.; Schumacher, R.K.; Akulov, A.; Thangavel, R.; Hernández-Restrepo, M.; Carnegie, A.J.; Cheewangkoon, R.; Wingfield, M.J.; Summerell, B.A.; Quaedvlieg, W. New and interesting fungi 2. *Fuse* **2019**, *3*, 57–134. [CrossRef] [PubMed]
6. Marin-Felix, Y.; Hernández-Restrepo, M.; Wingfield, M.J.; Akulov, A.; Carnegie, A.J.; Cheewangkoon, R.; Gramaje, D.; Groenewald, J.Z.; Guarnaccia, V.; Halleen, F. Genera of phytopathogenic fungi: GOPHY 2. *Stud. Mycol.* **2019**, *92*, 47–133. [CrossRef]
7. Huang, S.T.; Xia, J.W.; Zhang, X.G.; Sun, W.X.; Li, Z. Two new species of *Microdochium* from *Indocalamus longiauritus* in south-western China. *MycoKeys* **2020**, *72*, 93–108. [CrossRef]
8. Crous, P.W.; Cowan, D.A.; Maggs-Kölling, G.; Yilmaz, N.; Thangavel, R.; Wingfield, M.J.; Noordeloos, M.E.; Dima, B.; Brandrud, T.E.; Jansen, G.M.; et al. Fungal Planet description sheets: 1182–1283. *Pers. Mol. Phylogeny Evol. Fungi* **2021**, *46*, 313–528. [CrossRef]
9. Index Fungorum. Available online: http://www.indexfungorum.org/ (accessed on 27 March 2022).
10. Bánki, O.; Roskov, Y.; Döring, M.; Ower, G.; Vandepitte, L.; Hobern, D.; Remsen, D.; Schalk, P.; DeWalt, R.E.; Keping, M.; et al. Catalogue of Life Checklist (Version 2022-03-21). *Cat. Life* **2022**. [CrossRef]
11. Glynn, N.C.; Hare, M.C.; Parry, D.W.; Edwards, S.G. Phylogenetic analysis of EF-1 alpha gene sequences from isolates of *Microdochium nivale* leads to elevation of varieties Majus and Nivale to species status. *Fungal Biol.* **2005**, *109*, 872–880. [CrossRef]
12. Jewell, L.E.; Hsiang, T. Multigene differences between *Microdochium nivale* and *Microdochium majus*. *Botany* **2013**, *91*, 99–106. [CrossRef]

13. Domsch, K.H.; Gams, W. Fungi in agricultural soils. *J. Appl. Ecol.* **1972**, 160. [CrossRef]
14. Mouchacca, J.; Samson, R.A. Deux nouvelles espèces du genre *Microdochium* Sydow. *Rev. Mycol.* **1973**, *37*, 267–275.
15. Sutton, B.C. Eucalyptus microfungi: *Microdochium* and *Phaeoisaria* species from Brazil. *Nova Hedwig.* **1976**, *27*, 215–222.
16. Von, A.J.A. Notes on *Monographella* and *Microdochium*. *Trans. Br. Mycol. Soc.* **1984**, *83*, 373–374.
17. Harris, D.C. *Microdochium fusarioides* sp.nov. from oospores of *Phytophthora syringae*. *Trans. Br. Mycol. Soc.* **1985**, *84*, 358–361. [CrossRef]
18. Lombard, L.; Van, D.M.N.A.; Groenewald, J.Z.; Crous, P.W. Generic concepts in Nectriaceae. *Stud. Mycol.* **2015**, *80*, 189–245. [CrossRef]
19. Liang, J.M.; Li, G.S.; Zhao, M.Q.; Cai, L. A new leaf blight disease of turfgrasses caused by *Microdochium poae*, sp. nov. *Mycologia* **2019**, *111*, 265–273. [CrossRef]
20. Von Arx, J.A. Plant pathogenic fungi. *Nova Hedwig.* **1987**, *87*, 1–288.
21. Hiruma, K.; Kobae, Y.; Toju, H. Beneficial associations between Brassicaceae plants and fungal endophytes under nutrient-limiting conditions: Evolutionary origins and host–symbiont molecular mechanisms. *Curr. Opin. Plant Biol.* **2018**, *44*, 145–154. [CrossRef]
22. Mandyam, K.G.; Roe, J.; Jumpponen, A. Arabidopsis thaliana model system reveals a continuum of responses to root endophyte colonization. *Fungal Biol.* **2013**, *117*, 250–260. [CrossRef]
23. Mandyam, K.; Loughin, T.; Jumpponen, A. Isolation and morphological and metabolic characterization of common endophytes in annually burned tallgrass prairie. *Mycologia* **2010**, *102*, 813–821. [CrossRef] [PubMed]
24. Shadmani, L.; Jamali, S.; Fatemi, A. Biocontrol activity of endophytic fungus of barley, *Microdochium bolleyi*, against *Gaeumannomyces graminis* var. *tritici*. *Mycol. Iran.* **2018**, *5*, 7–14.
25. Chomnunti, P.; Hongsanan, S.; Aguirre-Hudson, B.; Tian, Q.; Peršoh, D.; Dhami, M.K.; Alias, A.S.; Xu, J.C.; Liu, X.Z.; Stadler, M. The sooty moulds. *Fungal Divers* **2014**, *66*, 1–36. [CrossRef]
26. Jiang, N.; Voglmayr, H.; Bian, D.R.; Piao, C.G.; Wang, S.K.; Li, Y. Morphology and Phylogeny of *Gnomoniopsis* (Gnomoniaceae, Diaporthales) from Fagaceae Leaves in China. *J. Fungi* **2021**, *7*, 792. [CrossRef] [PubMed]
27. Jiang, N.; Voglmayr, H.; Piao, C.G.; Li, Y. Two new species of *Diaporthe* (Diaporthaceae, Diaporthales) associated with tree cankers in the Netherlands. *MycoKeys* **2021**, *85*, 31–56. [CrossRef]
28. Cai, L.; Hyde, K.D.; Taylor, P.W.J.; Weir, B.; Waller, J.; Abang, M.M.; Zhang, Z.J.; Yang, Y.L.; Phoulivong, S.; Liu, Z.Y.; et al. A polyphasic approach for studying *Colletotrichum*. *Fungal Divers* **2009**, *39*, 183–204.
29. Crous, P.W.; Verkleij, G.J.M.; Groenewald, J.Z.; Houbraken, J. *Fungal Biodiversity*; CBS Laboratory Manuals Series 1; CBS-KNAW Fungal Biodiversity Centre Utrecht: Utrecht, The Netherlands, 2009.
30. Guo, L.D.; Hyde, K.D.; Liew, E.C.Y. Identification of endophytic fungi from *Livistona chinensis* based on morphology and rDNA sequences. *New Phytol.* **2000**, *147*, 617–630. [CrossRef]
31. Vilgalys, R.; Hester, M. Rapid genetic identification and mapping of enzymatically amplified ribosomal DNA from several *Cryptococcus* species. *J. Bacteriol.* **1990**, *172*, 4238–4246. [CrossRef]
32. White, T.J.; Bruns, T.; Lee, S.; Taylor, F.J.R.M.; Lee, S.H.; Taylor, L.; Shawe-Taylor, J. Amplification and direct sequencing of fungal ribosomal rna genes for phylogenetics. In *PCR Protocols: A Guide to Methods and Applications*; Innis, M.A., Gelfand, D.H., Sninsky, J.J., Eds.; Academic Press Inc.: New York, NY, USA, 1990; pp. 315–322. [CrossRef]
33. Liu, Y.J.; Whelen, S.; Hall, B.D. Phylogenetic Relationships among Ascomycetes: Evidence from an RNA polymerse II subunit. *Mol. Biol. Evol.* **1999**, *16*, 1799–1808. [CrossRef]
34. Sung, G.H.; Sung, J.M.; Hywel-Jones, N.L.; Spatafora, J.W. A multi-gene phylogeny of Clavicipitaceae (Ascomycota, Fungi): Identification of localized incongruence using a combinational bootstrap approach. *Mol. Phylogenet. Evol.* **2007**, *44*, 1204–1223. [CrossRef] [PubMed]
35. Nilsson, R.; Tedersoo, L.; Abarenkov, K.; Ryberg, M.; Kristiansson, E.; Hartmann, M.; Schoch, C.; Nylander, J.; Bergsten, J.; Porter, T.; et al. Five simple guidelines for establishing basic authenticity and reliability of newly generated fungal ITS sequences. *MycoKeys* **2012**, *4*, 37–63. [CrossRef]
36. Kumar, S.; Stecher, G.; Tamura, K. MEGA7: Molecular evolutionary genetics analysis version 7.0 for bigger datasets. *Mol. Phylogenet. Evol.* **2016**, *33*, 1870–1874. [CrossRef] [PubMed]
37. Katoh, K.; Rozewicki, J.; Yamada, K.D. MAFFT online service: Multiple sequence alignment, interactive sequence choice and visualization. *Brief. Bioinform.* **2017**, *20*, 1160–1166. [CrossRef]
38. Miller, M.A.; Pfeiffer, W.; Schwartz, T. The CIPRES science gateway: Enabling high-impact science for phylogenetics researchers with limited resources. In Proceedings of the 1st Conference of the Extreme Science and Engineering Discovery Environment. Bridging from the Extreme to the Campus and Beyond, Chicago, IL, USA, 16–20 July 2012; Association for Computing Machinery: San Diego, CA, USA, 2012; p. 8. [CrossRef]
39. Huelsenbeck, J.P.; Ronquist, F. MRBAYES: Bayesian inference of phylogeny. *Bioinformatics* **2001**, *17*, 754–755. [CrossRef]
40. Ronquist, F.; Huelsenbeck, J.P. MrBayes 3: Bayesian phylogenetic inference under mixed models. *Bioinformatics* **2003**, *19*, 1572–1574. [CrossRef]
41. Ronquist, F.; Teslenko, M.; Van, D.M.P.; Ayres, D.L.; Darling, A.; Höhna, S.; Larget, B.; Liu, L.; Suchard, M.A.; Huelsenbeck, J.P. MrBayes 3.2: Efficient Bayesian phylogenetic inference and model choice across a large model space. *Syst. Biol.* **2012**, *61*, 539–542. [CrossRef]

42. Stamatakis, A. RAxML Version 8: A tool for phylogenetic analysis and post-analysis of large phylogenies. *Bioinformatics* **2014**, *30*, 1312–1313. [CrossRef]
43. Nylander, J.A.A. *MrModelTest v. 2. Program Distributed by the Author*; Evolutionary Biology Centre, Uppsala University: Uppsala, Sweden, 2004.
44. Parkinson, V.O.; Sivanesan, A.; Booth, C. The perfect state of the rice leaf-scald fungus and the taxonomy of both the perfect and imperfect states. *Trans. Br. Mycol. Soc.* **1981**, *76*, 59–69. [CrossRef]
45. Samuels, G.J.; Hallett, I.C. *Microdochium stoveri* and *Monographella stoveri*, new combinations for Fusarium stoveri and Micronectriella stoveri. *Trans. Br. Mycol. Soc.* **1983**, *81*, 473–483. [CrossRef]
46. Jaklitsch, W.M.; Voglmayr, H. Phylogenetic relationships of five genera of Xylariales and *Rosasphaeria* gen. nov. (Hypocreales). *Fungal Divers* **2012**, *52*, 75–98. [CrossRef]
47. Hawksworth, D.L.; Crous, P.W.; Redhead, S.A.; Reynolds, D.R.; Samson, R.A.; Seifert, K.A.; Taylor, J.W.; Wingfield, M.J.; Abaci, O.; Aime, C.; et al. The Amsterdam declaration on fungal nomenclature. *IMA Fungus* **2011**, *2*, 105–111. [CrossRef] [PubMed]
48. Hsieh, H.M.; Ju, Y.M.; Rogers, J.D. Molecular phylogeny of *Hypoxylon* and closely related genera. *Mycologia* **2005**, *97*, 844–865. [CrossRef] [PubMed]
49. Læssøe, T.; Srikitikulchai, P.; Luangsa-ard, J.J.D.; Stadler, M. Theissenia reconsidered, including molecular phylogeny of the type species T *Pyrenocrata* and a new genus *Durotheca* (Xylariaceae, Ascomycota). *IMA Fungus* **2013**, *4*, 57–69. [CrossRef]

Journal of *Fungi*

MDPI

Article

Is *Hyperdermium* Congeneric with *Ascopolyporus*? Phylogenetic Relationships of *Ascopolyporus* spp. (*Cordycipitaceae*, *Hypocreales*) and a New Genus *Neohyperdermium* on Scale Insects in Thailand

Donnaya Thanakitpipattana [1,2], Suchada Mongkolsamrit [2], Artit Khonsanit [2], Winanda Himaman [3], Janet Jennifer Luangsa-ard [2,*] and Natapol Pornputtapong [1,*]

1 Department of Biochemistry and Microbiology, Center of Excellence for DNA Barcoding of Thai Medicinal Plants, Faculty of Pharmaceutical Sciences, Chulalongkorn University, Bangkok 10330, Thailand; 6176134833@student.chula.ac.th or donnaya.tha@biotec.or.th

2 Plant Microbe Interaction Research Team, National Center for Genetic Engineering and Biotechnology (BIOTEC), 113 Thailand Science Park, Phahonyothin Road, Khlong Nueng, Khlong Luang, Pathum Thani 12120, Thailand; suchada@biotec.or.th (S.M.); artit.kho@biotec.or.th (A.K.)

3 Forest Entomology and Microbiology Research Group, Forest and Plant Conservation Research Office, 61 Department of National Parks, Wildlife and Plant Conservation, Phahonyothin Road, Chatuchak, Bangkok 10900, Thailand; winandah@gmail.com

* Correspondence: jajen@biotec.or.th (J.J.L.-a.); natapol.p@chula.ac.th (N.P.); Tel.: +66-2-564-6700 (J.J.L.-a.); +66-2-218-8382 (N.P.)

Abstract: During surveys of insect pathogenic fungi (IPF) in Thailand, fungi associated with scale insects and plants were found to represent five new species of the genus *Ascopolyporus* in *Cordycipitaceae*. Their macroscopic features resembled both *Hyperdermium* and *Ascopolyporus*. Morphological comparisons with the type and known *Ascopolyporus* and *Hyperdermium* species and phylogenetic evidence from a multigene dataset support the appointment of a new species of *Ascopolyporus*. Moreover, the data also revealed that the type species of *Hyperdermium*, *H. caulium*, is nested within *Ascopolyporus*, suggesting that *Hyperdermium* is congeneric with *Ascopolyporus*. The specimens investigated here differ from other *Ascopolyporus* species by phenotypic characters including size and color of stromata. Phylogenetic analyses of combined LSU, *TEF1*, *RPB1* and *RPB2* sequences strongly support the notion that these strains are distinct from known species of *Ascopolyporus*, and are proposed as *Ascopolyporus albus*, *A. galloides*, *A. griseoperitheciatus*, *A. khaoyaiensis* and *A. purpuratus*. *Neohyperdermium* gen. nov. is introduced for other species originally assigned to *Hyperdermium* and *Cordyceps* occurring on scale insects and host plants as epiphytes, accommodating two new combinations of *Hyperdermium pulvinatum* and *Cordyceps piperis*.

Keywords: *Cordycipitaceae*; epiphyte; insect pathogenic fungi; multigene phylogeny; scale insects

Citation: Thanakitpipattana, D.; Mongkolsamrit, S.; Khonsanit, A.; Himaman, W.; Luangsa-ard, J.J.; Pornputtapong, N. Is *Hyperdermium* Congeneric with *Ascopolyporus*? Phylogenetic Relationships of *Ascopolyporus* spp. (*Cordycipitaceae*, *Hypocreales*) and a New Genus *Neohyperdermium* on Scale Insects in Thailand. *J. Fungi* **2022**, *8*, 516. https://doi.org/10.3390/jof8050516

Academic Editors: Cheng Gao and Lei Cai

Received: 6 April 2022
Accepted: 13 May 2022
Published: 17 May 2022

1. Introduction

Scale insects are a diverse group of sap-sucking insects in the superfamily *Coccoidea* of the order *Hemiptera*, associated with aphids (*Aphidoidea*) and whiteflies (*Aleyrodoidea*) [1,2]. These insects cause damage by sucking fluids from leaves, stems and other parts of host plants and excrete honeydew that favors sooty mold growth, which consequently decreases photosynthetic rates. They belong to seven families: *Antennulariellaceae*, *Capnodiaceae*, *Chaetothyriaceae*, *Coccodiniaceae*, *Euantennariaceae*, *Metacapnodiaceae* and *Trichomeriaceae* [3–5]. In addition, many groups of fungi are known to grow on various scale insects by covering the whole surface of the insect body and can be found in the phyla (a) *Basidiomycota*: *Septobasidiales* (*Septobasidium* and *Uredinella*), (b) *Chytridiomycota*: *Blastocladiales* (*Myiophagus*) and (c) *Ascomycota*: *Myriangiales* (*Myriangium*), *Pleosporales* (*Podonectria*), and especially in a large group of entomopathogens in the *Hypocreales* [6–9].

Hypocrealean fungi associated with armored (*Diaspididae*) and soft-scale insects (*Coccidae*) can be found in various genera within five families: (1) *Bionectriaceae* viz. *Clonostachys* Corda; (2) *Nectriaceae* viz. *Microcera* Desm. and *Fusarium* Link; (3) *Cordycipitaceae* viz. *Ascopolyporus* Möller, *Cordyceps* Fr. and *Hyperdermium* J.F. White, R.F. Sullivan, Bills and Hywel-Jones; (4) *Ophiocordycipitaceae* viz. *Ophiocordyceps* Petch; and (5) *Clavicipitaceae* viz. *Aschersonia* Mont., *Conoideocrella* D. Johnson, G.H. Sung, Hywel-Jones and Spatafora, *Dussiella* Pat., *Helicocollum* Luangsa-ard, Mongkols., Noisrip. and Thanakitp., *Hypocrella* Sacc., *Regiocrella* P. Chaverri and K.T. Hodge, *Orbiocrella* D. Johnson, G.H. Sung, Hywel-Jones and Spatafora and *Samuelsia* P. Chaverri and K.T. Hodge [9–21]. Among them, the most abundant and widespread members are found in *Clavicipitaceae* and *Cordycipitaceae*. The macromorphological characters of these genera in nature can be easily distinguished in each family. Scale insect pathogenic genera in *Clavicipitaceae,* such as *Conoideocrella, Hypocrella, Moelleriella* and *Orbiocrella*, possess diverse morphological characters, such as the formation of hard stromata, pulvinate, subglobose or hemispherical and ring-like stromata, as well as the presence of only superficial, cone-shaped perithecia, while in *Cordycipitaceae,* most have pulvinate, subglobose, hemispherical, soft stromata with crowded perithecia. Two different colors are found the upper and lower surface of stromata in some species of *Ascopolyporus* and *Hyperdermium* [11,12,14,16].

Ascopolyporus is an epiphytic fungal genus in *Cordycipitaceae* that produces stromata on the stems of living plants as biotrophs and infects scale insects as necrotrophs comprising only seven species [22]. *Ascopolyporus* species are commonly found in tropical forests where bamboo is present [23]. The type species of *Ascopolyporus, A. polychrous,* is a pathogen of bamboo scale insects that produces up to 4 cm large subglobose to polypore-like, bright rusty-red or white to yellow perithecial stromata, which are usually fertile only on the underside of the stroma [12,13,24]. In 2005, a new species of *Ascopolyporus, A. philodendrus,* was described by Bischoff et al. [14] on bamboo scale insects, and a new description for *A. villosus* was made. They considered that the morphology of perithecial stromata and the conidial states of *Ascopolyporus* resemble the scale insect pathogenic genus *Hyperdermium,* especially its type species, *H. caulium* [11,14]. Both of these species in the two genera share similar morphological characters, having large stromata, immersed perithecia, filiform ascospores and phialidic conidiogenous cells. The anamorph state is referred to as cylindrocarpon-like phialides, characterized by producing multiseptate conidia, a unique character in the *Cordycipitaceae.* Moreover, a species of *Cordyceps, C. piperis,* is also capable of parasitizing scale insects but differ by producing verticillium-like anamorph with aseptate conidia [11,12].

During our continuous survey of insect pathogenic fungi (IPF) in national parks and community forests in Thailand, we encountered hyperdermium-like specimens with differences in phenotypic characters including colors and sizes of stromata. These morphologically diverse specimens were preliminarily identified as members of the genus *Hyperdermium* and *Ascopolyporus.* The aims of this study are thus (1) to determine the phylogenetic relationship of these two genera and (2) to identify and describe new species of hyperdermium-like fungi on scale insects from Thailand by combining morphological characteristics and reconstructing their phylogeny based on sequence data of LSU, *TEF1, RPB1* and *RPB2* loci.

2. Materials and Methods

2.1. Collection and Isolation

The 63 epiphytic isolates in this study were collected from various localities in Thailand since June 1992, representing the first recorded collection from Khao Yai National Park, Nakhon Ratchasima Province. Thereafter, these specimens have been found throughout every region in Thailand, albeit not frequently, including the Ban Hua Thung community forest in Chiang Mai Province; Chiang Dao, Khao Soi Dao and Khlong Nakha wildlife sanctuaries; Kaeng Krachan and Khlong Lan national parks; and the Khao Chong wildlife development and conservation promotion station. Specimens were examined for fungal colonization from the stems and leaves of monocotyledonous and dicotyledonous plants. The specimens were collected and stored in plastic boxes before returning to the laboratory for isolation. Pure cultures were made from the isolation of the sexual morph following Luangsa-ard et al. [25]. The cultures and the voucher specimens were deposited in Thailand Bioresource Research Center (TBRC) and BIOTEC Bangkok Herbarium (BBH), Thailand, respectively.

2.2. Morphological Study

For obtaining morphological descriptions, all isolates were cultured on oatmeal agar (OA: oatmeal 60 g, agar 12.5 g, in 1 L distilled water, Difco) and potato dextrose agar (PDA: potato 200 g, dextrose 20 g, agar 15 g, in 1 L distilled water) for 14–20 days. Colony morphology was examined for color, size, shape and appearance. Fungal structures of teleomorph and anamorph states were mounted in lactophenol cotton blue solution, and their characters were investigated by light microscopy, as described by Mongkolsamrit et al. [26] and Khonsanit et al. [27]. Sections of the stroma on stems were prepared by using a freezing microtome (Slee Cryostat MEV, Mainz, Germany), and mounted in distilled water and in lactophenol cotton blue solution [28]. The Sixth Royal Horticultural Society (RHS) color chart was used to characterize the colors of fresh specimens and cultures [29]. Twenty to fifty individual length and width measurements were taken, and the amount of variability is provided as average $\pm$ standard deviation with absolute minima and maxima in parentheses.

2.3. DNA Extraction, PCR and Sequencing

The mycelial mass of fungi was obtained from cultures grown on PDA for 7 days at 25 °C. A modified CTAB protocol used for DNA extraction using polyvinylpyrrolidone instead of β-mercaptoethanol in CTAB buffer and increasing temperature in the incubation process from 60 °C to 65 °C was previously described by Thanakitpipattana et al. [30]. PCR was used to amplify the nuclear ribosomal large subunits (LSU), the region of the elongation factor 1-α (*TEF1*), and the largest and second-largest subunits of RNA polymerase II (*RPB1* and *RPB2*). The reaction mix was prepared in 25 μL volumes containing 1× Dream Taq Buffer (with included 20 mM $MgCl_2$), 0.4 M betaine, 200 μM dNTP mix, 0.5 μM of each primer, 1 Unit Dream Taq DNA polymerase (Thermo Scientific, Waltham, MA, USA), 50 ng of DNA template and Milli-Q water. PCR amplifications of four loci were carried out with the following primers: LROR and LR5 for LSU [31,32], 983F and 2218R for *TEF1* [33], CRPB1 and RPB1-Cr for *RPB1* [34], and RPB2-5F2 and RPB2-7Cr for RPB2 [35,36]. The PCR conditions were performed as follows: 94 °C for 3 min, followed by 35 cycles of denaturation at 94 °C for 1 min, annealing at a suitable temperature for 1 min, extension at 72 °C for 1 min and a final extension of 72 °C for 10 min. The annealing temperature of each gene was 50 °C for *RPB1* and *RPB2*, and 55 °C for *TEF1* and LSU. PCR products were purified and subsequently sequenced with PCR amplification primers.

2.4. Sequence Alignment and Phylogenetic Analyses

The newly generated sequences from the twelve strains in this study were assembled using BioEdit v. 7.2.5 [37] and then deposited in the GenBank database under the accession numbers of *TEF1* (OL322029-OL322040), LSU (OL322041-OL322052), *RPB1* (OL322053-OL322059) and *RPB2* (OL322060-OL322070) (Table 1). Sequences of each locus were aligned using MUSCLE 3.6 [38] together with other sequences of related taxa from previous studies for phylogenetic analyses (see Table 1), and manually refined to minimize gaps. The concatenated (LSU + *TEF1* + *RPB1* + *RPB2*) sequences were analyzed by maximum likelihood (ML) and Bayesian inference (BI), both on the CIPRES Science Gateway portal [39]. Maximum likelihood analysis was performed with RAxML-HPC2 on XSEDE v.8.2.12 with default parameters [40] using the GTRCAT substitution model with 1000 rapid bootstrap replicates. The program MrModeltest v.2.2 [41] was used to determine the model of evolution under the Akaike Information Criterion (AIC) implemented in PAUP v.4.0a169 [42], which selected SYM + G as the best nucleotide substitution model. The BI analysis was performed using MrBayes on XSEDE v.3.2.7a with default parameters [43]. The Markov Chain Monte Carlo (MCMC) searches were run for 5,000,000 generations with sampling every 1000 generations and a burn-in value of 10%. Nodes were considered as strongly supported with bootstrap and posterior probability values greater than 70% and 0.7, respectively.

Table 1. List of species and GenBank accession numbers of sequences used in this study. Bold accession numbers were generated for this study. The symbol "–" denotes no available data.

Species	Strain	GenBank Accession No.				References
		LSU	*TEF1*	*RPB1*	*RPB2*	
Akanthomyces aculeatus	HUA 186145	MF416520	MF416465	–	–	[44]
Akanthomyces attenuatus	CBS 402.78	AF339565	EF468782	EF468888	EF468935	[45]
Akanthomyces lecanii	CBS 101247	AF339555	DQ522359	DQ522407	DQ522466	[45]
Akanthomyces neoaraneogenus	GZU1032Lea	KX845704	KX845698	KX845700	KX845702	[46]
Akanthomyces sabanense	ANDES-F 1024	KC875225	KC633266	–	KC633249	[47]
Akanthomyces tuberculatus	OSC 111002	DQ518767	DQ522338	DQ522384	DQ522435	[45]
Ascopolyporus albus	**BCC48975**	**OL322048**	**OL322035**	**OL322056**	**OL322065**	**This study**
Ascopolyporus albus	**BCC48976**	**OL322049**	**OL322036**	**OL322057**	**OL322066**	**This study**
Ascopolyporus galloides	**BCC25446**	**OL322042**	**OL322029**	**OL322053**	**OL322060**	**This study**
Ascopolyporus galloides	**BCC47981**	**OL322043**	**OL322030**	**OL322054**	**OL322061**	**This study**
Ascopolyporus galloides	**BCC48704**	**OL322044**	**OL322031**	**OL322055**	**OL322062**	**This study**
Ascopolyporus griseoperitheciatus	**BCC22358**	**OL322050**	**OL322037**	–	**OL322067**	**This study**
Ascopolyporus griseoperitheciatus	**BCC25788**	**OL322051**	**OL322038**	**OL322058**	**OL322068**	**This study**
Ascopolyporus khaoyaiensis	**BCC43314**	**OL322052**	**OL322039**	–	**OL322069**	**This study**
Ascopolyporus khaoyaiensis	**BCC43741**	**OL322041**	**OL322040**	–	**OL322070**	**This study**
Ascopolyporus polychrous	**P.C. 546**	**DQ118737**	**DQ118745**	**DQ127236**	–	[15]
Ascopolyporus purpuratus	**BCC88388**	**OL322046**	**OL322033**	–	**OL322064**	**This study**
Ascopolyporus purpuratus	**BCC88389**	**OL322047**	**OL322034**	–	–	**This study**
Ascopolyporus purpuratus	**BCC88430**	**OL322045**	**OL322032**	**OL322059**	**OL322063**	**This study**
Ascopolyporus villosus	ARSEF 6355	AY886544	DQ118750	DQ127241	–	[14,15]
Beauveria bassiana	ARSEF 300	–	AY531924	HQ880831	HQ880903	[33,48]
Beauveria kipukae	ARSEF 7032	–	HQ881005	HQ880875	HQ880947	[48]
Beauveria staphylinidicola	ARSEF 5718	EF468836	EF468776	EF468881	–	[45]
Beauveria varroae	ARSEF 2694	–	HQ881004	HQ880874	HQ880946	[48]
Blackwellomyces cardinalis	OSC 93609	AY184962	DQ522325	DQ522370	DQ522422	[49,50]
Blackwellomyces cardinalis	OSC 93610	AY184963	EF469059	EF469088	EF469106	[45,50]
Blackwellomyces pseudomilitaris	BCC 1919	MF416534	MF416478	–	MF416440	[44]
Blackwellomyces pseudomilitaris	BCC 2091	MF416535	MF416479	–	MF416441	[44]
Cordyceps bifusispora	EFCC 5690	EF468806	EF468746	EF468854	EF468909	[45]
Cordyceps lepidopterorum	TBRC 7263	MF140699	MF140819	MF140768	MF140792	[51]
Cordyceps piperis	CBS 116719	AY466442	DQ118749	DQ127240	EU369083	[15,17]
Cordyceps takaomontana	BCC28612	FJ765252	FJ765268	–	–	[52]
Cordyceps tenuipes	ARSEF 5135	JF415980	JF416020	JN049896	JF416000	[53]
Engyodontium aranearum	CBS 309.85	AF339526	DQ522341	DQ522387	DQ522439	[49]
Gibellula leiopus	BCC 16025	MF416548	MF416492	MF416649	–	[44]
Gibellula pulchra	NHJ 10808	EU369035	EU369018	EU369056	EU369076	[17]
Gibellula sp.	NHJ 10788	EU369036	EU369019	EU369058	EU369078	[17]
Gibellula sp.	NHJ 13158	EU369037	EU369020	EU369057	EU369077	[17]
Hevansia arachnophila	NHJ 10469	EU369031	EU369008	EU369047	–	[17]
Hevansia cinerea	NHJ 3510	–	EU369009	EU369048	EU369070	[17]
Hevansia nelumboides	BCC 2093	MF416530	MF416473	–	MF416437	[44]
Hevansia novoguineensis	NHJ 4314	–	EU369012	EU369051	EU369071	[17]
Hyperdermium caulium	AF242354	AF242354	–	–	–	[11]
Hyperdermium pulvinatum	P.C. 602	DQ118738	DQ118746	DQ127237	–	[15]
Lecanicillium antillanum	CBS 350.85	AF339536	DQ522350	DQ522396	DQ522450	[45]
Lecanicillium psalliotae	CBS 363.86	AF339559	EF468784	EF468890	–	[45]

Table 1. *Cont.*

Species	Strain	GenBank Accession No.				References
		LSU	*TEF1*	*RPB1*	*RPB2*	
Lecanicillium psalliotae	CBS 532.81	AF339560	EF469067	EF469096	EF469112	[45]
Neotorrubiella chinghridicola	BCC 39684	MK632096	MK632148	MK632071	MK632181	[30]
Neotorrubiella chinghridicola	BCC 80733	MK632097	MK632149	MK632072	MK632176	[30]
Samsoniella aurantia	TBRC 7271	MF140728	MF140846	MF140791	–	[51]
Samsoniella aurantia	TBRC 7273	MF140726	MF140844	–	MF140816	[51]
Samsoniella inthanonensis	TBRC 7915	MF140725	MF140849	MF140790	MF140815	[51]
Samsoniella inthanonensis	TBRC 7916	MF140724	MF140848	MF140789	MF140814	[51]
Outgroup						
Flavocillium bifurcatum	YFCC 6101	MN576781	MN576951	MN576841	MN576897	[54]
Lecanicillium sp.	CBS 639.85	KM283801	KM283824	KM283843	KM283865	[54]

3. Results

3.1. Molecular Phylogeny

The combined four-gene dataset of 54 taxa consisted of 3404 bp (LSU 861 bp, *TEF1* 954 bp, *RPB1* 730 bp, *RPB2* 859 bp). *Flavocillium bifurcatum* and *Flavocillium* sp. in *Cordycipitaceae* were used as an outgroups. Phylogenetic tree topology obtained from ML was similar to the BI analysis. Therefore, only the ML tree is shown (Figure 1). Multigene phylogenetic analyses revealed that the sequenced strains comprise five novel species and are nested with the type and other species of *Ascopolyporus*, *A. polychrous* and *A. villosus*, as well as type species of *Hyperdermium*, *H. caulium*, within the *Ascopolyporus* clade, with strong support (81% ML bootstrap (MLBS) and 0.99 BI posterior probability (BIPP)), as shown in Figure 1. The type species *H. caulium* is clustered within this clade, suggesting that *Hyperdermium* is congeneric with *Ascopolyporus*, although with low internal bootstrap support because only LSU sequence data are available (<50 MLBS and <0.5 BIPP, data not shown).

Three of our new species are found in the pulvinate subclade showing irregularly subglobose to globose stromata, namely, *Ascopolyporus albus*, *A. galloides* and *A. griseoperitheciatus*, with 93% MLBS and 0.97 BIPP. Another subclade comprises both flattened and pulvinate stromata of two new and known species, including *Ascopolyporus khaoyaiensis*, *A. purpuratus*, *A. polychrous*, *A. villosus* and *H. caulium* (Figure 1). The *Ascopolyporus* clade is sister to the *Blackwellomyces* clade, which produces similar types of phialides and conidial arrangement as well as acremonium-like or lecanicillium-like anamorphs.

The position of *Hyperdermium pulvinatum* and *Cordyceps piperis*, on the other hand, is clearly distant from the *Ascopolyporus* clade, and these two species always clustered together separate from the type species of *Hyperdermium*, *H. caulium*. These two species form a basal clade to *Akanthomyces*, *Samsoniella*, *Beauveria* and *Cordyceps*. Based on their multigene phylogenetic position presented in this study, we propose to transfer these two species to the genus *Neohyperdermium*.

3.2. Taxonomy

Ascopolyporus Möller **emend**. Thanakitpipattana and Luangsa-ard.

Stromatal mass exceeding scale insect host. *Stroma* bulbous (lumpy or tuberous) or ungulate, flattened or pulvinate, fleshy or gall-like, polypore-like, white, yellowish white, purple to orange; sterile surface and fertile underneath the stroma. *Perithecia* semi-immersed to immersed, ovate to obclavate or cone-shaped. *Asci* hyaline, filiform. *Ascospores* hyaline, whole with septation or aseptate. Conidiogenous cells phialidic, solitary, slightly curved. *Conidia* hyaline, fusiform to subcylindrical, acerose, aseptate or 1–5 septate when mature, in chains or in sticky heads.

Typification: Ascopolyporus polychrous.

Habit and type host: On dead culms of bamboo, stems or leaf midrib of monocotyledonous and dicotyledonous plants.

Distribution: Argentina, Bolivia, Brazil, Colombia, Costa Rica, Ecuador, Peru, Thailand [23].

Ascopolyporus albus Mongkolsamrit, Thanakitpipattana and Luangsa-ard **sp. nov.** Figure 2.

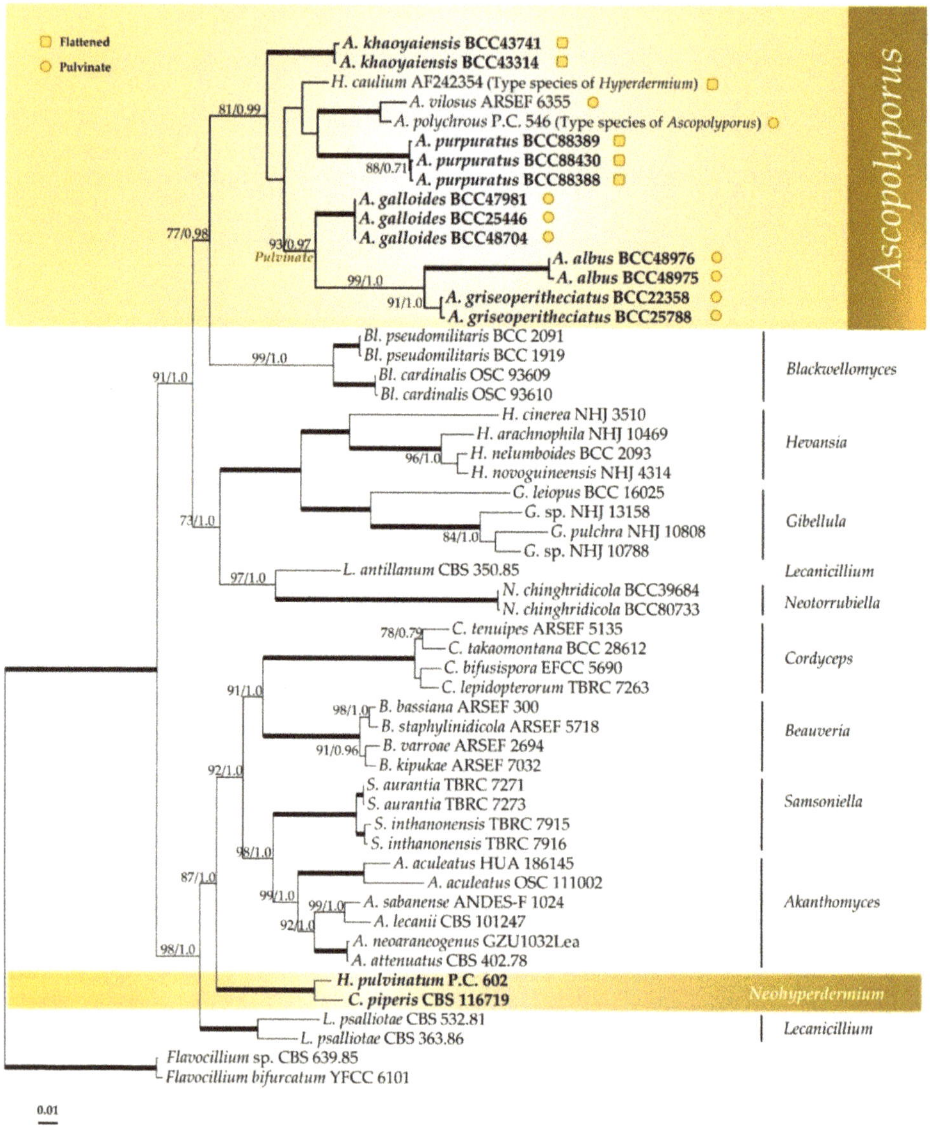

Figure 1. Phylogenetic reconstruction of *Ascopolyporus* and related genera in the *Cordycipitaceae* obtained from the combined LSU, *TEF1*, *RPB1* and *RPB2* sequence dataset based on maximum likelihood (RAxML) and Bayesian inference. Numbers on the nodes are ML bootstrap and Bayesian posterior probability values above 70% (MLBS) or 0.7 (BIPP). Thickened lines mean support for the two analyses was 100% (MLBS) or 1.0 (BIPP). ○ represents species with pulvinate stromata while □ represents species with flattened stromata.

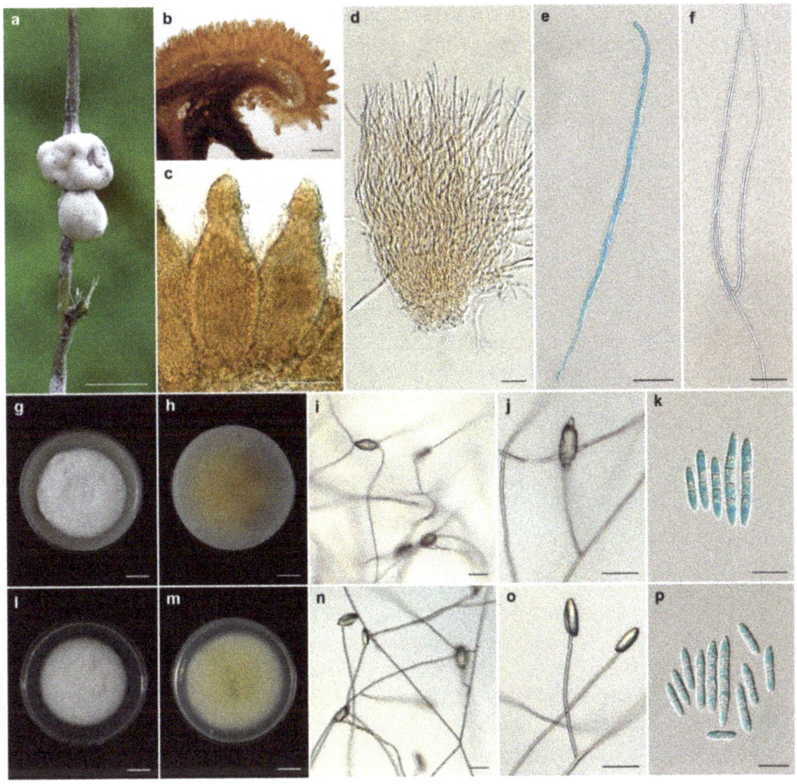

Figure 2. *Ascopolyporus albus*. (**a**) Stromata on living stem of bamboo (*Bambusae*); (**b**) cross-section through stroma showing perithecia (BBH30734); (**c**) perithecia; (**d**,**e**) asci. (**f**) ascospores; (**g**) colony obverse on OA; (**h**) colony reverse on OA; (**i**,**j**) phialide apex with conidial head on OA; (**k**) conidia on OA; (**l**) colony obverse on PDA; (**m**) colony reverse on PDA; (**n**,**o**) phialide and conidia on PDA; (**p**) conidia on PDA. Scale bars: (**g**,**h**,**l**,**m**) = 10 mm; (**a**) = 5 mm; (**b**) = 200 μm; (**c**) = 100 μm; (**d**,**f**,**j**,**n**,**o**) = 20 μm; (**e**,**k**,**p**) = 10 μm.

MycoBank: MB 841855.

Etymology: From the Latin "albus", referring to the white color of the fresh stromata.

Typification: Thailand, Chiang Mai Province, Chiang Dao Wildlife Sanctuary, Doi Chiang Dao Wildlife Research Station; 19°23′10.70″ N, 98°50′28.50″ E, on scale insects (*Coccidae; Hemiptera*), on living stem of bamboo (*Bambusae*), 17 August 2011, K. Tasanathai (K.T.), P. Srikitikulchai (P.S.), S. Mongkolsamrit (S.M.), A. Khonsanit (A.K.) (holotype BBH30734, ex-holotype culture BCC48975). GenBank: ITS = OL331502, LSU = OL322048, *TEF1* = OL322035, *RPB1* = OL322056, *RPB2* = OL322065.

Description: *Stromata* epibiotic, pulvinate, subglobose, globose, white (NN155C) to pinkish white (N155B), becoming dark brown when old, 3–6 mm wide, 2–3 mm thick. *Perithecia* semi-immersed, with slightly protruding orifices, obpyriform, 250–320 × 100–120 μm. *Asci* cylindrical up to 250 μm long and 3–4 μm wide, *Asci caps* 2–3 × 3–4 μm. *Ascospores* hyaline, whole, filiform, multiseptate, 95–135 × 1 μm.

Culture characteristics: Colonies on OA attaining a diameter of 3.5–4 cm in 14 days, slightly convex to the agar surface, white, reverse moderate orange yellow (164C). *Phialides* arising from aerial hyphae, solitary, cylindrical, slightly curved, up to 80 μm long, 1–2 μm wide. *Conidia* hyaline, enteroblastic, fusiform to acerose, early in development aseptate, becoming 1–4 septa, aggregated at the apex of the phialides, (8–)10–23(–28) × (2–)2.5–3 μm.

Colonies on PDA attaining a diameter of 3.5–4 cm in 14 days, slightly convex to the agar surface, white, reverse light yellow (162C). *Phialides* arising from aerial hyphae, solitary, cylindrical, slightly curved, up to 120 μm long, 1–2 μm wide. *Conidia* hyaline, enteroblastic, fusiform to acerose, early in development aseptate, developing 1–4 septa, aggregated at the apex of the phialides, (8–)10.5–21(–30) × (2–)2.5–3(–3.5) μm.

Habitat: On scale insects (*Coccidae; Hemiptera*), found on living stems of bamboo (*Bambusae*).

Additional specimen examined: Thailand, Chiang Mai Province, Chiang Dao Wildlife Sanctuary, Doi Chiang Dao Wildlife Research Station; 19°23′10.70″ N, 98°50′28.50″ E, on scale insects (*Coccidae; Hemiptera*), on the living stems, 17 August 2011, K.T., P.S., S.M., A.K. (BBH30734, BCC48976). GenBank: ITS = OL331503, LSU = OL322049, *TEF1* = OL322036, *RPB1* = OL322057, *RPB2* = OL322066.

Notes: *Ascopolyporus albus* significantly differs from other species in *Ascopolyporus* herein. The difference is in the color of stromata. *Ascopolyporus albus* produces white to pinkish white stromata (Figure 2), whereas other species produce very pale violet (91D) to yellowish white (158) with strong orange (25A) stromata. Based on *Ascopolyporus* species from Thailand, the perithecia of *A. albus* are semi-immersed, similar to those in *A. galloides*, *A. griseoperitheciatus* and *A. purpuratus*. The perithecial shape of *A. albus* differs from *A. galloides*, *A. griseoperitheciatus* and *A. purpuratus* by having an obpyriform shape, whereas perithecia in *A. galloides*, *A. griseoperitheciatus* and *A. purpuratus* are obclavate, obovoid and ovoid, respectively.

Ascopolyporus caulium (Berk. and M.A. Curtis) Thanakitp. and Luangsa-ard, **comb. nov.**

MycoBank: MB 842779.

≡ *Corticium caulium* Berk. and M.A. Curtis, J. Acad. nat. Sci. Philad. 2: 279. 1854.

≡ *Hypocrella caulium* (Berk. and M.A. Curtis) Pat., Bull. Soc. Mycol. France 30: 346. 1915.

≡ *Hyperdermium caulium* (Berk. and M.A. Curtis) P. Chaverri and K.T. Hodge, 2008.

= *Hypocrella camerunensis* Henn., Engler's Bot. Jahrb. 23: 540. 1897.

= *Hypocrella brasiliana* (Henn.) Mains, Mycopath. Myc. Appl. 11: 311. 1959.

≡ *Stigmatea brasiliana* Henn., Hedwigia 36: 230. 1897.

≡ *Hypocrella camerunensis* var. *brasiliana* Henn., Hedwigia 43: 85. 1904.

= *Hyperdermium bertonii* J.F. White, R.F. Sullivan, Bills and Hywel-Jones, Mycologia 92: 910. 2000.

≡ *Epichloë bertonii* Speg., An. Mus. Nac. Hist. Nat. Buenos Aires 31: 416. 1922.

Ascopolyporus galloides Khonsanit, Thanakitpipattana and Luangsa-ard **sp. nov.** Figure 3.

MycoBank: MB 841853.

Etymology: Refers to the character of the stromata, which look similar to plant galls.

Typification: Thailand, Nakhon Ratchasima Province, Khao Yai National Park, Mo Singto Nature Trail; 14°26′21.46″ N, 101°22′20.20″ E, on scale insects (*Coccidae; Hemiptera*), on the living stems of dicotyledonous plant, 5 July 2011, A.K., K.T., K. Sansatchanon (K.S.), P.S., S.M., W. Noisripoom (W.N.) (holotype BBH30629, ex-holotype culture BCC48704). GenBank: ITS = OL331509, LSU = OL322044, *TEF1* = OL322031, *RPB1* = OL322055, *RPB2* = OL322062.

Description: *Stromata* epibiotic, pulvinate, subglobose, hemispherical, upper surface white (NN155B) to yellowish white (158); lower surface strong orange (N25C), 1–7 mm wide. *Perithecia* semi-immersed, crowded, obclavate, 170–340 × 60–110 μm. *Asci* cylindrical, (129–)133–153.5(–175) × (3–)3.5–5(–6) μm. *Asci caps* 1–2 × 2.5–4 μm. *Ascospores* hyaline, whole, filiform, aseptate, (131–) 154.5–211.5(–216) × 0.5 μm.

Culture characteristics: Colony on OA attaining a diameter of 3.5–4 cm in 20 days, flat, slightly convex to the agar surface, white (158), reverse pale greenish yellow (2D). *Phialides* arising from aerial hyphae, solitary, cylindrical, slightly curved, 35–161 × 1–2 μm. *Conidia* hyaline, enteroblastic, fusiform to acerose, early in development aseptate, becoming 1–3 septa, aggregated at the apex of the phialides, (4–)6–22(–34) × (1.5–)2–2.5(–3) μm.

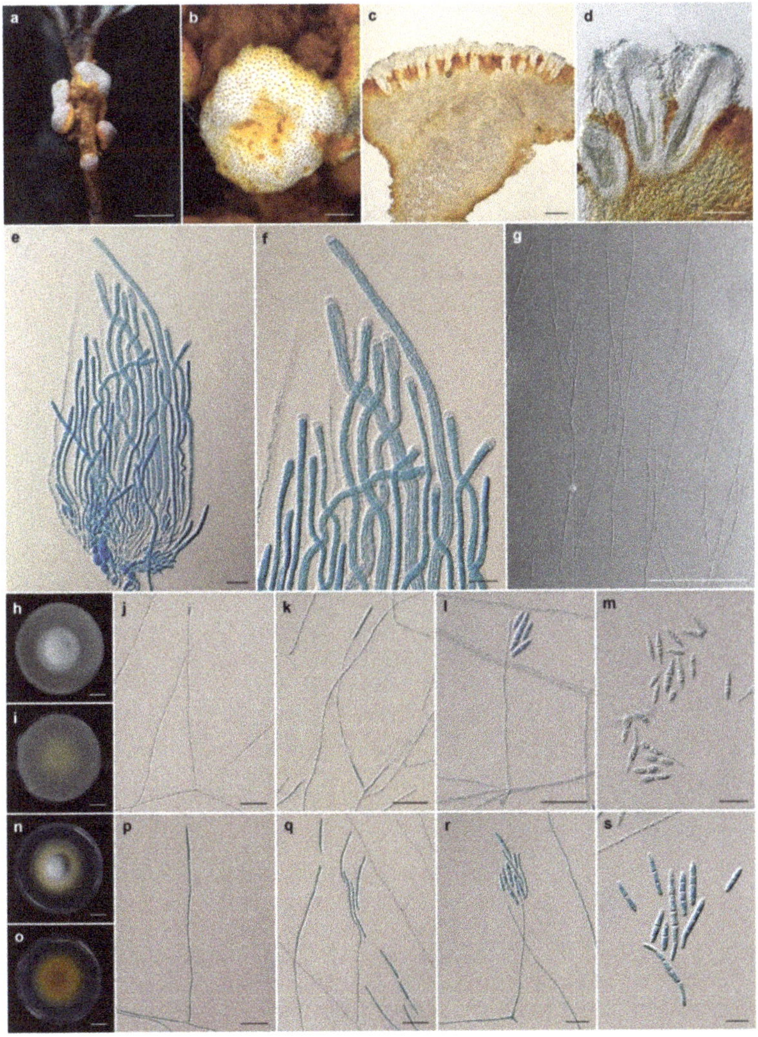

Figure 3. *Ascopolyporus galloides*. (**a**,**b**) Stromata on living stem of dicotyledonous plant (BBH48704); (**c**) cross-section through stroma showing perithecia; (**d**) perithecia; (**e**) asci; (**f**) asci-caps; (**g**) ascospores; (**h**) colony obverse on OA; (**i**) colony reverse on OA; (**j**–**l**) phialide apex with conidial head on OA; (**m**) conidia on OA; (**n**) colony obverse on PDA; (**o**) colony reverse on PDA; (**p**–**r**) phialide apex with conidial head on PDA; (**s**) conidia on PDA. Scale bars: (**h**,**i**,**n**,**o**) = 10 mm; (**a**) = 5 mm; (**b**) = 1 mm; (**c**) = 200 μm; (**d**,**g**) = 100 μm; (**j**,**k**,**l**,**q**,**r**) = 20 μm; (**e**,**f**,**m**,**p**,**s**) = 10 μm.

Colony on PDA attaining a diameter of 4 cm in 20 days, fluffy in the middle, flat to umbonate, white in the middle, pale orange yellow (23D), light yellow (11B), brilliant yellow (11A), reverse moderate orange (173C) in the middle, strong orange yellow (N163D), light yellow (14D). *Phialides* arising from aerial hyphae, solitary, cylindrical, slightly curved, 30–294 × 1–2 μm. *Conidia* hyaline, enteroblastic, fusiform to acerose, cylindrical, early in development aseptate, becoming 1–4 septa, aggregated at the apex of the phialides, (5–)8–16(–27) × (2–)2.5–3.5(–4) μm.

Habitat: On scale insects (*Coccidae*, *Hemiptera*), found on living stems of dicotyledonous plant.

Additional specimens examined: Thailand, Ranong Province, Khlong Nakha Wildlife Sanctuary, Khlong Nakha Nature Trail; 9°27′33″ N, 98°30′16″ E, on scale insects (*Coccidae; Hemiptera*), on living stems of dicotyledonous plant, 5 October 2004, B. Thongnuch (B.T.), D. Johnson (D.J.), K.T., S.M., W. Chaygate (W.C.) (BBH10163, BCC16408; BBH10176, BCC16419; BBH10177, BCC16420; BBH10178, BCC16421; BBH10179, BCC16422; BBH10180, BCC16423); Nakhon Nayok Province, Khao Yai National Park, Tat Ta Phu Waterfall Nature Trail; 14°26′21.46″ N, 101°22′20.20″ E, on scale insects (*Coccidae; Hemiptera*), on living stems of dicotyledonous plant, 24 August 2005, K.T. (BBH14835, BCC18980); Phetchaburi Province, Kaeng Krachan National Park, Ban Krang Camp Nature Trail; 12°54′05″ N, 99°37′48″ E, on scale insects (*Coccidae; Hemiptera*), on living stems of dicotyledonous plant, 14 November 2005, B.T., K.T., R. Ridkaew (R.R.), W.C. (BBH15034, BCC19720); Ranong Province, Khlong Nakha Wildlife Sanctuary, Khlong Nakha Nature Trail; 9°27′33″ N, 98°30′16″ E, on scale insects (*Coccidae; Hemiptera*), on living stems of dicotyledonous plant, 10 January 2006, B.T., K.T., L.N. Yen (L.N.Y.), L.T. Huyen (L.T.H.), PS, SM, WC (BBH16500, BCC20115); 12 January 2006, B.T., K.T., L.N.Y., L.T.H., P.S., S.M., W.C. (BBH16554, BCC20123); Nakhon Ratchasima Province, Khao Yai National Park, Bueng Phai Nature Trail; 14°26′21.46″ N, 101°22′20.20″ E, on scale insects (*Coccidae; Hemiptera*), on living stems of dicotyledonous plant, 5 July 2006, B.T., J. Luangsa-ard (J.J.L.), K.T., P.S., S.M., W.C. (BBH18631, BCC22237; BBH18632, BCC22238); Chanthaburi Province, Khao Soi Dao Wildlife Sanctuary, Withiphrai Nature Trail; 13°06′13″ N, 102°11′39″ E, on scale insects (*Coccidae; Hemiptera*), on living stems of dicotyledonous plant, 1 May 2007, B.T., K.T., R.R., S.M., W.C. (BBH19873, BCC25446; BCC25447, BCC25448); Nakhon Ratchasima Province, Khao Yai National Park, km. 33 Nature Trail; 14°26′21.46″ N, 101°22′20.20″ E, on scale insects (*Coccidae; Hemiptera*), on living stems of dicotyledonous plant, 8 August 2007, B.T., P. Puyngain (P.P.), W.C. (BBH22627, BCC26680), Trang Province, Khao Chong Wildlife Development and Conservation Promotion Station, 1.8 km. Nature Trail; 7°32′57″ N, 99°47′11″ E, on scale insects (*Coccidae; Hemiptera*), on living stems of dicotyledonous plant, 18 September 2007, B.T., K.T. (BBH23089, BCC27812); Nakhon Ratchasima Province, Khao Yai National Park, Mo Singto Nature Trail; 14°26′21.46″ N, 101°22′20.20″ E, on scale insects (*Coccidae; Hemiptera*), on living stems of dicotyledonous plant, 18 June 2009, K.T., P.S., R.R., S.M., T. Chohmee (T.C.) (BBH30139, BCC36656); 20 July 2009, K.T., P.S., R.R., S.M., T.C. (BBH27634, BCC37668); 23 July 2009, K.T., P.S., R.R., S.M. (BCC37879); Khao Yai National Park, km. 29 Nature Trail; 14°26′21.46″ N, 101°22′20.20″ E, on scale insects (*Coccidae; Hemiptera*), on living stems of dicotyledonous plant, 2 June 2011, A.K., K.T., K.S., P.S., S.M., W.N. (BBH30577, BCC47981); GenBank: ITS = OL331511, LSU = OL322043, *TEF1* = OL322030, *RPB1* = OL322054, *RPB2* = OL322061; Khao Yai National Park, Mo Singto Nature Trail; 14°26′21.46″ N, 101°22′20.20″ E, on scale insects (*Coccidae; Hemiptera*), on living stems of dicotyledonous plant, 3 August 2011, A.K., K.T., K.S., P.S., S.M., W.N. (BBH30683, BCC48951).

Notes: Based on the macromorphologies of the natural samples, the lower surface of stromata of *A. galloides* and *A. griseoperitheciatus* are orange and their perithecial layers are white and pale violet to light purplish gray, respectively. The perithecia in these two species are semi-immersed, but perithecia in *A. galloides* are obclavate, whereas those of *A. griseoperitheciatus* are obovoid. The colony color of *A. galloides* on PDA is pale orange yellow, light yellow and brilliant yellow, whereas *A. griseoperitheciatus* is white and produces a pale purplish pink pigment diffusing in the medium.

Ascopolyporus griseoperitheciatus Khonsanit, Thanakitpipattana and Luangsa-ard **sp. nov**. Figure 4.

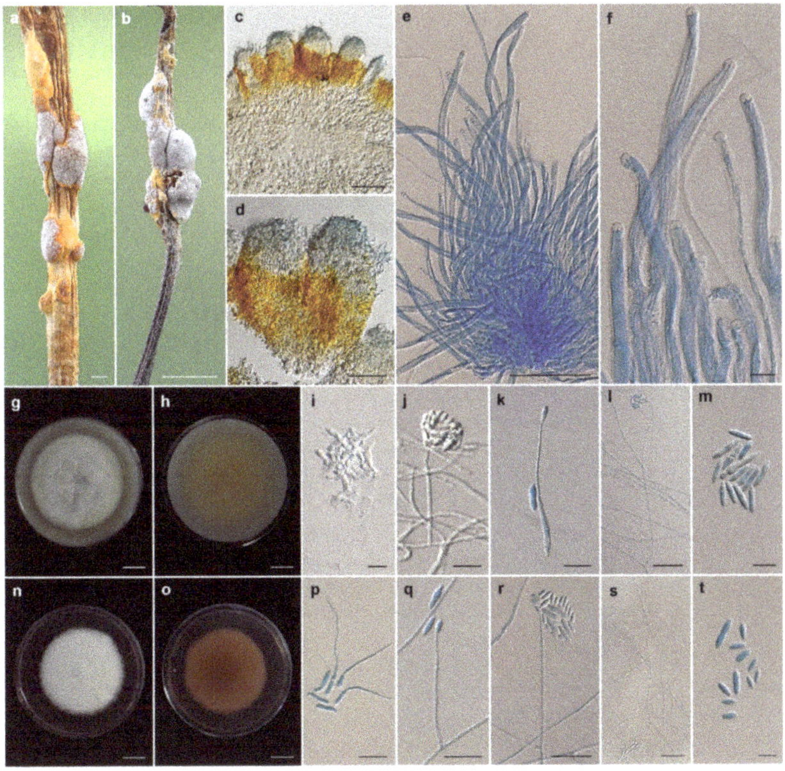

Figure 4. *Ascopolyporus griseoperitheciatus*. (**a**,**b**) Stromata on living stem of dicotyledonous plants (**a** BBH18679, **b** BBH30155); (**c**,**d**) perithecia; (**e**) asci; (**f**) asci-caps; (**g**) colony obverse on OA; (**h**) colony reverse on OA; (**i**) microsclerotium-like structure on OA; (**j**) phialide apex with conidial head; (**k**) conidium germination; (**l**) phialide and conidia on OA; (**m**) conidia on OA; (**n**) colony obverse on PDA; (**o**) colony reverse on PDA; (**p**) conidia germination; (**q**–**s**) phialide and conidia on PDA; (**t**) conidia on PDA. Scale bars: (**a**,**g**,**h**,**n**,**o**) = 10 mm; (**b**) = 5 mm; (**c**) = 100 µm; (**d**,**e**,**l**) = 50 µm; (**p**,**q**,**r**,**s**) = 20 µm; (**i**,**j**,**k**,**m**,**t**) = 10 µm; (**f**) = 5 µm.

MycoBank: MB 841854.

Etymology: From the Latin "griseo", referring to the gray color of the fresh stromata.

Typification: Thailand, Kamphaeng Phet Province, Khlong Lan National Park, Khlong Lan Waterfall; 16°07′50.20″ N, 99°16′36.30″ E, on scale insects (*Coccidae*; *Hemiptera*), on the living stems of dicotyledonous plant, 19 June 2006, B.T., J.J.L., K.T., P.S., R.R., S.M., W.C. (holotype BBH18679, ex-holotype culture BCC22358). GenBank: ITS = OL331507, LSU = OL322050, *TEF1* = OL322037, *RPB1* = *RPB2* = OL322067.

Description: *Stromata* epibiotic, irregularly pulvinate or subglobose, upper surface very pale violet (91D) to light purplish gray (N187D); lower surface vivid yellow (14C) to strong orange (25A), 3–7 mm wide. *Perithecia* semi-immersed, crowded, obovoid, 150–320 × 80–140 µm. *Asci* cylindrical, (150–)154–179(–193) × 4–5 µm. *Asci caps* 1.5–2 × 3–3.5 µm. *Ascospores* hyaline, whole, filiform, aseptate, extending the length of ascus.

Culture characteristics: Colony on OA attaining a diameter of 4 cm in 20 days, flat, slightly convex to the agar surface, white with light yellow green (150D), reverse pale orange yellow (23D). *Phialides* arising from aerial hyphae, solitary, cylindrical or acremonium-like, slightly curved, 50–250 × 1–2 µm. *Conidia* hyaline, enteroblastic, fusiform to acerose, early in development aseptate, becoming 1–2 septa, aggregated at the apex of the phialides, (6–)7–14(–19) × (1.5–)2–3 µm.

Colony on PDA attaining a diameter of 3.5–4 cm in 20 days, compact mycelium, slightly convex to the agar surface, white (158), pale purplish pink (56A) pigment diffusing in medium, reverse strong yellowish pink (31C). *Phialides* arising from aerial hyphae, solitary, cylindrical or acremonium-like, slightly curved, 43–265 × 1–2 µm. *Conidia* hyaline, enteroblastic, fusiform to acerose, early in development aseptate, becoming 1–2 septa, aggregated at the apex of the phialides, (4–)6–11(–17) × 2–3.5(–4) µm.

Habitat: On scale insects (*Coccidae*; *Hemiptera*), found on living stems of dicotyledonous plants.

Additional specimens examined: Thailand, Chanthaburi Province, Khao Soi Dao Wildlife Sanctuary, Withiphrai Nature Trail; 13°06′13″ N, 102°11′39″ E, on scale insects (*Coccidae*; *Hemiptera*), on living stems of dicotyledonous plant, 1 May 2007, B.T., K.T., R.R., S.M., W.C. (BBH19872, BCC25788); Nakhon Ratchasima Province, Khao Yai National Park, Mo Singto Nature Trail; 14°26′21.46″ N, 101°22′20.20″ E, on scale insects (*Coccidae*; *Hemiptera*), on living stems of dicotyledonous plant, 30 June 2010, A.K., K.T., K.S., P.S., R. Somnuk (R.S.), S.M. (BBH30155, BCC43315).

Notes: Our molecular phylogenetic study has shown that *A. griseoperitheciatus* is closely related to *A. albus*. However, *A. griseoperitheciatus* significantly differs from *A. albus* in having a perithecial layer on the upper surface of stromata that is very pale violet to light purplish gray, the lower surface of stromata is vivid yellow to strong orange, while in *A. albus*, the stromata are only white. Additionally, *A. griseoperitheciatus* produces a pale purplish pink pigment diffusing in PDA plates, whereas *A. albus* does not produce any pigment.

Ascopolyporus gollmerianus Henn., Hedwigia 41: 8. 1902.

Ascopolyporus khaoyaiensis Mongkolsamrit, Thanakitpipattana and Luangsa-ard sp. nov. Figure 5.

MycoBank: MB 841856.

Etymology: Named after Khao Yai National Park, where the type specimen was found.

Typification: Thailand, Nakhon Ratchasima Province, Khao Yai National Park, Mo Singto Nature Trail; 14°26′21.46″ N, 101°22′20.20″ E, on scale insects (*Coccidae*; *Hemiptera*), on the living stems of dicotyledonous plant, 5 A0ugust 2010, K.T., P.S., S.M., A.K., R.S., K.S. (holotype BBH30157, ex-holotype culture BCC43741). GenBank: ITS = OL331513, LSU = OL322041, *TEF1* = OL322040, *RPB2* = OL322070.

Description: *Stromata* epibiotic, flattened to convex, cylindrical to irregularly shaped, upper surface very pale violet (91D) to dark purple (59A); lower surface white to pale orange (20A), 3–25 mm wide, 1–3 mm thick. *Perithecia* semi-immersed, slightly protruding apices, narrow flask shaped, slightly protruding, obclavate, 300–360 × 100–120 µm. *Asci* cylindrical, up to 215 µm long, 3–4 µm wide, *Asci caps* 2–4 × 3–4 µm. *Ascospores* hyaline, whole, filiform, aseptate, 175–200 × 1 µm.

Culture characteristics: Colonies on OA attaining a diameter of 3.5 cm in 14 days, cottony, white, reverse moderate brown (165A). *Phialides* arising from aerial hyphae, solitary, cylindrical or acremonium-like, slightly curved, up to 60 µm long, 1–2 µm wide. *Conidia* hyaline, enteroblastic, fusiform to acerose, early in development aseptate, mostly becoming 1 septum, occasionally 2–3 septa, aggregated at the apex of the phialides, (5–)8–16(–20) × (1.5–)2–3 µm.

Colonies on PDA attaining a diameter of 3–4 cm in 14 days, cottony, white, pale orang in the middle of colony, reverse dark red (59A) in the middle, bright strong purplish red (60D) pigment diffusing in medium. *Phialides* arising from aerial hyphae, solitary, cylindrical or acremonium-like, up to 50 µm long, 1–2 µm wide. *Conidia* hyaline, enteroblastic, fusiform to acerose, early in development aseptate, mostly becoming 1 septum, occasionally 2–3 septa, aggregated at the apex of the phialides, (7–)9–16.5(–22) × (1.5–)2–3 µm.

Habitat: On scale insects (*Coccidae*; *Hemiptera*), found on living stems of dicotyledonous plant.

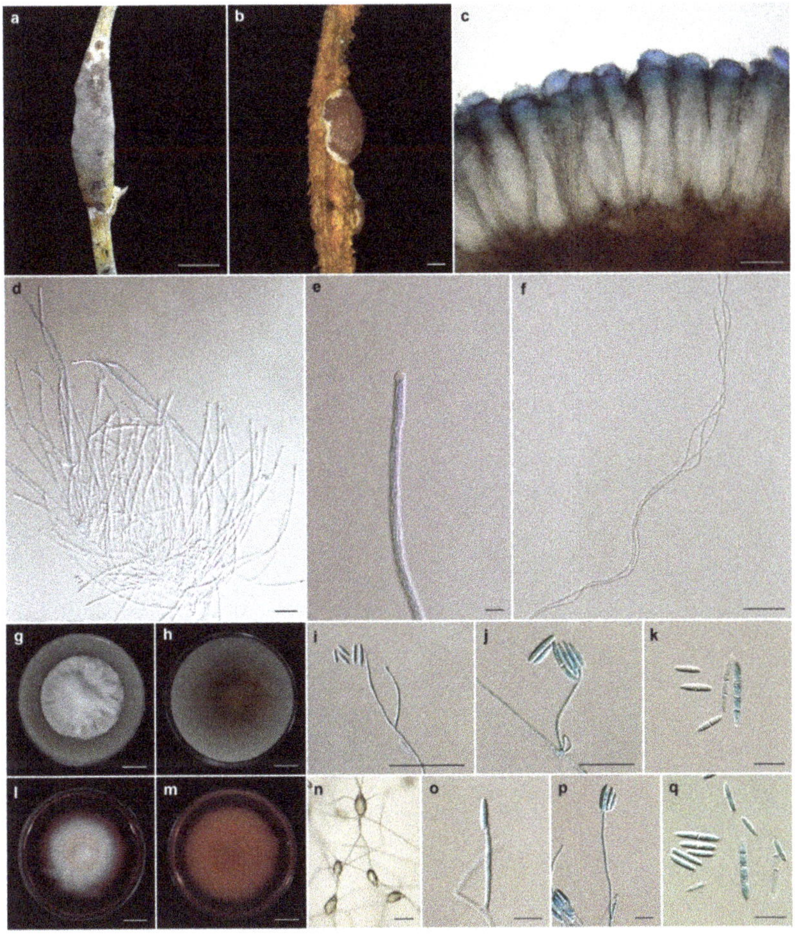

Figure 5. *Ascopolyporus khaoyaiensis*. (**a**,**b**) Stromata on living stem of dicotyledonous plants (**a** BBH30157, **b** BBH30154); (**c**) perithecia; (**d**) asci; (**e**) asci-caps; (**f**) ascospores; (**g**) colony obverse on OA; (**h**) colony reverse on OA; (**i**,**j**) phialide apex with conidial head on OA; (**k**) conidia on OA; (**l**) colony obverse on PDA; (**m**) colony reverse on PDA; (**n**–**p**) phialide apex with conidial head on PDA; (**q**) conidia on PDA. Scale bars: (**b**,**g**,**h**,**l**,**m**) = 10 mm; (**a**) = 5 mm; (**c**) = 100 μm; (**i**) = 50 μm; (**d**,**f**) = 20 μm; (**g**,**j**,**k**,**n**,**o**,**p**,**q**) = 10 μm; (**e**) = 5 μm.

Additional specimen examined: Thailand, Nakhon Ratchasima Province, Khao Yai National Park, Mo Singto Nature Trail; 14°26′21.46″ N, 101°22′20.20″ E, on scale insects (*Coccidae*; *Hemiptera*), on living stems of dicotyledonous plant, 30 June 2010, K.T., P.S., S.M., A.K., R.S., K.S. (BBH30154, BCC43314). GenBank: ITS = OL331512, LSU = OL322052, *TEF1* = OL322039, *RPB2* = OL322069.

Notes: The stromatal color of the natural samples of *A. khaoyaiensis* is similar to the purple stromata of *A. purpuratus*. However, perithecia in *A. khaoyaiensis* are immersed and obclavate, whereas perithecia in *A. purpuratus* are semi-immersed and ovoid. Asci of *A. khaoyaiensis* are shorter than those of *A. purpuratus* (up to 215 × 3–4 vs. 200–240 × 4–5 μm). Additionally, *A. khaoyaiensis* and *A. purpuratus* produce bright strong purplish red pigment diffusing in PDA plates.

Ascopolyporus möellerianus (Henn.) Möller, Phycomyc. Ascomyc. Bras.: 301. 1901.

Ascopolyporus philodendri J.F. Bisch. (as "philodendrus"), Mycologia 97(3): 711. 2005.

Ascopolyporus polychrous Möller, Bot. Mitt. Trop. 9: 300. 1901.

Ascopolyporus polyporoïdes Möller, Bot. Mitt. Trop. 9: 301. 1901.

Ascopolyporus purpuratus Mongkolsamrit, Thanakitpipattana, Himaman and Luangsa-ard **sp. nov**. Figure 6.

Figure 6. *Ascopolyporus purpuratus*. (**a**,**b**) Stromata on living midrib of leaves of dicotyledonous plant (BBH44511); (**c**) cross-section through stroma showing perithecia; (**d**) ovoid perithecia; (**e**) asci; (**f**) asci-caps; (**g**) ascospores; (**h**) colony obverse on OA; (**i**) colony reverse on OA; (**j**,**k**) phialide apex with conidial head on OA; (**l**) conidia on OA; (**m**) colony obverse on PDA; (**n**) colony reverse on PDA; (**o**,**p**) phialide apex with conidial head on PDA; (**q**) conidia on PDA. Scale bars: (**h**,**i**,**m**,**n**) = 10 mm; (**a**,**b**,**c**) = 1 mm; (**d**,**e**) = 100 μm; (**g**,**j**,**o**) = 20 μm; (**f**,**k**,**l**,**p**,**q**) = 10 μm.

MycoBank: MB 841857.

Etymology: Referring to the purple color of the fresh stroma.

Typification: Thailand, Nakhon Ratchasima Province, Khao Yai National Park, Pong Chang Chomrom Phoen (Nong Phakchi); 14°27′04.0″ N, 101°22′03.60″ E, on scale insects (*Coccidae*; *Hemiptera*), on the living stems and midrib of dicotyledonous leaves, 19 September 2018, J.J.L., K.T., D. Thanakitpipattana (D.T.), B. Sakolrak (B.S.), R.S., S.M., W.N., W. Himaman (W.H.), P.S. (holotype BBH44511, ex-holotype culture BCC88430). GenBank: ITS = OL331506, LSU = OL322045, *TEF1* = OL322032, *RPB1* = OL322059.

Description: *Stromata* epibiotic, flattened to convex, consisting of dense white mycelial mat, upper surface yellow (18D) to very pale purple-violet (75D), 5–12 mm long, 3–8 mm wide. *Perithecia* semi-immersed, crowded, ovoid, (300–)335–414(–420) × (100–)110–142(–150) μm.

Asci cylindrical, up to 240 × 2–4 μm. *Asci caps* 2–4 × 3–4 μm. *Ascospores* hyaline, whole, filiform, aseptate, (100–)131–190(–220) × 1–1.5 μm.

Culture characteristics: Colonies on OA attaining a diameter of 3–4 cm in 14 days, flat, white, reverse white. *Phialides* arising from aerial hyphae, solitary, cylindrical or acremonium-like, slightly curved, up to 40 μm long, 1–2 μm wide. *Conidia* hyaline, enteroblastic, fusiform to acerose, aseptate, aggregated at the apex of the phialides, (5–)6.5–18(–25) × (1.5–)2–2.5(–3) μm.

Colonies on PDA attaining a diameter of 2.5–3.5 cm in 14 days, slightly convex to the agar surface, vivid orange (28B) with white in the middle of colony, reverse strong reddish orange (34C), bright moderate reddish orange (N34D) pigment diffusing in medium. *Phialides* arising from aerial hyphae, solitary, cylindrical or acremonium-like, slightly curved, up to 55 μm long, 1–2 μm wide. *Conidia* hyaline, enteroblastic, fusiform to acerose, aseptate, aggregated at the apex of the phialides, (5–)7–13.5(–18) × 1.5–2.5(–3) μm.

Habitat: On scale insects (*Coccidae; Hemiptera*), found on living stem and midrib of leaves of dicotyledonous plants.

Additional specimen examined: Thailand, Phetchaburi Province, Kaeng Krachan National Park, Ban Krang Camp Nature Trail; 12°54′05″ N, 99°37′48″ E, on scale insects (*Coccidae; Hemiptera*), on the midrib of leaves, 14 November 2005, K.T., W.C., R.R., B.T. (BBH 15035, BCC 19721); Nakhon Ratchasima Province, Khao Yai National Park, Mo Singto Nature Trail; 14°26′21.46″ N, 101°22′20.20″ E, on scale insects (*Coccidae; Hemiptera*), on the twigs of tree, 23 July 2009, K.T., P.S., R.R., S.M. (BBH26373, BCC37880); Nakhon Ratchasima Province, Khao Yai National Park, Bueng Phai Nature Trail; 14°26′21.46″ N, 101°22′20.20″ E, on scale insects (*Coccidae; Hemiptera*), on the midrib of leaves and twigs of tree, 18 September 2018, J.J.L., K.T., D.T., B.S., R.S., S.M., W.N., W.H., P.S. (BBH44547, BCC88388); GenBank: ITS = OL331505, LSU = OL322046, *TEF1* = OL322033, RPB2 = OL322064, (BBH44551, BCC88389); GenBank: ITS = OL331504, LSU = OL322047, *TEF1* = OL322034.

Notes: *Ascopolyporus purpuratus* can be found on the midrib of leaves and living stems of dicotyledonous plants. In natural samples, the perithecia are pale yellow and/or very pale purple-violet. Additionally, *A. purpuratus* produces red pigment diffusing in PDA plates the same as *A. griseoperitheciatus* and *A. khaoyaiensis*. However, the mycelia of *A. purpuratus* are white in the middle of the colony with orange edges, whereas *A. griseoperitheciatus* produces only white mycelia and *A. khaoyaiensis* produces white mycelia that turn pale orange in the center of colony.

Ascopolyporus villosus Möller, Bot. Mitt. Trop. 9: 301. 1901.

Neohyperdermium Thanakitpipattana and Luangsa-ard, gen. nov.

MycoBank: MB 842780.

Etymology: Referring to the phenotypic similarity of the stromatal formation to *Hyperdermium*.

Typification: *Neohyperdermium piperis* (J.F. Bisch. and J.F. White) Thanakitpipattana and Luangsa-ard.

Description: *Stroma* epibiotic, flattened to pulvinate, white to yellow. Hosts are scale insects (*Coccoidea, Hemiptera*). *Perithecia* immersed, obpyriform, cymbiform to cone-shaped. *Asci* cylindrical, linear with enlarged refractive tip. Asexual morph verticillium-like.

Notes: This genus is a phylogenetically separate lineage from other scale insect pathogens in *Cordycipitaceae*, as shown in Figure 1. Two species are recognized in this genus that produce white to yellow stromata, immersed perithecia and a verticillium-like anamorph.

Neohyperdermium piperis (J.F. Bisch. and J.F. White) Thanakitpipattana and Luangsa-ard, **comb. nov**.

MycoBank MB 842782.

≡ *Torrubiella piperis* J.F. Bisch. and J.F. White, Studies in Mycology 50: 89–94. 2004.

≡ *Cordyceps piperis* (J.F. Bisch. and J.F. White) D. Johnson, G.H. Sung, J.F. Bisch. and Spatafora, Mycol. Res. 113(3): 284. 2009.

Description and illustration: See J.F. Bisch. and J.F. White (2004).

Typification: Panama, Barro Colorado Island, Lutz Creek, scale insect (*Coccoidea, Hemiptera*) on *Piper carrilloanum* (*Piperaceae*) August 2003, J.F. Bischoff and J.F. White, Jr., New York Botanical Garden (NY), culture ex-type CBS 116719.

Habitat: Scale insects.

Known distribution: Panama.

Note: *Neohyperdermium piperis* is closely related to *N. pulvinatum* and can be distinguished from *N. pulvinatum* in producing part-ascospores, whereas in *N. piperis*, the ascospores are whole with multiple septations and the conidia are aseptate.

Neohyperdermium pulvinatum (J.F. White et al.) Thanakitpipattana and Luangsa-ard, comb. nov.

MycoBank: MB 842783.

≡ *Hyperdermium pulvinatum* J.F. White et al., Mycologia 92(5): 908–918. 2000.

Description and illustration: See J.F. White et al. (2000).

Typification: Costa Rica, Guanacaste, Parque Nacional Guanacaste, Sector El Hacha, Puesto Los Almendros, on *Asteraceae*, 6 October 1998, J.F. White, G. Bills and S. Salas, RUTPP, culture ex-type ATCC MYA-69.

Habitat: Scale insects.

Known distribution: Costa Rica.

Note: *Neohyperdermium pulvinatum* is closely related to *N. piperis*, which can be distinguished by the type of ascospores and the presence of multiseptate conidia.

4. Discussion

The results of our multigene phylogenetic analyses show that our specimens were closely related to *Ascopolyporus polychorus*, *A. villosus* and *Hyperdermium caulium* (Figure 1). Importantly, the specimens in this study are clearly distinct species in *Ascopolyporus* because of the differences in the sizes, color, perithecial position and features of the stromata, which also overlap with morphological characters of some species previously treated as belonging to the genera *Hyperdermium* (*H. caulium*, *H. bertonii*, *H. pulvinatum*) and *Cordyceps* (*C. piperis*) in *Cordycipitaceae*, by producing flattened to pulvinate stromata and producing unique cylindrocarpon-like anamorph with multiseptate conidia [11,12,14,16]. The two genera, *Ascopolyporus* and *Hyperdermium*, differ only in the sizes and characters of ascomata [11,13,14]; *Hyperdermium* stromata are either flattened or pulvinate, whereas in *Ascopolyporus* sensu Möller, the stromata are subglobose to polypore-like. Based on these results, since the type species of *Hyperdermium*, *H. caulium*, is nested within *Ascopolyporus*, *Hyperdermium* is synonymized with *Ascopolyporus* and a new combination is proposed for *H. caulium*. The generic description of *Ascopolyporus* is therefore emended to include flattened to pulvinate stromata.

Our new species in *Ascopolyporus* are characterized by possessing flattened and pulvinate stromata, two groups that are supported as separate clades in phylogenetic analyses (Figure 1). The three pulvinate species (*Ascopolyporus albus*, *A. galloides* and *A. griseoperitheciatus*) have smaller stromata than previously described by Möller [24] and Bischoff et al. [14]. The two new species of *Ascopolyporus khaoyaiensis* and *A. purpuratus* have flattened stromata, and their sizes are in the same range as *A. caulium* (Table 2). All new *Ascopolyporus* species in this study possess semi-immersed perithecia with ostioles slightly protruding on the surface of the fertile cushion, whereas *A. polychrous* and *A. philodendrus* have completely immersed perithecia; *A. vilosus* does not produce perithecia on stromata [14].

Table 2. Morphological comparisons of *Ascopolyporus* and related species. NA, not applicable.

Name	Host	Stromata (mm)	Perithecia (μm)	Asci (μm)	Ascospores (μm)	Conidiogenous Cell (μm)	Conidia (μm)	References
Ascopolyporus albus	Scale insect, Epiphyte	pulvinate, subglobose to globose white to pinkish white 3–6	semi-immersed, obpyriform, 250–320 × 100–120	hyaline, cylindrical, up to 250 × 3–4	hyaline, filiform, whole, multiseptate, 95–135 × 1	solitary, slightly curved, cylindrical, up to 120 × 1–2	enteroblastic, fusiform to acerose, 1–4 septate, 8–30 × 2–3.5	This study
A. caulium	Scale insect, Epiphyte	crustose, subcircular yellow to orange, 5–100	cylindrical, 200–250 × 65–80	cylindrical to slightly fusiform, 100–160 × 8–9	filiform, multiseptate, extending to the length of ascus × 1 wide	phialidic, sparse layer	enteroblastic, cylindrical to fusiform, slightly truncate at end, aseptate: 5–7 × 1–1.5 1–5 septate: 15–30 × 1.5–3	Sullivan et al., 2000
A. galloides	Scale insect, Epiphyte	pulvinate, hemispherical, upper: white-yellowish white lower: strong orange 1–7	semi-immersed, obclavate, 170–340 × 60–110	hyaline, cylindrical, 129–175 × 3–6	hyaline, filiform, whole, aseptate, 131–216 × 0.5	solitary, slightly curved, cylindrical, 30–294 × 1–2	enteroblastic, fusiform to acerose, aseptate to 1–4 septate, 5–27 × 2–4	This study
A. griseoperitheciatus	Scale insect Epiphyte	pulvinate, irregular pulvinate to subglobose, 3–7 upper: very pale violet to light purplish gray lower: vivid yellow to orange	semi-immersed, obovoid, 150–320 × 80–140	hyaline, cylindrical, 150–193 × 4–5	hyaline, filiform, whole, aseptate, extending to the length of ascus	solitary, slightly curved, cylindrical, acremonium-like, 43–265 × 1–2	enteroblastic, fusiform to acerose, aseptate to 1–2 septate, 4–17 × 2–4	This study
A. khaoyaiensis	Scale insect, Epiphyte	flattened to convex, cylindrical to irregular shaped, very pale violet to dark purple 3–25	semi-immersed, obclavate, 300–360 × 100–120	hyaline, cylindrical, up to 215 × 3–4	hyaline, filiform, whole, aseptate, 175–200 × 1	solitary, slightly curved, cylindrical, acremonium-like, up to 50 × 1–2	enteroblastic, fusiform to acerose, aseptate to 1–3 septate, 7–22 × 1.5–3	This study

Table 2. *Cont.*

Name	Host	Stromata (mm)	Perithecia (μm)	Asci (μm)	Ascospores (μm)	Conidiogenous Cell (μm)	Conidia (μm)	References
A. philodendrus	Scale insect, Epiphyte	subglobose, upper: sterile, red-purple lower: fertile, white to tan 12–25	immersed, obclavate, 200–300 × 40–80	cylindrical, 90–140 × 3–5	filiform, length of ascus	simple, phialidic, 30–60 × 1–3	enteroblastic, subcylindrical, aseptate to 1–4 septate, 7–25 × 3–4	White et al., 2003
A. polychrous	Scale insect, Epiphyte	polypore-like, bright-rusty red or white to yellow 40	underside of stroma, immersed, narrow obclavate, up to 750	hyaline, cylindrical, 500 × 4	hyaline, filiform to spiroid, 300 × 1, disarticulate into part-spores, 6 × 1	NA	enteroblastic, oval, 1-multiseptate, 7–12 × 4–6	Möller, 1901; White et al., 2003
A. purpuratus	Scale insect, Epiphyte	flattened to convex, yellow to very pale, purple-violet, 5–12 × 3–8	semi-immersed, ovoid, 300–420 × 100–150	hyaline, cylindrical, up to 240 × 2–4	hyaline, filiform, whole, aseptate, 100–220 × 1–1.5	solitary, slightly curved, cylindrical, acremonium-like, up to 55 × 1–2	enteroblastic, fusiform to acerose, aseptate, 5–18 × 1.5–3	This study
A. villosus	Scale insect, Epiphyte	white to pale yellow, 12–25	not produced	NA	NA	phialidic	enteroblastic, subcylindrical, guttulate, aseptate to 1–4 septate, 10–22 × 2–5	White et al., 2003; Bischoff et al., 2005
Neohyperdermium piperis	Scale insect, Epiphyte	pulvinate, subglobose to cylindrical, white to yellow, 3–10 × 3–6	immersed, obpyriform to cymbiform, 175–290 × 40–80	cylindrical, 120–170 × 3–5	filiform, disarticulating into part-spores, 4–9 × 1–2	upright, verticillate, 150–400	hyaline, subcylindrical, rarely subglobose, aseptate, 3–5 × 1–2	Bischoff and White, 2004
N. pulvinatum	Scale insect, Epiphyte	pulvinate, white to tan, 3–6	cone-shaped, 150–250 × 100–130	linear, 150–240 × 5–7	filiform, multiseptate, 130–225 × 1	phialidic, hyaline, 1 to several septa, 40–100 × 1–3	enteroblastic, subcylindrical, slightly arcuate, aseptate: 14–16 × 2.5–3 1–5 septate: 22–30 × 2.5–4	Sullivan et al., 2000

Another species in *Hyperdermium* found in *Cordycipitaceae*, *H. pulvinatum*, did not cluster with type species *H. caulium*, which was congruent with Sung et al. [45], Kepler et al. [44] and Wang et al. [54], and is grouped together with *Cordyceps piperis* possessing pulvinate stromata that are white to yellow, producing aseptate, subcylindrical conidia on cultures (Table 2). These two species are proposed as new combinations in a new genus *Neohyperdermium*, as *Neohyperdermium pulvinatum* and *N. piperis*, which were described as epiphytes on scale insect pathogens in *Cordycipitaceae*.

The evolution and ecology of insect pathogenic fungi using insects and plants as the main source of nutrients remain not fully understood. Humber [55] suggested that the interaction between higher fungi and plants range from virulent pathogens to decomposer to mutualistic symbiosis. In *Hypocreales*, the genera *Aschersonia*, *Ascopolyporus*, *Conoideocrella*, *Dussiella*, *Hyperdermium*, *Hypocrella*, *Moelleriella*, *Regiocrella* and *Samuelsia* also utilize nutrients from the phloem of host plants through scale insects and white flies (*Coccidae* and *Aleyrodidae*) to continue their growth on plants [11,13–15]. Our new *Ascopolyporus* species are also found in this position, in which the scale insect attached to host plants was parasitized until it was consumed, but the fungus continues to utilize the nutrients that are being released through the stylet apparatus. The interactions occurred on the underside of fungal stroma, which is where the bridge for the exchange of nutrients between the fungus and plant exists (Figure 7).

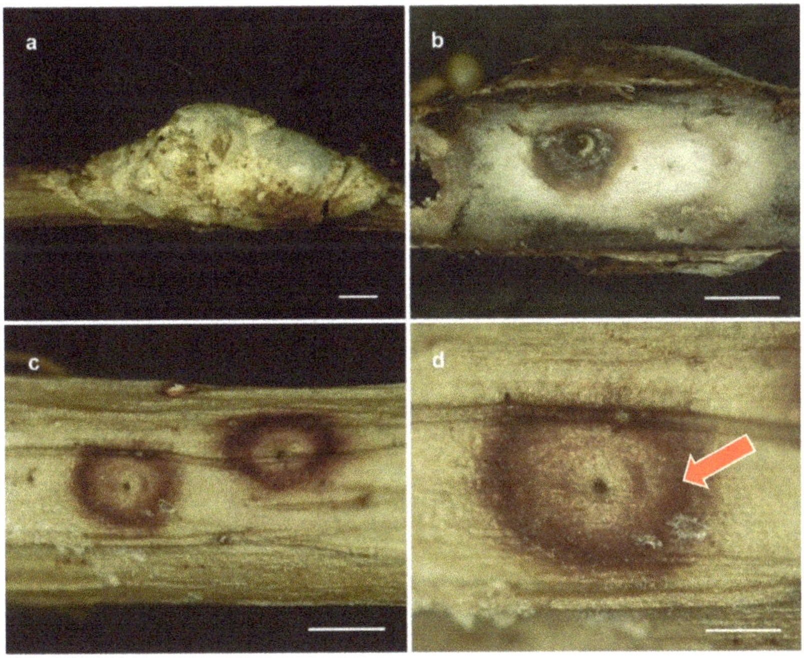

Figure 7. Photographs showing the interaction between fungus, plant and scale insect. (**a**) Fungal stroma on the stem of a dicotyledonous plant. (**b**) Underside of fungal stroma removed from the stem. (**c**,**d**) Underside of fungal stromata with ventage (arrow) from which the scale insect stylet entered the plant host. Scale bars = (**a**,**b**,**c**) = 1 mm; (**d**) = 0.05 mm.

Hypocrealean fungi are excellent producers of secondary metabolites which can be used to reduce the damage from insect fungi herbivores and phytopathogenic fungi [13,56]. However, no report has been made on the secondary metabolites produced from any of the reported species in *Ascopolyporus*, which should be a focus of future studies.

Residual Species of *Ascopolyporus*.

The remaining taxon could not be accommodated in the genus *Ascopolyporus* because its morphological description resembles other genera in *Clavicipitaceae* by producing paraphyses, which are not found in *Ascopolyporus* (*Cordycipitaceae*; *Hypocreales*), and molecular phylogenetic data are not available.

Ascopolyporus puttemansii Henn., Hedwigia 48: 6. 1908.

Key to *Ascopolyporus* species:

1a. Conidia aseptate 2
1b. Conidia aseptate to multiseptate 3
2a. Conidia oval . . . . *A. möllerianus*
2b. Conidia fusiform to acerose . . . . *A. purpuratus*
3a. Conidia 1–5 septate, cylindrical to fusiform, 5–30 × 1–3 μm . . . . *A. caulium*
3b. Conidia 1–4 septate 4
4a. Conidia subcylindrical 5
4b. Conidia fusiform to acerose 6
5a. Conidia subcylindrical, 7–25 × 3–4 μm . . . . *A. philodendrous*
5b. Conidia subcylindrical, guttulate, 10–22 × 2–5 μm . . . . *A. villosus*
6a. Ascospores hyaline, disarticulate into part-spores 7
6b. Ascospores hyaline, whole 8
7a. Ascospores filiform to spiroid, 6 × 1 μm . . . . *A. polychrous*
7b. Ascospores filiform, 8–15 μm . . . . *A. polyporoïdes*
8a. Ascospores multiseptate, 95–135 × 1 μm . . . . *A. albus*
8b. Ascospores aseptate 9
9a. Perithecia semi-immersed, obovoid . . . . *A. griseoperithciatus*
9b. Perithecia semi-immersed, obclavate 10
10a. Stromata pulvinate, hemispherical, 1–7 mm . . . . *A. galloides*
10b. Stromata flattened to convex, cylindrical to irregular shaped, 3–25 mm . *A. khaoyaiensis*

Key to *Neohyperdermium* species

1a. Ascospores filiform, multiseptate, whole . . . . *N. pulvinatum*
1b. Ascospores filiform, disarticulating into part-spores . . . . *N. piperis*

Author Contributions: Conceptualization, D.T., A.K., J.J.L.-a. and N.P.; Data curation, W.H. and N.P.; Formal analysis, D.T. and A.K.; Funding acquisition, D.T., S.M., J.J.L.-a. and N.P.; Investigation, D.T., S.M., A.K., W.H., J.J.L.-a. and N.P.; Methodology, D.T., S.M., A.K., W.H., J.J.L.-a. and N.P.; Resources, W.H. and N.P.; Supervision, D.T., S.M., J.J.L.-a. and N.P.; Writing—original draft, D.T., S.M., A.K., W.H. and J.J.L.-a.; Writing—review and editing, D.T., S.M., A.K., J.J.L.-a. and N.P. All authors have read and agreed to the published version of the manuscript.

Funding: This research was funded by a Graduate School Thesis Grant, Chulalongkorn University and Platform Technology Management Section, National Center for Genetic Engineering and Biotechnology (BIOTEC), grant no. P19-50231.

Institutional Review Board Statement: Not applicable.

Informed Consent Statement: Not applicable.

Data Availability Statement: Publicly available datasets were analyzed in this study. These data can be found in: Genbank, https://www.ncbi.nlm.nih.gov/genbank/ (accessed on 9 November 2021; MycoBank, https://www.mycobank.org/ (accessed on 9 November 2021); Index Fungorum, http://www.indexfungorum.org/Names/Names.asp (accessed on 5 December 2021).

Acknowledgments: The authors would like to thank Philip Shaw for editing the manuscript. We are also grateful to the anonymous reviewers for their valuable comments and suggestions.

Conflicts of Interest: The authors declare no conflict of interest.

References

1. Kondo, T.; Gullan, P.J.; Williams, D.J. Coccidology. The study of scale insects (*Hemiptera: Sternorrhyncha: Coccoidea*). *Cienc. Tecnol. Agropecuaria* **2008**, *9*, 55–61. [CrossRef]
2. Mansour, R.; Grissa-Lebdi, K.; Suma, P.; Mazzeo, G.; Russo, A. Key scale insects (*Hemiptera: Coccoidea*) of high economic importance in a Mediterranean area: Host plants, bio-ecological characteristics, natural enemies and pest management strategies—A review. *Plant Prot. Sci.* **2017**, *53*, 1–14.
3. Dhami, M.K.; Weir, B.S.; Taylor, M.W.; Beggs, J.R. Diverse honeydew-consuming fungal communities associated with scale insects. *PLoS ONE* **2013**, *8*, e70316.
4. Chomnunti, P.; Hongsanan, S.; Aguirre-Hudson, B.; Tian, Q.; Peršoh, D.; Dhami, M.K.; Alias, A.S.; Xu, J.; Liu, X.; Stadler, M.; et al. The sooty moulds. *Fungal Divers.* **2014**, *66*, 1–36. [CrossRef]
5. Hongsanan, S.; Tian, Q.; Hyde, K.; Chomnunti, P. Two new species of sooty moulds, *Capnodium coffeicola* and *Conidiocarpus plumeriae* in *Capnodiaceae*. *Mycosp* **2015**, *6*, 814–824. [CrossRef]
6. Henk, D.A.; Vilgalys, R. Molecular phylogeny suggests a single origin of insect symbiosis in the *Pucciniomycetes* with support for some relationships within the genus *Septobasidium*. *Am. J. Bot.* **2007**, *94*, 1515–1526. [CrossRef]
7. Dao, H.T.; Beattie, G.A.C.; Rossman, A.Y.; Burgess, L.W.; Holford, P. Four putative entomopathogenic fungi of armoured scale insects on Citrus in Australia. *Mycol. Prog.* **2016**, *15*, 47. [CrossRef]
8. Araújo, J.P.; Hughes, D.P. Diversity of entomopathogenic fungi: Which groups conquered the insect body? *Adv. Genet.* **2016**, *94*, 1–39.
9. Xu, X.-L.; Zeng, Q.; Lv, Y.-C.; Jeewon, R.; Maharachchikumbura, S.S.; Wanasinghe, D.N.; Hyde, K.D.; Xiao, Q.-G.; Liu, Y.-G.; Yang, C.-L. Insight into the systematics of novel entomopathogenic fungi associated with armored scale insect, kuwanaspis howardi (*Hemiptera: Diaspididae*) in China. *J. Fungi* **2021**, *7*, 628. [CrossRef]
10. Hywel-Jones, N.L.; Samuels, G.J. Three species of *Hypocrella* with large stromata pathogenic on scale insects. *Mycologia* **1998**, *90*, 36–46. [CrossRef]
11. Sullivan, R.F.; Bills, G.F.; Hywel-Jones, N.L.; White, J.F. *Hyperdermium*: A new clavicipitalean genus for some tropical epibionts of dicotyledonous plants. *Mycologia* **2000**, *92*, 908–918. [CrossRef]
12. Bischoff, J.F.; White, J.F. *Torrubiella piperis* sp. nov. (*Clavicipitaceae, Hypocreales*), a new teleomorph of the *Lecanicillium* complex. *Stud. Mycol.* **2004**, *50*, 89–94.
13. Koroch, A.; Juliani, H.; Bischoff, J.; Lewis, E.; Bills, G.; Simon, J.; White, J.F. Examination of plant biotrophy in the scale insect parasitizing fungus *Dussiella tuberiformis*. *Symbiosis* **2004**, *37*, 267–280.
14. Bischoff, J.F.; Chaverri, P.; White, J.F. Clarification of the host substrate of *Ascopolyporus* and description of *Ascopolyporus philodendrus* sp. nov. *Mycologia* **2005**, *97*, 710–717. [CrossRef] [PubMed]
15. Chaverri, P.; Bischoff, J.F.; Evans, H.C.; Hodge, K.T. *Regiocrella*, a new entomopathogenic genus with a pycnidial anamorph and its phylogenetic placement in the *Clavicipitaceae*. *Mycologia* **2005**, *97*, 1225–1237. [CrossRef] [PubMed]
16. Chaverri, P.; Liu, M.; Hodge, K. A monograph of the entomopathogenic genera *Hypocrella*, *Moelleriella*, and *Samuelsia* gen. nov. (*Ascomycota, Hypocreales, Clavicipitaceae*), and their aschersonia-like anamorphs in the Neotropics. *Stud. Mycol.* **2008**, *60*, 1–66. [CrossRef]
17. Johnson, D.; Sung, G.H.; Hywel-Jones, N.L.; Luangsa-Ard, J.J.; Bischoff, J.F.; Kepler, R.M.; Spatafora, J.W. Systematics and evolution of the genus *Torrubiella* (*Hypocreales, Ascomycota*). *Mycol. Res.* **2009**, *113*, 279–289. [CrossRef]
18. Fan, J.-H.; Xie, Y.-P.; Xue, J.-L.; Xiong, Q.; Jiang, W.-J.; Zhang, Y.-J.; Ren, Z.-M. The strain HEB01 of *Fusarium* sp., a new pathogen that infects brown soft scale. *Ann. Microbiol.* **2014**, *64*, 333–341. [CrossRef]
19. Luangsa-ard, J.J.; Mongkolsamrit, S.; Noisripoom, W.; Thanakitpipattana, D.; Khonsanit, A.; Wutikhun, T. *Helicocollum*, a new clavicipitalean genus pathogenic to scale insects (*Hemiptera*) in Thailand. *Mycol. Prog.* **2017**, *16*, 419–431. [CrossRef]
20. Deng, J.; Yu, Y.; Wang, X.; Liu, Q.; Huang, X. The ubiquity and development-related abundance dynamics of *Ophiocordyceps* fungi in soft scale insects. *Microorganisms* **2021**, *9*, 404. [CrossRef]
21. Szklarzewicz, T.; Michalik, K.; Grzywacz, B.; Kalandyk-Kołodziejczyk, M.; Michalik, A. Fungal Associates of Soft Scale Insects (*Coccomorpha: Coccidae*). *Cells* **2021**, *10*, 1922. [CrossRef] [PubMed]
22. Index Fungorum Database. Available online: http://www.indexfungorum.org/ (accessed on 5 December 2021).
23. Wood, M. Webwatch: Observing Mushrooms. Fungi Magazine 1, 2. Available online: https://mushroomobserver.org/ (accessed on 27 December 2021).
24. Möller, A. *Phycomyceten und Ascomyceten. Untersuchungen aus Brasilien. Botanische Mittheilungen aus den Tropen 9*; Fischer: Jena, Germany, 1901.
25. Luangsa-ard, J.J.; Tasanathai, K.; Thanakitpipattana, D.; Khonsanit, A.; Stadler, M. Novel and interesting *Ophiocordyceps* spp. (*Ophiocordycipitaceae, Hypocreales*) with superficial perithecia from Thailand. *Stud. Mycol.* **2018**, *89*, 125–142. [CrossRef] [PubMed]
26. Mongkolsamrit, S.; Noisripoom, W.; Thanakitpipattana, D.; Khonsanit, A.; Lamlertthon, S.; Luangsa-ard, J.J. New species in *Aciculosporium*, *Shimizuomyces* and a new genus *Morakotia* associated with plants in *Clavicipitaceae* from Thailand. *FUSE* **2021**, *8*, 27–37. [CrossRef] [PubMed]
27. Khonsanit, A.; Noisripoom, W.; Mongkolsamrit, S.; Phosrithong, N.; Luangsa-ard, J.J. Five new species of *Moelleriella* infecting scale insects (*Coccidae*) in Thailand. *Mycol. Prog.* **2021**, *20*, 847–867. [CrossRef]
28. Heritage, J.; Evans, E.G.V.; Killington, R. *Introductory Microbiology*; Cambridge University Press: Cambridge, UK, 1996.

29. The Royal Horticultural Society. *The Fifth Edition Published by The Royal Horticultural Society*; Vincent Square: London, UK, 2007.
30. Thanakitpipattana, D.; Tasanathai, K.; Mongkolsamrit, S.; Khonsanit, A.; Lamlertthon, S.; Luangsa-Ard, J.J. Fungal pathogens occurring on *Orthopterida* in Thailand. *Persoonia* **2020**, *44*, 140. [CrossRef]
31. Vilgalys, R.; Hester, M. Rapid genetic identification and mapping of enzymatically amplified ribosomal DNA from several *Cyptococcus* species. *J. Bacteriol. Res.* **1990**, *172*, 4238–4246. [CrossRef]
32. Rehner, S.A.; Samuels, G.J. Taxonomy and phylogeny of *Gliocladium* analysed from nuclear large subunit ribosomal DNA sequences. *Mycol. Res.* **1994**, *98*, 625–634. [CrossRef]
33. Rehner, S.A.; Buckley, E. A *Beauveria* phylogeny inferred from nuclear ITS and EF1-α sequences: Evidence for cryptic diversification and links to *Cordyceps* teleomorphs. *Mycologia* **2005**, *97*, 84–98. [CrossRef]
34. Castlebury, L.A.; Rossman, A.Y.; Sung, G.H.; Hyten, A.; Spatafora, J.W. Multigene phylogeny reveals new lineage for *Stachybotrys chartarum*, the indoor air fungus. *Mycol. Res.* **2004**, *108*, 864–872. [CrossRef]
35. Liu, Y.J.; Whelen, S.; Hall, B.D. Phylogenetic relationships among ascomycetes: Evidence from an RNA polymerase II subunit. *Mol. Biol. Evol.* **1999**, *16*, 1799–1808. [CrossRef]
36. O'Donnell, K.; Sarver, B.A.; Brandt, M.; Chang, D.; Noble-Wang, J.; Park, B.J.; Sutton, D.A.; Benjamin, L.; Lindsley, M.; Padhye, A.; et al. Phylogenetic diversity and microsphere array-based genotyping of human pathogenic *Fusaria*, including isolates from the multistate contact lens-associated U.S. keratitis outbreaks of 2005 and 2006. *J. Clin. Microbiol.* **2007**, *45*, 2235–2248. [CrossRef] [PubMed]
37. Hall, T. BioEdit: A user-friendly biological sequence alignment editor and analysis program for Windows 95/98/NT. *Nucleic Acids Symp. Ser.* **1999**, *41*, 95–98.
38. Edgar, R.C. MUSCLE: Multiple sequence alignment with high accuracy and high throughput. *Nucleic Acids Res.* **2004**, *32*, 1792–1797. [CrossRef]
39. Miller, M.A.; Pfeiffer, W.; Schwartz, T. The CIPRES science gateway: A community resource for phylogenetic analyses. In Proceedings of the 2011 TeraGrid Conference: Extreme Digital Discovery; Association for Computing Machinery: New York, NY, USA, 2011; pp. 1–8.
40. Stamatakis, A. RAxML version 8: A tool for phylogenetic analysis and post-analysis of large phylogenies. *Bioinformatics* **2014**, *30*, 1312–1313. [CrossRef] [PubMed]
41. Nylander, J. *MrModeltest v2. Program Distributed by the Author*; Uppsala University: Uppsala, Sweden, 2004.
42. Wilgenbusch, J.C.; Swofford, D. Inferring evolutionary trees with PAUP. *Curr. Protoc. Bioinform.* **2003**, *1*, 6.4.1–6.4.28. [CrossRef]
43. Ronquist, F.; Teslenko, M.; Van Der Mark, P.; Ayres, D.L.; Darling, A.; Höhna, S.; Larget, B.; Liu, L.; Suchard, M.A.; Huelsenbeck, J.P. MrBayes 3.2: Efficient Bayesian phylogenetic inference and model choice across a large model space. *Syst. Biol.* **2012**, *61*, 539–542. [CrossRef]
44. Kepler, R.M.; Luangsa-Ard, J.J.; Hywel-Jones, N.L.; Quandt, C.A.; Sung, G.H.; Rehner, S.A.; Aime, M.C.; Henkel, T.W.; Sanjuan, T.; Zare, R. A phylogenetically-based nomenclature for *Cordycipitaceae* (*Hypocreales*). *IMA Fungus* **2017**, *8*, 335–353. [CrossRef]
45. Sung, G.H.; Hywel-Jones, N.L.; Sung, J.M.; Luangsa-Ard, J.J.; Shrestha, B.; Spatafora, J.W. Phylogenetic classification of *Cordyceps* and the clavicipitaceous fungi. *Stud. Mycol.* **2007**, *57*, 5–59. [CrossRef]
46. Chen, W.; Han, Y.; Liang, Z.; Jin, D. *Lecanicillium araneogenum* sp. nov., a new araneogenous fungus. *Phytotaxa* **2017**, *305*, 29–34. [CrossRef]
47. Chiriví-Salomón, J.S.; Danies, G.; Restrepo, S.; Sanjuan, T. *Lecanicillium sabanense* sp. nov. (*Cordycipitaceae*) a new fungal entomopathogen of coccids. *Phytotaxa* **2015**, *234*, 63–74. [CrossRef]
48. Rehner, S.A.; Minnis, A.M.; Sung, G.H.; Luangsa-ard, J.J.; Devotto, L.; Humber, R.A. Phylogeny and systematics of the anamorphic, entomopathogenic genus *Beauveria*. *Mycologia* **2011**, *103*, 1055–1073. [CrossRef] [PubMed]
49. Spatafora, J.W.; Sung, G.H.; Sung, J.M.; Hywel-Jones, N.L.; White, J.F. Phylogenetic evidence for an animal pathogen origin of ergot and the grass endophytes. *Mol. Ecol.* **2007**, *16*, 1701–1711. [CrossRef] [PubMed]
50. Sung, G.H.; Spatafora, J.W. *Cordyceps cardinalis* sp. nov., a new species of *Cordyceps* with an east Asian-eastern North American distribution. *Mycologia* **2004**, *96*, 658–666. [CrossRef] [PubMed]
51. Mongkolsamrit, S.; Noisripoom, W.; Thanakitpipattana, D.; Wutikhun, T.; Spatafora, J.W.; Luangsa-Ard, J.J. Disentangling cryptic species with isaria-like morphs in *Cordycipitaceae*. *Mycologia* **2018**, *110*, 230–257. [CrossRef]
52. Tasanathai, K.; Thanakitpipattana, D.; Noisripoom, W.; Khonsanit, A.; Kumsao, J.; Luangsa-ard, J.J. Two new *Cordyceps* species from a community forest in Thailand. *Mycol. Prog.* **2016**, *15*, 28. [CrossRef]
53. Kepler, R.M.; Sung, G.H.; Ban, S.; Nakagiri, A.; Chen, M.J.; Huang, B.; Li, Z.; Spatafora, J.W. New teleomorph combinations in the entomopathogenic genus *Metacordyceps*. *Mycologia* **2012**, *104*, 182–197. [CrossRef]
54. Wang, Y.B.; Wang, Y.; Fan, Q.; Duan, D.E.; Zhang, G.D.; Dai, R.Q.; Dai, Y.D.; Zeng, W.B.; Chen, Z.H.; Li, D.D. Multigene phylogeny of the family *Cordycipitaceae* (*Hypocreales*): New taxa and the new systematic position of the Chinese cordycipitoid fungus *Paecilomyces hepiali*. *Fungal Divers.* **2020**, *103*, 1–46. [CrossRef]
55. Humber, R.A. Evolution of entomopathogenicity in fungi. *J. Invertebr. Pathol.* **2008**, *98*, 262–266. [CrossRef]
56. Torres, M.S.; Singh, A.P.; Vorsa, N.; White, J.F. An analysis of ergot alkaloids in the *Clavicipitaceae* (*Hypocreales*, *Ascomycota*) and ecological implication. *Symbiosis* **2008**, *46*, 11–19.

Journal of
Fungi

MDPI

Article

Diversity of *Cantharellus* (Cantharellales, Basidiomycota) in China with Description of Some New Species and New Records

Ming Zhang [1], Chao-Qun Wang [1], Man-Shui Gan [2], Yi Li [3], Shi-Cheng Shao [4], Wei-Qiang Qin [5], Wang-Qiu Deng [1] and Tai-Hui Li [1,*]

1 State Key Laboratory of Applied Microbiology Southern China, Guangdong Provincial Key Laboratory of Microbial Culture Collection and Application, Institute of Microbiology, Guangdong Academy of Sciences, Guangzhou 510070, China; zhangming@gdim.cn (M.Z.); wangcq@gdim.cn (C.-Q.W.); dengwq@gdim.cn (W.-Q.D.)
2 Bureau of Environmental Protection of Dinghai District, Zhoushan Municipality, Zhoushan 316000, China; ganmanshui2022@163.com
3 School of Food Science and Engineering, Yangzhou University, Yangzhou 225127, China; liyi062@yzu.edu.cn
4 CAS Key Laboratory of Tropical Plant Resources and Sustainable Use, Xishuangbanna Tropical Botanical Garden, Chinese Academy of Sciences, Mengla 666303, China; shaoshicheng@xtbg.org.cn
5 School of Marxism, Jishou University, Zhangjiajie 427000, China; qinweiq@163.com
* Correspondence: lith@gdim.cn

Abstract: *Cantharellus* is a well-known genus of edible mushrooms, belonging to the family Hydnaceae in the class Agaricomycetes. In this study, a phylogenetic overview of *Cantharellus* subg. *Cinnabarinus* and C. subg. *Parvocantharellus* in China is carried out with the description of four new species. Species description are based on morphological characters of basidiomata and phylogenetic analyses of multi-locus dataset of 28S + *tef*1 + *rpb*2. Among the new species, two species, *C. chrysanthus* and *C. sinocinnabarinus*, belong to *C.* subg. *Cinnabarinus* and two new species, *C. convexus* and *C. neopersicinus*, belong to *C.* subg. *Parvocantharellus*. Species delimitation characters of the new taxa are compared with closely related species. In addition, three new records of *Cantharellus* are reported for China: *C. albovenosus* and *C. citrinus* of subg. *Cinnabarinus* and *C. koreanus* of subg. *Parvocantharellus*. A key to the species of subg. *Cinnabarinus* in China was provided.

Keywords: chanterelle; East Asia; new species; phylogeny; taxonomy

Citation: Zhang, M.; Wang, C.-Q.; Gan, M.-S.; Li, Y.; Shao, S.-C.; Qin, W.-Q.; Deng, W.-Q.; Li, T.-H. Diversity of *Cantharellus* (Cantharellales, Basidiomycota) in China with Description of Some New Species and New Records. *J. Fungi* **2022**, *8*, 483. https://doi.org/10.3390/jof8050483

Academic Editors: Cheng Gao and Lei Cai

Received: 7 April 2022
Accepted: 29 April 2022
Published: 6 May 2022

1. Introduction

Cantharellus Fr. was firstly described by Fries [1] based on the type species *Cantharellus cibarius* Fr. Most *Cantharellus* species are popular edible mushrooms, especially beloved in Europe. *Cantharellus* is an ectomycorrhizal genus, forming symbiosis with various plants, such as the trees of Fagaceae, Pinaceae, Betulaceae, Salicaceae, Juglandaceae, Leguminosae, etc. [2–8]. Species in *Cantharellus* are widely distributed and are especially rich in subtropical to tropical zones [3,9,10]. Up to now, about 300 species of *Cantharellus* have been reported worldwide [7]. However, the species diversity is poorly known in Asia in the past decades, and many specimens were named after European or North American species [6,11–13]. In recent years, some new species were reported from Asia based on the combination of morphological characters and DNA phylogenetic analyses [6,7,11–16]. Recent phylogenetic studies demonstrated that *Cantharellus* is monophyletic and forms a sister relationship with *Craterellus* Pers. [3,4,7]. Species in *Cantharellus* were divided into seven subgenera based on multi-locus phylogenetic analyses in Buyck et al. [3], and a subsequent study in Cao et al. [7]. *Cantharellus* subg. *Cinnabarinus* Buyck & V. Hofst., typified by *C. cinnabarinus* (Schwein.) Schwein. was introduced for a monophyletic assemblage of mostly quite small, yellow, orange, pink or red species, sometimes mixed with lilac-purple or brownish tones in the pileus center, strongly veined in the lamellate hymenophore with principally thin-walled hyphal endings and abundant in clamp connections [3,17,18]. Species in subg. *Cinnabarinus* are widely distributed in Asia, Europe, North America, Australasia and Africa, and 16 species have been reported worldwide. In China, a large number of *Cantharellus* species have been reported, but only two species in the subg. *Cinnabarinus* were recorded, i.e., *C. cinnabarinus* and *C. phloginus*, by S.C. Shao & P.G. Liu. *Cantharellus cinnabarinus*, originally reported from North America, was recorded to be widely distributed in China [19–21]; *C. phloginus* was described as being from southwestern China [22].

In this study, a number of *Cantharellus* specimens were collected from China; further study proved that they represented eight distinct species, five of which belong to the subg. *Cinnabarinus* and three to the subg. *Parvocantharellus*. Four species are described below as new to science, which would make a contribution to understanding the species diversity of *Cantharellus* in China, and revealing the phylogenetic relationships of *Cantharellus* species.

2. Materials and Methods

2.1. Morphological Studies

Photographs of fresh basidiomata were taken in the field. Specimens were dried and deposited in the Fungarium of Guangdong Institute of Microbiology (GDGM). Descriptions of macro-morphological characters and habitats were obtained from photographs and field notes. The color codes followed Kornerup and Wanscher [23]. Microscopic observations were carried out on tissue sections stained with 5% aqueous KOH and 1% aqueous Congo red under a light microscope (Carl Zeiss Microscopy GmbH, Göttingen, Germany) with a magnification up to 1000×. For basidiospore descriptions, the notation (a–)b–c(–d) describes basidiospore dimensions, where the range b–c represented 90% or more of the measured values and 'a' and 'd' were the extreme values; L_m and W_m indicated the average length and width (±standard deviation) of the measured basidiospores, respectively; Q referred to the length/width ratio of an individual basidiospore and Q_m referred to the average Q value of all basidiospores ± sample standard deviation. All line-drawings of microstructures were made based on rehydrated materials.

2.2. DNA Extraction, PCR Amplification and Sequencing

Genomic DNA was extracted from the voucher specimens using the Sangon Fungus Genomic DNA Extraction kit (Sangon Biotech Co., Ltd., Shanghai, China) according to the manufacturer's instructions. Primer pairs LROR/LR7 [24], tef1F/tef1R and RPB2-5FCanth/RPB2-7cRCanth [3,25] were used to amplify the LSU, *tef1* and *rpb2* region, respectively. PCR reactions were performed in a total volume of 25 μL containing 0.5 μL template DNA, 11 μL sterile deionized water, 0.5 μL of each primer and 12.5 μL 2 × PCR mix [DreamTaqtm Green PCR Master Mix (2×), Fermentas, MA, USA]. Amplification reactions were performed in a Tprofessional Standard thermocycler (Biometra, Göttingen, Germany) under the following conditions: 95 °C for 4 min; then, 35 cycles of denaturation at 94 °C for 60 s, annealing at 53 °C (LSU)/50 °C (*tef*1)/52 °C (*rpb*2) for 60 s and extension at 72 °C for 60 s; with a final extension at 72 °C for 8 min. The PCR products were electrophoresed on 1% agarose gels and then send for sequencing on an ABI Prism® 3730 Genetic Analyzer (PE Applied Biosystems, Foster, CA, USA) at the Beijing Genomic Institute (BGI) using the same PCR primers. The raw sequences were assembled and checked with SeqMan implemented in Lasergene v7.1 (DNASTAR Inc., Madison, WI, USA). The newly generated sequences in this study were submitted to GenBank.

2.3. Phylogenetic Analyses

Sequences generated in this study and those downloaded from GenBank were combined and used for phylogenetic reconstruction. Detailed information of specimens included in this study was given in Table 1. Three sequence matrices, i.e., nrLSU, *tef1* and *rpb2*, were aligned separately with software MAFFT v6.853 using the E-INS-i strategy [26] and then manually adjusted in MEGA 6 [27]. The ambiguous aligned regions and introns of the two protein-coding genes of *tef1* and *rpb2* were retained in the final analyses.

Maximum Likelihood (ML) analyses were inferred using RAxML v7.2.6 [28], and all parameters were kept as defaults except for choosing GTRGAMMAI as the model; statistical supports were obtained using rapid non-parametric bootstrapping with 1000 replicates. Bayesian Inference (BI) phylogenies were inferred using MrBayes 3.2.6 [29]; the best models of the multi-locus datasets were searched via the PartitionFinder 2 [30] for each locus, i.e., K80 + I + G, SYM + I + G and SYM + I + G for 28S, *tef1* and *rpb2*, respectively. BI analysis using 4 chains were conducted by setting generations to 20 million and stoprul command with the value of stopval set to 0.01; trees were sampled every 1000 generations, the first 25% generations were discarded as burn-in and posterior probabilities (PP) were then calculated from the posterior distribution of the retained Bayesian trees. *Cantharellus cibarius* Fr. was selected as the outgroup based on recent studies [3,13]. The phylogenetic trees were visualized using FigTree v1.4.23.

Table 1. Specimen information used in this study. Sequences newly generated in this study are in bold; HT, NT and ET refer to holotype, neotype and epitype, respectively.

Taxa	Voucher	Locality	GenBank Accession No.			Reference
			LSU	*tef1*	*rpb2*	
Cantharellus afrocibarius	BB 96.236	Zambia	KF294669	JX192994	KF294747	[3]
C. afrocibarius	BB 96.235 (HT)	Zambia	KF294668	JX192993	KF294746	[3]
C. albovenosus	1690 (HT)	South Korean	–	KY271942	–	[11]
C. albovenosus	1713	South Korean	–	MW124387	–	[11]
C. albovenosus	**GDGM85853**	**China**	**OM978952**	**ON119062**	**ON119006**	**Present study**
C. albovenosus	**GDGM85846**	**China**	**OM978950**	**ON119060**	**ON119004**	**Present study**
C. albovenosus	**GDGM85142**	**China**	**OM978949**	**ON119059**	**ON229082**	**Present study**
C. albovenosus	**HMAS279296**	**China**	**OM978948**	**ON119066**	**ON119010**	**Present study**
C. albovenosus	**HMAS279284**	**China**	**ON212414**	**ON119064**	**ON119008**	**Present study**
C. albovenosus	**HMAS279292**	**China**	**ON212412**	**ON119065**	**ON119009**	**Present study**
C. albovenosus	**HMAS279262**	**China**	**OM978947**	**ON119063**	**ON119007**	**Present study**
C. albovenosus	**GDGM85852**	**China**	**OM978951**	**ON119061**	**ON119005**	**Present study**
C. albus	HKAS107045 (HT)	China	MT782540	MT776015	MT776012	[12]
C. albus	GDGM81399	China	MZ605074	MZ613977	MZ614022	[13]
C. albus	GDGM81064	China	MZ605073	MZ613976	MZ614021	[13]
C. appalachiensis	GRSM77088	USA	DQ898690	–	DQ898748	[31]
C. appalachiensis	BB 07.123	USA	KF294635	GQ914979	KF294711	[3]
C. aurantinus	GDGM46278 (HT)	China	MZ766517	MZ766560		[13]
C. aurantinus	GDGM46279	China	MZ766518	MZ766561	MZ766571	[13]
C. aurantinus	GDGM81899	China	MZ766520	MZ766563	MZ766573	[13]
C. aurantinus	GDGM84974	China	MZ766521	MZ766564	MZ766572	[13]
C. austrosinensis	GDGM81303	China	MZ605084	MZ613986	MZ614029	[13]
C. austrosinensis	GDGM81249 (HT)	China	MZ605082	MZ613983	MZ614027	[13]
C. austrosinensis	GDGM80616	China	MZ605081	MZ613982	MZ614026	[13]
C. austrosinensis	GDGM81381	China	MZ605086	MZ613988	MZ614031	[13]
C. austrosinensis	GDGM81379	China	MZ605085	MZ613987	MZ614030	[13]
C. austrosinensis	GDGM81985	China	MZ605087	MZ613989	MZ614032	[13]
C. chrysanthus	**GDGM45166**	**China**	**OM978959**	**ON119074**	**ON119011**	**Present study**
C. chrysanthus	**GDGM45937**	**China**	**OM978960**	**ON119075**	**ON119012**	**Present study**
C. chrysanthus	**GDGM85298**	**China**	**OM978975**	**ON119089**	**ON119025**	**Present study**
C. chrysanthus	**GDGM85305**	**China**	**OM978976**	**ON119090**	**ON119026**	**Present study**
C. chrysanthus	**GDGM53485**	**China**	**OM978962**	**ON119077**	**ON119014**	**Present study**
C. chrysanthus	**GDGM80220 (HT)**	**China**	**OM978970**	**ON119083**	**ON119019**	**Present study**
C. chrysanthus	**GDGM82511**	**China**	**OM978973**	**ON119087**	**ON119023**	**Present study**
C. chrysanthus	**GDGM82516**	**China**	**OM978974**	**ON119088**	**ON119024**	**Present study**
C. chrysanthus	**GDGM80436**	**China**	**OM978971**	**ON119084**	**ON119020**	**Present study**

Table 1. *Cont.*

Taxa	Voucher	Locality	GenBank Accession No.			Reference
			LSU	*tef*1	*rpb*2	
C. chrysanthus	**GDGM80202**	**China**	**OM978965**	**ON119080**	**ON119016**	**Present study**
C. chrysanthus	**GDGM80204**	**China**	**OM978966**	**ON119081**	**ON119017**	**Present study**
C. chrysanthus	**HMAS279434**	**China**	**ON212413**	**ON119091**	**ON229079**	**Present study**
C. chrysanthus	**GDGM80438**	**China**	–	**ON119085**	**ON119021**	**Present study**
C. chrysanthus	**GDGM82473**	**China**	**OM978972**	**ON119086**	**ON119022**	**Present study**
C. chrysanthus	**GDGM77035**	**China**	**OM978964**	**ON119079**	**ON229081**	**Present study**
C. chrysanthus	**GDGM60524**	**China**	**OM978963**	**ON119078**	**ON119015**	**Present study**
C. chrysanthus	**GDGM80217**	**China**	**OM978969**	**ON119082**	**ON119018**	**Present study**
C. chrysanthus	**GDGM49628**	**China**	**OM978961**	**ON119076**	**ON119013**	**Present study**
C. chrysanthus	**GDGM87950**	**China**	**OM978968**	–	**ON119027**	**Present study**
C. chrysanthus	**GDGM87951**	**China**	**OM978967**	–	**ON119028**	**Present study**
C. cibarius	GE 07.025	France	KF294658	GQ914949	KF294736	[3]
C. cibarius	BB 07.300	Slovakia	KF294641	GQ914950	KF294718	[3]
C. cinnabarinus	BB 04.263 (NT)	USA	–	GQ914983	–	[32]
C. cinnabarinus	BB 07.053	USA	KF294630	GQ914984	KF294705	[32]
C. cinnabarinus	BB 07.001	USA	KF294624	GQ914985	KF294698	[32]
C. citrinus	1691 (HT)	South Korean	–	MW124385	–	[16]
C. citrinus	1715	South Korean	–	MW124388	–	[16]
C. citrinus	1710	South Korean	–	MW124386	–	[16]
C. citrinus	1711	South Korean	–	MW124384	–	[16]
C. citrinus	**GDGM86140**	**China**	**OM978955**	**ON119070**	**ON119032**	**Present study**
C. citrinus	**GDGM86141**	**China**	**OM978956**	**ON119071**	**ON119033**	**Present study**
C. citrinus	**GDGM80825**	**China**	–	**ON119069**	**ON119031**	**Present study**
C. citrinus	**GDGM86142**	**China**	**OM978957**	**ON119072**	**ON119034**	**Present study**
C. citrinus	**GDGM80724**	**China**	**OM978954**	**ON119068**	**ON119030**	**Present study**
C. citrinus	**GDGM86143**	**China**	**OM978958**	**ON119073**	**ON119035**	**Present study**
C. citrinus	**GDGM80723**	**China**	**OM978953**	**ON119067**	**ON119029**	**Present study**
C. coccolobae	1064_RC. 14_24	Guadeloupe	KX857088	KX857020	KX856992	[33]
C. coccolobae	1065_RC. 11_25 (HT)	Guadeloupe	KX857089	KX857021	KX856993	[33]
C. congolensis	1645/BB16.044	Saharan Africa	KX857102	KX857075	KX857006	[33]
C. congolensis	1676/BB16.123	Saharan Africa	KX857106	KX857078	KX857010	[33]
C. aff. *congolensis*	BB 06.176	Madagascar	KF294606	–	KF294680	[3]
C. aff. *congolensis*	BB 06.197	Madagascar	KF294608	–	KF294683	[3]
C. convexus	**GDGM54841**	**China**	**OM978940**	**ON119052**	**ON119036**	**Present study**
C. convexus	**GDGM70307 (HT)**	**China**	**OM978941**	**ON119053**	**ON119037**	**Present study**
C. corallinus	1083_JJ_MO_CANT_2	USA	–	KX857031	–	[34]
C. corallinus	1086_JJ_MO_CANT_5	USA	–	KX857034	–	[34]
C. corallinus	FLAS_F_61106	USA	–	MK045368	–	[34]
C. curvatus	BRNM:825749 (HT)	South Korea		MW124390		[16]
C. cyphelloides	TNS F-61721 (HT)	Japan	NG059027	–	–	[35]
C. decolorans	BB 08.278 (HT)	Madagascar	KF294654	GQ914968	KF294731	[3]
C. fistulosus	DT_43	Tanzania	JQ976965	JX192997	–	[3]
C. friesii	AH44798	Spain	KR677522	KX828831	KX828752	[36]
C. friesii	VDKO 1165	Africa	–	KX834408	KX881922	[5]
C. galbanus	GDGM86249 (HT)	China	ZM766516	MZ766568	MZ766577	[13]
C. garnierii	BB 09.024	New Caledonia	KX857085	KX857017	KX856989	[34]
C. garnierii	BB 09.283	New Caledonia	KX857087	KX857019	KX856991	[34]
C. garnierii	BB 09.033	New Caledonia	KX857086	KX857018	KX856990	[34]
C. garnierii	RF33	New Caledonia	AY392768	–		[37]
C. garnierii	RF32	New Caledonia	AY392767	–		[37]
C. koreanus	1697	South Korea	–	KY271940	–	[11]
C. koreanus	1689 (HT)	South Korea	–	KY271941	–	[11]
C. koreanus	**GDGM85306**	**China**	**OM978978**	**ON119093**	**ON229077**	**Present study**
C. koreanus	**GDGM79233**	**China**	**OM978977**	**ON119092**	**ON229078**	**Present study**
C. koreanus	1693	South Korea	–	–		Unpublished
C. koreanus	1694	South Korea	–	–		Unpublished

Table 1. *Cont.*

Taxa	Voucher	Locality	GenBank Accession No.			Reference
			LSU	*tef*1	*rpb*2	
C. koreanus	1696	South Korea	–	–		Unpublished
C. luteolus	GDGM60393 (HT)	China	ZM766515	MZ766566	MZ766575	[13]
C. luteolus	GDGM86247	China	MZ766513	MZ766567	MZ766576	[13]
C. luteolus	GDGM44258	China	ZM766514	MZ766566	MZ766570	[13]
C. luteovirens	GDGM81079	China	MZ605092	MZ613994	MZ614036	[13]
C. luteovirens	GDGM80672 (HT)	China	MZ605090	MZ613992	MZ614035	[13]
C. luteovirens	GDGM80680	China	MZ605091	MZ613993	–	[13]
C. minioalbus	GDGM78910	China	MZ605098	MZ613999	MZ614043	[13]
C. minioalbus	GDGM78901 (HT)	China	MZ605097	MZ613998	MZ614042	[13]
C. minioalbus	GDGM78916	China	MZ605100	MZ614001	MZ614045	[13]
C. minor	BB 07.057	USA	KF294632	JX192979	KF294707	[3]
C. minor	BB 07.002	USA	KF294625	JX192978	KF294699	[3]
C. neopersicinus	**GDGM85145-1**	**China**	**OM978942**	**ON119054**	**ON119039**	**Present study**
C. neopersicinus	**GDGM85145-2**	**China**	**OM978945**	**ON119055**	**ON119040**	**Present study**
C. neopersicinus	**GDGM85145-3**	**China**	**OM978946**	**ON119056**	**ON119041**	**Present study**
C. neopersicinus	**GDGM87366-1 (HT)**	**China**	**OM978943**	**ON119057**	**ON119042**	**Present study**
C. neopersicinus	**GDGM87366-2**	**China**	**OM978944**	**ON119058**	**ON119043**	**Present study**
C. phloginus	**GDGM79007-1**	**China**	**OM978979**	**ON119094**	**ON119044**	**Present study**
C. phloginus	**GDGM79007-2**	**China**	**OM978980**	**ON119095**	**ON119045**	**Present study**
C. phloginus	SSC99 (HT)	China	–	KF801096	–	[22]
C. phloginus	SSC98	China	–	KF801095	–	[22]
C. phloginus	Yuan14468	China	–	MW999424.	–	[7]
C. phloginus	Yuan14490	China	–	MW999425	–	[7]
C. phloginus	**GDGM82514**	**China**	–	**ON119096**	–	**Present study**
C. pseudominimus	JV 00.663	Portugal	KF294657	JX192991	KF294735	[3,10]
C. romagnesianus	AH44218	Spain	KX828807	KX828836	KX828757	[36]
C. roseofagetorum	AH44789	Georgia	KX828812	KX828839	KX828760	[36]
C. sinocinnabarinus	**GDGM83229**	**China**	**OM978983**	**ON119098**	**ON119047**	**Present study**
C. sinocinnabarinus	**GDGM83238**	**China**	**OM978985**	**ON119101**	**ON119051**	**Present study**
C. sinocinnabarinus	**GDGM83023**	**China**	**OM978981**	**ON119097**	**ON119050**	**Present study**
C. sinocinnabarinus	**GDGM83232**	**China**	–	**ON119100**	**ON119049**	**Present study**
C. sinocinnabarinus	**GDGM83027**	**China**	**OM978982**	–	**ON119046**	**Present study**
C. sinocinnabarinus	**GDGM83230 (HT)**	**China**	**OM978984**	**ON119099**	**ON119048**	**Present study**
C. sinocinnabarinus	HKAS58243	China	JF906727	–	–	[20]
C. sinominor	GDGM80788	China	MZ605105	MZ614004	MZ614048	[13]
C. sinominor	GDGM80842 (HT)	China	MZ605107	MZ614006	MZ614050	[13]
C. sinominor	GDGM80885	China	MZ605108	MZ614007	MZ614051	[13]
C. aff. subcyanoxanthus	BB 98.014	Tanzania	KF294615	JX192973	KF294689	[3]
C. tabernensis	BB 07.119	USA	KF294634	GQ914976	KF294709	[3]
C. tabernensis	BB 07.056 (ET)	USA	KF294631	GQ914974	KF294706	[3,38]
C. texensis	341/O7.120	USA	JN940601	GQ914987	KF294710	[3]
C. texensis	BB 07.018	USA	KF294626	GQ914988	KF294701	[3]
C. xanthocyaneus	1751	Congo	MT006309	MT002277	–	[39]
C. xanthocyaneus		Congo	MT006310	MT002278	–	[39]
C. zangii	GDGM82389	China	MZ605110	MZ614009	MZ614053	[13]
C. zangii	GDGM82393	China	MZ605111	MZ614010	MZ614054	[13]
C. zangii	GDGM82374	China	MZ605109	MZ614008	MZ614052	[13]

3. Results

3.1. Molecular Phylogeny

For phylogenetic analyses, a total of 152 sequences were newly produced in this study, containing 49 nrLSU, 51 *tef1* and 52 *rpb2*, and 185 reliable sequences were downloaded from the GenBank database based on previous studies [3,13]. The combined dataset (LSU + *tef1* + *rpb2*) contained 2892 characters (1311, 707 and 874 for LSU, *tef1* and *rpb2*, respectively), of which 2013 were conserved and 708 were parsimony-informative. ML and BI analyses of the concatenated data set resulted in almost identical topologies, and no strongly-supported conflicts between ML and BI analyses were discovered; thus, only the tree inferred from ML analysis was displayed (Figure 1). Our phylogenetic analyses indicated that members of *C.* subg. *Cinnabarinus* formed a highly support monophyletic group (MLB/BPP = 100%/1.0). Five well-supported clades in the subg. *Cinnabarinus* were identified based on samples newly collected from China, including two new species, two species newly recorded in China and a known species in China. Besides, three well-supported clades in the subg. *Parvocantharellus* were firstly discovered in China, containing two new species and a newly recorded species from China.

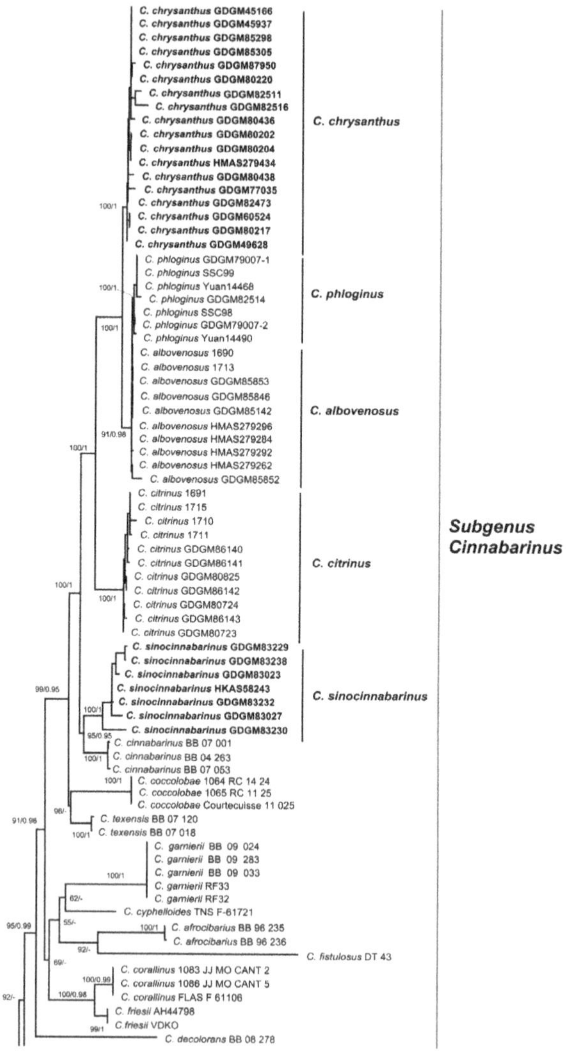

Figure 1. *Cont.*

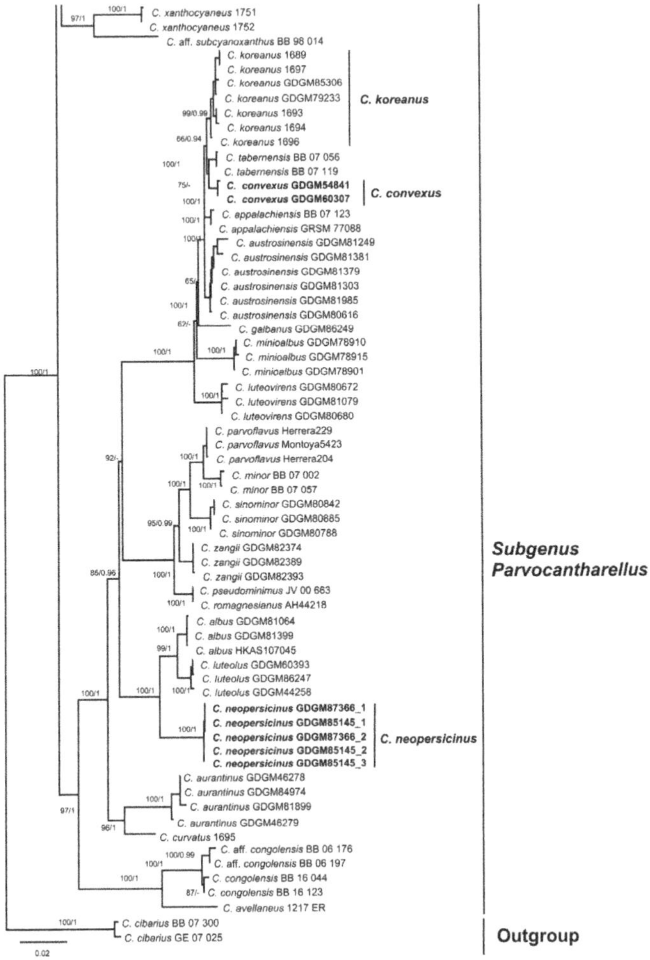

Figure 1. Phylogenetic tree of representative species of *Cantharellus* inferred from LSU-*tef1*-*rpb2* dataset by means of both ML and BI methods. *Cantharellus cibarius* Fr. served as outgroup. Bootstrap Supports (BS > 50%) and Bayesian Posterior Probabilities (BPP > 0.90) are shown on the supported branches. Bold names represent new species.

3.2. *Taxonomy*

3.2.1. *Cantharellus* subgen. *Cinnabarinus* Buyck & V. Hofst.

Cantharellus chrysanthus Ming Zhang, C.Q. Wang & T.H. Li sp. nov.; Figures 2 and 3.

MycoBank: MB843657.

GenBank: OM978970 for LSU, ON119083 for *tef1* and ON119019 for *rpb2*.

Etymology—refers to the color of pileus similar to the yellow chrysanthemum flower.

Diagnosis—This species is characterized by its orange to orange-yellow pileus, pinkish white to orange white hymenophore, thin-walled pileipellis terminal hyphae, broadly ellipsoid basidiospores (7.5–9 × 5–6.5 μm) and long basidia up to 100 μm.

Type—CHINA. Guangdong Province, Shaoguan City, Ruyuan town, Nanling National Natural Reserve, alt. 500 m, 10 June 2020, Ming Zhang (GDGM80220).

Figure 2. Basidiomata of *Cantharellus chrysanthus*. (**a**,**b**) GDGM80220, holotype. (**c**) GDGM60524. (**d**) GDGM80438. (**e**) GDGM82516. (**f**) GDGM80217. (**g**) GDGM49628. (**h**) GDGM80436. (**i**) GDGM60334. (**j**) GDGM80202. (**k**) GDGM45937. (**l**) GDGM85298. (**m**) GDGM82473. Bars = 2 cm.

Basidiomata small-sized. Pileus 20–60 mm broad, convex, with involute margin when young, then gradually to nearly applanate or broadly infundibuliform with depressed center and inflexed to straight, irregularly undulate or slightly cracked at maturity; surface dry or hygrophanous, glabrous or finely subtomentose, orange (5A7–6A7) to deep orange (5A8–6A8) when young, slightly fading to orange yellow to yellow (3A7–4A7) when mature. Context yellowish white to orange white (4A2–6A2), 1–2 mm thick in the center of the pileus, sharply attenuate towards margin, unchanging when exposed. Hymenophore decurrent, subdistant, composed of bifurcate, 2–3 mm high venose folds, particularly towards pileus margin, pinkish white (7A2–10A2), but in some specimens yellowish white to orange white, unchanging when bruised. Stipe 20–60 × 3–14 mm, central, cylindrical or slightly tapering towards base, solid, glabrous or finely pubescent, concolorous with pileus or paler, unchanging when handled. Odor fruity and pleasant. Taste mild.

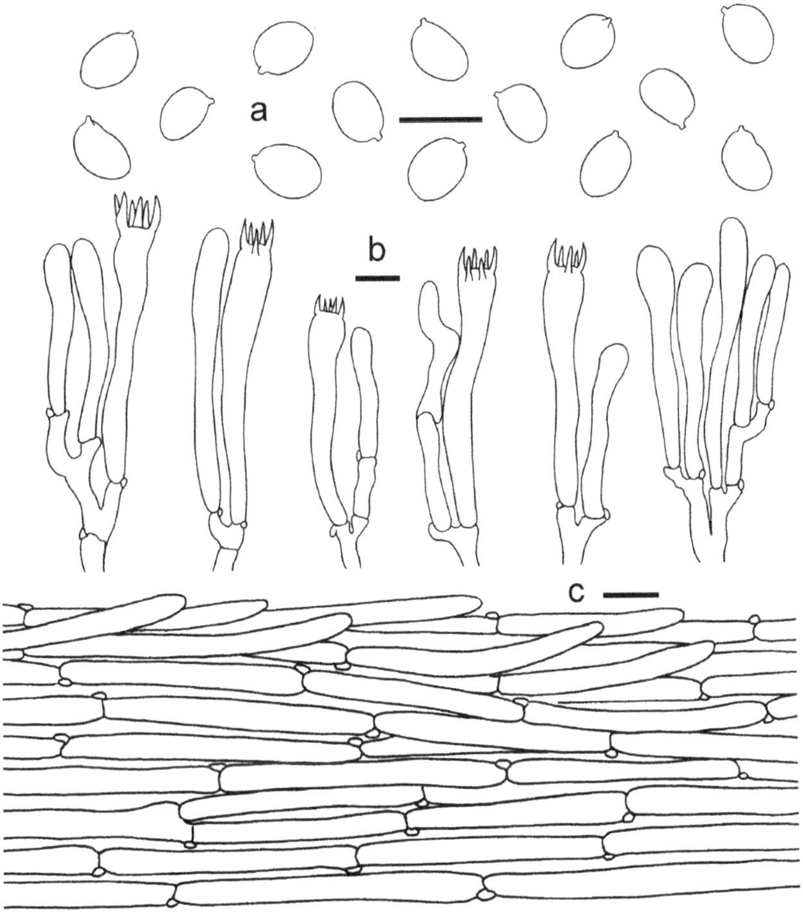

Figure 3. *Cantharellus chrysanthus* (GDGM80220, Holotype). (**a**) Basidiospores. (**b**) Basidia, basidiola and elements of the subhymenium. (**c**) Pileipellis. Bars: (**a**,**b**) = 10 μm; (**c**) = 20 μm.

Basidiospores 7.5–9 × 5–6.5 μm, $L_m \times W_m$ = 8.45(±0.47) × 5.98(±0.42) μm, Q = (1.25) 1.28–1.6(1.64), Q_m = 1.42 ± 0.1; broadly elliptical to subglobose, smooth, guttulate, thin-walled. Basidia 55–100 × 7–11 μm, 2–6-spored, narrowly clavate, colorless to hyaline in KOH; sterigmata 6–10 μm long. Pileipellis a cutis with long, repent and occasionally interwoven hyphae, subcylindrical cells that are 6–12 μm wide, thin-walled. Stipitipellis a cutis of cylindrical, parallel hyphae, 3–8 μm wide. Clamp connections abundant in all tissues.

Habitat and distribution—Solitary or scattered under Fagaceae trees mixed with other broadleaf trees in subtropical forests. Known from southern and southwestern China.

Additional specimen examined—China. Guangdong Province, Shaoguan City, Ruyuan town, Nanling National Natural Reserve, alt. 500 m, 7 June 2017, Ming Zhang (GDGM49628); same location, alt. 500 m, 21 July 2017, Ming Zhang (GDGM60524); same location, alt. 500 m, 9 June 2020, Ming Zhang (GDGM80436, GDGM80438); same location, alt. 500 m, 10 June 2020, Ming Zhang (GDGM80202, GDGM80204, GDGM80217, GDGM80220,); Huizhou city, Xiangtoushan National Natural Reserve, alt. 550 m, 17 May 2016, Ting Li (GDGM45937); Hunan Province, Rucheng town, Jiulongjiang National Forest Park, alt. 300 m, 4 September 2016, Ming Zhang (GDGM53485); Zhejiang province, Jinhua city, Wuyi

Town, 23 August 2015, Tai-Hui Li (GDGM45166); Hangzhou City, Laohushan, 15 July 2021, Bao-Juan Ling (GDGM85298, GDGM85305); Qingyuan Town, Baishanzu National Natural Reserve, alt. 29 July 2020, Tai-Hui Li (GDGM82473); Longquan City, Fengyangshan National Natural Reserve, 25 August 2016, Rui-Lin Zhao (ZRL20161616, HMAS279434); Quzhou City, Kaihua County, He Tian township, Chi Keng village, 24 May 2021, Yi Li (GDGM87950); Quzhou City, Kaihua County, Shengtangou Scenic Spot, 30 May 2021, Yi Li (GDGM87951); Anhui Province, Huangshan City, Huangshan scenic spot, 11 August 2020, Ming Zhang (GDGM82511), same location, 13 August 2020, Ming Zhang (GDGM82516); Guizhou Province, Guiyang City, Longli County, Guanyin Village, bought from a wild mushroom market, 2 August 2019, alt. 1000 m, Yong He (GDGM77035).

Notes—*Cantharellus chrysanthus* is different from other *Cantharellus* species by the combined features of the orange to orange-yellow pileus, the pinkish white to orange white hymenophore, the thin-walled terminal hyphae of pileipellis, the broadly ellipsoid basidiospores (7.5–9 × 5–6.5 μm) and the long basidia up to 100 μm.

Phylogenetically, *C. chrysanthus* is related to *C. albovenosus* and *C. phloginus* in the analyses of the multi-locus datasets. However, *C. albovenosus* differs in its orange to reddish orange pileus with tomentoum or fibrilla, white to orange white and better-developed hymenophore, orange to reddish orange stipe, smaller basidiospores (7–8.5 × 5–6 μm) and shorter basidia (48–63 × 7–9 μm) [11]; *Cantharellus phloginus*, reported from southwest China, differs in its pastel red to pastel pink pileus and stipe, pale yellow to light yellow hymenophore, larger basidiospores [6.8–9.5 (–12) × 5–7 μm] and shorter basidia (60–95 × 8–10 μm) [22].

Cantharellus sinocinnabarinus Ming Zhang, S.C. Shao & T.H. Li sp. nov.; Figures 4 and 5.

Figure 4. *Cont.*

Figure 4. Basidiomata of *Cantharellus sinocinnabarinus*. (**a**,**c**) GDGM83230. (**b**) GDGM83232. (**d**) GDGM83229. (**e**) GDGM832296. (**f**) GDGM83027. (**g**) GDGM83238. (**h**) HKAS58243. Bars = 2 cm.

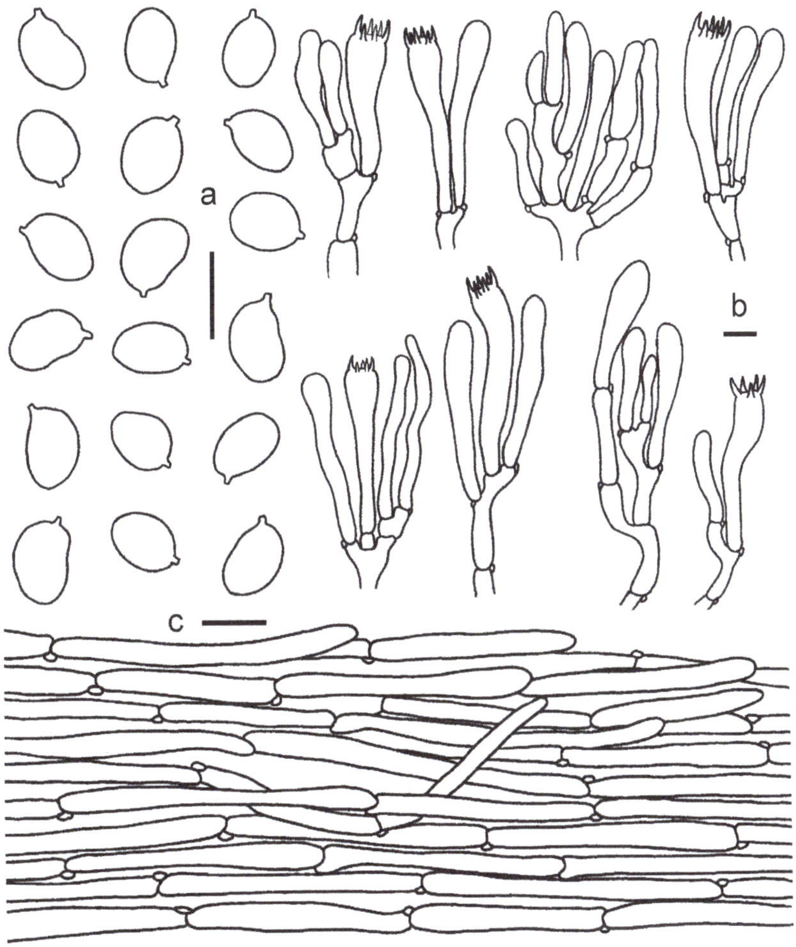

Figure 5. *Cantharellus sinocinnabarinus*. (**a**) Basidiospores. (**b**) Basidia, basidiola and elements of the subhymenium. (**c**) Pileipellis. Bars: (**a**,**b**) = 10 μm; (**c**) = 20 μm.

MycoBank: MB843658.
GenBank: OM978984 for LSU, ON119099 for *tef1* and ON119048 for *rpb2*.

Etymology—Refers to the species distributed in China and is similar to *C. cinnabarinus* in morphology.

Diagnosis—This species is characterized by its small basidiomata, reddish orange to yellowish red pileus covered with white minute fibrils, yellowish orange to orange hymenophore and elongate elliptical basidiospores measuring (6.5–) 7–8 (9) × (4.5) 5–6 μm.

Type—China. Yunnan Province, Lijiang City, Yulong County, Jiuhe Village, 1 September 2020, alt. 2400 m, Ming Zhang (GDGM83230).

Basidiomata small-sized. Pileus 5–15 mm broad, applanate with a depressed center, not perforate; margin slightly incurved when young, applanate to reflexed with age; surface dry, orange, reddish orange to yellowish red (6A7–8A7), locally with white minute fibrils. Context thin, 0.5–1.5 mm thick, fleshy to fibrous, yellowish orange to reddish orange, unchanging when bruised. Hymenophore subdecurrent, with a clearly delimitation from stipe surface; lamellate ridges subdistant to close, well-developed, 1–2 mm high, appropriately bifurcate, with low interconnected low venose folds, particularly at pileus margin, yellowish orange to orange (4A7–6A7), unchanging when bruised. Stipe 10–15 mm long, 1–2.5 mm thick, subcylindrical, slightly tapering downward, glabrous or with obscure white minute fibrils, hollow, concolorous with pileus. Odor pleasant.

Basidiospores (100/4/4) (6.5)7–8(9) × (4.5)5–6 μm, $L_m \times W_m$ = 7.47(±0.5) × 5.21(±0.39) μm, Q = (1.25)1.27–1.6(1.67), Q_m = 1.43 ± 0.09; elliptical to elongate elliptical. Basidia 50–75 × 10–12 μm, clavate, with 4–8 sterigmata. Pileipellis a cutis, composed of procumbent hyphae; hyphae 4–13 μm in diam., colorless, thin-walled. Hymenophoral trama composed of cylindrical hyphae 5–10 μm in diam. Stipitipellis a cutis, composed of procumbent, branched hyphae; hyphae 4–12 μm in diam., mostly 7 μm in diam. Cystidia absent. Clamp connections common.

Habitat and distribution—Gregarious on soil in subalpine mixed forest dominated by *Cyclobalanopsis delavayi* (Franch.) Schott. and *Pinus yunnanensis* Franch. Currently known from southwest China.

Additional specimens examined—China. Yunnan Province, Jianchuan County, Qianshi Mountain, 7 September 2009, alt. 2491 m, Yu23 (HKAS58243); Lijiang City, Yulong County, Jiuhe Village, 1 September 2020, alt. 2400 m, Ming Zhang (GDGM83229, GDGM83232, GDGM83027), Li-Qiang Wu (GDGM83238).

Notes—*Cantharellus sinocinnabarinus* can be easily recognized in the field by its small reddish orange basidiomata. Morphologically, *C. sinocinnabarinus* is similar to *C. cinnabarinus*, *C. persicinus* R.H. Petersen and *C. texensis*. However, the latter three species were all originally reported from North America; *C. cinnabarinus* and *C. persicinus* differ in their larger basidiomata (pileus up to 40 mm), thicker-walled hyphae of pileipellis terminal cells, and different sizes of basidiospores (6.7–7.57 × 3.82–4.68 μm for *C. cinnabarinus*, and 10.2–11.9 × 6.3–7.2 μm for *C. persicinus*) [32]; *C. texensis* differs in its robust basidiomata and longer but narrower basidiospores (8–8.95 × 3.7–4.3 μm), with a larger Q value (1.8–2.2) [32].

Shao et al. [20] has described a specimen (HKAS58243) under the name *C. cinnabarinus* on the basis of the LSU sequence, which is geographically close to *C. sinocinnabarinus* in southwest China. In this study, the specimen (HKAS58243) was re-examined; the morphological features and molecular phylogenetic analyses all demonstrated that it is actually *C. sinocinnabarinus*.

In the multi-locus phylogentic trees, specimens of *C. sinocinnabarinus* formed a well-supported independent terminal branch (BS = 100%, BPP = 1.0) in the subg. *Cinnabarinus*, and are closely related to *C. cinnabarinus*. However, they can be easily distinguished by the morphological features and large genetic distance.

Cantharellus albovenosus Buyck, Antonín & Ryoo, in Antonín, Hofstetter, Ryoo, Ka and Buyck, Mycol. Progr. 16(8): 757 (2017); Figures 6 and 7.

Basidiomata small-sized. Pileus 20–55 mm broad, convex at first, then broad applanate with a depressed centre, subinfundibuliform when mature or old; margin inflexed to straight when young, then undulate; surface tomentose when young, then glabrescent and radially (innately) fibrillose to finely striate and rugulose, orange, deep orange to reddish orange (5A6–7A6, 5A8–7A8), then pallescent to light orange at margin. Hymenophore decurrent, with a clearly delimitation from stipe surface; lamellate ridges, subdistant to distant, relatively well-developed, 1–1.5 mm high, appropriately bifurcate and interconnected with low veined folds, particularly towards pileus margin, white to orange white (5A2–6A2), unchanging when bruised. Stipe 25–50 × 2.5–9 mm, cylindrical and slightly clavate to bulbose at base, finely tomentose when young, then glabrous or with finely longitudinally fibrillose, concolorous with pileus, orange to reddish orange, sometimes paler to light orange in some specimens. Context white, orangish under pileipellis, solid, becoming hollow-fibrous in stipe. Odor spicy. Taste mild.

Basidiospores 7–8.5 × 5–6 µm, $L_m \times W_m$ = 7.9(±0.48) × 5.5(±0.34) µm, Q = (1.33)1.4–1.5(1.54), Q_m = 1.44 ± 0.05; ellipsoid to subglobose, thin-walled, sometimes with granulose contents. Basidia 48–63 × 7–9 µm, 2–6-spored, clavate, sometimes subcapitate. Hymenial trama hyphae cylindrical to subinflated, sometimes irregular, thin-walled, 3–8 µm wide. Pileipellis a cutis composed of cylindrical, rarely subinflated, thin-walled, 4–10 µm wide hyphae; terminal cells 37–87 × 5–8 µm, adpressed, cylindrical, clavate or subfusoid. Stipitipellis a cutis of cylindrical, parallel, thin-walled, clamped, 3–7 µm wide hyphae.

Habitat and distribution—Scattered or gregarious on soil under mixed forest dominated by Fagaceae trees. Known to be from eastern China and Korea.

Specimens examined—China. Jiangsu Province, Nanjing City, Purple Mountain, 19 June 2021, alt. 150 m, Zi-Hang Zhang (GDGM85846); same location, 28 June 2021, Zi-Hang Zhang (GDGM85852, GDGM85853); Anhui Province, Huangshan National Scenic Area, 26 August 2021, alt. 1400 m, Chen-Jie Jiang (GDGM85142). Zhejiang Province, Lishui City, Jingning Town, Wangdongyang Alpine Wetland Nature Reserve 22 September 2016, Rui-Lin Zhao (HMAS279296, HMAS279292); same location, 23 September 2016, Rui-Lin Zhao (HMAS279262, HMAS279284).

Figure 6. Basidiomata of *Cantharellus albovenosus*. (**a**,**b**) GDGM85852. (**c**,**d**) GDGM85846. (**e**) GDGM85142. Bars = 2 cm.

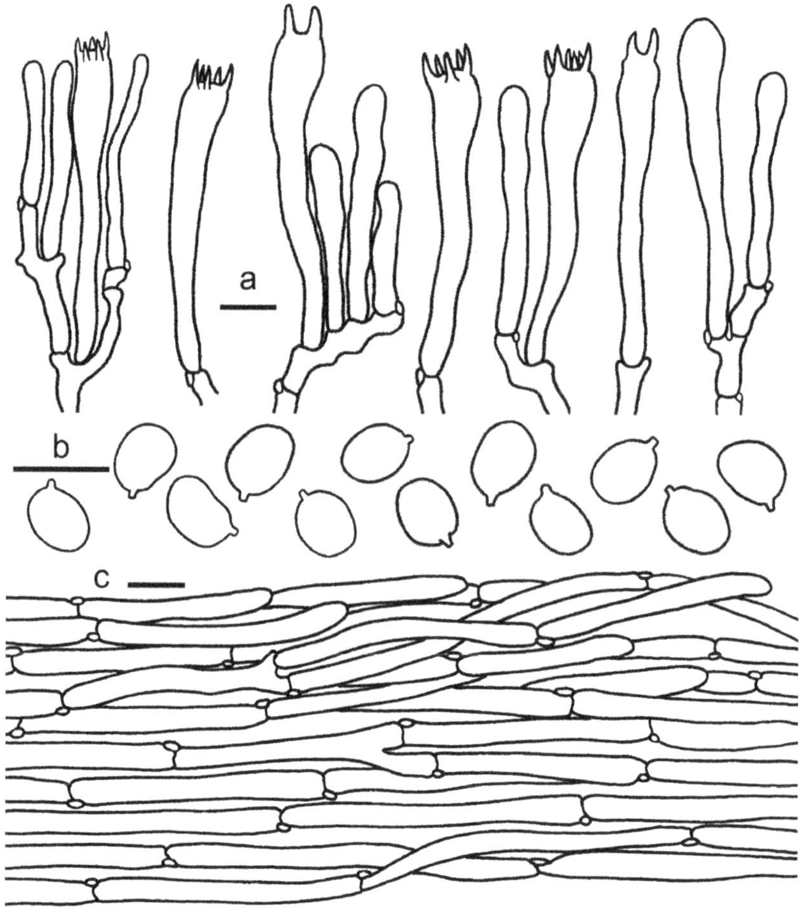

Figure 7. *Cantharellus albovenosus*. (**a**) Basidia, basidiola and elements of the subhymenium. (**b**) Basidiospores. (**c**) Pileipellis. Bars: (**a**,**b**) = 10 μm; (**c**) = 20 μm.

Notes—*Cantharellus albovenosus*, recently reported from South Korea, is characterized by the combined features of the orange to reddish orange pileus, white to orange white and relatively well-developed lamellate hymenophore, the orange to reddish orange stipe, and the ellipsoid to nearly globose basidiospores (7–8.5 × 5–6 μm) [11]. Phylogenetically, *C. albovenosus* and *C. phloginus* clustered together in an almost similar phylogenetic position, and cannot be separated in our multi-locus phylogenetic tree (Figure 1). Morphologically, *C. phloginus* can be distinguished by its pastel red to pastel pink pileus and stipe, pale yellow to yellowish orange hymenophore and large basidiospores [6.8–9.5 (−12) × 5–7 μm] [22]. Ecologically, *C. albovenosus* is known from subtropical regions of South Korea and eastern China; meanwhile, *C. phloginus* is currently only known from tropical regions of southwest China. The distinguishable morphological features and different growth habits supported them as two distinct species, but some more effective molecular markers are needed to distinguish the two species.

Cantharellus citrinus Buyck, R. Ryoo & Antonín, in Buyck, Hofstetter, Ryoo, Ka and Antonín, MycoKeys 76: 35 (2020); Figures 8 and 9.

Basidiomata small-sized. Pileus 15–45 mm broad, convex, with involute margin when young, then gradually to broadly infundibuliform with depressed center, irregularly undulate or slightly cracked margin when old; surface dry or hygrophanous, glabrous or finely subtomentose, greenish yellow, light yellow, yellow to yellowish orange (1A4–4A4, 1A7–4A7). Context yellowish white, 1 mm thick in the center of the pileus, sharply attenuate towards margin, unchanging when exposed. Hymenophore decurrent, subdistant, composed of bifurcate, less than 1 mm high veined folds, particularly towards pileus margin, white to yellowish white (1A2–3A2), unchanging when bruised. Stipe 15–30 × 3–5 mm, central, cylindrical or slightly tapering towards base, hollow, glabrous, concolorous with pileus or paler, unchanging when handled. Odor fruity and pleasant. Taste mild.

Basidiospores 7–9 × 5–6(6.5) μm, $L_m \times W_m$ = 7.77(±0.47) × 5.29(±0.40) μm, Q = (1.17)1.23–1.6(1.64), Q_m = 1.47 ± 0.11; broadly elliptical to subglobose, smooth, guttulate, thin-walled. Basidia 55–65 × 7–8 μm, 4–6-spored, narrowly clavate, colorless to hyaline in KOH; sterigmata 5–10 μm long. Pileipellis a cutis with long, repent and occasionally interwoven hyphae, subcylindrical cells that are 5–15 μm wide, thin-walled. Stipitipellis a cutis of cylindrical, parallel hyphae, 5–10 μm wide. Clamp connections abundant in all tissues.

Habitat and distribution—Gregarious on soil under mixed forests in southwest China. Known from southwest China and Korea.

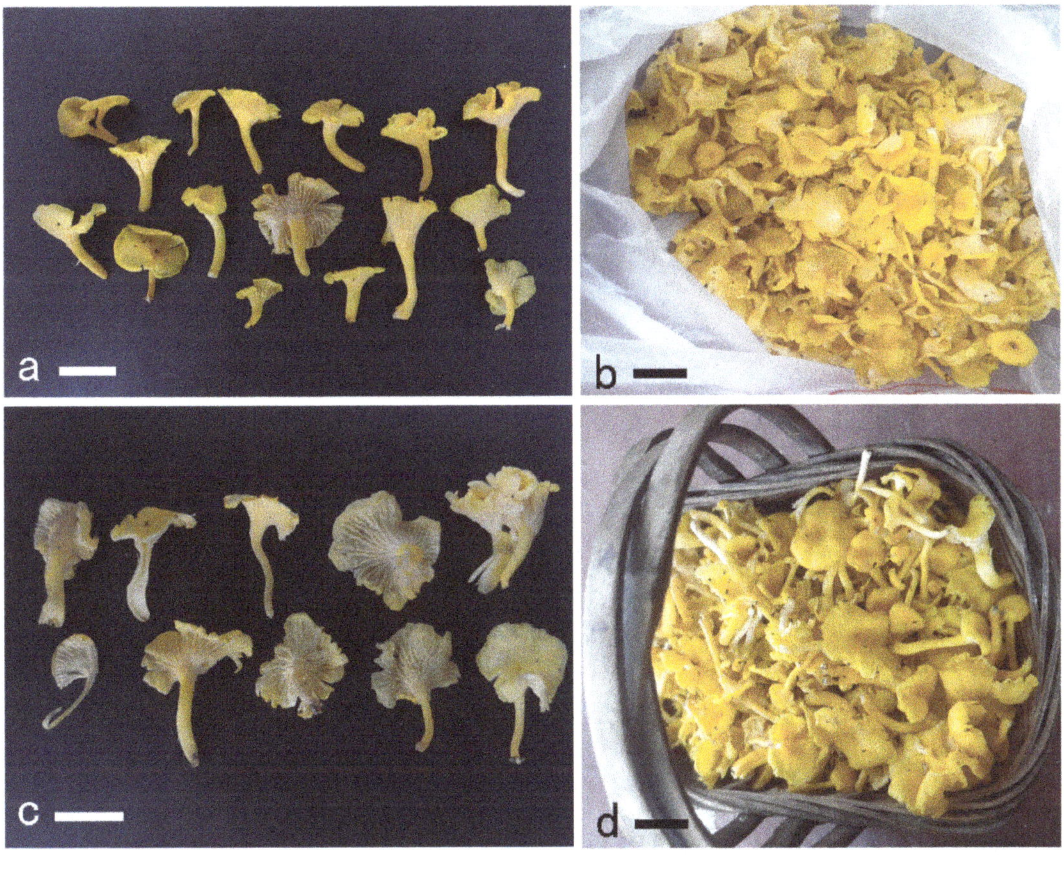

Figure 8. Basidiomata of *Cantharellus citrinus*. (**a**,**b**) GDGM86143. (**c**) GDGM86141. (**d**) GDGM80723. Bars = 2 cm.

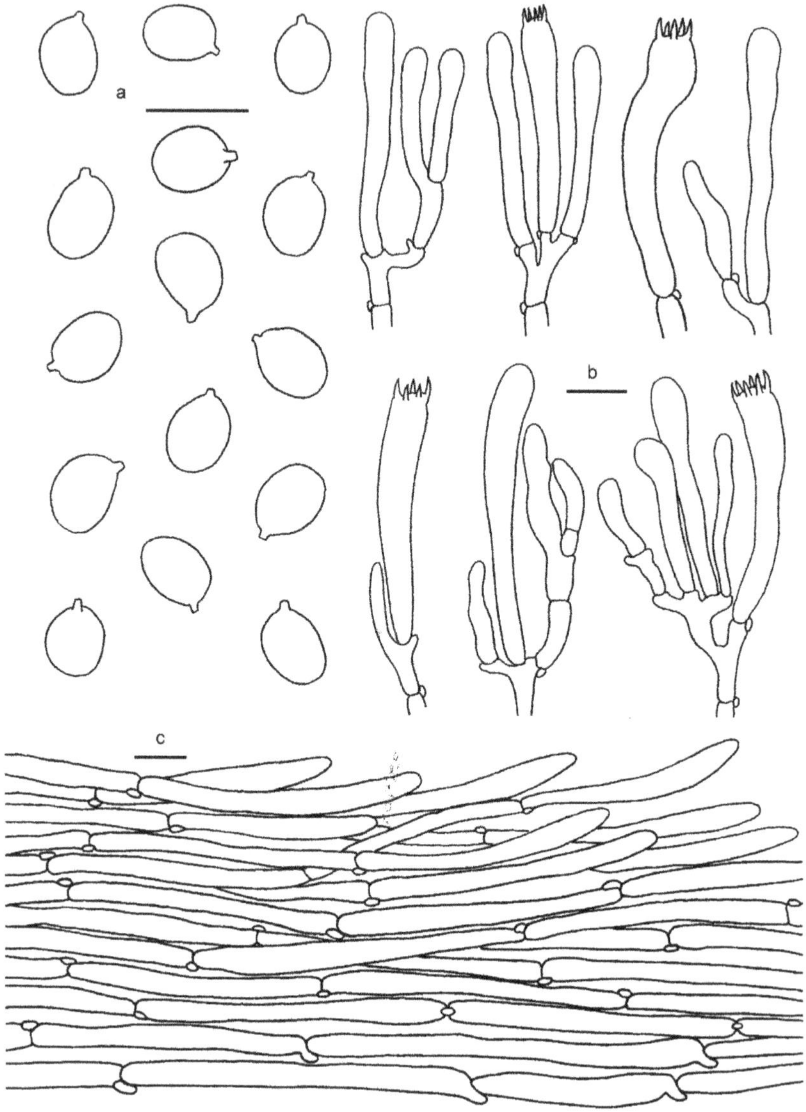

Figure 9. *Cantharellus citrinus*. (**a**) Basidiospores. (**b**) Basidia, basidiola and elements of the subhymenium. (**c**) Pileipellis. Bars: (**a**,**b**) = 10 μm; (**c**) = 20 μm.

Specimen examined—China. Guizhou Province, Guiyang City, Longli County, Guanyin Village, bought from a wild mushroom market, 1 July 2020, alt. 1000 m, Ming Zhang (GDGM80825); Same location, 16 June 2020, Ting Li (GDGM80724, GDGM80723); 7 July 2021, Ming Zhang (GDGM86140, GDGM86141, GDGM86142, GDGM86143).

Notes—*Cantharellus citrinus*, recently reported from Korea [11], is characterized by its small basidiomata, greenish yellow to yellowish orange pileus, white to yellowish white hymenophore strongly bifurcate at pileus margin, glabrous and hollow stipe, and broadly elliptical to subglobose basidiospores [7–9 × 5–6 (6.5) μm]. In the multi-locus phylogentic tree, samples of *C. citrinus* formed a well-supported monophyletic terminal clade, and can be easily distinguished from other *Cantharellus* species.

Morphologically, *C. citrinus* might be easily identified as a species in the subg. *Parvocantharellus* by the small basidioma with a greenish yellow to yellowish orange pileus, and similar to *C. galbanus* Ming Zhang, C.Q. Wang & T.H. Li and *C. luteovirens* Ming Zhang, C.Q. Wang & T.H. Li. However, *C. galbanus*, recently reported from tropical China, differs in its smaller basidiomata, relatively well-developed hymenophore, and smaller basidiospores (6–7.5 × 4.8–5.5 μm) [13]; *C. luteovirens*, recently reported from subtropical China, differs in its yellow to yellowish-orange pileus, yellowish white to pale yellow hymenophore and smaller basidiospores (6–7.5 × 4.5–6 μm) [13].

Cantharellus phloginus S.C. Shao & P.G. Liu, in Shao, Buyck, Tian, Liu and Geng, Mycoscience 57(2): 146 (2016); Figures 10 and 11.

Basidiomata small to medium-sized. Pileus 20–60 mm broad, applanate with a concave center, margin incurved at first, then becoming applanate or slightly reflexed with age, glabrous, pastel red to pastel pink (7A4–11A4); Context 2–3 mm thick, white, with pinkish hues under pileipellis, unchanging when bruised; Hymenophore decurrent, well-developed, lamellate ridges with anastomosing veins, forking towards pileus margin, pale yellow to light yellow (3A3–4A3), unchanging when touched. Stipe 20–40 × 4–8 mm, central, solid, subcylindrical, or slightly tapering towards base, glabrous, concolorous with pileus or paler to pinkish with yellowish hues, unchanging when handled. Odor fruity. Taste pleasant.

Basidiospores 6.8–9.5 (–12) × 5–7 μm, $L_m \times W_m$ = 8.49(±1.09) × 5.71(±0.69) μm, Q = (1.33)1.36–1.6(1.7), Q_m = 1.49 ± 0.18; broadly ellipsoid to subglobose, smooth, guttulate. Basidia 60–95 × 8–10 μm, 2–6-spored, narrowly clavate, colorless to hyaline in KOH; sterigmata 3–7 μm long. Hymenophoral trama composed of cylindrical interwoven hyphae 3–13 μm in diam. Pileipellis a subcutis, composed of long, repent, branched, and slightly interwoven hyphae, with subcylindrical cells in 3–13 μm wide, thin-walled. Clamp connections abundant in all tissues.

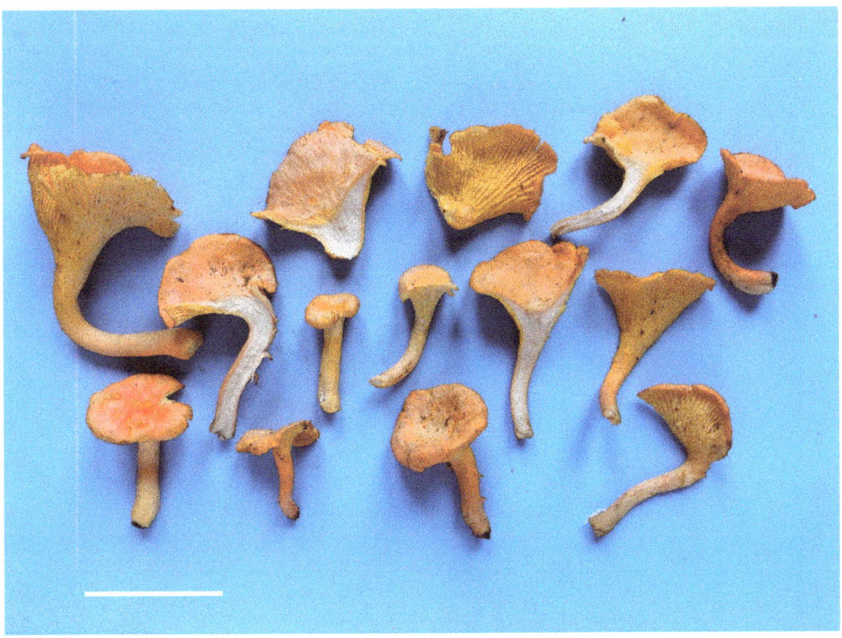

Figure 10. Basidiomata of *Cantharellus phloginus* (GDGM79007). Bar = 5 cm.

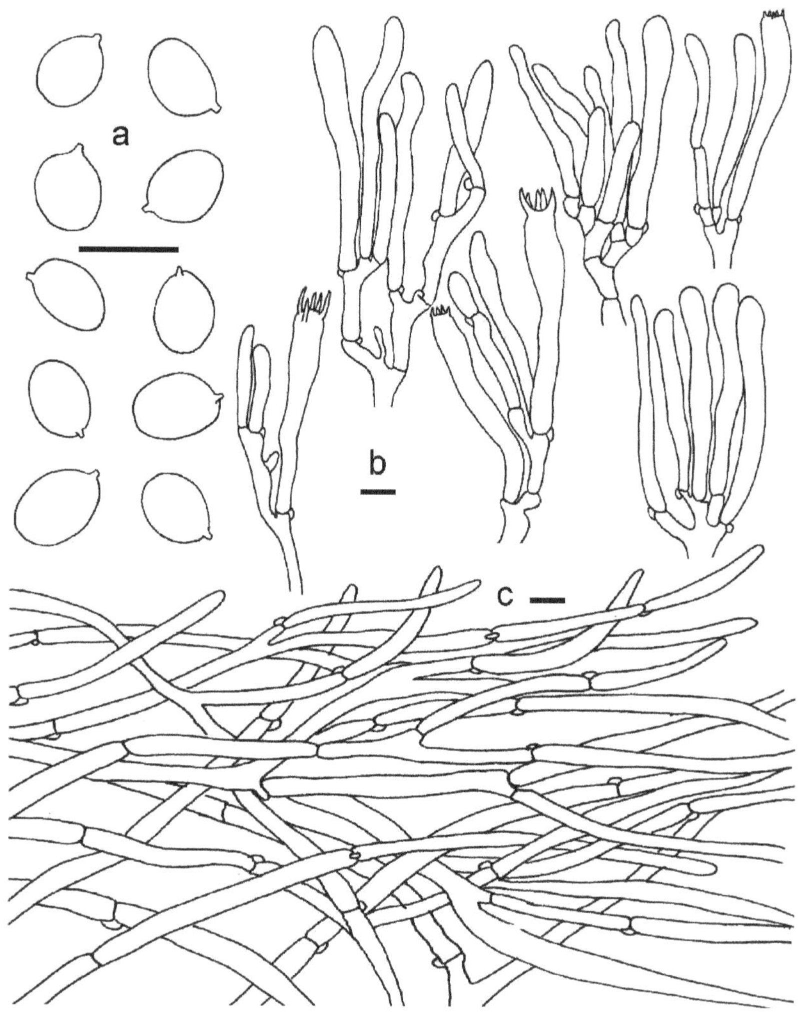

Figure 11. *Cantharellus phloginus*. (**a**) Basidiospores. (**b**) Basidia, basidiola and elements of the subhymenium. (**c**) Pileipellis. Bars: (**a**,**b**) = 10 μm; (**c**) = 20 μm.

Habitat and distribution—Gregarious or caespitose under mixed forests, dominated by *Pinus* sp. and *Castanopsis* in the tropical forest. Currently known to be southwest China.

Specimens examined—China. Yunnan Province, Puer City, alt. 1500 m, 26 August 2009, S.C. Shao 98 (HKAS58208, holotype); Puer City, bought from a mushroom market, alt. 1500 m, 28 September 2019, Ming Zhang (GDGM79007).

Notes—*Cantharellus phloginus*, recently reported from southwest China, is characterized by its pastel red to pastel pink pileus and stipe, pale yellow to yellowish orange, well-developed hymenophore, and ellipsoid basidiospores [6.8–9.5 (–12) × 5–7 μm] [22]. Morphologically, *C. phloginus* is similar to *C. cinnabarinus* and *C. texensis* Buyck & V. Hofst with the pinkish red pileus color. However, *C. cinnabarinus* differs in its small basidiomata, reddish pink pileus, small basidiospores [(6.4) 6.7–7.5 (8.1) × (3.7) 3.8–4.6 (5.2) μm] and thick-walled pileipellis [32]; *C. texensis* differs in its slender basidiomata, reddish pink pileus, relatively well developed hymenophore, small basidiospores [8–8.95 (9.4) × (3.3) 3.7–4.3 μm], and thinner-walled pileipellis that is faintly covered with zebroid incrustation [32]. Ecologically, *C. phloginus* occurs under trees of *Pinus* sp. and *Castanopsis* sp. in tropical regions of southwest China, while *C. cinnabarinus* and *C. texensis* occur on sandy loam in oak-pine forests in temperate regions of North America [32].

3.2.2. *Cantharellus* subgen. *Parvocantharellus* Eyssart. & Buyck

Cantharellus convexus Ming Zhang & T.H. Li sp. nov.; Figures 12 and 13.

MycoBank: MB843659.

GenBank: OM978941 for LSU, ON119053 for *tef1* and ON119037 for *rpb2*.

Etymology— "*convexus*" refers to the convex of the pileus center.

Diagnosis—This species can be easily distinguished from others in *Cantharellus* by its small basidiomata, yellowish white pileus, distant and well-developed lamellate hymenophore with or without bifurcate low veins and smaller basidiospores at 6–7 × 4.5–5 μm.

Type—China. Guangdong Province, Shaoguan City, Nanling National Nature Reserve, alt. 800 m, 29 July 2017, Ming Zhang (GDGM70307).

Basidiomata small-sized. Pileus 5–12 mm broad, convex when young, then gradually to nearly applanate with a central shallow depression at maturity; surface dry, tomentosus, mostly yellowish white, pale yellow to pale orange (2A2, 2A3–5A3), but in some specimens can be yellowish brown to brown, with a deeper center to olive brown to yellowish brown (4E5–5E5); margin wavy, incurved when young, decurved to slightly upturned at maturity, unchanging when handled. Context yellowish white, thin, unchanging when exposed. Hymenophore decurrent, lamellate ridges distant, relatively well developed, occasionally forking towards pileus margin, with or without bifurcate low veins between ridges, yellowish white to pale yellow (2A2–4A2, 2A3–4A3), unchanging when bruised. Stipe 10–20 × 1.5–3 mm, central, cylindrical or slightly tapering towards base, glabrous or faintly scaly, concolorous with pileus or paler, unchanging when handled. Odor not distinct.

Basidiospores (50/2/2) 6.0–7.0 × 4.5–5.0 μm, $L_m \times W_m$ = 5.71(±0.64) × 4.87(±0.49) μm, Q = (1)1.1–1.27(1.37), Q_m = 1.17 ± 0.07, broadly ellipsoid to subglobose, smooth, guttulate. Basidia 32–50 × 7–9 μm, 4–6-spored, narrowly clavate, colorless to hyaline in KOH, sterigmata 3–7 μm long. Hymenophoral trama irregular, composed of colorless and branched hyphae, 5–22 μm wide, septate, thin-walled. Pileipellis a cutis with long, repent, branched, and usually interwoven hyphae consisting of subcylindrical cells in 3–15 μm wide, thin-walled; terminal cells appressed to suberect, mostly cylindrical, up to 110 μm long, 5–15 μm wide. Stipitipellis a cutis of cylindrical, parallel hyphae, 3–10 μm wide; terminal cells clavate or cylindrical. Clamp connections abundant in all tissues.

Figure 12. Basidiomata of *Cantharellus convexus*. (**a**) GDGM70307. (**b**) GDGM54841. Bars = 2 cm.

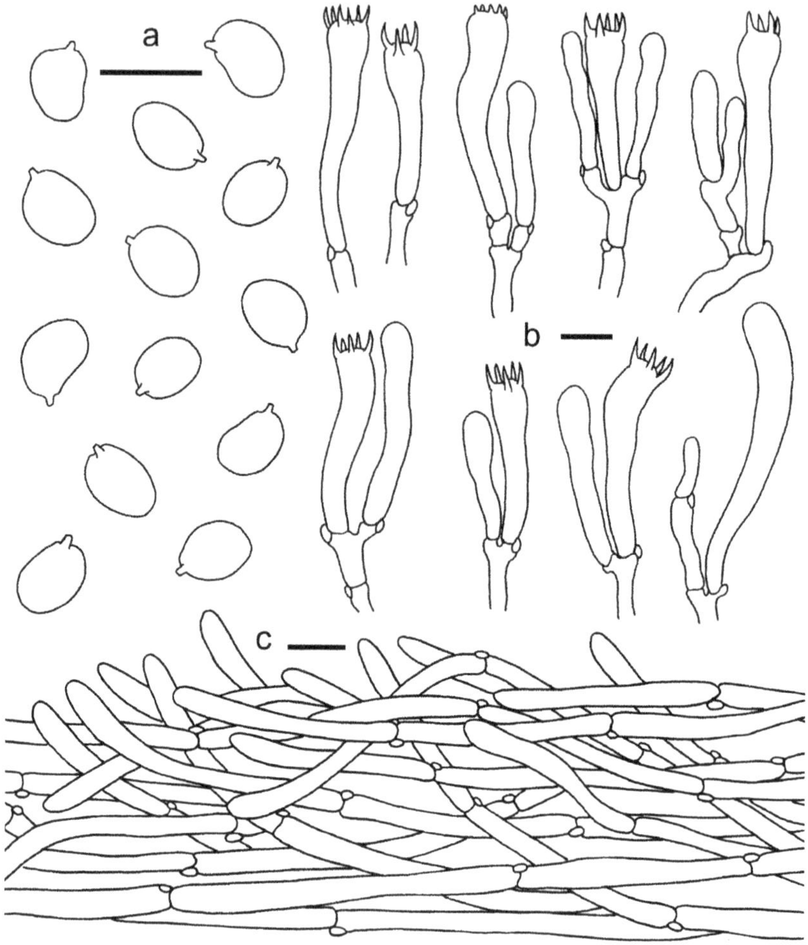

Figure 13. *Cantharellus convexus*. (**a**) Basidiospores. (**b**) Basidia, basidiola and elements of the subhymenium. (**c**) Pileipellis. Bars: (**a**,**b**) = 10 μm; (**c**) = 20 μm.

Habitat and distribution—Gregarious or scattered under broadleaf forests (dominated by Fagaceae trees) in subtropical China. Currently known from Guangdong and Hunan Province, Southern China.

Additional specimens examined—China. Hunan Province, Chenzhou City, Sanjiangkou Town, Jiulongjiang National Forest Park, under *Castanopsis hystrix* mixed with other broadleaf trees, alt. 200 m, 3 August 2017, Ming Zhang (GDGM54841).

Notes—*Cantharellus convexus* is characterized by its small basidiomata, convex pileus covered with fibrillose scales, distant and well-defined lamellate hymenophore without anastomosis between the folds, broad elliptic to subglobose basidiospores and thin-walled hyphae of the pileipellis. These traits taxonomically enable the placement of *C. convexus* into subg. *Parvocantharellus*.

Phylogenetically, two specimens of *C. convexus* formed an isolated lineage in subg. *Parvocantharellus*, and are closely related to *C. tabernensis*. A BLAST result of ITS sequence in the GenBank database also demonstrated that the similarity between *C. convexus* and *C. tabernensis* (JN944012, O7.064) is 93.7%. However, *C. tabernensis*, originally reported from North America, differs in its more robust basidiomata, dull orange-yellow to yellowish-brown pileus, vivid orange-yellow hymenophore and stipe and larger basidiospores (6–9 × 4.4–5.9 μm) [40]. Additionally, *C. tabernensis,* currently only known from Texas, Louisiana and Mississippi in North America, occurs in well-drained (sandy) soil in mixed woods, and near to *Pinus elliottii* Engelm. Meanwhile, *C. convexus* was found in broadleaf forests in southern China, close to Fagaceae trees. Another North America species, *C. appalachiensis,* also demonstrates a close relationship with *C. convexus*. However, *C. appalachiensis* differs in its larger and more robust basidiomata, with a drab yellow to dull brown pileus applanate with the center depressed, surface locally dull-grayish due to aggregate minute fibrils and with larger basidiospores (6.6–8.9 × 4.4–5.9 μm) [41,42].

Morphologically, *C. convexus* is similar to *C. austrosinensis* Ming Zhang, C.Q. Wang & T.H. Li, *C. koreanus* Buyck, Antonín & Ryoo and *C. luteovirens*. However, *C. austrosinensis* differs in its pastel yellow to greyish-yellow pileus, usually with a greyish-orange to brownish-orange center, broader basidiospores (6–8 × 4.8–6 μm) and strictly associated with coniferous trees (*Pinus massoniana*) [13]; *C. koreanus*, originally described from the temperate region of the Republic of Korea, differs in its dirty yellow-brown to pale brown pileus usually with a brown to dark brown center and larger basidiospores [6–8 (–9) × 4.2–5.5 (–6.5) μm] [11]; *C. luteovirens* differs in its yellow to orange pileus, greyish-yellow to greyish-orange hymenophores, broadly ellipsoid to subglobose basidiospores (7–8 × 5.2–6.5 μm) and is currently only found be associated with *Acacia* trees [13].

Cantharellus neopersicinus Ming Zhang, T.H. Li & X.Y. Chen sp. nov. Figures 14 and 15.

Figure 14. Basidiomata of *Cantharellus neopersicinus*. (**a–e**) GDGM87366. (**f,g**) GDGM85145. Bars: (**a,b,d–f**) = 2 cm; (**c,g**) = 5 cm.

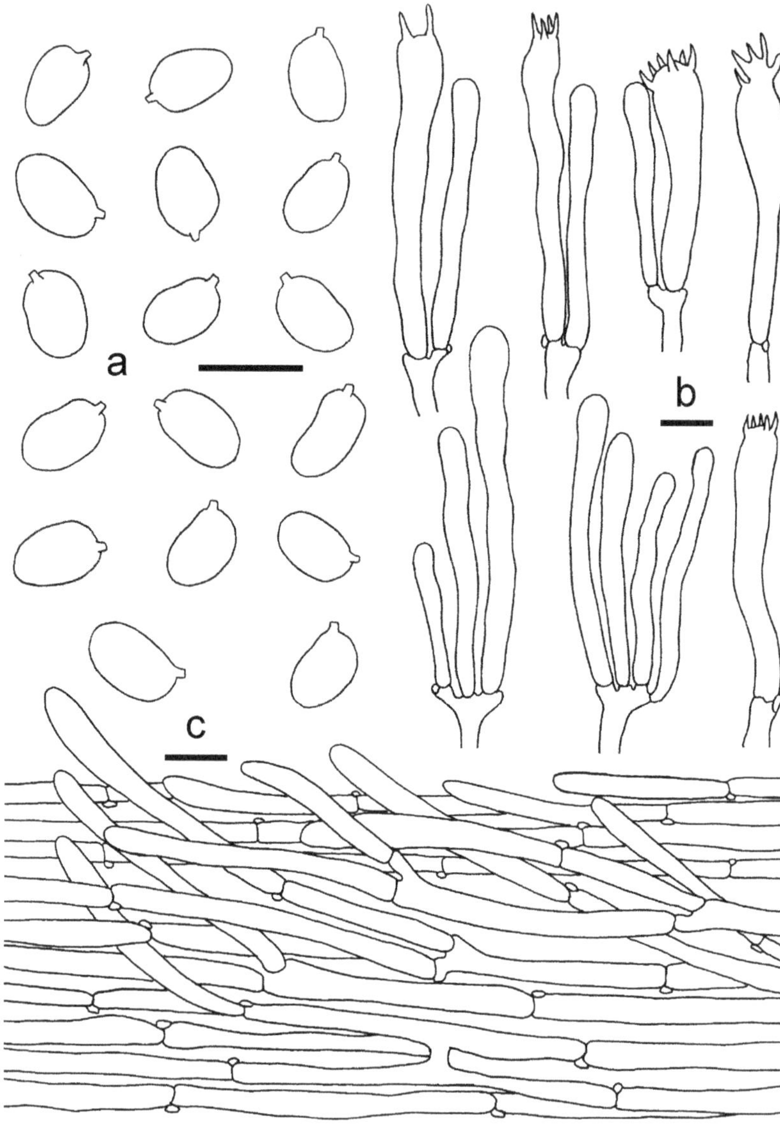

Figure 15. *Cantharellus neopersicinus.* (**a**) Basidiospores. (**b**) Basidia, basidiola and elements of the subhymenium. (**c**) Pileipellis. Bars: (**a**,**b**) = 10 µm; (**c**) = 20 µm.

MycoBank: MB843660

GenBank: OM978943 for LSU, ON119057 for *tef1* and ON119042 for *rpb2*

Etymology—refers to the color similar to *Cantharellus persicinus.*

Diagnosis—The pastel red to pink pileus, white to pinkish hymenophore with strongly bifurcate low veins and ellipsoid to subglobose [(6–)7–8.5(–9) × (4–)4.5–5.5(–6) µm], make *C. neopersicinus* easily distinguished from other species in the subg. *Parvocantharellus.*

Type—China. Guangdong Province, Leizhou City, Fangcha Village, under Eucalyptus robusta, alt. 105 m, 16 October 2021, Xiu-Yuan Chen (GDGM87366).

Basidiomata small-sized. Pileus 15–45 mm broad, convex when young, then gradually to nearly applanate with a central shallow depression at maturity; surface dry, glabrous, pastel red, pastel pink to pink (8A4–12A4); margin incurved when young, reflexed with age, wavy, sometimes irregularly split; unchanging when touched. Context thin, reddish white or pinkish (8A2–12A2), unchanging when exposed. Hymenophore decurrent, but clearly demarcated with stipe, lamellate ridges close to subdistant, poorly-developed, strongly forking towards pileus margin, with bifurcate low veins between ridges, white to pinkish, unchanging when bruised. Stipe 15–40 × 3–8 mm, central, cylindrical or slightly tapering towards base, hollow, glabrous, concolorous with pileus, unchanging when handled. Odor fruity. Taste mild.

Basidiospores (50/2/2) (6–)7–8.5(–9) × (4–)4.5–5.5(–6) μm, $L_m \times W_m$ = 7.78(±0.64) × 4.871(±0.46) μm, Q = (1.2)1.4–1.77(2), Q_m = 1.6 ± 0.15, ellipsoid to subglobose, smooth, guttulate. Basidia 45–62 × 7–9 μm, 4–6-spored, narrowly clavate, colorless to hyaline in KOH, sterigmata 3–7 μm long. Hymenophoral trama irregular to subregular, composed of colorless and branched hyphae, 8–16 μm wide, septate, thin-walled. Pileipellis a cutis with long, repent to suberect, branched, and slightly interwoven hyphae, subcylindrical cells in 8–15 μm wide, thin-walled; terminal cells appressed, mostly cylindrical, up to 100 μm long, 5–15 μm wide. Stipitipellis a cutis of cylindrical, parallel hyphae, 3–8 μm wide, terminal cells cylindrical. Clamp connections abundant in all tissues.

Habitat and distribution—Gregarious or scattered under *Eucalyptus robusta* Smith in tropical China. Currently known from Guangdong Province, Southern China.

Additional specimens examined—China. Guangdong Province, Leizhou City, Fangcha Village, alt. 105 m, 25 October 2021, Xiu-Yuan Chen (GDGM85145).

Notes—*Cantharellus neopersicinus* is characterized by its small basidiomata, pastel red to pink pileus, poorly-developed lamellate hymenophore with strongly bifurcate low veins and ellipsoid to subglobose basidiospores [(6–) 7–8.5 (–9) × (4–) 4.5–5.5 (–6) μm]. Phylogenetic analyses based on multi-locus datasets demonstrated that *C. neopersicinus* was well nested into the subg. *Parvocantharellus*, formed a well-supported terminal clade, and was closely related to *C. albus* S.P. Jian & B. Feng and *C. luteolus*. However, *C. albus*, recently reported from China, can be easily distinguished by its white basidiomata slightly changing to yellowish when bruised, a spicy taste and smaller basidiospores (5.5–7.5 × 4.5–6 μm) [12,13]; *C. luteolus* differs in its small basidiomata, yellow to orange pileus, greyish-yellow to greyish-orange hymenophore and oval to subglobose basidiospores (7–8 × 5.2–6.5 μm) [13].

Morphologically, the pastel red to pink pileus color is easily reminiscent of the species *C. cinnabarinus*, *C. coccolobae* Buyck, P.-A. Moreau & Courtec., *C. phloginus* and *C. persicinus*. However, the former three species belong to the subg. *Cinnabarinus*, and can be easily distinguished from *C. neopersicinus* by the genetic distances. Besides, *C. cinnabarinus* differs in its cinnabar red to bright orange pileus, thick-walled hyphal terminal cells of pileipellis and smaller basidiospores (6.7–7.57 × 3.82–4.68 μm) [32]. *Cantharellus coccolobae* differs in its salmon orange hymenophore, white stipe context partly changing to yellowish when cut, large basidiospores [(7.9) 8.3–9.3 (9.8) × (4.8) 5.3–5.9 (6) μm], longer basidia up to 120 μm and the thick-walled hyphae of the pileipellis. Additionally, *C. coccolobae* was reported to be strictly associated with *Coccoloba* trees, while *C. neopersicinus* is under *Eucalyptus* trees [33]. *Cantharellus phloginus* is redescribed in this study and differs in its darker pileus color, pale yellow to light yellow hymenophore, white context and larger basidiospores [6.8–9.5 (–12) × 5–7 μm]. *Cantharellus persicinus*, originally reported from North America, differs in its more robust basidiomata, larger basidiospores (9.6–10.9 × 6.3–7.1 μm), and thick-walled cells of pileipellis. In addition, *C. persicinus* is reported to be associated with oaks or eastern hemlock [32,43,44].

Cantharellus koreanus Buyck, Antonín & Ryoo, in Antonín, Hofstetter, Ryoo, Ka and Buyck, Mycol. Progr. 16(8): 755 (2017); Figures 16 and 17.

Basidiomata small-sized. Pileus 15–40 mm broad, convex at first, then gradually applanate with slightly an umbilicate centre; margin involute at first, undulate; surface dry, glabrous or finely tomentose-fibrillose at centre, mostly pale yellow to light yellow (1A3–4A3,1A4–4A4), olive brown to light brown (4D4–5D4) at centre, with obscurely sulcate at margin. Hymenophore with lamellate ridges; ridges broadly adnate to subdecurrent, with a clearly delimitation from the stipe surface, well-developed, bifurcate and with interconnected low veins, up to 1 mm high, yellowish white (2A2–4A2), unchanging when bruised. Stipe10–40 mm long, 2–5 mm thick, subcylindrical to cylindrical, slightly enlarged downward, but sometimes tapering towards base, glabrous or with faintly scaly, hollow, concolorous with pileus, darker and more somber than lamellae ridges. Odor fruity. Taste mild.

Basidiospores 5–8 × (4–) 4.5–6 μm, $L_m \times W_m$ = 7.05(±0.51) × 5.192(±0.34) μm, Q = (1.08)1.2–1.45(1.6), Q_m = 1.36 ± 0.097, ellipsoid, broadly ellipsoid, thin-walled. Basidia 40–70 × 8–12 μm, 4–6-spored, narrowly clavate, sometimes subcapitate, thin-walled, clamped. Hymenophoral trama composed of clavate, subcylindrical, subregular, branched, thin-walled, clamped hyphae 5–12 μm wide. Pileipellis a cutis, composed of cylindrical, thin-walled hyphae, 5–15 μm wide; terminal cells clavate, fusoid to cylindrical, up to 100 μm long. Stipitipellis a cutis of cylindrical, parallel, branched, thin-walled hyphae 2–9 μm wide. Clamp connections abundant in all tissues.

Habitat and distribution—Gregarious or scattered under broadleaf forests (dominated by *Fagaceae* trees) in subtropical regions of China. Known from Hunan Province, China and Korea.

Specimens examined—China, Hunan Province, Zhangjiajie City, Zhangjiajie National Forest Park, alt. 1200 m, 17 July 2020, Wei-Qiang Qin (GDGM79233); same location, alt. 1100 m, 5 July 2021, Wei-Qiang Qin (GDGM85306).

Notes—*Cantharellus koreanus*, recently reported from Korea, is firstly reported from China in this study. It is characterized by the small basidiomata, the dirty yellow-brown to pale brown pileus with a brown to dark brown center, the well-development hymenophoral ridges with yellow tinge, and the ellipsoid to broadly ellipsoid basidiospores 6–8 (–9) × 4.2–5.5 (–6.5) μm in Antonín et al. [11] and 5–8 × (4–) 4.5–6 μm in this study.

Figure 16. Basidiomata of *Cantharellus koreanus*. (**a**,**b**) GDGM79233. (**c**,**d**) GDGM85306. Bars = 2 cm.

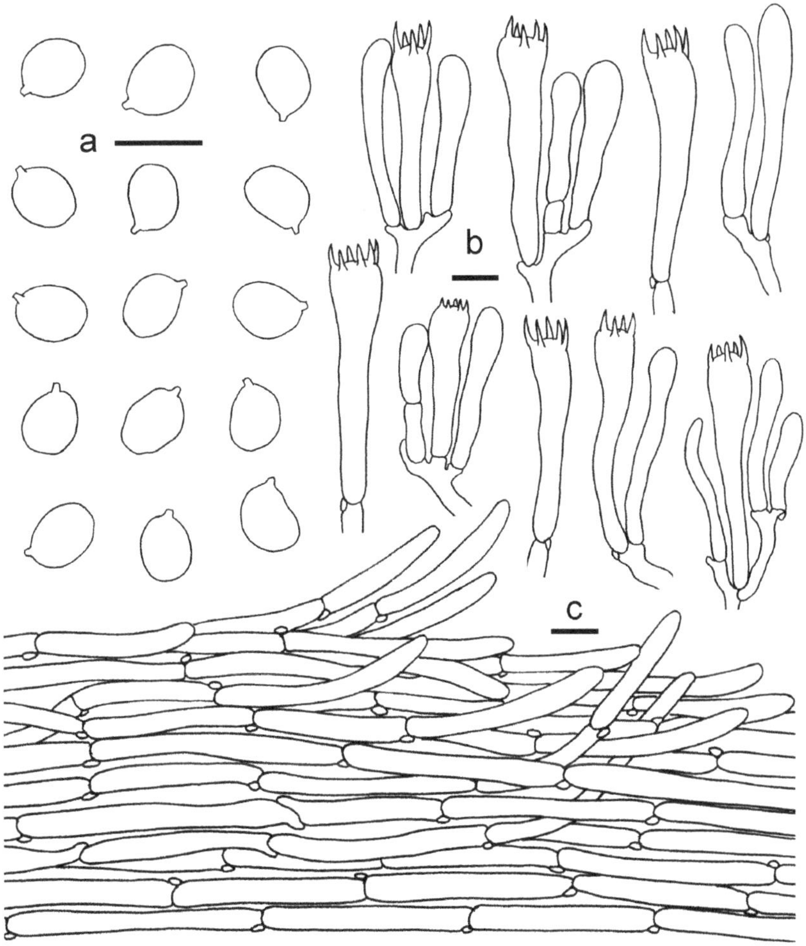

Figure 17. *Cantharellus koreanus*. (**a**) Basidiospores. (**b**) Basidia, basidiola and elements of the subhymenium. (**c**) Pileipellis. Bars: (**a**, **b**) = 10 μm; (**c**) = 20 μm.

Phylogenetically, *C. koreanus* is closely related to *C. appalachiensis*, *C. austrosinensis* and *C. tabernensis*. Indeed, *C. koreanus* is similar to *C. appalachiensis*, *C. austrosinensis* and *C. tabernensis* in morphology. However, *C. appalachiensis* differs in its larger and more robust basidiomata (pileus up to 50 mm broad), drab yellow to dull brown pileus, narrower basidia (5.5–9 μm in diam.), shorter and slightly thickened end cell of pileipellis, narrower hyphae of hymenophoral trama, and association with oaks and other hardwoods [41,45,46]; *C. austrosinensis* differs in its smaller basidiomata, pastel yellow to greyish-yellow pileus with a greyish-orange to brownish-orange center, shorter and narrower basidia (50–55 × 7–9 μm), interwoven hyphae of pileipellis, and symbiosis with coniferous trees [13]; *C. tabernensis* differs in its dull orange yellow to yellowish brown pileus, vivid orange yellow hymenophore and stipe, shorter and narrower basidia (35–55 × 5–8 μm), and distribution in North America [40,42,46].

In addition, several species were recently reported from China, and are also similar to *C. koreanus* in morphology, such as *C. galbanus*, *C. luteolus* Ming Zhang, C.Q. Wang & T.H. Li, *C. luteovirens* and *C. sinominor* Ming Zhang, C.Q. Wang & T.H. Li [13], but they can be easily separated from each other by the large genetic distances.

3.3. Key to Species of Subgenus Cinnabarinus in China

1 Basidiomata with pastel red or reddish orange tinge..2
1′ Basidiomata without red tinge..4
2 Pileus: small, always <20 mm broad...***C. sinocinnabarinus***
2′ Pileus: relatively large, usually >20 mm wide...3
3 Basidiospores: 7–8.5 × 5–6 μm...***C. albovenosus***
3′ Basidiospores: 6.8–9.5 (−12) × 5–7 μm...***C. phloginus***
4 Pileus: greenish yellow to yellowish orange, hymenophore white to yellowish white; basidiospores: 7–9 × 5–6(6.5) μm..***C. citrinus***
4′ Pileus: orange to orange-yellow, hymenophore pinkish white to orange white; basidiospores: 7.5–9 × 5–6.5 μm, basidia up to 100 μm....................................***C. chrysanthus***

4. Discussion

In this study, the species diversity of C. subg. *Cinnabarinus* from China were examined. Five species were identified based on morphological characters and multi-locus phylogenetic analyses, containing two new species *C. chrysanthus* and *C. sinocinnabarinus*, two newly recorded species *C. albovenosus* and *C. citrinus* to China, and a known species, *C. phloginus*. In addition, three species belonging to the subg. *Parvocantharellus* were firstly discovered from China, including two new species *C. convexus* and *C. neopersicinus*, and a new recorded species, *C. koreanus*.

In the past, the knowledge of species diversity of *Cantharellus* in China was poor and the specimens with large and yellow to orange basidiomata were mostly misidentified as the type species of the genus *C. cibarius*; meanwhile, specimens with small and yellow to orange red basidiomata were often inaccurately treated as *C. minor* Peck or *C. cinnabarinus*. However, a recent study proved that the distribution of *C. cibarius* is limited to northeast China, and the so-called "*C. cibarius*" reported from southwest China is actually *C. yunnanensis* W.F. Chiu [8]; meanwhile, the specimens labeled as "*C. minor*" in China were also proven to be misidentified, several new species with small basidiomata have been reported from China, and the distribution of *C. minor* with correctly identified specimens has not been found in China [13]. *Cantharellus cinnabarinus* was widely reported in China [19,21], but those photos of *C. cinnabarinus* used in the two literatures look like *C. albovenosus*; the correctly identified specimens of *C. cinnabarinus* in China have not been found in the present study. However, three morphologically similar species were discovered. The specimen HKAS58243 from southwest China, firstly identified as *C. cinnabarinus* in Shao et al. [20], was proven to be a native species of *C. sinocinnabarinus* in the present study. In addition, *C. sinocinnabarinus* seems to be restricted to subalpine habitats, and prefers symbiosis with *Cyclobalanopsis delavayi* and *Pinus yunnanensis*. The other two species, *C. albovenosus* and *C. phloginus*, are easily misidentified as *C. cinnabarinus* by their small basidiomata and reddish pileus color. However, *C. albovenosus*, recently reported from Korea, has been also found in eastern China, and *C. phloginus* seems to be restricted to tropical to subtropical regions in southwest China. Thus, we speculate that the specimens of "*C. cinnabarinus*" in Anhui, Guangdong, Jiangsu and Zhejiang provinces could be *C. albovenosus*, the distribution of "*C. cinnabarinus*" from tropical to subtropical regions of southwest China could be *C. phloginus* and the collections of "*C. cinnabarinus*" from subalpine regions of southwest China could be *C. sinocinnabarinus*.

Cantharellus neopersicinus, newly discovered in this study, is a remarkable species in *Cantharellus*. Morphologically, *C. neopersicinus* can be easily identified as a member of subg. *Cinnabarinus* or subg. *Cantharellus*, due to its pastel red to pink pileus and white to pinkish hymenophore; however, phylogenetic analyses demonstrated that it belongs to the subg. *Parvocantharellus*, which makes it the first species reported from China with pastel red to pink tinge in the subg. *Parvocantharellus*. Ecologically, *C. neopersicinus* is distributed in tropical areas of southern China, and currently, the only known symbiosis is with *Eucalyptus robusta*.

Cantharellus subg. *Parvocantharellus*, mainly composed of small-sized species, was suggested to be a monophyletic group, and closely related to the subg. *Cinnabarinus* [3]. However, in the present study, the subgenus was proven to be paraphyletic or polyphyletic; two species of *C. cyanoxanthus* R. Heim ex Heinem. and *C. subcyanoxanthus* Buyck, Randrianj. & Eyssart formed an isolated clade in the multi-locus phylogenetic tree, and could represent a separate generic clade. The result is similar to previous studies [13,16].

Species in the two subgenera are difficult to separate in morphology because most species share similar characteristics of small basidiomata, abundant clamps and thin-walled hyphal ends at the pileus surface. However, they formed two separate clades in the multi-locus phylogenetic trees, and can be easily distinguished by molecular phylogenetic evidence. In addition, the species in subg. *Cinnabarinus* mostly own distinct orange, pink or red tinge, and can be distinguished from subg. *Parvocantharellus*. In future work, more detailed morphological observations are needed to provide new evidences for distinguishing the two subgenera.

Author Contributions: Conceptualization, M.Z. and T.-H.L.; methodology, M.Z. and C.-Q.W.; performing the experiment, M.Z.; phylogenetic analysis, M.Z. and Y.L.; validation, M.Z., C.-Q.W., Y.L., M.-S.G., S.-C.S., W.-Q.Q., W.-Q.D. and T.-H.L.; writing—original draft preparation, M.Z.; writing—review and editing, C.-Q.W., Y.L., S.-C.S. and T.-H.L.; visualization, M.Z.; supervision, T.-H.L.; project administration, M.Z.; funding acquisition, M.Z. All authors have read and agreed to the published version of the manuscript.

Funding: This research was funded by the National Natural Science Foundation of China (Nos. 32070020 and 32170010), the Science and Technology Planning Project of Guizhou Province, China [No. Qian Ke He Fu Qi (2019) 4007], and Biodiversity Survey, Observation and Evaluation Project (2019–2023) of the Ministry of Ecology and Environment of China.

Institutional Review Board Statement: Not applicable.

Informed Consent Statement: Not applicable.

Data Availability Statement: Publicly available datasets were analyzed in this study. These data can be found here: (https://www.ncbi.nlm.nih.gov/; https://www.mycobank.org/page/Release%20 names; accessed on 20 March 2022).

Acknowledgments: The authors sincerely thank the editors and anonymous reviewers for their efforts and contributions towards this manuscript. Sincere acknowledgments are expressed to the curators of the HMAS and HKAS for loan of the research specimens; to B. Bart (Muséum National d'Histoire Naturelle) for providing the sequences of *Cantharellus koreanus*; to W.Z. Ma (Kunming Institute of Botany, Chinese Academy of Sciences) and Z. Du (Institute of Microbiology, Chinese Academy of Sciences) for their assistance during the loan of specimens; to X.Y. Chen, Z.H. Zhang, and Y. He for the collaboration in the field.

Conflicts of Interest: The authors declare no conflict of interest.

References

1. Fries, E.M. *Systema Mycologicum, Sistens Fungorum Ordines, Genera et Species*; Ex Offificina Berlingiana: Lundae, Sweden, 1821; Volume 1.
2. Kumari, D.; Upadhyay, R.C.; Reddy, M.S. *Cantharellus pseudoformosus*, a new species associated with *Cedrus deodara* from India. *Mycoscience* **2011**, *52*, 147–151. [CrossRef]
3. Buyck, B.; Kauff, F.; Eyssartier, G.; Couloux, A.; Hofstetter, V. A multilocus phylogeny for worldwide *Cantharellus* (Cantharellales, Agaricomycetidae). *Fungal Divers.* **2014**, *64*, 101–121. [CrossRef]
4. Henkel, T.W.; Wilson, A.W.; Amie, M.C.; Dierks, J.; Uehling, J.K.; Roy, M.; Schimann, H.; Wartchow, F.; Mueller, G.M. *Cantharellaceae* of Guyana II: New species of *Craterellus*, new South American distribution records for *Cantharellus guyanensis* and *Craterellus excelsus*, and a key to the neotropical taxa. *Mycologia* **2014**, *106*, 307–322. [CrossRef] [PubMed]
5. De Kesel, A.D.; Amalfi, M.; Kasongo Wa Ngoy, B.; Yorou, N.S.; Raspé, O.; Degreef, J.; Buyck, B. New and interesting *Cantharellus* from tropical Africa. *Cryptog. Mycol.* **2016**, *37*, 283–327. [CrossRef]
6. Ogawa, W.; Endo, N.; Fukuda, M.; Yamada, A. Phylogenetic analyses of Japanese golden chanterelles and a new species description, *Cantharellus anzutake* sp. nov. *Mycoscience* **2018**, *59*, 153–165. [CrossRef]

7. Cao, T.; Hu, Y.P.; Yu, J.R.; Wei, T.Z.; Yuan, H.S. A phylogenetic overview of the *Hydnaceae* (Cantharellales, Basidiomycota) with new taxa from China. *Stud. Mycol.* **2021**, *99*, 100121. [CrossRef] [PubMed]
8. Shao, S.C.; Liu, P.G.; Wei, T.Z.; Herrera, M. New insights into the taxonomy of the genus *Cantharellus* in China: Epityfication of *C. yunnanensis W.F. Chiu* and the first record of *C. cibarius* Fr. *Cryptog. Mycol.* **2021**, *42*, 25–37. [CrossRef]
9. Corner, E.J.H. *A Monograph of Cantharelloid Fungi*; Oxford University Press: Oxford, UK, 1966.
10. Buyck, B.; Kauff, F.; Cruaud, C.; Hofstetter, V. Molecular evidence for novel *Cantharellus* (Cantharellales, Basidiomycota) from tropical African miombo woodland and a key to all tropical African chanterelles. *Fungal Divers.* **2013**, *58*, 281–298. [CrossRef]
11. Antonín, V.; Hofstetter, V.; Ryoo, R.; Ka, K.H.; Buyck, B. New *Cantharellus* species from the Republic of Korea. *Mycol. Prog.* **2017**, *16*, 75–759. [CrossRef]
12. Jian, S.P.; Dai, R.; Gao, J.U.N.; Feng, B. *Cantharellus albus*, a striking new species from Southwest China. *Phytotaxa* **2020**, *470*, 133–144. [CrossRef]
13. Zhang, M.; Wang, C.Q.; Buyck, B.; Deng, W.Q.; Li, T.H. Multigene Phylogeny and Morphology Reveal Unexpectedly High Number of New Species of *Cantharellus Subgenus Parvocantharellus* (Hydnaceae, Cantharellales) in China. *J. Fungi* **2021**, *7*, 919. [CrossRef] [PubMed]
14. An, D.Y.; Liang, Z.Q.; Jiang, S.; Su, M.S.; Zeng, N.K. *Cantharellus hainanensis*, a new species with a smooth hymenophore from tropical China. *Mycoscience* **2017**, *58*, 438–444. [CrossRef]
15. Zhang, Y.Z.; Liang, Z.Q.; Xie, H.J.; Wu, L.L.; Xue, R.; Zeng, N.K. *Cantharellus macrocarpus* (Cantharellaceae, Cantharellales), a new species from tropical China. *Phytotaxa* **2021**, *484*, 170–180. [CrossRef]
16. Buyck, B.; Hofstetter, V.; Ryoo, R.; Ka, K.-H.; Antonín, V. New *Cantharellus* species from South Korea. *MycoKeys* **2020**, *76*, 31–47. [CrossRef] [PubMed]
17. Eyssartier, G.; Buyck, B. Note nomenclaturale et systématique sur le genre *Cantharellus*. *Doc. Mycol.* **2001**, *31*, 55–56.
18. Buyck, B.; Henkel, T.W.; Dentinger, B.T.M.; Séné, O.; Hofstetter, V. Multigene sequencing provides a suitable epitype, barcode sequences and a precise systematic position for the enigmatic, *African Cantharellus miniatescens*. *Cryptog. Mycol.* **2016**, *37*, 269–282. [CrossRef]
19. Wu, X.L.; Dai, Y.C.; Li, T.H.; Yang, Z.L.; Song, B. *Fungi of Tropical China*; Science Press: Beijing, China, 2011.
20. Shao, S.C.; Tian, X.F.; Liu, P.G. Two species with intercontinental disjunct distribution of the genus *Cantharellus*. *J. Yunnan Agric. Univ.* **2012**, *27*, 150–155.
21. Yang, Z.L.; Wu, G.; Li, Y.C.; Wang, X.H.; Cai, Q. *Common Edible and Poisonous Mushrooms of Southwestern China*; Science Press: Beijing, China, 2021.
22. Shao, S.C.; Buyck, B.; Tian, X.F.; Liu, P.G.; Geng, Y.H. *Cantharellus phloginus*, a new pink-colored species from southwestern China. *Mycoscience* **2016**, *57*, 144–149. [CrossRef]
23. Kornerup, A.; Wanscher, J.H. *Taschenlexikon der Farben*, 3rd ed.; Muster-Schmidt Verlag: Göttingen, Germany, 1981.
24. Vilgalys, R.; Hester, M. Rapid genetic identification and mapping of enzymatically amplified ribosomal DNA from several *Cryptococcus* species. *J. Bacteriol.* **1990**, *172*, 4238–4246. [CrossRef]
25. Morehouse, E.A.; James, T.Y.; Ganley, A.R.D.; Vilgalys, R.; Berger, L.; Murphy, P.J.; Longcore, J.E. Multilocus sequence typing suggests the chytrid pathogen of amphibians is a recently emerged clone. *Mol. Ecol.* **2003**, *12*, 395–403. [CrossRef]
26. Katoh, K.; Rozewicki, J.; Yamada, K.D. MAFFT online service: Multiple sequence alignment, interactive sequence choice and visualization. *Brief Bioinform.* **2019**, *20*, 1160–1166. [CrossRef] [PubMed]
27. Tamura, K.; Stecher, G.; Peterson, D.; Filipski, A.; Kumar, S. MEGA6: Molecular Evolutionary Genetics Analysis version 6.0. *Mol. Biol. Evol.* **2013**, *30*, 2725–2729. [CrossRef] [PubMed]
28. Stamatakis, A. RAxML-VI-HPC: Maximum likelihood-based phylogenetic analyses with thousands of taxa and mixed models. *Bioinformatics* **2006**, *22*, 2688–2690. [CrossRef] [PubMed]
29. Ronquist, F.; Teslenko, M.; van der Mark, P.; Ayres, D.L.; Darling, A.; Höhna, S.; Larget, B.; Liu, L.; Suchard, M.A.; Huelsenbeck, J.P. MrBayes 3.2: Effificient bayesian phylogenetic inference and model choice across a large model space. *Syst. Biol.* **2012**, *61*, 539–542. [CrossRef] [PubMed]
30. Lanfear, R.; Frandsen, P.B.; Wright, A.M.; Senfeld, T.; Calcott, B. PartitionFinder 2: New methods for selecting partitioned models of evolution for molecular and morphological phylogenetic analyses. *Mol. Biol. Evol.* **2017**, *34*, 772–773. [CrossRef]
31. Moncalvo, J.M.; Nilsson, R.H.; Koster, B.; Dunham, S.M.; Bernauer, T.; Matheny, P.B.; McLenon, T.; Margaritescu, S.; Weiß, M.; Garnica, S.; et al. The cantharelloid clade: Dealing with incongruent gene trees and phylogenetic reconstruction methods. *Mycologia* **2006**, *98*, 937–948. [CrossRef]
32. Buyck, B.; Cruaud, C.; Couloux, A.; Hofstetter, V. *Cantharellus texensis* sp. nov. from Texas, a Southern lookalike of *C. cinnabarinus* revealed by tef-1 sequence data. *Mycologia* **2011**, *103*, 1037–1046. [CrossRef]
33. Buyck, B.; Moreau, P.-A.; Courtecuisse, R.; Kong, A.; Roy, M.; Hofstetter, V. *Cantharellus coccolobae* sp. nov. and *Cantharellus garnieri* two tropical members of *Cantharellus* subg. *Cinnabarinus*. *Cryptog. Mycol.* **2016**, *37*, 391–403. [CrossRef]
34. Buyck, B. Special issue: *Cantharellus*. *Cryptog. Mycol.* **2016**, *37*, 255–258. [CrossRef]
35. Suhara, H.; Kurogi, S. *Cantharellus cyphelloides* (Cantharellales), a new and unusual species from a Japanese evergreen broad-leaved forest. *Mycol. Prog.* **2015**, *14*, 55. [CrossRef]
36. Olariaga, I.; Moreno, G.; Manjón, J.L.; Salcedo, I.; Hofstetter, V.; Rodríguez, D.; Buyck, B. *Cantharellus* (Cantharellales, Basidiomycota) revisited in Europe through a multigene phylogeny. *Fungal Divers.* **2017**, *83*, 263–292. [CrossRef]

37. Ducousso, M.; Contesto, C.; Cossegal, M.; Prin, Y.; Rigault, F.; Eyssartier, G. *Cantharellus garnierii* sp. nov. from nickel mine maquis in New Caledonia. *Cryptog. Mycol.* **2004**, *25*, 115–125.
38. Buyck, B.; Hofstetter, V. The contribution of tef-1 sequences to species delimitation in the *Cantharellus cibarius* complex in the southeastern USA. *Fungal Divers.* **2011**, *49*, 35–46. [CrossRef]
39. Buyck, B.; Ndolo Ebika, S.T.; De Kesel, A.; Hofstetter, V. Tropical African Cantharellus Adans.: Fr. (Hydnaceae, Cantharellales) with lilac-purplish tinges revisited. *Cryptog. Mycol.* **2020**, *41*, 161–177. [CrossRef]
40. Feibelman, T.P.; Bennett, J.W.; Cibula, W.G. *Cantharellus tabernensis*: A new species from the Southeastern United States. *Mycologia* **1996**, *88*, 295–301. [CrossRef]
41. Ryvarden, L.; Petersen, R. Notes on cantharelloid fungi IV. Two new species of *Cantharellus*. *Sven. Bot. Tidskr.* **1971**, *65*, 399–405.
42. Buyck, B.; Lewis, D.P.; Eyssartier, G.; Hofstetter, V. *Cantharellus quercophilus* sp. nov. and its comparison to other small, yellow or brown American chanterelles. *Cryptog. Mycol.* **2010**, *31*, 17–33.
43. Petersen, R.H. *Notes on Clavarioid Fungi. XIX.* Colored illustrations of selected taxa, with comments on *Cantharellus*. *Nova Hedwig.* **1985**, *42*, 151–160.
44. Kuo, M. *Cantharellus persicinus*. Retrieved from the MushroomExpert.Com. 2015. Available online: http://www.mushroomexpert.com/cantharellus_persicinus.html (accessed on 20 March 2022).
45. Bigelow, H.E. The cantharelloid fungi of New England and adjacent areas. *Mycologia* **1978**, *70*, 707–756. [CrossRef]
46. Montoya, L.; Herrera, M.; Bandala, V.M.; Ramos, A. Two new species and a new record of yellow *Cantharellus* from tropical Quercus forests in eastern Mexico with the proposal of a new name for the replacement of *Craterellus confluens*. *MycoKeys* **2021**, *80*, 91–114. [CrossRef]

Journal of *Fungi*

MDPI

Article

A Phylogenetic and Taxonomic Study on *Phellodon* (Bankeraceae, Thelephorales) from China

Chang-Ge Song [1], Yuan-Yuan Chen [2], Shun Liu [1], Tai-Min Xu [1], Xiao-Lan He [3], Di Wang [3] and Bao-Kai Cui [1,*]

1 Institute of Microbiology, School of Ecology and Nature Conservation, Beijing Forestry University, Beijing 100083, China; changgesong@126.com (C.-G.S.); liushun2017@bjfu.edu.cn (S.L.); fungitaiminx@bjfu.edu.cn (T.-M.X.)
2 College of Forestry, Henan Agricultural University, Zhengzhou 450002, China; cyuan091@163.com
3 Sichuan Institute of Edible Fungi, Sichuan Academy of Agricultural Sciences, Chengdu 610066, China; xiaolanhe1121@aliyun.com (X.-L.H.); wang.di19881213@163.com (D.W.)
* Correspondence: cuibaokai@bjfu.edu.cn; Tel./Fax: +86-10-6233-6309

Abstract: In this study, phylogenetic analyses of *Phellodon* from China were carried out based on sequences from the internal transcribed spacer (ITS) regions, the large subunit of nuclear ribosomal RNA gene (nLSU), the small subunit of nuclear ribosomal RNA gene (nSSU), the largest subunit of RNA polymerase II (RPB1), and the second largest subunit of RNA polymerase II (RPB2), combined with morphological characters of the collected specimens in China. The fruiting bodies of the specimens were used to observe their characteristics, and three new species of *Phellodon* are discovered. *Phellodon crassipileatus* is characterized by its pale brown to dark brown pileal surface, tomentose pileal margin, white spines, and the presence of clamp connections in generative hyphae of pileal surface, context, and stipe. *Phellodon griseofuscus* is characterized by its dark brown to black pileal surface, white to pale brown pileal margin, the presence of both simple septa and clamp connections in generative hyphae of spines, and moderately long basidia. *Phellodon perchocolatus* is characterized by its woody and broad pileus, brown to greyish brown pileal surface when fresh, tomentose pileal margin when young, which becomes glabrous with age, and the presence of both simple septa and clamp connections in the generative hyphae of the spines. This is the first time both single and multi-genes analysis is used in such a phylogenetic and taxonomic study on *Phellodon*, which can provide the basis for the phylogenetic study of the genus.

Keywords: ectomycorrhizal fungi; molecular phylogeny; morphology; stipitate hydnoid fungi

Citation: Song, C.-G.; Chen, Y.-Y.; Liu, S.; Xu, T.-M.; He, X.-L.; Wang, D.; Cui, B.-K. A Phylogenetic and Taxonomic Study on *Phellodon* (Bankeraceae, Thelephorales) from China. *J. Fungi* **2022**, *8*, 429. https://doi.org/10.3390/jof8050429

Academic Editors: Lei Cai and Cheng Gao

Received: 23 March 2022
Accepted: 20 April 2022
Published: 22 April 2022

1. Introduction

Phellodon P. Karst. was established by Petter Adolf Karsten and typified by *P. niger* (Fr.) P. Karst [1]. The genus, together with *Hydnellum* P. Karst. and *Sarcodon* Quél. ex P. Karst. were stipitate hydnoids, and they were affiliated to Bankeraceae of Thelephorales. All of the three genera belong to ectomycorrhizal fungi, which are associated with broad-leaved or coniferous trees in forest ecological systems [2–4]. Ectomycorrhizal fungi are symbionts of trees in forests, which can reflect the conservation state of forest ecosystems [5]. They can connect plant roots to soil by promoting the decomposition of organic matter in soil and the absorption of organic and inorganic elements by host plants [6]. Therefore, they are of great significance to the growth of plants and the material circulation of ecosystems. During the second half of the 20th century, the numbers of most species of stipitate hydnoid fungi have declined [7], and many species have been included in national red lists [8]. This is most likely ascribed to habitat loss due to forestry operations, such as massive logging, the disappearance of old *Picea* forests on calcareous soils, deciduous forest transformed into coniferous forest, direct effects of air pollutants, and forest soil acidification [7,9]. In addition, sulfur and nitrogen depositions and soil acidification also contribute to the decline of stipitate hydnoid fungi [7,9,10]. In recent decades, the number of stipitate hydnoid fungi has dropped significantly, which reflects that we need to pay more attention to protecting them [4]. Meanwhile, discovering new species of stipites hydnoid fungi is also of great significance for helping us to further recognize and protect them.

Macro-morphologically, species of *Phellodon*, *Hydnellum*, and *Sarcodon* are relatively similar in having single to concrescent basidiomata and spines. However, the three genera can be distinguished by the color of their basidiospores. Traditionally, species in *Hydnellum* and *Sarcodon* have brown basidiospores, while species in *Phellodon* have white basidiospores [4]. While in a recent comprehensive study, Larsson et al. [11] suggested that basidiospore size can distinguish the *Hydnellum* and *Sarcodon*, species in *Hydnellum* have basidiospore lengths in the range 4.45−6.95 μm while the corresponding range for *Sarcodon* is 7.4−9 μm. Species in *Phellodon* are characterized by solitary to gregarious or concrescent, stipitate basidiomata, hydnoid hymenophore, and echinulate basidiospores [12], and often occur in forests of Fagaceae and Pinaceae [2,6].

In 1881, Karsten divided the genus *Hydnellum* into two parts: the white toothed and the dark toothed, and the former was named *Phellodon* [13]. Banker [13] revised all of the Hydnaceae found in the continent of North America and its adjacent areas, which included *Hydnellum*, *Phellodon*, and *Sarcodon*, and 10 species of *Phellodon* were described based on morphological features. Species of *Phellodon* were described only based on morphological characteristics in the past few decades, which resulted in the lack of molecular basis for taxonomic studies of the genus [11,13–25]. Morphological and phylogenetic studies were used to identify the genus in recent years. Parfitt et al. [4] carried out a systematic study of *Hydnellum* and *Phellodon* based on molecular and morphological analyses, which identified the taxonomic status of the known *Phellodon* species from Britain. Ainsworth et al. [26] revealed the cryptic taxa of the genera *Hydnellum* and *Phellodon* based on the combination of molecular and morphological analysis. Moreover, Baird et al. [27] reevaluated the species of stipitate hydnums from the southern United States, and 41 distinct taxa of *Hydnellum*, *Phellodon*, and *Sarcodon* were determined. At the same time, they described 10 species of *Phellodon*. They provided phylogenetic analyses on *Phellodon* based on ITS sequences, which provided a morphological and molecular basis for taxonomic and phylogenetic studies of the genus. Furthermore, *Bankera fuligineoalba* (J.C. Schmidt) Pouzar, the typified species of *Bankera* Coker and Beers, was recombined in *Phellodon* in their study, which suggests that the genus *Bankera* has already been combined into *Phellodon*. In recent years, the genus has been studied in China. Mu et al. [12] described *Phellodon subconfluens* H.S. Yuan and F. Wu in Liaoning Province based on morphological characters and molecular data. Later, Song et al. [28] described four species of *Phellodon*, *P. atroardesiacus* B.K. Cui and C.G. Song, *P. cinereofuscus* B.K. Cui, and C.G. Song, *P. stramineus* B.K. Cui, and C.G. Song and *P. yunnanensis* B.K. Cui, and C.G. Song, based on morphological characters and ITS sequences data from southwestern China [28].

During the investigations of stipitate hydnoid fungi from China, abundant fruiting bodies were obtained, and three undescribed species of *Phellodon* were discovered. To confirm the affinity of the undescribed species corresponding to *Phellodon*, phylogenetic analyses were carried out based on ITS and ITS + nLSU + nSSU + RPB1 + RPB2 sequences. The new species were described based on the combination of morphological and phylogenetic analysis.

2. Materials and Methods

2.1. Morphological Studies

Methods of specimen collection and preservation followed the methods of Wang [29]. The specimens used in this study were collected during the annual growing season of macrofungi. At the same time, the specimen information, host trees, ecological habits, location, altitude, collector, date were recorded, and the photos of the fruiting bodies and growth environment were taken. Then, the specimens were dried and bagged in time for preservation. After that, the specimens were registered and deposited at the herbarium of the Institute of Microbiology, Beijing Forestry University (BJFC). Macromorphological descriptions were based on the field notes and measurements of herbarium specimens. Microscopic characteristics, measurements, and drawings were made from slide preparations stained with Cotton Blue and Melzer's reagent and observed at magnifications up to ×1000 under a light microscope (Nikon Eclipse E 80i microscope, Nikon, Tokyo, Japan) following Liu et al. [30]. Basidiospores were measured from sections cut from the spines. The following abbreviations are used: IKI, Melzer's reagent; IKI–, neither amyloid nor dextrinoid; KOH, 5% potassium hydroxide; CB, Cotton Blue; CB–, acyanophilous; L, mean spore length (arithmetic average of all spores); W, mean spore width (arithmetic average of all spores); Q, variation in the L/W ratios between the specimens studied; n (a/b), and number of spores (a) measured from given number (b) of specimens. A field Emission Scanning Electron Microscope (FESEM) Hitachi SU-8010 (Hitachi, Ltd., Tokyo, Japan) was used to film the spore's morphology. Sections were studied at up to 2200 times magnification, according to the method by Sun et al. [31].

2.2. Molecular Study and Phylogenetic Analysis

DNA extraction, amplification, and sequencing: the CTAB rapid plant genome extraction kit (Aidlab Bio technologies Co., Ltd., Beijing, China) was used to obtain PCR products from dry specimens, and for polymerase chain reaction (PCR), according to the manufacturer's instructions with some modifications [32]. The primer pairs ITS5/ITS4, LR0R/LR7, NS1/NS4, AF/Cr, and 5F/7Cr were used to amplify ITS, nLSU, nSSU, RPB1, and RPB2 sequences [28]. The PCR process for ITS was as follows: initial denaturation at 95 °C for 3 min, followed by 35 cycles at 94 °C for 40 s, 56 °C for 45 s and 72 °C for 1 min and a final extension of 72 °C for 10 min. The PCR process for nLSU and nSSU was as follows: initial denaturation at 94 °C for 1 min, followed by 35 cycles at 94 °C for 30 s, 50 °C for 1 min and 72 °C for 90 s and a final extension of 72 °C for 10 min. The PCR process for RPB1 and RPB2 was as follows: initial denaturation at 94 °C for 2 min, 9 cycles at 94 °C for 45 s, 60 °C for 45 s, followed by 36 cycles at 94 °C for 45 s, 53 °C for 1 min, 72 °C for 90 s and a final extension of 72 °C for 10 min. The PCR products were purified and sequenced in Beijing Genomics Institute, China, with the same primers. All newly generated sequences were submitted to GenBank and are listed in (Table 1). Moreover, other sequences in the dataset for phylogenetic analysis were downloaded from GenBank (http://www.ncbi.nlm.nih.gov/genbank/php, accessed on 15 October 2021).

New sequences generated in this study were aligned with additional sequences downloaded from GenBank (Table 1) using ClustalX [33] and manually adjusted in BioEdit [34]. The sequences of *Amaurodon aquicoeruleus* Agerer and *A. viridis* (Alb. and Schwein.) J. Schröt. were used as the outgroups, according to Mu et al. [12].

Maximum parsimony (MP) analysis followed and was applied to the sequence datasets using PAUP* version 4.0b10 [35], and the congruences of the 5-gene (ITS, nLSU, nSSU, RPB1, and RPB2) were evaluated with the incongruence length difference (ILD) test [36]. Gaps in the alignments were treated as missing data. Maxtrees were regular to 5000, branches of zero length were collapsed, and all parsimonious trees were saved. Clade might be assessed using a bootstrap (BS) analysis with 1000 replicates [37]. Descriptive tree statistics tree length (TL), consistency index (CI), retention index (RI), rescaled consistency index (RC), and homoplasy index (HI) were calculated for each maximum parsimonious tree generated.

Maximum likelihood (ML) analysis was conducted with RAxML-HPC252 on Abe through the Cipres Science Gateway (www.phylo.org, accessed on 18 October 2021), which referred to 100 ML searches, and the program estimated all model parameters. The maximum likelihood bootstrap (ML-BS) values were performed with a rapid bootstrapping with 1000 replicates. Phylogenetic trees were viewed using FigTree v1.4.2 (http://tree.bio.ed.ac.uk/software/figtree/, accessed on 18 October 2021).

Table 1. A list of species, specimens and GenBank accession numbers of sequences used in this study.

Species	Specimen No.	Locality	GenBank Accession No.				
			ITS	nLSU	nSSU	RPB1	RPB2
Amaurodon aquicoeruleus	UK 452	Australia	AM490944	AM490944	-	-	-
A. viridis	TAA 149664	Russia	AM490942	AM490942	-	-	-
Hydnellum atrospinosum	Yuan 6520	China	MW579912	-	MW579912	-	-
H. atrospinosum	Yuan 6495	China	MW579938	MW579885	MW579911	-	-
H. suaveolens	ELarsson 139-09	Norway	MK602734	MK602734	-	-	-
H. suaveolens	ELarsson 8-14	Sweden	MK602735	MK602735	-	-	-
Phellodon alboniger	REB-70	USA	KC571749	-	-	-	-
P. alboniger	REB-57	USA	JN135206				
P. atratus	CL-72	Canada	MK281471	-	-	-	-
P. atratus	DAVFP 28189	Canada	HQ650766	-	-	-	-
P. atroardesiacus	Cui 18449	China	MZ221189	**MZ225598**	**MZ225636**	-	-
P. atroardesiacus	Cui 18457	China	MZ225577	**MZ225599**	**MZ225637**	-	-
P. atroardesiacus	Cui 18458	China	MZ225633	**MZ225600**	**MZ225638**	-	-
P. atroardesiacus	Cui 18459	China	MZ225634	**MZ225601**	**MZ225639**	-	-
P. atroardesiacus	Cui 16951	China	MZ225632	**MZ225597**	**MZ225635**	**MZ343209**	**MZ343197**
P. brunneoolivaceus	REB-166	USA	KC571752	-	-	-	-
P. cinereofuscus	Cui 14231	China	MZ225579	-	-	-	-
P. cinereofuscus	Cui 16940	Australia	MZ225580	**MZ225602**	**MZ225640**	**MZ343210**	**MZ343198**
P. cinereofuscus	Cui 16944	China	MZ225581	**MZ225603**	**MZ225641**	**MZ343211**	**MZ343199**
P. cinereofuscus	Cui 16945	China	MZ225582	**MZ225604**	**MZ225642**	-	-
P. cinereofuscus	Cui 16962	China	MZ225583	**MZ225605**	**MZ225643**	**MZ352084**	**MZ343200**
P. cinereofuscus	Cui 16963	China	MZ225584	**MZ225606**	**MZ225644**	**MZ352085**	**MZ343201**
P. confluens	WAT 28574	UK	EU622361	-	-	-	-
P. confluens	E00 186901	UK	EU622362	-	-	-	-
P. crassipileatus	Cui 18532	China	**OL449267**	**OL439037**	**OL439027**	-	-
P. crassipileatus	Cui 18533	China	**OL449268**	**OL439038**	**OL439028**	-	-
P. ellisianus	REB-264	USA	KC571757	-	-	-	-
P. ellisianus	REB-407	USA	KC571759	-	-	-	-
P. fibulatus	REB-168	USA	JN135205	-	-	-	-
P. fibulatus	REB-34	USA	KC571761	-	-	-	-
P. fuligineoalbus	REB-271	USA	KC571760	-	-	-	-
P. fuligineoalbus	REB-285	USA	JN135196	-	-	-	-
P. fuligineoalbus	SL8	-	EU622316	-	-	-	-
P. griseofuscus	Cui 18544	China	**OL449265**	**OL439035**	**OL439025**	**OL456229**	**OL449087**
P. griseofuscus	Cui 18561	China	**OL449266**	**OL439036**	**OL439026**	-	-
P. melaleucus	LH4	UK	EU622368	-	-	-	-
P. melaleucus	E00219373	UK	EU622369	-	-	-	-
P. melaleucus	Cui 18614	China	**OL449262**	**OL439032**	**OL439022**	**OL456228**	-
P. melaleucus	Cui 18620	China	**OL449263**	**OL439033**	**OL439023**	-	-
P. melaleucus	Cui 18623	China	**OL449264**	**OL439034**	**OL439024**	-	-
P. mississippiensis	MS-1	USA	JN247563	-	-	-	-
P. mississippiensis	MS-3	USA	JN247564	-	-	-	-
P. niger	REB-46	USA	JN135202	-	-	-	-
P. niger	REB-282	USA	KC571766	-	-	-	-
P. cf. *nothofagi*	MES-175	Chile	MH930224	-	-	-	-
P. perchocolatus	Cui 18534	China	**OL449259**	**OL439029**	**OL439020**	**OL456227**	-
P. perchocolatus	Cui 18536	China	**OL449260**	**OL439030**	-	-	-
P. perchocolatus	Cui 18540	China	**OL449261**	**OL439031**	**OL439021**	-	-
P. putidus	REB-8	USA	JN135200	-	-	-	-
P. secretus	0097	Russia	MG597404	-	-	-	-
P. sinclairii	PDD 89028	New Zealand	GU222291	-	-	-	-
P. stramineus	Cui 16942	China	MZ225585	**MZ225607**	**MZ225645**	**MZ352086**	-
P. stramineus	Cui 16943	China	MZ225586	**MZ225608**	**MZ225646**	**MZ352087**	**MZ343202**
P. stramineus	Cui 16956	China	MZ225587	**MZ225609**	**MZ225647**	**MZ352088**	**MZ343203**
P. stramineus	Cui 16959	China	MZ225588	**MZ225610**	**MZ225648**	**MZ352089**	**MZ343204**
P. stramineus	Cui 16961	China	MZ225589	**MZ225611**	**MZ225649**	**MZ352090**	**MZ343205**
P. stramineus	Cui 16964	China	MZ225590	**MZ225612**	**MZ225650**	**MZ352091**	-
P. subconfluens	Yuan 11123	China	MK677464	-	-	-	-
P. subconfluens	Yuan 11150	China	MK677465	-	-	-	-
Phellodon sp.1	REB-83	USA	KC571747	-	-	-	-
Phellodon sp.1	REB-325	USA	KC571748	-	-	-	-
P. tomentosus	SL70	UK	EU622381	-	-	-	-
P. tomentosus	LH22	UK	EU622382		-	-	-
P. yunnanensis	Cui 14292	China	MZ225591	-	-	-	-
P. yunnanensis	Cui 14294	China	MZ225592	-	-	-	-
P. yunnanensis	Cui 17097	China	MZ225593	**MZ225613**	**MZ225651**	-	**MZ343206**
P. yunnanensis	Cui 17129	China	MZ225594	**MZ225614**	**MZ225652**	-	**MZ343207**
P. yunnanensis	Cui 17131	China	MZ225595	**MZ225615**	**MZ225653**	-	**MZ343208**
P. violascens	2359-QFB-25626	Canada	KM406977	-	-	-	-
Sarcodon imbricatus	JRova 1408292	Sweden	MK602746	MK602746	-	-	-
S. imbricatus	ELarsson 384-10	Norway	MK602747	MK602747	-	-	-
S. squamosus	OF 177452	Norway	MK602768	MK602768	-	-	-
S. squamosus	OF 295554	Norway	MK602769	MK602769	-	-	-

New sequences are shown in bold.

MrModeltest2.3 [38,39] was used to determine the best-fit evolution model for the combined dataset for Bayesian inference (BI). BI was performed using MrBayes 3.2.6 on

Abe through the Cipres Science Gateway (www.phylo.org, accessed on 19 October 2021) with 2 independent runs, beginning from random trees with 4 simultaneous independent Chains, performing 2 million replicates, sampling 1 tree for every 100 generations. The burn-in was set to discard 25% of the trees. The remaining ones were used to construct a majority rule consensus and calculate the Bayesian posterior probabilities (BPP) of the clades.

Branches that received bootstrap support for maximum parsimony (MP), maximum likelihood (ML), and Bayesian posterior probabilities (BPP) greater than or equal to 50% (MP and ML) and 0.95 (BPP) were regarded as prominently supported.

3. Results

3.1. Phylogenetic Analyses

The dataset of ITS included 73 sequences representing 32 taxa. The ITS dataset had an aligned length of 873 characters, of which 376 characters were constant, 46 were variable and parsimony-uninformative, and 451 were parsimony-informative. Maximum parsimony analysis yielded 516 equally parsimonious trees (TL = 1516, CI = 0.547, RI = 0.839, RC = 0.459, HI = 0.453), and 1 of the maximum parsimonious trees is shown in Figure 1. The best fit model selected for these three partitions of ITS sequences was GTR + G for ITS1, JC for 5.8 s, and HKY + G for ITS2. BI resulted in a similar topology with an average standard deviation of split frequencies = 0.007630 to MP analysis. The MP topology is shown with MP (≥75%), ML (≥75%), and BPP (≥0.95) supported values at the nodes (Figure 1).

In the ITS based phylogenetic tree (Figure 1), the three new species *P. crassipileatus*, *P. griseofuscus*, and *P. perchocolatus* formed distinct well-supported lineages distant from other species of *Phellodon*.

The combined ITS + nLSU + nSSU + RPB1 + RPB2 dataset included sequences from 73 fungal samples representing 32 taxa. The combined dataset had an aligned length of 5639 characters including gaps (873 characters for ITS, 1379 characters for nLSU, 1097 characters for nSSU, 1203 characters for RPB1, 1087 characters for RPB2), of which 4599 characters were constant, 192 were variable and parsimony-uninformative, and 848 were parsimony-informative. Maximum parsimony analysis yielded 12 equally parsimonious trees (TL = 2195, CI = 0.643, RI = 0.866, RC = 0.557, HI = 0.357), and 1 of the maximum parsimonious trees is shown in Figure 2. The best fit model selected for the combined ITS + nLSU + nSSU + RPB1 + RPB2 sequence dataset was GTR + I+G with equal frequency of nucleotides. BI resulted in a similar topology with an average standard deviation of split frequencies = 0.008906 to MP analysis. The MP topology is shown with MP (≥75%), ML (≥75%), and BPP (≥0.95) supported values at the nodes (Figure 2).

The ITS + nLSU + nSSU + RPB1 + RPB2 based phylogenetic tree (Figure 2) produced a topology similar to that generated by the ITS based phylogenetic tree, and confirmed the affinities of the three new species within *Phellodon*.

3.2. Taxonomy

Phellodon crassipileatus B.K. Cui and C.G. Song, sp. nov., Figure 3a, Figure 4a,b, and Figure 5.

MycoBank: 843670

Diagnosis—This species is characterized by its pale brown to dark brown pileal surface, thick pileus, tomentose pileal margin, white spines, and the presence of clamp connections in generative hyphae of pileal surface, context, and stipe.

Etymology—*crassipileatus* (Lat.): refers to the thick pileus.

Holotype—CHINA. Sichuan Province, Pingwu County, Bazi, on the ground of forest dominated by trees of *Quercus* sp., alt. 1190 m, 18 September 2020, Cui 18533 (BJFC 035394).

Fruiting body—Basidiomata annual, centrally or eccentrically stipitate, solitary or gregarious, with a fenugreek odor when dry. Pileus infundibuliform, up to 6.5 cm in diam, 2 cm thick at the center. Pileal surface pale brown to dark brown when fresh, becoming dark brown upon drying, azonate, tomentose at the margin; pileal margin blunt or irregular, white when fresh, becoming cream upon drying, up to 1.2 cm wide. Spines soft, white when fresh, becoming fragile, cream to clay-buff upon drying, up to 3 mm long. Context vinaceous grey, tough, up to 6 mm thick. Stipe brown to dark brown in the outer layer, fuscous in the inner layer, cylindrical, glabrous, up to 1.5 cm long, 1 cm in diameter.

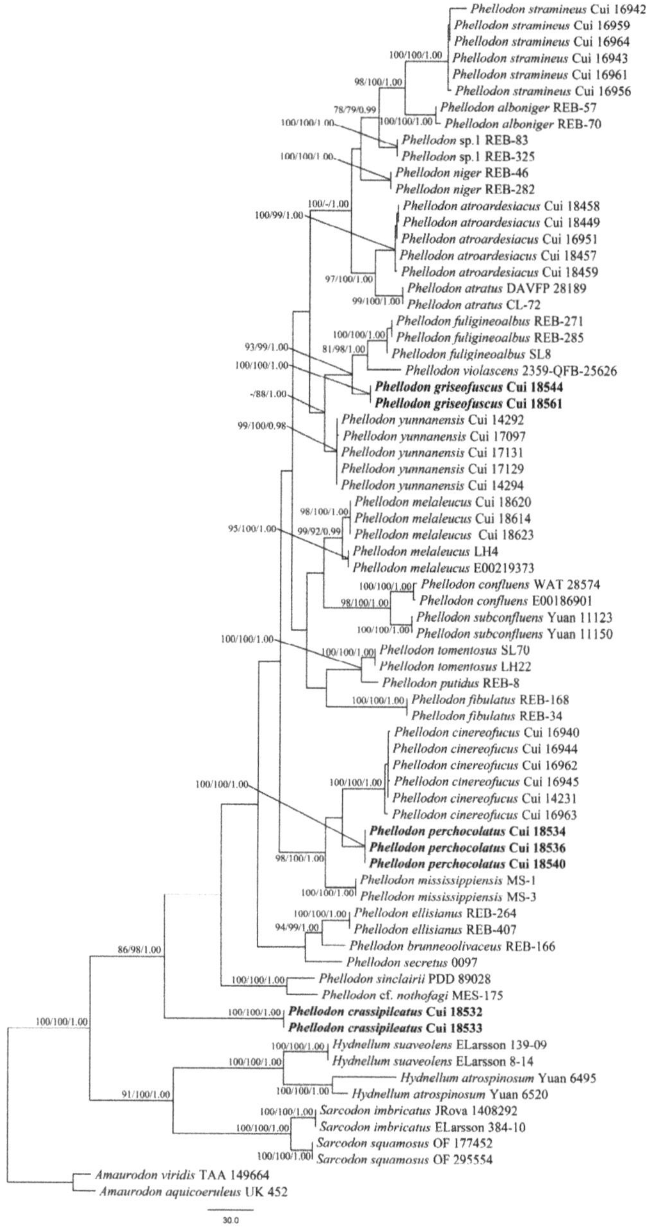

Figure 1. Maximum parsimony (MP) phylogram of the *Phellodon* species based on ITS sequences data. The supported branches are labeled with parsimony bootstrap values higher than 75%, maximum likelihood bootstrap values higher than 75%, and Bayesian posterior probabilities more than 0.95. Bold names = New species.

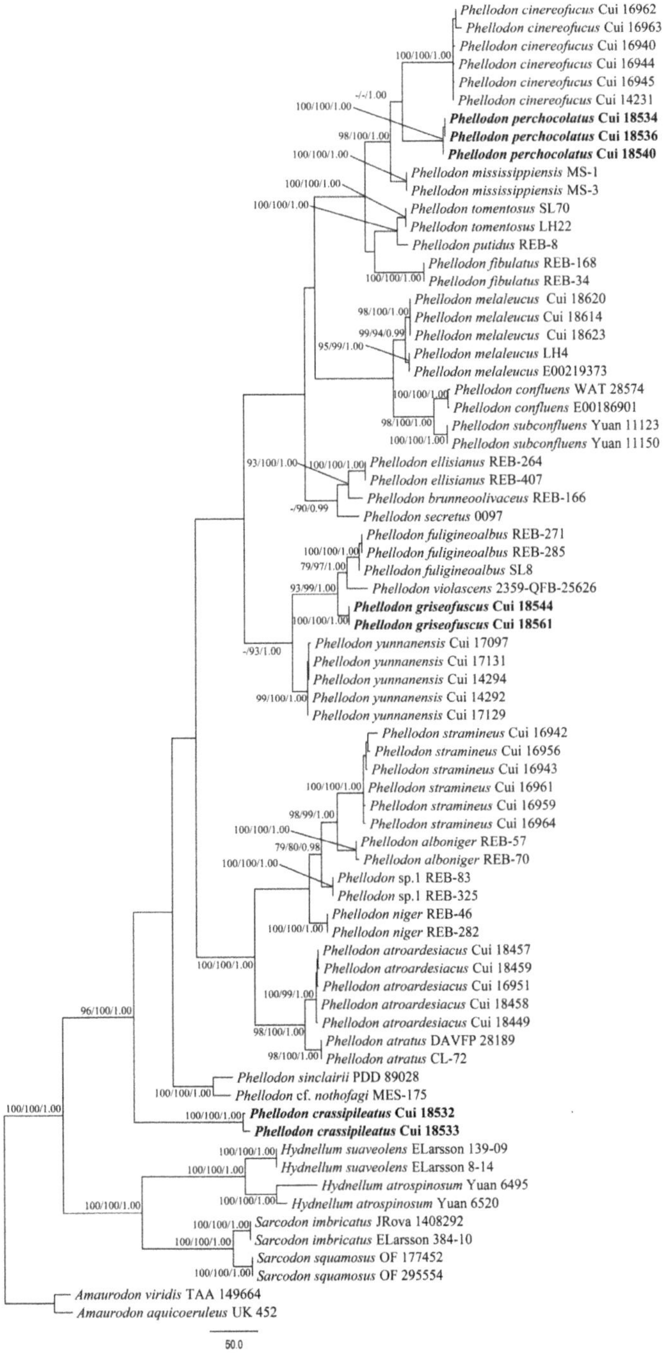

Figure 2. Maximum parsimony (MP) phylogram of the *Phellodon* species based on the combined ITS + nLSU + nSSU + RPB1 + RPB2 sequences data. The supported branches are labeled with parsimony bootstrap values higher than 75%, maximum likelihood bootstrap values higher than 75%, and Bayesian posterior probabilities more than 0.95. Bold names = New species.

Hyphal structure—Hyphal system monomitic; generative hyphae mostly with simple septa, occasionally with clamp connections; all the hyphae IKI–, CB–; tissues turned to olive green in KOH. Generative hyphae in pileal surface pale brown, thick-walled, rarely branched, mostly with simple septa, occasionally with clamp connections, parallel, 2–6 µm in diameter. Generative hyphae in context clay-buff to pale brown, thick-walled, occasionally branched, mostly with simple septa, occasionally with clamp connections, 2–5 µm in diameter. Generative hyphae in spines clay-buff to pale brown, thin-walled, branched, with simple septa, more or less parallel along the spines, 2–4 µm in diameter. Generative hyphae in stipe clay-buff to brown, thick-walled, rarely branched, mostly bearing simple septa, occasionally with clamp connections, parallel along the stipe, 2–6 µm in diameter.

Cystidia—Cystidia and other sterile hyphal elements absent.

Basidia—Basidia clavate, bearing four sterigmata and a basal simple septum, 22–45 × 4–7 µm; sterigmata, 1.5–5 µm; basidioles similar to basidia in shape, but slightly smaller.

Spores—Basidiospores subglobose to globose, hyaline, thin-walled, echinulate, IKI–, CB–, (3.5–) 4–5 × 4–5 µm, L = 4.67 µm, W = 4.19 µm, Q = 1–1.25 (n = 60/2, without the ornamentation).

Additional specimen (paratype) examined—China, Sichuan Province, Pingwu County, Bazi, on the ground of forest dominated by trees of *Quercus* sp., alt. 1190 m, 18 September 2020, Cui 18532 (BJFC 035393).

Ecological habits—*P. crassipileatus* was found on the ground of forest dominated by trees of *Quercus* sp., under a humid monsoon-climate in the northern subtropical region.

Figure 3. Basidiomata of *Phellodon* species. (**a**,**b**) *P. crassipileatus*, (**c**,**d**) *P. griseofuscus*, and (**e**,**f**) *P. perchocolatus*. Scale bars: 2 cm.

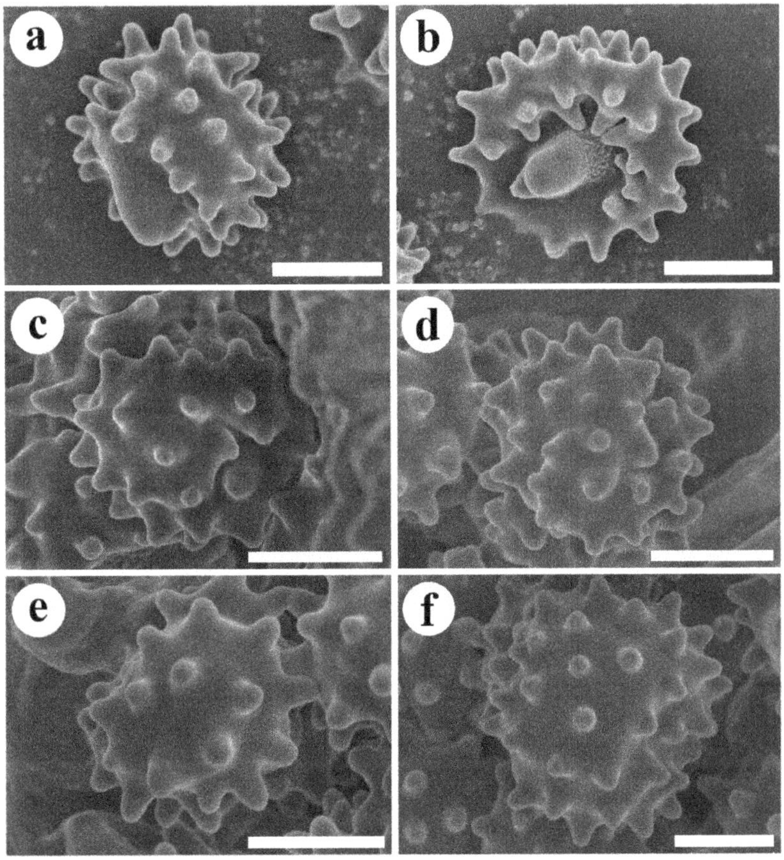

Figure 4. SEM of basidiospores of *Phellodon* species. (**a**,**b**) *P. crassipileatus*, (**c**,**d**) *P. griseofuscus*, and (**e**,**f**). *P. perchocolatus*. Scale bars: 1.5 μm.

Phellodon griseofuscus B.K. Cui and C.G. Song, sp. nov., Figure 3b, Figure 4c,d, and Figure 6.

MycoBank: 843671.

Diagnosis—This species is characterized by its dark brown to black pileal surface, white to pale brown pileal margin, short spines, generative hyphae with both simple septa and clamp connections in spines, and moderately long basidia.

Etymology—*griseofuscus* (Lat.): refers to the pale brown to dark brown or blackish basidiomata.

Holotype—China, Sichuan Province, Jiuzhaigou County, Jiuzhaigou Nature Reserve, on the ground of forest dominated by trees of *Pinus* sp., alt. 2400 m, 20 September 2020, Cui 18561 (BJFC 035422).

Fruiting body—Basidiomata annual, centrally or eccentrically stipitate, solitary or gregarious, with strong odor when dry. Pileus infundibuliform, up to 4 cm in diameter, 5 mm thick at the center. Pileal surface pale brown to dark brown or black when fresh and becoming dark grey to mouse-grey upon drying, azonate, fibrillose; margin blunt or irregular, white to pale brown when fresh, vinaceous grey with age, becoming fuscous upon drying, up to 3 mm wide. Spines soft, white when young, brown with age when fresh, becoming fragile, pale mouse-grey upon drying, up to 1 mm long. Context dark grey, tough, up to 2 mm thick. Stipe fuscous in the outer layer, fuscous to black in the inner layer, cylindrical, glabrous, up to 1.5 cm long, 0.6 cm in diameter.

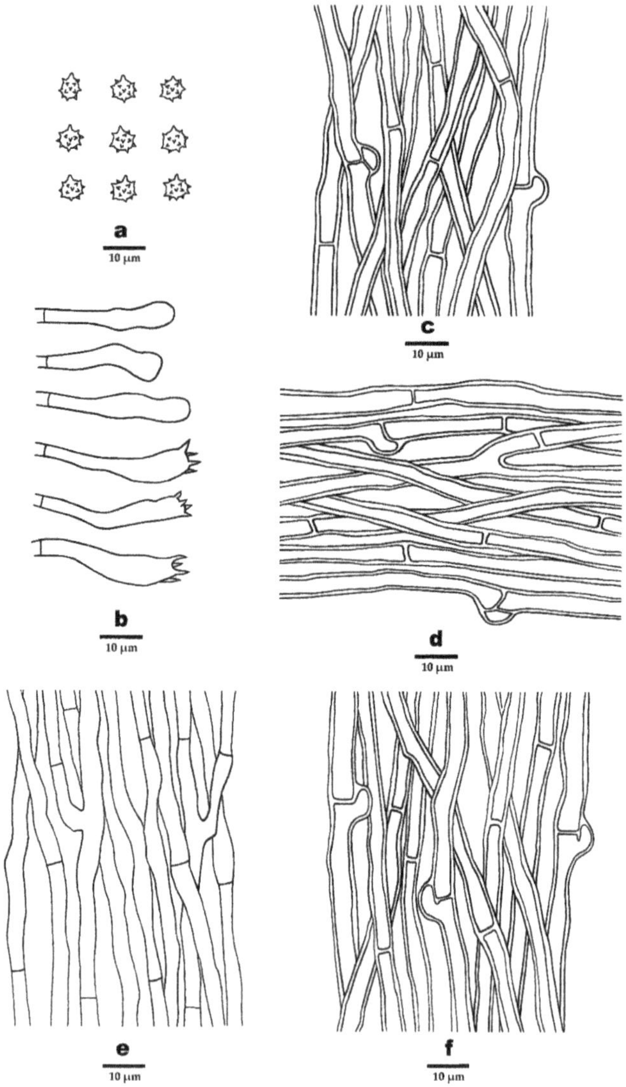

Figure 5. Microscopic structures of *P. crassipileatus* (drawn from the holotype). (**a**) Basidiospores, (**b**) Basidia and basidioles, (**c**). Hyphae from pileal surface, (**d**). Hyphae from context, (**e**). Hyphae from spines, and (**f**) Hyphae from stipe.

Hyphal structure—Hyphal system monomitic; generative hyphae mostly with simple septa, occasionally with clamp connections; all the hyphae IKI–, CB–; tissues turned to olive green in KOH. Generative hyphae in pileal surface greyish brown, thick-walled, rarely branched, with simple septa, parallel, 3–6 μm in diameter. Generative hyphae in context pale brown, thick-walled, occasionally branched, with simple septa, parallel, 3–5 μm in diameter. Generative hyphae in spines clay-buff, thin-walled, branched, mostly with simple septa, occasionally with clamp connections, more or less parallel along the spines, 2–4 μm in diameter. Generative hyphae in stipe greyish brown, slightly thick-walled, rarely branched, bearing simple septa, subparallel along the stipe, 2–6 μm in diameter.

Cystidia—Cystidia and other sterile hyphal elements absent.

Basidia—Basidia clavate, bearing four sterigmata and a basal simple septum, 22–55 × 5–6 μm; sterigmata, 1.5–5μm; basidioles similar to basidia in shape, but slightly smaller.

Spores—Basidiospores subglobose to globose, hyaline, thin-walled, echinulate, IKI–, CB–, 4–5 × 3.5–4.5 μm, L = 4.4 μm, W = 4 μm, Q = 1–1.25 (n = 60/2, without the ornamentation).

Additional specimen (paratype) examined—China, Sichuan Province, Jiuzhaigou County, Jiuzhaigou Nature Reserve, on the ground of forest dominated by trees of *Pinus* sp., alt. 2400 m, 19 September 2020, Cui 18544 (BJFC 035405).

Ecological habits—*P. griseofuscus* was found on the ground of forest dominated by trees of *Pinus* sp., under the humid climate of the plateau. This species grows in well-watered bryophytes, which are often interspersed with pine needles.

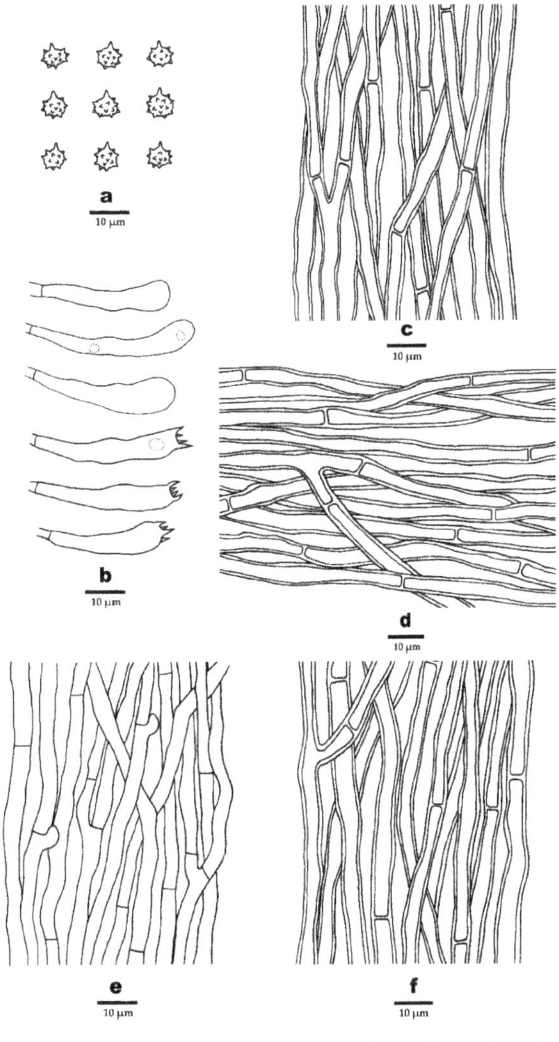

Figure 6. Microscopic structures of *P. griseofuscus* (drawn from the holotype). (**a**) Basidiospores, (**b**) Basidia and basidioles, (**c**) Hyphae from pileal surface, (**d**) Hyphae from context, (**e**) Hyphae from spines, and (**f**) Hyphae from stipe.

Phellodon perchocolatus B.K. Cui and C.G. Song, sp. nov., Figure 3c, Figure 4e,f, and Figure 7.

MycoBank: 843672

Diagnosis—This species is characterized by its woody and broad pileus, brown to greyish brown pileal surface when fresh, tomentose pileal margin when young and glabrous after mature, and the presence of both simple septa and clamp connections in generative hyphae of spines.

Etymology—*perchocolatus* (Lat.): refers to the brown to greyish brown pileal surface.

Holotype—China, Sichuan Province, Pingwu County, Bazi, on the ground of forest dominated by trees of *Quercus* sp., alt. 1190 m, 18 September 2020, Cui 18536 (BJFC 035397).

Fruiting body—Basidiomata annual, centrally or eccentrically stipitate, solitary or gregarious, with a fenugreek odor when dry. Pileus infundibuliform, woody, up to 9 cm in diam, 6.5 mm thick at the center. Pileal surface brown to greyish brown when fresh, becoming fuscous to black upon drying, zonate, tomentose when young, glabrous after the mature; margin blunt or irregular, white when fresh, becoming buff upon drying, up to 5 mm wide. Spines soft, white when fresh, and becoming fragile, pinkish buff to olivaceous buff upon drying, up to 3 mm long. Context vinaceous grey to greyish brown, tough, up to 3 mm thick. Stipe dark brown to fuscous in the outer layer, fuscous in the inner layer, cylindrical, glabrous, up to 4.8 cm long, 1.9 cm in diameter.

Hyphal structure—Hyphal system monomitic; generative hyphae mostly with simple septa, occasionally with clamp connections; all the hyphae IKI–, CB–; tissues turned to olive-green in KOH. Generative hyphae in pileal surface olivaceous-buff, slightly thick-walled, rarely branched, with simple septa, interwoven, 3–7 μm in diameter. Generative hyphae in context olivaceous-buff, thick-walled, occasionally branched, with simple septa, parallel, 2–6 μm in diameter. Generative hyphae in spines olivaceous-buff, thin-walled, branched, mostly with simple septa, occasionally with clamp connections, more or less parallel along the spines, 2–4 μm in diameter. Generative hyphae in stipe clay-buff, thick-walled rarely branched, bearing simple septa, interwoven in the outer layer, parallel along the stipe in the inner layer, 3–8 μm in diameter.

Cystidia—Cystidia and other sterile hyphal elements absent.

Basidia—Basidia clavate, bearing four sterigmata and a basal simple septum, 16–36 × 5–7 μm; sterigmata, 1–4 μm; basidioles similar to basidia in shape, but slightly smaller.

Spores—Basidiospores subglobose to globose, hyaline, thin-walled, echinulate, IKI–, CB–, 4–5 (–5.5) × (3.5–) 4–4.5 (–5) μm, L = 4.7 μm, W = 4.3 μm, Q = 1–1.25 (n = 90/3, without the ornamentation).

Additional specimens (paratypes) examined—China, Sichuan Province, Pingwu County, Bazi, on the ground of forest dominated by trees of *Quercus* sp., alt. 1190 m, 18 September 2020, Cui 18534 (BJFC 035394) & Cui 18540 (BJFC 035401).

Ecological habits—*P. griseofuscus* was found in forest dominated by trees of *Quercus* sp., under a humid monsoon-climate in the northern subtropical region.

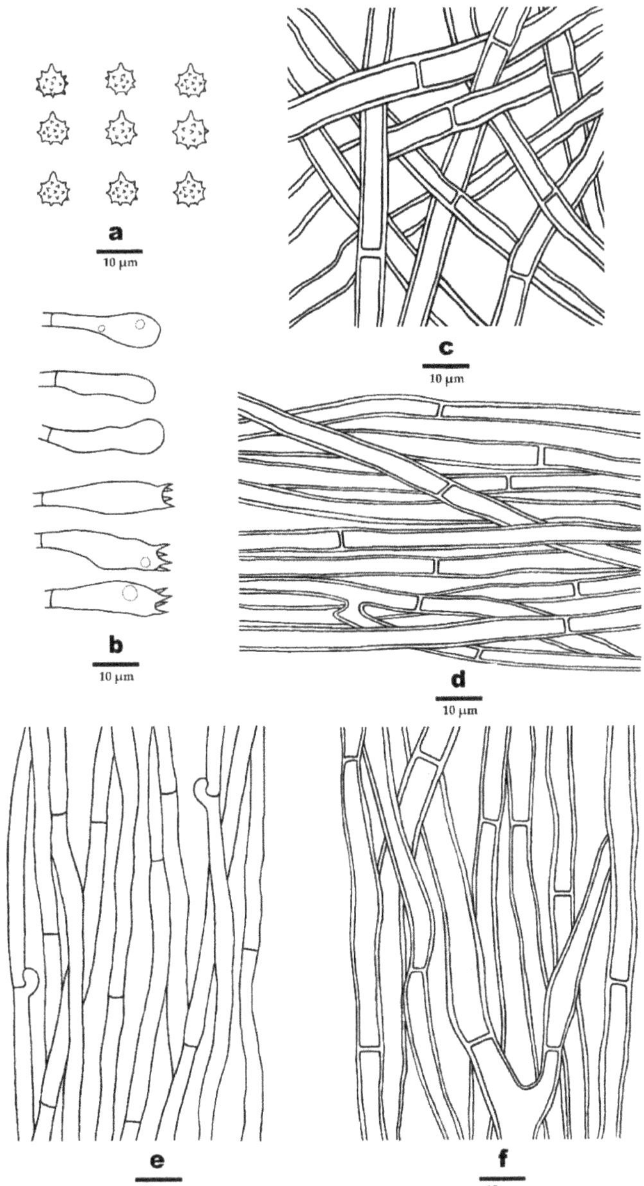

Figure 7. Microscopic structures of *P. perchocolatus* (drawn from the holotype). (**a**) Basidiospores, (**b**) Basidia and basidioles, (**c**). Hyphae from pileal surface, (**d**) Hyphae from context, (**e**) Hyphae from spines, and (**f**) Hyphae from stipe.

Key to species of *Phellodon* from China

1. Pileal surface colored straw buff ---------- *P. stramineus*
1. Pileal surface differently colored ---------- 2
2. Pileal surface blackish blue to dark grey ---------- *P. atroardesiacus*
2. Pileal surface different colored---------- 3
3. Tissues change color in KOH---------- 4
3. Tissues unchanged in KOH---------- *P. subconfluens*
4. Clamp connections exist---------- 5
4. Clamp connections absent ---------- *P. cinereofuscus*
5. Clamp connections exist in spines ---------- 6
5. Clamp connections not exist in spines ---------- 7
6. Spines brown after mature---------- *P. griseofuscus*
6. Spines white after mature ---------- *P. perchocolatus*
7. Pileal surface tomentose and azonate ---------- *P. crassipilieatus*
7. Pileal surface glabrous and zonate---------- *P. yunnanensis*

4. Discussion

In this study, phylogenetic analyses of *Phellodon* were conducted based on the ITS sequences and the combined ITS + nLSU + nSSU + RPB1 + RPB2 sequences to confirm the affinities of the new species and reveal the relationships of *Phellodon* species.

Phellodon crassipileatus formed a single lineage different from other species of *Phellodon* in our phylogenetic analyses (Figures 1 and 2). Morphologically, *P. crassipileatus* is similar to *P. griseofuscus* in having infundibuliform and dark brown pileus. However, *P. griseofuscus* can be distinguished from *P. crassipileatus* by its fibrillose pileus, brown spines after maturity, presence of clamp connections in generative hyphae of spines, and longer basidia (22–55 × 5–6 µm).

Phellodon griseofuscus is closely related to *P. violascens* (Alb. and Schwein.) A.M. Ainsw. and *P. fuligineoalbus* (J.C. Schmidt) R.E. Baird in our phylogenetic analyses (Figures 1 and 2). *Phellodon violascens* is similar to *P. griseofuscus* in having solitary or gregarious basidiomata and larger basidiospores measuring 4.5–5.4 × 4.3–4.5 µm [22]. However, *P. violascens* differs from *P. griseofuscus* by its white to flesh brown basidiomata and lack of clamp connections. *Phellodon fuligineoalbus* differs from *P. griseofuscus* by its yellow-white or light brown basidiomata, lack of clamp connections, and larger basidiospores measuring 4–6 × 4–5 µm [40]. *Phellodon yunnanensis* B.K. Cui and C.G. may be confused with *P. griseofuscus* in having white to brown spines. However, *P. yunnanensis* can be distinct from *P. griseofuscus* by its smaller basidiospores measuring 3.5–4.5 × 3–4 µm [28] and shorter clavate basidia measuring 24–27 × 6–7 µm [28].

Phellodon perchocolatus and *P. cinereofuscus* B.K. Cui and C.G. Song clustered together and then grouped with *P. mississippiensis* R.E. Baird, L.E. Wallace and G. Baker, forming a high supported lineage (98% MP, 100% ML, 1.00 BPP) in our phylogenetic trees (Figures 1 and 2). Morphologically, *P. cinereofuscus* is similar to *P. perchocolatus* in having infundibuliform basidiomata and white pileal margin. However, *P. cinereofuscus* differs from *P. perchocolatus* by its reddish brown to cinnamon brown pileal surface, glabrous basidiomata and lack of clamp connections [28]. *Phellodon mississippiensis* is similar to *P. perchocolatus* in having solitary or gregarious basidiomata, and subglobose to globose basidiospores. However, *P. mississippiensis* can be distinguished from *P. perchocolatus* by its white, light orange to light brown pileal surface and shorter basidia measuring 16–22 × 5–6 µm [27]. Moreover, tissues in *P. mississippiensis* turned light to dark brown in KOH while turned to olive green in *P. perchocolatus*.

5. Conclusions

This study not only fills in the blank of multiple gene fragments of *Phellodon*, but also enriches the species diversity of the genus, which will promote the taxonomy and phylogeny of the genus. This is the first step to infer the phylogeny of *Phellodon* on the basis of multiple genes rather than ITS sequences. Therefore, this study provides a basis for further research on *Phellodon*. However, only a few species of *Phellodon* with available multiple genes could be used for the analyses, which limited the systematic study of the genus. For the time being, the best gene marker for the identification of *Phellodon* is ITS, while more samples with more gene markers including TEF, RPB1, and RPB2 are needed to further investigate the species diversity and phylogenetic relationships of *Phellodon* species.

Author Contributions: B.-K.C. designed the research; B.-K.C., X.-L.H., D.W., Y.-Y.C., S.L., T.-M.X., and C.-G.S. prepared the samples; C.-G.S., S.L., and T.-M.X. conducted the molecular experiments and analyzed the data; C.-G.S. and B.-K.C. drafted the manuscript. All authors have read and agreed to the published version of the manuscript.

Funding: The research is supported by the National Natural Science Foundation of China (Nos. 31870008, 31900017) and Beijing Forestry University Outstanding Young Talent Cultivation Project (No. 2019JQ03016).

Institutional Review Board Statement: Not applicable.

Informed Consent Statement: Not applicable.

Data Availability Statement: The data and results of this study are available upon reasonable request. Please contact the main author of this publication.

Conflicts of Interest: The authors declare no conflict of interest. The funders had no role in the design of the study; in the collection, analyses, or interpretation of data; in the writing of the manuscript, or in the decision to publish the results.

References

1. Karsten, P.A. Enumeratio boletinearum et polyporearum fennicarum, systemate novo dispositarum. *Rev. Mycol.* **1881**, *3*, 16–19.
2. Stalpers, J.A. The aphyllophoraceous fungi I. keys to the species of the Thelephorales. *Stud. Mycol.* **1993**, *35*, 1–168.
3. Pegler, D.N.; Roberst, P.J.; Spooner, B.M. *British Chanterelles and Tooth Fungi*; Royal Botanic Gardens: Kew, UK, 1997.
4. Parfitt, D.; Ainsworth, A.M.; Simpson, D.; Rogers, H.J.; Boddy, L. Molecular and morphological discrimination of stipitate hydnoids in the genera *Hydnellum* and *Phellodon*. *Mycol. Res.* **2007**, *3*, 761–777. [CrossRef] [PubMed]
5. Van der Heijden, M.; Klironomos, J.; Ursic, M.; Moutoglis, P.; Streitwolf-Engel, R.; Boller, T.; Wiemken, A.; Sanders, I. Mycorrhizal fungal diversity determines plant biodiversity, ecosystem variability and productivity. *Nature* **1998**, *396*, 69–72. [CrossRef]
6. Erland, S.; Taylor, A.F.S. Resupinate ectomycorrhizal fungal genera. In *Ectomycorrhizal Fungi Key Genera in Profile*; Cairney, J.W.G., Chambers, S.M., Eds.; Springer: Berlin/Heidelberg, Germany, 1999; pp. 347–363.
7. Arnolds, E. Decline of ectomycorrhizal fungi in Europe. *Agric. Ecosyst. Environ.* **1991**, *35*, 209–244. [CrossRef]
8. Termorshuizen, A.J.; Schaffers, A.P. The decline of sporocarps of ectomycorrhizal fungi in stands of *Pinus sylvestris* L. in The Netherlands: Possible causes. *Nova Hedwig.* **1991**, *53*, 267–289.
9. Arnolds, E. The fate of hydnoid fungi in The Netherlands and Northwestern Europe. *Fungal Ecol.* **2010**, *3*, 81–88. [CrossRef]
10. Holec, J.; Kučera, T. Hydnoid fungi of the family Bankeraceae—Their assemblages and vegetation ecology in Central Europe, Czech Republic. *Fungal Ecol.* **2018**, *32*, 40–48. [CrossRef]
11. Larsson, K.H.; Svantesson, S.; Miscevic, D.; Kljalg, U.; Larsson, E. Reassessment of the generic limits for *Hydnellum* and *Sarcodon* (*Thelephorales, Basidiomycota*). *MycoKeys* **2019**, *54*, 31–47. [CrossRef]
12. Mu, Y.H.; Wu, F.; Yuan, H.S. Hydnaceous fungi of China 7. Morphological and molecular characterization of *Phellodon subconfluens* sp. nov. from temperate, deciduous forests. *Phytotaxa* **2019**, *414*, 280–288. [CrossRef]
13. Banker, H.J. A contribution to a revision of the North American Hydnaceae. *Mem. Torrey Bot. Club* **1906**, *12*, 99–194.
14. Coker, W.C.; Beers, A.H. *The Stipitate Hydnums of the Eastern United States*; University of North Carolina Press: Chapel Hill, NC, USA, 1951; p. 211.
15. Harrison, K.A. New or little known north American stipitate hydnums. *Can. J. Bot.* **1964**, *42*, 1205–1233. [CrossRef]
16. Harrison, K.A. A new species of *Phellodon* possessing clamp connections. *Can. J. Bot.* **1972**, *50*, 1219–1221. [CrossRef]
17. Maas Geesteranus, R.A. The stipitate hydnums of the Netherlands. *Fungus* **1958**, *28*, 48–61.
18. Maas Geesteranus, R.A. Notes on the hydnums. *Persoonia* **1960**, *1*, 341–384.
19. Maas Geeseranus, R.A. Hyphal structures in the hydnums. *Persoonia* **1962**, *2*, 476.
20. Maas Geesteranus, R.A. Hydnaceous fungi of the eastern old world. *Verh. Kon. Ned. Akad. Wetensch. Afd. Natuurk* **1971**, *60*, 176.

21. Maas Geesteranus, R.A. The terrestrial hydnums of Europe. *Verh. Kon. Ned. Akad. Wetensch. Afd. Natuurk* **1975**, *65*, 127.
22. Hrouda, P. Hydnaceous fungi of the Czech Republic and Slovakia. *Czech Mycol.* **1999**, *51*, 99–155. [CrossRef]
23. Hrouda, P. Bankeraceae in Central Europe. 1. *Czech Mycol.* **2005**, *57*, 57–78. [CrossRef]
24. Hrouda, P. Bankeraceae in Central Europe. 2. *Czech Mycol.* **2005**, *57*, 279–297. [CrossRef]
25. Niemellä, T.; Kinnunen, J.; Renvall, P.; Schigel, D. *Phellodon secretus* (Basidiomycota), a new hydnaceous fungus from northern pine woodlands. *Karstenia* **2003**, *43*, 37–44. [CrossRef]
26. Ainsworth, A.M.; Parfitt, D.; Rogers, H.J.; Boddy, L. Cryptic taxa within European species of *Hydnellum* and *Phellodon* revealed by combined molecular and morphological analysis. *Fungal Ecol.* **2010**, *3*, 65–80. [CrossRef]
27. Baird, R.E.; Wallace, L.E.; Baker, G. Stipitate hydnums of the southern United States 1: *Phellodon mississippiensis* sp. nov. *Mycotaxon* **2013**, *123*, 183–191. [CrossRef]
28. Song, C.G.; Ji, X.; Liu, S.; He, X.L.; Cui, B.K. Taxonomy and molecular phylogeny of *Phellodon* (Thelephorales) with descriptions of four new species from Southwest China. *Forests* **2021**, *12*, 932. [CrossRef]
29. Wang, M. Taxonomy and Phylogeny of Hydnoid Fungi in Hericiaceae from China. Master's Thesis, Beijing Forestry University, Beijing, China, 2018.
30. Liu, S.; Han, M.L.; Xu, T.M.; Wang, Y.; Wu, D.M.; Cui, B.K. Taxonomy and phylogeny of the *Fomitopsis pinicola* complex with descriptions of six new species from east Asia. *Front. Microbiol.* **2021**, *12*, 644979. [CrossRef]
31. Sun, Y.F.; Costa-Rezende, D.H.; Xing, J.H.; Zhou, J.L.; Zhang, B.; Gibertoni, T.B.; Gates, G.; Glen, M.; Dai, Y.C.; Cui, B.K. Multi-gene phylogeny and taxonomy of *Amauroderma* s. lat. (*Ganodermataceae*). *Persoonia* **2020**, *44*, 206–239. [CrossRef]
32. Stöger, A.; Schaffer, J.; Ruppitsch, W. A rapid and sensitive method for direct detection of *Erwinia amylovora* in symptomatic and asymptomatic plant tissues by polymerase chain reaction. *J. Phytopathol.* **2006**, *154*, 469–473. [CrossRef]
33. Thompson, J.D.; Gibson, T.J.; Plewniak, F.; Jeanmougin, F.; Higgins, D.G. The Clustal_X windows interface: Flexible strategies for multiple sequence alignment aided by quality analysis tools. *Nucleic Acids Res.* **1997**, *25*, 4876–4882. [CrossRef]
34. Hall, T.A. Bioedit: A user-friendly biological sequence alignment editor and analysis program for Windows 95/98/NT. *Nucleic Acids Symp. Ser.* **1999**, *41*, 95–98.
35. Swofford, D.L. *PAUP*: Phylogenetic Analysis Using Parsimony (*and Other Methods)*; Version 4.0b10; Sinauer Associates: Sunderland, MA, USA, 2002.
36. Farris, J.S.; Källersjö, M.; Kluge, A.G.; Kluge, A.G.; Bult, C. Testing significance of incongruence. *Cladistics* **1995**, *10*, 315–319. [CrossRef]
37. Felsenstein, J. Confidence intervals on phylogenetics: An approach using bootstrap. *Evolution* **1985**, *39*, 783–791. [CrossRef] [PubMed]
38. Posada, D.; Crandall, K.A. Modeltest: Testing the model of DNA substitution. *Bioinformatics* **1998**, *14*, 817–818. [CrossRef] [PubMed]
39. Nylander, J.A.A. *MrModeltest v2. Program. Distributed by the Author*; Evolutionary Biology Center, Uppsala University: Uppsala, Sweden, 2004.
40. Baird, R.E.; Wallace, L.E.; Baker, G.; Scruggs, M. Stipitate hydnoid fungi of the temperate southeastern United States. *Fungal Divers.* **2013**, *62*, 41–114. [CrossRef]

Journal of *Fungi*

MDPI

Article

Morphological and Molecular Evidence Reveal Eight New Species of *Gymnopus* from Northeast China

Jiajun Hu [1,2], Guiping Zhao [2], Yonglan Tuo [2], Gu Rao [2], Zhenhao Zhang [2], Zhengxiang Qi [2], Lei Yue [2], Yajie Liu [2], Tong Zhang [3], Yu Li [1,*] and Bo Zhang [2,*]

1 School of Life Science, Northeast Normal University, Changchun 130024, China; hujjfungi@163.com
2 Engineering Research Centre of Edible and Medicinal Fungi, Jilin Agricultural University, Ministry of Education, Changchun 130118, China; guipingz6@163.com (G.Z.); tuoyonglan66@163.com (Y.T.); raogufungi@126.com (G.R.); zzhzz34@163.com (Z.Z.); qzx7007@163.com (Z.Q.); yuelei02@126.com (L.Y.); lyj117@163.com (Y.L.)
3 Department of Pesticide and Pesticide Machinery, Jilin Province Agricultural Technology Extension Station, Changchun 130000, China; zt_fungi@163.com
* Correspondence: yuli966@126.com (Y.L.); zhangbofungi@126.com (B.Z.)

Abstract: *Gymnopus* is a widely distributed genus consisting of about 300 species thus far, including *Gymnopus fusipes* as a generic type. A total of nine species from China belong to the sect. *Levipedes*, including eight new species—*Gymnopus longisterigmaticus*, *Gymnopus longus*, *Gymnopus macrosporus*, *Gymnopus striatus*, *Gymnopus changbaiensis*, *Gymnopus tomentosus*, *Gymnopus tiliicola*, and *Gymnopus globulosus*—which were delimited and proposed based on morphological and molecular evidence; and one new record from Jilin Province, China—*Gymnopus erythropus*. Detailed descriptions and illustrations are presented, as well as comparisons to similar species. Overall, our results broaden the morphological characterization of the genus. The pileipellis of sect. *Levipedes* typically takes on the "*Dryophila* structure", while, in our findings, pileipellis terminal hyphae inflated to spherical to prolate were observed, in addition to extremely long basidia sterigma. The phylogenies inferred from the ITS and nLSU dataset supported the *Gymnopus*, which was defined by Oliveira et al. as a monophyletic genus, and the novel species as separate lineages within. A key to all species described in this study is also provided.

Keywords: *Gymnopus* sect. *Levipedes*; *Gymnopus erythropus* complex; new species; Northeast China; phylogenetic analysis

Citation: Hu, J.; Zhao, G.; Tuo, Y.; Rao, G.; Zhang, Z.; Qi, Z.; Yue, L.; Liu, Y.; Zhang, T.; Li, Y.; et al. Morphological and Molecular Evidence Reveal Eight New Species of *Gymnopus* from Northeast China. *J. Fungi* **2022**, *8*, 349. https://doi.org/10.3390/jof8040349

Academic Editors: Lei Cai and Cheng Gao

Received: 21 February 2022
Accepted: 25 March 2022
Published: 28 March 2022

1. Introduction

The genus *Gymnopus* (Pers.) Roussel belongs to the family Omphalotaceae, according to Antonín and Noordeloos [1]. There is a long taxonomic research history on this genus, beginning with its proposal by Persoon in 1801 as a tribe of *Agaricus* L. [2]. Later, Fries [3] established *Agaricus* trib. *Collybia* Fr., transferring the species of *Agaricus* trib. *Gymnopus* Pers. into it accordingly. This perspective was widely accepted by other mycologists, until Staude [4] established the genus *Collybia* (Fr.) Staude. Singer [5–7] divided the genus *Collybia* into nine sections—sect. *Striipedes* (Fr.) Quél., sect. *Dictyoplocae* (Mont.) Sing., sect. *Iocephalae* Sing. ex Halling, sect. *Levipedes* (Fr.) Quél., sect. *Vestipedes* (Fr.) Quél., sect. *Subfulmosae* Sing., sect. *Cystidiatae* Sing., sect. *Ixotrma* Sing., and sect. *Collybia* Sing.—in his book, *The Agaricales in Modern Taxonomy*. It is the embryonic form of the modern taxonomy of the genus *Collybia* (*Gymnopus*). Based on their research [8–10], Halling, Antonín and Noordeloos pointed out that the genus *Collybia* had a problematic and controversial taxonomy and, thus, lacked a clear definition; then Antonín et al. [11] shifted the section and species into *Gymnopus* and *Rhodocollybia* Singer, leaving three species in the genus *Collybia*. The members of *Gymnopus*, in the conception of Antonín and Noordeloos [1,12], are mainly characterized by basidiomata, usually collybioid, marasmioid, and gymnopoid, stipes only rarely arising from the sclerotia, which is a white spore print with smooth basidiospores that are commonly ellipsoid to oblong, typically the presence of clamp connections, a cutis-type pileipellis, ixocutis or similar to a trichoderm, terminal elements mostly coralloid to diverticulate, and usually encrusted pigments.

With the advent and development of molecular technologies, the phylogenetic analysis of marasmioid and collybioid fungi based on sequences of nuclear ribosomal DNA is just beginning to help clarify generic and infrageneric circumscriptions. Moncalvo et al. [13] pointed out that the genus *Gymnopus* was multiphyletic. Mata et al. [14], who found similar conclusions, stated that *Gymnopus* is more closely related to *Marasmiellus* Murrill. The type specimen of *Marasmiellus*, *Marasmiellus juniperinus* Murrill, was confirmed within *Gymnopus* sect. *Levipedes* (Fr.) Antonín, Halling and Noordel. [12]. More recently, Oliveira et al. [15] redefined the genus *Gymnopus* more strictly standard based on their combined ITS + nLSU phylogenetic analysis. In its conception, the key features of the genus *Gymnopus* are collybioid (rarely tricholomatoid or marasmioid) basidiomata, free, emarginate, or adnate lamellae that are usually crowded, an insititious stipe or not, usually with a strigose base; a white spore print, basidiospores ellipsoid to short-oblong, inamyloid; cheilocystidia usually present, or a variety of cheilocystidia, a cutis or ixocutis pileipellis with radially arranged cylindrical hyphae or interwoven more akin to a trichoderm or ixotrichoderm, made up of irregular coralloid terminal elements ("*Dryophila* structures")—often incrusted, diverticulate hyphal elements, mixed with broom cells and coralloid hyphae; and clamp connections present in all tissues. As a result, *Gymnopus* sect. *Vestipedes* (Fr.) Antonín, Halling and Noordel. is segregated and placed within *Marasmiellus* s. str, and *Gymnopus* sect. *Perforanita* (Singer) R.H. Petersen is considered a new independent genus *Paragymnopus* J.S. Oliveira. In addition, some *Gymnopus* species were transferred to two new genera [16].

Most species of *Gymnopus* sect. *Levipedes* (Quél.) Halling have smooth, polished, or pubescent stipe; pileipellis mostly as an entangled trichoderm (never radially oriented), composed of inflated, often lobed elements or coralloid ("*Dryophila* structures"); non-dextrinoid trama and elements, with some species turning green in alkali [1,17]. *Gymnopus erythropus* (Pers.) Antonín, Halling and Noordel. is one of the most confusing species in this section. This species was named by Persoon as *Agaricus erythropus* Pers. [2], and then transferred to *Gymnopus* [11]. However, Persoon had a broad conception of this species. Singer [18] selected a neotype from the herbarium of Persoon, labeled *Agaricus erythropus*, which was confirmed to be a *Mycena* (Pers.) Roussel species later on. There remained more specimens labeled *A garicus erythropus* in the herbarium of Persoon, until Jansen [19] studied the material and found that one specimen fit well with the current concept of *Collybia erythropus* (Pers.) P. Kumm. (≡ *Gymnopus erythropus*); considering this, Singer's choice was rejected. Prior to the current study, only two red stipe species had been reported in this section *Gymnopus erythropus*, and *Gymnopus fagiphilus* (Velen.) Antonín, Halling, and Noordel. These two species are morphologically very similar, with the clear distinguishing factor being the tomentose stipe. Specifically, *Gymnopus erythropus* have a smooth stipe, while those with a dense tomentose at the base of the stipe are *Gymnopus fagiphilus*.

Approximately 300 species have been validly published in the genus *Gymnopus* [20], with most species having been reported from Europe and America. However, research on *Gymnopus* in China is lacking. Teng [21] was the first to report a *Gymnopus* species in China; based on the genus *Collybia*, four taxa were reported. Later, Tai [22] reported eight taxa. Deng [23], a preliminary study on the resources of the genus *Gymnopus* in Southern China, reported 19 taxa. Recently, three new species and 11 new records were recorded from China [23–29]. Until now, 24 species of *Gymnopus* s. str. have been recognized from China.

This paper aims to describe and illustrate nine species of *Gymnopus* sect. *Levipedes*—eight species new to science, and one new record from Jilin Province, China—based on morphology and molecular studies.

2. Materials and Methods

2.1. Sampling and Morphological Studies

The studied specimens were photographed in situ. The size of the basidiomata was measured when fresh. After examination and description of the fresh macroscopic characteristics, the specimens were dried in an electric drier at 40–45 °C.

Descriptions of the macroscopic characteristics were based on field notes and photographs, with the colors corresponding to the *Flora of British fungi: Color identification chart* [30]. The dried specimens were rehydrated in 94% ethanol for microscopic examination, and then mounted in 3% potassium hydroxide (KOH), 1% Congo red (0.1 g Congo red dissolved in 10 mL distilled water), and Melzer's reagent (1.5 g potassium iodide, 0.5 g crystalline iodine and 22 g chloral hydrate dissolved in 20 mL distilled water) [31]; they were then examined with a Zeiss Axio lab. A1 microscope at magnifications up to 1000×. All measurements were taken from the sections mounted in the 1% Congo red. For each specimen, a minimum of 40 basidiospores, 20 basidia, 20 cheilocystidia, and 20 widths of pileipellis were measured from two different basidiocarps. When reporting the variation in the size of the basidiospores, basidia, cheilocystidia, and width of the pileipellis, 5% of the measurements were excluded from each end of the range, and are given in parentheses. The basidiospores measurements are given as length × width (L × W). Q denotes the variation in the ratio of L to W among the studied specimens, and Qm denotes the average Q value of all the basidiospores ± standard deviation. "I" refers to the number of lamellulae between every two complete lamellae, and "L" refers to the number of complete lamellae. The specimens examined are deposited in the Herbarium of Mycology of Jilin Agricultural University (HMJAU).

2.2. DNA Extraction, PCR Amplification, and Sequencing

The total DNA was extracted from dried specimens by using the NuClean Plant Genomic DNA Kit (Kangwei Century Biotechnology Company Limited, Beijing, China), according to the manufacturer's instructions. Sequences of the internal transcribed spacer (ITS) region, and nuclear large ribosomal subunits (nLSU) were used for phylogenetic analysis. The ITS sequence was amplified by using the primer pair ITS1-F (CTT GGT CAT TTA GAG GAA GTA A) and ITS4-B (CAG GAG ACT TGT ACA CGG TCC AG) [32], and the nLSU sequence was amplified by using the primer pair LROR (GTA CCC GCT GAA CTT AAG C) and LR7 (TAC TAC CAC CAA GAT CT) [33,34]. PCR reactions (25 μL) contained 8 μL 2 × EasyTaq® PCR SuperMix (TransGen Biotech Co., Ltd., Beijing, China), 1 μL 10 μM primer L, 1 μL 10 μM primer R, 3 μL DNA solution, and 12 μL dd H_2O. The reaction programs were as follows: for the ITS, initial denaturation at 94 °C for 4 min, followed by 30 cycles at 94 °C for 1 min, 54 °C for 1 min and 72 °C for 1 min, and a final extension of 72 °C for 10 min [35]; for the nLSU, initial denaturation at 95 °C for 3 min, followed by 30 cycles at 94 °C for 30 s, 47 °C for 45 s, and 72 °C for 90 s, and a final extension of 72 °C for 10 min [36]. The PCR products were visualized via UV light after electrophoresis on 1% agarose gels stained with ethidium bromide and purified by using the Genview High-Efficiency Agarose Gels DNA Purification Kit (Gen-View Scientific Inc., Galveston, TX, USA). The purified PCR products were then sent to Sangon Biotech Limited Company (Shanghai, China) for sequencing, using the Sanger method. The new sequences were deposited in GenBank (http://www.ncbi.nlm.nih.gov/genbank (accessed on 17 November 2021); see Table 1).

2.3. Data Analysis

Based on the results of BLAST and morphological similarities, the sequences obtained and related to these samples were collected and are listed in Table 1. We used a dataset of ITS and nLSU resign comprising sequences from this study, with 49 representative sequences showing the highest similarity to *Gymnopus* spp. This dataset included all *Gymnopus* s. str. section (sect. *Androcacei* (Kühner) Antonín and Noordel., sect. *Levipedes* (Quél.) Halling, sect. *Impudicae* (Antonín and Noordel.) Antonín and Noordel., and sect. *Gymnopus*

(Pers.) Roussel) to further explore the relationships of the newly sequenced Chinese specimens within the genus *Gymnopus*. Moreover, the species within this genus and those in allied genera, including *Lentinula* Earle, *Rhodocollybia* Singer, *Mycetinis* Earle, *Marasmiellus* Murrill, *Collybiopsis* (J. Schröt.) Earle, and *Paragymnopus* J.S. Oliveira were included. The sequences of *Marasmius* sect. *Globulares* Kühner, *Marasmius stenophyllus* Mont., *Marasmius aurantioferrugineus* Hongo, *Marasmius brunneospermus* Har. Takah., *Marasmius maximus* Hongo, and *Marasmius nivicola* Har. Takah., were selected as the outgroup taxa [15].

For the dataset, each gene region was aligned by using ClustalX [37], MACSE V2.03 [38], or MAFFT 7.490 [39], and then manually adjusted in BioEdit 7.0.5.3 [40]. The datasets first were aligned, and then the ITS and nLSU sequences were combined with Phylosuit V1.2.2 [41]. The best-fit evolutionary model was estimated by using Modelfinder [42]. Following the models, Bayesian inference (BI) algorithms were used to perform the phylogenetic analysis. Specifically, BI was calculated with MrBayes 3.2.6 with a general time-reversible DNA substitution model and a gamma distribution rate variation across the sites [43]. Four Markov chains were run for two runs from random starting trees for two million generations until the split deviation frequency value was <0.01; the trees were sampled every 100 generations. The first 25% of the sampled trees were discarded as burn-in, while all remaining trees were used to construct a 50% majority consensus tree and for calculating the Bayesian posterior probabilities (BPPS). RaxmlGUI 2.0.5 [44] was used for maximum likelihood (ML) analysis, along with 1000 bootstraps (BS) replicates, using the GTRGAMMA algorithm to perform a tree inference and search for the optimal topology [45]. Then the FigTree v1.3.1 was used to visualize the resulting trees.

Table 1. Voucher/specimen numbers, country, and GenBank accession numbers of the specimens included in this study. Sequences produced in this study are in bold.

Scientific Name	Country	Voucher/Specimen Numbers	GenBank Accession Numbers ITS	LSU	References
Collybiopsis dichroa	USA	TENN56726	AY256702		[46]
Co. filamentipes	USA	TENN-F-065861	MN897832	MN897832	[47]
Co. furtiva	USA	SFSU: DED4425	DQ450031	AF042650	[47]
Co. hasanskyensis	Russia	TENN-F-060730	MN897829		[47]
Co. juniperina	USA	TENN59540	AY256708		[14]
Co. melanopus	Indonesia	SFSU: A.W. Wilson 54	NR_137539	NG_060624	[48]
Co. melanopus	China	LF1758	KU529307		[23]
Co. mesoamericana	Costa Rica	TENN 058613	NR_119583	KY019632	[49,50]
Co. minor	USA	TENN-F-059993	MN413334	MW396880	[47]
Co. parvula	USA	TENN-F-059993	MN413334		Unpublished
Co. stenophylla	USA	TENN59449	DQ450033		[46]
Gymnopus alkalivirens	USA	TENN51249	DQ450000		[46]
G. alliifoetidissimus	China	GDGM76695	MT023344	MT017526	[25]
G. alpinus	Latvia	CB16251	JX536168		[51]
G. androsaceus	Russia	TENN-F-59594	KY026663	KY026663	[50]
G. androsaceus	France	CBS239.53	MH857174	MH868713	[52]
G. aquosus	Czech Republic	BRNM665362	JX536172		[51]
G. aurantiipes		AWW118	AY263432	AY639410	[48]
G. bicolor		AWW116	AY263423	AY639411	[48]
G. biformis	USA	TENN58541	DQ450054		[48]
G. brunneigracilis		AWW01	AY263434	AY639412	[48]
G. changbaiensis	**China**	**HMJAU60300**	**OM030272**	**OM033387**	**this study**
G. changbaiensis	**China**	**HMJAU60301**	**OM030273**	**OM033388**	**this study**
G. changbaiensis	**China**	**HMJAU60302**	**OM030274**	**OM033389**	**this study**

Table 1. *Cont.*

Scientific Name	Country	Voucher/Specimen Numbers	GenBank Accession Numbers ITS	LSU	References
G. collybioides	Costa Rica	TENN58020	AF505772		[46]
G. confluens	Sweden	TENN50524	DQ450044		[46]
G. confluens	USA	TENN55695	DQ450050		[46]
G. cylindricus	Costa Rica	TENN-058097	NR_119464		[49]
G. densilamellatus	Republic of Korea	BRNM714984	KP336686	KP336695	[36]
G. dryophilus	Czech Republic	BRNM695586	JX536143		[51]
G. dryophilus	Germany	BRNM737691	JX536139		[51]
G. dryophilus	China	HMAS290095	MK966542		Unpublished
G. dryophilus	Japan	Duke31	DQ480099		[46]
G. dryophioides	Republic of Korea	BRNM781447	MH589967	MH589985	[53]
G. dysodes	USA	TENN59141	AF505778		[46]
G. erythropus	Czech Republic	BRNM714784	JX536136		[51]
G. erythropus	USA	JFA12910	DQ449998		[46]
G. erythropus	Austria	TENN59329	AF505786		[46]
G. erythropus	**China**	**HMJAU60313**	**OM030281**	**OM033395**	**this study**
G. erythropus	**China**	**HMJAU60315**	**OM030280**	**OM033396**	**this study**
G. fagiphilus	Czech Republic	BRNM707079	JX536129		[51]
G. fusipes	Austria	TENN59300	AF505777		[46]
G. fusipes	France	TENN59217	AY256710	AY256710	[14]
G. globulosus	**China**	**HMJAU60307**	**OM030269**	**OM033406**	**this study**
G. globulosus	**China**	**HMJAU60308**	**OM030270**	**OM033407**	**this study**
G. globulosus	**China**	**HMJAU60308**	**OM030271**	**OM033408**	**this study**
G. hybridus	Italy	BRNM695773	JX536177		[51]
G. inexpectatus	Italy		EU622905	EU622906	[54]
G. inusitatus	Spain	SCM B-4058	JN247553	JN247557	[51]
G. junquilleus	USA	TENN55224	NR_119582		[49]
G. lanipes	Spain	BRNM670686	JX536137		[51]
G. longisterigmaticus	**China**	**HMJAU60288**	**OM030282**	**OM033403**	**this study**
G. longisterigmaticus	**China**	**HMJAU60289**	**OM030283**	**OM033404**	**this study**
G. longisterigmaticus	**China**	**HMJAU60290**	**OM030284**	**OM033405**	**this study**
G. longus	**China**	**HMJAU60291**	**OM030285**	**OM033400**	**this study**
G. longus	**China**	**HMJAU60292**	**OM030286**	**OM033401**	**this study**
G. longus	**China**	**HMJAU60293**	**OM030287**	**OM033402**	**this study**
G. macrosporus	**China**	**HMJAU60294**	**OM030266**	**OM033397**	**this study**
G. macrosporus	**China**	**HMJAU60295**	**OM030267**	**OM033398**	**this study**
G. macrosporus	**China**	**HMJAU60296**	**OM030268**	**OM033399**	**this study**
G. ocior	Czech Republic	BRNM699795	JX536166		[51]
G. pallipes	China	GDGM81513	MW582856		[24]
G. ramulicola	China	GDGM44256	KU321529		[27]
G. similis	Republic of Korea	BRNM766739	KP336692	KP336699	[36]
G. similis	China	GDGM78308	MT023352	MT017530	[25]
G. striatus	**China**	**HMJAU60297**	**OM030263**	**OM033384**	**this study**
G. striatus	**China**	**HMJAU60298**	**OM030264**	**OM033385**	**this study**
G. striatus	**China**	**HMJAU60299**	**OM030265**	**OM033386**	**this study**
G. tiliicola	**China**	**HMJAU60305**	**OM030275**	**OM033393**	**this study**
G. tiliicola	**China**	**HMJAU60306**	**OM030276**	**OM033394**	**this study**
G. tiliicola	**China**	**HMJAU60307**	**OM030277**	**OM033392**	**this study**
G. tomentosus	**China**	**HMJAU60303**	**OM030278**	**OM033390**	**this study**
G. tomentosus	**China**	**HMJAU60304**	**OM030279**	**OM033391**	**this study**

Table 1. *Cont.*

Scientific Name	Country	Voucher/Specimen Numbers	GenBank Accession Numbers ITS	LSU	References
Letinula aciculospora	Costa Rica	TENN37996	AY016443		[55]
L. boryana	Brazil	TENN58368	AY016440		[55]
L. edodes	China	STCL125	AF031183		[56]
Marasmius aurantioferrugineus	Republic of Korea	BRNM714752	FJ904962	MK278334	[57]
M. brunneospermus	Republic of Korea	KPM-NC0005011	FJ904969	FJ904951	[57]
M. maximus	Republic of Korea	BRNM714570	FJ904976	FJ904958	[57]
M. nivicola	Republic of Korea	KPM-NC0006038	FJ904973	FJ904955	[57]
Marasmiellus ramealis	Sweden	TENN50324	DQ450030		[46]
Mycetinis. alliaceus	Russia	TENN-F-55630	KY696784	KY696752	[58]
My. curraniae	New Zealand	PDD95301	KY696778		[58]
My. opacus	USA	TENN-F-59451	KY696755		[58]
My. scorodonius	Switzerland	TENN-F-59451	KY696725		[58]
Paragymnopus foliiphilus	USA	TENN-F-68183	KY026705	KY026705	[50]
P. perforans	Sweden	TENN-F-50319	KY026625	KY026625	[50]
P. pinophilus	USA	TENN-F-69207	KY026725	KY026725	[50]
Rhodocollyba butyracea	Sweden	TENN53580	AY313293		[46]
R. butyracea	China	HFJAU0269	MN258680		Unpublished
R. maculata	Dominican Republic	TFB11720	KT205402		[59]
R. maculata	USA	TENN59459	AY313296		[46]

3. Results

3.1. Phylogenetic Analyses

In the dataset, 143 sequences derived from two gene loci (ITS and nLSU) from 92 samples were used to build phylogenetic trees; 50 of these were newly generated, with 25 ITS sequences and 25 nLSU sequences. The phylogenetic construction performed via ML and BI analysis for the two combined datasets showed a similar topology. The combined ITS and nLSU dataset represented 63 taxa and 2600 characters after being trimmed. The Bayesian analysis was run for two million generations and resulted in an average standard deviation of split frequencies of 0.004989. The same dataset and alignment were analyzed by using the ML method. In the phylogenetic tree, six clades corresponding to *Gymnopus*, *Rhodocollybia*, *Lnetinula*, *Marasmiellus*, *Marasmius*, *Mycetines*, *Collybiopsis*, and *Paragymnopus* were revealed (Figures 1 and 2). Twenty-one sampled specimens formed eight new species and were clustered in a clade comprising the species of *Gymnopus* sect. *Levipedes* (Figure 2). At the same time, two sampled specimens—clustered with *Gymnopus erythropus* with strong support—were confirmed as new records from Jilin Province, China.

The phylogeny inferred from the dataset of ITS and nLSU region recovered *Gymnopus* s. str. as a monophyletic genus divided into four clades, sect. *Androcacei* clade, sect. *Levipedes* clade, sect. *Impudicae* clade, and sect. *Gymnopus* clade, formed a sister clade to *Rhodocollybia*, *Paragymnopus*, and *Lentinula* (Figure 1). The sect. *Levipedes* clade was mainly divided into three clades, the red stipe species formed an independent clade, and the *Gymnopus dryophilus* complex species formed an independent clade. These two clades mentioned above are near the species *Gymnopus alkalivirens* (Singer) Halling that turns green in KOH, representing *Gymnopus* sect. *Levipedes* subsect. *Alkalivirentes* Antonín and Noorde.

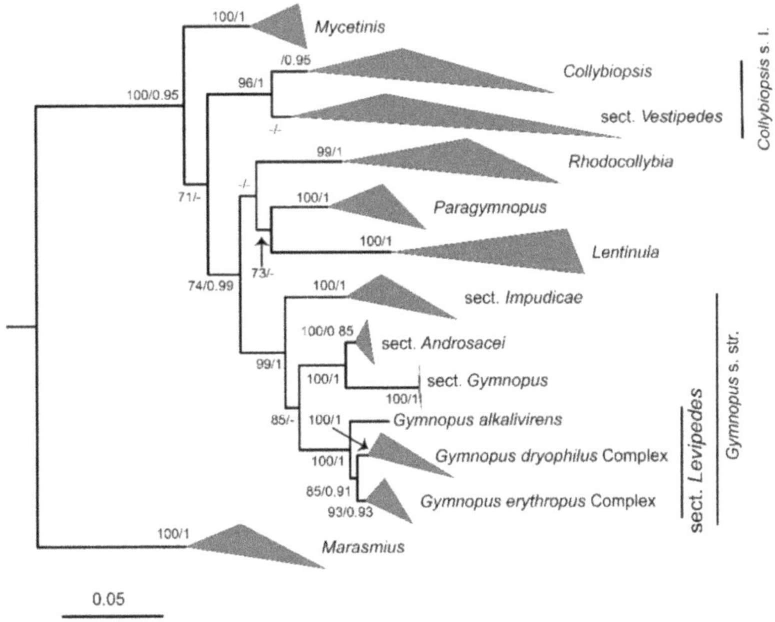

Figure 1. Bayesian 50% majority-rule consensus tree from the ITS and nLSU analyses. Support values at the nodes consist of BPPS $\geq$ 0.90 and BS $\geq$ 70; unsupported nodes under BPPS 0.5 are collapsed. The major clade is simplified, representing genus-level groups, as depicted in the figure. The outgroup consists of members of *Marasmius*.

3.2. Taxonomy

Gymnopus longisterigmaticus J.J. Hu, B. Zhang and Y. Li sp. nov.

Figures 3a and 4

MycoBank: MB 842333

Etymology: The epithet "*longisterigmaticus*" refers to the extremely long sterigmata of the basidia.

Diagnosis: This species is distinguished from closed species by pileus brown at the center, light brown to yellow towards the margin, margin light yellow to yellowish white, stipe reddish brown, covered with white to light reddish brown density hairs at base, basidia four-spored, sterigmata extremely long, pileipellis wider than *Gymnopus longus* and *Gymnopus macrosporus*, branched, pigment yellowish brown incrusting in pileipellis, and larger basidiospores.

Type: China. Jilin Province: Yanbian Korean Autonomous Prefecture, Antu County, Erdaobaihe Town, 42.39° N, 128.11° E, 4 September 2018, Jia-Jun Hu and Bo Zhang, HMJAU 60288, holotype (GenBank accession no.: ITS = OM030282, nLSU = OM033403).

Basidiomata small-to-medium-sized, scattered to gregarious. Pileus convex to applanate, 1.5–3.2 cm diameter, smooth, hygrophanus, brown at the center, light brown to yellow towards the margin, margin light yellow to yellowish white, entire. Context thin, fleshy, light reddish brown, odorless. Stipe center, cylindrical, 3.2–5.0 cm long and 0.2–0.3 cm wide, reddish brown, smooth, covered with white to light reddish brown density hairs at base, fistulose, fibrous. Lamellae subfree to adnate, white to light yellow, I = 1–3, L = 15–18, crowded.

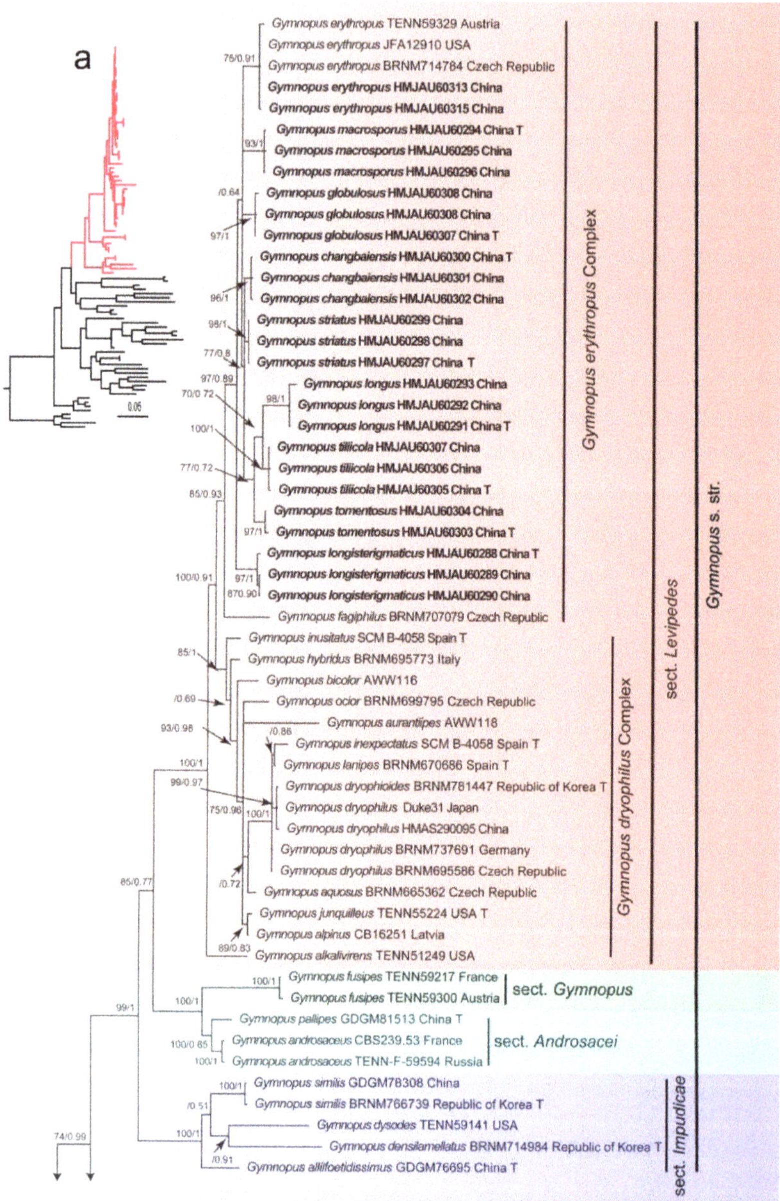

Figure 2. *Cont.*

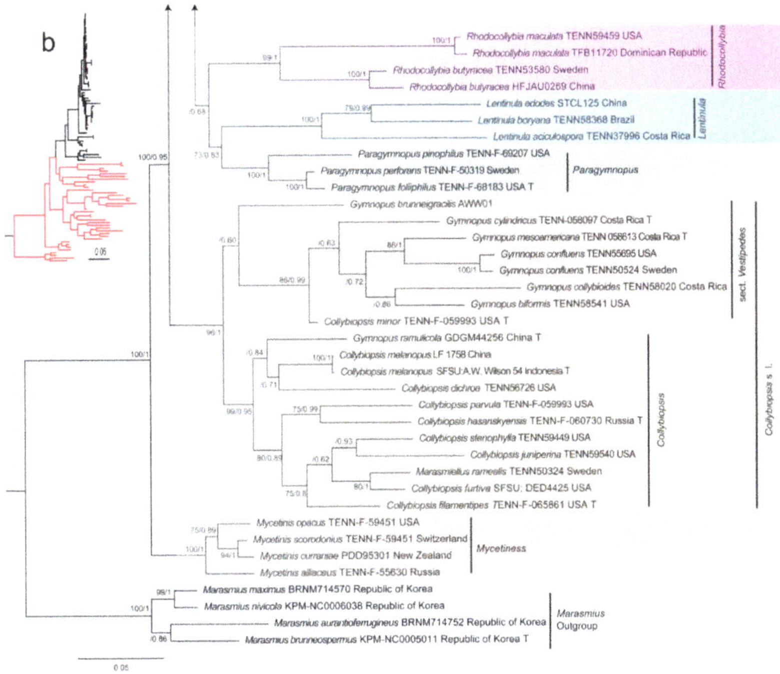

Figure 2. Maximum likelihood phylogenetic tree generated from the ITS and nLSU dataset. Bootstraps values (BS) ≥ 70% from ML analysis and Bayesian posterior probabilities (BPPS) ≥ 0.90 are shown on the branches. Newly sequenced collections are indicated in bold, and the type specimens are denoted by (T).

Basidiospores elliptic, (6.2) 6.7–9.0 × (3.0) 3.1–4.3 (5.0) μm, Q = (1.40) 1.67–2.25 (2.26), Qm = 1.93 ± 0.20, smooth, hyaline, inamyloid, thin-walled. Basidia clavate, (18) 19–27 (28) × (5) 6–10 μm, four-spored, thin-walled, smooth, hyaline; sterigmata extremely long, up to 40 μm. Cheilocystidia abundant, clavate, with obtuse on the top, or branched, (16) 18–27 × (4) 5–8 (9) μm, thin-walled, smooth, hyaline. Pileipellis a cutis, made up of irregularly branched hyphae, inflated, 10–27 (35) μm wide, hyaline to light yellow, smooth or pigment yellowish brown incrusting in pileipellis, thin-walled, clamps present.

Ecology: Grows on the deciduous layer or rotten branches in coniferous and broad-leaved mixed forest.

Distribution: China (Jilin Province)

Other specimen examined: China. Jilin Province: Yanbian Korean Autonomous Prefecture, Antu County, Erdaobaihe Town, 42.39° N, 128.11° E, 13 September 2019, Jia-Jun Hu and Bo Zhang, HMJAU 60289 (GenBank accession no.: ITS = OM030283, nLSU = OM033404); Yanbian Korean Autonomous Prefecture, Antu County, Erdaobaihe Town, 42.39° N, 128.11° E, 13 September 2019, Jia-Jun Hu and Bo Zhang, HMJAU 60290 (GenBank accession no.: ITS = OM030284, nLSU = OM033405).

Note: Morphologically, *Gymnopus longisterigmaticus* is similar to *Gymnopus erythropus* and *Gymnopus fagiphilus* with its reddish brown stipe. However, *Gymnopus longisterigmaticus* differs from *Gymnopus erythropus* with its light reddish brown density hairs on the stipe, extremely long sterigmata of basidiomata (up to 40 μm), different shape of cheilocystidia—cheilocystidia of *Gymnopus erythropus* is clavate to subclavate or somewhat flexuous, coralloid at apex sometimes [1], while clavate of *Gymnopus longisterigmaticus*, and quite larger basidiospores [(6.2) 6.7–9.0 × (3.0) 3.1–4.3 (5.0) μm].

Gymnopus longisterigmaticus and *Gymnopus fagiphilus* are both covered with hairs on the stipe, but the lamellae of *Gymnopus longisterigmaticus* is white to light yellow, while that of *Gymnopus fagiphilus* is pinkish brown to pinkish yellow; on the other hand, *Gymnopus longisterigmaticus* has extremely long sterigmata of the basidia and lack of chaulocystidia. Moreover, the different shape and size of cheilocystidia can differentiate *Gymnopus longisterigmaticus* from *Gymnopus fagiphilus*. The cheilocystidia of *Gymnopus fagiphilus* is usually irregularly clavate, often with lobed apex or with short-to-long rostrum, sometimes very slender and lageniform and quite larger [15–40 (60) × 4.0–8.0 (10) µm] [1], while cheilocystidia of *Gymnopus longisterigmaticus* is clavate, branched or obtuse.

Figure 3. Fresh basidiomata of *Gymnopus* species: (**a**) *Gymnopus longisterigmaticus* (Holotype, HMJAU 60288), (**b**) *Gymnopus longus* (Holotype, HMJAU 60291), (**c**) *Gymnopus macrosporus* (Holotype, HMJAU 60294), (**d**) *Gymnopus tiliicola* (Holotype, HMJAU 60304), (**e**) *Gymnopus globulosus* (Holotype, HMJAU 60308), (**f**) *Gymnopus changbaiensis* (HMJAU 60300), (**g**) *Gymnopus striatus* (Holotype, HMJAU 60297), (**h**) *Gymnopus erythropus* (HMJAU 60315), and (**i**) *Gymnopus tomentosus* (Holotype, HMJAU 60303). Scale bars = 1 cm.

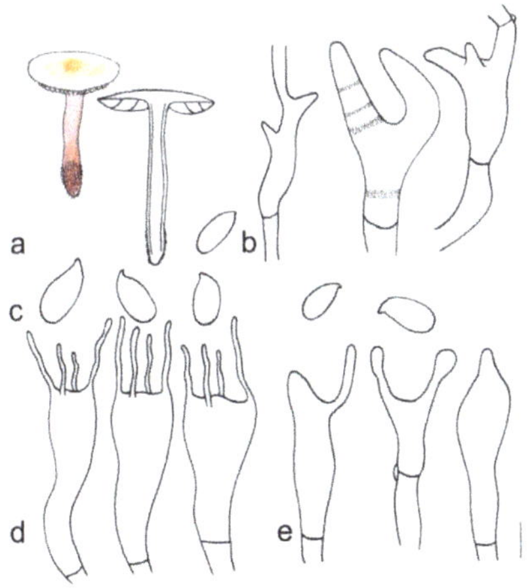

Figure 4. Morphological characteristics of *Gymnopus longisterigmaticus* (HMJAU 60288, holotype): (**a**) basidiomata, (**b**) pileipellis, (**c**) basidiospores, (**d**) basidia, and (**e**) cheilocystidia. Scale bars: 1 cm (**a**), 25 µm (**b**), and 5 µm (**c–e**).

Gymnopus longus J.J. Hu, B. Zhang and Y. Li sp. nov.

Figures 3b and 5

MycoBank: MB 842334

Etymology: The epithet "*longus*" refers to the extremely long sterigmata of the basidia.

Diagnosis: *Gymnopus longus* can be easily differentiated from closely-related species *Gymnopus fagiphilus* by its pileus reddish brown, stipe reddish brown, with brown farinose on the upper part and white to light reddish brown tomentose at the base, basidia 2- or 4-spored, sterigmata extremely long, and smaller basidiospores.

Type: China. Jilin Province: Yanbian Korean Autonomous Prefecture, Antu County, Erdaobaihe Town, 42.39° N, 128.11° E, 4 September 2018, Jia-Jun Hu, Bo Zhang, and Gui-Ping Zhao, HMJAU 60291, holotype (GenBank accession no.: ITS = OM030285, nLSU = OM033400).

Basidiomata small-to-medium-sized, scattered to gregarious. Pileus 1.7–3.7 cm in diameter, convex to applanate or revolute, smooth, hygrophanus, reddish brown at the center, towards margin light reddish brown to brown; margin white to light yellow or light brown, entire. Context thin, fleshy, light reddish brown, odorless. Stipe center, cylindrical to clavate, 3.7–4.3 cm long and 0.3–0.6 cm wide, reddish brown, with brown farinose on the upper part, and white to light reddish brown tomentose at the base, hollow, filiform. Lamellae adnate, white to light yellow, I = 5–7, L = 19–24, crowded. Spores print unknown.

Basidiospores (5.6) 6.0–8.0 × (3.0) 3.1–4.1 (4.9) µm, Q = (1.27) 1.47–2.19 (2.58), Qm = 1.8 ± 0.24, oblong, smooth, hyaline, thin-walled, inamyloid. Basidia (19) 20–28 (29) × 6–9 µm, two- or four-spored, hyaline, thin-walled, clavate; sterigmata extremely long, up to 33 µm long. Cheilocystidia (21) 22–29 (30) × 5–7 µm, mass, clavate, with obtuse on the top, hyaline, thin-walled, smooth. Pileipellis a translation between a cutis and a trichoderm, made up of irregularly interwoven, repent or ascending, inflated hyphae with inflated and irregularly branched terminal elements, hyaline to light brown, (6) 7–13 (15) µm wide, smooth or pigment yellowish brown incrusting in pileipellis.

Ecology: Grows on the deciduous layer or rotten branches in coniferous and broad-leaved mixed forest.

Other specimen examined: China. Jilin Province: Yanbian Korean Autonomous Prefecture, Antu County, Erdaobaihe Town, 42.39° N, 128.11° E, 4 September 2018, Jia-Jun Hu, Bo Zhang, and Gui-Ping Zhao, HMJAU 60292 (GenBank accession no.: ITS = OM030286, nLSU = OM033401); Yanbian Korean Autonomous Prefecture, Antu County, Erdaobaihe

Town, 42.39° N, 128.11° E, 31 August 2020, Jia-Jun Hu, Bo Zhang, and Gui-Ping Zhao, HMJAU 60293 (GenBank accession no.: ITS = OM030287, nLSU = OM033402).

Note: *Gymnopus longus* is closed to *Gymnopus erythropus*, *Gymnopus fagiphilus*, and *Gymnopus longisterigmaticus* in morphological, because of the red pileus and stipe. However, *Gymnopus longus* differs from *Gymnopus erythropus* by being covered with brown farinose on the upper part, white to light reddish brown tomentose at the base, slight thin basidiospores, smaller Qm [1], and extremely long sterigmata (up to 33 μm long).

A deeper color pileus, covered with brown farinose on the stipe, smaller basidiospores, clavate with obtuse cheilocystidia, and pileipellis a translation between a cutis and a trichoderm differs *Gymnopus longisterigmaticus* from *Gymnopus longus*. *Gymnopus longus* differs from *Gymnopus fagiphilus* by a farinose stipe, deep color pileus and stipe, white lamellae, smaller basidiospores, lack of caulocystidia, and uncoralloid pileipellis [1].

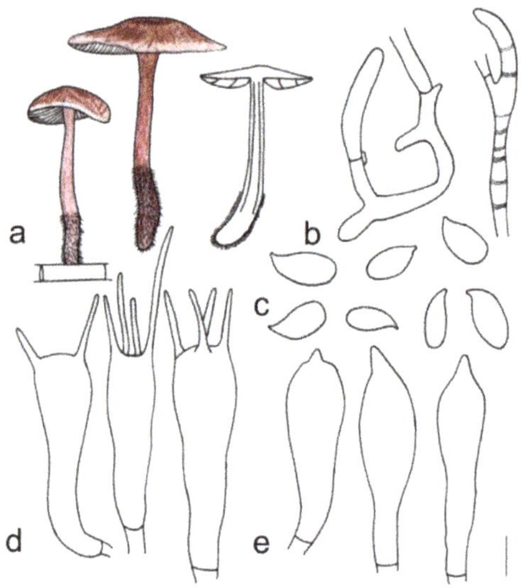

Figure 5. Morphological characteristics of *Gymnopus longus* (HMJAU 60291, holotype): (**a**) basidiomata, (**b**) pileipellis, (**c**) basidiospores, (**d**) basidia, and (**e**) cheilocystidia. Scale bars: 1 cm (**a**), 25 μm (**b**), and 5 μm (**c**–**e**).

Gymnopus macrosporus J.J. Hu, B. Zhang and Y. Li sp. nov.

Figures 3c and 6

MycoBank: MB 842335

Etymology: the epithet "*macrosporus*" refers to the big basidiospores of this species.

Diagnosis: This species is distinguished from closed species by a convex to applanate pileus that is deep reddish brown at the center and reddish brown to yellowish brown toward the margin, with the margin beige to light yellow, striped; a deep reddish brown to reddish brown stipe with smooth, light yellow to light reddish brown tomentose at the base, coralloid pileipellis, bigger basidiospores, and extremely long basidia sterigmata.

Type: China. Jilin Province: Yanbian Korean Autonomous Prefecture, Antu County, Erdaobaihe Town, 42.39° N, 128.11° E, 4 September 2018, Jia-Jun Hu and Bo Zhang, HMJAU 60294, holotype (GenBank accession no.: ITS = OM030266, nLSU = OM033397).

Basidiomata small-to-medium-sized, gregarious. Pileus convex to applanate, 1.2–4.6 cm diameter, smooth, hygrophanus, deep reddish brown at the center, reddish brown to yellowish brown towards margin; margin beige to light yellow, striped, entire, wavy. Context thin, fleshy, light reddish brown to light yellowish brown, odorless. Stipe center, cylindrical, 7.8–9.5 cm long and 0.2–0.5 cm wide, deep reddish brown to reddish brown, smooth, fistulose, fibrous, and light yellow to light reddish brown tomentose at the base. Lamellae adnexed to adnate or near free, light yellow, I = 1–3, L = 13–17, crowded.

Basidiospores elliptic, (6.0) 6.8–7.9 (8.3) × (3.0) 3.1–4.2 (4.3) μm, Q = (1.63) 1.67–2.32 (2.37), Qm = 1.88 ± 0.18, smooth, hyaline, inamyloid, thin-walled. Basidia clavate, 20–29 × 6–9 μm, two- or four-spored, thin-walled, smooth, hyaline; sterigmata extremely long, up to 32 μm. Cheilocystidia abundant, clavate, with obtuse on the top, 20–28 (30) × 5 (6)–9 μm, thin-walled, smooth, hyaline. Pileipellis a cutis, made up of irregular branched or weakly coralloid hyphae, inflated, 10–27 (35) μm wide, hyaline to light yellow, smooth, thin-walled, clamps present.

Ecology: Grows on the deciduous layer or rotten branches in coniferous and broad-leaved mixed forest.

Distribution: China (Jilin Province)

Other specimen examined: China. Jilin Province: Yanbian Korean Autonomous Prefecture, Antu County, Erdaobaihe Town, 42.39° N, 128.11° E, 4 September 2018, Jia-Jun Hu and Bo Zhang, HMJAU 60295 (GenBank accession no.: ITS = OM030267, nLSU = OM033398); Yanbian Korean Autonomous Prefecture, Antu County, Erdaobaihe Town, 42.39° N, 128.11° E, 13 September 2019, Jia-Jun Hu and Bo Zhang, HMJAU 60296 (GenBank accession no.: ITS = OM030268, nLSU = OM033399).

Note: *Gymnopus macrosporus* is morphologically similar to *Gymnopus longisterigmaticus* and *Gymnopus longus* because of its reddish brown, tomentose stipe, and long sterigmata of basidia. *Gymnopus macrosporus* differs from *Gymnopus longus* due to its pileus in a darker color—pileus deep reddish brown at center, reddish brown to yellowish brown towards margin; margin beige to light yellow, striped characteristics, and smooth texture on the upper part of the stipe, coralloid and without pigment incrusting pileipellis. These two *Gymnopus* species have a similar basidiospore size; however, the Qm of *Gymnopus macrosporus* is larger than *Gymnopus longus*. *Gymnopus longisterigmaticus* differs in smooth, pale color, and unstriped pileus; pileipellis a bit wider and pigment yellowish brown incrusting in pileipellis, and it has bigger basidiospores [(6.2) 6.7–9.0 × (3.0) 3.1–4.3 (5.0) μm].

Figure 6. Morphological characteristics of *Gymnopus macrosporus* (HMJAU 60294, holotype): (**a**) basidiomata, (**b**) pileipellis, (**c**) basidiospores, (**d**) basidia, and (**e**) cheilocystidia. Scale bars: 1 cm (**a**), 25 μm (**b**), and 5 μm (**c**–**e**).

Gymnopus striatus J.J Hu, B. Zhang and Y. Li sp. nov.

Figures 3g and 7

MycoBank: MB 842336

Etymology: the epithet "*striatus*" refers to the striped stipe of this species.

Diagnosis: This species is distinguished from closed species by a cinnamon pileus, with a lighter color toward the margin and a white to light yellow margin, striped; yellow to light brown lamellae, a deep reddish brown stipe, longitudinal striped stipe up to 1/3 covered with yellow to light brown hairs (from the base upward), short sterigmata of basidia, and smaller basidiospores.

Type: China. Jilin Province: Yanbian Korean Autonomous Prefecture, Antu County, Erdaobaihe Town, 42.39° N, 128.11° E, 9 September 2019, Jia-Jun Hu, Gui-Ping Zhao, and Bo Zhang, HMJAU 60297, holotype (GenBank accession no.: ITS = OM030263, nLSU = OM033384).

Basidiomata small-to-medium-sized, gregarious. Pileus convex to applanate, depressed when old, 2.3–4.1 cm diameter, smooth, hygrophanus, cinnamon at the center, brown to light brown towards margin; margin white to light yellow, striped, entire, wavy. Context thin, fleshy, light yellowish brown, odorless. Stipe center, cylindrical to clavate, 5.5–7.0 cm long and 0.3–0.8 cm wide, deep reddish brown to reddish brown, smooth in the upper part, longitudinal striped, covered with yellow to light brown hairs up to 1/3 (from the base upwards), fistulose, fibrous. Lamellae adnate, yellow to light brown, I = 3–9, L = 17–23, crowded.

Basidiospores elliptic, 6.0–8.0 (9.0) × 3.0–4.0 μm, Q = (1.50) 1.58–2.50 (2.60), Qm = 2.01 ± 0.25, smooth, hyaline, inamyloid, thin-walled. Basidia clavate, 20 (21)–34 (37) × 5–10 μm, two- or four-spored, thin-walled, smooth, hyaline. Cheilocystidia abundant, clavate, with obtuse on the top, (17) 20–30 × 4–8 (10) μm, thin-walled, smooth, hyaline. Pileipellis a cutis, made up of irregular branched or weakly coralloid hyphae, inflated, 10–30 (35) μm wide, hyaline to light yellow, smooth, thin-walled, clamps present.

Ecology: Grows on the deciduous layer or rotten branches in coniferous and broad-leaved mixed forest.

Distribution: China (Jilin Province)

Other specimen examined: China. Jilin Province: Yanbian Korean Autonomous Prefecture, Antu County, Erdaobaihe Town, 42.39° N, 128.11° E, 18 September 2020, Jia-Jun Hu, Gui-Ping Zhao, and Bo Zhang, HMJAU 60298 (GenBank accession no.: ITS = OM030264, nLSU = OM033385); Yanbian Korean Autonomous Prefecture, Antu County, Erdaobaihe Town, 42.39° N, 128.11° E, 18 September 2020, Jia-Jun Hu, Gui-Ping Zhao, and Bo Zhang, HMJAU 60299 (GenBank accession no.: ITS = OM030265, nLSU = OM033386).

Note: *Gymnopus striatus* is easily confused with *Gymnopus longisterigmaticus*, *Gymnopus longus*, and *Gymnopus macrosporus* due to their highly similar morphology. However, *Gymnopus striatus* differs from those three species by its deeper color lamellae, longitudinal stripes on the stipe and stripes on the margin of pileus, bigger Qm, and short basidia sterigmata. *Gymnopus striatus* can be easily differentiated from *Gymnopus fagiphilus* by its deeper color pileus, uniform colored and longitudinally striped stipe, lack of caulocystidia, uncoralloid cheilocystidia, without pigment incrusting in pileipellis, and smaller basidiospores.

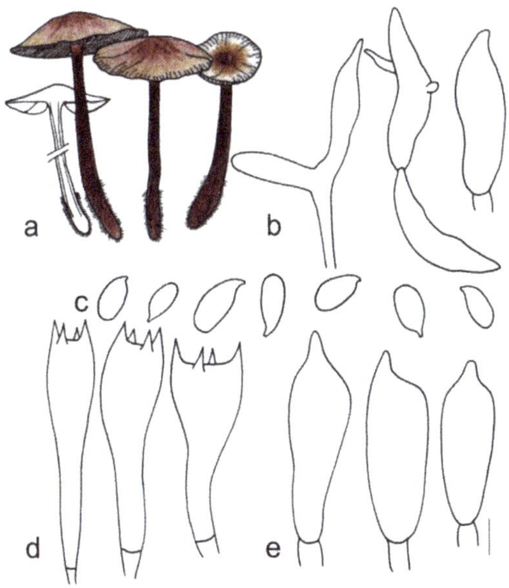

Figure 7. Morphological characteristics of *Gymnopus striatus* (HMJAU 60297, holotype): (**a**) basidiomata, (**b**) pileipellis, (**c**) basidiospores, (**d**) basidia, and (**e**) cheilocystidia. Scale bars: 1 cm (**a**), 25 µm (**b**), and 5 µm (**c**–**e**).

Gymnopus changbaiensis J.J. Hu, B. Zhang and Y. Li sp. nov.

Figures 3f and 8

MycoBank: MB 842337

Etymology: the epithet "*changbaiensis*" refers to Mt. Changbai, the location of the holotype.

Diagnosis: This species is distinguished from closed species by a reddish brown pileus and depressed when mature at the center, light pink to white outwards and margin, striped; fresh to pink lamellae, and a reddish brown stipe up to 1/3 covered with light yellow to brown hairs (from the base upwards), short sterigmata of basidia, lack of caulocystidia, uncoralloid cheilocystidia and smaller basidiospores.

Type: China. Jilin Province: Baishan City, Changbai Korean Autonomous County, Wangtian'e Scenic Spot, 41.56° N, 127.95° E, 17 September 2020, Jia-Jun Hu, Gui-ping Zhao, and Bo Zhang, HMJAU 60300, holotype (GenBank accession no.: ITS = OM030272, nLSU = OM033387).

Basidiomata small-to-medium-sized, gregarious. Pileus hemispherical, deep reddish brown when young, convex or slightly depressed sometimes when mature, 2.1–3.4 cm diameter, smooth, hygrophanus, reddish brown at the center, light pink towards margin; margin white to light pink, striped, entire. Context thin, fleshy, light yellowish brown, odorless. Stipe center, cylindrical, 4.2–5.3 cm long and 0.2–0.3 cm wide, deep reddish brown to reddish brown, smooth in the upper part, covered with light yellow to brown hairs up to 1/3 (from the base upwards), fistulose, fibrous. Lamellae adnate, fresh to pink, I = 1–5, L = 19–24, crowded.

Basidiospores elliptic, (5.8) 6.0–8.1 (9.0) × 3.0–4.1 (4.2) µm, Q = (1.41) 1.53–2.40 (2.50), Qm = 1.98 ± 0.24, smooth, hyaline, inamyloid, thin-walled. Basidia clavate, (19) 20–29 (32) × 5–8 µm, two- or four-spored, thin-walled, smooth, hyaline. Cheilocystidia abundant, clavate, with obtuse on the top, (23) 24–34 (39) × (5) 6–7 (9) µm, thin-walled, smooth, hyaline. Pileipellis a cutis, made up of irregular branched or weakly coralloid hyphae, inflated, 8–23 (25) µm wide, hyaline to light yellow, smooth, thin-walled, clamps present.

Ecology: Grows on the deciduous layer or rotten branches in coniferous and broad-leaved mixed forest.

Distribution: China (Jilin Province)

Other specimen examined: China. Jilin Province: Baishan City, Changbai Korean Autonomous County, Wangtian'e Scenic Spot, 41.56° N, 127.95° E, 9 September 2019, Jia-Jun Hu, Gui-ping Zhao, and Bo Zhang, HMJAU 60301 (GenBank accession no.: ITS = OM030273, nLSU = OM033388); Baishan City, Changbai Korean Autonomous County, Wangtian'e Scenic Spot, 41.56° N, 127.95° E, 9 September 2019, Jia-Jun Hu, Gui-ping Zhao, and Bo Zhang, HMJAU 60302 (GenBank accession no.: ITS = OM030274, nLSU = OM033389).

Note: *Gymnopus changbaiensis* is significantly related to *Gymnopus fagiphilus* and *Gymnopus striatus* based on its reddish brown and tomentose stipe, and short basidia sterigmata. *Gymnopus changbaiensis* can be distinguished from *Gymnopus fagiphilus* by its lighter and depressed pileus, denser and fresh to pink lamellae, and, in terms of microscopic characteristics, smaller basidiospores, uncoralloid cheilocystidia, and lack of caulocystidia. *Gymnopus changbaiensis* differs from *Gymnopus striatus* by its pale color, striped, and depressed pileus, fresh-to-pink lamellae, non-striped stipe, a bit longer cheilocystidia, coralloid pileipellis.

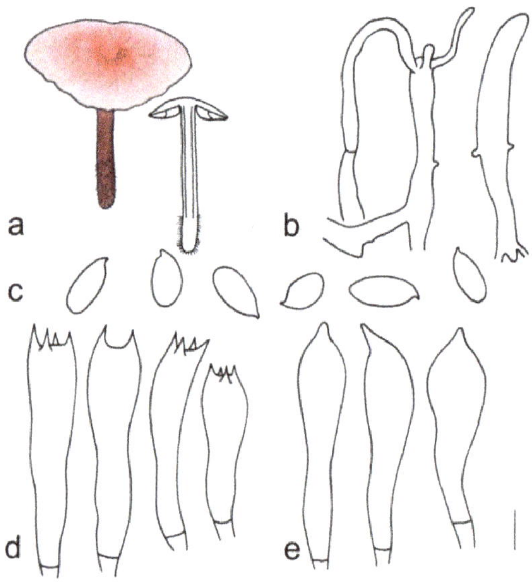

Figure 8. Morphological characteristics of *Gymnopus changbaiensis* (HMJAU 60300, holotype): (**a**) basidiomata, (**b**) pileipellis, (**c**) basidiospores, (**d**) basidia, and (**e**) cheilocystidia. Scale bars: 1 cm (**a**), 25 μm (**b**), and 5 μm (**c**–**e**).

Gymnopus tomentosus J.J. Hu, B. Zhang and Y. Li sp. nov.

Figures 3i and 9

MycoBank: MB 842338

Etymology: the epithet "*tomentosus*" refers to the tomentose margin of pileus.

Diagnosis: This species is distinguished from closed species by a near white pileus with a tomentose margin, yellowish green lamellae, and a reddish brown stipe up to 1/4 covered with reddish brown hairs (from the base upwards), smaller basidiospores, clavate cheilocystidia, and inflated to bulbous pileipellis.

Type: China: Jilin Province, Jiaohe City, Lafa Mountain National Forest Park Red Leaf Valley Scenic Spot, 43.71° N, 127.08° E, 7 September 2019, Jia-Jun Hu, Gui-ping Zhao, and Bo Zhang, HMJAU 60303, holotype (GenBank accession no.: ITS = OM030278, nLSU = OM033390).

Basidiomata small-to-medium-sized, scattered. Pileus convex, 1.6–3.0 cm diameter, smooth, tan at the center, light brown towards margin; margin white, tomentose, entire. Context thin, fleshy, white to light yellow, odorless. Stipe center, cylindrical, 3.3–4.3 cm long and 0.2–0.5 cm wide, blackish green at apex, reddish brown below, covered with reddish brown hairs up to 1/4 (from the base upwards), fistulose, fibrous. Lamellae adnexed, yellowish green, I = 3–7, L = 19–25, crowded.

Basidiospores elliptic, (6.0) 6.2–8.2 (9.0) × 3.0–4.1 (4.2) µm, Q = (1.50) 1.59–2.33 (2.40), Qm = 1.92 ± 0.23, smooth, hyaline, inamyloid, thin-walled. Basidia clavate, 20–30 (31) × 5–8 µm, two- or four-spored, thin-walled, smooth, hyaline. Cheilocystidia abundant, clavate, with obtuse on the top sometimes, (20) 22–30 (32) × 5–7 µm, thin-walled, smooth, hyaline. Pileipellis a cutis, made up of irregular branched to weakly coralloid or bulbous hyphae, inflated, 10–18 (21) µm wide, light brown, smooth, thin-walled, clamps present.

Ecology: Grows on the deciduous layer in broad-leaved forest.

Distribution: China (Jilin Province)

Note: The reddish brown and tomentose stipe makes *Gymnopus tomentosus* similar to *Gymnopus fagiphilus*, *Gymnopus longisterigmaticus*, *Gymnopus longus*, *Gymnopus macrosporus*, *Gymnopus striatus*, and *Gymnopus changbaiensis*. However, its white-to-pale-yellow pileus with a tomentose margin and inflated bulbous terminal hyphae of the pileipellis differentiates *Gymnopus tomentosus* from *Gymnopus longisterigmaticus*, *Gymnopus longus*, *Gymnopus macrosporus*, *Gymnopus striatus*, and *Gymnopus changbaiensis*. *Gymnopus tomentosus* can be distinguish from *Gymnopus fagiphilus* by its near-white pileus with a tomentose margin, coralloid-to-bulbous pileipellis, smaller basidiospores, and lack of caulocystidia [1].

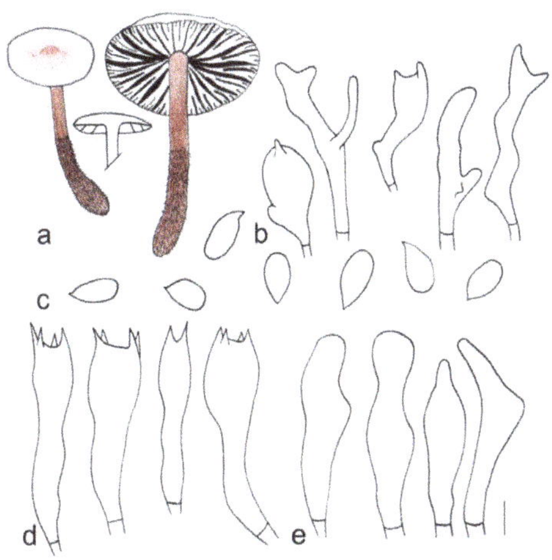

Figure 9. Morphological characteristics of *Gymnopus tomentosus* (HMJAU 60303, holotype): (**a**) basidiomata, (**b**) pileipellis, (**c**) basidiospores, (**d**) Basidia, and (**e**) cheilocystidia. Scale bars: 1 cm (**a**), 25 µm (**b**), and 5 µm (**c**–**e**).

Gymnopus tiliicola J.J. Hu, B. Zhang and Y. Li sp. nov.

Figures 3d and 10

MycoBank: MB 842339

Etymology: the epithet "*tiliicola*" refers to this species grows at the base of *Tilia* sp.

Diagnosis: This species is distinguished from closed species by a deep rose-red pileus that is pale color outward, light pink to fresh lamellae, with a deep reddish brown and smooth stipe that is longitudinally striped, grows at the base of *Tilia* sp., uncoralloid cheilocystidia, two- or four-spored basidia, and a bit bigger basidiospores.

Type: China. Jilin Province: Yanbian Korean Autonomous Prefecture, Antu County, Erdaobaihe Town, 42.39° N, 128.11° E, 13 September 2019, Jia-Jun Hu, Gui-Ping Zhao, and Bo Zhang, HMJAU 60304, holotype (GenBank accession no.: ITS = OM030275, nLSU = OM033392).

Basidiomata medium-to-large-sized, gregarious. Pileus convex, 3.0–6.7 cm diameter, smooth, deep rose-red at the center, yellowish pink towards margin; margin white to light yellow, striped, entire, wavy. Context thin, fleshy, white to pink, odorless. Stipe center, cylindrical, 2.2–4.5 cm long and 0.3–0.7 cm wide, deep reddish brown, smooth, fistulose, fibrous. Lamellae adnexed to adnate, light pink to fresh, I = 1–3, L = 19–24, crowded.

Basidiospores elliptic, (6.0) 6.9–8.0 (8.2) × (3.0) 3.1–4.0 (4.2) μm, Q = (1.70) 1.75–2.26 (2.33), Qm = 1.93 ± 0.17, smooth, hyaline, inamyloid, thin-walled. Basidia clavate, 20–30 × 6–8 μm, two- or four-spored, thin-walled, smooth, hyaline. Cheilocystidia abundant, clavate, with obtuse on the top, (20) 21–27 (28) × 5–7 μm, thin-walled, smooth, hyaline. Pileipellis a cutis, made up of irregular branched to weakly coralloid hyphae, inflated, (5) 6–15 (17) μm wide, light brown, smooth, thin-walled, clamps present.

Ecology: Grows at the base of *Tilia* sp.

Distribution: China (Jilin Province)

Other specimen examined: China. Jilin Province: Yanbian Korean Autonomous Prefecture, Antu County, Erdaobaihe Town, 42.39° N, 128.11° E, 31 August 2020, Jia-Jun Hu, Gui-Ping Zhao, and Bo Zhang, HMJAU 60305 (GenBank accession no.: ITS = OM030277, nLSU = OM033393); Yanbian Korean Autonomous Prefecture, Antu County, Erdaobaihe Town, 42.39° N, 128.11° E, 27 August 2021, Jia-Jun Hu, Gui-Ping Zhao, and Bo Zhang, HMJAU 60304 (GenBank accession no.: ITS = OM030276, nLSU = OM033394).

Note: Morphologically, the rose-red to dark red pileus and stipe make *Gymnopus tiliicola* closed to *Gymnopus erythropus*. *Gymnopus tiliicola* differs from *Gymnopus erythropus* in a lighter color and striped pileus, light pink to fresh and denser lamellae. Besides, *Gymnopus tiliicola* grows at the base of *Tilia* sp., while *Gymnopus erythropus* grows on the deciduous layer or rotten branches. In regard to microfeatures, *Gymnopus tiliicola* differs from *Gymnopus erythropus* by a weakly coralloid pileipellis, uncoralloid cheilocystidia, bigger basidiospores, and two- or four-spored basidia.

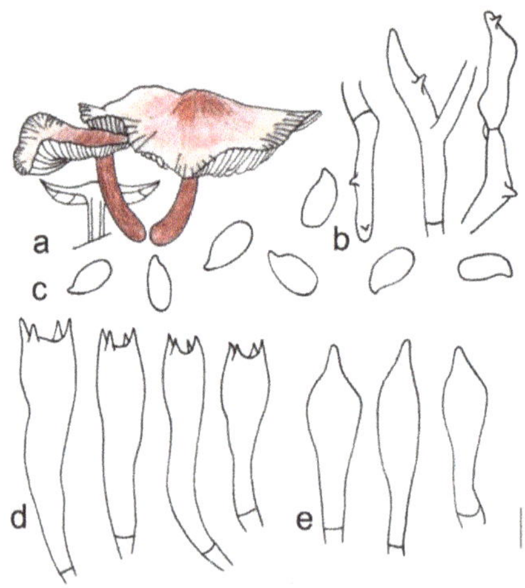

Figure 10. Morphological characteristics of *Gymnopus tiliicola* (HMJAU 60304, holotype): (**a**) basidiomata, (**b**) pileipellis, (**c**) basidiospores, (**d**) basidia, and (**e**) cheilocystidia. Scale bars: 1 cm (**a**), 25 μm (**b**), and 5 μm (**c**–**e**).

Gymnopus globulosus J.J. Hu, Y.L. Tuo, B. Zhang and Y. Li sp. nov.

Figures 3e and 11

MycoBank: MB 842340

Etymology: the epithet "*globulosus*" refers to pileipellis terminal hyphae inflated to spherical to prolate.

Diagnosis: This species is distinguished from closed species by a convex to applanate pileus that is deep reddish brown at the center, lighter-colored outwards, and light yellow margin, striped; white to light yellowish green lamellae, with a reddish brown and smooth stipe, pileipellis two layers and the terminal hyphae inflated to spherical to prolate, and a bit bigger basidiospores.

Type: China. Jilin Province: Tonghua City, Ji'an County, Wunvfeng National Forest Park, 41.28° N, 126.14° E, 28 August 2019, Yong-Lan Tuo and Jia-Jun Hu, HMJAU 60307, holotype (GenBank accession no.: ITS = OM030269, nLSU = OM033406).

Basidiomata medium-sized, gregarious. Pileus convex to applanate, 4.5–5.5 cm diameter, smooth, deep reddish brown at the center, yellowish brown towards margin; margin white to light yellow, striped, entire, wavy. Context thin, fleshy, brown, odorless. Stipe center, clavate, 4.8–6.0 cm long and 0.6–0.8 cm wide, deep reddish brown, paler at apex, smooth, fistulose, fibrous. Lamellae adnexed to adnate, white to light yellowish green, I = 1–3, L = 9–15, crowded.

Basidiospores elliptic, (6.8) 7.0–8.8 (9.0) × (3.1) 3.3–4.2 (4.8) μm, Q = (1.63) 1.75–2.20 (2.26), Qm = 1.93 ± 0.16, smooth, hyaline, inamyloid, thin-walled. Basidia clavate, (23) 25–32 (33) × 6–9 (11) μm, two- or four-spored, thin-walled, smooth, hyaline. Cheilocystidia abundant, clavate, with obtuse on the top, (22) 24–38(39) × 5–9 (10) μm, thin-walled, smooth, hyaline. Pileipellis layered, the upper layer inflated to spherical to prolate hyphae, 15–33 (47) μm wide, brown, smooth, thin-walled; down layer made up of branched and inflated hyphae, pigment light brown to brown incrusting in pileipellis, thin-to-thick-walled.

Ecology: Grows on rotten wood.

Distribution: China (Jilin Province)

Other specimen examined: China. Jilin Province: Tonghua City, Ji'an County, Wunvfeng National Forest Park, 41.28° N, 126.14° E, 3 September 2021, Yong-Lan Tuo and Jia-Jun Hu, HMJAU 60308 (GenBank Accession no.: ITS = OM030270, nLSU = OM033407).

Note: In terms of morphology, *Gymnopus globulosus* resembles *Gymnopus erythropus* and *Gymnopus tiliicola* in its red to dark red pileus and stipe. However, *Gymnopus globulosus* is distinguishable from *Gymnopus erythropus* due to its deeper-colored pileus, light yellowish green lamellae, which is light yellow of *Gymnopus erythropus*. In terms of microfeature, the pileipellis of *Gymnopus erythropus* is between a cutis and a trichoderm, while the pileipellis of *Gymnopus globulosus* is layered, with the upper layer inflated to spherical to prolate hyphae and the down layer made up of branched and inflated hyphae, and bigger basidiospores. *Gymnopus globulosus* differs from *Gymnopus tiliicola* with its deeper-colored pileus, light yellowish green lamellae, grows on rotten wood, pileipellis two layers and the terminal hyphae inflated to spherical to prolate, and bigger basidiospores.

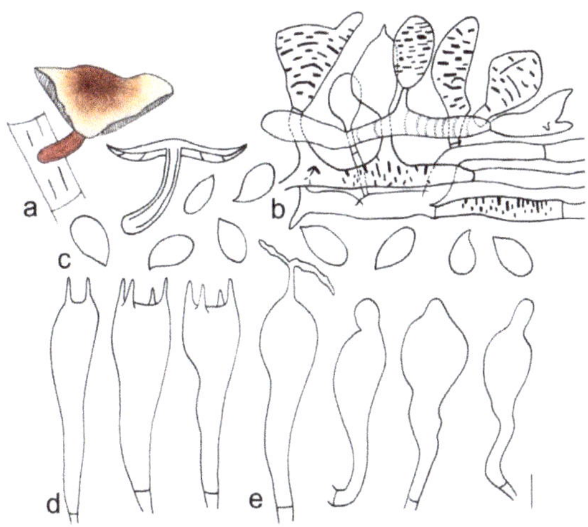

Figure 11. Morphological characteristics of *Gymnopus globulosus* (HMJAU 60307, holotype): (**a**) basidiomata, (**b**) pileipellis, (**c**) basidiospores, (**d**) basidia, and (**e**) cheilocystidia. Scale bars: 1 cm (**a**), 25 μm (**b**), and 5 μm (**c**–**e**).

New record from Jilin Province, China

Gymnopus erythropus (Pers.) Antonín, Halling and Noordel.

Figures 3h and 12

Basidiomata small-to-medium-sized, scattered to gregarious. Pileus convex to applanate, 1.1–3.2 cm diameter, smooth, hygrophanus, reddish brown to brown at the center, light reddish brown to yellowish brown towards margin; margin beige to light yellow, entire, wavy sometimes. Context thin, fleshy, light brown, odorless. Stipe center, cylindrical, 4.1–10.0 cm long and 0.2–0.5 cm wide, deep reddish brown to light reddish brown, paler at apex, smooth, covered with scattered light yellow to brown hairs hairy at base, fistulose, fibrous. Lamellae adnate, fresh-pink, I = 3–5, L = 14–27, crowded.

Basidiospores elliptic, (5.0) 6.0–8.2 (10.0) × (2.1) 3.0–5.0 (6.0) μm, Q = (1.20) 1.48–2.33 (3.00), Qm = 1.87 ± 0.27, smooth, hyaline, inamyloid, thin-walled. Basidia clavate, (17) 21–33 (38) × (4) 5–9 (10) μm, two- or four-spored, thin-walled, smooth, hyaline. Cheilocystidia abundant, clavate, with obtuse on the top, (15)21–33 (39) × (3) 4–8 (9) μm, thin-walled, smooth, hyaline. Pileipellis a cutis, made up of irregular branched or weakly coralloid hyphae, inflated, (6) 8–20 (20) μm wide, hyaline to light yellow, smooth, thin-walled, clamps present.

Ecology: Grows on the deciduous layer or rotten branches in coniferous and broad-leaved mixed forest.

Distribution: China (Jilin Province)

Specimen examined: China. Jilin Province: Baishan City, Changbai Korean Autonomous County, Wangtian'e Scenic Spot, 41.56° N, 127.95° E, 8 September 2019, Jia-Jun Hu, Gui-ping Zhao, and Bo Zhang, HMJAU 60309; Baishan City, Changbai Korean Autonomous County, Wangtian'e Scenic Spot, 41.56° N, 127.95° E, 8 September 2019, Jia-Jun Hu, Gui-ping Zhao, and Bo Zhang, HMJAU 60315 (GenBank Acc. no.: ITS = OM030280, nLSU = OM033395); Baishan City, Fusong County, Lushuihe Town, 42.53° N, 127.80° E, 8 September 2019, Jia-Jun Hu, Gui-ping Zhao, and Bo Zhang, HMJAU 60313 (GenBank Acc. no.: ITS = OM030281, nLSU = OM033396); Yanbian Korean Autonomous Prefecture, Antu County, Edaobaihe Town, 42.39° N, 128.11° E, 4 September 2018, Jia-Jun Hu and Bo Zhang, HMJAU 60310; HMJAU 60311; HMJAU 60312; Liaoning Province: Jinzhou City, Yi County, Mt. Yiwulv, 24 September 2013, Di Wang, HMJAU 28892; Jinzhou City, Yi County, Mt. Yiwulv, 25 September 2013, Di Wang, HMJAU 28839.

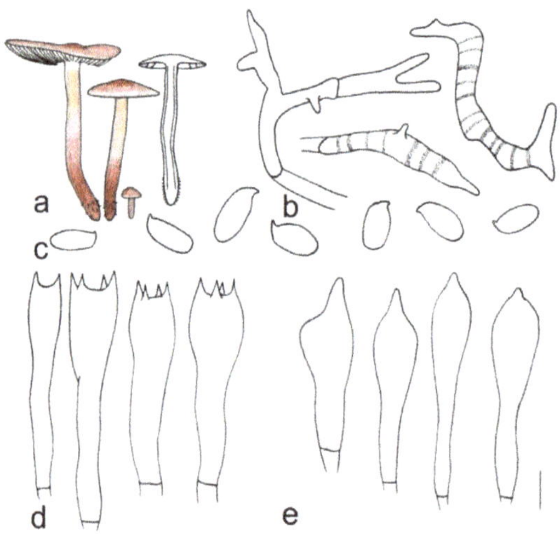

Figure 12. Morphological characteristics of *Gymnopus erythropus* (HMJAU 60315). (**a**) Basidomata, (**b**) pileipellis, (**c**) basidiospores, (**d**) basidia, and (**e**) cheilocystidia. Scale bars: 1 cm (**a**), 25 μm (**b**), and 5 μm (**c**–**e**).

Table 2. Macrocharacteristics comparison between our new species, *Gymnopus erythropus* and *Gymnopus fagiphilus*.

Scientific Name	Pileus	Lamellae	Stipe
G. longisterigmaticus	1.5–3.2 cm diameter, smooth, brown at center, light brown to yellow towards the margin, margin light yellow to yellowish white, entire	Lamellae subfree to adnate, white to light yellow, I = 1–3, L = 15–18, crowded	3.2–5.0 × 0.2–0.3 cm, reddish brown, smooth, covered with white to light reddish brown density hairs at base
G. longus	1.7–3.7 cm diameter, smooth, reddish brown at center, towards margin light reddish brown to brown; margin white to light yellow or light brown	Adnate, white to light yellow, I = 5–7, L = 19–24, crowded	3.7–4.3 cm × 0.3–0.6 cm, reddish brown, with brown farinose on the upper part, and white to light reddish brown tomentose at the base
G. macrosporus	1.2–4.6 cm diameter, smooth, deep reddish brown at center, reddish brown to yellowish brown towards margin; margin beige to light yellow, striped, wavy	Adnexed to adnate or near free, light yellow, I = 1–3, L = 13–17, crowded	7.8–9.5 × 0.2–0.5 cm, deep reddish brown to reddish brown, smooth, and light yellow to light reddish brown tomentose at the base
G. striatus	Depressed when old, 2.3–4.1 cm diameter, smooth, cinnamon at center, brown to light brown towards margin; margin white to light yellow, striped, wavy	Adnate, yellow to light brown, I = 3–9, L = 17–23, crowded	5.5–7.0 cm × 0.3–0.8 cm, deep reddish brown to reddish brown, longitudinal striped, covered with yellow to light brown hairs up to 1/3 (from the base upwards), fistulose, fibrous
G. changbaiensis	2.1–3.4 cm diameter, smooth, hygrophanus, reddish brown at the center, light pink towards margin; margin white to light pink, striped	Adnate, fresh to pink, I = 1–5, L = 19–24, crowded	4.2–5.3 cm × 0.2–0.3 cm, deep reddish brown to reddish brown, covered with light yellow to brown hairs up to 1/3 (from the base upwards)
G. tomentosus	1.6–3.0 cm diameter, smooth, tan at the center, light brown towards margin; margin white, tomentose	Adnexed, yellowish green, I = 3–7, L = 19–25, crowded	3.3–4.3 cm × 0.2–0.5 cm, blackish green at apex, reddish brown below, covered with reddish brown hairs up to 1/4 (from the base upwards)
G. tiliicola	3.0–6.7 cm diameter, smooth, deep rose-red at the center, yellowish pink towards margin; margin white to light yellow, striped	Adnexed to adnate, light pink to fresh, I = 1–3, L = 19–24, crowded	2.2–4.5 cm × 0.3–0.7 cm, deep reddish brown, smooth
G. globulosus	4.5–5.5 cm diameter, smooth, deep reddish brown at the center, yellowish brown towards margin; margin white to light yellow, striped	Adnexed to adnate, white to light yellowish green, I = 1–3, L = 9–15, crowded	4.8–6.0 cm × 0.6–0.8 cm, deep reddish brown, paler at apex, smooth
G. erythropus	1.1–3.2 cm diameter, smooth, hygrophanus, reddish brown to brown at the center, light reddish brown to yellowish brown towards margin; margin beige to light yellow	Adnate, fresh-pink, I = 3–5, L = 14–27, crowded	4.1–10.0 cm × 0.2–0.5 cm, deep reddish brown to light reddish brown, paler at apex, smooth, covered with scattered light yellow to brown hairs hairy at base
G. fagiphilus	(7) 15–25 (35) mm broad, when moist slightly translucently striate at margin, yellow-brown or reddish brown, paler at margin	Moderately distant, L = 18–21, I = 3–7 (15), free or narrowly adnate, rarely adnate, pinkish brown or pinkish cream, darker with age or tinged gray	20–40 (70) × 1–3 mm, orange-brown to red-brown, sometimes paler at apex, dark red-brown towards base, covered with fine, white or yellow hairs up to 2/3 of length (from base upwards)

Note: The description of *Gymnopus fagiphilus* is based on Antonín and Noordeloos [1].

Table 3. Microcharacteristics comparison between our new species, *Gymnopus erythropus* and *Gymnopus fagiphilus*.

Scientific Name	Pileipellis	Q	Qm	Basidiospores	Basidia	Cheilocystidia	Caulocystidia
G. longisterigmaticus	Cutis, irregularly branched hyphae, inflated, 10–27 (35) µm wide, hyaline to light yellow, smooth or pigment yellowish brown incrusting in pileipellis	(1.40) 1.67–2.25 (2.26)	1.93 ± 0.20	(6.2) 6.7–9.0 × (3.0) 3.1–4.3 (5.0) µm	Clavate, (18) 19–27 (28) × (5) 6–10 µm, four-spored; sterigmata extremely long, up to 40 µm	Clavate, with obtuse on the top, or branched, (16) 18–27 × (4) 5–8 (9) µm	None
G. longus	A translation between a cutis and a trichoderm, made up of irregularly interwoven, repent or ascending inflated hyphae with inflated and irregularly branched terminal elements, hyaline to light brown, (6) 7–13 (15) µm wide, smooth or pigment yellowish brown incrusting in pileipellis	(1.27) 1.47–2.19 (2.58)	1.80 ± 0.24	(5.6) 6.0–8.0 × (3.0)3.1–4.1 (4.9) µm	(19) 20–28 (29) × 6–9 µm, two- or four-spored, clavate; sterigmata extremely long, up to 33 µm long	Clavate, (21) 22–29 (30) × 5–7 µm, with obtuse on the top	None
G. macrosporus	Cutis, made up of irregular branched or weakly coralloid hyphae, inflated, 10–27 (35) µm wide, hyaline to light yellow, smooth	(1.63) 1.67–2.32 (2.37)	1.88 ± 0.18	(6.0) 6.8–7.9 (8.3) × (3.0) 3.1–4.2 (4.3) µm	Clavate, 20–29 × 6–9 µm, two- or four-spored, thin-walled; sterigmata extremely long, up to 32 µm	Clavate, with obtuse on the top, 20–28 (30) × 5 (6)–9 µm	None
G. striatus	Cutis, made up of irregular branched or weakly coralloid hyphae, inflated, 10–30 (35) µm wide, hyaline to light yellow, smooth	(1.50) 1.58–2.50 (2.60)	2.01 ± 0.25	6.0–8.0 (9.0) × 3.0–4.0 µm	Clavate, 20 (21)–34 (37) × 5–10 µm, two- or four-spored	Clavate, with obtuse on the top, (17) 20–30 × 4–8 (10) µm	None
G. changbaiensis	Cutis, made up of irregular branched or weakly coralloid hyphae, inflated, 8–23 (25) µm wide, hyaline to light yellow	(1.41) 1.53–2.40 (2.50)	1.98 ± 0.24	(5.8) 6.0–8.1 (9.0) × 3.0–4.1(4.2) µm	Clavate, (19) 20–29 (32) × 5–8 µm, two- or four-spored	Clavate, with obtuse on the top, (23) 24–34 (39) × (5) 6–7 (9) µm	None
G. tomentosus	Cutis, made up of irregular branched to weakly coralloid or bulbous hyphae, inflated, 10–18 (21) µm wide, light brown	(1.50) 1.59–2.33 (2.40)	1.92 ± 0.23	(6.0) 6.2–8.2 (9.0) × 3.0–4.1 (4.2) µm	Clavate, 20–30 (31) × 5–8 µm, two- or four-spored	Clavate, with obtuse on the top sometimes, (20) 22–30 (32) × 5–7 µm	None
G. tiliicola	Cutis, made up of irregular branched to weakly coralloid hyphae, inflated, (5)6–15(17) µm wide, light brown, smooth	(1.70) 1.75–2.26 (2.33)	1.93 ± 0.17	(6.0) 6.9–8.0 (8.2) × (3.0) 3.1–4.0 (4.2) µm	Clavate, 20–30 × 6–8 µm, two- or four-spored	Clavate, with obtuse on the top, (20) 21–27 (28) × 5–7 µm	None
G. globulosus	Layered, the upper layer inflated to spherical to prolate hyphae, 15–33 (47) µm wide, brown, smooth, thin-walled; down layer made up of branched and inflated hyphae, pigment light brown to brown incrusting in pileipellis, thin-to-thick-walled	(1.63) 1.75–2.20 (2.26)	1.93 ± 0.16	(6.8) 7.0–8.8(9.0) × (3.1) 3.3–4.2 (4.8) µm	Clavate, (23) 25–32 (33) × 6–9 (11) µm, two- or four-spored	Clavate, with obtuse on the top, (22) 24–38 (39) × 5–9 (10) µm	None
G. erythropus	Cutis, made up of irregular branched or weakly coralloid hyphae, inflated, (6) 8–20 (20) µm wide, hyaline to light yellow, smooth	(1.20) 1.48–2.33 (3.00)	1.87 ± 0.27	(5.0) 6.0–8.2 (10.0) × (2.1) 3.0–5.0 (6.0) µm	Clavate, (17) 21–33 (38) × (4) 5–9 (10) µm, two- or four-spored	Clavate, with obtuse on the top, (15) 21–33 (39) × (3) 4–8 (9) µm, thin-walled, smooth, hyaline	None
G. fagiphilus	Cutis with transitions to a trichoderm, made up of irregularly shaped, 4.0–15 (25) µm–wide coralloid elements ("Dryophila-structure"); pigment brown-yellow, incrusting in pileipellis	1.7–2.3	2.1	(6.0) 7.0–9.0 × (3.0) 3.5–4.5 µm	21–31 × 6.0–8.5 µm, 4-spored	15–40 (60) × 4.0–8.0 (10) µm, irregularly clavate, often with lobed apex or with short to long, up to 10 µm long rostrum, sometimes very slender lageniform	20–80 (120) × 4.0–12 µm, subcylindrical or sublageniform, numerous

Note: The description of *Gymnopus fagiphilus* is based on Antonín and Noordeloos [1].

Key to the species reported in this study
1 Stipe covered with dense hairs at the base 2
1 Stipe smooth, or covered with sparse hairs at the base .... *Gymnopus erythropus*
2 Basidia sterigmata extremely long 3
2 Basidia sterigmata short 5
3 Stipe smooth in upper part 4
3 Stipe covered with brown farinose on the upper part .... *Gymnopus longus*
4 Pileus pale color, stipe color uneven .... *Gymnopus longisterigmaticus*
4 Pileus dark color, stipe color uniform .... *Gymnopus macrosporus*
5 Growing on the deciduous layer or rotten branches 6
5 Grows at the base of *Tilia* sp. .... *Gymnopus tiliicola*
6 Pileus pale color, near white .... emphGymnopus tomentosus
6 Pileus deep color 7
7 Stipe covered with longitudinally stripes .... *Gymnopus striatus*
7 Stipe without longitudinally stripes 8
8 Pileipellis a cuits, typically "*Dryophila* type" .... *Gymnopus changbaiensis*
8 Pileipellis layered, hyphae inflated to spherical to prolate .... *Gymnopus globulosus*

4. Discussion

4.1. New Sights on Morphological Characteristics

The genus *Gymnopus* is geographically widely distributed; however, in China, its species diversity is poorly known. Moreover, in China, only three species were originally described with molecular evidence. One of these is *Gymnopus ramulicola* T.H. Li and S.F. Deng [27] from Hainan Province, China; the second one is *Gymnopus alliifoetidissimus* T.H. Li and J.P. Li [25] from Guangdong Province, China; and the third is *Gymnopus pallipes* J.P. Li and Chun Y. Deng [25] from Guangdong and Guizhou Province, China. In our study, eight new species of *Gymnopus* from China are described as new species. They are well-supported by molecular phylogenetic and morphological evidence. Our newly recognized and delimited species are distributed in the broad-leaved and mixed forests, and occur in early autumn in Northeast China. The species we described here are hardly seen in the wild mushroom market; thus, their edibility is not yet known.

The description of these new species also broadens the morphological characterization of the genus *Gymnopus*. In the previous study, the pileipellis of the species in this genus was a cuits to trichoderm. Moreover, the pileipellis in the species of sect. *Levipedes* was an entangled, not radially oriented trichoderm of inflated, often lobed or coralloid elements of the "*Dryophila* type" [1,17]. In this study, the pileipellis of *Gymnopus globulosus* was divided into two layers, with the upper layer comprising hyphae inflated to spherical to prolate, differing from that of all known species in the genus, while the second layer was typical of the "*Dryophila* type". To our knowledge, the sterigmata of the basidia are usually not too long; however, the species *Gymnopus longistrigmaticus*, *Gymnopus longus*, and *Gymnopus macrosporus* had extremely long sterigmata, up to 40 μm. Thus, the structure of extremely long basidia sterigmata is traceable in our species. In addition, all the species described from this study are detailed compared in macro- and micro-features (Tables 2 and 3).

4.2. Phylogenetic Relationships of Gymnopus s.l. with Related Genera

Phylogenetic analyses of the species of *Gymnopus* s.l. and the related genera presented in this study confirmed that the genus *Gymnopus* defined by Antonín and Noordeloos, as well as Halling, is not monophyletic in a strongly supported clade. Similar results were observed with our phylogenetic analysis. Our results, thus, support the finding of Oliveira et al., promoting sect. *Perforanita* to the genus level, *Paragymnopus*, and share a close affinity with *Lentinula*. Moreover, sect. *Vestipedes* was clearly separated from *Gymnopus* s. str [12,15], and were closed to *Marasmiellus*, *Collybiopsis*, and *Rhodocollybia*. However, in their study, the species of *Gymnopus* sect. *Vestipedes* was involved with

Marasmiellus; therefore, Oliveira et al. [15] proposed to transfer *Gymnopus* sect. *Vestipedes* to *Marasmiellus* and redefined the genus *Gymnopus* more strictly.

However, in our phylogenetic analyses, a different result was obtained. In our results, sect. *Vestipedes* did not group into one clade with *Marasmiellus* to form an independent clade, forming a sister clade to genus *Collybiopsis*. Furthermore, the taxonomic status of *Collybiopsis minor* R.H. Petersen still needs to be clarified; in our study, *C. minor* was separated far away from *Collybiopsis*, while being clustered with sect. *Vestipedes* within a single clade.

Some species of sect. *Vestipedes* and genus *Marasmiellus* have been proposed for transfer to other genera in recent years. *Gymnopus cylindricus* J.L. Mata and *Gymnopus brunneigracilis* (Corner) A.W. Wilson, Desjardin and E. Horak were suggested to be switched into *Marasiellus*. The type species of *Marasmiellus*, *Marasmiellus juniperinus*, and some other species within the genus, were advised to be relocated to *Collybiopsis* [47]. Thus, the boundaries between *Gymnopus*, *Marasmiellus*, and *Collybiopsis* would be more blurred, especially between *Marasmiellus* and *Collybiopsis*, as well as if these species were transferred to *Collybiopsis*; then it would be multiphyletic, with *Rhodocollybia*, *Paragymnopus*, and *Lentinula* would becoming synonyms of *Gymnopus*.

4.3. Nova Suggestions of Phylogenetic Relationships within Gymnopus s. str.

In our phylogenetic results, the genus *Gymnopus*, which was defined by Oliveira et al. [15], was mainly divided into four clades. Sect. *Levipedes*, sect. *Gymnopus*, and sect. *Androsacei* are somewhat more closely related, whereas they are distant from the sect. *Impudicae*. Before 2010, both sect. *Impudicae* and sect. *Levipedes* were subsections below the same section. However, from our results, sect. *Impudicae* and sect. *Levipedes* are more distantly related, probably due to similar environments, causing a similar appearance. In addition, the genus *Mycetines* and sect. *Impudicae*, with a strong odor, are not closely related to each other, and this is consistent with the result that they have a different pileipellis structure of pileus.

Thus, sect. *Levipedes* being split into two sections was supported by the phylogenetic analysis. Sect. *Levipedes* subsect. *Levipedes* was also divided into two subclades: one is *Gymnopus dryophilus* complex, a subclade (defined here as/*dryophila*) that includes all the *Gymnopus dryophilus* complex species reported around the world (characterized by a *Gymnopus dryophilus*–like appearance and arises in early spring or later in the autumn). From the result, the East Asia sequences of *Gymnopus dryophilus* were not clustered with the European sequences, while they were clustered with the new species—*Gymnopus dryophiloides*—that Antonín, Ryoo and Ka reported from Korea in 2020. Antonín et al. [51] do not accept *Gymnopus lanipes* (Malençon and Bertault) Vila and Llimona as a separate species and consider it to be a variant of *Gymnopus dryophilus*. From our phylogenetic result, it is clear that *Gymnopus lanipes* clusters with *Gymnopus inexpectatus*, which Consiglio, Vizzini, Antonín and Contu described from Europe, which, if *Gymnopus lanipes* is not considered an independent species, then *Gymnopus dryophioides* and *Gymnopus inexpectatus* should equally be treated as *Gymnopus dryophilus*. Moreover, *Gymnopus erythropus* complex, a subclade (defined here as/*erythropus*), includes *Gymnopus erythropus*, *Gymnopus fagiphilus*, and our new species (characterized by a red to reddish brown color, a smooth or scattered-to-dense tomentose stipe, and occurring in early autumn). The above results imply the need for a deeper and more extensive study on sect. *Levipedes*.

Based on the current study's findings, we increased the species diversity of the genus *Gymnopus* from China. However, probably due to the lacking of species sampling or the inadequate genetic variation in the DNA loci in our study, the deep phylogenetic relationships within the genus *Gymnopus* and between the related genera—*Lentinula*, *Rhodocollybia*, *Mycetinis*, *Collybiopsis*, etc.—remain unresolved. Thus, in future work, more species of this genus and similar genera will be discovered, which will provide new evidence and, thus, lead to a deeper understanding of the relationships within and among these genera.

Author Contributions: Conceptualization, Y.L. (Yu Li), B.Z. and J.H.; experimental design and methodology, Y.L. (Yajie Liu), B.Z., J.H. and Y.T.; performance of practical work, G.R., Y.L. (Yajie Liu), Y.T., J.H., L.Y., Y.L. (Yu Li) and G.Z.; statistical analyses, J.H., Y.T., Z.Z., T.Z. and Z.Q.; validation, J.H. and B.Z.; writing—original draft preparation, J.H.; writing—review and editing, B.Z.; supervision, Y.L. (Yajie Liu) and B.Z.; project administration, B.Z.; funding acquisition, B.Z. All authors have read and agreed to the published version of the manuscript.

Funding: This study was funded by the Natural Science Foundation of China (No. 31970020), National Key R&D of Ministry of Science and Technology (2018YFE0107800, 2019YFD1001905-33), Jilin Province Science and Technology Development Plan Project (20190201256JC, 20190201026JC), Scientific and Technological Tackling Plan for the Key Fields of Xinjiang Production and Construction Corps (No. 2021AB004), and "111" program (No. D17014).

Institutional Review Board Statement: Not applicable.

Informed Consent Statement: Not applicable.

Data Availability Statement: Not applicable.

Conflicts of Interest: The authors state no conflict of interest.

References

1. Antonín, V.; Noordeloos, M.E. *A Monograph of Marasmioid and Collybioid Fungi in Europ*; IHW Verl: Eching, Germany, 2010.
2. Persoon, C.H. *Synopsis Methodica Fungorum 1*; Apud Henricum Dieterich: Göttingen, Germany, 1801. (In Latin)
3. Fries, E.M. *Systema Mycologicum: Sistens Fungorum Ordines, Genera Et Species, Huc Usque Cognitas*; Ex officina Berlingiana: Lund, Berling, Germany, 1821; Volume 1. (In Latin)
4. Staude, F. *Die Schwämme Mitteldeutschlands, Insbesondere Des Herzogtums Coburg*; Dietz: Coburg, Germany, 1857; Volume 1. (In German)
5. Singer, R. *The Agaricales in Modern Taxonomy*, 2nd ed.; J. Cramer: Vaduz, Germany, 1962.
6. Singer, R. *The Agaricales in Modern Taxonomy*, 3rd ed.; J. Cramer: Vaduz, Germany, 1975.
7. Singer, R. *The Agaricales in Modern Taxonomy*, 4th ed.; Koeltz Botanical Books: Koenigstein, Germany, 1986.
8. Halling, R.E. The Genus Collybia (Agaricales) in the Northeastern United States and Adjacent Canada. *Mycol. Mem.* **1983**, *8*, 1–148.
9. Antonín, V.; Noordeloos, M.E. *A Monograph of Marasmius, Collybia and Related Genera in Europe. Part 1: Marasmius, Setulipes, and Marasmiellus*; Libri Botanici; IHW-Verlag: Eching, Germany, 1993; Volume 8, pp. 1–229.
10. Antonín, V.; Noordeloos, M.E. *A Monograph of Marasmius, Collybia and Related Genera in Europe. Part 2: Collybia, Gymnopus, Rhodocollybia, Crinipellis, Chaetocalathus, and Additions to Marasmiellus*; Libri Botanici; IHW-Verlag: Eching, Germany, 1997; Volume 17, pp. 1–256.
11. Antonín, V.; Halling, R.; Noordeloos, M. Generic Concepts within the Groups of Marasmius and Collybia Sensu Lato. *Mycotaxon* **1997**, *63*, 359–368.
12. Wilson, A.W.; Desjardin, D.E. Phylogenetic Relationships in the Gymnopoid and Marasmioid Fungi (Basidiomycetes, Euagarics Clade). *Mycologia* **2005**, *97*, 667–679. [CrossRef] [PubMed]
13. Moncalvo, J.M.; Vilgalys, R.; Redhead, S.A.; Johnson, J.E.; James, T.Y.; Catherine Aime, M.; Hofstetter, V.; Verduin, S.J.W.; Larsson, E.; Baroni, T.J.; et al. One Hundred and Seventeen Clades of Euagarics. *Mol. Phylogenet. Evol.* **2002**, *23*, 357–400. [CrossRef]
14. Mata, J.L.; Hughes, K.W.; Petersen, R.H. Phylogenetic Placement of Marasmiellus juniperinus. *Mycoscience* **2004**, *45*, 214–221. [CrossRef]
15. Oliveira, J.J.; Vargas-Isla, R.; Cabral, T.S.; Rodrigues, D.P.; Ishikawa, N.K. Progress on the Phylogeny of the Omphalotaceae: Gymnopus S. Str., Marasmiellus S. Str., Paragymnopus gen. nov. and Pusillomyces gen. nov. *Mycol. Prog.* **2019**, *18*, 713–739. [CrossRef]
16. Petersen, R.H.; Hughes, K.W. Two New Genera of Gymnopoid/Marasmioid Euagarics. *Mycotaxon* **2020**, *135*, 1–95. [CrossRef]
17. Halling, R.E. Notes on Collybia V. Gymnopus Section Levipedes in Tropical South America, with Comments on Collybia. *Brittonia* **1996**, *48*, 487–494. [CrossRef]
18. Singer, R. Type Studies in Basidiomycetes. X. *Pers. Mol. Phylogeny Evol. Fungi* **1961**, *2*, 1–62.
19. Jansen, A. *Het Geslacht Collybia (De Fungi Van Nederland)*; Stichting Uitgeverij Koninklijke Nederlandse Natuurhistorische Vereniging: Zeist, Nederlands, 1991. (In Dutch)
20. Kirk, P.; Cannon, P.; Minter, D.; Stalpers, J. *Dictionary of the Fungi*, 10th ed.; CABI International: Wallingford, UK, 2008.
21. Teng, S.Q. *Fungi of China*; Science Press, Academic Sinica: Beijing, China, 1963.
22. Tai, F.L. *Sylloge Fungorum Sinicorum*; Science Press, Academic Sinica: Beijing, China, 1979. (In Latin)
23. Deng, S.F. Taxonomy of Gymnopus and Preliminary Study of Marasmiaceae Resource in South China. Master's Thesis, South China Agricultural University, Guangzhou, China, 2016.
24. Li, J.-P.; Song, B.; Feng, Z.; Wang, J.; Deng, C.-Y.; Yang, Y.-H. A New Species of Gymnopus Sect. Androsacei (Omphalotaceae, Agaricales) from China. *Phytotaxa* **2021**, *521*, 1–14. [CrossRef]

25. Li, J.-P.; Li, Y.; Li, T.-H.; Antonín, V.; Hosen, I.; Song, B.; Xie, M.-L.; Feng, Z. A Preliminary Report of Gymnopus Sect. Impudicae (Omphalotaceae) from China. *Phytotaxa* **2021**, *497*, 263–276. [CrossRef]
26. Mešić, A.; Tkalčec, Z.; Deng, C.-Y.; Li, T.-H.; Pleše, B.; Ćetković, H. Gymnopus Fuscotramus (Agaricales), a New Species from Southern China. *Mycotaxon* **2011**, *117*, 321–330. [CrossRef]
27. Deng, S.-F.; Li, T.-H.; Jiang, Z.-D.; Song, B. Gymnopus Ramulicola sp. nov., a Pinkish Species from Southern China. *Mycotaxon* **2016**, *131*, 663–670. [CrossRef]
28. Wang, X.S.; Bao, J.S.; Bao, H.; Feng, J. Macrofungal Diversity in Hanwula National Nature Reserve, Inner Mongolia. *Mycosystema* **2020**, *39*, 695–706. [CrossRef]
29. Sun, Y.L.; Luo, Y.; Bau, T. Notes on Basidiomycetes of Jilin Province (X). *J. Fungal Res.* **2021**, *19*, 139–147. [CrossRef]
30. Royal Botanic Garden, E. *Flora of British Fungi: Colour Identification Chart*; HM Stationery Office: Edinburgh, UK, 1969.
31. César, E.; Bandala, V.M.; Montoya, L.; Ramos-Fernández, A. A New Gymnopus Species with Rhizomorphs and Its Record as Nesting Material by Birds (Tyrannideae) in the Subtropical Cloud Forest from Eastern Mexico. *MycoKeys* **2018**, *42*, 21–34. [CrossRef]
32. Gardes, M.; Bruns, T.D. Its Primers with Enhanced Specificity for Basidiomycetes-Application to the Identification of Mycorrhizae and Rusts. *Mol. Ecol.* **1993**, *2*, 113–118. [CrossRef]
33. Cubeta, M.A.; Echandi, E.; Abernethy, T.; Vilgalys, R. Characterization of Anastomosis Groups of Binucleate Rhizoctonia Species Using Restriction Analysis of an Amplified Ribosomal Rna Gene. *Phytopathology* **1991**, *81*, 1395–1400. [CrossRef]
34. Vilgalys, R.; Hester, M. Rapid Genetic Identification and Mapping of Enzymatically Amplified Ribosomal DNA from Several Cryptococcus Species. *J. Bacteriol.* **1990**, *172*, 4238–4246. [CrossRef]
35. Coimbra, V.R.M.; Pinheiro, F.G.B.; Wartchow, F.; Gibertoni, T.B. Studies on Gymnopus Sect. Impudicae (Omphalotaceae, Agaricales) from Northern Brazil: Two New Species and Notes on G. Montagnei. *Mycol. Prog.* **2015**, *14*, 110. [CrossRef]
36. Ryoo, R.; Antonín, V.; Ka, K.-H.; Tomšovský, M. Marasmioid and Gymnopoid Fungi of the Republic of Korea. 8. Gymnopus Section Impudicae. *Phytotaxa* **2016**, *286*, 75–88. [CrossRef]
37. Thonpson, J. The Clustal X Windows Interface: Flexible Strategies for Multiple Sequence Alignment Aided by Quality Analysis Tools. *Nucl. Acids Res.* **1997**, *24*, 4876–4882. [CrossRef]
38. Ranwez, V.; Douzery, E.J.P.; Cambon, C.; Chantret, N.; Delsuc, F. Macse V2: Toolkit for the Alignment of Coding Sequences Accounting for Frameshifts and Stop Codons. *Mol. Biol. Evol.* **2018**, *35*, 2582–2584. [CrossRef]
39. Katoh, K.; Standley, D.M. Mafft Multiple Sequence Alignment Software Version 7: Improvements in Performance and Usability. *Mol. Biol. Evol.* **2013**, *30*, 772–780. [CrossRef]
40. Hall, T. Bioedit: A User-Friendly Biological Sequence Alignment Editor and Analysis Program for Windows 95/98/Nt. *Nucleic Acids Symp. Ser.* **1999**, *41*, 95–98.
41. Zhang, D.; Gao, F.; Jakovlić, I.; Zhou, H.; Zhang, J.; Li, W.X.; Wang, G.T. Phylosuite: An Integrated and Scalable Desktop Platform for Streamlined Molecular Sequence Data Management and Evolutionary Phylogenetics Studies. *Mol. Ecol. Resour.* **2020**, *20*, 348–355. [CrossRef]
42. Kalyaanamoorthy, S.; Minh, B.Q.; Wong, T.K.F.; Von Haeseler, A.; Jermiin, L.S. Modelfinder: Fast Model Selection for Accurate Phylogenetic Estimates. *Nat. Methods* **2017**, *14*, 587–589. [CrossRef]
43. Ronquist, F.; Huelsenbeck, J.P. Mrbayes 3: Bayesian Phylogenetic Inference under Mixed Models. *Bioinformatics* **2003**, *19*, 1572–1574. [CrossRef]
44. Edler, D.; Klein, J.; Antonelli, A.; Silvestro, D. Raxmlgui 2.0: A Graphical Interface and Toolkit for Phylogenetic Analyses Using Raxml. *Methods Ecol. Evol.* **2021**, *12*, 373–377. [CrossRef]
45. Vizzini, A.; Antonín, V.; Sesli, E.; Contu, M. Gymnopus Trabzonensis sp. nov. Omphalotaceae and Tricholoma Virgatum var. Fulvoumbonatum var. nov. Tricholomataceae, Two New White-Spored Agarics from Turkey. *Phytotaxa* **2015**, *226*, 119–130. [CrossRef]
46. Mata, J.L.; Hughes, K.W.; Petersen, R.H. An Investigation of/Omphalotaceae (Fungi: Euagarics) with Emphasis on the Genus Gymnopus. *SYDOWIA-HORN-* **2006**, *58*, 191–289.
47. Petersen, R.H.; Hughes, K.W. Collybiopsis and Its Type Species, Co. Ramealis. *Mycotaxon* **2021**, *136*, 263–349. [CrossRef]
48. Wilson, A.; Desjardin, D.; Horak, E. Agaricales of Indonesia: 5. The Genus Gymnopus from Java and Bali. *Sydowia* **2004**, *56*, 137–210.
49. Schoch, C.L.; Robbertse, B.; Robert, V.; Vu, D.; Cardinali, G.; Irinyi, L.; Meyer, W.; Nilsson, R.H.; Hughes, K.; Miller, A.N.; et al. Finding Needles in Haystacks: Linking Scientific Names, Reference Specimens and Molecular Data for Fungi. *Database* **2014**, *2014*, bau061. [CrossRef]
50. Petersen, R.H.; Hughes, K.W. Micromphale Sect. Perforantia (Agaricales, Basidiomycetes); Expansion and Phylogenetic Placement. *MycoKeys* **2016**, *18*, 1–122. [CrossRef]
51. Antonín, V.; Sedlák, P.; Tomšovský, M. Taxonomy and Phylogeny of European Gymnopus Subsection Levipedes (Basidiomycota, Omphalotaceae). *Persoonia* **2013**, *31*, 179–187. [CrossRef]
52. Vu, D.; Groenewald, M.; De Vries, M.; Gehrmann, T.; Stielow, B.; Eberhardt, U.; Al-Hatmi, A.; Groenewald, J.Z.; Cardinali, G.; Houbraken, J.; et al. Large-Scale Generation and Analysis of Filamentous Fungal DNA Barcodes Boosts Coverage for Kingdom Fungi and Reveals Thresholds for Fungal Species and Higher Taxon Delimitation. *Stud. Mycol.* **2019**, *92*, 135–154. [CrossRef]

53. Ryoo, R.; Antonín, V.; Ka, K.H. Marasmioid and Gymnopoid Fungi of the Republic of Korea. 8. Gymnopus Section Levipedes. *Mycobiology* **2020**, *48*, 252–262. [CrossRef]
54. Vizzini, A.; Consiglio, G.; Antonin, V.; Contu, M. A New Species within the Gymnopus Dryophilus Complex (Agaricomycetes, Basidiomycota) from Italy. *Mycotaxon* **2008**, *105*, 43–52. [CrossRef]
55. Mata, J.L.; Petersen, R.H.; Hughes, K.W. The Genus Lentinula in the Americas. *Mycologia* **2001**, *93*, 1102–1112. [CrossRef]
56. Hibbett, D.S.; Hansen, K.; Donoghue, M.J. Phylogeny and Biogeography of Lentinula Inferred from an Expanded Rdna Dataset. *Mycol. Res.* **1998**, *102*, 1041–1049. [CrossRef]
57. Antonín, V.; Ryoo, R.; Shin, H.-D. Marasmioid and Gymnopoid Fungi of the Republic of Korea. 2. Marasmius Sect. Globulares. *Persoonia* **2010**, *24*, 49–59. [CrossRef]
58. Petersen, R.H.; Hughes, K.W. An Investigation on Mycetinis (Euagarics, Basidiomycota). *MycoKeys* **2017**, *24*, 1–139. [CrossRef]
59. Mata, J.L.; Ovrebo, C.L.; Baroni, T.J.; Hughes, K.W. New Species of Neotropical Rhodocollybia. *Mycotaxon* **2016**, *131*, 235–245. [CrossRef]

Journal of
Fungi

MDPI

Article

New Species and New Records of *Otidea* from China Based on Molecular and Morphological Data

Yu-Yan Xu †, Ning Mao †, Jia-Jia Yang and Li Fan *

College of Life Science, Capital Normal University, Xisanhuanbeilu 105, Haidian, Beijing 100048, China; 2190801004@cnu.edu.cn (Y.-Y.X.); 2210801013@cnu.edu.cn (N.M.); 2210802122@cnu.edu.cn (J.-J.Y.)
* Correspondence: fanli@mail.cnu.edu.cn
† These authors contributed equally to this work.

Abstract: Species of genus *Otidea* previously reported in China are mainly distributed in the northeast, northwest and southwest regions of China, but the species diversity of *Otidea* in north China is not very clear. In this study, newly collected *Otidea* specimens from northern China and some herbarium specimens deposited in three important Chinese fungus herbaria (HMAS, HKAS, HMJAU) were studied using morphological and phylogenetic methods. The internal transcribed spacers of the nrDNA (ITS), the nrRNA 28S subunit (nrLSU), the translation elongation factor 1-alpha (*tef1-α*), and the second largest subunit of RNA polymerase II (*rpb2*), were employed to elucidate the phylogenetic relationships between *Otidea* species. Results identified 16 species of *Otidea*, of which seven new species are described, namely *O. aspera*, *O. cupulata*, *O. filiformis*, *O. khakicolorata*, *O. parvula*, *O. plicara* and *O. purpureobrunnea*. *Otidea bicolor* and *O. pruinosa* are synonymized as *O. subpurpurea*. Two species, *O. mirabilis* and *O. nannfeldtii*, are being reported for the first time in China. The occurrence of *O. bufonia*, *O. leporina* and *O. onotica* are confirmed by molecular data in China.

Keywords: Ascomycota; Pyronemataceae; phylogeny; seven new taxa; taxonomy

Citation: Xu, Y.-Y.; Mao, N.; Yang, J.-J.; Fan, L. New Species and New Records of *Otidea* from China Based on Molecular and Morphological Data. *J. Fungi* **2022**, *8*, 272. https://doi.org/10.3390/jof8030272

Academic Editors: Cheng Gao and Lei Cai

Received: 31 January 2022
Accepted: 6 March 2022
Published: 8 March 2022

1. Introduction

The genus *Otidea* (Pers.) Bonord. (Pyronemataceae, Pezizales), with *O. onotica* (Pers.) Fuckel as the type species, was established in the mid-19th century [1,2]. The genus is characterized by the following criteria: epigeous, cup- to ear-like apothecia, split to the base on one side (less often entirely), and being stipitate or not; or as in a single species, hypogeous with enclosed ascomata, subcylindrical, nonamyloid, operculate asci, ellipsoid to fusoid, biguttulate ascospores, and filiform paraphyses [3–5]. *Otidea* species are considered to form ectomycorrhizae with both broad leaf and conifer trees [3,4], and are widely distributed across the temperate regions of Europe, North America and Asia in the northern hemisphere [4–10]. The taxonomy of *Otidea* species are mainly based on the morphological characteristics studied in work before 2006 [5,6,11–17]. Since then, taxonomists began to introduce molecular methods into the taxonomy and identification of *Otidea* species. Molecular techniques have revolutionized phylogenetics and species delimitation of *Otidea* [3,4,10,18–25]. In 2009, a whole new set of morphological and histochemical features was also introduced in *Otidea* by Harmaja [15] and further by Hansen and Olariaga [3] and Olariaga et al. [4]. These features have made it possible to recognize the species morphologically to a large extent and have since been employed. *Otidea* species in Europe (c. 32 accepted species) and North America (c. 14 accepted species) have been comprehensively and systematically studied in recent years [3,4,10,21,22].

China has a huge temperate area in the northern hemisphere and likely has diverse *Otidea* species. However, only a few taxonomic works focus on this genus, and about a quarter of Chinese *Otidea* species are not supported by molecular evidence [6,8,16–18,23–29]. Currently, a total of 24 *Otidea* species are reported in China, including 17 native species and seven known species originally described from Europe and/or North America.

During our investigation of fungal resources in northern China since 2017, many apothecia of the genus *Otidea* were collected. On the bases of these new collections and some of herbarium specimens deposited in three important Chinese fungus herbaria (HMAS, HKAS, HMJAU), we recognized seven species and two records new to China based on both morphological examination and molecular analysis. Also, both *O. bicolor* W.Y. Zhuang & Zhu L. Yang, and *O. pruinosa* Ekanayaka, Q. Zhao & K.D. Hyde were conspecific with *O. subpurpurea* W.Y. Zhuang. Our aim in this paper is to describe and illustrate these new species and new records and to synonymize *O. bicolor* and *O. pruinosa* as *O. subpurpurea*. Molecular data for some known species existing in China are additionally provided.

2. Materials and Methods

2.1. Morphological Studies

Fresh specimens were collected and photographed in the field from the Shanxi and Hebi provinces, as well as Beijing, China. The specimens were dried and deposited in the BJTC (Herbarium, Biology Department, Capital Normal University) and the HSA (Herbarium Institute of Edible Fungi, Shanxi Academy of Agricultural Science, Taiyuan, China). Other specimens were studied from the HMAS (Herbarium Mycologicum Academiae Sinicae, Institute of Microbiology, Chinese Academy of Sciences), HKAS (Herbarium of Cryptogams at the Kunming Institute of Botany, Chinese Academy of Sciences) and HMJAU (Herbarium of Mycology, Jilin Agricultural University). Macroscopic characteristics were recorded from fresh specimens. Standardised color values matching the described colour were taken from ColorHexa (http://www.colorhexa.com/, accessed on 30 January 2022). Microscopic characteristics were observed in thin sections of dry specimens mounted in 5% KOH and Melzer's reagent [30]. The dimensions for ascospores are given using notation of the form (a-)b-c(-d). The range b-c contains a minimum of 90% of the measured values. The extreme values, i.e., a and d, are given in the parentheses. L_m and W_m indicate the average ascospore length and width for the measured ascospores, respectively. Q is used to represent length/width ratio of a ascospore in side view and Q_m represents average Q of all specimens. The number of populations that the statistics are based on is indicated by n.

2.2. DNA Extraction, PCR Amplification, Sequencing

Herbarium specimens were crushed by shaking for 30 s at 30 Hz 2–4 times (Mixer Mill MM 301, Retsch, Haan, Germany) in a 1.5 mL tube together with one 3 mm diameter tungsten carbide ball, and total genomic DNA was extracted using the modified CTAB method [31]. The following primers were used for PCR amplification and sequencing: ITS1f/ITS4 [31,32] were used for the internal transcribed spacers of the nrDNA (ITS1-5.8S-ITS2 = ITS), LR0R/LR5 [33] for the nrDNA 28S subunit (nrLSU), EF1-983F/EF1-2218R [34] for the translation elongation factor 1-alpha (*tef1-α*), and RPB2-Otidea6F/RPB2-Otidea7R and fRPB2-7cF/fRPB2-11aR [3] for the RNA polymerase II second largest subunit (*rpb2*), respectively. PCRs were performed in 50 μL reactions containing 4 μL DNA template, 2 μL of per primer (10 μM), 25 μL 2× Master Mix (Tiangen Biotech Co., Beijing, China), and 17 μL ddH_2O. PCR reactions were performed as follows: for the ITS gene: initial denaturation at 94 °C for 3 min, followed by 35 cycles at 94 °C for 30 s, 56 °C for 45 s, 72 °C for 1 min, and a final extension at 72 °C for 10 min; for the nrLSU gene: initial denaturation at 94 °C for 4 min, followed by 35 cycles at 94 °C for 30 s, 55 °C for 45 s, 72 °C for 1 min, and a final extension at 72 °C for 10 min; for the *tef1-α* gene: initial denaturation at 94 °C for 3 min, followed by 35 cycles at 94 °C for 30 s, 60 °C for 45s, 72 °C for 1min, and a final extension at 72 °C for 10 min; for the *rpb2* gene: initial denaturation at 94 °C for 3 min, followed by 10 cycles (including denaturation) at 94 °C for 30 s, annealing temperature started at 62 °C (decreased by 1 °C per cycle, until to 52 °C) for 45 s and extension at 72 °C for 1 min, then followed by 30 cycles at 94 °C for 35 s, 55 °C for 45 s, 72 °C for 1 min, and a final extension at 72 °C for 10 min. The PCR products were sent to Beijing Zhongkexilin Biotechnology Co., Ltd. (Beijing, China) for purification, sequencing, and editing. The newly generated sequences were assembled and edited using SeqMan (DNA STAR package,

DNAStar Inc., Madison, WI, USA) with generic-level identities for sequences confirmed via BLAST queries of GenBank.

2.3. Sequence Alignment and Phylogenetic Analyses

A total of 730 sequences from 283 collections of *Otidea* were used in the molecular phylogenetic analyses. The detail information about them is provided in Supplementary Table S1, including the geographic origin and accession numbers. Sequences of all DNA regions generated in this study were deposited in GenBank. The sequences obtained from GenBank are based on published literature or selected by using BLASTn search to find similar matches with taxa in *Otidea*. Two datasets were assembled for this study. Dataset I (ITS/nrLSU) and datasets II (ITS/nrLSU/*tef1-α*/*rpb2*) contained the backbone species and all phyloclades of *Otidea*, which were used to infer the phylogenetic status of Chinese *Otidea* species of the genus *Otidea*. The taxa *Monascella botryosa* Guarro & Arx and *Warcupia terrestris* Paden & J.V. Cameron were selected as outgroups. The ITS, nrLSU, *tef1-α* and *rpb2* sequences were respectively aligned using the MAFFT v.7.110 online program under the default parameters [35], and manually adjusted to allow for maximum sequence similarity in Se-Al version.2.03a [36]. Ambiguously aligned regions of each sequence were detected and excluded using Gblocks 0.91b [37] before the phylogenetic analyses. Unsampled gene regions were coded as missing data and all introns of *tef1-α* and *rpb2* were excluded because of the alignment difficulty. To examine the conflict among topologies with maximum likelihood (ML), separate single-gene analyses were conducted. Alignments were concatenated using SequenceMatrix v1.7.8 [38] and are provided in Supplementary Files S2 and S3. We conducted maximum likelihood (ML) and Bayesian inference (BI) analyses on the two datasets.

Maximum likelihood (ML) analyses of the two datasets were carried out using RAxML 8.0.14 [39] with all parameters kept to the default settings using a GTRGAMMAI model. The ML bootstrap replicates (1000) were computed in RAxML using a rapid bootstrap analysis searching for the best scoring ML tree. Bayesian inference (BI) analyses were performed with MrBayes v3.1.2 [40] based on the best substitution models for each gene region as determined by MrModeltest 2.3 [41]. The GTR + I+G model was the best model for ITS, nrLSU and *rpb2*, whereas the best model for *tef1-α* was the SYM + I+G model. Two independent executions of four chains were conducted: 3,485,000 for ITS/nrLSU and 765,000 for the ITS/nrLSU/*tef1-α*/*rpb2* datasets. Markov chain Monte Carlo generations were conducted using the default settings and sampled every 100 generations. The temperature value was lowered to 0.20, burn-in was set to 0.25, and the program was automatically stopped as soon as the average standard deviation of split frequencies reached below 0.01. A 50% majority-rule consensus tree was constructed. Clades with a bootstrap support (BS) $\geq$ 70% and a Bayesian posterior probability (PP) $\geq$ 0.95 were considered as significantly supported [42,43]. All phylogenetic trees were viewed with TreeView32 [44].

3. Results

3.1. Phylogenetic Analyses

No topological incongruence was detected when the four genes were analyzed individually. Dataset I (ITS/nrLSU) contained 528 sequences from 51 species, including 93 novel sequences the two genes from Chinese collections, and four from the outgroups (*Monascella botryosa* and *Warcupia terrestris*). The dataset had an aligned length of 1363 characters (551 bp from ITS and 812 bp from nrLSU), of which 647 were constant, 716 were variable, and 609 of these variable sites were informative. ML and BI analyses yielded similar tree topologies. Only the tree inferred from the ML analyses is shown (Figure 1). The species of *Otidea* formed a monophyletic clade with high support values (BS = 100%, PP = 1.00). A total of 10 clades were recognized in the two-gene phylogram, which is consistent with Olariaga et al. [4] and Hansen and Olariaga [3]. Newly obtained Chinese *Otidea* specimens were nested in six clades: *O. bufonia-onotica*, *O. formicarum*, *O. leporina*, *O. cantharella*, *O. alutacea*, and *O. platyspora* clade (Figure 1). A total of 16 species were recognized.

In the *O. bufonia-onotica* clade, 24 Chinese specimens were clearly placed in eight well-supported clades, represented by five known species and three new species. The known species are *O. bufonia* (Pers.) Boud., *O. subpurpurea*, *O. mirabilis* Bolognini & Jamoni, *O. onotica* and *O. brevispora* (W.Y. Zhuang) Olariaga & K. Hansen. The three new species are respectively described as *O. filiformis*, *O. cupulata* and *O. purpureobrunnea* in this study. It is interesting that the sequences from the type specimens of *O. bicolor* and *O. pruinosa* fall into the clade of *O. subpurpurea* and shared 98.87–99.84% similarity in the ITS region, which implied they may be conspecific, although they have some difference in apothecial color. Fortunately, we borrowed the type specimens of these three species for observation and research. We thought that they should be conspecific and formally synonymized *O. bicolor* and *O. pruinosa* in this study.

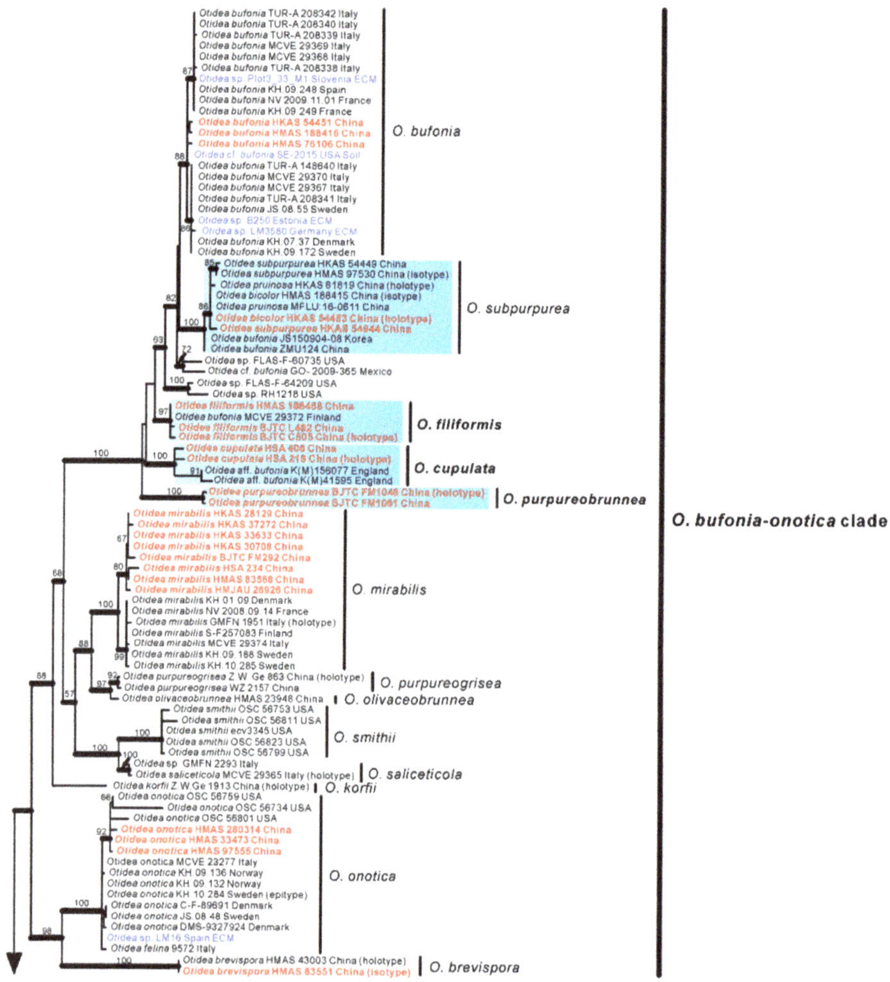

Figure 1. *Cont.*

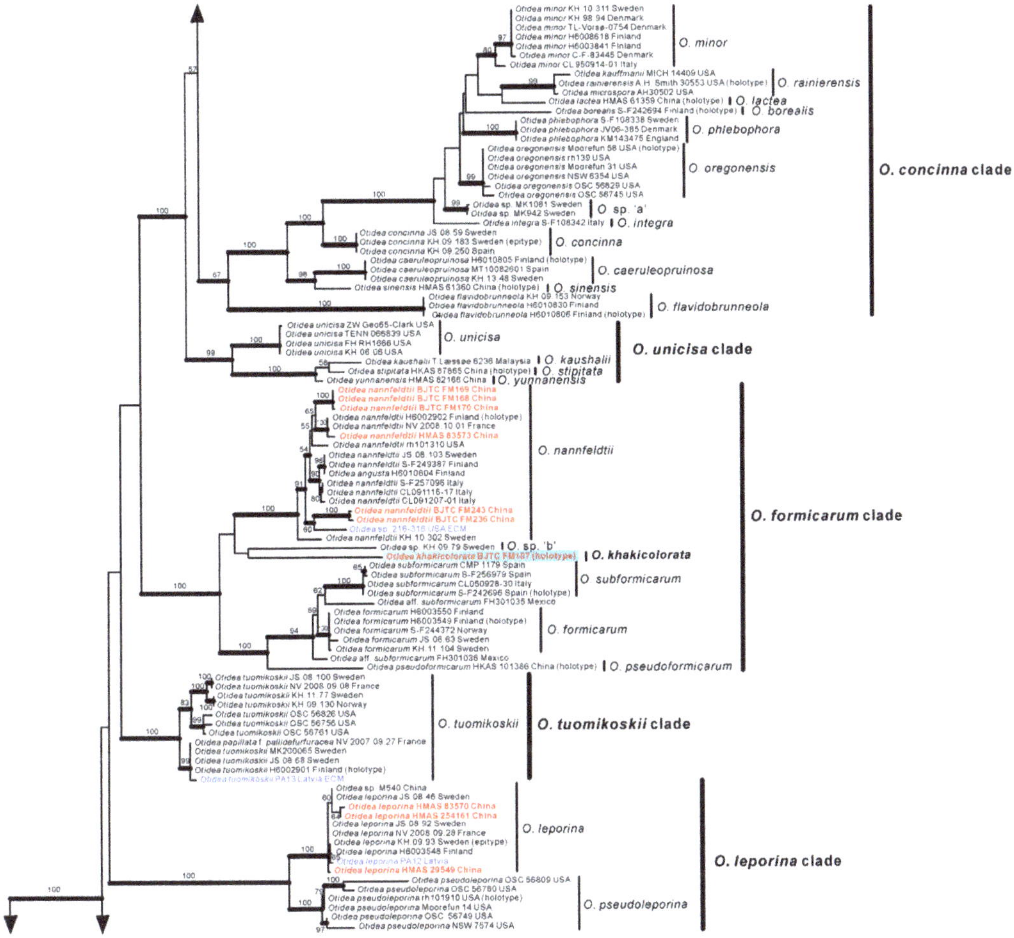

Figure 1. *Cont.*

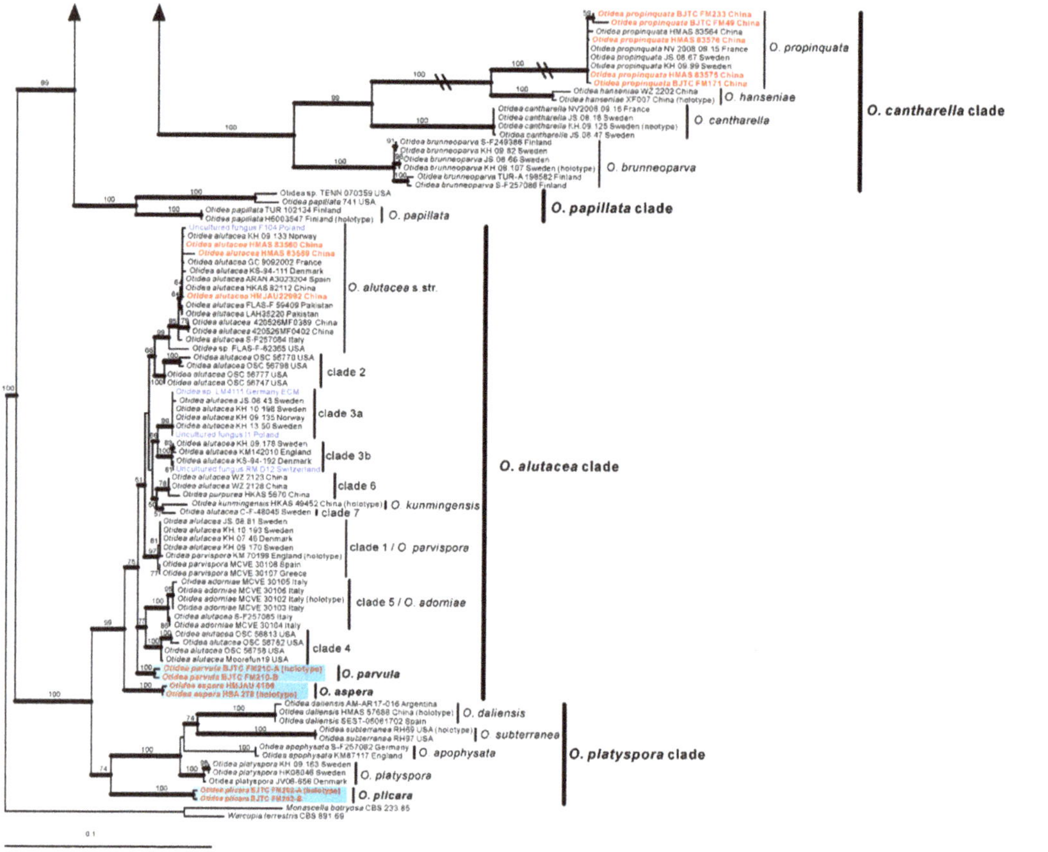

Figure 1. Phylogenetic tree generated from maximum likelihood analysis based on ITS and nrLSU sequences, showing the phylogenetic relationships of *Otidea*. *Monascella botryosa* and *Warcupia terrestris* are the outgroups. Maximum likelihood bootstrap support values (≥50%) are indicated above the nodes as BS. Thick black branches received Bayesian posterior probabilities (BPP) ≥ 0.95. Novel sequences are printed in bold red. Mycorrhizal or environmental sequences are printed in blue. The new species are in bold font and highlighted by blue boxes.

In the *O. formicarum* clade, seven Chinese specimens were clearly placed in two well supported clades. One of clades corresponds to *O. nannfeldtii* Harmaja, marking the first time it is discovered in China. The other one is a new species described as *O. khakicolorata* in this study. In the *O. leporina* clade and *O. cantharella* clade, the Chinese specimens were identified as the previously known species, *O. leporina* (Batsch) Fuckel and *O. propinquata* (P. Karst.) Harmaja, respectively. In the *O. alutacea* clade, seven Chinese collections were clearly placed into three well supported clades, represented by the known species *O. alutacea* (Pers.) Massee and two new species described as *O. parvula* and *O. aspera* in this paper. In the *O. platyspora* clade, the two Chinese collections formed a distinct clade with high evidential support, which was described as *O. plicara* in this study.

In order to further verify the phylogenetic positions of the seven new species and whether *O. bicolor* and *O. pruinosa* are conspecific with *O. subpurpurea*, we performed a further phylogenetic analysis based on the four-gene dataset II. This dataset II does not include sequences from seven confirmed known species, namely *O. bufonia*, *O. mirabilis*, *O. onotica*, *O. nannfeldtii*, *O. leporina*, *O. propinquata* and *O. alutacea*. Dataset II (ITS/nrLSU/*tef1-α*/*rpb2*) contained 501 sequences from 49 species, including 73 novel sequences these four genes from Chinese collections, and 8 from the outgroups (*M. botryosa* and *W. terrestris*). The dataset had an aligned length of 4400 characters (551 bp from ITS, 812 bp from nrLSU, 1383bp from *tef1-α* and 1654 bp from *rpb2)*, of which 2456 were constant, 1854 were variable, and 1753 of these variable sites were informative. ML and BI analyses yielded similar tree topologies. Only the tree inferred from the ML analysis is shown (Figure 2).

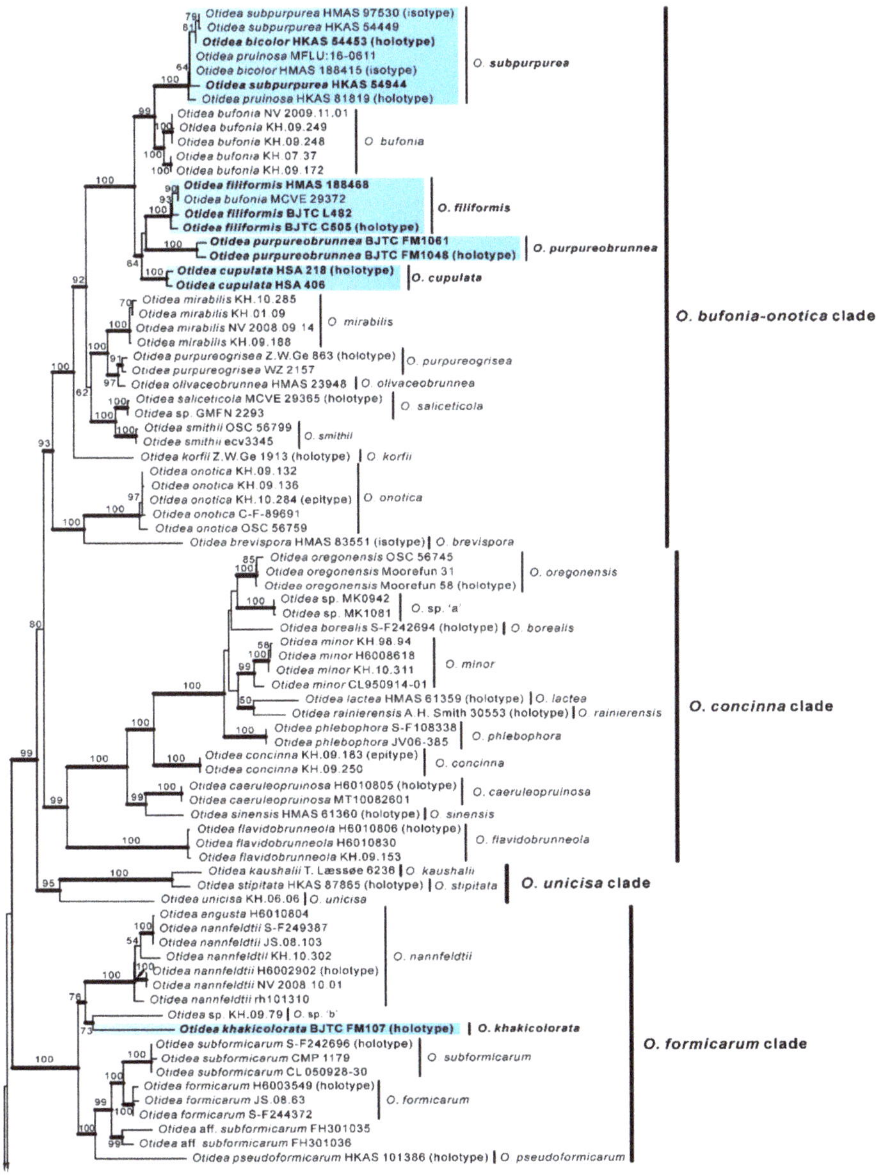

Figure 2. *Cont.*

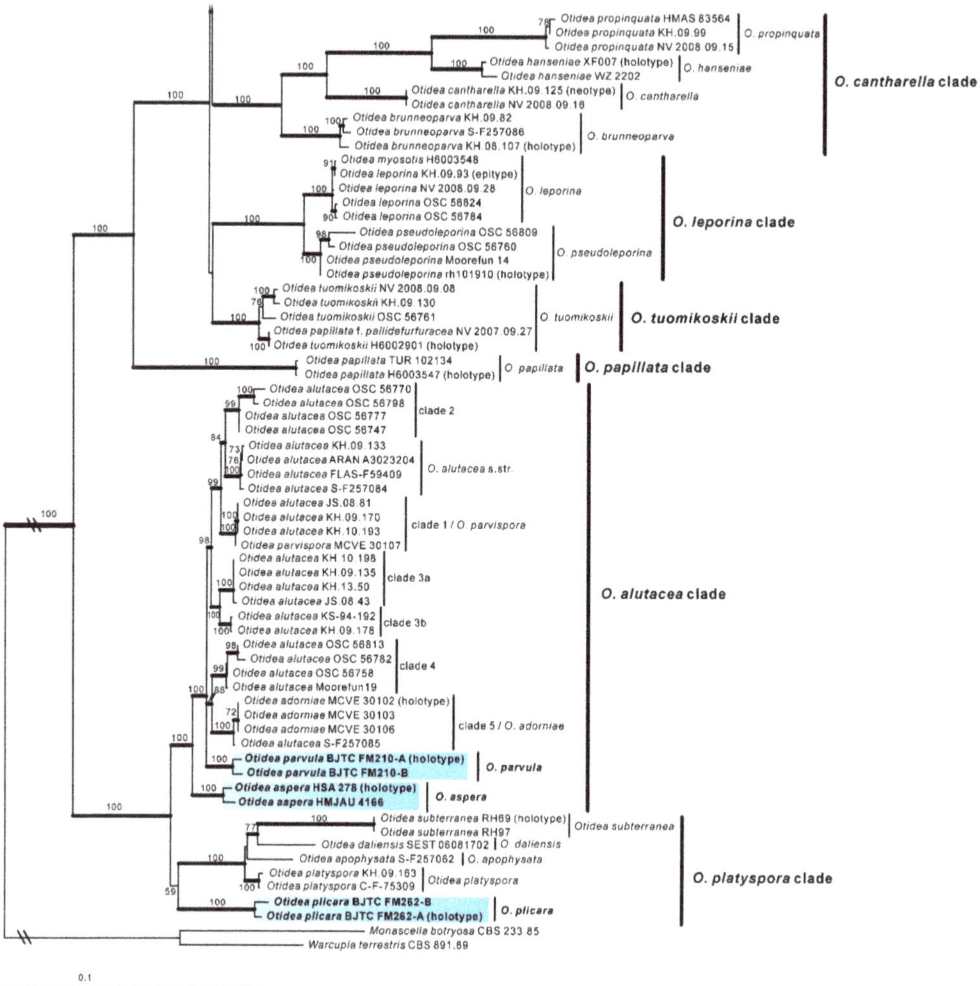

Figure 2. Phylogenetic tree generated from a maximum likelihood analysis based on ITS, nrLSU, *tef1-α* and *rpb2* sequences, showing the phylogenetic relationships of *Otidea*. *Monascella botryosa* and *Warcupia terrestris* are the outgroups. Maximum likelihood bootstrap support values (≥50%) are indicated above the nodes as BS. Thick black branches received Bayesian posterior probabilities (BPP) ≥ 0.95. The new species are in bold font and highlighted by blue boxes.

The species of *Otidea* formed a monophyletic clade with high support values (BS = 100%, PP = 1.00). A total of 10 clades were recognized in the four-gene phylogram, which is consistent with Hansen and Olariaga [3]. Similar to the two-gene phylogram results, the specimens from China formed seven apparently independent clades with high support values, representing seven new species. The type sequences of *O. bicolor*, *O. pruinosa* and *O. subpurpurea* clustered into a clade with high support values (BS = 100%, PP = 1.00), indicating that they are indeed the same species, and *O. bicolor* and *O. pruinosa* are placed in synonymy as *O. subpurpurea* in the taxonomy section below. Compared with the two-gene tree phylogram (Figure 1), the 10 clades identified in the four-gene tree are consistent, but inside the *O. bufonia-onotica* clade, the *O. concinna* clade, and the *O. alutacea* clade, the topological positions of some species are slightly different (e.g., *O.*

purpureobrunnea and *O. cupulata* in the *O. bufonia-onotica* clade), which may be due to the fact that these species have a lower support value between them.

3.2. Taxonomy

Based on our phylogenies and morphological data, a total of seven new species, a known species, and two new records of *Otidea* from China were described and illustrated here.

Otidea aspera L. Fan & Y.Y. Xu, sp. nov. (Figure 3).

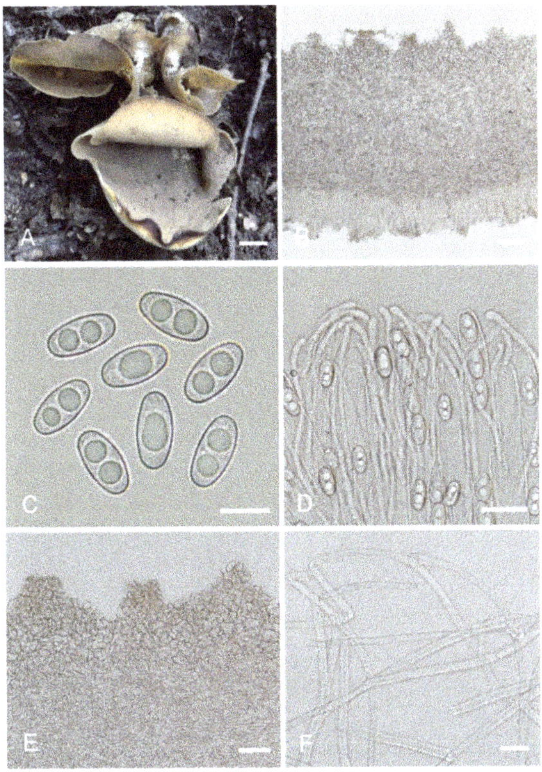

Figure 3. *Otidea aspera* (HSA 278) (**A**) apothecia, (**B**) anatomy of apothecium, (**C**) ascospores, (**D**) asci and paraphyses, (**E**) ectal excipulum in water, and (**F**) basal mycelium. Scale bars: (**A**) = 1 cm, (**B**) = 100 μm, (**C**) = 10 μm; (**D**) = 20 μm; (**E**) = 50 μm; (**F**) = 10 μm.

MycoBank: MB843176.

Etymology: *aspera*, referring to the rough receptacle surface.

Holotype: China. Shanxi Province, Jiaocheng County, Pangquangou National Nature Reserve, HaoJiagou, alt. 1800m, in the mixed forest dominated by *Pinus tabuliformis* Carrière and *Quercus mongolica* Fisch. ex Ledeb., 29 August 2018, J.Z. Cao, LH278 (HSA 278).

Saprobic on soil. Apothecia solitary or caespitose in nature, 30–45 mm high, 25–60 mm wide, initially ear-shaped, then expanding and sometimes becoming irregularly ear-shaped or shallowly to deeply cup-shaped, sometimes elongated on one side or obconical, split, stipitate. Hymenium surface greyish yellow (#fbe8c5) to light brown (#baab98) when fresh, ochre brown (#a87832) when dry, subsmooth. Receptacle surface pale yellow (#fafad2) to yellowish brown (#b5a27f) when fresh, slightly hygrophanous, pale whitish ochre (#c8b99f) when dry, finely warty. Stipe 5–12 × 4–8 mm. Basal tomentum and mycelium white. Apothecial section 600–900 μm thick. Ectal excipulum of *textura angularis*, 100–150 μm thick, cells thin walled, hyaline to pale brown, 11–28 × 8–20 μm. Medullary excipulum of *textura intricata*, 300–500 μm thick, hyphae 3–10 μm wide, sometimes slightly swollen, thin to thick walled, septate, hyaline to light brown. Subhymenium c. 50–120 μm thick, visible as a brown zone of densely arranged cylindrical to swollen cells. Paraphyses septate, straight to slightly curved, of uniform width or slightly enlarged at the apices to 3–4.7 μm wide, without or rarely with 1–2 low notches. Asci 150–200 × 9–13 μm, 8-spored, unitunicate, cylindrical, hyaline, long pedicellate, arising from croziers, non-amyloid, ascospores released from an eccentric split at the apical apex. Ascospores ellipsoid to slightly subfusoid, inequilateral, with two large guttules, sometimes only with one big guttule, smooth, hyaline, (12–) 12.8–15 (–15.5) × (5.8–) 6.5–7.5 (–8) μm ($L_m \times W_m$ = 14 × 7 μm, Q = 1.8–2.2, Q_m = 2, n = 50). Receptacle surface with warts, 50–80 μm high, formed by short, fasciculate, hyphoid hairs, of 5–7 subglobose to elongated cells, constricted at septa, 6–13 μm wide. Resinous exudates absent to scarce. Basal mycelium of 2–5 μm wide, septate, hyaline to pale brown hyphae, smooth, unchanged in KOH, smooth, turning yellow in MLZ.

Other materials examined: China. Inner Mongolia Autonomous Region, Daqinggou National Nature Reserve, on the broad-leaved woodland, 24 August 2005, Tolgor Bau (HMJAU 4166).

Notes: *Otidea aspera* is diagnosed by the combination of the stipitate, broadly ear shaped to cup-shaped, greyish yellow to light brown hymenium, pale-yellow to yellowish-brown receptacle surface, ellipsoid to slightly subfusoid ascospores and straight to slightly curved paraphyses. *Otidea aspera* and *O. parvispora* have comparable apothecia color, however *O. parvispora* differs from *O. aspera* by the smaller ascospores ((11.0–) 11.5–13.0 × 5.0–6.5 μm) and shorter asci. DNA analysis showed that *O. aspera* shared less than 93.39% ITS sequence similarity with other *Otidea* species. Phylogenetic analyses revealed that the sequences of *O. aspera* were grouped into an independent clade with a strong support value (Figures 1 and 2). These supported the erection of the new species.

Otidea cupulata L. Fan & Y.Y. Xu, sp. nov. (Figure 4).

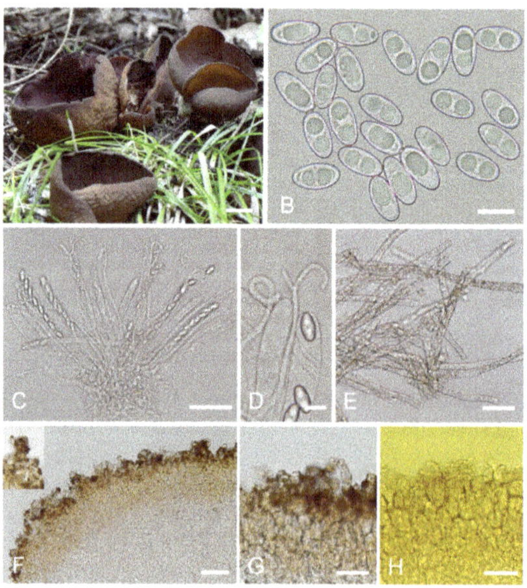

Figure 4. *Otidea cupulata* (HSA 218) (**A**) apothecia, (**B**) ascospores, (**C**) asci and paraphyses, (**D**) paraphyses, (**E**) basal mycelium, and (**F**) ectal and medullary excipulum in water, (**G**) ectal excipulum in water, (**H**) ectal excipulum in MLZ. Scale bars: (**A**) = 1 cm, (**B**) = 10 μm, (**C**) = 50 μm, (**D**) = 10 μm, (**E**) = 25 μm, (**F**) = 50 μm, (**G,H**) = 30 μm.

MycoBank: MB843177.

Etymology: *cupulata*, referring to the apothecia shape of the fungus.

Holotype: China. Shanxi Province, Jiaocheng County, Pangquangou Township, Badaogou valley, alt. 2200m, on soil in mixed forest of *Larix* sp. and *Betula* sp., 28 August 2018, L.J. Guo, LH218 (HSA 218).

Saprobic on soil. Apothecia gregarious or caespitose in nature, 25–40 mm high, 15–50 mm wide, initially ear-shaped, soon expanding, becoming broadly ear-shaped or deeply cup-shaped, split, often broader above, margin sometimes lobate, stipitate. Hymenium surface dark orange brown (#734a12) to dark purple brown (#4d282d) when fresh, and usually with bluish-lilaceous shades, gray ochraceous brown (#37290e) when dry, subsmooth. Receptacle surface yellowish brown (#a1805f) to brown (#8a6660) when fresh, slightly hygrophanous, surface with shallow wrinkles, dark brown (#261600) when dry, furfuraceous. Stipe 5–10 × 5–7 mm. Basal tomentum and mycelium whitish to grayish yellow (#c6cbac). Apothecial section 900–1200 μm thick. Ectal excipulum of *textura angularis*, 80–110 μm thick, cells thin walled, brownish, 10–25 × 8–22 μm. Medullary excipulum of *textura intricata*, 500–700 μm thick, hyphae 3.5–10 μm wide, sometimes slightly swollen, thin to slightly thick walled, septate, hyaline to light brown. Subhymenium c. 60–100 μm thick, visible as a brown zone, of densely arranged cylindrical to swollen cells, with scattered brown resinous exudate at septa. Paraphyses septate, curved to hooked of uniform width or slightly enlarged at the apices to 2.6–4.2 μm wide, without or rarely with 1–2 low notches, sometimes forked near the apex. Asci 150–200 × 8–11 μm, 8-spored, unitunicate, cylindrical, hyaline, long pedicellate, arising from croziers, non-amyloid, ascospores release from an eccentric split at the apical apex. Ascospores ellipsoid to slightly subfusoid, inequilateral, with two large guttules, sometimes with only one big guttule, smooth, hyaline, (12–) 12.5–15 (–15.5) × (6–) 6.5–7.4 (–7.9) μm ($L_m \times W_m$= 13.9 × 7 μm, Q = 1.9–2.1, Q_m = 2, n = 50). Receptacle surface with low warts, 25–45 μm high, formed by short, fasciculate, hyphoid hairs, of 3–4 subglobose to elongated cells, constricted at septa, 6–11 μm wide, sometimes with a gelatinous sheath. Resinous exudates abundant on the outer surface, dark brown, partly dissolving and converting into small particles in MLZ, entirely dissolving and turning bright yellow in KOH. Basal mycelium of 3–6.5 μm wide, septate, hyaline to pale brown hyphae, bright yellow in KOH, with abundant, very small, irregularly, brown, resinous exudates on the surface, dissolving and turning bright yellow in KOH, unchanged in MLZ.

Other materials examined: China. Shanxi Province, Jiaocheng County, Pangquangou Township, Badaogou Valley, alt. 1800m, on soil under *Larix* sp., 6 September 2018, L.J. Guo, LH 406 (HSA 406).

Notes: *Otidea cupulata* is recognized by the stipitate, broadly cup-shaped, dark-orange-brown to dark-purple-brown hymenium, yellowish-brown to brown receptacle surface, ellipsoid to subfusoid ascospores, forked or notched paraphyses, and furfuraceous receptacle surface. The forked paraphyses is rarely found in other species in the O. bufonia-onotica clade. Several species in the O. bufonia-onotica clade are similar to *O. cupulata* in apothecial shape and color, including *O. bufonia*, *O. filiformis*, *O. mirabilis*, and *O. olivaceobrunnea* Harmaja, but *O. bufonia* can be distinguished by its distinctly narrowly fusoid ascospores, the presence of hyphae with striate resinous exudates in the medullary excipulum, and resinous exudates of the ectal excipulum that does not turn bright yellow in KOH. *Otidea filiformis* differs in its apothecia with pinkish shades, distinctly narrowly fusoid ascospores, unenlarged and curved to hooked paraphyses, and higher warts (40–75 μm) on the receptacle surface. *Otidea mirabilis* differs by having purple to lilaceous-bluish shades on the receptacle surface and narrowly fusoid ascospores (Q_m = 2.1–2.3). *Otidea olivaceobrunnea* can be separated by its olive-brown hymenium and wider ascospores (14–17 × 8–8.5 μm). Four species (*O. purpureogrisea* Pfister, F. Xu & Z.W. Ge, *O. purpureobrunnea*, *O. simithii* Kanouse and *O. subpurpurea*) have more or less bluish or lilaceous shades on apothecia that resembles that of *O. cupulata*. However, *O. purpureogrisea* is distinguished by its ear-shaped apothecia, dark-purple-brown to purple-gray receptacle surface, and the resinous exudate in the ectal excipulum turning amber and brown in KOH. *Otidea purpureobrunnea* is distinguished by its grayish-purple-brown to dark-purple-brown receptacle surface and mostly smooth basal mycelium. *Otidea simithii* differs in having typically narrower, ear-shaped apothecia, and resinous exudates of the ectal excipulum that does not turn bright yellow in KOH. *Otidea subpurpurea* in having smaller ascospores (9–12 × 4.5–6 μm) and lilac to purplish receptacle surface.

In the two-gene phylogenetic tree (Figure 1), two specimens (K(M)41595 and K(M)156077) from England fall into the clade of *O. cupulata*, which are noted to belong to an unnamed taxon related to *O. bufonia* and *O. subpurpurea* by Parslow et al. [10]. Regrettably, they did not further describe the morphology of these two specimens. According to the ITS sequence similarity analysis (ITS: 98%) and phylogenetic analyses (Figure 1), they may be conspecific with *O. cupulata*. Final confirmation requires morphological observation of these two specimens.

Otidea filiformis C.L. Hou, Y.Y. Xu & H. Zhou, sp. nov. (Figure 5).

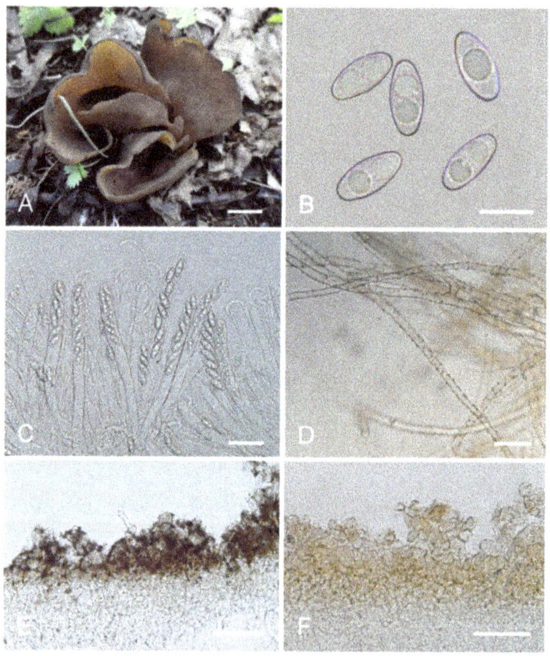

Figure 5. *Otidea filiformis* (BJTC C505) (**A**) apothecia, (**B**) ascospores, (**C**) asci and paraphyses, (**D**) basal mycelium, (**E**) ectal excipulum in water, and (**F**) ectal excipulum in KOH. Scale bars: (**A**) = 1 cm, (**B**) = 10 μm, (**C**) = 30 μm, (**D**) = 20 μm, (**E**,**F**) = 50 μm.

MycoBank: MB843178.

Etymology: *filiformis*, referring to the filiform paraphyses in the hymenium.

Holotype: China. Beijing City, Huairou District, Sunshanzi Village, alt. 770m, on soil in mixed forest of *Populus* sp. and *Larix* sp., 28 August 2020, G.Q. Chen C505 (BJTC C505).

Saprobic on soil. Apothecia gregarious or caespitose in nature, 15–55 mm high, 15–55 mm wide, initially ear-shaped, soon expanding, becoming shallowly or deeply cup-shaped, split, margin sometimes lobate, sessile or shortly stipitate, regular or sometimes undulate in the margin. Hymenium surface yellowish brown (#b19461) to ochre yellow (#cd9575) with pinkish shades, sometimes with brown spots or stains when fresh, margin dark brown when bruised, gray brown (#321f15) when dry, subsmooth. Receptacle surface yellowish brown (#b19461) to orange brown (#967059) when fresh, slightly hygrophanous, dark brown (#321f15) when dry, furfuraceous to finely warty. Stipe not well developed. Basal tomentum and mycelium whitish to pale brown (#dccdbf). Apothecial section 800–1000 μm thick. Ectal excipulum of *textura angularis*, 75–110 μm thick, cells thin walled, brown, 10–35 × 7–26 μm. Medullary excipulum of *textura intricata*, 400–500 μm thick, hyphae 4–10 μm wide, sometimes slightly swollen, thin to thick walled, septate, hyaline to light brown, without resinous exudates. Subhymenium ca. 75–100 μm thick, visible as a brown zone of densely arranged cylindrical to swollen cells, with scattered brown resinous exudate at septa. Paraphyses septate, curved to hooked of uniform width at the apices to 2–3 μm wide, without or with a low notch. Asci 140–175 × 10–14 μm, 8-spored, unitunicate, cylindrical, hyaline, long pedicellate, arising from croziers, non-amyloid, ascospores released from an eccentric split at the apical apex. Ascospores narrowly fusoid, narrowed at both ends, inequilateral, with two large guttules, sometimes only with one big guttule, smooth, hyaline, (12.5–) 13–14.5 (–15) × (6–) 6.5–7 (–7.3) μm ($L_m \times W_m$ = 13.5 × 6.8 μm, Q = 1.9–2.1, Q_m = 2, n = 50). Receptacle surface with broad conical warts, 40–75 μm high, formed by short, fasciculate, hyphoid hairs, of 6–7 subglobose to elongated cells, constricted at septa, 6–10 μm wide. Resinous exudates abundant on the outer surface, dark yellow brown, partly dissolving and converting into small particles in MLZ, partially dissolving and turning yellowish brown in KOH. Basal mycelium of interwoven, 3–6 μm wide, septate, hyaline to pale brown hyphae, turning yellow in KOH, with abundant small, regularly arranged, spheroid, pale brown, resinous exudates, partly dissolving in KOH, unchanged in MLZ.

Other materials examined: China. Hebei Province, Chicheng County, Dahaituo nature reserve, alt. 1640m, on soil under *Betula platyphylla* Suk., 22 August 2019, J.Q. Li L482 (BJTC L482); China. Inner Mongolia Autonomous Region, Balinyou Banner, Saihanwula Nature Reserve, on soil in mixed forest of *Larix gmelinii* (Rupr.) Kuzen. and *Betula platyphylla* Suk., 2 September 2008, T.Z. Liu, H.M. Zhou & C. Sun 3858 (HMAS 188468).

Notes: *Otidea filiformis* diagnosed by the combination of yellowish-brown to ochre-yellow apothecia, sometimes with brown spots or stains, narrowly fusoid ascospores, uniform width, narrow paraphyses (≤3 μm) and basal mycelium with abundant spheroid, pale brown, resinous exudates. *Otidea bufonia* and *O. mirabilis* are similar to *O. filiformis* in apothecia color and ascospore shape. *Otidea bufonia* differs from *O. filiformis* in having the longer ascospores (12–) 13–16.5 (–18) × 6–7.5 (–8) μm, and brown striate exudates on some hyphae of the medullary excipulum. *O. mirabilis* differs in dark-brown apothecia, purple to lilaceous-bluish shades on the receptacle surface and biflabellate crystal-like exudates in the medullary excipulum. The apothecia of *Otidea korfii* Pfister, F. Xu & Z.W. Ge, *O. olivaceobrunnea* and *O. saliceticola* Cartabia, M. Carbone & P. Alvarado also have brown tones. *Otidea korfii* is distinguished from *O. filiformis* by its ear-shaped apothecia, olivaceous brown receptacle surface, ellipsoid to broadly ellipsoid bigger ascospores (14.5–17 × 6.5–9 μm), resinous exudate of the ectal excipulum dissolving in MLZ and smooth basal mycelium. *Otidea olivaceobrunnea* differs from *O. filiformis* in olive-brown hymenium, ellipsoid ascospores. *Otidea saliceticola* differs in pale alutaceous-greyish hymenium surface, dark brown receptacle surface, wider ascospores (14–15 × 7.5–8 μm) with low Q value of 1.75–1.87.

Phylogenetic analyses revealed that the sequences of *O. filiformis* were grouped into an independent clade with a strong support value (Figures 1 and 2). DNA analysis showed that *O. filiformis* shared less than 95.88% ITS sequence similarity with other *Otidea* species. These supported the erection of the new species. One Finnish collection (MCVE 29372) identified as *O. bufonia* by Carbone et al. [22] is clustered into the *O. filiformis* clade with strong support values in our phylogenetic trees (Figures 1 and 2), and it shared more than 98% ITS similarity with our *O. filiformis* and less than 94.6% with *O. bufonia*. This evidence showed MCVE 29372 is more closely related to *O. filiformis*, and whether it is conspecific with *O. filiformis* need further morphological identification.

Otidea khakicolorata L. Fan & Y.Y. Xu, sp. nov. (Figure 6).

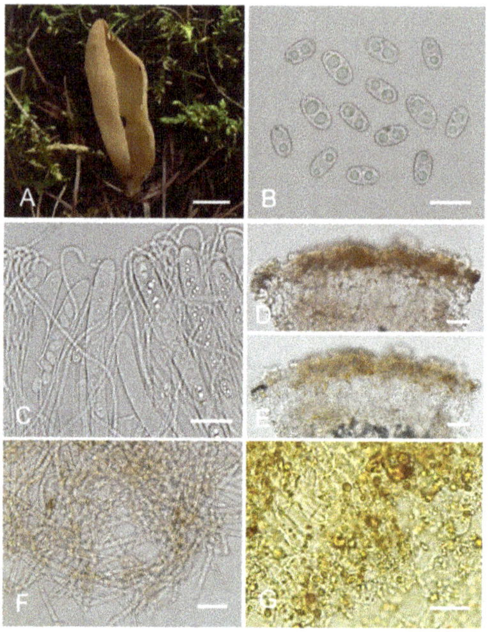

Figure 6. *Otidea khakicolorata* (BJTC FM107) (**A**) apothecia, (**B**) ascospores, (**C**) asci and paraphyses, (**D**) ectal excipulum in KOH, (**E**) ectal excipulum in water, (**F**) basal mycelium, and (**G**) amber drops on the outermost ectal excipulum cells in Melzer's reagent. Scale bars: (**A**) = 0.5 cm, (**B**) = 10 μm, (**C**) = 20 μm, (**D**,**E**) = 25 μm, (**F**,**G**) = 20 μm.

MycoBank: MB843179.

Etymology: *khakicolorata*, referring to the khaki color of apothecia.

Holotype: China. Shanxi Province, Ningwu County, Guancen Mountain, Dashidong Forest Farm, alt. 2200m, among moss under coniferous forest dominated by *Picea wilsonii* Mast. and *Larix principis-rupprechtii* Mayr, 24 August 2017, X.Y. Yan YXY170824 (BJTC FM107).

Saprobic on soil. Apothecia solitary or gregarious in nature, 15–25 mm high, 5–8 mm wide, narrowly long ear-shaped, margin rounded, split, sessile or sub-stipitate. Hymenium surface khaki (#ca9a67) to pale ochre (#c08649) when fresh, yellowish ochre(#e3c57f) when dry, subsmooth. Receptacle surface concolorous with hymenium when fresh, slightly hygrophanous, dark reddish brown (#8c5738) when dry, furfuraceous. Warts absent or very low. Stipe, if present, very short. Basal tomentum and mycelium whitish to grayish whitish (#e7e2d9). Apothecial section 500–800 μm thick. Ectal excipulum of *textura angularis*, 70–120 μm thick, cells thin walled, brown, 10–25 × 6–20 μm. Medullary excipulum of *textura intricata*, 250–500 μm thick, hyphae 3.5–8 μm wide, sometimes slightly swollen, thin to slightly thick walled, septate, hyaline to light brown. Subhymenium ca. 60–90 μm thick, visible as a yellowish-brown zone, of densely arranged cylindrical to swollen cells. Paraphyses septate, curved to hooked, of uniform width at the apices, 2.5–3.7 μm wide, without notch. Asci 150–180 × 8.5–11 μm, 8-spored, unitunicate, cylindrical, hyaline, long pedicellate, arising from croziers, non-amyloid, ascospores released from an eccentric split at the apical apex. Ascospores ellipsoid, sometimes slightly inequilateral, with two large guttules, smooth, hyaline, (8.5–) 9–10 (–10.5) × (4.5–) 5–6 (–6.5) μm ($L_m \times W_m$= 9.5 × 5.5 μm, Q= 1.6–1.9, Q_m= 1.75, n = 50). Receptacle surface with broadly conical warts, 30–40 μm high, formed by hyphoid hairs, of 2–5 subglobose to elongated cells, constricted at septa, 6–10 μm wide, sometimes with a gelatinous sheath. Resinous exudates abundant on the outer surface, yellow brown to dark brown, partly dissolving into amber drops in MLZ, turning reddish brown to dark reddish brown in KOH. Basal mycelium of 3–5.5 μm wide, septate, hyaline to pale-brown hyphae, unchanged in KOH, smooth or with little resinous exudates on the surface, partly dissolving in MLZ, and partly dissolving and more slowly in KOH.

Notes: *Otidea khakicolorata* is characterized by khaki to pale-ochre, long, narrowly ear-shaped apothecia, small ascospores and resinous exudates on the ectal excipulum turning reddish brown in KOH. *Otidea nannfeldtii* and *O. khakicolorata* share similar apothecia shape and the reaction of the resinous exudate in the ectal excipulum and basal mycelium in MLZ and KOH, but *O. nannfeldtii* can be distinguished by an ochre to orangish-ochre hymenium surface with pink tones, higher warts (45–85 μm) on the apothecial outer surface, medullary excipulum of textura intricata differentiated into two parts, and relatively bigger ascospores ((9–) 9.5–10.5 (–11.5) × 5.5–6.5 (–7) μm). *Otidea khakicolorata* is phylogenetically close to *O. nannfeldtii*; however, they are separated by a low support value (Figures 1 and 2). DNA analyses showed that *O. khakicolorata* shared less than 91% similarity in ITS sequence with *O. nannfeldtii*.

Otidea parvula L. Fan & Y.Y. Xu, sp. nov. (Figure 7).

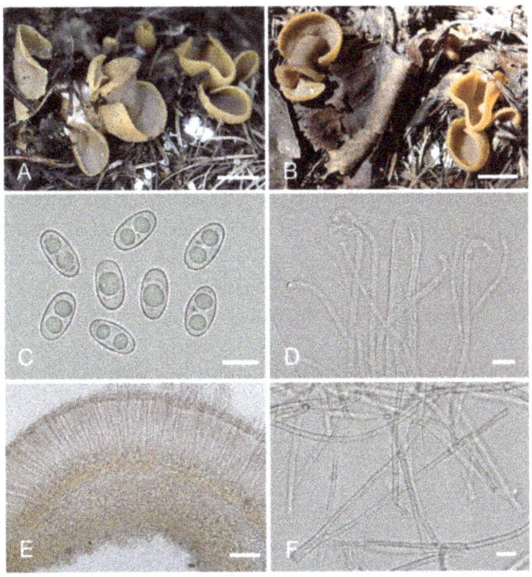

Figure 7. *Otidea parvula* (BJTC FM210-A) (**A**,**B**) apothecia, (**C**) ascospores, (**D**) paraphyses and asci, (**E**) anatomy of apothecium, and (**F**) basal mycelium. Scale bars: (**A**,**B**) = 1 cm, (**C**,**D**) = 10 μm, (**E**) = 100 μm, (**F**) = 10 μm.

MycoBank: MB843180.

Etymology: *parvula*, referring to the small apothecia of the fungus.

Holotype: China. Shanxi Province, Jiaocheng County, Guandi Mountain, Pangquangou National Nature Reserve, alt. 2000m, on soil in the mixed forest dominated by *Picea wilsonii* Mast., 7 September 2017, J.Z. Cao, Cao170803 (BJTC FM210-A).

Saprobic on soil. Apothecia gregarious to caespitose in nature, 8–15 mm high, 5–13 mm wide, initially narrowly to broadly ear-shaped, margin rounded, then expanding and sometimes becoming irregularly ear-shaped or cup-shaped, split, stipitate, or sessile. Hymenium surface pale whitish ochre (#ffffed) to ochre yellow (#b39a7f) when fresh, pale ochre brown (#c2a461) when dry, subsmooth. Receptacle surface pale ochre (#d7c498) to orangish ochre (#dba35e) when fresh, hygrophanous, light yellowish brown (#d7c498) when dry, furfuraceous. Stipe 3–7 × 3–5 mm. Basal tomentum and mycelium white. Apothecial section 500–700 μm thick. Ectal excipulum of *textura angularis*, 70–120 μm thick, cells thin walled, pale brown, 9–24 × 7–20 μm. Medullary excipulum of *textura intricata*, 80–160 μm thick, hyphae 3–7 µm wide, sometimes slightly swollen, thin to thick walled, septate, hyaline to light brown. Subhymenium c. 50–80 µm thick, visible as a brown zone, of densely arranged cylindrical to swollen cells, with scattered brown resinous exudate at septa. Paraphyses septate, bent to curved, sometimes straight, of uniform width or slightly enlarged at the apices, 2.5–4.5 µm wide, sometimes with 1–2 notches near the apex. Asci 150–200 × 10.5–13.5 µm, 8-spored, unitunicate, cylindrical, hyaline, long pedicellate, arising from croziers, non-amyloid, ascospores released from an eccentric split at the apical apex. Ascospores ellipsoid, slightly inequilateral, with two large guttules, sometimes with only one big guttule, smooth, hyaline, (12.5–) 13–15.5 (–16) × (6.5–) 6.8–8 (–8.6) µm ($L_m \times W_m$ = 14.3 × 7.5 µm, Q = 1.7–2.1, Q_m = 1.9, n = 50). Receptacle surface with low warts, 30–50 µm high, formed by short, fasciculate, hyphoid hairs, of 5–6 subglobose to elongated cells, constricted at septa, 5–10 µm wide. Resinous exudates absent. Basal mycelium of 2.5–5 µm wide, septate, hyaline to pale brown hyphae, smooth, unchanged in KOH, turning yellow in MLZ.

Other materials examined: China. Shanxi Province, Jiaocheng County, Guandi Mountain, Pangquangou National Nature Reserve, alt. 2000m, on soil in the mixed forest dominated by *Picea wilsonii*, 7 September 2017, J.Z. Cao, Cao170803 (BJTC FM210-B).

Notes: *Otidea parvula* is easily recognized by the stipitate, irregularly ear-shaped or cup-shaped, small, pale-whitish-ochre, ochre-yellow to orangish-ochre apothecia, straight to bent paraphyses and furfuraceous receptacle surface. *Otidea parvula* and *O. adorniae* Agnello, M. Carbone & P. Alvarado are somewhat similar in the color of the apothecia, but *O. adorniae* differs in its larger apothecia and smaller ascospores (11.8 × 6.4 µm). *Otidea parvula* and *O. parvispora* (Parslow & Spooner) M. Carbone, Agnello, Kautmanová, Z.W. Ge & P. Alvarado have the highest ITS sequence similarity of 96%, but upon examination of the phylogenetic tree (Figures 1 and 2), they don't seem to be closely related, and *O. parvispora* is easily distinguished by its pale hymenium and smaller ascospores ((11.0–) 11.5–13.0 × 5.0–6.5 µm).

Otidea plicara L. Fan & Y.Y. Xu, sp. nov. (Figure 8).

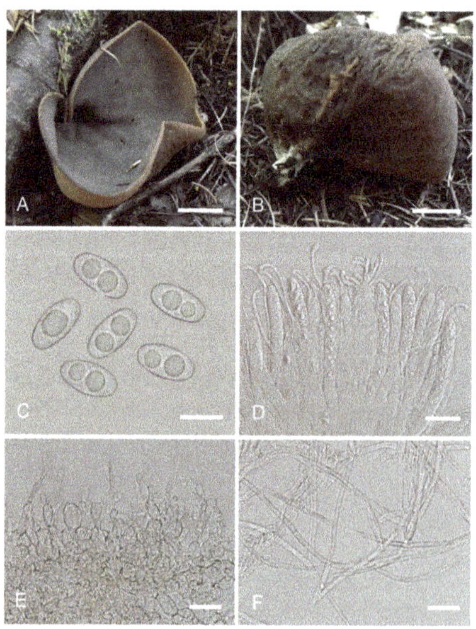

Figure 8. *Otidea plicara* (BJTC FM262-A) (**A**,**B**) apothecia, (**C**) ascospores, (**D**) paraphyses and asci, (**E**) ectal excipulum in water, and (**F**) basal mycelium. Scale bars: (**A**,**B**) = 1 cm, (**C**) = 10 µm, (**D**,**E**) = 30 µm, (**F**) = 20 µm.

MycoBank: MB843181.

Etymology: *plicara*, referring to the small fold on the apothecia.

Holotype: China. Shanxi Province, Jiaocheng County, Guandi Mountain, Badaogou valley, alt. 2000m, on soil in the mixed forest dominated by *Picea wilsonii*, 7 September 2017, J.Z. Cao, Cao170855 (BJTC FM262-A).

Saprobic on soil. Apothecia solitary or gregarious in nature, 15–28 mm high, 12–42 mm wide, initially spoon-shaped or broadly ear-shaped, soon expending, in the end almost deeply cup-shaped, often broader above, margin entire, with a small fold on the apothecia, seemingly spilt yet not split, stipitate. Hymenium surface pale greyish brown (#6f6a61) to dark brown (#665856), margin yellow ochre (#837050) when fresh, when dry becoming slightly lighter but dull, light brown (#ab9876), subsmooth. Receptacle surface dark reddish brown (#7c6052) when fresh, slightly hygrophanous, pale ochre brown (#a48e6a) when dry, finely furfuraceous, wrinkle veined at the base. Stipe 8–15 × 5–8 mm. Basal tomentum and mycelium abundant, white to pale cream (#f1f9ed). Apothecial section 650–1000 µm thick. Ectal excipulum of *textura angularis*, 60–100 µm thick, cells thin walled, brown, 11–34 × 9–28 µm. Medullary excipulum of *textura intricata*, 300–500 µm thick, hyphae 3–7.5 µm wide, sometimes slightly swollen, thin walled, septate, hyaline to light brown. Subhymenium ca. 80–130 µm thick, visible as a yellowish-brown zone. Paraphyses septate, curved to hooked, usually enlarged at the apices, 3.5–5.5 µm wide at apex, 2–3 µm below. Asci 160–220 × 10–14 µm, 8-spored, unitunicate, cylindrical, hyaline, long pedicellate, arising from croziers, non-amyloid, ascospores released from an eccentric split at the apical apex. Ascospores ellipsoid, sometimes slightly inequilateral, with one to two large guttules, smooth, hyaline, (12.5–) 13.5–16 (–17) × (6–) 6.5–8 (–8.5) µm ($L_m \times W_m$ = 14.7 × 7.4 µm, Q = 1.8–2.2, Q_m = 2, n = 50). Receptacle surface with hyphoid hairs, 50–80 µm long, of 3–6 ovoid or subglobose to elongated cells, constricted at septa, 4–9 µm wide. Resinous exudates absent. Basal mycelium of interwoven, 2.5–6 µm wide, septate, hyaline to pale brown hyphae, unchanged in KOH, smooth, turning yellow in MLZ.

Other materials examined: China. Shanxi Province, Jiaocheng County, Guandi Mountain, Badaogou Scenic Area, alt. 2000m, on soil in the mixed forest dominated by *Picea wilsonii* Mast., 7 September 2017, J.Z. Cao, Cao170855 (BJTC FM262-B).

Notes: *Otidea plicara* is characterized by greyish-brown to dark-brown, stipitate, rarely split, deeply cup-shaped apothecia, small ascospores, enlarged paraphyses and the lack of resinous exudates on the ectal excipulum and basal mycelium. Macroscopically, *Otidea apophysata* (Cooke & W. Phillips) Sacc. and *O. platyspora* Nannf. have similar apothecial shape and color to *O. plicara*, but *O. apophysata* can be distinguished by the larger ascospores (20–24.5 × 9–11 µm) and frequently branched paraphyses. *Otidea platyspora* can be distinguished by split apothecia and larger ascospores (18–22 × (9.5–)10.5–12 µm). DNA analyses showed that *O. plicara* shared less than 92% similarity in ITS sequence with other species of *Otidea*. Phylogenetic analyses revealed that the sequences of *O. plicara* were grouped into an independent clade with a strong support value (Figures 1 and 2). These supported the erection of the new species.

Otidea purpureobrunnea L. Fan & Y.Y. Xu, sp. nov. (Figure 9).

MycoBank: MB843182.

Etymology: *purpureobrunnea*, referring to the purple-brown tone of apothecia.

Holotype: China. Shanxi Province, Qinshui County, Tuwo Township, Shangwoquan Village, alt. 1200m, on soil under *Quercus* sp., 25 August 2020, H. Liu 1065 (BJTC FM1048).

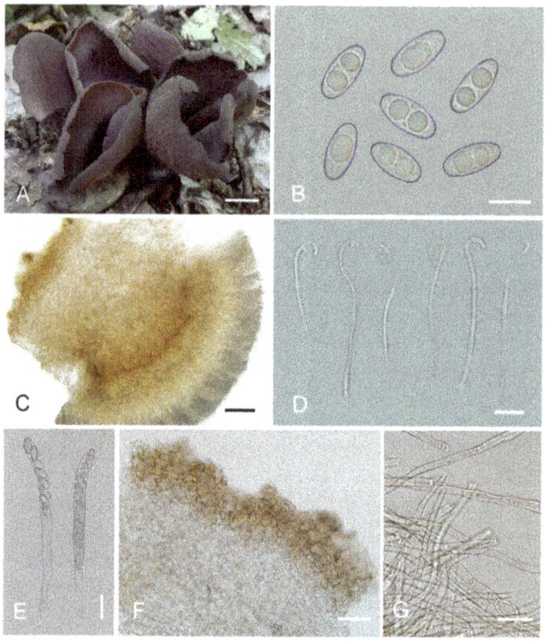

Figure 9. *Otidea purpureobrunnea* (BJTC FM1048) (**A**) apothecia, (**B**) ascospores, (**C**) anatomy of apothecium, (**D**) paraphyses, (**E**) asci, (**F**) ectal and medullary excipulum in water, and (**G**) basal mycelium. Scale bars: (**A**) = 1 cm, (**B**) = 10 μm, (**C**) = 100 μm, (**D**,**E**) = 20 μm, (**F**) = 50 μm, (**G**) = 20 μm.

Saprobic on soil. Apothecia gregarious to caespitose in nature, 25–55 mm high, 50–80 mm wide, initially ear shaped, soon expanding, becoming broadly ear shaped or deeply cup shaped, often elongated on one side, split, margin sometimes lobate, stipitate or sessile. Hymenium surface ochraceous brown (#804618), grayish purple (#5e4f5f) to purple brown (#39242f) when fresh, gray brown to dark brown (#3e2c1c) when dry, subsmooth. Receptacle surface grayish purple brown (#816e71) to dark purple brown (#483131) when fresh, sometimes partly dark yellow brown, slightly hygrophanous, some apothecia with shallowly wrinkled, dark brown (#492615) when dry, furfuraceous to finely warty. Stipe 5–10 × 4–8 mm. Basal tomentum and mycelium whitish to pale brown (#dccdbf). Apothecial section 900–1300 μm thick. Ectal excipulum of *textura angularis*, 80–120 μm thick, cells thin walled, brownish, 13–33 × 7–26 μm. Medullary excipulum of *textura intricata*, 500–900 μm thick, formed of loosely woven cylindrical to slightly swollen thin-walled hyphae, 4.5–11 μm wide, septate, hyaline to light brown, with brown resinous exudates at septa. Subhymenium c. 100–150 μm thick, visible as a brown zone, of densely arranged cylindrical to swollen cells, with scattered brown resinous exudate at septa. Paraphyses septate, curved to hooked, a few curved, sometimes forming a coil or helix, of the same width or often enlarged at the apices, 3.5–5 μm wide, 2–3.3 μm below, sometimes with 1–2 notches, or with an obvious bulge near the apex. Asci 140–190 × 8.5–15 μm, 8-spored, unitunicate, cylindrical, hyaline, long pedicellate, arising from croziers, non-amyloid, ascospores released from an eccentric split at the apical apex. Ascospores ellipsoid to slightly subfusoid, inequilateral, with two large guttules, sometimes with only one big guttule, smooth, hyaline, (12.5–) 13–15 (–15.5) × (6–) 6.5–7 (–7.5) μm ($L_m \times W_m$= 14 × 6.5 μm, Q= 1.9–2.3, Q_m= 2.1, n = 50). Receptacle surface with broad conical warts, 35–60 μm high, formed by short, fasciculate, hyphoid hairs, of 2–5 subglobose to elongated cells, constricted at septa, 6–13 μm wide. Resinous exudates abundant on the outer surface, yellow brown to dark brown, partly dissolving into particles in MLZ, entirely dissolving and turning yellow in KOH. Basal mycelium of 3.5–6 μm wide, septate, hyaline to pale brown hyphae, turning yellow in KOH, mostly smooth, a few with very small, spheroid, pale-brown, resinous exudates, dissolving in KOH, partially dissolving in MLZ.

Other materials examined: China. Shanxi Province, Qinshui County, Tuwo Township, Shangwoquan Village, alt. 1200m, on soil under *Quercus* sp., 25 August 2020, H. Liu 1079 (BJTC FM1061).

Notes: *Otidea purpureobrunnea* is characterized by the stipitate, broadly ear-shaped to cup-shaped, grayish-purple to purple-brown apothecia, ellipsoid to slightly subfusoid ascospores, paraphyses enlarged at the apices, with 1–2 notches or an obvious bulge and smooth basal mycelium. Similar to *O. purpureobrunnea*, the apothecia of *O. bufonia*, *O. cupulata*, *O. mirabilis*, *O. purpurea*, *O. purpureogrisea*, *O. smithii*, and *O. subpurpurea* all have some purple tones, but *Otidea bufonia* differs in its fusoid ascospores, the presence of hyphae with striate resinous exudates in the medullary excipulum, resinous exudates of the ectal excipulum not turning bright yellow in KOH, and in having abundant resinous exudates on the basal mycelium. *Otidea mirabilis* differs by having fusoid ascospores, resinous exudates of the ectal excipulum that do not turn bright yellow in KOH and when present, biflabellate, crystal-like exudates in the medullary excipulum. *Otidea purpurea* and *O. subpurpurea* are easily distinguished by the obviously smaller spores (*O. subpurpurea*: 9–12 × 4.5–6 μm; *O. purpurea*: 8–10 × 4.5–6 μm). *Otidea purpureogrisea* is distinguished by the purple-gray tone of the receptacle surface near the base and resinous exudates of the ectal excipulum turning amber in MLZ and turning brown in KOH. *Otidea smithii* is distinguished by typically narrower, ear-shaped apothecia, relatively shorter ascospores (12–14 × 6–7.5 μm) with a lower Q_m value (1.9–2), and resinous exudates of the ectal excipulum not turning bright yellow in KOH. For a comparison with *O. cupulata* see under that species below.

Phylogenetic analyses revealed that *O. purpureobrunnea* and *O. filiformis* are grouped together with a low support value (Figure 2), but *O. filiformis* is easy to distinguish from *O. purpureobrunnea* by its apothecia without purple tones, fusoid ascospores, same width and narrow paraphyses (≤3 μm), as well as its basal mycelium with abundant spheroid, pale brown, resinous exudates. DNA analysis showed that *O. purpureobrunnea* shared less than 94.53% similarity in its ITS sequence with *O. filiformis*. These indicate that they are two different species.

Otidea subpurpurea W.Y. Zhuang, Mycologia Montenegrina 10: 238 (2007).

Holotype: China, Yunnan Province, Kunming City, Kunming Institute of Botany, alt. 1980m, 8 October 2005, Z.L. Yang 4602, (HKAS 49443); Isotype (HMAS 97530).

= *Otidea bicolor* W.Y. Zhuang & Zhu L. Yang, Mycotaxon 112: 35 (2010).

Holotype: China, Yunnan Province, Kunming City, Heilongtan Park, 16 August 2008, Z.L. Yang 5156, (HKAS 54453); Isotype (HMAS 188415).

= *Otidea pruinosa* Ekanayaka, Q. Zhao & K.D. Hyde, Fungal Diversity 87: 130 (2017).

Holotype: China, Yunnan Province, Kunming City, Xishan Scenic Area, 15 September 2012, T. Guo 617, (HKAS 81819).

Materials examined: China, Yunnan Province, Kunming City, Kunming Institute of Botany, alt. 1980m, 9 October 2005, Z.L. Yang 4602, (HKAS 49443); Isotype (HMAS 97530). China, Yunnan Province, Kunming City, Heilongtan Park, 16 August 2008, Z.L. Yang 5156, (HKAS 54453); Isotype (HMAS 188415). ibid., (HKAS 54449). China, Yunnan Province, Kunming City, Xishan Scenic Area, 15 September 2012, T. Guo 617, (HKAS 81819).

Notes: *Otidea bicolor*, *O. pruinosa* and *O. subpurpurea* are highly similar species. In fact, previous scholars have also noticed the phenomenon that the type sequences of the three species are clustered together [22,25]; however, due to the unavailability of specimens, this issue has not been formally addressed. In this study, we examined the type specimens and obtained multiple loci sequences from them. DNA analyses revealed that *O. pruinosa*, *O. bicolor*, and *O. subpurpurea* share high sequence similarity (ITS: >98.87%; nrLSU: >99.53%; *tef1-α*: >99.72%; *rpb2*: >99.45%). We performed morphological observation on these type specimens and found that there was no obvious difference in microscopic features. The reaction of the resinous exudate in the ectal excipulum and basal mycelium in MLZ and

KOH are also the same. Although the receptacle surface of *O. bicolor* and *O. subpurpurea* is purplish in tint when fresh, the receptacle surface of *O. pruinosa* is without a purplish tint [23,28,29], but that may be influenced by its habitat. *Otidea pruinosa* is proposed as a new species because of receptacle surface with pruinose, but we found a similar granulate on the surface of dry specimens of *O. bicolor* and *O. subpurpurea*. In addition, phylogenetic analyses based on the two-gene and four-gene datasets also confirmed that they represent the same species, so here we formally treat *O. bicolor* and *O. pruinosa* as synonyms of *O. subpurpurea*. The sequence from ZMU124 (label as *O. bufonia*) from Guizhou province of China grouped with *O. subpurpurea* with a high support value (Figure 1), indicating that *O. subpurpurea* seems widely distributed in southwest China. Similarly, the sequence from JS150904-08 from Korea named *O. bufonia* [45] was also grouped into this clade (Figure 1). We checked the original morphological description by Jin et al. [45] and found that its ascospores size does not conform to *O. bufonia*, but instead to *O. subpurpurea*. This indicates that *O. subpurpurea* also occurs in Korea.

Otidea mirabilis Bolognini & Jamoni in Jamoni, Funghi e Ambiente 85–86: 56 (2001). (Figure 10).

Figure 10. Two new record species from China. (**A**) *Otidea mirabilis* (HSA 234), (**B**) *Otidea nannfeldtii* (BJTC FM236). Scale bars: (**A**,**B**) = 1 cm.

Habitat: on soil under mixed forest of *Larix principis-rupprechtii* and *Betula* sp.

Distribution: Known from the northeast, northern, northwest and southwest regions of China.

Materials examined: China, Yunnan Province, Jingdong County, Ailao Mountain, Xujiaba Village, alt. 2500m, 24 August 1994, M. Zang, 12389 (HKAS 28129). China, Gansu Province, Wudu County, Liangshui Town, Gongba River Beach, alt. 2600m, 11 July 1996, M.S. Yuan, 2213 (HKAS 30708). China, Sichuan Province, Hongyuan County, Kangle Town, alt. 3400m, 19 August 1998, M.S. Yuan, 3433 (HKAS 33633). China, Jilin Province, Fusong County, Songjiang River, 19 August 2000, M.S. Yuan, 4725 (HKAS 37272). China, Xinjiang Autonomous Region, Jimusa'er, alt. 1700m, 1 August 2003, W.Y. Zhuang & Y. Nong, 4657 (HMAS 83568). China, Inner Mongolia Autonomous Region, Chifeng City, Baiyin Aobao National Nature Reserve, 2 August 2013, Tolgor Bau (HMJAU 26926). China, Shanxi Province, Jiaocheng County, Guandi Mountain, Shanshui Village, alt. 1800m, 8 September 2017, J.Z. Cao, CAO170863 (BJTC FM292). China, Shanxi Province, Jiaocheng County, Pangquangou Nature Reserve, alt. 2100m, 28 August 2018, H. Liu, LH234 (HSA 234).

Notes: The occurrence of *O. mirabilis* is confirmed in China based on morphological and DNA evidence in this study. Olariaga et al. [4] showed already that *O. mirabilis* occur in China using nrLSU sequences from two Chinese collections in GenBank (identified as *O. leporina* by Zhuang [17] and Liu and Zhuang [18], but by morphology it has not been previously confirmed, as Olariaga et al. did not study those two collections morphologically. It is interesting that two distinct clades were revealed, one comprising Chinese specimens, and another comprising specimens from Europe. The Chinese specimens shared 98.46–99.84% ITS sequence similarity and the European ones had 99.54–99.85% similarity, while the similarities between the two proveniences were 97–98.5%. However, we found no significant morphological differences between the Chinese and European specimens, which probably resulted from the geographic distance.

Otidea nannfeldtii Harmaja, Karstenia 15: 31 (1976). (Figure 10).

Habitat: on soil under mixed forest of *Larix principis-rupprechtii.*

Distribution: Known in northern China and northwest China.

Materials examined: China, Shanxi Province, Ningwu County, Guancen Mountain, Qiuqiangou Village, alt. 2100m, on soil under *L. principis-rupprechtii* Mayr, 25 August 2017, X.Y. Yan, YXY170836 (BJTC FM168); ibid., X.Y. Yan, YXY170837 (BJTC FM169); ibid., X.Y. Yan, YXY170838 (BJTC FM170). China, Shanxi Province, Jiaocheng County, Guandi Mountain, Pangquangou Nature Reserve, alt. 2000m, 6 September 2017, J.Z. Cao, CAO170829 (BJTC FM236); ibid., J.Z. Cao, CAO170836 (BJTC FM243). China, Xinjiang Autonomous Region, Jimusa'er, alt. 1700m, 1 August 2003, W.Y. Zhuang & Y. Nong, 4655 (HMAS 83573).

Notes: The occurrence of *O. nannfeldtii* in China is first confirmed based on molecular and morphological evidence. *Otidea nannfeldtii* is originally described in Europe, and also reported from North America [4]. Before this study, there are no DNA data that support the existence of this species in China.

4. Discussion

Temperate China is surely rich in *Otidea* species. Nine species are added to this genus by this study. A total of 31 species is thus recorded in this huge country currently. Of these species, 27 species are supported by morphological and molecular data, but four species (*O. cochleata* (L.) Fuckel, *O. purpurea* (M. Zang) Korf & W.Y. Zhuang, *O. smithii*, *O. tianshuiensis* J.Z. Cao, L. Fan & B. Liu) still lack DNA evidence. Compared to the records from the continents of Europe (c. 32 accepted species) and North America (c. 14 accepted species), more studies of this large and widely distributed temperate fungal group in China are needed. From the present point of view, the *Otidea* species is widely distributed in the southwest and northern regions of China. Four species, *O. alutacea*, *O. bufonia*, *O. mirablilis*, and *O. onotica*, are widely distributed and are often encountered in the wild. So far, almost no *Otidea* species have been reported from south-central China and east China, which also have abundant forest resources, so it is necessary to investigate fungal resources in these regions in the future.

Key to species of *Otidea* in the study

1. Apothecia entire——————————————————————————2
1. Apothecia split——————————————————————————3
2. Apothecia broadly cup-shaped, ochre brown to reddish brown, with abundant resinous exudates on basal mycelium and ectal excipulum, ascospore lenth > 18 µm———————————————————————*O. propinquata*
2. Apothecia deeply cup-shaped, with a small fold, pale greyish brown to dark brown, ectal excipulum and basal mycelium without resinous exudates, ascospore lenth ≤17 µm———————————*O. plicara*
3. Apothecia long, narrowly ear-shaped——————————————————4
3. Apothecia cup-shaped or broadly ear-shaped——————————————6
4. Apothecia khaki to pale ochre, receptacle surface with warts of 30–40 µm high, ascospores (8.5–) 9–10 (–10.5) × (4.5–) 5–6 (–6.5) µm ——————————*O. khakicolorata*
4. Apothecia cinnamon brown or yellowish ochre to brown, receptacle surface with warts of 45–85 µm high——————————————————————————5
5. Ascospore length < 12 µm, resinous exudates of the ectal excipulum turning reddish brown in KOH——————————————————————*O. nannfeldtii*
5. Ascospore length > 12 µm, resinous exudates of the ectal excipulum turning yellowish reddish grey heterogeneous drops in KOH————————————*O. leporina*
6. Apothecia pale yellow, yellowish brown, ochraceous yellow, ochre orange————7
6. Apothecia brown, dark brown, dark reddish brown or purple brown——————10
7. Apothecia small-sized (<1.5cm), ascospores (12.5–) 13–15.5 (–16) × (6.5–) 6.8–8 (–8.6) µm——————————————————————————*O. parvula*
7. Apothecia big-sized (>2cm)——————————————————————8
8. Resinous exudates of the ectal excipulum and basal mycelium absent———*O. asperior*
8. Resinous exudates of the ectal excipulum and basal mycelium present——————9
9. Hymenium light yellow to ochraceous yellow, often with pink tones, ascospore length < 11 µm——————————————————————*O. brevispora*
9. Hymenium yellow to dull yellow, ascospore length > 11 µm——————*O. onotica*
10. Resinous exudates of the ectal excipulum and basal mycelium absent——*O. alutacea*
10. Resinous exudates of the ectal excipulum and basal mycelium present——————11
11. Apothecia yellowish brown, without purple tones——————————*O. filiformis*
11. Apothecia yellowish brown to dark brown, or purple brown, with purple to lilaceous-bluish tones——————————————————————————12
12. Ascospores narrowly fusoid——————————————————————13
12. Ascospores ellipsoid to slightly subfusoid————————————————14
13. Receptacle surface mostly without purple tones, medullary excipulum with striate exudates covering some hyphae————————————————————*O. bufonia*
13. Receptacle surface strikingly purple-violaceous (fresh); medullary excipulum without, or rarely with flabellate crystal-like exudates, forming cross-like aggregates——————————————————————————*O. mirabilis*
14. Ascospore length <12 µm——————————————————*O.subpurpurea*
14. Ascospore length >12 µm——————————————————————15
15. Receptacle surface with purple tones, grayish purple brown to dark purple brown, basal mycelium smooth, or with a few pale brown resinous exudates——————————————————————*O. purpureobrunnea*
15 Receptacle surface without purple tones, yellowish brown to brown, basal mycelium with abundant resinous exudates ——————————————————*O. cupulata*

Supplementary Materials: The following supporting information can be downloaded at: https://www.mdpi.com/article/10.3390/jof8030272/s1, Table S1: Information on sequences used in molecular phylogenetic analyses for Otidea; File S1: Aligned, concatenated dataset of Otidea, consisting of two partitions (ITS, nrLSU).; File S2: Aligned, concatenated dataset of Otidea, consisting of four partitions (ITS, nrLSU, tef-α, rpb2).

Author Contributions: L.F. conceived and designed the study; Y.-Y.X. and N.M. wrote the manuscript; Y.-Y.X. conducted phylogenetic analyses and morphological observations; N.M. and J.-J.Y. conducted the experiments. All authors have read and agreed to the published version of the manuscript.

Funding: This study was supported by the National Natural Science Foundation of China (No. 31750001) and the Beijing Natural Science Foundation (No. 5172003).

Institutional Review Board Statement: Not applicable.

Informed Consent Statement: Not applicable.

Data Availability Statement: The sequencing data were submitted to GenBank.

Acknowledgments: J.Z. Cao is appreciated for collecting specimens and providing valuable suggestions. Many thanks to the curator Lei Cai and staff Zhuo Du (HMAS), to the curator Tao Deng and staff Wen-Zhang Ma (HKAS), and to the curator Tolgor Bau (HMJAU) for loan of specimens used in this study.

Conflicts of Interest: The authors declare no conflict of interest.

References

1. Bonorden, H.F. *Handbuch der Allgemeinen Mykologie als Anleitung zum Studium Derselben: Nebst Speciellen Beiträgen zur Vervolkommung Dieses Zweiges der Naturkunde;* Schweizerbart: Stuttgart, Germany, 1851.
2. Fuckel, L. Symbolae mycologicae. *Jahrbücher des Nassauischen Vereins für Naturkunde* **1870**, *23–24*, 1–459.
3. Hansen, K.; Olariaga, I. Species limits and relationships within Otidea inferred from multiple gene phylogenies. *Pers. Mol. Phylogeny Evol. Fungi* **2015**, *35*, 148–165. [CrossRef]
4. Olariaga, I.; Van Vooren, N.; Carbone, M.; Hansen, K. A monograph of *Otidea* (Pyronemataceae, Pezizomycetes). *Pers. Mol. Phylogeny Evol. Fungi* **2015**, *35*, 166–229. [CrossRef] [PubMed]
5. Kanouse, B.B. Studies in the Genus *Otidea*. *Mycologia* **1949**, *41*, 660–677. [CrossRef]
6. Cao, J.Z.; Fan, L.; Liu, B. Some species of *Otidea* from China. *Mycologia* **1990**, *82*, 734–741. [CrossRef]
7. Peterson, E.T. Systematics of the Genus *Otidea* in the Pacific Northwest. Master's Thesis, Oregon State University, Corvallis, OR, USA, 1998.
8. Zhuang, W.Y. *Flora Fungorum Sinicorum Vol. 48—Pyronemataceae;* Science Press: Beijing, China, 2014; pp. 92–118.
9. Parslow, M.; Spooner, B.M. The British species of *Otidea*: Overview and the large-spored species. *Mycosystema* **2013**, *32*, 347–365.
10. Parslow, M.; Suz, L.M.; Spooner, B.M. The British species of *Otidea* (3): Taxa present in Britain. *Ascomycete.org* **2019**, *11*, 256–284. [CrossRef]
11. Nannfeldt, J.A. On *Otidea caligata*, *O. indivisa* and *O. platyspora* (Discomycetes, Operculatae). *Ann. Bot. Fenn.* **1966**, *3*, 309–318.
12. Harmaja, H. *Flavoscypha*, a new genus of the Pezizales for *Otidea cantharella* and *O. phlebophora*. *Karstenia* **1974**, *14*, 105–108. [CrossRef]
13. Harmaja, H. New species and combinations in the genera *Gyromitra*, *Helvella* and *Otidea*. *Karstenia* **1976**, *15*, 29–32. [CrossRef]
14. Harmaja, H. Studies on the Pezizales. *Karstenia* **1986**, *26*, 41–48. [CrossRef]
15. Harmaja, H. Studies in *Otidea* (Pezizales). *Karstenia* **2009**, *48*, 33–48. [CrossRef]
16. Korf, R.P.; Zhuang, W.Y. Some new species and new records of discomycetes in China. *Mycotaxon* **1985**, *22*, 483–514.
17. Zhuang, W.Y. Notes on *Otidea* from Xinjiang, China. *Mycotaxon* **2005**, *94*, 365–370.
18. Liu, C.Y.; Zhuang, W.Y. Relationships among some members of the genus *Otidea* (Pezizales, Pyronemataceae). *Fungal Divers.* **2006**, *23*, 181–192. [CrossRef]
19. Smith, M.E.; Healy, R.A. *Otidea subterranea* sp. *nov.*: *Otidea* goes below ground. *Mycol. Res.* **2009**, *113*, 858–866. [CrossRef]
20. Parslow, M.; Spooner, B.M. The British species of *Otidea* (2), *O. alutacea* and related taxa. *Ascomycete.org* **2015**, *7*, 295–302. [CrossRef]
21. Carbone, M.; Cartabia, M.; Alvarado, P. *Otidea saliceticola* (Pezizales) a new species from the Italian Alps. *Ascomycete.org* **2017**, *9*, 215–224. [CrossRef]
22. Carbone, M.; Agnello, C.; Kautmanová, C.; Ge, Z.W.; Alvarado, P. Phylogenetic and morphological studies in *Otidea alutacea* and *O. bufonia* clades (Pezizales), with the new species *Otidea adorniae*. *Ascomycete.org* **2019**, *11*, 117–126. [CrossRef]
23. Hyde, K.D.; Norphanphoun, C.; Abreu, V.P.; Bazzicalupo, A.; Chethana, K.W.T.; Clericuzio, M.; Dayarathne, M.C.; Dissanayake, A.J.; Ekanayaka, A.H.; He, M.-Q.; et al. Fungal diversity notes 603–708: Taxonomic and phylogenetic notes on genera and species. *Fungal Divers.* **2017**, *87*, 1–235. [CrossRef]
24. Hyde, K.D.; Chaiwan, N.; Norphanphoun, C.; Boonmee, S.; Camporesi, E.; Chethana, K.W.T.; Dayarathne, M.C.; de Silva, N.I.; Dissanayake, A.J.; Ekanayaka, A.H.; et al. Mycosphere notes 169–224. *Mycosphere* **2018**, *9*, 271–430. [CrossRef]
25. Xu, F.; Ge, Z.-W.; LoBuglio, K.F.; Pfister, D.H. *Otidea* species from China, three new species with comments on some previously described species. *Mycol. Prog.* **2018**, *17*, 77–88. [CrossRef]
26. Zhuang, W.Y.; Korf, R.P. Some new species and new records of discomycetes in China. III. *Mycotaxon* **1989**, *35*, 297–312.
27. Harmaja, H. A note on *Otidea* (Pezizales, Fungi). *Phytotaxa* **2009**, *2*, 49–50. [CrossRef]
28. Zhuang, W.-Y. Taxonomic assessment of some pyronemataceous fungi from China. *Mycotaxon* **2010**, *112*, 31–46. [CrossRef]
29. Zhuang, W.Y.; Yang, Z.L. Some pezizalean fungi from alpine areas of southwestern China. *Mycologia Montenegrina* **2007**, *10*, 235–249.

30. Dring, D.M. Techniques for microscopic preparation. In *Methods in Microbiology*; Booth, C., Ed.; Academic Press: New York, NY, USA, 1971; Volume 4.
31. Gardes, M.; Bruns, T.D. Its primers with enhanced specificity for basidiomycetes—Application to the identification of mycorrhizae and rusts. *Mol. Ecol.* **1993**, *2*, 113–118. [CrossRef]
32. White, T.J.; Bruns, T.; Lee, S.; Taylor, J. Amplification and direct sequencing of fungal ribosomal RNA genes for phylogenetics. In *PCR Protocols*; Innis, M.A., Gelfand, D.H., Sninsky, J.J., White, T.J., Eds.; Academic Press: San Diego, CA, USA, 1990; pp. 315–322.
33. Vilgalys, R.; Hester, M. Rapid genetic identification and mapping of enzymatically amplified ribosomal DNA from several *Cryptococcus species*. *J. Bacteriol.* **1990**, *172*, 4238–4246. [CrossRef]
34. Rehner, S.A.; Buckley, E. A Beauveria phylogeny inferred from nuclear ITS and EF1-α sequences: Evidence for cryptic diversification and links to Cordyce psteleomorphs. *Mycologia* **2005**, *97*, 84–98. [CrossRef]
35. Katoh, K.; Standley, D.M. Mafft Multiple Sequence Alignment Software Version 7: Improvements in Performance and Usability. *Mol. Biol. Evol.* **2013**, *30*, 772–780. [CrossRef]
36. Rambaut, A. Estimating the rate of molecular evolution: Incorporating non-contemporaneous sequences into maximum likelihood phylogenies. *Bioinformatics* **2000**, *16*, 395–399. [CrossRef] [PubMed]
37. Castresana, J. Selection of Conserved Blocks from Multiple Alignments for Their Use in Phylogenetic Analysis. *Mol. Biol. Evol.* **2000**, *17*, 540–552. [CrossRef] [PubMed]
38. Vaidya, G.; Lohman, D.J.; Meier, R. Sequence Matrix: Concatenation software for the fast assembly of multi-gene datasets with character set and codon information. *Cladistics* **2011**, *27*, 171–180. [CrossRef] [PubMed]
39. Stamatakis, A. RAxML version 8: A tool for phylogenetic analysis and post-analysis of large phylogenies. *Bioinformatics* **2014**, *30*, 1312–1313. [CrossRef]
40. Ronquist, F.; Huelsenbeck, J.P. MrBayes 3: Bayesian phylogenetic inference under mixed models. *Bioinformatics* **2003**, *19*, 1572–1574. [CrossRef]
41. Nylander, J. *MrModeltest 2.2*; Evolutionary Biology Centre, University of Uppsala: Uppsala, Sweden, 2004.
42. Hillis, D.M.; Bull, J.J. An Empirical Test of Bootstrapping as a Method for Assessing Confidence in Phylogenetic Analysis. *Syst. Biol.* **1993**, *42*, 182–192. [CrossRef]
43. Alfaro, M.E.; Zoller, S.; Lutzoni, F. Bayes or Bootstrap? A Simulation Study Comparing the Performance of Bayesian Markov Chain Monte Carlo Sampling and Bootstrapping in Assessing Phylogenetic Confidence. *Mol. Biol. Evol.* **2003**, *20*, 255–266. [CrossRef]
44. Page, R.D. *TreeView*; Glasgow University: Glasgow, Scotland, 2001.
45. Jin, S.L.; Sun, Y.C.; Kim, C.; Hyany, B.L. Twelve Undescribed Species of Macrofungi from Korea. *Korean J. Mycol.* **2016**, *44*, 233–239. [CrossRef]

Journal of
Fungi

MDPI

Article

New Species of *Mallocybe* and *Pseudosperma* from North China

Ning Mao †, Yu-Yan Xu †, Tao-Yu Zhao, Jing-Chong Lv and Li Fan *

College of Life Science, Capital Normal University, Xisanhuanbeilu 105, Haidian, Beijing 100048, China; 2210801013@cnu.edu.cn (N.M.); 2190801004@cnu.edu.cn (Y.-Y.X.); 2200802110@cnu.edu.cn (T.-Y.Z.); 2200802057@cnu.edu.cn (J.-C.L.)

* Correspondence: fanli@mail.cnu.edu.cn

† These authors contributed equally to this work.

Abstract: Within the family Inocybaceae, many species of *Mallocybe* and *Pseudosperma* have been reported, but there are only a few reports on these two genera from north China. In this study, six collections of *Mallocybe* and 11 collections of *Pseudosperma* were studied by morphological and phylogenetic methods. Phylogenetic analyses based on sequence data from three or two different loci (ITS, LSU, and *rpb2* for *Mallocybe;* ITS and LSU for *Pseudosperma*) are performed to infer species relationships within genera *Mallocybe* and *Pseudosperma*, respectively. Results indicate that eight species of *Mallocybe* and *Pseudosperma* are found in Shanxi province, north China; two new species of *Mallocybe*, *M. depressa* and *M. picea*, are described. Overall, six species belong to *Pseudosperma*, of which three are new: *P. gilvum*, *P. laricis* and *P. pseudoniveivelatum*.

Keywords: inocybaceae; multigene; phylogeny; taxonomy

Citation: Mao, N.; Xu, Y.-Y.; Zhao, T.-Y.; Lv, J.-C.; Fan, L. New Species of *Mallocybe* and *Pseudosperma* from North China. *J. Fungi* **2022**, *8*, 256. https://doi.org/10.3390/jof8030256

Academic Editors: Lei Cai and Cheng Gao

Received: 6 February 2022
Accepted: 1 March 2022
Published: 2 March 2022

Publisher's Note: MDPI stays neutral with regard to jurisdictional claims in published maps and institutional affiliations.

1. Introduction

Inocybaceae Jülich (Basidiomycota, Agaricales) is an ecologically important fungal family, and is estimated to contain 1050 species [1]. These ectomycorrhizal fungi of Inocybaceae form a mutually symbiotic association with as many as 23 families of vascular plants [2]. Inocybaceae was initially considered by many researchers to include only one or two genera [3–9]. Recently, Matheny et al. [2] revised Inocybaceae to include seven genera based on a six-locus phylogeny, namely *Auritella* Matheny & Bougher, *Inocybe* (Fr.) Fr., *Inosperma* (Kühner) Matheny & Esteve-Rav., *Mallocybe* (Kuyper) Matheny, Vizzini & Esteve-Rav., *Nothocybe* Matheny & K.P.D. Latha, *Pseudosperma* Matheny & Esteve-Rav., and *Tubariomyces* Esteve-Rav. & Matheny.

The genus *Mallocybe* was originally described as a subgenus of *Inocybe*. It is elevated to the genus level by Matheny et al. [2], with *Mallocybe terrigena* (Fr.) Matheny, Vizzini & Esteve-Rav. As the type species. The species of this genus are very widely distributed, reported in Africa, Asia, Australia, Europe, New Zealand, and North America [2]. Approximately 56 species are recorded in Index Fungorum [www.indexfungorum.org/Names/Names.asp (accessed on 5 February 2022)]. This genus is mainly characterized by a coarsely fibrillose or woolly-squamulose and often flattened pileus, which becomes noticeably dark upon the application of 5% potassium hydroxide; adnate lamellae; a short stipe; necropigmented basidia; short cheilocystidia (<50 μm long); and the absence pleurocystidia [2,7,10–14]. The genus *Pseudosperma* was originally included in *Inocybe* section *Rimosae* sensu stricto (= clade *Pseudosperma*) [15–17], and traditionally placed in the subgenus *Inosperma*. Now, it is one of the seven genera in Inocybaceae [2]. There are 93 records listed in Index Fungorum, and approximately 70 species are accepted according to Matheny et al. [2]. The species of this genus are characterized by fibrillose or rarely squamulose, often rimose pileus; furfuraceous to furfuraceous–fibrillose stipe, distinctly pruinose stipe apex; adnexed to sinuate lamellae; hyaline or not necropigmented basidia; cylindrical to clavate cheilocystidia; absent pleurocystidia; and spermatic odor [2,17–19].

During an investigation of Inocybaceae fungi in Shanxi (north China), some fruitbodies of *Mallocybe* and *Pseudosperma* were collected. Subsequent morphological examination and molecular analyses showed they represented eight species, including five undescribed species. The aim of this study is to improve the knowledge of the genus *Pseudosperma* and *Mallocybe* by adding descriptions of five new species, and provide the DNA data of three previously described *Pseudosperma* species from China.

2. Materials and Methods

2.1. Morphological Studies

Collections were obtained and photographed in the field from Shanxi and Hebei province in China, dried in a fruit drier at 40–50 °C, and deposited in the herbarium of Capital Normal University, Beijing, China (BJTC) and the Herbarium Institute of Edible Fungi, Shanxi Academy of Agricultural Science, Taiyuan, China (HSA). Standardized color values were obtained from ColorHexa [http://www.colorhexa.com.asp (accessed on 5 February 2022)]. Microscopic characteristics were observed in sections obtained from dry specimens mounted in 3% KOH, Congo Red, or Melzer's reagent [20]. The term '[n/m/p]' means n. basidiospores from m. basidiomata of p collections. Dimensions of basidiospores are given using the following format '(a–)b–c(–d)', where the range 'b–c' represents at least 90% of the measured values, and 'a' and 'd' are the most extreme values. L_m and W_m indicate the average basidiospore length and width (± standard deviation) for the measured basidiospore, respectively. 'Q' refers to the length/width ratio of basidiospores in side-view; 'Q_{av}' refers to the average Q of all basidiospores ± standard deviation.

2.2. DNA Extraction, PCR Amplification, Sequencing

A small amount of basidiomata material (20−30 mg) was crushed by shaking for 45 s at 30 Hz 2−4 times (Mixer Mill MM301, Retsch, Haan, Germany) in a 1.5 mL tube, together with a 3 mm diam tungsten carbide ball. Total genomic DNA was extracted from the powdered basidiomata using NuClean Plant Genomic DNA Kit (CWBIO, Beijing, China), following the manufacturer's instructions. Primers ITS1F and ITS4 were employed for the ITS [21,22], while LR0R and LR5 for LSU [23], and bRPB2-6F and bRPB2-7R2 for the *rpb2* were used [16,24]. Polymerase chain reactions (PCR) for the ITS region, LSU region, and *rpb2* gene were performed in 25 μL reaction containing 2 μL DNA template (concentration: 12−20 ng/μL), 1 μL primer (10 μM) each, 12.5 μL of 2 × Master Mix [Tiangen Biotech (Beijing) Co., Beijing, China], 8.5 μL ddH2O.

PCR reactions were implemented as follows: an initial denaturation at 94 °C for 5 min, then to 35 cycles of the following denaturation at 94 °C for 30 s, annealing at 52 °C for 45 s (ITS), 60 s (LSU and *rpb2*), 72 °C for 1 min; and a final extension at 72 °C for 10 min. The PCR products were sent to Beijing Zhongkexilin Biotechnology Co. Ltd. (Beijing, China) for purification and sequencing. The newly generated sequences were assembled and edited using SeqMan (DNA STAR package; DNAStar Inc., Madison, WI, USA) with generic-level identities for sequences confirmed via BLAST queries of GenBank. These sequences of *Mallocybe* and *Pseudosperma* were mainly selected from those used by previous studies [2,12,14–19,25–32]. The accession numbers of all sequences employed are provided in Supplementary Tables S1 and S2.

2.3. Sequence Alignment and Phylogenetic Analyses

For this study, two datasets were assembled. Dataset I (ITS/LSU/*rpb2*) was used to investigate the phylogenetic placement of the *Mallocybe* species. *Pseudosperma triaciculare* Saba & Khalid and *P. breviterincarnatum* (D.E. Stuntz ex Kropp, Matheny & L.J. Hutchison) Matheny & Esteve-Rav. were selected as outgroup taxon. Dataset II (ITS/LSU) was used to investigate the phylogenetic placement of the *Pseudosperma* species. *Mallocybe velutina* Saba & Khalid and *M. africana* Aïgnon, Yorou & Ryberg were selected as outgroup taxon. The sequences of each marker were independently aligned in MAFFT v.7.110 [33] under default parameters, and edited by BioEdit 1.8.1. Maximum Likelihood (ML) and Bayesian Inference (BI) analyses were conducted on the resulting concatenated dataset.

Maximum Likelihood (ML) was performed using RAxML 8.0.14 [34] by running 1000 bootstrap replicates under the GTRGAMMAI model (for all partitions). Bayesian Inference (BI) analyses was performed with MrBayes v3.1.2 [35] based on the best substitution models (GTR + I + G for ITS and LSU; GTR + G for *rpb2*) determined by MrModeltest 2.3 [36]. A total of two independent runs with four Markov chains were conducted for 10 M generations under the default settings. Average standard deviations of split frequency (ASDSF) values were far lower than 0.01 at the end of the runs. Trees were sampled every 100 generations after burn-in (25% of trees were discarded as the burn-in phase of the analyses, set up well after convergence), and a 70% majority-rule consensus tree was constructed.

Trees were visualized with TreeView32 [37]. Bootstrap values (BS) $\geq$ 70% and Bayesian Posterior Probability values (BPP) $\geq$ 0.99 were considered significant [38,39].

3. Results

3.1. Phylogenetic Analyses

In this study, 36 sequences of ITS, LSU and *rpb2* were newly generated from our collections. Dataset I (ITS/LSU/*rpb2*) contained 122 sequences from 39 species, including 15 novel sequences of all three genes from our collections. *P. triaciculare* and *P. breviterincarnatum* were selected as the outgroup. The length of the aligned dataset was 2175 bp after exclusion of poorly aligned sites, with 620 bp for ITS, 884 bp for LSU, and 671 bp for *rpb2*. The topologies of ML and BI phylogenetic trees obtained in this dataset were practically the same, therefore only the tree inferred from the ML analyses is shown (Figure 1). The *Mallocybe* species formed a monophyletic lineage with strong support (MLB = 93%, BPP = 1.00). The sequences of our six collections formed two independent clades, which were respectively recognized and described as two new species: *Mallocybe depressa* and *Mallocybe picea*. *M. depressa* was sister to *M. velutina* Saba & Khalid with high supports, implying that they are closely related to each other. Another species *M. picea* was sister to *M. arthrocystis* (Kühner) Matheny & Esteve-Rav., with strong support (MLB = 96%, BPP = 1.00), and then grouped with *M. multispora* (Murrill) Matheny & Esteve-Rav. and *M. unicolor* (Peck) Matheny & Esteve-Rav. without supported data.

Dataset II (ITS/LSU) contained 1409 total characters (539 from ITS, 870 from LSU, gaps included) and included of 123 samples of 65 taxa. Since the topologies of ML phylogenetic trees is similar to that of the BI phylogenetic tree, only the tree inferred from the ML analyses is shown (Figure 2). A total of 21 sequences newly generated from our collections were resolved as six strong support clades, which indicated that they were six distinct species. Of them, the sequences of five collections clustered well with the authentic sequence of *Pseudosperma bulbosissimum* (Kühner) Matheny & Esteve-Rav., *P. rimosum* (Bull.) Matheny & Esteve-Rav., and *P. solare* Bandini, B. Oertel & U. Eberh., showing their identities with these three species, respectively. The remaining sequences of our collections formed three independent clades, which was recognized and described as three new species *Pseudosperma gilvum*, *Pseudosperma laricis*, and *Pseudosperma pseudoniveivelatum*. *P. gilvum* was sister to *P. citrinostipes* Y.G. Fan & W.J. Yu. *P. laricis* was closely grouped with *P. huginii* Bandini & U. Eberh and *P. arenicola* (R. Heim) Matheny & Esteve-Rav. with lower supports. *P. pseudoniveivelatum* was sister to *P. notodryinum* (Singer, I.J.A. Aguiar & Ivory) Matheny & Esteve-Rav.

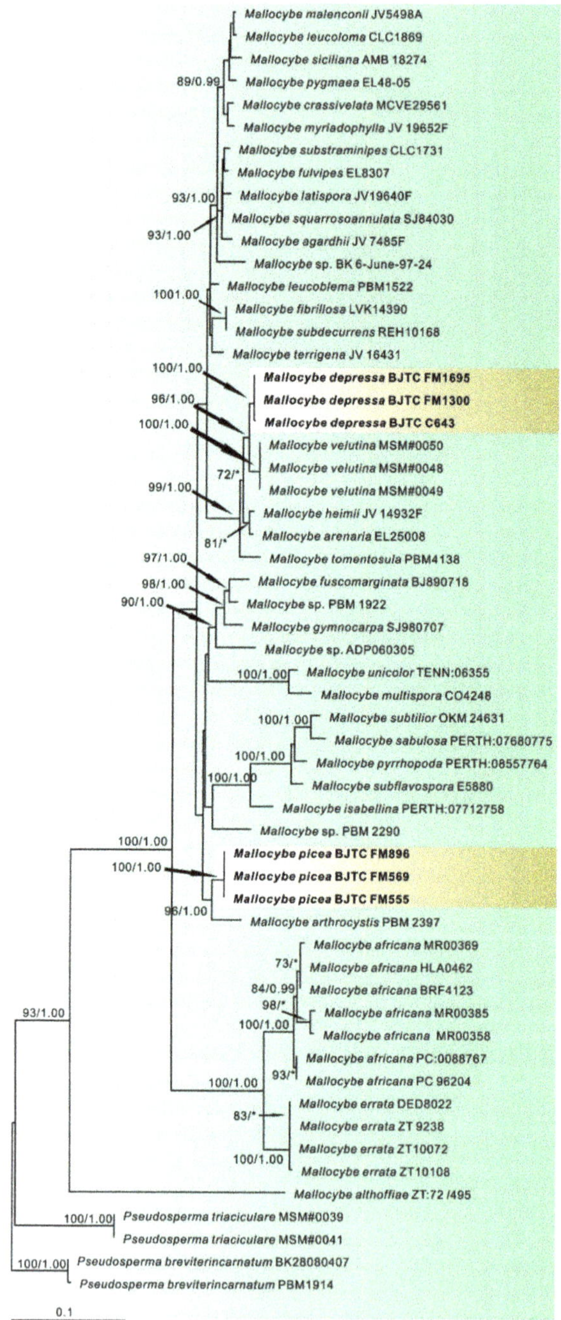

Figure 1. Phylogeny derived from Maximum Likelihood analysis of the combined (ITS/LSU/*rpb2*) dataset of *Mallocybe* and related genera in the family Inocybaceae. *Pseudosperma triaciculare* and *P. breviterincarnatum* were employed to root the tree as an outgroup. Numbers representing likelihood bootstrap support (BS $\geq$ 70%, left) and significant Bayesian posterior probability (BPP $\geq$ 0.99, right) are indicated above the nodes. New sequences are highlighted in black bold.

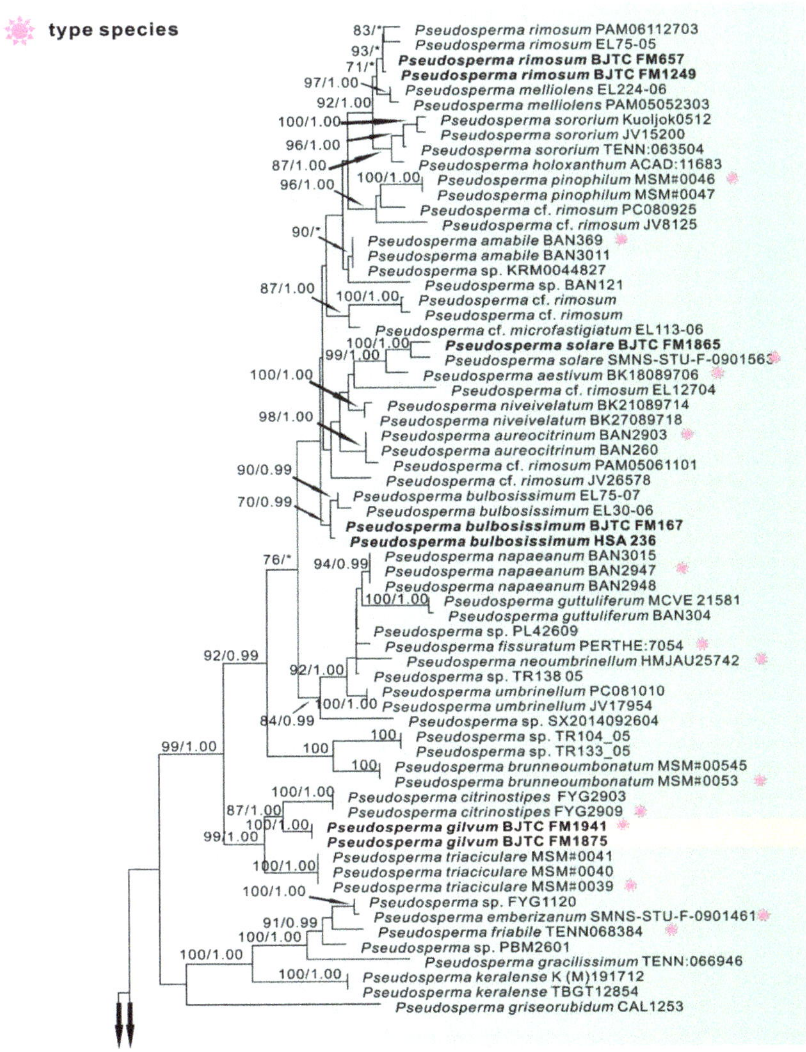

Figure 2. *Cont.*

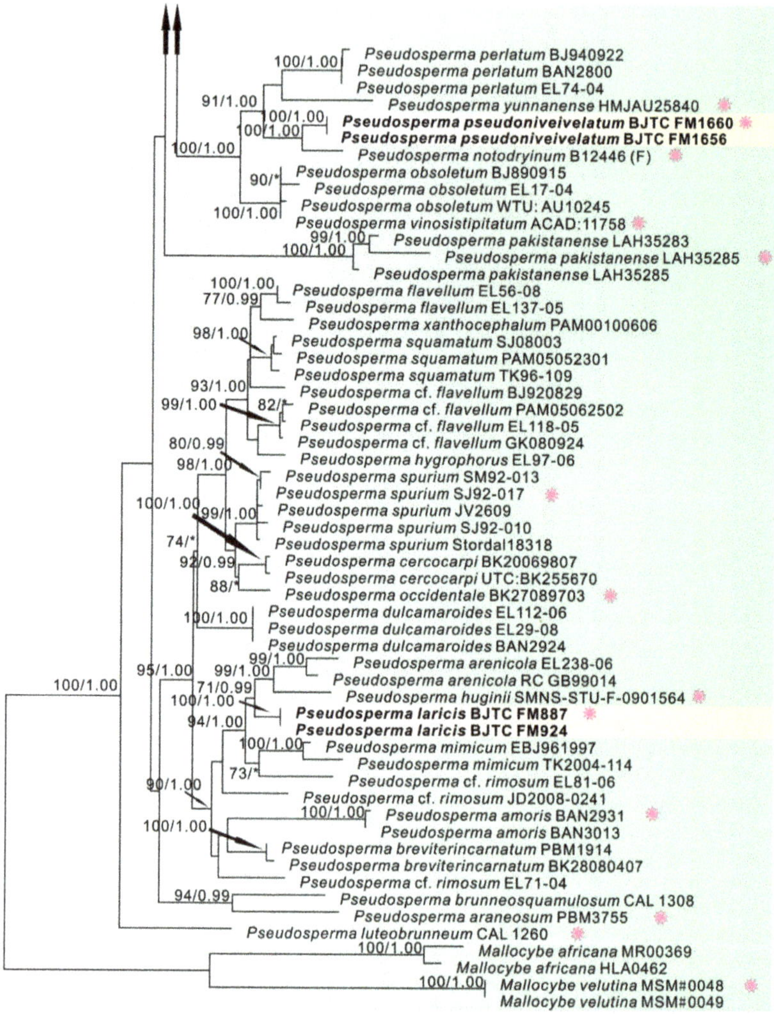

Figure 2. Phylogeny derived from Maximum Likelihood analysis of the ITS/LSU sequences from *Pseudosperma* and related genera in the family Inocybaceae. *Mallocybe velutina* and *M. africana* were employed to root the tree as an outgroup. Numbers representing likelihood bootstrap support (BS $\geq$ 70%, left) and significant Bayesian posterior probability (BPP $\geq$ 0.99, right) are indicated above the nodes. New sequences are highlighted in black bold. The red symbol represents the type species.

3.2. *Taxonomy*

Mallocybe depressa L. Fan, H. Zhou & N. Mao, sp. Nov. (Figures 3C and 4)

Figure 3. Basidiomata *of Mallocybe* and *Pseudosperma.* (**A**,**B**). *Mallocybe picea* (BJTC FM555, holotype), (**C**) *Mallocybe depressa* (BJTC C643), (**D**) *Pseudosperma laricis* (BJTC FM887, holotype), (**E**–**G**) *Pseudosperma pseudoniveivelatum* (BJTC FM1660, holotype), (**H**–**J**). *Pseudosperma gilvum* (BJTC FM1941, holotype). Scale bars: (**A**–**J**) = 10 mm.

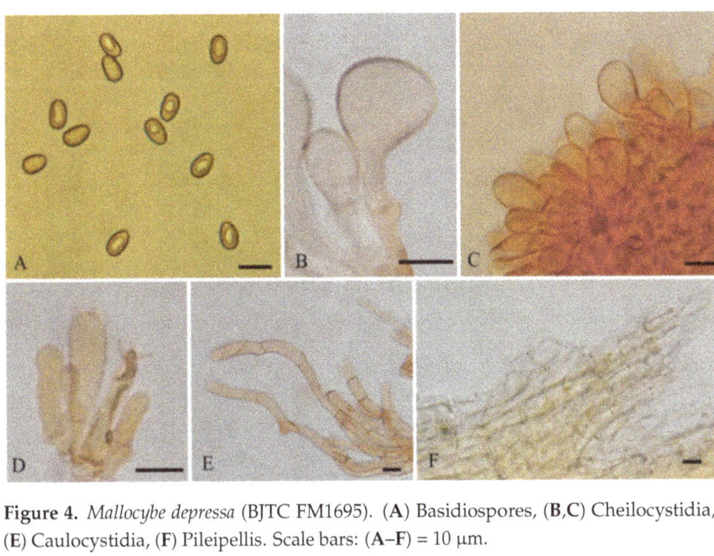

Figure 4. *Mallocybe depressa* (BJTC FM1695). (**A**) Basidiospores, (**B**,**C**) Cheilocystidia, (**D**) Basidia, (**E**) Caulocystidia, (**F**) Pileipellis. Scale bars: (**A**–**F**) = 10 μm.

MycoBank: MB843127

Diagnosis: *Mallocybe depressa* is characterized by its golden yellow to yellowish-brown pileus, central depression of pileus when old, pileus margin splitting when mature, amygdaloid, and subamygdaloid to subcylindrical basidiospores, clavate to broadly clavate, and septate cheilocystidia, and usually grow in coniferous forest dominated by *Pinus* sp. It is most similar to *M. velutina,* but differs by its narrower basidiospores and broadly clavate cheilocystidia.

Etymology: *depressa*, refers to the depression in the center of the pileus with age.

Holotype: China. Shanxi Province, Taiyuan City, Xishan Forest Park, 37°82.39′ N, 112°46.99′ E, alt. 1100 m, 22 July 2021, on the ground in coniferous forest dominated by *Pinus* sp., J.Z. Cao CF1014 (BJTC FM1695).

Description—Pileus 10–35 mm wide, convex to plano-convex at young age, then applanates to uplifted, with a shallow depression at the center; margin initially decurved, becoming flattened and splitting with age; surface dry, strongly fibrous towards the margin, squamulose at the center, dark brown (#4e3000) around the disc, golden yellow (#ff9f00) to yellowish-brown (#cd7f00) elsewhere. Lamellae regular, adnate, brown (#915b25) to dark brown (#68421b) when mature, 1–2 tiers of lamellulae and concolorous with lamellae. Stipe 20–32 × 2.5–5 mm, central, equal with a slightly swollen apex and base, longitudinally fibrillose downwards the stipe, yellowish-brown (#cd7f00) to orange-brown (#9a4d00). Context pale yellow brown. Odor unrecorded.

Basidiospores [60/2/2] (7–)7.5–9(–11) × 4–5 μm, $L_m \times W_m$ = 8.33 (± 0.72) × 4.65 (± 0.34), Q = (1.4–)1.6–1.9(–2.2) (Q_{av} = 1.79 ± 0.16), smooth, amygdaloid, subamygdaloid to subcylindrical, sometimes ellipsoid, thick-walled, yellowish-brown. Basidia with yellowish necropigment, 20–30 × 6–8 μm, clavate, four-spored, occasionally two-spored. Cheilocystidia 15–36 × 9–12(–16) μm, often in clusters, septate, clavate to broadly clavate, occasionally balloon-shaped, apices rounded to obtuse, hyaline, thin-walled. Pleurocystidia absent. Caulocystidia only near the apex, 20–58 × 6–15 μm, clavate to elongate clavate, hyaline or pale yellow. Pileipellis a cutis, composed of parallel arranged of yellowish-brown to brown, cylindrical hyphae, often septate, 3–14 μm wide, thin-walled. Stipitipellis a cutis, composed of parallel, compactly arranged, hyaline, cylindrical hyphae, 4–12 μm wide, thin-walled. Clamp connections abundant in all tissues.

Habitat: In groups on the ground in coniferous forest dominated by *Pinus* sp., Hebei province and Shanxi province, China.

Additional specimens examined: China. Shanxi Province, Taiyuan City, Jinci Park, 38°57.18′ N, 113°30.52′ E, alt. 1370 m, 20 August 2020, on the ground in coniferous forest dominated by *Pinus* sp., H. Liu LH1266A (BJTC FM1300). Hebei Province, Chicheng county, Yanshan Mountains, 38°57.18′ N, 113°30.52′ E, alt. 947 m, 26 August 2020, on the ground in coniferous forest dominated by *Pinus* sp., H. Zhou 130732MFBPC643 (BJTC C643).

Notes: *Mallocybe velutina* is sister to *M. depressa* in our phylogenetic analyses (Figure 1). Morphologically, *M. velutina* differs from *M. depressa* by its pileus center fulvous, margin light yellow, larger, and broader basidiospores (9.0 × 5.4 μm on average) [13]. Molecular analyses reveal that *M. velutina* shares less than 96.04% similarity in ITS sequence with *M. depressa*, supporting their separation. *Inocybe caesariata* (Fr.) P. Karst. is recorded in Shanxi province, China [40]. It is similar to the new species by its pileus color and size. However, it can be differentiated from *M. depressa* by its pileus not splitting, white lamellae edge, and ellipsoid basidiospores. Another collected species *M. picea* is distinguished from *M. depressa* based on its larger basidiospores (10.44 × 5.69 μm on average) and broadly clavate to balloon-shaped cheilocystidia.

Mallocybe picea L. Fan & N. Mao, sp. nov. (Figure 3A,B and Figure 5)

MycoBank: MB843129

Diagnosis: *Mallocybe picea* is characterized by its flattened and splitting pileus margin when mature, broadly clavate to balloon-shaped cheilocystidia, and usually grows in coniferous forest dominated by *Picea asperata*. It is most similar to *M. arthrocystis* but differs by its slightly broader basidiospores and often splitting pileus margin.

Etymology: *picea*, refers to the habitat of the species amongst forest of Picea.

Holotype: China. Shanxi Province, Wutai County, Wutai Mountain, 38°57.13′ N, 113°29.58′ E, alt. 2038 m, 25 July 2019, on the ground in coniferous forest dominated by *Picea asperata* Mast., L.J. Guo GLJM004 (BJTC FM555).

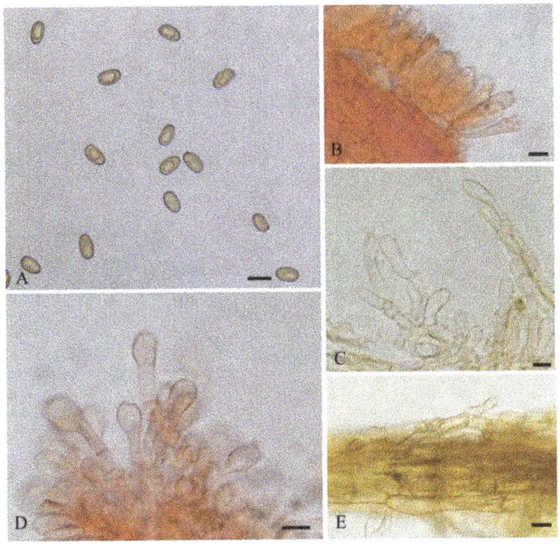

Figure 5. *Mallocybe picea* (BJTC FM555) (**A**) Basidiospores, (**B**) Basidia, (**C**) Caulocystidia, (**D**) Cheilocystidia, (**E**) Pileipellis. Scale bars: (**A–E**) = 10 µm.

Description—Pileus 20–55 mm wide, hemispherical to broadly convex at young age, becoming plano-convex to applanate with age, with distinctly umbo; margin turned down when young, then flattened and splitting; surface dry, uniformly grayish velutinous when young, fibrillose to tomentose, earthy yellow (#e1a95f), yellowish-brown (#cd8526), sometimes dark brown (#764d16) towards the margin when mature. Lamellae regular, adnate, subdistant, yellowish-brown (#cc8526) to dark brown (#764d16), 2–3 tiers of lamellulae and concolorous with lamellae. Stipe 21–45 × 4–8 mm, hollow, central, equal, or sometimes slight widening at base, longitudinally fibrillose downwards the stipe, with white tomentose hyphae at the base, earthy yellow (#e1a95f) to yellowish-brown (#cd8526), paler white (#f5f5f5) at base. Context yellowish white. Odor unrecorded.

Basidiospores [100/2/3] (9–)9.5–11.5(–12) × (5–)5.2–6(–6.5) µm, $L_m \times W_m$ = 10.44 (±0.69) × 5.69 (±0.35), Q = 1.5–2.2 (Q_{av} = 1.84 ± 0.16), smooth, subamygdaloid to subcylindrical, cylindrical, thick-walled, yellowish-brown. Basidia with yellowish necropigment, (24–)28–39 × 8–10 µm, cylindrical to clavate, four-spored, rarely two-spored; sterigmata 2–5 µm long. Cheilocystidia 18–35 × 9–16 µm, often in clusters, septate, broadly clavate to balloon-shaped, apices rounded to subcapitate, hyaline, thin-walled. Pleurocystidia absent. Caulocystidia only near the apex, 22–50 × 7–10 µm, clavate to cylindric, hyaline or pale yellow. Pileipellis a cutis, composed of dense layers of repent hyphae; hyphae cylindrical, often septate, 6–15 µm wide and with yellowish-brown to brown intracellular or parietal pigment, thin-walled. Stipitipellis a cutis, made up of parallel, compactly arranged, thin-walled, cylindrical hyphae, 3.5–12 µm wide, hyaline or pale brown in KOH. Clamp connections abundant in all tissues.

Habitat: In groups on the ground in coniferous forest dominated by *Picea asperata*, Shanxi province, China.

Additional specimens examined: China. Shanxi Province, Wutai County, Wutai Mountain, 38°57.52′ N, 113°31.9′ E, alt. 1910 m, 25 July 2019, on the ground in coniferous forest dominated by *Picea asperata*, H. Liu LH636 (BJTC FM569). Ibid, 38°57.18′ N, 113°30.52′ E, alt. 2013 m, 27 August 2019, on the ground in coniferous forest dominated by *P. asperata*, Y. Shen SYM078 (BJTC FM896).

Notes: *Mallocybe picea* and *M. arthrocystis* are not only closely related phylogenetically, but also morphologically very similar. *Mallocybe arthrocystis* is originally reported from France and distinguished from *M. picea* by its pileus margin not splitting, slightly narrower basidiospores of 9.5–1.2 × 4.5–5.5 μm [12]. Molecular analyses also revealed that *M. arthrocystis* shares less than 90.23% similarity in ITS sequence with *M. picea*, supporting their separation. The species *Inocybe dulcamara* (Pers.) P. Kumm. is easily confused with *M. picea* in morphology, which is reported from China, but classified into *Mallocybe* by Fan and Tolgor [41]. *Inocybe dulcamara* differs from *M. picea* by its longer basidia 30–60 × 8–12 μm and white context in cap [12].

Pseudosperma gilvum L. Fan & N. Mao, sp. nov. (Figure 3H,J and Figure 6)

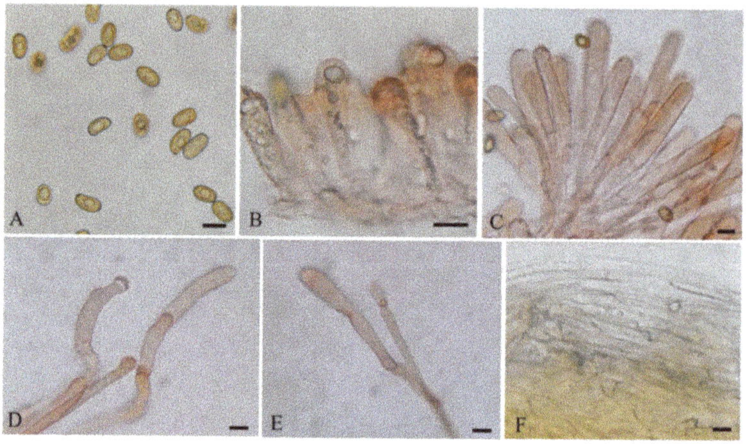

Figure 6. *Pseudosperma gilvum* (BJTC FM1941) (**A**) Basidiospores, (**B**) Basidia, (**C**) Cheilocystidia, (**D**,**E**). Caulocystidia, (**F**) Pileipellis. Scale bars: (**A**–**F**) = 10 μm.

MycoBank: MB843130

Diagnosis: *Pseudosperma gilvum* is characterized by convex to broadly convex pileus with subacute or obtuse umbo, mostly subphaseoliform, subcylindrical to cylindrical basidiospores, cylindrical, clavate to broadly clavate cheilocystidia. It is most similar to *P. triaciculare* but differs by its narrower basidiospores and paler pileus color.

Etymology: *gilvum*, Latin indicating light yellow, refers to the color of the pileus and stipe.

Holotype: China. Shanxi Province, Wenshui County, Lvliang Mountains, 37°28.28′ N, 111°34.22′ E, alt. 1750 m, 30 July 2021, on the ground in coniferous and broad-leaved mixed forest dominated by *Pinus* sp., L. FAN CF1115 (BJTC FM1941).

Description—Pileus 30–45 mm wide, convex to broadly convex with subacute or obtuse umbo; margin decurved or straight, not splitting; surface dry, fibrillose-rimulose, presence of a pale velipellis coating over the disc, light yellow (#ffffd4), becoming yellowish-brown (#ffc000) in some places with age, background pallid to cream white. Lamellae regular, adnate to sinuate, pale white (#f2f2f2) to grayish white (#e6e6e6) when young, becoming yellowish-brown (#ffb31a) with age, 1–2 tiers of lamellulae and concolorous with lamellae. Stipe 47–104 × 4–6 mm, solid, central, nearly terete, base slightly swollen, covered with whitish tomentum at young age, longitudinally fibrillose, pale yellow (#ffffd4) to yellowish brown (#ffab00), pale white at apex and base. Context white. Odor unrecorded.

Basidiospores [70/2/2] 10.5–12.5(–14) × (5.5–)6–7 μm, $L_m \times W_m$ = 11.40 (± 0.80) × 6.34 (±0.39), Q = 1.7–2.0 (Q_{av} = 1.80 ± 0.12), smooth, mostly subphaseoliform, subcylindrical to cylindrical, occasionally ellipsoid, slightly thick-walled, yellowish-brown to reddish-brown. Basidia 30–40 × 9–11 μm, clavate to broadly clavate, occasionally rounded-swollen at apex, primarily with four spored, rarely two spored, often with oily inclusions, hyaline in KOH. Cheilocystidia 25–75 × 9–14 μm, often in clusters, septate, cylindrical, clavate to broadly clavate, sometimes ovoid or subfusiform, often catenate with much shorter elements below the terminal element, hyaline to pale brown, thin-walled.

Pleurocystidia absent. Caulocystidia only near the apex, 20–85 × 8–12 μm, clavate to cylindric, similar to cheilocystidia, hyaline or pale yellow. Pileipellis a cutis, composed of parallel, compactly arranged, thin-walled, hyaline or yellowish-brown, cylindrical hyphae, 4–12.5 μm wide. Stipitipellis a cutis, composed of compactly hyphae, 4–10 μm wide, hyaline or pale brown in KOH. Clamp connections abundant in all tissues.

Habitat: Scattered or in groups on the ground in mixed coniferous and broad-leaved forest dominated by *Pinus* sp., Shanxi Province, China.

Additional specimens examined: China. Shanxi Province, Pu County, Wulu Mountain, 36°33.34′ N, 113°30.52′ E, alt. 1910 m, 28 July 2021, on the ground in coniferous and broad-leaved mixed forest dominated by *Pinus* sp., N. Mao MNM275 (BJTC FM1875).

Notes: *Pseudosperma gilvum* is clustered with *P. citrinostipes* and *P. triaciculare* Saba & Khalid, and forms a distinct monophyletic group. This indicates that the three species are phylogenetically closely related to each other. However, there are clear differences in morphology among them. *Pseudosperma citrinostipes* has brownish yellow or straw yellow to golden yellow pileus, mostly ellipsoid spores, subfusiform, utriform to lageniform cheilocystidia and a different phylogenetic position (Figure 2) [19], that separates it well from our new species. *Pseudosperma triaciculare* can be distinguished by its darker pileus (brownish-orange to fulvous) with radially rimose margin, broader basidiospores (9.0 × 5.4 μm on average) and slightly smaller cheilocystidia (23–54 × 9–16 μm) [17]. Molecular analyses also reveal that *P. gilvum* shares less than 89.20% similarity in ITS sequence with *P. triaciculare*, supporting their separation. Both *P. brunneoumbonatum* Saba & Khalid and *P. gilvum* are presumed to be associated with *Pinus*, and have similar pileus shape. However, *P. brunneoumbonatum* differs from *P. gilvum* by its strongly rimose pileus margin and lager basidiospores (12.5 × 7.5 μm on average) [17]. We also collected some specimens of *P. bulbosissimum* (Kühner) Matheny & Esteve-Rav from Shanxi province, China. Its pileus is pale yellow, then ochraceous to reddish brown, and covered with fibrillose-rimose, similar to that of *P. gilvum*. The basidiospores of *P. bulbosissimum*, however, are remarkably larger (12–15 × 6–8 μm) [15].

Pseudosperma laricis L. Fan & N. Mao, sp. nov. (Figures 3D and 7)

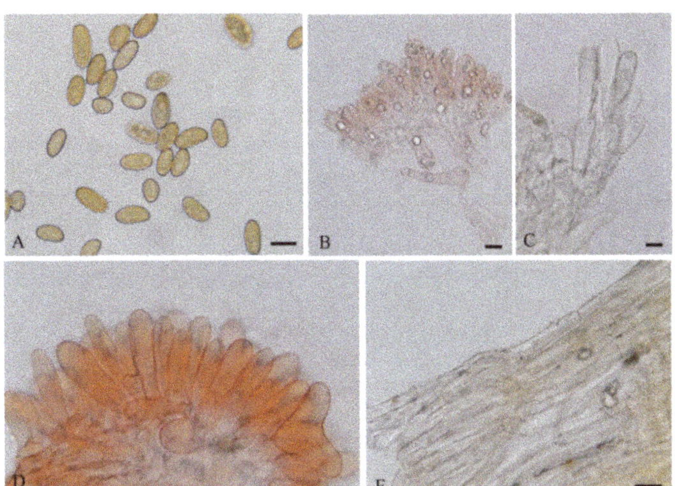

Figure 7. *Pseudosperma laricis* (BJTC FM887) (**A**) Basidiospores, (**B**) Basidia, (**C**) Caulocystidia, (**D**) Cheilocystidia, (**E**) Pileipellis. Scale bars: (**A–E**) = 10 μm.

MycoBank: MB843131

Diagnosis: *Pseudosperma laricis* is characterized by the pileus surface with fibrillose and strongly rimose, subcylindrical to cylindrical basidiospores and an ecological association with *Larix principis-rupprechtii*. It is most similar to *P. rimosum*, but differs by its narrower basidiospores and orange brown or brown pileus.

Etymology: *laricis*, refers to the habitat of the species amongst forest of *Larix*.

Holotype: China. Shanxi Province, Wutai County, Wutai Mountains, 38°57.7′ N, 113°30.16′ E, alt. 2075 m, 27 August 2019, on the ground in coniferous forest dominated by *Larix principis-rupprechtii* Mayr, Y. Shen SYM069 (BJTC FM887).

Description—Pileus 20–35 mm wide, convex, broadly convex or plane with subacute or obtuse umbo; margin at first incurved, then straight to somewhat wavy, not splitting; surface dry, smooth at the umbo, fibrillose and strongly rimose cracked towards center, yellowish-orange (#e6a800) to orange brown (#cc8400) or brown (#d18e4a), background pale white. Lamellae regular, adnate, grayish white (#e6e6e6) when young, becoming yellowish-brown (#ffb31a) to brown (#d18e4a) with age, 1–3 tiers of lamellulae and concolorous with lamellae. Stipe 51–65 × 4–6 mm, hollow, central, equal, longitudinally fibrillose, yellowish brown (#ffdf80) in different intensity, grayish white (#e6e6e6) at apex of the stipe. Context white. Odor unrecorded.

Basidiospores [85/2/2] (10–)11–14(–17) × (5–)5.5–7(–7.5) µm, $L_m \times W_m$ = 12.77 (± 1.38) × 6.16 (± 0.55), Q = 1.8–2.4 (Q_{av} = 2.07 ± 0.12), smooth, subcylindrical to cylindrical, slightly thick-walled, yellowish-brown to reddish-brown. Basidia 30–40 × 10–12 µm, clavate to broadly clavate, rounded-swollen at apex, generally with four spored, rarely two spored, often with oily inclusions, hyaline in KOH. Cheilocystidia 25–57 × 9–20 µm, often in clusters, mostly cylindrical, clavate to broadly clavate, rarely ovoid or fusiform, hyaline to pale brown, thin-walled. Pleurocystidia absent. Caulocystidia only near the apex, 17–47 × 8–11 µm, clavate or cylindrical, similar to cheilocystidia, often catenate with much shorter elements below the terminal element, hyaline or pale yellow. Pileipellis a cutis, composed of parallel, compactly arranged, thin-walled, yellowish-brown, cylindrical hyphae, with 4–13 µm wide. Stipitipellis a cutis, composed of parallel, compactly hyphae, 3–9 µm wide, hyaline or pale brown in KOH. Clamp connections abundant in all tissues.

Habitat: Scattered or in groups on the ground in coniferous forest dominated by *L. principis-rupprechtii*, Shanxi province, China.

Additional specimens examined: China. Shanxi Province, Wutai County, Wutai Mountains, 38°58.25′ N, 113°31.13′ E, alt. 1900 m, 27 August 2019, on the ground in coniferous forest dominated by *L. principis-rupprechtii*, H. Liu LH842 (BJTC FM924).

Notes: *Pseudosperma laricis* is clustered with *P. arenicola* (R. Heim) Matheny & Esteve-Rav. and *P. mimicum* (Massee) Matheny & Esteve-Rav. However, *P. arenicola* is distinguished by its stipe solid, often deeply buried in sand, and a broad host range, including species in *Salicaceae* and *Pinaceae*, and *P. mimicum* by its larger pileus (>65 mm) and ellipsoid basidiospores [15]. *Pseudosperma rimosum* (Bull.) Matheny & Esteve-Rav. is similar to *P. laricis*, and we also collected the fruit-bodies of *P. rimosum* from Shanxi Province, north China. It differs from the new species by its ellipsoid basidiospores (9.5–12.5 × 6–7 µm) [15]. Another species, *P. gilvum*, is easily distinguished from *P. laricis* by its paler color pileus and smaller basidiospores (11.4 × 6.34 µm on average).

Pseudosperma pseudoniveivelatum L. Fan & N. Mao, sp. nov. (Figure 3E–G and Figure 8)

MycoBank: MB843132

Diagnosis: *Pseudosperma pseudoniveivelatum* is characterized by pileus surface with a distinct pale grayish to pale white velipellis, pileus margin splitting with age, ellipsoid basidiospores, broadly clavate to papillate or utriform cheilocystidia. It is most similar to *P. niveivelatum*, but differs by its smaller basidiospores and darker (yellowish-brown to brown) pileus.

Etymology: *pseudoniveivelatum*, refers to this species is similar to *P. niveivelatum*.

Holotype: China. Shanxi Province, Qinshui County, Lishan Mountains, 35°29.48′ N, 112°4.12′ E, alt. 1690 m, 7 July 2021, on the ground in coniferous and broad-leaved mixed forest dominated by *Quercus* sp., N. Mao MNM232 (BJTC FM1660).

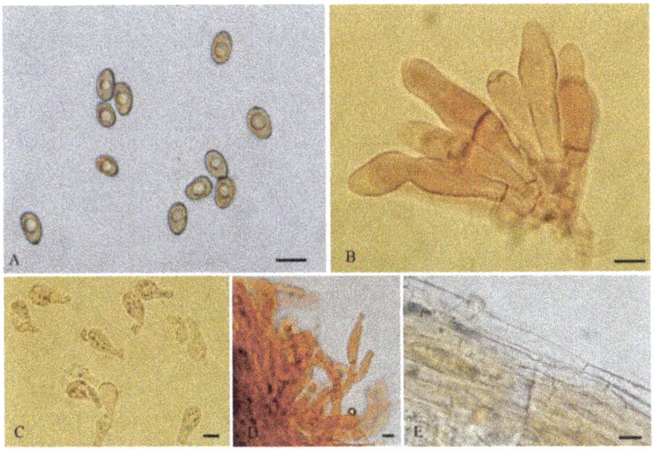

Figure 8. *Pseudosperma pseudoniveivelatum* (BJTC FM1660) (**A**) Basidiospores, (**B**) Cheilocystidia, (**C**) Basidia, (**D**) Caulocystidia, (**E**) Pileipellis. Scale bars: (**A–E**) = 10 μm.

Description—Pileus 20–45 mm wide, conical to conical-convex at first, then broadly convex to plane-convex with obtuse umbo; margin decurved or straight, becoming splitting with age; surface dry, with a distinct pale greyish to pale white velipellis, indistinctly fibrillose-rimulose, yellowish-brown (#cd9900) to brown (#8b4513), sometimes dark brown (#5e2f0d) at the center, background cream white. Lamellae regular, crowded, adnate, pale white (#ffffff) or yellowish white (#ffffe7) when young, later yellowish-brown (#ffbf00) to ochraceous (#a5682a), 1–2 tiers of lamellulae and concolorous with lamellae. Stipe 33–75 × 4–9 mm, solid, central, cylindrical, equal, or base slightly swollen, covered with whitish tomentum for a long time, later longitudinally fibrillose, pale orange (#ffae1a) to pale brownish (#997654). Context white. Odor unrecorded.

Basidiospores [70/2/2] (8.5–)9.5–11(–12) × (5–)5.5–7(–7.5) μm, $L_m \times W_m$ = 10.19 (± 0.77) × 6.22 (± 0.55), Q = 1.4–1.8 (Q_{av} = 1.63 ± 0.10), smooth, mostly ellipsoid, occasionally broadly ellipsoid or subcylindrical, slightly thick-walled, yellowish-brown to reddish-brown. Basidia 23–33 × 9–12.5 μm, clavate to broadly clavate, often rounded-swollen at apex, primarily four-spored, occasionally two-spored, usually with oily inclusions, hyaline in KOH. Cheilocystidia 30–65 × 7–16 μm, often in clusters, mostly clavate, broadly clavate to papillate or utriform, sometimes cylindrical, hyaline, thin-walled. Pleurocystidia absent. Caulocystidia only near the apex, 32–95 × 9–25 μm, in clusters, broadly clavate or utriform, at times with apices tapered or papillate, similar to cheilocystidia but larger, hyaline or pale yellow. Pileipellis a cutis, composed of parallel, compactly arranged, thin-walled, hyaline or yellowish-brown, cylindrical hyphae, 4–15 μm wide, with some encrustations, septate. Stipitipellis a cutis, composed of compactly hyphae, 5–14 μm wide, hyaline or pale brown in KOH. Clamp connections abundant in all tissues.

Habitat: Scattered or in groups on the ground in mixed coniferous and broad-leaved forests dominated by *Quercus* sp., north China, south China, and Europe.

Additional specimens examined: China. Shanxi Province, Qinshui County, Lishan Mountains, 35°29.14′ N, 112°1.20′ E, alt. 1660 m, 7 July 2021, on the ground in coniferous and broad-leaved mixed forest dominated by *Quercus* sp., N. Mao MNM224 (BJTC FM1656).

Notes: *Pseudosperma niveivelatum* is easily confused with *P. pseudoniveivelatum* in morphology, due to the presence of a white, abundant velipellis in *P. niveivelatum* that covers the pileus. However, this species has pale brown or yellow hues pileus, larger basidiospores (13.9 × 6.4 µm on average) and a different phylogenetic position (Figure 2) that separates it well from our new species [29]. *Pseudosperma notodryinum* is sister to *P. pseudoniveivelatum* in our phylogenetic analyses (Figure 2), implying that they have a close relationship. However, there are clear differences in the morphology. *P. notodryinum* can be distinguished by its darker pileus (yellow-ocher to rich yellowish-fuscous) and narrower basidiospores (9–12 × 5–6 µm) [42]. Molecular analyses revealed that *Pseudosperma notodryinum* shares less than 91.88% similarity in ITS sequence with *P. pseudoniveivelatum*, supporting their separation. A total of three species reported in China in previous studies, *P. obsoletum* (Quadr.) Valade, *P. perlatum* (Cooke) Matheny & Esteve-Rav. and *P. yunnanense* (T. Bau & Y.G. Fan) Matheny & Esteve-Rav., are all easily confused with *P. pseudoniveivelatum* in morphology [41,43]. However, *P. obsoletum* differs from *P. pseudoniveivelatum* by its gray brown to pinkish gray pileus, the absence of velipellis and narrower basidiospores (9–13 × 5–6 µm) [29]; it differs from *P. perlatum* by its larger basidiomata (pileus 35–100 mm, stipe 80–120 × 8–13 mm), and pileus color without yellow tinges [15]; and *P. yunnanens* by its fibrillose with densely squamules stipe and slightly smaller and narrower basidiospores (9–10.5 × 5–6 µm) [43]. Moreover, eleven ITS sequences respectively labelled '*P. obsoletum*', '*Inocybe obsoleta*' and '*P.* aff. *perlatum*' are conspecific to the new species *P. pseudoniveivelatum* since they clustered together with *P. pseudoniveivelatum* in ITS tree (not shown), and have more than 98.66% similarity in ITS region. Of them, two (MT072905, MG367271) are respectively from Inner Mongolia in northern China, Hainan Province in southern China, and nine from Europe (UDB035861, UDB015340, MW355002, MG367270, HG937630, JF908256, MZ410669, JX625280, JQ994477). These show that the new species *P. pseudoniveivelatum* is distributed in both northern and southern China and in Europe.

4. Discussion

Shanxi Province is located in north China, where the climate ranges from subtropical to cold temperate. Our analyses revealed two species of *Mallocybe* and six species of *Pseudosperma* in this region, i.e., *M. depressa*, *M. picea*, *P. bulbosissimum*, *P. gilvum*, *P. laricis*, *P. pseudoniveivelatum*, *P. rimosum* and *P. solare*. They all are associated with coniferous forests. *Pseudosperma bulbosissimum* is the most commonly encountered species, which is distributed in the central and northern regions in Shanxi Province. *Pseudosperma gilvum* is found in both central and southern regions. The remaining species probably have distribution limitations: *P. pseudoniveivelatum*, *P. rimosum*, and *P. solare* are distributed in the southern region, *M. depressa* are distributed in the central region, and *M. picea* and *P. laricis* are distributed in the northern region.

In China, the species diversity of the two genera of *Mallocybe* and *Pseudosperma* are scarce. A total of five species are reported in *Mallocybe* and 13 species in *Pseudosperma* [19,40–46]. With the exception of *P. citrinostipes*, *P. neoumbrinellum* (T. Bau & Y.G. Fan) Matheny & Esteve-Rav. and *P. yunnanense*, and the new species described in this study, the remaining seven species all need to be reexamined and verified with molecular data. They are *M. heimii* (Bon) Matheny & Esteve-Rav. [= *Inocybe heimii* Bon], *M. leucoloma* (Kühner) Matheny & Esteve-Rav. [= *Inocybe leucoloma* Kühner], *M. terrigena* (Fr.) Matheny, Vizzini & Esteve-Rav. [= *Inocybe terrigena* (Fr.) Kühner], *P. avellaneum* (Kobayasi) Matheny & Esteve-Rav. [= *I. avellanea* Kobayasi], *P. obsoletum* [= *I. obsoleta* Romagn.], *P. perlatum* [= *I. perlata* (Cooke) Sacc.], and *P. sororium* (Kauffman) Matheny & Esteve-Rav. [= *I. sororia* Kauffman] [19,40–46].

Key to the species of *Mallocybe* from China

1. Annulus present	*M. terrigena*
1. Annulus absent	2
2. Pileus applanate to uplifted, with a shallow depression at the right	*M. depressa*
2. Pileus plano-convex or applanate, with distinctly umbo or indistinctly umbo	3
3. Habitat not associated with *Picea*	*M. heimii*
3. Habitat associated with *Picea*	4
4. Basidiospores subamygdaloid to subcylindrical, cylindrical, length > 9 μm	*M. picea*
4. Basidiospores ellipsoid to subphaseoliform, cylindrical, length < 9 μm	*M. leucoloma*

Key to the species of *Pseudosperma* from China

1. Basidiomata uniformly brown (including pileus, lamellae, stipe)	*P. neoumbrinellum*
1. Basidiomata not uniformly brown	2
2. Pileus color with pinkish tinges	3
2. Pileus color without pinkish tinges	4
3. Pileus yellowish buff, grayish brown to pale pinkish beige, basidiospores 10–15 × 5.5–7.5 μm	*P. sororium*
3. Pileus gray brown to pinkish gray, basidiospores 9–13 × 5–6 μm	*P. obsoletum*
4. Pileus surface with a distinct pale white or gray white velipellis	5
4. Pileus surface without velipellis	7
5. Stipe surface fibrillose with densely squamules	*P. yunnanense*
5. Stipe surface fibrillose without squamules	6
6. Basidiospores mostly subphaseoliform, subcylindrical to cylindrical, $L_m \times W_m = 11.40 \times 6.34$ μm	*P. gilvum*
6. Basidiospores mostly ellipsoid, $L_m \times W_m = 10.19 \times 6.22$ μm	*P. pseudoniveivelatum*
7. Pileus color with yellow tinges	8
7. Pileus color without yellow tinges	9
8. Habitat associated with *Larix*	*P. laricis*
8. Habitat not associated with *Larix*	10
9. Pileus brownish to dark brown	*P. perlatum*
9. Pileus white or pallid ivory	*P. bulbosissimum*
10. Basidiospores small, length < 10 μm	*P. avellaneum*
10. Basidiospores large, length > 10 μm	11
11. Cheilocystidia missing (sub)capitate	*P. rimosum*
11. Cheilocystidia (sub)capitate	12
12. Stipe surface with lemon yellow fibrils	*P. citrinostipes*
12. Stipe surface with whitish to dingy whitish rough fibres or glabrous	*P. solare*

Supplementary Materials: The following supporting information can be downloaded at: https://www.mdpi.com/article/10.3390/jof8030256/s1, Table S1: Information on sequences used in molecular phylogenetic analyses for *Mallocybe*; Table S2: Information on sequences used in molecular phylogenetic analyses for *Pseudosperma*.

Author Contributions: L.F. conceived and designed the study; N.M. and Y.-Y.X. wrote the manuscript; N.M. conducted phylogenetic analyses and morphological observations; Y.-Y.X., T.-Y.Z. and J.-C.L. conducted the experiments. All authors have read and agreed to the published version of the manuscript.

Funding: This study was supported by the National Natural Science Foundation of China (No. 31750001) and the Beijing Natural Science Foundation (No. 5172003).

Institutional Review Board Statement: Not applicable.

Informed Consent Statement: Not applicable.

Data Availability Statement: All sequence data are available in NCBI GenBank following the accession numbers in the manuscript.

Acknowledgments: J.Z. Cao was appreciated for collecting specimens and providing valuable suggestions.

Conflicts of Interest: The authors declare no conflict of interest.

References

1. Matheny, P.B.; Kudzma, L.V. New species of *Inocybe* (Agaricales) from eastern North America. *J. Torrey Bot. Soc.* **2019**, *146*, 213–235. [CrossRef]
2. Matheny, P.B.; Hobbs, A.M.; Esteve-Raventós, F. Genera of Inocybaceae: New skin for the old ceremony. *Mycologia* **2019**, *112*, 83–120. [CrossRef] [PubMed]
3. Massee, G. A monograph of the genus *Inocybe*, Karsten. *Ann. Bot.* **1904**, *18*, 459–504. [CrossRef]
4. Heim, R. *Le Genre Inocybe, Encylopédie Mycologique 1*; Paul Lechevalier & Fils: Paris, France, 1931; 429p.
5. Kühner, R.; Romagnesi, H. *Flore Analytique des Champignos Supérieurs*; Masson et Cie: Paris, France, 1953; 556p.
6. Horak, E. Synopsis generum Agaricalium (Die Gattungstypen der Agaricales). *Beiträge Zur Kryptogamenflora Der Schweiz* **1967**, *13*, 1–741.
7. Kuyper, T.W. A revision of the genus *Inocybe* in Europe 1. Subgenus *Inosperma* and the smooth-spored species of subgenus *Inocybe*. *Persoonia* **1986**, *3*, 1–247.
8. Singer, R. *The Agaricales in Modern Taxonomy*, 4th ed.; Koeltz Scientific Books: Koenigsten, Germany, 1986; 981p.
9. Horak, E. *Röhrlinge und Blätterpilze in Europa*; Spektrum Akademischer Verlag: Heildelberg, Germany, 2005; 575p.
10. Stangl, J. Die Gattung *Inocybe* in Bayern. *Hoppea* **1989**, *46*, 5–388.
11. Jacobsson, S. Key to *Inocybe*. In *Funga Nordica: Agaricoid, Boletoid and Cyphelloid Genera*; Knudsen, H., Vesterholt, J., Eds.; Nordsvamp: Copenhagen, Denmark, 2008; pp. 868–906.
12. Cripps, C.L.; Larsson, E.; Horak, E. Subgenus *Mallocybe* (*Inocybe*) in the Rocky Mountain alpine zone with molecular reference to European arctic-alpine material. *N. Am. Fungi* **2010**, *5*, 97–126.
13. Saba, M.; Khalid, A.N. *Mallocybe velutina* (Agaricales, Inocybaceae), a new species from Pakistan. *Mycoscience* **2020**, *61*, 348–352. [CrossRef]
14. Aïgnon, H.; Naseer, A.; Matheny, P.B.; Yorou, N.S.; Ryberg, M. *Mallocybe africana* (Inocybaceae, Fungi), the first species of *Mallocybe* described from Africa. *Phytotaxa* **2021**, *478*, 49–60. [CrossRef]
15. Larsson, E.; Ryberg, M.; Moreau, P.A.; Mathiesen, Å.D.; Jacobsson, S. Taxonomy and evolutionary relationships within species of section *Rimosae* (*Inocybe*) based on ITS, LSU and mtSSU sequence data. *Persoonia* **2009**, *23*, 86–98. [CrossRef]
16. Matheny, P.B. Improving phylogenetic inference of mushrooms with RPB1 and RPB2 nucleotide sequences (*Inocybe*; Agaricales). *Mol. Phylogenet. Evol.* **2005**, *35*, 1–20. [CrossRef] [PubMed]
17. Saba, M.; Haelewaters, D.; Pfister, D.H.; Khalid, A.N. New species of *Pseudosperma* (Agaricales, Inocybaceae) from Pakistan revealed by morphology and multi-locus phylogenetic reconstruction. *MycoKeys* **2020**, *69*, 1–31. [CrossRef] [PubMed]
18. Bandini, D.; Oertel, B. Three new species of the genus *Pseudosperma* (Inocybaceae). *Czech Mycol.* **2020**, *72*, 221–250. [CrossRef]
19. Yu, W.J.; Chang, C.; Qin, L.W.; Zeng, N.K.; Wang, S.X.; Fan, G.Y. Pseudosperma citrinostipes (Inocybaceae), a new species associated with Keteleeria from southwestern China. *Phytotaxa* **2020**, *450*, 8–16. [CrossRef]
20. Dring, D.M. Techniques for microscopic preparation. In *Methods in Microbiology*; Booth, C., Ed.; Academic Press: New York, NY, USA, 1971; Volume 4.
21. White, T.J.; Bruns, T.; Lee, S.; Taylor, J. Amplification and direct sequencing of fungal ribosomal RNA genes for phylogenetics. In *PCR Protocols*; Innis, M.A., Gelfand, D.H., Sninsky, J.J., White, T.J., Eds.; Academic Press: San Diego, CA, USA, 1990; pp. 315–322.
22. Gardes, M.; Bruns, T.D. ITS primers with enhanced specificity for basidiomycetes—Application to the identification of mycorrhizae and rusts. *Mol. Ecol.* **1993**, *2*, 113–118. [CrossRef]
23. Vilgalys, R.; Hester, M. Rapid genetic identification and mapping of enzymatically amplified ribosomal DNA from several *Cryptococcus* species. *J. Bacteriol.* **1990**, *172*, 4239–4246. [CrossRef]
24. Matheny, P.B.; Wang, Z.; Binder, M.; Curtis, J.M.; Lim, Y.W.; Nilsson, R.H.; Hughes, K.W.; Hofstetter, V.; Ammirati, J.F.; Schoch, C.L.; et al. Contributions of *rpb2* and *tef1* to the phylogeny of mushrooms and allies (Basidiomycota, Fungi). *Mol. Phylogenet. Evol.* **2007**, *43*, 430–451. [CrossRef]
25. Ryberg, M.; Nilsson, R.H.; Kristiansson, E.; Töpel, M.; Jacobsson, S.; Larsson, E. Mining metadata from unidentified ITS sequences in GenBank: A case study in *Inocybe* (Basidiomycota). *BMC Evol. Biol.* **2008**, *8*, 50. [CrossRef]
26. Matheny, P.B. A phylogenetic classification of the Inocybaceae. *McIlvainea* **2009**, *18*, 11–21.
27. Ariyawansa, H.A.; Hyde, K.D.; Jayasiri, S.C.; Buyck, B.; Chethana, K.T.; Dai, D.Q.; Dai, Y.C.; Daranagama, D.A.; Jayawardena, R.S.; Lücking, R.; et al. Fungal diversity notes 111–252—Taxonomic and phylogenetic contributions to fungal taxa. *Fungal Divers.* **2015**, *75*, 27–274. [CrossRef]
28. Ryberg, M.; Larsson, E.; Jacobsson, S. An evolutionary perspective on morphological and ecological characters in the mushroom family Inocybaceae (Agaricomycotina, Fungi). *Mol. Phylogenet. Evol.* **2010**, *55*, 431–442. [CrossRef] [PubMed]
29. Kropp, B.R.; Matheny, P.B.; Hutchison, L.J. *Inocybe* section *Rimosae* in Utah: Phylogenteic affinities and new species. *Mycologia* **2013**, *105*, 728–747. [CrossRef] [PubMed]
30. Horak, E.; Matheny, P.B.; Desjardin, D.E.; Soytong, K. The genus *Inocybe* (Inocybaceae, Agaricales, Basidiomycota) in Thailand and Malaysia. *Phytotaxa* **2015**, *230*, 201–238. [CrossRef]
31. Pradeep, C.K.; Vrinda, K.B.; Varghese, S.P.; Korotkin, H.B.; Matheny, P.B. New and noteworthy species of Inocybe (Agaricales) from tropical India. *Mycol. Prog.* **2016**, *15*, 24. [CrossRef]
32. Brugaletta, E.; Consiglio, G.; Marchetti, M. Inocybe siciliana, una nuova specie del Sottogenere *Mallocybe*. *Riv. Micol.* **2018**, *60*, 195–209.

33. Katoh, K.; Standley, D.M. MAFFT multiple sequence alignment software version 7: Improvements in performance and usability. *Mol. Phylogenet. Evol.* **2013**, *30*, 772–780. [CrossRef]
34. Stamatakis, A. RAxML version 8: A tool for phylogenetic analysis and post-analysis of large phylogenies. *Bioinformatics* **2014**, *30*, 1312–1313. [CrossRef]
35. Ronquist, F.; Huelsenbeck, J.P. MrBayes 3: Bayesian phylogenetic inference under mixed models. *Bioinformatics* **2003**, *19*, 1572–1574. [CrossRef]
36. Nylander, J. *MrModeltest*, Version 2.2; Computer Software; Evolutionary Biology Centre, University of Uppsala: Uppsala, Sweden, 2004.
37. Page, R.D. *TreeView*; Glasgow University: Glasgow, UK, 2001.
38. Hillis, D.M.; Bull, J.J. An empirical test of bootstrapping as a method for assessing confidence in phylogenetic analysis. *Syst. Biol.* **1993**, *42*, 182–192. [CrossRef]
39. Alfaro, M.E.; Zoller, S.; Lutzoni, F. Bayes or bootstrap? A simulation study comparing the performance of Bayesian Markov chain Monte Carlo sampling and bootstrapping in assessing phylogenetic confidence. *Mol. Phylogenet. Evol.* **2003**, *20*, 255–266. [CrossRef]
40. Du, F.; Liu, B.; Li, Z.Y. Investigation Report on Wild Eating Mushrooms and Poisonous Mushrooms in Shanxi (Part 2). *J. Shanxi Univ.* **1980**, 71–78. [CrossRef]
41. Fan, G.Y.; Tolgor, B. Taxonomy in *Inocybe* subgen. *Mallocybe from China. J. Fungal Res.* **2016**, *14*, 129–141.
42. Matheny, P.B. *Key to Species of Inocybaceae from Eastern North America—v7*; University of Tennessee: Knoxville, TN, USA, 2020; pp. 1–23.
43. Tolgor, B.; Fan, G.Y. Three new species of *Inocybe* sect. *Rimosae from China. Mycosystema* **2018**, *37*, 693–702.
44. Zheng, G.Y.; Bi, Z.S.; Li, C.; Li, T.H. A preliminary report on the Taxonomy of Inocybe in Guangdong Province. *J. Shanxi Univ.* **1985**, *3*, 65–75.
45. Song, G.; Wang, L.Y.; Yang, W.S. Taxonomy of higher fungi in Helanshan Mountains II. *Yinshan Acad. J.* **1994**, *12*, 30–44.
46. Tolgor, B. List of Agarics and Boletoid Fungi from Eastern Inner Mongolia. *J. Fungal Res.* **2012**, *10*, 20–30.

MDPI
St. Alban-Anlage 66
4052 Basel
Switzerland
Tel. +41 61 683 77 34
Fax +41 61 302 89 18
www.mdpi.com

Journal of Fungi Editorial Office
E-mail: jof@mdpi.com
www.mdpi.com/journal/jof

www.ingramcontent.com/pod-product-compliance
Lightning Source LLC
Chambersburg PA
CBHW061503220926
51850CB00004B/158
9783036564494